Tuning the A-Series Engine

The definitive manual on tuning for performance or economy

3rd Edition

David Vizard

Haynes Publishing

First published in 1985 by G. T. Foulis & Company
as *Tuning BL's A-Series Engine*
Reprinted 1985 (twice), 1986, 1987, 1988 and 1989
Second edition published 1989
Reprinted 1990, 1991, 1993, 1995 (twice), 1996, 1997, and 1998
Third edition published 1999 by Haynes Publishing
as *Tuning the A-Series Engine*
Reprinted 2000, 2001, 2002, 2003, 2004, 2006, 2008, 2009, 2011, 2012, 2013, 2014 and 2015 (twice)

Published by Haynes Publishing
Sparkford, Yeovil, Somerset BA22 7JJ, UK
Tel: 01963 440635
Int. tel: +44 1963 440635
Website: www.haynes.co.uk

Haynes North America, Inc.
861 Lawrence Drive, Newbury Park,
California 91320 USA

A catalogue record for this book is available from the British Library

ISBN 1 85960 620 2

Library of Congress catalog card number
98-7540

Printed in the USA.

Dedications

This, the third edition of '*Tuning the A-Series Engine*' is dedicated to the Memory of David Taber (1936–1995) for setting examples we can all look up to.

To my mother, Eileen Vizard who, single-handedly ran my business affairs and looked after my home in California, thus freeing me to do the vital six months final research for this book. Without her efforts this book would never have come into existence.

A special 'thank you' must also go to Daphne Abbley, whose efforts toward producing the first edition will be understood by typists the world over. Whilst working under difficult circumstances, she typed and edited grammatically the first draft, then incorporated my editing into the second draft. She then laid out most of the charts, sorted out the chapters to put them into some sort of logical order, typed the photo captions together with a few other small, but important, jobs. All in all, this amounted to over a quarter of a million words. Since then this book has gone through a second and now this third update. Updates can often be almost as big a job as writing the book in the first place, as it proved in this instance. Here I would like to thank my wife Josephine for her typing efforts at what was a very busy time for us.

About the author in brief

David Vizard is a man who successfully wears many hats which makes it very difficult to 'pigeon hole' him for those who, for what ever reason, feel a need to do so. Educated in England and trained in the aerospace industry his work in the field of automotive research has brought him recognition amongst millions of enthusiasts worldwide. In short, he is a university lecturer on the theory of high performance engines, a skilled machinist/fabricator and engine builder, a prolific writer, innovator and inventor, a driver of race winning performance, crew chief, team manager, marketing advisor, and consultant engineer. With the contents of this book backed by such credentials you will be hard pressed to find such easy to understand, functional quality high tech information in any other book of this kind.

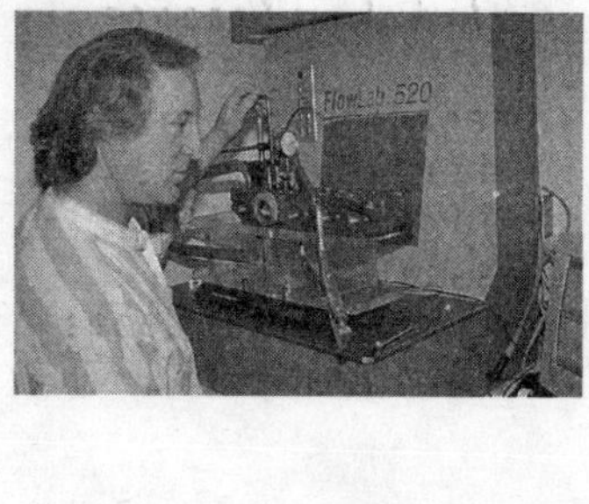

David Vizard engaged in one of his common pastimes – modifying and air flow testing a cylinder head

Contents

Contents

Acknowledgements

In writing this book, it is my fondest wish that it will be accepted as the bible for Series 'A' engine enthusiasts. Although over 95 per cent of the experimental work and test data contained and referred to in this book was generated by myself or my company (now A.P.T. in the US) there are a number of other engine men who have been particularly involved in the Series 'A' and who have reputations which are second to none in the high performance and tuning industry. I asked my friends amongst these to 'vet' the book and ensure that the content had achieved the desired standard and authority as judged by some of the best brains in the business. These guest technical contributors were: Keith Dodd, Richard Longman, Steve Harris, Keith Ripp, Paul Ivey and Dave Mountain, and for each I have written a short introduction which follows this page. Apart from our guest Technical Contributors many other people read and commented on certain parts of this book; people who were essentially extra-special experts in certain areas as well as being in many cases very much at home with the Series 'A' engine.

A special 'thank you' has to go to my Irish colleague Shaun Wall. Shaun is not only a Mini expert but also a journalist, so I was very pleased to have him go though the book to edit it not only for technical accuracy, but also to check the 'sense' and the grammar, so as to produce a book which though very technical, is easily understandable.

Various chapters were also edited by Pete Dugdale late of Iskenderian, Geoff Kershaw of Turbotechnics, and Graham Hickman. To say their time and effort was appreciated would be an understatement. To this list I must also add the people who gave me other valuable help. John Chappe (Advanced Products), Ian Cox (Piper), Eddie Perk (Piper), Ian Elliott (Austin/Rover), Clive Richardson (Austin Rover), Tom Oliver and Phil Hepworth, Jan O'dor (Janspeed), Bill Quinne (Manx Racing), Dave Bulman, Steve Roberts (Midas Cars), Fred Hadley (Omega), David Gardiner (A.E. Autoparts), Jack Field (Cosworth), Nolan Pitts, Tony Beadle (*Street Machine* magazine), Dimitri Elgin (Elgin Cams), Robin Chan (Contact Developments), Martin Goodall (Weber U.K.), Alan Jeffery, John Mowatt (John Mowatt), Mike Abramson (Seven Enterprises), Malcolm Riches (Oselli Engineering), and Mike Parry of Race Engine Technique.

Although this book only took six months to write, it was years in the making and researching, and over those years, literally hundreds of people have, in some way or another, contributed something to the information contained in this book, and although there is not room for those hundreds of names, their help has been valued all the same.

The Guest Technical Contributors

Richard Longman

Richard Longman's association with Series 'A' and, more particularly, Minis, dates back to the mid 60s. His fame then was mostly as a driver, and it was his exploits driving a Mini while he worked at Downton Engineering that put him on the map. Since he started racing in 1966 he has won well over 150 races, of which over 140 were with engines built by himself. As far as lap records are concerned, Richard has set them so often that they are just too numerous to document.

In 1971 he started his own business which, over the years, has steadily grown, both in size and reputation. His schooling, under the auspicies of the late Daniel Richmond at Downton Engineering, obviously stood Richard in good stead. Not only was he taught by one of the best, but he also proved to be a very apt pupil, as his subsequent career proves, and there are few people that would argue that when it comes to piloting Minis at incredible speed, Richard has few, if any, equals.

Those of us who are grossly absorbed with Series 'A', to the exclusion of all else, may be forgiven for thinking that Richard deals only in the Series 'A' engine. The fact that Richard is an engine development man as well as a driver, has brought him a reputation of success in the competition sphere which many of the big automobile manufacturers have been eager to capitalize on. Not only has he been heavily involved with British Leyland and the Austin/Rover Group, but also Ford, Jaguar and several other well-known companies.

Richard's input into this book has been of tremendous

value because it has been based on his successes doing it all: developing, building and driving Series 'A' powered vehicles on famous road courses all over the Western world.

As far as championship wins are concerned, Richard has captured three, which is no mean feat because in England the opposition in road racing is traditionally very stiff. Even more to his credit though, is the fact that Richard's engines have powered various vehicles and drivers to more than 20 championship wins in various places in the world.

The author of this book is one of Richard's greatest fans, and would sum him up by saying, "He's a quiet man who doesn't talk much, but when he does, you'll be that much wiser for listening."

Keith Ripp

Keith Ripp is a dynamic character who likes to play dirty ... in the dirt of Autocross and Rallycross, that is. Since starting racing in 1966, he has built a reputation for success in Autocross and Rallycross that is second to no-one. Keith has won four Autocross championships and was three times British Rallycross champion. Most of his success has been achieved in Minis. When Keith first started in serious competition there seemed to be none in his area of London that was able to supply parts or technical information that enabled him to make his car go faster. Since he and his group of helpers were more competitive than most, Keith figured it might be as well to capitalize on this, and so Ripspeed International was established. Since 1971, this company has grown to be, perhaps, the biggest retailer of speed equipment in Europe. Keith's philosophy seems to be the same in both business and racing: if you are going to do it, put all you've got into it. As a result, his very successful company has spawned many successful competition cars. In an effort to be as successful as possible, Keith has explored almost every conceivable avenue for getting extra power from his Rallycross Minis.

With something in the region of 300 class wins to his credit, Keith's dedication and perseverance has been well-rewarded, but these successes haven't always come easily. Autocross and Rallycross cars tend to be most competitive when they have big torquey engines, and as a result, Keith has run the gambit of large Mini engines, having tried everything from 1275 to 1600cc to power these cars.

Keith's editing on this book has been from the point of view of a man who knows what it takes to win as far as Rallycross is concerned. His race car was truly a test bed for the speed equipment manufacturers' parts that he sold from his retail speed equipment shops. As a successful racer, Keith knows what works and what doesn't; which parts will live and which won't.

Although personally involved in Rallycross and Autocross, other forms of racing and competition most certainly are not ignored at Ripspeed. Probably due to the now international status of his company, Ripspeed were, in 1982, appointed as the official Austin/Rover Group, MG Metro Challenge parts supplier for Europe.

Steve Harris

Although Steve Harris has been a race champion twice, set lap records and won many races, his forte is the building and development of high performance engines.

Steve started Steve Harris Motor Engineering in 1976 and to date his engines have well over 200 wins to their credit. Steve learnt his trade the classical way. In 1964 he started an apprenticeship at Downton Engineering, working under Daniel Richmond on the experimental and development aspects of the Series 'A' engine. Of course, a lot of the work being done there then was for Leyland (then called BMC). While there, Steve was responsible for preparing the Group 11 Mini Cooper 'S' for Gordon Spice which ran so successfully. When Richard Longman started Richard Longman & Co., Steve went into the company as a partner. In 1975 Steve was asked to run Downton Engineering after Daniel Richmond's untimely death. In the hope of greater things to come, he accepted this. This, however, proved not to be the case and in 1976 Steve started his own company. Though small, his company has a keen following of racers who appreciate his work. He spends a lot of time on his flow bench and dynamometer, developing successful engine specifications for many classes of racing. Steve employs a tenacious and methodical approach to his engine building and it is this approach that his most loyal customers feel is the quality to which he owes his success. Sure, Steve Harris's is a small company in terms of people, and it may always be small if it's kept on such a personal level with the customers it deals with, but in stature it is growing progressively.

Steve's contribution to this book is the confirmation or addition of numerous little details that lesser engine builders tend to pass up as irrelevant. The knowledge of such details is brought about by 20 years of personal, successful, experience with almost as many years working with some of the best that the rest of the industry has to offer.

Keith Dodd

Keith Dodd is Mr Mini Spares Centre, a company which also owns and runs Competition Silencers. Keith started in the motor trade 21 years ago, working within a BMC distributor network. This is about the time the first Coopers arrived on the scene and these sparked Keith's enthusiasm. Since then he has worked either within the Leyland network or at his own business, almost exclusively on Minis. In the late-60s and early-70s, Keith became well-known for his acute knowledge as a parts man; a walking parts computer not only on standard parts but also special tuning parts and performance after-market parts. In other words, if there was a part available to do the job, Keith would know what it would do, how it would do it and where it was obtainable.

After setting up the first special tuning dealership in London, Keith realized that a specialist parts business was needed that could cater for the people who owned and raced Minis, simply because, at the time (1974) most BMC/Leyland businesses seemed unable to cater for the enthusiast or racer in sufficient depth to give them a wholly satisfactory service. Many parts men were doing the job just as a job and seemed too baffled to understand the vehicles' peculiarities, the change points for different modifications and models, etc. As a result of this, in 1975 Mini Spares Centre of London was started. The growth of this company is almost legendary. Not only does it now supply vast numbers of original Leyland spares but also has manufactured to original drawings those parts which Leyland no longer choose to supply.

Keith's interest in the Mini and the Series 'A' engine is wide. He covers everything from restorations to racing. As you can probably guess, Keith's input into this book took the form of part number checking, interchangeability factors, introduction dates, availability confirmation and so on and so forth. In fact most of the facts that relate to which, when and where.

David Mountain

David Mountain is the newcomer to the scene. His enthusiasm, dedication and willingness to discover new things, either from his own efforts or learning from others, could well carry him to the top of his profession as a race engine builder. His heroes are the men he shares these pages with. Dave started his career in 1979, building engines for friends as a hobby. He read everything he could about making engines go fast. In the same year he commenced work at Swiftune Engineering who at the time were building some of the most successful Rallycross

Mini engines. In 1982 he began working for himself, setting up a company called Mountune Race Engines.

In order to be a successful race engine builder, Dave believes you must have a healthy and progressive attitude towards the subject. He believes in a logical approach to everything he does, and if there is any doubt after the application of copious quantities of logic, try it all the same. He will always consider and look at what other people do who are working in the same field.

Dave believes there is almost always some degree of merit in any intelligent modification, but that some engine builders, professional and otherwise, have such delicate egos that they find it difficult to accept that someone else may have come up with a good idea, so said idea is rejected out of hand. This, he feels, has hampered the progress of many potentially good engine builders. It's a trap he doen't intend to fall into. Well, has this philosophy worked for him? It would seem so. Not only has Dave in his short career, built a number of Rallycross and Autocross championship-winning engines, but also since the formation of his own company in 1982, he has virtually dominated 850 Minicross racing. In 1983 alone, he took first, second and third in the Brands Hatch Championship, first, second and third in the Lydden Hill Championship, and first, second, third and fourth in the Northampton Championship. This, plus numerous other wins from engines from Dave's company suggests he is going the right way. Dave's deep knowledge of racing 850s in this class really helped fill out this book in this area.

Paul Ivey

If you ask Paul Ivey what he does for a living, he'll simply tell you he's an engineer. That's a gross understatement that should be followed by 'race engine builder and dyed-in-the-wool performance enthusiast'. Paul runs several companies, namely Speed Sport Conversions, Race Engine Components and Specialized Valves. All these companies cater for the performance needs of the more dedicated and serious Series 'A' engine builder.

Paul started in business for himself in a general sort of way in the early-50s and during 1957 he started racing. He won his first event at Mallory Park in an Elva Courier sports car and he scored many wins and places in both circuit racing and hill climb events, driving a variety of machines from Lister Jaguars to F2 Coopers. As time went by, driving took a back seat to engine building. If you talk to Paul, he claims he learnt most of his skills as an engine builder by working with the late Daniel Richmond, who he, along with many other people, still claims is the Father of modern Series 'A' tuning.

Paul joined Downton Engineering after doing his National Service in the RAF and it was while working at Downton Engineering that he helped work on the first Austin/Morris conversion that Downton ever produced.

Paul started Speed Sport Conversions around 1961. Race Engine Components followed some time later, and Specialized Valves came into being as a limited company in 1980.

When it came to Series 'A' engines in general, and Mini engines in particular, Paul's forte was exotica. Dry sump Minis, 8-port fuel injected Series 'A' engines, Lotus twin-cam powered Minis, four valve per cylinder engines, etc. You name it and if it's in any way exotic, the chances are Paul has dabbled in it at one time or another.

During the time Paul has been building race engines, he has lost count of the number of wins they have scored.

Paul's input into this book is simple: what gets results and what doesn't; what's practical and what isn't.

Introduction – where this book will take you

Picking up a book like this from the bookshelf, one that very obviously deals with a technical subject, you may wonder if this book is for you. When putting it together, I was very careful to make sure the contents could be understood by a wide audience. For instance, it was necessary for me to make sure the reader did not have to have the skills of a typical Formula 1 race mechanic or the university learning of an engine designer. **The most basic assumption I made about the reader was that he or she would be of normal intelligence, and having made that assumption, I then went on to write the book for people whose background may stem from all walks of life, i.e. butcher, baker, candlestick maker, lawyer, painter, builder and even the undertaker!** In taking this viewpoint, it was

If you want to build an engine for 850 Mini-Xing, this book will tell you how. It will also tell you ...

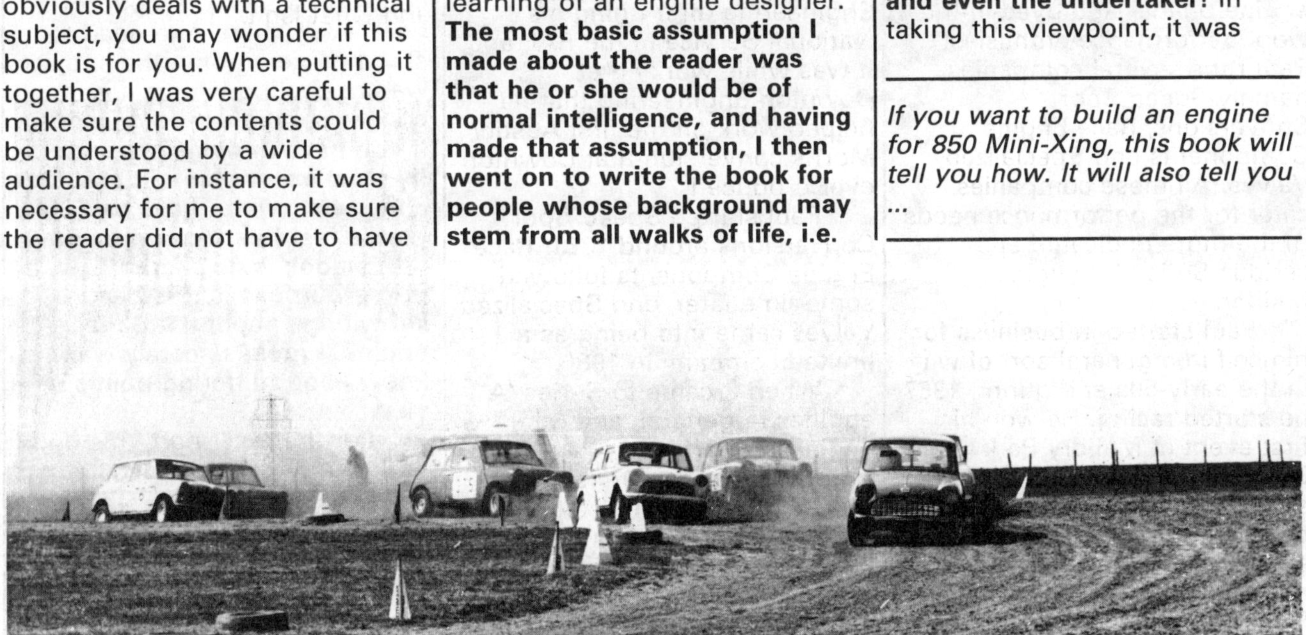

... how to build a supercharged drag racing engine, or ...

... an engine for drag racing a Mini.

If circuit racing is your cup of tea, then the ingredients to build a Metro racer are within these pages, as also are ...

necessary to understand that some people, due to lack of exposure to the subject, would know very little about high performance engines and what it takes to make them tick. On the other hand, I was also expecting this book to be read by people who are professional engine builders, even if only to compare notes, and they too may find new information within these covers.

Many people who read this book will undoubtedly be experts on the Series 'A' engine. No doubt some more expert than I. Catering for such people is not easy, but I've tried to do so not by the complexity of the information this book conveys, but rather by the depth of the information. In fact, throughout the entire text, it has been my intention to put things in such a manner that they can be simply and easily understood by all.

Should the reader be an expert in one particular area or another, then I hope I have put the text down in such a manner so as not to bore him by covering ground that he is already familiar with. Of course there may be many experts who read this book who have views contrary to those I've put forth here. Well, there's no need to stand on ceremony: the subject we are dealing with is a vast conglomeration of complexities. **If you have findings supported by solid experimental data, that contradict what I have said, please feel free to let me know – this book is no sacred cow.** Its prime purpose is to spread knowledge about the Series 'A' engine. I most sincerely want to know about differing points of view.

For the most part, this book

... the ingredients for a 140 HP hillclimb Mini that you can drive on the road.

contains data – largely my own – accumulated from experiments, tests and experience, but there are times when experimental data is insufficiently well defined to allow absolutely concrete conclusions to be drawn. In these instances I have put forward what I hope to be at least an intelligent guess, and of course we all know that the one single word for an intelligent guess is - 'opinion'. Well, where it's an opinion, I have tried to make it quite clear that that is the case.

The chapter heading did pose a question which I have not really got to, namely 'Where Is This Book Likely To Take You?' Well, if you are a beginner in the field of tuning engines for high performance or economy, whichever your requirement may be, then hopefully this book will put you fair and square into the expert class. **Of course it does make the assumption that whatever job you do, as described in this book, you do it properly.** Remember, I am giving you the information to do the job here; the workmanship you have to supply for yourself . . .

If it's a big engine you are after, up around the 1600 cc mark, just as Stud Nicolaou uses in the Ripspeed autocross Mini, then all the information you will need is right here.

If you are inexperienced at building engines, then there may be another commodity that you are short of, namely judgement. It is all too easy, when building your first high performance engine to go over the top. It's an unfortunate fact of life that most performance enthusiasts are power-hungry, and they see the big number at the top of the power scale as the only one of any importance. This number may be good to brag about whilst you are propping up a bar and talking to your buddies, but it's not necessarily the most satisfying number in terms of how the car feels when you are driving it. Until you know exactly what your end product is going to be like, from experience, it is best to be on the cautious side. For instance you may think you want a full race cam in there to get maximum hp for your street burner. Well if that's what you think, take two paces back and use a middle-of-the-road high performance street cam. If you think you would like to spend 200 hours grinding away on a

Of course some potential engine builders may want a highly refined, but high performance engine that's in keeping with some of the world's most refined kit cars. This book has it, as it also ...

cylinder head to use the hugest valves possible and you are prepared to risk going into the water jacket, don't. Just be more conservative and grind a cylinder head you know won't break into the water jacket. You'll like the results a lot better because you'll get to drive the engine that much sooner. Don't fall into the trap of thinking that bigger is better. Try and override the impulse of thinking that if some's good, more's better and too much is just right, because more than likely it will not work that way.

If you take these hints in the spirit in which they are

. . . has the ingredients to build a car that can better 80mpg at 30 mph, just like the author's economy Mini project car.

given, then, **even as a novice, you will, if you absorb the information within these pages, very quickly acquire the skills of tuning the Series 'A' engine that have taken some professionals 20 years to obtain.**

If you consider yourself to be fairly well versed in the art of extracting more from the Series 'A' engine, in other words you consider yourself an expert, what does this book have to offer? Well, first and foremost, it will hopefully tell you something you didn't know, simply because it is not a product of just one mind, but many. There are probably thousands of little facts within these pages that you already knew, but over the years you have just never had time to record them, write them down, corrolate them or look at the inferences they may have. Hopefully, in writing this book, I have done a lot of your engine building 'housekeeping' for you. Next time you want to find out how many cc a combustion chamber should have, you won't have to go through all the calculations; it's all in here for you. Next time you need to make a cam choice, you won't have to search through all those oily pieces of paperwork that you left lying around your tool box somewhere; it will be in here for you, and so on. You should also be able to equate and compare your own findings and experiences with modifying the Series 'A' engine with that of many others. That in itself should give you new targets, and even new ideas.

1 The History of the A-Series

Without going into the most minute detail, the history of the Series 'A' engine is of interest as it shows how the engine has developed over the years. Usually development takes place mostly as a result of competition. That's not competition on the track, but competition within the market place. This is competition manufacturers have to face if they are to stay in business. Such competition is brought about largely by the fact that it is necessary for a manufacturer to continue to supply what its general public customers want. If it doesn't, another manufacturer will simply step in. As much as anything then, the pressures that caused the evolution of the Series 'A' engine to the stage where we see it today were social.

The first Series 'A' engine to be introduced to the public became available in 1952 as an 803cc pushrod, overhead valve unit, and although it bears very obvious family ties with the current production Series 'A' engine, it was, in fact, a very different animal. Its 2.28 inch bore, along with a 3 inch stroke, meant that it was very much a long stroke unit with a bore/stroke ratio of 1.316:1. The small bore meant only small valves could be accommodated in the cylinder head, and the relatively long stroke, together with a crankshaft which was just a shade on the spindly side, meant that the rpm potential of the engine was somewhat limited. Sure, this engine can be modified, but it should be treated with a great deal of care because it is a delicate unit.

Even in the early days of performance tuning, when rpm was restricted by what are now seen as rather antiquated crankshaft designs, the crankshaft of the 803 engine was often described as 'a bent piece of wire with some journals ground on'.

This engine, which was used in the Austin A30 from 1952 to 1956, also found its way into the Morris Minor and occupied the engine bay of that vehicle from 1953 to 1956. If you have one of these early engines and you are contemplating modifying it, do so, but do bear in mind that its rpm potential is very limited. Of course most people wanting to modify these early engines would be those restoring vehicles of that era, in which case you would probably want

The search for power was on even before the advent of the Series 'A'. This is the twin-carb Derrington conversion, later popular on the Series 'A', but here seen fitted to its forerunner, the Series 'E'. (Ray Newell)

the engine to look absolutely standard on the outside. This being the case, it is best to tune these engines for torque output. In other words, improving the breathing efficiency by virtue of small internal mods is fine, but camshafts with significantly extended valve timing should be considered out of court.

The biggest difference between the Austin and Morris 803 engines, in fact virtually the only difference, is that the Austin version used a fixed choked Solex/Zenith carburettor, which cannot remotely be considered a performance item, whereas the Morris version used an SU carburettor and, as a result, produced 3 more hp. Even though it wasn't the world's best engine to modify, the early Series 'A' engine in 803cc form did inspire quite a few companies to produce suitable speed equipment. In fact, many companies that were to become big names in the tuning world started off by turning their hands to 803 modifying.

In 1957 the 'A' Series engine grew to 948cc, principally by virtue of an increase in bore size to 2.477 inch: up nearly 3⁄16 inch on the 803 engine. Along with the increase in bore size also went some toughening-up of the bottom end, to the point where the engine was good for a relatively reliable 6500rpm. The 948 engine found its way into the A30 which then became the A35 and the Morris Minor which became the Morris Minor 1000, but BMC as it was then, also had new models planned and the 948 engine was used to power the all-new Austin Healey Sprite and Austin A40 Farina. Such companies as **Alexander Engineering, Speedwell, Aquaplane** and a little later, **Downton Engineering** all began turning their attention in earnest to this engine. Factory involvement with the high-performance aspects of this engine, through a competition department, also meant that speed equipment began to acquire a form of respectability.

During the years 1957 and 1958 a new, relatively cheap, form of single-seater racing was conceived, planned and brought to fruition. Its intention was to train and encourage young hopefuls with their eyes set on the future Formula 1 championships. The new formula was known as 'Formula Junior'.

John Cooper was a name already firmly entrenched in Formula 1 racing, but at this time he became deeply involved with not only the running of a Formula Junior team, but also the manufacture, design and selling of Cooper Formula Junior cars. Of course many of the early Formula Junior cars used modified Series 'A' engines. A great many of these were based on standard units but there were quite a few 'specials': built with different bore/stroke ratios, special cams, EN40B nitrided cranks, special rods, forged pistons, etc. In all, it seems a few dozen of these special A-type engines were built and utilized in Formula Juniors.

As you can imagine, Formula Junior racing gave quite a boost to the 'A' Series engine as a performance power plant. As the end of the decade approached, sophisticated performance equipment began to appear and at least two companies went to the extent of manufacturing aluminium cylinder heads for the Series 'A', though many people thought at the time that it was unfortunate that these alloy heads were still of 5-port design.

1959 saw a change in the Series 'A' engine and its introduction in a car that probably did more to promote it as a performance engine than anything before, or since. De-stroking the 948 engine down to 848cc, a change in crankshaft pattern to allow a revised clutch design to be installed, and the mounting of the unit on top of a gearbox so as to allow transverse installation in a car produced the basic power unit for the new model. As most of us know, the car was the 850 Mini.

As far as engine parts were concerned, the 850 engine shared much of the 948cc unit's components. The cylinder head and valve sizes were the same; the carburation of both Austin and Morris vehicles was now by SUs, rather than the fixed jet Solex/Zenith type carbs (thank goodness!). With the Mini's lightness, compared with other vehicles utilizing the Series 'A' engine, it easily out-performed BMC's other Series 'A'-powered small cars. In fact, the only car that could outpace the Mini in a straight line at that particular time was the Sprite, and much to everyone's embarrassment, the new four-seater Mini saloon car could out-corner, with ease, the Austin Healey Sprite!

With the introduction of the 850 Mini, the performance equipment business in England changed into high gear. The Mini's capable chassis just asked for more hp. Most of the vehicles that preceded the Mini could easily be embarrassed by a little more power situation simply because contemporary chassis were not up to coping with much in the way of extra power. On the other hand, the Mini had, for that era at least, tenacious road holding, and the 34hp it was supposedly endowed with was far less than it could constructively use. While many of the tuning firms were getting to grips with 850 Minis and finding ways and means of making them perform better, John Cooper, with his inside contacts, was taking a different approach, though with the same end result in view. Rather than tell the story myself, it might be just as well to quote John's own words so you can see just how the next phase of Series 'A' performance got off the ground . . .

Let me set the scene: it's 1959 and now, in John's own words: 'Because of my associations with BMC and the Series 'A' engine, I managed to borrow a prototype Mini before the car was announced, and went to the Italian Grand Prix in it. Roy Salvadori and I drove the thing. In fact, Salvadori beat old Reg Parnell and his Aston Martin over one of the mountain passes. Anyway, we got there. I saw old Lampredi who had been Chief Engineer for Ferrari, but by then he had left Ferrari and was working at Fiat. He said, 'What's that thing you've got there?' I replied, 'It's the new Austin 7' though I really didn't know what it was going to be called. He asked, 'Can I try it?' and promptly went off along the roads around Monza, disappearing for about half-an-hour. When he came back, he said 'You know, John, if that car wasn't so darned ugly, I'd shoot myself', then added, 'This is the future car.' He was so right . . .

'It was then that I got the idea, and I went and saw Alec Issigonis and sort of suggested, how about a special Mini, one a bit better and faster for the boys, you know. So Alec and I saw Sir George Harriman, or George Harriman as he was then, and suggested that we build a Mini with a bit more steam, better brakes and so on, and he said, 'Well, take one away' so I took one away. Producing more hp from the engine was no problem. We had all that technology to hand. It was easy to get another 20hp. We made up a remote gearchange and stuck it on the car and, as luck would have it, the brakes were no problem because Lockheed at the time were developing a disc brake to go in a Mini wheel, purely as an engineering exercise, to prove to the world that if a disc brake could go in a Mini wheel, it would go into any; we got disc brakes on it just like that [snapping his fingers], you see.'

Well, any student of automotive history will know that what John Cooper had started there, was, in fact, the Mini Cooper, and in 1961, the 997 Mini Cooper was introduced. The bore of the 997 was slightly smaller than that of the 948 engine, and the stroke at 81.24mm (3.2 inch) was quite a bit longer. On the face of it, a strange move, particularly as the Ford Motor Company in Europe had just brought out a range of engines embodying the latest thinking: big bore and short stroke. At the time, short stroke engines were becoming all the rage because of their obvious assets for high performance. They allowed bigger valves to be installed and short strokes meant less piston inertia problems at high rpm. So here was the new 997 Mini Cooper with what was very much a long stroke engine. In fact its bore/stroke ratio of 1.302 was almost as long as the old 803 engine. When questioned on the subject, John Cooper said: 'Basically speaking, it tied up with BMC's production facilities. Don't forget that the engine was made at Longbridge on a production line and had to be compatible with the existing equipment for making engines. I think it was all tied up with the Churchill boring machines. In other words, it was obviously not the best way to go, but it was the way they chose to do it for production simplicity. I personally can't tell you any more than that. That's the way they wanted to do it.'

The 997 Cooper was essentially a performance road car and it would seem there was no real thought of producing it as a machine with maximum competition potential. However, as one would expect, the 997 Cooper was used for racing and successfully so at that, but because it had really been intended as a high performance road car it fell short in many areas. Stuart Turner, who ran BMC's Competition Department, really wanted something more purposeful for the rally team. This, coupled to the fact that the Mini had obvious competition potential, soon brought about the suggestion that the performance potential of the basic Series 'A' design should be plumbed in more depth, so they went ahead and built the 'real thing'. Well the real thing was, of course, the 1071 Cooper 'S', introduced in March of 1963. Much of the technology that went into the 1071 Cooper 'S' stemmed from the special 1100cc Formula Junior engines built some years earlier for John Cooper. Although externally very similar to all the Series 'A's that had gone before, the 1071 'S' engine was vastly different under the surface. The engine was much more of a big bore, short stroke unit. In fact this was the first over-square Series 'A' engine with a bore/stroke ratio of 0.967:1. The nitrided steel crank featured bigger main bearings than the Series 'A' engines that had gone before and the big bores allowed the use of much larger valves. The cylinder head was designed from the outset to be conducive to good performance, and the net result was a Series 'A' engine with incredibly high rpm potential. As most of us have subsequently found out, the rpm potential far outstrips the other capabilities of the engine, thus making the 'S' engine a very 'safe' engine to modify for high performance.

The 'S' engine did, in fact, start a whole new family of engines in the Series 'A' line. A group of engines that were subsequently to become commonly known in public circles as the big bore engines. Of course this leaves all engine types prior to the introduction of the 'S' engine to become known as small bore units. Although the 'S' engine was the big news of 1963, it wasn't the only significant factor in Series 'A' evolution. Amongst the small bore engines, as they can now be referred to, several major changes were made.

Firstly there was the introduction of the 998 cc engine. This utilized the same 3 inch stroke that the 948 engine had used, but it had a bore of some 1.65mm (65 thousandths of an inch) larger diameter. This engine was initially used in the then newly introduced Riley Elf and Wolseley Hornet. **In essence the 998 engine could be viewed as an overbored, stroked 850, as the cylinder head design was pretty much the same as the 948 and 850 engines that had gone before.** A number of bottom end design changes were made, the most significant being connecting rods with fully floating piston pin design.

Along with the introduction of this engine, an 1100cc engine was also produced. This unit was a stroked version of the 998, utilized the same bore, but with an 83.8mm (3.3 inch) stroke. It was used to power the new hydro-elastic suspended, front-wheel-drive, transverse-engined, 1100 saloon.

With all these changes going on, it didn't surprise anyone when the 997 Cooper was discontinued in January of 1964, and in March of that year, a Cooper with the 998 engine was introduced. On the face of it, only 1cc different, but let us not forget that the 998cc engine has a much shorter stroke and bigger bore than the old 997. Essentially the bottom end of the 998cc engine was as per the unit used in the Riley Elf and Wolseley Hornet, the only significant difference being that D-section, raised crown, high compression pistons were used rather than the dished pistons of the lesser performance models. Along with these raised crown pistons, a new open chamber cylinder head was used which sported a larger intake valve than had previously

Mike Cannon's Mini displays a typical rear wheel-lifting trait when provoked hard during a hillclimb.

An S.C.C.A. Midget powering through turn 6 at Riverside Raceway in Southern California.

been used on any of the small bore engines. The design of this new (12G 295) cylinder head bore a great resemblance to the Cooper 'S' cylinder head, although the valve sizes involved were smaller.

Along with the introduction of the 998 Cooper, came the MG Midget MkII and Sprite MkIII 1100cc engine. This engine used the same high performance cylinder head as the 998 Cooper and from engine numbers with the prefix 10CC had 2 inch main bearings as for the 'S'. (The earlier engine number prefix was 10CG.)

At least on the face of things it looked like the pattern of engines had settled down, and really few people expected anything new on the Series 'A' front for quite some time, especially in view of the intense amount of activity there had been in developing the Cooper 'S' engine. Well anyone who had made that assumption was mistaken, for in March of 1964 the 970 and 1275 Cooper 'S' engines were introduced. To be honest the 970 looked like it was something of a 'homologation-special'. It was built in substantial numbers solely to justify its existence in International 1000cc saloon car racing where it did exceptionally well. The real killer in the Mini range was, of course, the 1275 Cooper 'S'. This engine/chassis combination was instantly hailed as the world's greatest fun car to drive, a position which it was to retain virtually unchallenged for at least a decade, and even now, 20 years after its introduction, a well set up 'S' is still difficult to beat in terms of sheer driving pleasure and fun.

Technically the 970 and 1275 'S' engines differed little from the 1071 engine. The 970 engine size was achieved by destroking the engine to 62mm (2.44 inch). With its short stroke and relatively big bore this diminutive 'S' can be made to wind some horrendously high rpm without doing itself a mischief. To get the pistons to come to the top of the bore on a 970 'S' it was necessary for the designer of the engine to use a slightly longer rod together with a block about ¼ inch shorter. For the 1275cc engine, the opposing viewpoint was taken: i.e., take one 1071 engine and increase the stroke to produce a 1275.

To say that the 'S' engine, especially the 1275 'S', was a fully sorted engine when it was introduced, would have been quite a way from the truth. It was, for a production engine of its type, quite radical in many ways and produced a lot of hp. Such problems as valve seat cracking, cylinder head overheating and block flexure had to be cured, and as time went by, so they were to a large extent. However, the 'S' engine in all forms was an expensive engine

Even in standard trim the Mini provided a competitive mount in such events as this local rally.

to produce. The nitrided EN40B crank which was almost bullet-proof, had to be paid for dearly. However, engineers at the factory readily accepted that the 1275 formula for the Series 'A' was a good, sound one and so, in the late '60s, the non-'S' 1275 engine was introduced. This engine had the same bore and stroke as the 'S' but it used much cheaper materials and was less refined in many areas. Gone was the EN40B nitrided crank and in its place was a crank of more conventional material. However, to make up for some of the loss in strength by going to a cheaper material, the crankshaft was given big ends ⅛ inch bigger. Early engines with the engine number prefix 12G had big end journals the same as the 'S' at 1.625 inch, but the cranks were in EN16. Some of these engines were supplied in Austin 1100s in 1969 as an option. The material used in the rods was down-graded, although this has subsequently never proved to be a problem, thus indicating that the Cooper 'S' rods were overly strong in the first instance.

In the cylinder head department, a 12G 940 casting, very similar to the Cooper 'S' casting, was utilized, but the design was refined slightly. **Valve sizes went down just a little (33mm inlet, 29mm exhaust) and, in so doing, the cracking problem that the 'S' cylinder head had, was virtually eliminated.** An engine of very similar specification did, in fact, find its way into the last Cooper 'S' models. The standard 1275 GT crank was given a tuftriding treatment and this produced a crank adequate for all uses including competition. On the last of the 'S's, that's the MkII 'S', the intake valve size used was larger than those normally found in 1275 non-'S' engines. Essentially the intake valve size was that of the original Cooper 'S' (35.6mm), but exhaust valve size remained that of the 1275 non-'S' engines (29mm).

By now, largely taking the place of the 1275 'S' was the new Mini 1275GT Clubman although, unlike the 'S', the 1275 engine in the Clubman sported no real performance equipment. It was, in fact, the basic 1275 as could be found in any transverse-engined 1300 saloon.

With the demise of the Cooper 'S' in 1971, no twin carb transverse engine was seen in production. Only the 1275 Midget and Sprite engines in the Leyland range used twin carbs. When the Sprite and Midget ceased production in '69 and '74 (for Series 'A' engine versions), this left British Leyland with no real high performance Series 'A' engines in their range, but toward the end of the '70s the introduction of the Metro was looming into view. When the Metro did finally hit the market, it did so with a bang.

A lot of refinement work had been done on the Series 'A' engine. Some of it introduced prior to the introduction of the

The latest in performance from the factory: Leyland's Turbo MG Metro.

Metro. For instance, the engine block of the Metro had been upgraded by stiffening, but these modified blocks had been used on the production line for as much as 18 months before the Metro had actually been announced. Specificationwise at least, the Metro engine was very obviously an engine to power a non-sporting saloon, but the significance here is that the engine power output had been substantially increased along with its economy potential while detracting little, if any, in other areas. The new Series 'A' engine was to become known as the 'A+'. With rolled radius (supposedly stronger) crank, stiffer blocks and improved induction and exhaust systems and revised cam timings, the new Metro engine did, on a single carburettor, produce substantially more hp than the old 1275 engine. Ignoring British Leyland's figures, many independent dyno tests have shown that the standard Metro engine is capable of a solid 62hp.

1982 saw the re-emergence of the MG marque, and along with this, the Series 'A' engine gathered some more hp. By utilizing some of the tricks known to the performance industry for many years, Leyland managed to boost the power of a standard 1275 engine, still on a single carburettor, to 72hp. This gave the aerodynamic MG Metro a very creditable performance, but the MG Metro wasn't

A Turbo Metro engine on a dyno at Leyland. Over 30 years development separates this engine from the original 803cc unit.

Another approach to blown power. This is a 1275 Series 'A' fitted with Shorrock C75B supercharger to produce 105hp at 5,800rpm. (Colin Moore)

Leyland's ace in the hole. Late in '82 the turbo Metro made its debut. A substantial amount of re-engineering went into the Series 'A+' engine to make it withstand the rigours of 7lb of boost. Engineered as a drop-in package, the turbocharger on the uprated 'A+' engine endowed the Metro with 93hp: enough power to propel this aerodynamic machine at speeds around 110mph, thus making it the fastest MG saloon ever to be produced!

At 93hp in turbocharged form, the Series 'A' engine is a far cry from the original 803cc, 25hp engine that saw the light of day over 30 years ago.

2 What it takes to make power

If you were to read through this book without realizing that in almost every instance the goal I am trying to achieve is more hp, you would think (and rightly so) that I was obsessed with airflow. All the way through the chapter on ram charging, air filters, carburettors, induction systems, cylinder heads and exhaust, the airflow potential of the various parts is, sooner or later, brought into the discussion.

I assure you this is no obsession, as you will see when we delve into the detail of how an engine's performance is increased.

Let's start off with a very basic assumption. Why do engines have carburettors? The simple answer is that they are there to mix fuel and air. Why is it necessary to mix fuel and air? Again, the simple answer to that is, fuel will not burn on its own; it needs oxygen to burn. The oxygen is acquired from the air drawn into the engine. For a given amount of air intake, the engine can effectively burn a given amount of fuel. It is the action of the burning fuel causing the gases to heat up which increases the pressure in the cylinder, thus pushing the piston down the bore on the power stroke. The greater the amount of air drawn into the cylinder, the greater amount of available oxygen there is to burn fuel. The more fuel that can be burned, the greater the amount of heat generated and therefore the higher are the pressures generated in the cylinders. Greater pressures mean higher hp by virtue of higher torque. The bottom line then is that the more air the engine can inhale, the better off we will be from the point of view of extracting power. Breathing efficiency boils down to utilizing components which will allow air to flow as efficiently as possible into the engine, hence the apparent obsession with airflow capability. Any time a component does not flow air effectively it will rob the engine of some of its potential as a power producer.

It would be wrong to think that airflow is the sole key to horsepower. It just happens to be one of the most important ones. Like any complex device, if we fail to produce results in one area then the overall result can be less than hoped for. There are in fact many other things that we have to consider and it is best when discussing various topics to refer to the power production flow chart Fig. 2-1 A. In this chart, we go through major factors that have to be considered in the development of high output from any engine. Not just the Series 'A'.

Let's start at the top of the chart. Here we have the atmosphere which supplies the oxygen on which the engine runs. In passing this into the engine we have to move on to the step 2 which involves maximizing the air density. In this context it is difficult to cool the air below that of the prevailing (ambient) temperature of the surrounding atmosphere. However it is worth putting in the effort to see that the air isn't heated any more than necessary, thereby maximizing its density.

The next step involves minimizing intake flow restrictions. Here's where our flow bench work on the induction system, intake port and to an extent the combustion chamber, takes place. Also in terms of induction effectiveness we have to maximize pressure wave tuning. For 'A' Series engines with siamese port heads this takes a little different route to the norm. However if you intend employing one of the more exotic heads having seven or eight ports or even four valves per cylinder, then intake pressure wave tuning can account for substantial amounts of horsepower. If it's done properly a power increase in the region of 10 to 15% is not by any means an optimistic estimate. Next we come to our first box. Everything within this box labelled 'friction' really pertains for the most part to the long block assembly. Everything within this is affected by friction so must be dealt with on an overall basis.

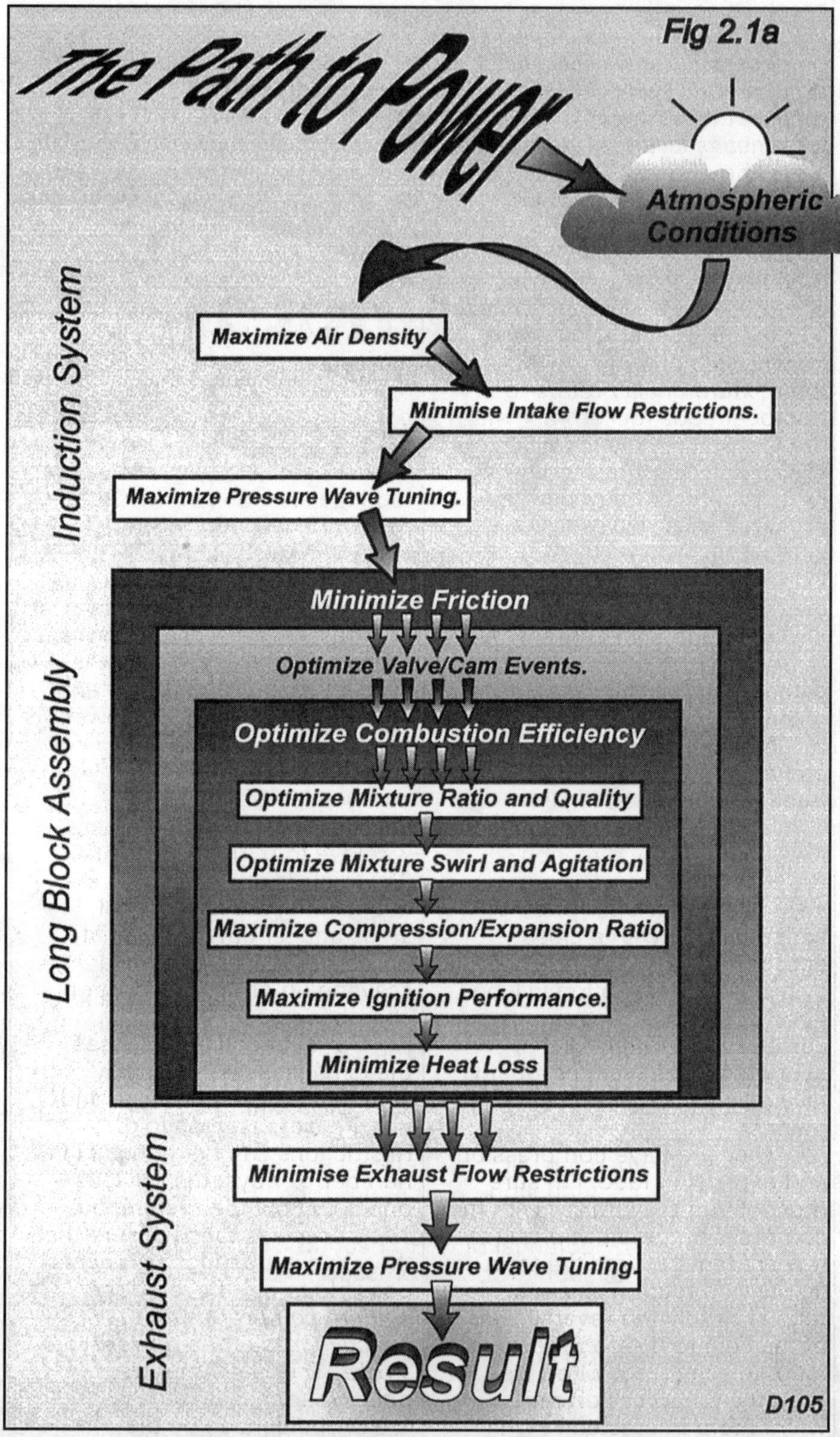

The next box within the friction box is the valve/cam event box. Again this is very much a factor which has to be dealt with simultaneously on both the intake and exhaust. To be successful all the events have to happen at the appropriate time in relation to the crankshaft rotation. This makes cam events more of an overall factor rather than something pertinent just to the intake.

Now we're getting down to the real core of the process of the heat engine, that of combustion efficiency. In the combustion efficiency box there are five factors which have to be dealt with, and failure to optimize any one of these means reduced torque which in turn means less power. In some areas the output may only be affected minimally, in other areas ill effects could be substantial. For instance the mixture ratio has to be closely controlled within narrow limits and the mixture quality has to be *what the engine wants*. Here the mixture quality refers to a homogeneous fuel/air mix with the fuel sufficiently atomized for the engine's combustion requirements but not over atomized so as to cut volumetric efficiency. Establishing the best mixture quality for power or economy is a major issue with five-port 'A' Series engines so it will be dealt with in detail in the carburation chapter.

Next we have mixture swirl and agitation. Fortunately, the traditional 'A' Series-shaped combustion chamber and a typical port entry direction produces very good mixture swirl. If the block assembly is built correctly with a close approach of the piston to the cylinder head, agitation will be good and very effective combustion will, given that everything else works properly, take place in a fast and efficient manner.

Then we have compression and expansion ratio. I'm sure most of you are familiar with the compression ratio but probably few are familiar with the term 'expansion ratio'. The expansion ratio is simply the reverse side of the coin to the compression ratio. In actual fact when we raise the compression ratio we're more interested in what it does to the expansion ratio as it has a significant effect on our optimal valve event timing, and the ratio of the cylinder head's intake and exhaust valve sizes. Again we'll go into sufficient detail in the chapter to arm you with the working knowledge needed.

Next we come down of maximizing ignition performance. Here we have to insure the engine has more than just an adequate spark. Once a more than adequate spark is established any extra spark output will really serve no measurable purpose in terms of increased engine output. However anything short of adequate will cost power!

The last factor in the combustion efficiency box is the minimization of heat loss. This is an area which has been given too little attention of the past, mostly because the materials to improve it were not really available. They are now. So we'll look at minimizing the combustion heat loss from the chamber so that more heat is available to expand the air and thus deliver more horsepower.

We have now passed through the friction and combustion efficiency box and into the exhaust system. Here we have to minimize exhaust flow restriction while maintaining cross sectional areas and lengths appropriate for effective pressure wave scavenging of the combustion chamber.

For all practical purposes the exhaust system actually starts at the exhaust valve and not the exhaust manifold. Restrictions can be reduced in a number of ways, the obvious one is to flow the system, but another one is to make sure that the valve is open for as long as possible to get the job done *without compromising the effectiveness of the power stroke*. This is an important point to appreciate as it is greatly influenced by the compression and expansion ratio. *Understanding the implications here can make the difference between 105 horsepower per litre and 140!*

Last we come to the pressure wave tuning on the exhaust. This is a very important factor. It's not commonly realized that exhaust pressure wave tuning has a far greater effect on engine output than the intake. A normal intake system with one port per cylinder is difficult to tune over a band width of more than 400 or so rpm wide. However, if the exhaust system is correctly done it can be made to effectively scavenge the combustion chamber of spent charge and pull in a greater charge over a power band as wide as 4,000 rpm – sometimes more. The following pages deal with all of these things and hopefully, by the time you reach the end of this book, you'll have a better understanding of how they all interact and how they fit together to make what an experienced engine builder calls 'the right engine combination to make horsepower'.

Well, so far things look simple enough. Now let's move onto the subject of torque and hp. I mentioned these openly earlier on as if you would, naturally, know exactly what they were. In fact they are terms that are bandied around quite casually, but few people outside of professional engineers or engine builders really know what the exact definition of torque and hp is. Let's deal with torque first, as this is the simpler of the two.

Torque, simply, is the turning effort that an engine is capable of. It's no different from the torque you exert on a spanner when you tighten up a nut. Use a torque wrench to tighten that nut and the torque wrench will tell you how much torque is being applied to the nut. The torque you exert on that nut is a function of the force that you apply to the end of the wrench, multiplied by the radius about which you are applying it. That will be the length of the torque wrench in this particular instance. Just to run through an example, if you

For a rally car, all-out hp is not the main criterion. Since most rallies cover a fair distance from start to finish, a rally engine needs to have power but with high reliability.

pull on a 2 foot torque wrench with a force of 50lb, you will apply 2 x 50lb ft. That works out to 100lb ft. You'd probably feel quite impressed at your own prowess if I were to tell you that you can, with a suitably long lever, exert far more torque than a Series 'A' or, indeed, any engine. You may conclude from this you have more power than a typical engine. Doesn't sound right does it? The key factor there was that you selected the lever arm length. The lever arm length in an engine is, to all intents and purposes, half the stroke length i.e. the distance from the crank centre to the big end centre line, and the force the engine exerts at that radius is due to the pressures in the cylinder. The stroke length can't be changed, at least not without redesigning the engine, and the pressures in the cylinder are limited by the amount of air drawn into the cylinder.

Ultimately an engine's torque output is dictated by how much air is trapped in the cylinder and how much it is compressed. The more air we can cram into the cylinder, then all other things being equal, the more torque the engine will deliver. Excluding the effects of shock wave ramming, an engine reaches the limit of breathing in normally aspirated form at a 100% volumetric efficiency: a fancy technical term for saying 100% breathing efficiency has been achieved.

Excluding the effects of compression and valve event timing, the only way to get more torque from the engine will be to force air into the cylinder by means of a compressor (turbocharger or supercharger). This doesn't increase the volumetric efficiency to over 100% because our engine now comprises of a cylinder, plus a compressor, so we have to consider the volumetric efficiency of the assembly as a whole. What a compressor does do is allow us to 'overfill' a cylinder of a given size, assuming, of course, the compressor is big enough to be able to do this. The net result is that for a given size of engine, the torque output can be increased.

At this point then, you should now have some understanding of torque and how it's generated, but how about hp? In actual fact, this is directly related to the torque output of an engine by rpm. Looking at it in simple mathematic terms, power is nothing more than torque times rpm. Now you will notice I use the term 'power', not hp. At this point in time, we haven't actually attached units to it. If there were no such thing as horses, hp wouldn't exist, but power would. The constant used would be different if we were going to measure things in terms of elephant or mouse power.

So that you know what hp is, let's look at the mathematical definition, this being:

$$\frac{\text{work done per minute}}{33{,}000}$$

In this equation, work done is equal to force x distance moved in 1 minute. In the case of an engine, there is a force generated, due to the engine's output, at, say, the edge of the flywheel. This force times the radius of the flywheel of course will be equal to the torque output of the engine. However, the distance moved by the force at the edge of the flywheel will, in fact, be the circumference of the flywheel.

Now for a bit of juggling of numbers here. We know that force times the radius of the flywheel is torque, so we can say that the torque times Pi (that will give us the distance around the edge of the flywheel) will equal the work done in one revolution of the engine. To get the amount of work done per minute, we simply multiply by rpm and to convert that to hp, we divide by our factor of 33,000.

Let's just quickly run

through that again, but in a more compact form:

$$hp = \frac{2\ Pi\ x\ torque\ x\ rpm}{33{,}000}$$

To find the circumference of the flywheel it will be necessary to use 2 Pi because the torque measurement involves the radius, not the diameter of the flywheel, *i.e.* force x radius.

Just to make things a little easier to use, everything can be simplified. The 2 and the Pi are constants, and so is the 33,000 if we divide through the equation with 2 x Pi, we get:

$$hp = \frac{torque\ x\ rpm}{5252}$$

There may be times when you may have a hp curve and you want to find what the relevant torque curve for it is, in which case

$$torque = \frac{hp\ x\ 5252}{rpm}$$

If you have followed what I've said so far about air consumption and how hp is measured, then you've got hold of the basics of understanding about power production and measurement. However, you probably now know enough to be dangerous rather than enlightened, so let's take things a few steps further.

These days, many motorists are becoming power conscious, and not surprisingly, performance enthusiasts as a group head the list. Understandably people like you and me want to know how much power the engine in our car may be putting out. We really need to know not only as a matter of interest but also to determine how our engine compares with other, similar engines. If we are involved in competition, this comparison becomes even more of a requirement as we simply need to know whether or not we have built a competitive engine.

The increasingly widespread use of both chassis and engine dynos since the late '60s has meant that verifying a vehicle's power figures is within the grasp of every enthusiast who cares to make the effort. Because of this it is understandable that we would expect the practice to spread of making comparisons of one dyno tested engine with another.

On the face of it, this is a simple method of revealing the comparative success of any tuning efforts that may have been applied to the engine. After all, if the fellow down the road has modified his 1275 Sprite and it subsequently gives 70hp at the wheels while yours gives 75, your engine must be more powerful than his, right? Well, if you believe that, you could be in for a big disappointment. It *may* have more power, but almost equally it may not. Misconceptions about an engine's potential power output are widespread and if we add to this the practice which some speed equipment companies (no names mentioned) have of generating the best possible figures for advertising purposes, then we can see that there is a distinct possibility that real world hp figures may not measure up to what many think their engines produce. On top of this, one can add the psychological factor of wanting to believe the bigger numbers

Finding the power to be competitive in 850 Mini-X is relatively simple, but finding the power to stay out front requires that you know every trick in the book.

rather than the smaller ones. It's easy to see that those people who deal in real numbers are going to run into severe problems caused by inflated hp numbers that are bandied around by those who do not think along the same lines.

To put things into perspective, let's go back to the point where the development of hp begins, right there in the cylinder, and see where potential losses occur and why there is such a big difference between various sources of quoted hp.

Heat Engine

An internal combustion engine is so named because it burns fuel and air internally, that is, within the working cylinders. Just by way of interest, an example of an external combustion engine might help you to see why we call the petrol/gasoline engine, an internal combustion engine. The most obvious example of an external combustion engine is the steam engine. Combustion of the fuel takes place outside of the cylinder, in the furnace that fires the boiler. Anyway, since there are very few steam Series 'A' engines around, I'll skip anything further on that subject!

Going back to the internal combustion engine, we find that heat produced by the burning of the fuel causes the gas in the cylinder to expand. It is the expansion of the gas pushing the piston down the cylinder that is the means by which power is developed. To burn the fuel efficiently it has to be mixed in certain well defined proportions. Although there are various valid reasons for working above and below the chemically correct mixture of fuel and air, we can say that to all intents and purposes, burning the fuel at the chemically correct mixture will give good results. So that you can get a better understanding of the situation, refer to **Fig. 2.1**. This shows the approximate proportions of fuel and air consumed in a minute by an engine of around 100hp.

We know that heat in the cylinder causes the gas to expand and this, in turn, pushes the piston down, but, unfortunately, we cannot utilize the whole potential of the heat energy delivered during the burning cycle as, due to the very design of the engine, heat is lost from many areas. If you refer to **Fig. 2.2** you will see where most heat losses occur. It is essential at this point to realize that heat energy is directly related to hp. Assuming no losses of any sort occur, it takes 778 B.T.U.s (British Thermal Units) of heat energy to develop 1hp. Looking at it the other way around, it takes 1hp of mechanical energy to produce 778 B.T.U.s of heat energy. Unfortunately for us, it's more practical to convert mechanical energy into heat energy than it is the other way around. At the end of the day we find that, out of the potential

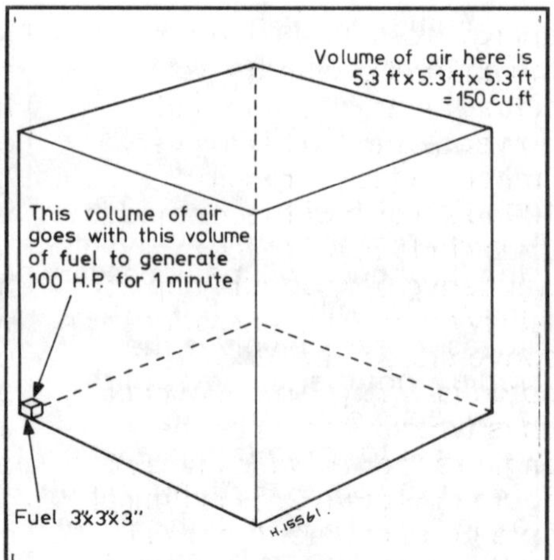

Fig. 2.1 Relationship between fuel/air ratio and power

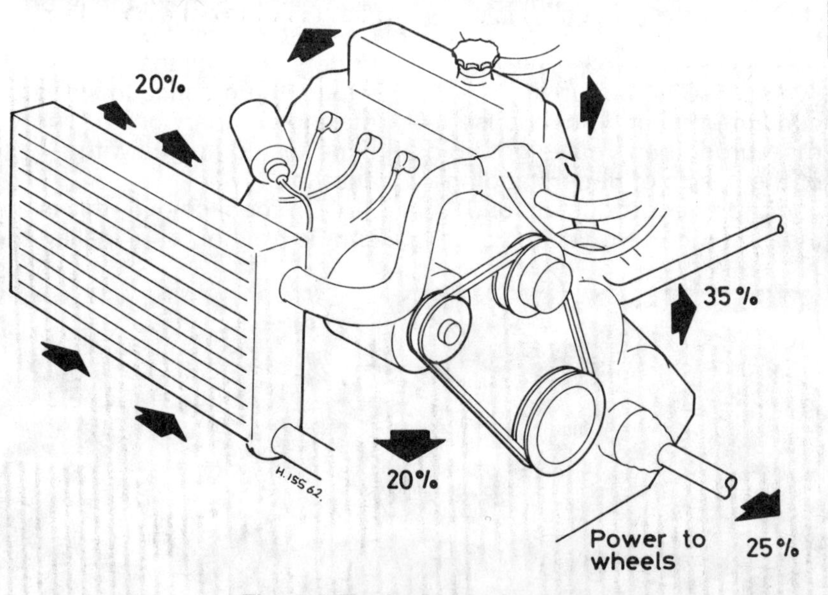

Fig. 2.2 Typical heat loss

If the heat generated by the burning fuel is represented by 100%, then a typical heat loss equation is as shown here. 20% of the heat is lost to the coolant, which in turn rejects it to the cooling air. 20% is lost by radiation from the engine itself. 35% of the heat is rejected in the exhaust. Only 25% is turned into mechanical power to drive the vehicle. But the losses don't stop there. Friction in the transmission and rolling resistance turn some of the mechanical energy back into heat.

energy from the fuel burned, the amount of energy we actually extract in terms of power at the flywheel is very limited. In fact, for every 100hp's worth of fuel burned in the cylinders, a *GOOD* engine will derive only about 25hp at the flywheel. This rate of energy conversion of the fuel's potential energy into flywheel hp is what is known as the engine's thermal efficiency. A figure of around 25% is typical for a good engine. A normal road engine is often around 18% thermal efficient.

Examining the cylinder pressures that occur within the engine, we find that the theoretical power developed by the cylinder pressures is more than that seen at the flywheel. The difference between these two numbers is a measure of the engine's mechanical efficiency, that is, its loss of power due to friction. Unfortunately, hp that is lost to friction is turned directly back into heat, heat which is subsequently carried away by the cooling system of the engine.

When the engine is modified, attempts are made to minimize all these losses on one hand, and improve the rate at which the engine consumes air on the other. As has already been demonstrated, the more air and fuel mixture that can be passed through the engine in a given time, the more the power output will be, so long as none of the other inefficiencies have unduly increased along with it. If this state of affairs is achieved, then the engine will show more power and the place we determine whether or not power has been increased is on the dynamometer, a device for measuring hp.

In case you are not aware, there are generally two types of dyno in common use, an engine dyno and a chassis dyno. The engine dyno tests the engine while it is out of the car: the output from the engine's flywheel being transmitted directly to an absorption unit. A chassis dyno measures the hp at the driving wheels of the car. This is done simply by driving the car onto a set of rollers which are then coupled up to the absorption unit. Whether it's an engine dyno or a chassis dyno, a dynamometer itself is merely a device for applying a braking load to the engine, hence the often used term '*Brake* Horse Power' (bhp). The braking load against which the engine can hold a steady rpm is measured on a torque arm, and the resultant gauge reading from the torque arm is a direct measurement of the engine's torque output in whatever units the dyno may be calibrated in. In fact, virtually all dynos read out directly in terms of torque. The torque output of the engine is then converted into hp by taking into account the rpm at which the engine is running, or in the case of a chassis dyno, the rpm at which the rollers are running. The formula given earlier will work for any dyno. However, many dynos have odd length torque arms so they end up with a constant in the equation which the manufacturers often hope will make the calculation of hp easier. On virtually all engine dynos, the formula for calculating the hp will be printed somewhere on the dyno.

As far as chassis dynos are concerned, the most important data for them to read out is hp

This cutaway of a Turbo MG Metro engine shows some of the intricacies involved in forced induction. The whole point of forced induction is to get the engine to consume more air so it can burn more fuel to make the extra power for the performance required.

at the driving wheels. The torque at the driving wheels can be varied depending upon what gear the vehicle is in, so torque figures measured on a chassis dyno do not necessarily relate to the torque output of the engine. The only time they truly equate to the torque output of the engine would be after correction using the gear ratios.

The figure so produced is best described as the effective torque, as it takes into account all transmission losses. In other words it would only be the true torque output of the engine if the car had 100% efficient transmission, which, of course, none do.

With all but the most sophisticated engine dynos, the calculations necessary to produce hp figures are carried out manually. With rolling road dynos, the torque figures at the rollers will, as has been stated before, be dependent on the gear ratios of the car, as well as the engine's power output. For this reason making comparisons of the torque output of one vehicle against another is very difficult. To get around this situation, chassis dynos convert the torque figures to hp (with varying degrees of accuracy) within the dyno itself and simply read out in hp only.

At this point you can see that for the purposes of making comparisons, there are a number of possible places for errors to creep in on a chassis dyno. Firstly the dyno must measure the torque accurately otherwise the final power figure will be in error. Secondly, the rev counter that is used to determine the rpm involved must be accurate, otherwise a proportionate error will be produced. In the case of a chassis dyno, the method used to multiply the torque and rpm factors must also be accurate, otherwise this will produce yet another error source. Of course it's easy to fall into the trap of thinking that if all these factors are right, then the dyno figures will also be right. Well, this need not necessarily be the case . . . I'm sure most of you are aware that automatic transmissions usually function with a torque converter between the engine and gearbox. The torque converter is not too dissimilar to the typical absorption unit of a dyno, assuming, of course, we are talking about a water dyno. Torque converters are designed from the outset to give some form of torque multiplication. Dyno absorption units are designed to give no torque multiplication. **However, there's no guarantee that some torque multiplication doesn't creep in. If this happens, then the torque figure measured can be higher than it really is.** As a result, the power figure finally arrived at can be in error.

Just running through possible sources of error in the dyno, we find that for a typical good engine dyno, the combined accuracy of the torque gauge and rev counter are typically within 2%. Assuming there is no torque multiplication in the absorption unit, this means that if we have an engine which is really producing exactly 100hp it could indicate on a high reading dyno, as high as 102, or on a low reading dyno, 98. So immediately we have a 4hp error possible in a situation where the power hasn't actually changed. On a chassis dyno we have much greater problems accounting for the frictional losses in the heavily loaded rollers, plus whatever system was used to carry out the measurement and computations. As a result, the chassis dyno may offer no better than about 5% accuracy, and some can be considerably more in error than this depending upon their age and how well they have been serviced. Again, an engine which delivers, say, exactly 100hp at the wheels, could, on a high reading dyno, read 105, and on a low reading one, 95. However, equipment errors aren't the sole cause of discrepancy in numbers. The conditions under which the figures have been taken have a significant effect on the numbers which we end up with.

Rating Standards

No doubt when you have been looking through car catalogues, information sheets, or new car specs, you have seen hp figures quoted. Some will be quoted as DIN figures; others as SAE gross or net. Maybe you have wondered why they are different. Well, it's because they have been obtained under different test conditions. At one time, many manufacturers used to quote power outputs under conditions that gave the most flattering results, just so they could advertise the highest figures in their promotional literature. Very often an engine's power output would be quoted without fan, dynamo, air cleaner or silencers. Also the carburettor and ignition timing may have been optimized at each point in the rpm range so as to produce the best possible power curve. Obviously a rating under such circumstances would represent a gross rating, that is, the engine would be driving nothing in terms of ancillaries and would have the best chance of a high power output due to an unrestricted induction and exhaust system.

Such a rating, however, does not bear any resemblance to the engine's installed power output, so such figures are unrealistic. Things don't stop there. Some manufacturers (again I am not quoting any names) have in the past used blueprinted engines from which to quote their catalogue hp figures, and this can stretch the margin of credibility between what the advertized literature stated and what was actually achieved in the vehicle, to extreme limits. A case that comes to mind from some years back on an American-produced V8 engine, showed that the manufacturer claimed 280hp from the engine, whereas the engine pulled straight from the vehicle and put onto an engine dyno, complete with its exhaust system and ancillaries, would often show no more than 195–200hp. Just imagine how difficult it would be for an engine tuner to say he had tuned the engine to give 250hp. Anyone believing the literature would immediately say, 'You have just tuned 30hp out', whereas in essence he has built an extra 50 or so into the typical engine.

Air Density

Because dynos are becoming more widespread and people are more concerned with the power output they are buying, these big discrepancies have become less of a problem over the last few years, but discrepancies still do occur, depending on the type of rating used, and the correction factor involved in quoting the hp. Having read this far, you should be aware by now that an engine's hp is proportional to the weight of air it inhales in a given time. If the air temperature goes up, the air becomes less dense, i.e. it weighs less and as a result, the hp comes down. If the barometric pressure drops, the weight of air taken into the engine is reduced and so the engine's power, again, drops. If the air becomes humid, the atmosphere now contains a proportion of gaseous water. This means that space is now used up in the intake manifold and cylinder by water vapour, so the total amount of oxygen drawn into the engine is reduced. As a result, the power drops.

So that direct comparisons between one engine and another can be made on different days and under different atmospheric conditions, a number of corrective formulae have been developed which allow the engine tester to correct the power to a given set of conditions. Unfortunately not all manufacturers or dynamometer operators correct their figures to the same set of standard conditions. If all these factors were taken into account, we would find that SAE figures most often used in America and still quite often used in England, show figures from 5%–10% higher than the DIN power figures which are commonly used in Europe and are beginning to be used in the U.K.

In Italy, yet another set of conditions are specified by CUNA, and this usually results in power figures appearing to be 2%–3% higher than the DIN power figures. So with all these different standards, we find that the numbers produced can vary quite considerably, though the power output of the engine itself hasn't actually changed.

All the modifications to this 1275 race engine were aimed at getting it to consume more air so it could burn more fuel to make more hp.

No doubt when you pick up this book, you will notice that there are literally dozens of power curves in it, and along the side of each graph it does, in many instances, say 'Corrected hp'. Since I have now pointed out that there are many correction standards, you are probably wondering what these power figures are actually corrected to. Well, throughout the book, unless otherwise stated, the power figures are corrected to a standard correction factor most often used in the performance industry, namely to 29.93 inch of Mercury for the barometric pressure (that's sea-level air pressure) and an air temperature of 68° Fahrenheit (20°C). Along with this, the air is also assumed to be dry. These correction factors give the most flattering power figures of any correction factor normally used. If you want to correct to the more often used SAE figures, then multiply the figures you see by 0.966. If you want to correct to DIN figures, then multiply by 0.971.

Relating Chassis and Engine Dyno Results

One of the principal factors that confront any rolling road or chassis dyno operator, is relating the power figures that he sees at the wheels of his customer's car with the flywheel hp. Inevitably, customers can relate to flywheel hp because they are the numbers they are confronted with most in technical specifications. Most enthusiasts like to know how much flywheel hp their engine puts out related to the driving wheel hp. Many dyno operators will simply say that a third of the engine's power output is lost in the transmission. Although in most cases this assessment is totally erroneous, it is quoted for a very good reason. In the old days, when factory hp numbers were at their most erroneous (due to the exaggeration factor) testing a standard car on a typical rolling road dyno tended to produce power figures equal to about two-thirds of the manufacturer's advertised output. For instance, a car which was supposed to develop 60 flywheel hp would produce a true 40 at the driving wheels. The problem here is that the 60 quoted hp very often wasn't, so that one-third loss ratio was based on numbers that really didn't apply in the first place. In all probability, the so-called 60hp engine may have been, in reality, nearer 52–55.

When this one-third loss is applied to competition cars, the situation gets out of hand really quickly. An example is a V8 powered special which produced 240hp at the wheels. Applying the one-third loss principle, this supposedly relates to an engine delivering 360hp at the flywheel. Two questions you can ask yourself here are, what sort of transmission is it that is so inefficient that it loses no less than 120hp? That's more than a good many Series 'A' engines put out in the first place! Secondly, if it loses this power, it can only be converted to heat, and 120hp is equivalent to about 90 kilowatts (the conversion equivalent is 746 watts to the hp). If you consider that a large household electric fire of the double bar variety is typically rated at 3kw, you can see that the inefficiencies of this transmission if the one-third loss applies, would be equal to having the heat output of no less than 30 big, electric fires underneath the car. This means the transmission alone would require a radiator about the size of a typical Cooper 'S' radiator to keep it cool.

Such was not the case, therefore that much power could not be dissipated through losses in the transmission. This poses the question, how much hp does get lost in the transmission of a typical vehicle? We will not concern ourselves with automatic transmissions and transmissions big enough to handle V8 engines, because it simply doesn't apply to Series 'A' technology, so we'll talk about smaller vehicles that do.

The first step is to figure out how much hp is actually being developed at the flywheel. This is going to depend on the conditions in which the engine is operating. **Remember, when typical hp figures are quoted, they are corrected for the air temperature and barometric conditions that exist outside the engine compartment.** The correction factors I mentioned earlier are about the conditions we would expect to see on a day early in an English summer. Unfortunately the conditions in the engine compartment which, after all, is where the engine lives, do not even remotely resemble that. To give you an idea of just how much the hp changes between the 'as dynoed under ideal conditions' and the 'as-installed conditions', let's consider what happens to a 100hp engine when we take the air temperature up to a typical engine compartment temperature. Also, another factor to take into account is that there are very few parts of the world which are actually at sea level. On average, our vehicles are probably 500 feet above sea level. If we take these factors into account, an engine which gave a 'corrected' 100hp on the engine dyno would, in the engine compartment of a typical Series 'A' powered car, produce approximately 88hp: that's 88 installed hp. So we can see straight away that the 100hp engine that we think we have, is, under installed conditions, only delivering 88hp. So far then, we are down 12hp on the quoted 100, not because we have lost 12hp but rather because it was never made in the first place due to the lower air density environment the engine is working within.

Moving along, the next point of power reduction is, in fact, a real loss: the gearbox. Whatever power is fed into the gearbox you can always guarantee less will come out. Of those gearboxes used in conjunction with Series 'A' engines, the transverse gearboxes tend to be less efficient than gearboxes used in inline situations. It's possible that much of this inefficiency in transverse gearboxes is due to the fact that there are more gears and they have to run more deeply immersed in oil than gearboxes for inline installations. With manual gearboxes, used with the Series 'A' engine, losses can range between about 12–18hp, although a carefully built box may lose less than 10hp. Let's say for our particular example that the gearbox loses 18hp. This means our engine power output which was assumed to be 100 is now down to 70. From the gearbox the power has to be fed through the rest of the transmission to the wheels, and through the wheels and tyres, into the rollers. It's only at this point that it gets read as hp. Let's assume we are talking about a transverse engined vehicle, so transmission losses are, in effect, just gearbox losses, so we have already taken those into account. The power going into the wheels, then, is the same power that's coming out of the gearbox.

At this stage, the next point where power will be lost is in the tyres. Tyres set up a force opposing motion known as 'rolling resistance'. As speed increases, so rolling resistance becomes greater at a rate faster than the increase in speed. On a chassis dyno (rolling road) the fact that the tyre is on a roller rather than a flat, road-like surface, affects rolling resistance. When on a roller, tyre wall distortion and contact pressure are higher, so losses are greater. It's not out of the question to lose 6–10hp in the tyres. Tyres inflated hard, lose a lot less than soft tyres. All smooth, slick tyres absorb a lot less on the rollers than tyres with knobbly tread patterns. If you should find your vehicle is off the mark check to see if tyre absorption is the culprit by making a tyre change. Anyway, getting back on track, let's assume that 6hp goes up in tyre heat. That 70hp is now reading out as 64. If we now apply the often-used one-third formula, we come back to 96hp, and that's pretty close to the 100 that the manufacturer originally quoted, but it's only by luck. In actual fact that vehicle does only have 88hp because, remember, the initial 12 didn't materialize. However, the point I am really trying to make here, is that trying to ascertain what the flywheel hp is from rolling road figures, is by and large, a fool's game. The possible exception to this is the latest high tech, rolling road dynos, which have a facility for checking transmission and tyre losses by inertial or motoring means. If you want numbers, real or imaginary, to talk to the guys about next time you are propping up a bar, remember the number from the rollers is almost worthless in terms of relating it back to flywheel hp.

If you are using a chassis dyno to monitor your vehicle, be sure the test conditions are repeated every time. It only needs a couple of P.S.I. change in tyre pressure to make you think you have either done well on your last modification or that it was worthless, depending on which way these pressures go. Also you should make note of the atmospheric conditions. Certainly in England it doesn't affect power readings that much and correction factors up to plus or minus 4% are about as broad a span as you can expect to see. **On the other hand, if**

Fig. 2.3 Factors to check to get comparable numbers from rolling road dyno tests

1 Always use the same dyno, and ask if any recalibration was done since you last put your car on it.
2 Use same type of tyres in same sort of conditions as previous test.
3 Use same tyre pressure.
4 Have bonnet open for every test, or closed for every test.
5 Always take the power curve either going up the rpm range or coming down, but never both.
6 This one is very important: make sure the rev counter holds steady for at least five seconds before taking the horsepower reading. If the horsepower needle flickers, take the average of the top and bottom reading.
7 Note weather conditions and correct to standard temperature and pressure with the following formula:

$$\frac{29.93}{p} \times \sqrt{\frac{460+Y}{520}}$$

where P is the barometric pressure in inches of mercury and Y is degrees Farenheit.
Multiply the dyno readings by the correction factor.

you are in the South West of U.S.A., dyno correction factors can be as high as 20%, so if you don't correct the power figures back to some standard, those rolling road figures will be totally useless. If you do want to compare numbers from a rolling road dyno, and it should be the same dyno all the time, then refer to **Fig. 2.3** for things you should do to give you the most consistent numbers.

Riding Your Own Dyno

If this book is anything like a prime example, it would seem that anything to do with high performance cars inevitably gets surrounded by numbers. Some of those numbers are important to us as performance enthusiasts and others possibly not. Right now we are looking at probably what is the most important number, concerning our high performance engine and that is the hp. As has been said before, hp numbers are often exaggerated by those people who have to sell hp to make a living and compete against other people doing likewise. Mostly stretching the truth about an engine's hp output is something they can get away with, simply because most enthusiasts do not have a dyno to check out these numbers. If you really get down to the nitty gritty, the truth of the matter is that many people selling engines, don't have dynos either, and so the numbers quoted are often nothing more than inspired guesses, based on the performance of some of their competitors who possibly do have a dyno.

The thing is, although I have talked about dynos at length, you don't necessarily need one to check out the peak power output of your vehicle's engine. It can be done in numerous ways, and basically the equipment you need is simple and straightforward. In fact many of you may already have the necessary equipment to check out your engine's power output. To measure hp it's necessary to absorb the engine's developed power in one form or another. An engine or chassis dyno does just this. Usually the power is absorbed into an absorption unit such as a water wheel housed in a case or, as many rolling roads do, into an eddy current brake. You can measure the hp of your own engines by doing exactly the same thing. The power developed by the engine has to be absorbed by something. There are two simple possibilities here. We can either use wind resistance to absorb the hp of the car, or a combination of wind resistance and the mass of the car. If wind resistance alone is chosen as the absorption medium, then hp output will be derived by means of some computations involving the vehicle's maximum speed. If wind resistance plus the mass of the car is used, then another avenue is open to us, namely this: it is fairly obvious that the more hp a vehicle has, the faster it will accelerate, and by measuring the speed at the end of a set distance, the hp of the engine can, within reasonable limits, be computed by virtue of the speed and weight of the vehicle. By comparison, the top speed method is fine if the relevant data about the car's various permanent characteristics such as aerodynamics, frontal area, rolling losses, etc., are known. It's also fine if the top speed of the car isn't so high that it requires the length of Bonneville Salt Flats to attain. Even if it's not that fast, it's worth remembering that anything over about 70 is speed ticket country in most parts of the world. Even though the Series 'A' engine may be small, it could be installed in vehicles which could top out in excess of 130mph. Rushing around public roads, trying to establish accurately what that top speed is, can be dangerous in the first instance and can cost you your licence in the second. On the other hand, the acceleration method does not involve such high speeds, and is not so sensitive to the aerodynamic attributes of the car. It is more dependent on the power to weight ratio of the vehicle, so high speeds are not involved until extremely high power to weight ratios are achieved. Except for the most exotically modified Series 'A' engine, that situation is not likely to be one that will bother us.

Power Formulae

Not surprisingly, the distance used in the following method for measuring the engine's power output will be the quarter-mile. This distance makes sense, not only because it provides a good balance between acceleration and top speed, but also because there are many drag strips dotted about all over the world and, lastly, and probably most importantly, who ever worked out these equations based all their calculations on a quarter mile, and I don't feel inclined to juggle them around to work on some other, less convenient, distance! Apart from that, if the distance used was much shorter, wheelspin would have a big effect on the outcome. If it was much longer, the wind resistance factor would play too much of a dominant role.

Using the speed reached from a standing start at the end of a quarter-mile, the formula to work out hp is shown in **Fig. 2.4.** All you need in order to come up with your vehicle's hp

Fig. 2.4 HP deduction formula

$$HP = \left(x \sqrt{\frac{S}{225}} \right) W$$

Where S = Speed in mph at the end of the quarter-mile.
W = Start line weight in pounds.
X = Vehicle index number from Fig. 2.5.

is the weight of the vehicle as it leaves the start line, plus an accurate speed at the end of the quarter-mile. To work out the formula, it's best you have an electronic calculator with a log function on it.

Okay, now for the technique. First weigh your car on a public weighbridge at the weight you intend to run it down the quarter-mile. Next, on a motorway, freeway or any conveniently marked piece of road, check the calibration of your speedometer. Inevitably, speedos, like hp numbers, tend to be designed to exaggerate the performance of the vehicle. If the speed you input into your formula is in error, your hp reading will also be in error. A few mph makes a much bigger difference than being a few pounds out on the weight. Of course if you are going to run at the dragstrip, the speed ticket you get there will be more than sufficiently accurate for the job in hand.

Once you have a means of establishing the mph at the end of a quarter-mile, then you are in business. If you are doing your runs on a piece of public road which may not be anything like as level as most dragstrips, it's best to do the runs in both directions to eliminate the effect of both wind and gradient.

Now we come to the application of the formula. As you can see from **Fig. 2.4,** everything is included except the X factor, that is the vehicle index number – the whole key to coming up with the right hp figures. The number varies slightly from vehicle to vehicle. The more hp the vehicle has in relation to its frontal area/drag coefficient, the lower the index number becomes. Also, another factor affecting the index number is that big, torquey engines tend to reduce this index number, whereas high revving, small engines tend to require a higher number to come up with the right answer. Turbocharged engines, with their potentially high torque capability, tend to pull the vehicle index number down. If you are applying the formulae here to a *well developed* turbocharged engine which is producing 90 or more hp per litre, then the index number can come down by about 0.008 on whatever index number is shown in the list of **Fig. 2.5.**

Working It Out

So much for all the whys and wherefores. Now let's run through an example so you can see how it's done. The particular vehicle I am going to choose is a 1435 Mini, shown somewhere in the pages of this book. It was a particularly favourite Mini of mine. I ran this car through the clocks at a dragstrip and it netted a speed at the end of the quarter of 89.9mph. The startline weight on the car was 1696lb, including me and the fuel it was carrying. The track was almost at sea-level and the weather was cool, so the atmospheric conditions were near to standard. Running the numbers through the formula given in **Fig. 2.4,** we have: S = 89.9; W = 1696lb.

Fig. 2.5 Vehicle index numbers for HP calculations

850 Mini	*0.345*
1000 Mini	*0.342*
1275–1430 Mini	*0.340*
1275–1430 Sprite/Midget	*0.330*
1098 Sprite/Midget	*0.332*
948 Sprite	*0.334*
Morris Minor 1000	*0.342*
Austin A30	*0.344*
Austin A35	*0.342*
Austin A40	*0.340*
1100 saloon (sedan) all types	*0.343*
1275 saloon (sedan) all types	*0.341*
1275 Morris Marina	*0.342*
1000 Mini Metro	*0.338*
1275 Mini Metro	*0.334*
1275 Turbo Metro	*0.326*

From **Fig. 2.5** we have the X factor which is equal to 0.34. The formula is:

$$HP = \left(0.34 \sqrt{\frac{S}{225}} \right) W$$

Substituting the value for S, we have

$$\frac{89.9}{225} = 0.3995$$

Finding the power from this 1400 street/drag racing Mini was not really a problem, but putting the power down on the ground through street tyres was.

It's now necessary to take the 0.34 root of this, i.e. $0.34 \sqrt{0.93995}$. To do this you use an electronic calculator (or log tables) and first find the log to the base 10 of 0.3995. This equals 0.39842. Now divide this by the index which in our example is 0.34. Working out this example, we have:

$$\frac{0.39842}{0.34} = 1.17183$$

Now take the anti-log of this number. This should come out to 0.06732. Right, we've got the tricky bit worked out. The last step is to multiply this figure by the car's weight, W. So we have 0.06732 x 1696lb. Working this out, we have an answer of 114.2: therefore, according to the formula, this 1435 Mini cranked out 114.2hp. On a real live dyno, the engine made 112hp. So the formula isn't too bad.

Although I mentioned it before, it's worth stressing once again that the mph input must be accurate, otherwise the power figures can be off by quite a way. If, in the previous example, the speed had been off by 3mph, *i.e.* 92.9mph instead of 89.9, then the hp produced by the calculations would have been 125.7hp. That's almost 14hp out and far enough out that a guess would have been good enough. An error in speed input has much more effect on the final answer than errors in weight. Errors in weight produce a corresponding error in hp, e.g. 2% error in weight produces 2% error in hp.

Calculating Dragstrip Potential

Up to this point we have simply used a car's performance over the quarter-mile to relate to its hp. If you are involved in drag racing, then it is the time the car takes to cover the quarter-mile, not the speed, that counts. For a car to reach a given speed at the end of a quarter-mile requires the input of a certain amount of energy. The speed at the end of a quarter-mile is a less variable factor than is the E.T. (Elapsed Time) it takes to cover the standing quarter. However, if you are looking for fast standing quarter performance figures, then it is possible to calculate, bearing in mind such things as traction, gearing, etc., the elapsed time your vehicle will require to cover the quarter-mile with a given power output. For this, use the formula in **Fig. 2.6**. This equation is handy in many respects. If your car won't manage anywhere near the quarter-mile time this formula predicts, then you can assume that the gearing you are employing is too high, or the suspension is not working properly, or the tyres aren't giving sufficient grip. The last factor translated into real-world terms, means that you could have a front-wheel-drive car, *i.e.* a Mini. Unfortunately, standing starts with front-wheel-drive cars are never spectacular in terms of acceleration, only tyre smoke, as the weight transfers right off the driving wheels. The net result is that the initial start from the line is a lot slower than in an equivalent rear-wheel-drive car.

Fig. 2.6 ET (elapsed time) formula

This formula can be used for calculating a vehicle's quarter-mile time, assuming the vehicle is set up correctly as far as suspension and tyres are concerned.

$$ET \text{ (in seconds)} = 3\sqrt{\frac{W \times 192 \times X}{HP \times 0.318}}$$

If you have calculated your engine's hp from one of the formulae given earlier, then the formula to predict a quarter-mile elapsed time will be as shown in **Fig. 2.6**. An example worked out on a Mini racer looks something like this:

$$3\sqrt{\frac{1260lb \times 192 \times 0.34}{115.5hp \times 0.318}}$$

Working out everything in the cube root bracket reduces the expression to $3\sqrt{2239.4}$. To get the cube root of the number, we take its log which is equal to 3.350, then divide it by 3. This gives 1.1167. The answer, when this number is anti-logged, is 13.1.

So the Mini racer in question should have run 13.1 seconds for the quarter. In reality, the best time the car ever ran was 14.2 seconds, so is the formula out? Not really. The 13.1 figure reflects what a vehicle with a power to weight ratio as per our example should do if everything is in its favour, and one of these factors is that it's rear-wheel-drive. Remember, with a front-wheel-drive vehicle, it's only possible to use gears just so low before uncontrollable wheelspin ensues. A typical gear such as this, still isn't low enough for the best quarter-mile time. As for wheelspin, for a Mini, this adds between 0.8 to 1.2 seconds to the quarter-mile time. A lightweight 1275 Sprite on a good set of soft slicks, close ratio gears, and well set up rear suspension, should be able to run in the low 13s!

3 The A-Series; its performance prospects

Many are no doubt reading this book with a view to extracting as much hp as possible from their Series 'A', but probably the majority of you are reading it to find out how to achieve more hp and economy, or at least not damage the existing economy during efforts to get more power. I expect there will also be a proportion of readers who are more interested in getting better economy from the engine, so long as it doesn't hurt the power.

Because in this instance it happens to be an easy subject to deal with, I am going to look at the economy aspect first. Basically it's all good news. Virtually irrespective of which Series 'A' engine you have, the potential economy properties are carried throughout the range, right from the early 803 engine to the current turbo 1275s. Essentially the Series 'A' engine has what it takes to produce good fuel economy. So long as you stick with the fuel economy recommendations made in this book, you will like the results. Really the only factor you have to remember is that the smaller the engine is, basically the more economical it's going to be, but of course that will be at the expense of performance. On the other hand, if you are one of those who is looking for hp for road use, then the whole situation has to be re-evaluated because each engine within the Series 'A' group has its strong and weak points. So I'll take things roughly in chronological order and look at what you need to know about your particular engine before you take the plunge with any modifications.

The little Grandfather of all Series 'A' engines, the 803, is characterized by having the longest stroke in relation to its bore for any Series 'A' engine. In other words, for high performance use, its bore/stroke ratio is the worst of any derivative produced. On top of this, it's hampered by having a crankshaft that is on the fragile side, and won't take high rpm for any length of time. Because the bore is so small at 2.280 inch, the cylinder head can only accommodate very small valves and the net result is that even if the engine were strong enough to tune to any extent, the cylinder head could never be made to breathe adequately to produce any really good power figures. As a result of these inadequacies, the 803 engine should be considered as an engine capable of benefiting from mild tuning only. When the 948 engine was introduced, from the performance point of view at least, the Series 'A' engine took a giant step forward. The 948 engine had the same 3 inch stroke the 803 had, but it sported a bore which was almost ¼ inch larger than its predecessor. This, together with general beefing up of the engine, suddenly put the Series 'A' engine well and truly into the hotrod category. With an rpm potential of 6,500 reliably and 7,000 if you weren't too bothered about having to change the crank once in a while, the 948 engine became THE engine to modify in the U.K. Even though it was significantly better than the 803, this engine still suffered from having an inadequate bore size, and ultimately this bore size will limit the hp of the engine simply because the room available for valves is limited. Still, to put things into perspective, if every trick in the book is employed, but a stock block, stock head formula is adhered to, then even with the standard stroke, it is possible to get about 98–100 hp from a 948 engine, bored +60 over, which brings it just under the 1-litre limit.

Because of its extensive use in racing in the late '50s and early '60s, special high quality steel cranks were made for the 948 engine by the factory, but these cranks are about as rare as hens teeth. Of course, if you are modifying one of these engines today, it is not necessary to have an original, factory steel crank unless the rules and regulations being raced under, call for it. Several good crankshaft manufacturing companies can supply cranks to order for any of the Series 'A' engines.

The 850 engine, introduced in 1959, was essentially a destroked 948 with a 68.25mm (2.687 inch, 2¹¹⁄₁₆th) stroke and a 62.15mm (2.447 inch) bore, it was the closest thing yet to a 'square' bore/stroke ratio Series 'A' engine. Unfortunately in achieving that near-squareness a capacity reduction occurred, thus reducing its potential to produce hp. We have to remember the old adage 'There's no substitute for cubic inches' was as true then as it is now. In terms of rpm capabilities, the early 850 engines were pretty good, but early cranks had a problem in as much as they suffered from lack of torsional stiffness. A torsional vibration set up in the crank could cause them to break if they turned more than 6500rpm with any regularity. Early cranks, often known as the thin tail cranks, were replaced with a crank having a thicker tail for the primary gear to run-on around about 1963. With this change, the 850 engine really did acquire some rpm potential with rpm capabilities up to 7000 or more if the bottom end was suitably prepared. Although an 850 could produce a fairly good power output per litre, its biggest stumbling block was the fact that it was only 850cc and this ultimately limited its total power capability. On the other hand, there are so many 850s around, especially in breakers' yards, that they are cheap to acquire and within reason, cheap to modify. As a result, there are many classes of competition which cater specifically for people who wish to race 850 Minis. Of the cheaper formulae, two that spring to mind are 850 Minicross and 850 Minirods. Both formulae seek to limit the modifications allowed so that relatively cheap, yet reliable competition engines can be built.

The 997, with its 81.3mm (3.2 inch) stroke, along with a bore size smaller than that of an 850 engine, is not a good engine to modify. Sure, it can be modified to produce more power, but its rpm potential is very much limited. Unless you are hell-bent on an authentic restoration job, the least expensive way to get worthwhile power from a 997 engine is to replace it with another engine. Any transverse engine other than the 850 is a better power producer than the 997. This doesn't mean to say the engine can't produce power, but it's always going to be two steps behind everything else, and on top of that, you have the perpetual worry that at any moment it could break if the rpm go too high.

The 997's replacement, the 998 and its bigger brother, the 1098, were much better engines. They utilized a bigger bore at 64.57mm (2.542 inch), the 1098 merely being a long stroke version of the 998. In fact the introduction of the 998 transverse engine, saw the return of a 76.2mm (3 inch) stroke engine, a popular stroke length for the Series 'A'. The 1098 engine was simply a stroked version of the 998, and although it doesn't have the rpm potential of the 998, this is more than made up for by the fact it has an extra 100cc.

With the stroke length of 83.8mm (3.3 inch), the 1100 engine is the longest stroked version of the Series 'A' engine. However, in spite of this long stroke and the fact that in transverse form it only ever had 1¾ inch main bearings, the transverse 1100 is capable of fairly high rpm. Indeed, an engine with a fully prepared bottom end is capable of hanging together for short bursts up to 8000rpm, but if reliability is to be counted for anything, revs have to be kept way under this on a long-term basis.

The 1098 engine was actually made in transverse and inline forms. The early inline engines shared, with the transverse engine, the 1¾ inch main bearings, though later engines had a 2 inch main bearing crank. These engines, used in Midgets and Sprites, could withstand the rigours of competition in spite of their long stroke because of the added strength the bigger main bearings gave. However, the 1098 engine, as much as any Series 'A' engine before, suffered from the fact that it had a long stroke in relation to its bore, and for its capacity the valve sizes that could be used were still on the small side. The introduction of the big bore range of engines, namely the 'S' engines and their subsequent derivatives, tended to put the situation nearer to the ideal.

The first 'S' engine was, of course, the 1071 unit. With a 70.6mm (2.78 inch) bore and a stroke of 68.26mm (2.687 inch), this engine was a high winder, and the subsequent 970 'S' with an even shorter stroke was, in fact, one of the few virtually square engine configurations that were factory built. Still, the really popular engine was to be the 1275 and even though it was tending towards being a long stroke engine again, the extra capacity more than made up for this deficiency. The advantage of the bigger bore of the 'S' engine was that bigger valves could be accommodated and, let's face it, this engine needed bigger valves. All through the Series 'A' range, the valve sizes the engine had, in relation to what it really needed, were always lacking. The 'S' engine with its relatively large valves, was still far off having adequate valve size or breathing capability, except maybe in 970 form. Still, the purpose of the 'S' engine was to make the most of the Series 'A' engine, and this it did. A superior crankshaft with bigger main bearings, together with stronger rods, allowed the Series 'A' engine to turn some very high rpm. In fact, the 'S' engine proved to be one of those almost unburstable units. The big problem though, was that it was expensive not only to produce but also to modify. If anything did break, then it was sure to cost two to three times as much as the equivalent part in a non-'S' specification unit. Of course Leyland did find that the 1275 configuration was a successful formula and so a 1275 non-'S' engine was introduced.

This engine was essentially a cheaper version of the 'S' unit. Cheaper in terms of production and cheaper for the end user to buy. Generally, critical material specifications were reduced. The most significant being those of crank and rods, but to make up for a potential loss in crankshaft strength BL's engineers increased the rod journal size by ⅛ inch. By comparison with the Cooper 'S' engines, valve sizes were reduced slightly to enhance reliability and the material they were made of was of a less exotic specification. Although on paper the 1275 engine looks a lot less of a high performance unit, very little of the tuning potential was lost by comparison with the 'S' engine. All the internal components were capable, though to a slightly lesser degree, of taking race engine stresses, but at a fraction of the cost of anything which had the 'S' label. This same 1275 configuration was also perpetuated through the inline engine series, and 1275 engines found their way into Sprites, Midgets, Marinas, etc.

It's probably fair to say that the most successful Series 'A' engine is, in fact, the 1275 and Leyland have probably recognized this since there has been more detailed development on this engine than any preceding Series 'A' engine. The 1275 engine has evolved into the Series 'A+' engine which is used in the Metro. The 'A+' engine does, in fact, share virtually total interchangeability with previous engines, but improved crankshafts, stiffer blocks, better cylinder head castings all make the 'A+' engine a much more refined unit, and one well worth using as a basis for a high performance engine.

Last but not least we have the 1275 turbo engine. This engine contains lots of heavy duty parts, many of which will inevitably find their way to production lines feeding more mundane Series 'A' powered vehicles. For the engine modifier at least, the situation concerning performance practicality of standard parts has never been so good.

If you are building a high performance engine, consider your needs carefully. Very often many enthusiasts who are building an engine from scratch do so on an engine obtained from the breaker's yard. Remember, interchangeability of engines from one size to another is exceptionally good. Of course this doesn't apply so much from inline to transverse, as the blocks are slightly different in the rear main bearing cap area, but since blocks of either type are so plentiful, as are the relevant cranks, there shouldn't be any need to worry about how to convert one to the other.

If you are selecting an engine which can be modified as it's rebuilt and then installed in the vehicle, choose your engine carefully as there are many criteria by which you should judge which engine is best for you. Take a look at the **engine selection chart**, and weigh up the pros and cons very carefully before choosing which engine to go for.

'Engine Selection Chart Overleaf'

Fig. 3.1 Engine selection chart

	Bore	Stroke	B/S ratio	Cost for road use	Used parts availability	Power per litre	Max power capability *	RPM potential	Competition potential	Cost for race preparation**	New parts availability	Capacity potential compared with other engines in the group
803 IL	2.280 in 57.9 mm	3.00 in 76.2 mm	1.315	Low	Fair	Poor	Very Poor	Poor	Nil	N.A.	Poor	Very Poor
848 T	2.477 in 62.92 mm	2.687 in 68.25 mm	1.085	Low	Very Good	Good	Poor	Good	Good	Can be very low to high	Very Good	Low
948 IL	2.477 in 62.92 mm	3.00 in 76.2 mm	1.211	Low	Good	Good	Average	Good	Good	Generally low but can be high	Average	Fair
997 T	2.457 in 62.42 mm	3.21 in 81.5 mm	1.306	Average	Average	Average	Average	Average	Average	Average	Average	Fair
998 T	2.542 in 62.93 mm	3.00 in 76.2 mm	1.180	Low	Very Good	Good	Average to Good	Good	Very Good	Can be low to high	Very Good	Fair
1098 T	2.542 in 62.93 mm	3.296 in 83.72 mm	1.330	Low	Very Good	Average to Good	Good	Average	Average to Good	Generally Low	Very Good	Good
1098 IL	2.542 in 62.93 mm	3.296 in 83.72 mm	1.330	Low	Good	Average to Good	Good	Good (big mains type)	Good	Generally Low	Very Good	Good
970 'S' T	2.780 in 70.61 mm	2.439 in 62.0 mm	0.877	Medium to High	Poor	Very Good	Good	Exceptionally Good	Very Good	High	Poor to Average	Fair
1071 'S' T	2.780 in 70.61 mm	2.693 in 68.4 mm	0.967	Medium to High	Poor	Very Good	Very Good	Very Good	Very Good	High	Poor to Average	Fair
1275 'S' T	2.780 in 70.61 mm	3.200 in 81.3 mm	1.15	Medium to High	Average	Good	Very Good	Good	Good	Medium to High	Good	Very Good
1275 IL	2.780 in 70.61 mm	3.200 in 81.3 mm	1.15	Low to Medium	Good	Good	Very Good	Good	Good	Medium to High	Very Good	Very Good
1275 T Non 'S'	2.780 in 70.61	3.200 in 81.3 mm	1.15	Low to Medium	Very Good	Good	Very Good	Good	Good	Medium to High	Very Good	Very Good
1275 A+ T	2.780 in 70.61 mm	3.200 in 81.3 mm	1.15	Medium	Good	Good	Very Good	Good	Good	Medium to High	Very Good	Very Good
1275 Turbo T	2.780 in 70.61 mm	3.200 in 81.3 mm	1.15	Medium to High	Average	Very Good	Exceptionally Good	Good	Good	High	Good	Good

T Denotes transverses installation. IL denotes inline installation.
**Engine capacity is the main factor governing total power output, therefore a small engine is always at a disadvantage to a bigger one.*
*** Any race engine can be expensive, this assumes average case.*

4 Workshop Practice

I am going to make this a brief chapter, but it's nonetheless important, ***It's possible to apply all the technology in the world and still end up with a dog of a motor if the workmanship and building techniques that went into the unit were dubious.*** Some of the things I am going to say here may sound obvious, but I am mentioning them because believe it or not, I have seen people build engines, making just the mistakes I am going to advise you not to.

For instance, when it comes to engine building, cleanliness is the order of the day; do not attempt to build an engine in the garage where you have just ground your cylinder head, at least not until you have been through that garage with a vacuum cleaner and cleaned it as spotlessly as you would your living-room. Except for grinding heads, grit and engines **do not** go together. If you are working in a garage that has a concrete floor, remember it will give off dust, dust that is nothing more than fine grit. Don't mix it with oil and engines. Your best plan is to seal the floor. If you can't afford proper sealing paint, then ordinary household paint, thinned down, is a good temporary measure.

If you've just torn your engine down, the chances are all the tools you used will be dirty: these all need to be washed off and thoroughly dried before putting together a new motor. Of course many builders work in the evenings and weekends, doing their engines piecemeal. First they get the block, crank or whatever done, and these parts may be lying idle for some time whilst other parts are being worked on. Don't leave these parts lying around. As each component is reconditioned or modified ready for use, oil its machined surfaces, paint the cast surfaces (if it's a block) and then pack said parts in sealed plastic bin liners so that they stay free of moisture and grit.

When it comes to parts cleaning, one of the worst components to handle is the block. Getting blocks clean can be a pretty lengthy job, but however long it takes, it's time well spent. In this respect, Series 'A' enthusiasts in the U.S.A. are somewhat better catered for than they are in England and most other countries. Many U.S. engine reconditioners have highly caustic hot tanks, boil-out tanks or jet-spray cleaners which really get blocks down to the bare metal, many using powerful chemicals to reduce the block to bare cast iron by removing rust and everything else. Cleaners such as this are not so common in Europe. If your local motor machine shop does have a block cleaning facility, then by all means make use of it as it not only gets the outside of the block clean, but also scours the oilways clean. However, don't assume the oilways will be perfectly clean, no matter what type of caustic cleaning is used. If you make the assumption that the oil galleries will have debris in them, you will more than likely be right. In view of this it is vital to pull out the brass plugs and tap the block so that screw-in gallery plugs can be installed. Then, after the tapping operation has been done on all the galleries concerned, use a rifle barrel brush to go through the galleries to get them spotlessly clean.

If cleaning facilities are limited, then it's still entirely practical to get the block completely clean by using a good de-greasing agent and a hose pipe. I know at first thought, putting water onto a block, especially after it has been machined, sounds like sacrilege, but it is of far lesser consequence to have a few spots of surface rust on a block than it is to have grit in the oilways. If you have to wash the block 'by hand', then here are a few pointers: firstly, pull **all** the plugs and go through the block with a stiff brush and rifle brushes to get everything clean. Use a de-greaser such as Jizer, Gunk or whatever the equivalents are in whatever country you happen to reside. After you are sure that all grease, muck and grit has been loosened off, use a hosepipe to wash off the block. Once the block has been washed, pop around to your local service station and use the airline to blow the water off the block. Bring the block back and immediately paint all the cast surfaces. When the block has been completely painted, that is inside and out (red primer on the inside and red primer plus a colour on the outside) then the block can be washed through with paraffin/kerosene to stop it going rusty. Once the excess of this is wiped off, the oil plugs can be smeared with a suitable sealer, such as a silicone, and installed.

Cranks need a similar treatment. After they have been ground, they need to have all the oil drillings in them thoroughly brushed out. If it's an old crankshaft, a lot of deposits can reside in those drillings. Not only can they starve the engine of oil, but if they do loosen and break free, and they surely will, a considerable amount of debris can go through the bearings. This will not do the bearings or the crankshaft journal any good.

Hopefully I have made the point on cleanliness. When it does actually come time to put the engine together, a steel or linoleum-topped bench beats a wooden bench any day. Wooden benches tend to have grit embedded in the surface. If you have a wooden-topped bench, do not build your engine directly on it. Buy an off-cut of suitably tough floor covering and put it on the top of the bench. Also get yourself a stack of paper kitchen rolls. I find that on an average build I will go through at least two, and usually three or four rolls of kitchen paper.

Although it might seem like a luxury at first, having a proper engine stand is most decidedly an asset. It allows you to get around the engine much easier. Of course the way the engine is

Although it may seem like a luxury item, an engine stand is worth its weight in gold.

bolted up to the stand will mean that at least some components can't be installed. For instance, some of my engine stands use the alternator bracket mounting bolts. This means I can't actually put the alternator on before taking the engine off the stand, but it's a small point. The rest of the engine build carries on uninhibited.

If you are doing the job of assembling the engine properly, then it's necessary to do a 'dry build'. The definition of a dry build is an engine build where all clearances are checked and adjusted as necessary. For instance, the connecting rods will all vary in length slightly and so will the piston pin-to-crown heights, and for that matter, the stroke will vary on each crank throw. This means that each piston could stop at a different distance from the top of the block. Ideally we need all the pistons to come up to the top of the block and stop in the same place, which could be flush, 5 thousandths down, 10 thousandths down, 2 thousandths out, or whatever is deemed to be the best figure for the particular application concerned. Doing this will allow precisely the desired compression ratio to be achieved for each cylinder. As we shall see, this may vary from cylinder to cylinder.

Assuming the job in hand is a total rebuild, and not just an ordinary engine freshen-up, then it's not necessary to put rods back in the same order as removal. Of course, if it's a total rebuild, pistons will be new, so will rings, bores and crank journals, so there is no question of a necessity of returning parts to original locations. If you are just 'freshening' the engine by replacing things like piston rings and bearings, then of course everything not renewed must go back from whence it came; however, for our total rebuild, the selection of parts for a cylinder is open. This means that at this stage we can select the shortest rod to go with the tallest pin-to-crown height of the piston and install that, if necessary, in a bore with the shortest stroke. This means that we can, by parts juggling, get the piston crown to deck height as close as possible on all cylinders. From there on it's a question of measuring each one and deciding how much each piston has to be machined to give them all the same crown-to-deck height clearance. Once the figure for each piston is decided on, the next thing that has to be done is to decide how much has to come off the top of the block, and really without doing a dry build it's impossible to tell. Of course when you are doing a dry build it's not necessary to put the rings on the pistons, so the assembly can be done pretty rapidly. It's also not necessary to put circlips into pistons for a dry build, so that speeds things up even more.

If you are building an engine which has very shallow combustion chambers, as is the case with a small displacement high compression engine, then sometimes it's necessary to check valve-to-piston clearance. Again, this can be done in the dry build. To check this piston clearance, the cam must be timed-in properly as detailed in a later chapter on cams and valve trains. Suffice to say here that the cam timing can be done and corrected as necessary during the dry build.

Once cam timing has been properly set, the cylinder head, with its valves and valve train, can be installed. I usually do this without the cylinder head gasket and this puts the valves approximately the thickness of the gasket closer to the pistons. Before actually putting the cylinder head on, place modelling clay (Plasticine) on the piston crowns. The head should then be assembled onto the block, together with the valve train. I usually check that the pistons are half-way down the bores at this assembly point. The tappets are then sequentially adjusted. Of course, on the first cylinder you adjust the tappets, you will find whether or not the valves hit the pistons. It's necessary to turn the engine over very slowly and feel if there is any resistance due to valve-to-piston contact. The most likely position for it to occur is 15–25° or so off the T.D.C. point. If any resistance is felt, don't turn the crank any further. Just stop, pull the head off and check the thickness of the compressed Plasticine. Obviously if the Plasticine has been totally flattened out, it's because there is insufficient valve-to-piston clearance. The obvious remedy for this is to have valve pockets machined into the pistons. The easy way to find out where the cutouts must be is to make up a centre punch from an old valve, making sure the point is truly concentric with the shaft, and then systematically install the 'valve' into each guide and lightly centre punch the pistons. This can then be used by the machine shop as a reference point for machining the cutouts.

Another factor that can be checked out at this point, although it can also be done during the final build, is the setting up of the valve train. If a lot of material has been skimmed from the head or the block, it may be necessary to compensate by either putting packing under the rocker shaft pillars or shortening the pushrods. Of the two, shortening the pushrods is preferable. However, you really need a lathe with a very slow speed plus a form tool for this job. As very few motor machine shops are geared to do this work, the price charged is often quite high. Also during the dry build you will get an opportunity to check whether valve spring coil binding is occurring. With the cam profiles that are coming into favour in the late' 80s, the amount of lift given is much higher than many of the outdated profiles previously used for the Series 'A' engine, some of which were designed right back in the '50s. Modern high lift cams can, especially if used with high ratio rockers, lift the valves sufficiently to cause traditional Leyland (Austin Rover) style valve springs to coil bind well before full lift is reached. Obviously if this situation occurs, the rocker or some part of the valve train will flex or the rocker will actually break. The dry build allows you to establish such criteria and rectify as necessary prior to a proper engine build.

It is unlikely that many readers will have their own fully equipped automotive machine shop and therefore do all the work on their engines themselves. The only people who are likely to do this are professional engine builders with a machine shop, or people who work in motor machine shops. This means that the rest of us have to go along to a motor machine shop for jobs such as boring, honing, crank grinding, etc. **If you are doing your own engine building you, not the machine shop, are responsible for checking the machined parts that go into your engine. If you don't you are not a good engine builder.** If you installed a crank that's too tight and subsequently ruined your engine, don't expect the machine shop to re-imburse you other than for regrinding the crank. As an engine builder, you should check every single component that goes into your engine. This means that any mis-machined part should be discovered before the final build. If there is any doubt about any of the components you are putting into the engine, it is a good idea to remind yourself that it is cheaper to fix a problem now than later on when it can cause a major catastrophe within the engine. Remember, doing the job properly is half the battle towards getting good results from the engine.

5 Performance Filtration

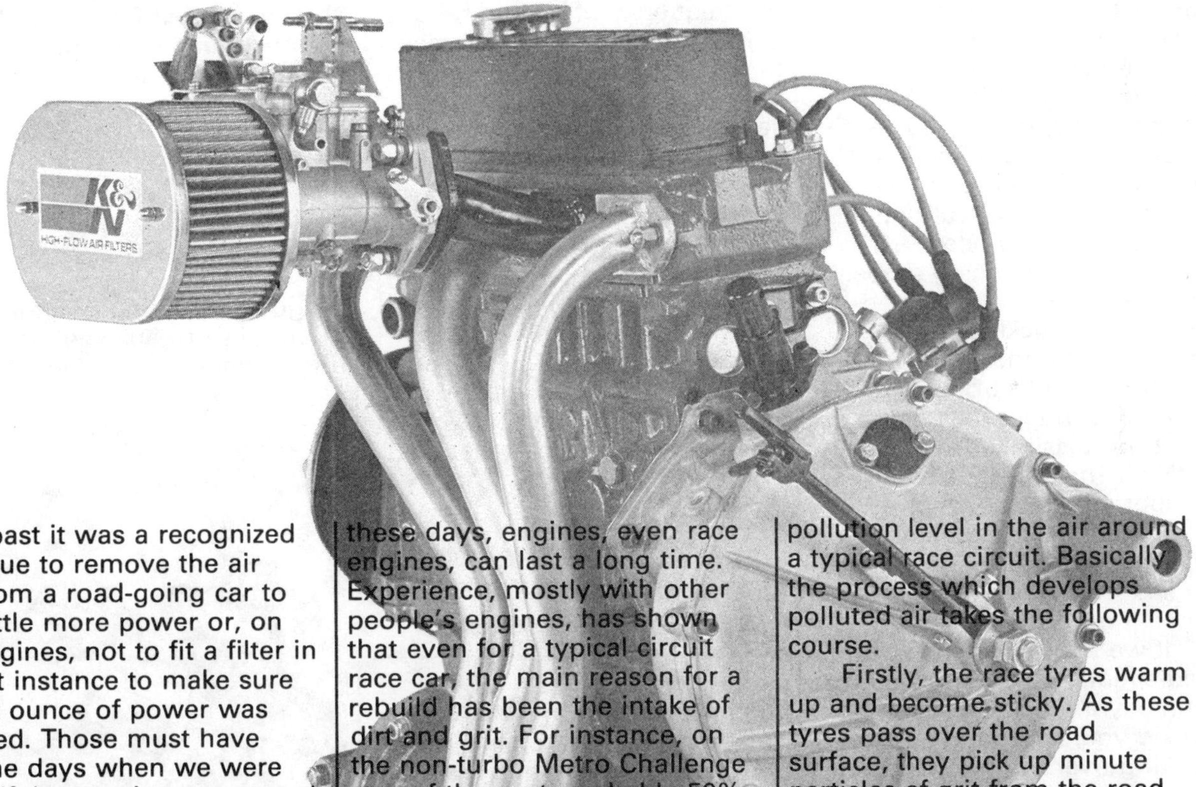

In the past it was a recognized technique to remove the air filter from a road-going car to get a little more power or, on race engines, not to fit a filter in the first instance to make sure the last ounce of power was extracted. Those must have been the days when we were less skilful as engine tuners and when air filter manufacturers were less skilful at building air filters, because that era has most certainly gone.

These days, it is not only practical but positively beneficial from every aspect to utilize a filtration system to supply clean air to the engine. The advantage of employing an effective air filter is that it will protect what is probably one of the most expensive engines you will ever build. Modern oil technology being what it is these days, engines, even race engines, can last a long time. Experience, mostly with other people's engines, has shown that even for a typical circuit race car, the main reason for a rebuild has been the intake of dirt and grit. For instance, on the non-turbo Metro Challenge cars of the past, probably 50% of the rebuilds required were due to pitted valve seats, scored bores and worn rings. All these maladies could have been avoided had a decent air filter been fitted. It is easy to fall into the trap of thinking that because a car is being raced on a circuit, the air it will be breathing will be relatively clean. In fact, **nothing could be further from the truth.** Remember, this is the age of the sticky compound tyre, and it's the use of such tyres which causes a high grit pollution level in the air around a typical race circuit. Basically the process which develops polluted air takes the following course.

Firstly, the race tyres warm up and become sticky. As these tyres pass over the road surface, they pick up minute particles of grit from the road. This happens mostly when the car is in or just exiting a corner. As the car builds up speed down the straight, so these particles are centrifuged off the tyres. Each particle of grit is usually covered in or mixed with, particles of sticky rubber. This grit is thrown up into the air and can then be ingested into an engine. As it goes through the engine, so the intense combustion temperatures cause the rubber to melt resulting in the production of an excellent substitute for grinding paste. Do I need to say more?

During a ten lap race at Silverstone, it would not be uncommon for about a teaspoonful of this tyre rubber/ grit compound to go through an engine. Ingesting this much dirt can quite easily wipe out 3% - 5% of the engine's hp. Is it any wonder that engines can get a shade on the smokey side after half-a-dozen races?

As far as road engines are concerned, a similar situation can exist, even in relatively damp climates such as that of Britain. Dust from the road is disturbed and put up into the air by passing vehicles. Since the early '70s, road tyres have become far more refined in terms of their wet road performance, and a lot of this wet road grip has been obtained by advancing rubber technology. Although a road-going tyre's stickiness is far removed from that of a race tyre, the net effect of the amount of rubber/dust in the air is not too dissimilar to a race track's, simply because the number of vehicles involved is far greater. So the argument for having a good filtration system applies just as much to a road car as it does a race car.

If we are talking about off-road events, then a filter is most definitely a must. However, do not kid yourself that a nylon stocking over the ram pipes of your engine's Weber is going to do any good. All it will succeed in doing is to reduce the engine's hp; typically by 6-8hp! The fine dust that is going to cause the damage, will pass right through that nylon stocking and chew up your engine's rings, bores and valve seats, just the same. Sure, the nylon will stop big rocks that could break things instantly rather than grind them away at a lesser rate, but that's all it will do. **It is foolhardy to run even one off-road event without an air filter.** It's no good saying, 'I'll just do this one event and put the air filter on next week' because the chances are you will already have ruined your motor during the one event when it was run filterless.

I think the point I am trying to make is probably getting across, but if not, let me summarize: if a correctly sized, high efficiency filter is installed on your engine, it will allow the engine to maintain its maximum power output over a considerably longer period. It will save you money because the frequency of engine rebuilds will be reduced. A typical race tuned Series 'A' engine could, breakages aside, run a whole season without really requiring a rebuild, whereas without a filter the engine may require two total rebuilds during the season to maintain its competitiveness.

Hopefully having convinced you that a filter is a necessary item on your engine, I will now move along and have a look at what's on the market to establish just exactly what an efficient filter is.

Filter Performance

It is fortunate for me that I live most of the time in California. Fortunate, because this area of the world is one of the great centres of off-road racing. It is also a very dry and dusty area. Such conditions place great demands on filters, so finding a type that will work really effectively amounts to the same thing as adding hp to the engine. With dust conditions as bad as they are in this area of the world, you could not expect a poorly filtered engine in an off-road car to last more than 5 or 10 miles before it ground itself to uselessness. 500-1000 miles off-road races mean that engines have to stay in one piece a long time, and it also means that filters have to do their job. It is with this background in mind that I pass on my findings on various filters, and from this experience **I can quite categorically say that the type of filter you need to use on your engine must be effective in three distinct areas. Firstly, it must filter; secondly it must be capable of holding a lot of dirt without necessarily impeding the airflow to any great extent. In other words it must have a low clog rate, and thirdly, it must be able to flow as much air as possible for a given size of filter.**

Airflow

As good a starting point as any in an effort to analyse the effectiveness of available filters,

Like many of the standard filter cases, the ordinary Metro item is anything but high performance. A free-flowing filter will often add about 5 hp to the engine's output.

is the air flow bench. By simply flowing an extensive variety of air filter elements, and calculating the flow rate per square inch, it is possible to get an accurate idea of their relative efficiency. A point I need to mention here is that for filter testing a pressure drop of 1.5 inch of water is often used by performance orientated filter manufacturers; this represents about the maximum allowable pressure drop for a high performance system. As far as filtering media is concerned, the most popular types of element I have tested were paper, foam or foam derivatives, plain wire, which is subsequently oiled to catch the particles, a combination of wire and wool, and lastly wire and cotton.

Let's start off with the good old paper filter element, as there are probably a hundred of these for every one of all the other types put together. This being the case, it's important to understand just how all the other filters lie in relation to the paper element filter. **Firstly, without a shadow of doubt, the paper element filter does filter out the dirt from the air very effectively.** This is a point which motor manufacturers are very keenly aware of. Manufacturers do not skimp on element filtering capability, since it is the reputation of their engine at stake, and they will have to bear the cost of replacing the engine (under warranty) should it suffer accelerated wear rates.

As far as airflow is concerned, the paper element ranges from pretty good to absolutely abysmal, depending upon which brand is used. If you refer to the chart, **Fig. 5.1**, you will see various major brands tested in an 'as new', straight out of the box, condition. These figures give the airflow capability of the outer surface area of the filter, or to put it another way, the installed filter area. I need to make this point clear because it's important to understand that a manufacturer can increase the airflow of the element simply by

For this chapter, the author drew heavily on flow bench and desert testing. Not only were Leyland filtration systems tried, but also a variety from other manufacturers such as this Ford item shown here. Some worked ...

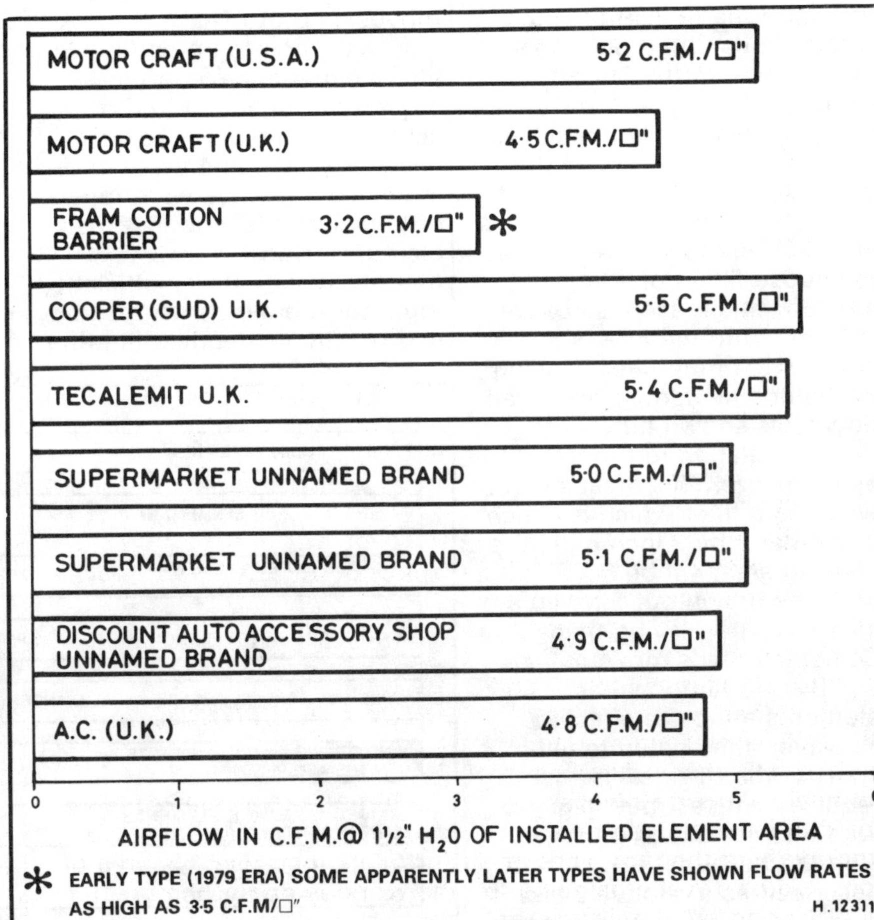

Fig. 5.1 Paper air filter element performance

Paper element filters certainly aren't all the same as regards air flow performance. This chart shows how things were in 1986 but (and it's a big 'but') things can change quite drastically overnight. The big problem is that one of the critical factors is the type of brand of filter paper used. There are several companies who supply paper to the filter manufacturers. As with almost anything in the motor industry, pricing is cut-throat. If a filter manufacturer changes from one paper supplier to another, the air flow performance may change, depending upon the efficiency of the new supplier's paper. Big name brands such as Motorcraft, A.C.,Tecalemit or Cooper, tend to be more consistent because they either have more stringent specifications for the paper they use, or they manufacture their own.

using a greater number of pleats, but you and I as end users, have no control over this. Obviously if a manufacturer puts more material into the element, the final cost will be more. Looking at it another way, there are many companies which make cheap-skate filters. They are cheap because the amount of element material in them is less, and therefore there is less total paper area for the air to pass through. The golden rule for a manufacturer is that within limits, the more paper that is put into the filter element, the higher the resulting airflow capability will be. For you and me as end users, the golden rule is that the bigger the element is, the more air it is likely to flow so we need to choose filters of adequate size. For this reason, since we are limited to the filter size which will fit the filter housing or the engine compartment, all flow tests shown here were done in relation to the installed element size. This means that if we have a filter which is 6 inch in diameter by 3 inch high, the filtering area will be 6 x Pi x 3, NOT the drawn out filter area of all the pleats. Having made that point clear, let's move on.

By and large, paper elements are quite effective. However, they begin to fall short of the racer's requirements as they begin to load up with dirt. A paper element has what can be best described as a relatively fast clog rate. In other words, **as the dirt begins to fill the pores in the wax paper, so the flow capability of the filter element as a whole, drops off very rapidly.** It is for this reason that performance drops off, unless the paper element is changed regularly. Additionally clogging can, depending on how the carburettor float bowl is vented, affect the mixture strength. A clogged filter, to a greater or lesser degree, causes the mixture to become richer and this is the case whichever type of filter you have. However, if a filter clogs faster, then the change in mixture occurs over a lower mileage.

As you can see from **Fig. 5.1**, the performance of paper elements can vary quite considerably. What I now intend to do is to look at the performance of other filtering media in relation to the best paper element I tested, namely the Cooper element. So this moves us on to **Fig. 5.2**.

Foam and foam derivative filter elements have found a degree of popularity in the accessory market for some years. My personal experience indicates there is a wide range of effectiveness amongst foam elements. Some filter manufacturers use insufficient foam for the elements, resulting in woefully inadequate filtering capability. An easy check for this type of filter is to hold the element up to the light. If you can see light through it, the chances are the foam is not thick enough. On the other hand, there are foam filters around which really do get the job done. A filter which is popular – particularly in the U.K. – is manufactured under the name 'Filtron'. This item is marketed in U.K. by such companies as **Performance Spares Centre, Cosworth Engineering** and a few others. In the U.K. another company whose products are rapidly gaining popularity is Pipercross. Though, I imagine, not up to 'desert terrain' filtration capabilities, their filters would be O.K. for most circuit race applications or where loading with wet debris would be unlikely.

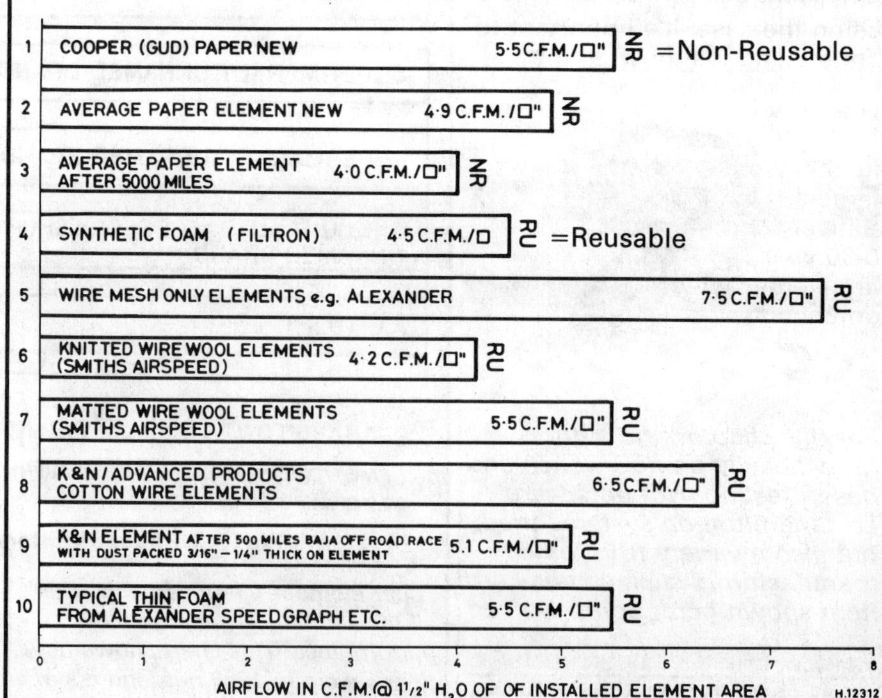

Fig. 5.2 Air filter element type comparison

The highest flowing type of element shown here is the wire type, but ... it doesn't filter very well, so there's not a lot of point in using it. The next alternative on the list is the K & N element. Its performance both new and really loaded with dust speaks for itself!

In the U.S.A., a limited number of shops throughout the country have Filtron elements. Generally they are easily recognizable by the fact that the element itself is bright green in colour. **Flow testing showed that a foam element was generally less effective on a per square inch basis than a typical paper element filter,** as depicted by Test 4 in **Fig. 5.2**. Of course, those elements having a thinner foam than typically used by a Filtron unit did flow more. Bear in mind though, they do not filter, so they can't really justify their existence on an engine in the first place. Although the Filtron element flowed less than the best paper element, it did score by having a significantly lower clog rate. In other words, a Filtron unit can hold a lot more dirt before flow begins to drop off significantly. On the debit side, because they flow less than a good paper filter to start with, a foam filter must be made larger to equal the same as-new performance as a paper element filter.

Moving along to Test 5 of **Fig. 5.2**, we have the wire element filter. These were relatively common up to about 20-30 years ago. Usually the wire element filter worked in conjunction with an oil bath. As such, it was reasonably effective, although still less effective than its modern counterpart, the paper element. Most wire filters sold on the accessory market, make do without the benefit of the oil bath, and in essence, it is the oil bath which is the key to a half-way decent filtering action from the wire element. As such, these wire elements do not filter anything like as effectively as the paper or foam element filters just mentioned. In fact, unless conditions are almost dust-free, the wire element filter should be looked upon only as a means of keeping large particles out of the engine. Even when these filters are oiled, a significant percentage of the small, engine-damaging particles, can still pass right through the element. As far as airflow is concerned, the chart **Fig. 5.2**, shows where, on the flow efficiency scale, the wire element is situated. As you can see, it is capable of flowing a lot of air, but that is a pointless exercise if it doesn't filter very well.

... and some didn't. This particular wire mesh filter did not stand up to the rigours of desert filtration and its flow rating was only average.

Filtration systems like this are a disaster, not because the element material itself is inefficient, but simply because there is insufficient area. This engine here would be better off with a standard filter. It would make more power and live longer.

Tests 6 and 7 show the airflow performance of the wire/wool type element. I must admit to not having tested this type of filtering element in harsh, off-road conditions in California, so I will have to stick my neck out and make an assessment based on what I think is considerable experience. Firstly, the wire/wool concept appears to have a lot more potential to filter the air than a plain, wire element. Of course it will probably perform best with oil applied to the element so as to make it tacky to catch the particles and hold them. Bearing in mind that this is only an educated guess, it would be fairly safe to assume that the wire/wool element is superior to the plain wire element, but probably inferior to the paper element. As far as I know the only company to manufacture this wire/wool type element was **Smiths Filters** and they sold them under the name of 'Speed Filters'. You will notice that in **Fig. 5.2** there are two flow ratings for this wire/wool type element. Basically, this stems from the fact the Speed Filters are made by one of two methods. Firstly the element can be 'knitted' out of the wire/wool material, and in so doing it produces a much more densely packed element than the second case, where the element is simply wound around in an ad lib form, and matted together. My guess is that the knitted element has the best chance of filtering effectively, but as such, its airflow capability is down.

Next on the list is the wire and cotton combination as produced by **K & N Engineering** in U.S.A. and **K & N (Europe)** in U.K. Like many of the other filters, this type requires element oiling for it to be effective as a filtering medium. As far as effective filtering goes, this filter has proved time and time again in off-road races that it's just

An adequately sized, efficient, air filter such as this K & N really does produce good results. The Superflow 300 flow bench showed that the filter caused no loss in flow at the airflow levels required by a 125 hp Series 'A' engine.

about the best available. It has also proved capable of satisfying the stringent needs of the aircraft industry, the mining industry, the trucking industry, as well as the requirements of most of the top racers.

Tests have shown that this filter's resistance to clogging is quite amazing, and I use that term guardedly. In its 'as new' form, the element topped the list in flow efficiency in terms of flow per square inch for a *high filtration filter*, but of even greater significance is the fact that a K & N filter, packed with 3/16 inch of dust, sucked up from a dusty Baja 500 off-road race, still out-flowed most brands of new, paper element filters. Put another way, if a standard K & N filter is used instead of a paper element filter, a typical road-going car, would need a filter cleaning service at intervals of about 100,000 miles!

Mentioning cleaning leads me onto another subject. Like anything, filters cost money. Some of the filters discussed are throw-away items, whereas others are permanent filters. If you study the chart, any filter which is designated 'RU' means it's Re-Usable. Any one designated 'NR' is Non Re-usable. You may pay a little more for a permanent filter, but at about the time your car's second or third service point falls due, you will break even on the cost of the filter. However, if you have chosen a filter which has a very low clog rate, the chances are, you have also saved some fuel, and if you have chosen a filter which really does filter the air, the chances are, you have saved on engine wear, **all of which means a good filter can be a real money saver.**

Those filters using a foam element, sandwiched between two grids of metal, such as the Speedograph filter or anything similar, are difficult to clean in as much as re-assembling them can be a tedious and arduous job. The Filtron/Cosworth/ Performance Spares Centre filters are simple to clean as the foam element can be pulled off its wire frame and easily washed. It takes only some suitable solvents, plus a tap, to get these elements back to as good as new. The same goes for the K & N filter. However, care must be exercised with the K & N as the filter must be washed from the inside out, otherwise dirt that is on the outside of the filter can be transposed to the inside, and once it's on the inside, it's going to find its way into the engine . . .

So much for elements. Out of all the elements tested, the one most popular amongst high performance enthusiasts and professional racers alike is the K & N element, and it's one that I personally recommend. Quite honestly, it's simply the best I have tested to date. However, there is one point I should make: do not fall into the trap of over-oiling the K & N filter. If you are using the squeeze-bottle/tube to oil the filter, run a bead of oil along each rib. If you have oiled it just right, it should take 15-20 minutes for the oil to 'wick' through to the other side. If the filter has soaked up all the oil in just 2 or 3 minutes, the chances are you have over-oiled it, and **when a K & N is over-oiled, the airflow drops dramatically.** Apart from this cautionary note, there is nothing else you really need concern yourself with as far as the K & N filter goes. It's made to very high standards and does its job well.

Cases

Having an adequately sized element is half the battle towards getting an efficient filtering system, but the effectiveness of the element can be completely offset by bad case design. Excluding the path of the air entry into the

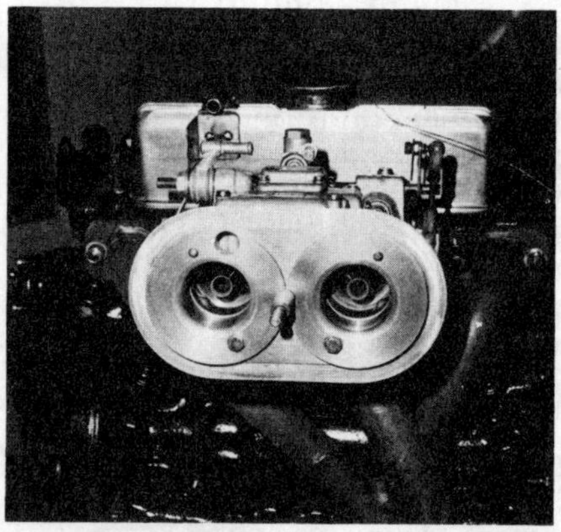

In the days when air filters on race engines weren't fashionable, I was running my engines with big air filters to save expensive engines which I could barely afford in the first place. This is the backplate of one such installation. Note the built-in ram pipes. The engine produced the same hp with or without a big air filter element in place.

carburettor mouth for a moment, the only really critical thing about the case design for an air filter on a high performance vehicle, is just how big an air filter element that case will hold. As I said earlier, with filtering systems it's very much a question of the bigger the better. **A reduction in the size of the filter element can only be successfully achieved if a highly efficient element is used.** Judging from some of the filter cases I have looked at and tested, it would appear that many filter manufacturers are barely aware of this situation, or if they are aware of it, they do not heed the situation to any worthwhile degree. In many instances, K & N filter cases flow tested were smaller than a lot of the other brands tested. However, they stood on the fact that the element used is highly efficient and therefore it can successfully be used as a smaller element case combination.

Basically an engine requires between 1.4 and 1.8 cubic feet of free air per minute for every hp it develops. That air must pass through a filter, so the filter must be big enough to allow it to pass through without any significant pressure drop or resistance to airflow. Knowing this figure, we can make an estimate of just how big a filter must be for a given engine power output.

Fig. 5.3 gives the applications of popular K & N filters, for use on Series 'A' engines. This chart shows, essentially, what a given filter can be installed on, without actually costing the engine any power, due to filter flow losses.

An important feature the design of the air filter case must take into account is the proximity of the lid of the case to the carburettor mouth. If too close, the air flow into the carb will be restricted. If just the right distance, it can, under certain conditions, aid airflow

Fig. 5.3 K&N Filters for 'A' Series Engines
SU units with 5⅞ inches diameter (elements measure 6 inches O.D.)

Application	APT Filter part number and old K&N number	K&N Europe filter part number	Stub Stack Cast	Stub Stack APT Billet
MINI				
1¼ inch HS2	*SD22-318	56-1630	SS50	MSS2
1½ inch HS4/HIF38	*SD23-318	56-9128	SS51	MSS4
1¾ inch HS6	*SD24-318	56-9132	SS52	MSS6
1¾ inch HIF6/HIF 44	*SD27-332 (Offset 4½ x 7 oval 1¾ deep for mini speedo cable clearance) also SD25-318	56-9142	SS53	MSS44
1¼ inch HS2 x 2	SD21-318 (most Mini's)	56-9121	SS50	MSS2
1¼ inch HS2 x 2 L/H Dr.	SDX-318	56-9122	SS50	MSS2
1¼ inch HS2 x 2	*SD22-318 (some)	56-1630	SS50	MSS2
1½ inch HS4 x 2	SD3-318 (most)	56-1390	SS51	MSS4
1½ inch HS4 x 2	SD23-318 (some)	56-9128	SS51	MSS4
1¾ inch HS6 x 2	SD4-318	56-1400	SS52	MSS6
*Modified engines over 80bhp use 2½ inch element suffix No. 317 e.g. single 1¾ inch HS6 SD24-317. Twin units e.g. SD23 318 adequate for 125bhp. *Indicates offset hole. 318 elements will cope with 95bhp with reduced cleaning intervals (10,000 miles rather than 25-30,000 miles).*				
28/36 DCD	56-1640	56-1640		
40/42 DCNF	56-1110	–		
40/45/48 DCOE/DHLA (see note #1)	Large oval 5½ x 9 x 3¼ 15-1360 (modify b/head on Mini) Small oval 4½ x 7 x 3¼ 56-1350	56-1630 56-9265	40 DCOE-WSS 40/16 45 DCOE-WSS 45/16 45 DCOE-SHSS-4512	
46/48 IDA	15-1210 (alter bonnet)	56-1210		
Split Webers 40/45/48	No APT no.	K&N no. See end of list		
40 DCOE Stub Stack				
45 DCOE Stub Stack				
40 DHLA Stub Stack				
METRO				
1½ inch HS4	SD3-318 or 317	56-1390/56-9094	SS51	
1¾ inch HIF 6/HIF44	SD25-318 or 317	56-9135/56-9136	SS53	
40/45/48/DOCE/DHLA	56-1355 (L/oval 2½ inches deep)	56-9110		
(using 4 inch manifold)	16, 26 and 39mm ram pipes available	–		
SPRITE/MIDGET etc.				
1¼ inch HS2 x 2 – 948 to 1275	SD22-318 SD21-318 R/H Dr. Spridget	56-1630	SS50	
1½ inch HS4 x 2 – 948 to 1275	SD3-318	56-1390	SS51	
1¾ inch HS6 x 2 – 948 to 1275	SD4 -318	56-1400	SS52	
New alloy units incorporating Stub Stacks in base plate.				
1¼ inch SU	MS 109S (Sprite)	56-9057		
1¼ inch SU	MS 109M (Midget)	56-9055		
MINOR (948, 1098)				
1¼ inch HS2	SD21-318	56-9121	SS50	
1½ inch HS4	SD3-318	56-1390	SS51	
1100 1300 mainly as Mini				
Cone shape elements, with Stub Stack, for special Mini applications.				
1¼ inch HS2	SD33	56-9335		
1½ inch HS4, HIF38	SD36	56-9320		
1¾ inch HS6	SD38	56-9327		
1¾ inch HIF6/HIF44	SD40	56-9330		

Notes:-

1) 40mm Webers must use the same filter as per 45/48's if the slide in type of K&N ram pipes are used.

2.) SU Stub Stacks K&N Europe no.
SS50 = 85-5038
SS51 = 85-5039
SS52 = 85-5040
SS53 = 85-5041

3.) Weber/Dellorto Ram Stacks

	K&N Europe no.	APT no.
DCOE 40's	16mm = 85-5042	WSS40/16
	26mm = 85-5043	WSS40/26
	39mm = 85-5041	WSS40/39
DCOE 45's	16mm = 85-5045	WSS45/16
	26mm = 85-5046	WSS45/26
	39mm = 85-5047	WSS45/39
DHLA 40's	16mm = 85-5029	
	26mm = 85-5030	
	39mm = 85-5031	
DHLA 45's	16mm = 85-5032	SHSS-4512
	26mm = 85-5033	
	39mm = 85-5034	

4.) A LHD Mini with twin SU's requires a special pair of offset filters. APT part No. for twin 1¼s is SDX-318 and for twin 1½s the cone filters part No. SD 36 should be used.

into the engine. This phenomenon is by no means common, so don't count on achieving it. Basically the best plan is to make sure the case lid never gets too close to the carburettor, thus avoiding carburettor shrouding. So how close is *too close*? Well, this actually depends on the amount of flare on the end of the intake trumpet. If the carburettor does possess a trumpet, the bigger the flare on the end, the closer the case lid can approach the carburettor mouth before restriction occurs. A good rule of thumb is to assume that the case must be no closer to the mouth of the carburettor than the bore of the carb. That is an absolutely rock-bottom figure, unless you have access to a flow bench to check out your proposed combination. If you can select a filter case that has more space than this, then so much the better.

Once the case lid has been spaced from the mouth of the carburettor by the diameter of the carb, then it is largely beyond the sphere of influence. An example here will serve to illustrate the point: if your engine is equipped with a 40mm Weber, make sure the air filter lid is 40mm away from the end of the carburettor ram pipe. If the air filter case is fitted with a snorkel which many original equipment air filters are, but few accessory ones are so equipped, then the intake diameter of the snorkel should be at least equal to 50% greater area than the venturis of the carb. Again, that's a rock-bottom figure, and for best results, figure on 4-5 times the venturi area for good effect.

Ram pipes are very important to the function of a side draft carb. Try and avoid using any air filtration system that cannot be used with the carb's ram pipes. Ram pipes are available even for the shortest filters.

Standard Filter Cases - Just How Good?

By and large, the standard filter cases on most Leyland vehicles are designed with a view to keeping the noise level of the induction system down. However, of late, more attention has been paid to making efficient air filtration systems. The MG Metro, for instance, has a system which is more than adequate for the standard power output of the engine. In fact it's good for power outputs up to 85 and maybe even 90hp, and on an engine such as this, because the case design is essentially good, it is easy to make a small but worthwhile improvement in the system by simply installing a K & N replacement element into the standard case. When a single carb system is employed on a Series 'A' engine, the normal air filter setup involves a 90° elbow being bolted to the carb. On the smaller carbs, 1¼ inch and 1½ inch, this elbow represents a prime restriction to airflow. On H.I.F.6s the elbow is less of a restriction simply because it has a much larger passageway. This is not to say it is more efficient in its design; it's just I am pointing out that because it's bigger, it flows more, and therefore it represents less of an impediment to the production of power than do the elbows on smaller carburettors. For this reason, any filter case which uses the 90° elbow, other than the H.I.F.6, will cause a loss in power due to the restriction at that elbow. Although the case design may carry an adequately sized filter, the elbow can be restricting the airflow.

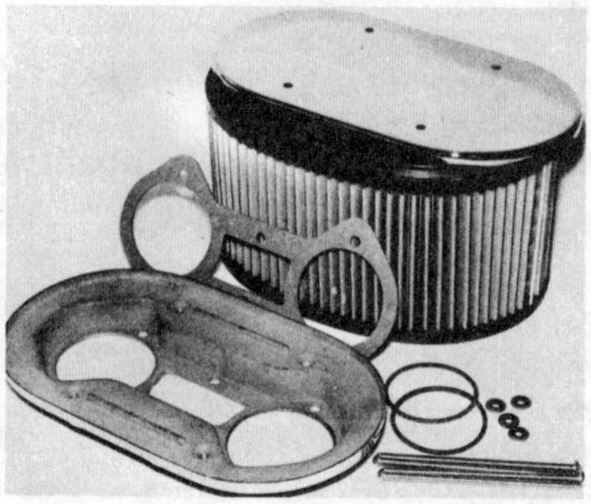

If your Series 'A' is equipped with a 46 or 48 IDA, you will need the K & N filter shown here. It's about the only one made for this type of carb and it flows sufficient air for about 400 hp. Wouldn't you love to have a Series 'A' that will run that one to the limit?

On some filter cases, such as the ordinary Metro, it is the snorkel itself which provides the major restriction to flow, and very often, in a situation like this, boring a series of holes of, say ½ inch to ¾ inch diameter, around the edge facing away from the passenger compartment, can considerably

Here's the filter setup for an 850 Mini racer employing one barrel of a single sidedraft Weber.

The term 'stub stack' is one that is often bandied about in the air filter business. Ever wondered what it is? Take a look inside this filter element and you will see that a stub stack is really nothing more than a flow bench-designed, stubby, short stack for the mouth of the carb. Hotter versions more nearly resemble a short ram pipe.

Seen here in its dismantled state, the latest high performance filter for transverse engines makes use of . . .

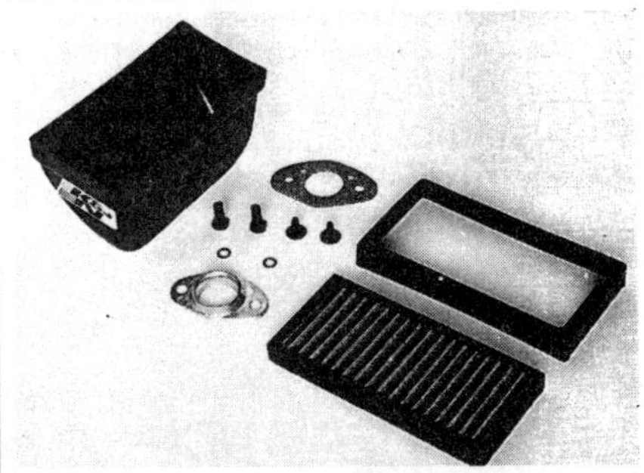

. . . a flat pad filter. By using an upswept design in conjunction with the flat pad filter, bulkhead clearance problems associated with transverse engines are all but eliminated. The K & N element assures high air flow capability.

improve performance of the filtering system as a whole. When doing this, be sure to round off the corners from the holes bored in the case, otherwise whistling will occur. With many of the twin carb systems, such as those on the Cooper 'S', the single snorkel into the case can limit hp. Also,

The air filter case on the MG Metro is one of the best to come out of Leyland for a long time. On engines up to about 80hp it works very well, and I thoroughly recommend it in installations where noise is a factor. However it won't fit all Series 'A' vehicles by any means, so check whether or not the installation is feasible before you buy.

The air filter setup on this beautifully restored 1275 'S' looks neat and functional, but airflow is restricted by the single pipe which picks up air in the area of the exhaust manifold down between the two carbs. Installing a bigger pipe here helps hp, but the system still has other restrictions.

Here's what a high performance filter looks like installed on a Mini. Nothing really spectacular, but it performs well. The only snag is, it's a little on the noisy side.

The engine builder's intention when intalling these beautifully formed ram pipes was to increase air flow. Unfortunately, he promptly undid all the good work in that respect by covering the mouths with nylon gauze. The resulting arrangement neither filters nor flows air. On the other hand ...

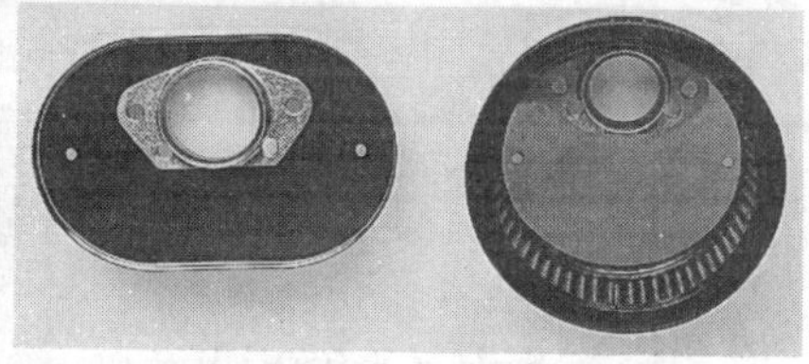

The latest 'stub stack' design from K & N Europe is seen here. I've dyno tested these and found that an improvement can be made if the radius in the throat is extended right around the full 180 degrees.

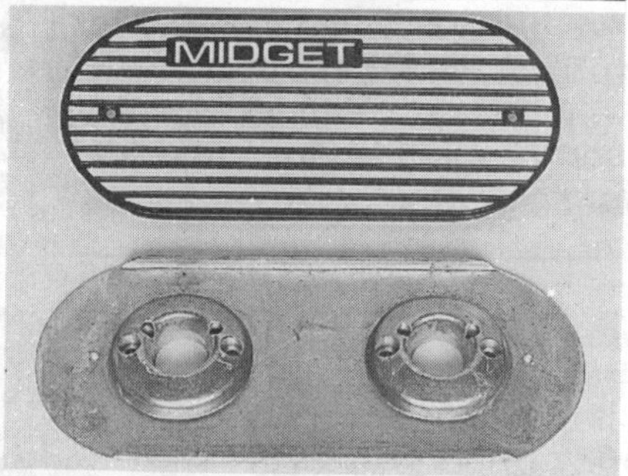

Series 'A' sport cars are well catered for in terms of high performance filters. This unit from K & N Europe, of which examples are available for both the Sprite and Midget, uses a large oval filter element plus a back plate which houses these two integral 'stub stacks'.

... these K&N filters, which house ram pipes, provide very clean air with no measurable air flow loss.

another aspect of case design which is often overlooked is the location at which the snorkel picks up air. **Remember, many of the standard cases are designed to feed hot air to the engine. That's good for economy, but not for hp.** However, having said it's good for economy, let me clarify that. It's good for economy in the first 6-10 miles; after that it's no real big advantage so long as the carb is calibrated for the cooler air it's likely to pick up if a cool air source is used.

Carb Air Entry

The way the air flows into the carb mouth can be quite critical on some carbs. On the smaller SUs, principally 1¼ inch and 1½ inch, a radiused entry makes a substantial difference to the airflow, and on something like an 850 Mini, a radiused intake to the carb, can boost the power by 5%-7% on an otherwise standard motor. In the following chapter on ram charging, I deal in detail with how best to get the air into the carb. For that chapter, a lot of tests were done on the effectiveness of various shapes of ram pipe. Suffice to say at this stage that **if you are installing an accessory-type air filter, you should be looking at one that is capable of having some sort of ram pipe installed within the filter case,** otherwise you will be throwing away a chunk of performance unnecessarily. As I have said, 1¼ inch and 1½ inch are quite critical in this respect, so, for that matter, are 1¾ inch SUs of the HS type. On the H.I.F. carb, we see a slightly different situation. The H.I.F. range of carb have a small, truncated radius at the mouth and are less sensitive to a ram pipe or a radiused entry than other carbs, but even so, they still respond to having a full radiused entry at this point.

Moving onto Webers, we find that the 28/36 Weber carb responds also to having a radiused entry, but really it only produces results in terms of hp on engines that are above the 50-55hp. Sidedraft Webers and Dellortos function best when they have a proper ram pipe. Removing the ram pipe can cause a dramatic reduction in airflow and a reduction in the main jet signal strength. All this has a tendency not only to reduce power output, but also to cut drivability. Let's face it, the whole object of putting on a sophisticated carb, is to allow the engine to breathe well and perform better. Cancelling out that extra breathing and performance by removing the ram pipes, is not really the way to go.

For virtually all the range of K & N filters available, Advanced Products (Warrington) Ltd. can supply special length ram pipes or stub stacks (very short ram pipes for shallow filters).

Calibration Changes

To facilitate calibration of the carb while dyno tuning, many engines are dynoed without a

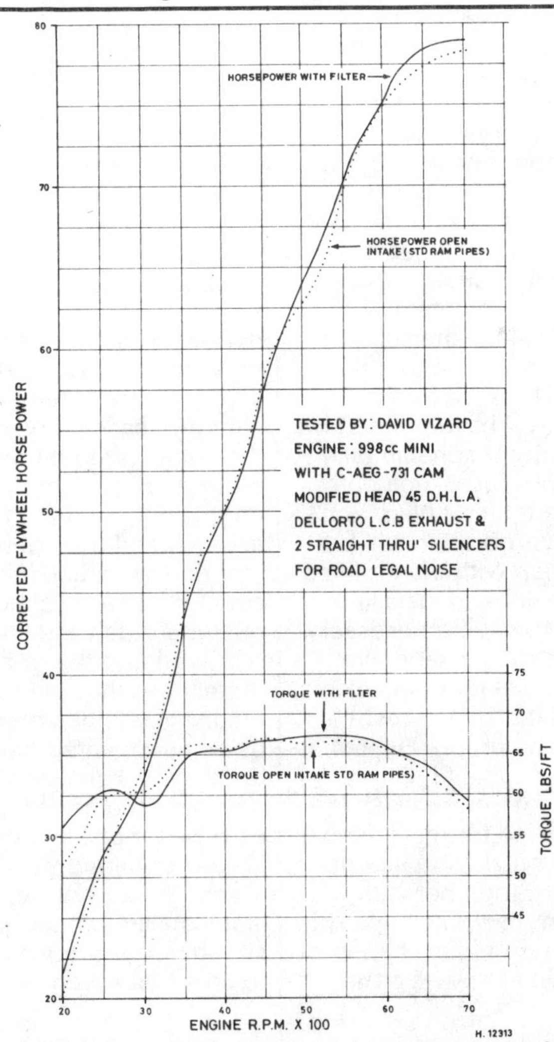

Fig. 5.4 Filter/ram pipe power and torque comparison

Here is the positive side to a well designed filter setup – more power. The unit used is shown in the accompanying photograph. This proves that extra power and extended engine life are not mutually exclusive.

filter for most part of their setting up. Using the filter recommendations so far passed along, the amount of drop in airflow going into the engine due to the presence of the filter, is limited to a virtually insignificant amount usually not detectable by the flow bench. However, experience has shown that on occasions it is still necessary to recalibrate the carb simply because certain effects associated with shock waves take place. When the air filter case is not present, shock waves simply emanate out into the atmosphere, but with the case lid present, the situation can be different. It would appear that certain shock wave reflections can come from the case lid, causing a change in carb calibration requirements. Sometimes the presence of the lid can actually enhance the power output of the engine to a slight degree. Sometimes it causes a small reduction in power at certain points of the rpm range. Usually the solution to this problem is to either move the case lid closer or further away. Anyway, I have pointed out that this phenomenon can take place, so if you are dynoing your engine without an air filter, don't assume the carburation will be exactly spot-on with the air filter, as it was previous to installing the air filter. It may be necessary to just juggle the jets and air correctors to dial it right back in.

Now for the sort of results obtainable. The graph, **Fig. 5.4,** is for a 1000cc engine I built many years ago, long before I knew of K & N filters, and for this I used a filter system employing a large paper element together with a case made in my workshop. This had internal ram pipes shaped to produce good airflow. **On the dyno it was found that, in this particular instance, the case lid aided power.** Whenever the air filter and case lid were installed, the engine would make 1-1½hp more than without. In fact it was on this engine that I first experienced the phenomenon mentioned earlier. Initially I concluded carb calibration was out when the case lid was removed and that replacing it merely put the calibration spot on. However, this proved not to be so as repeated back-to-back tests always showed that whatever the calibration of the carb was, optimum power was always obtained with the filter assembly on, not off. Of course just installing an air filter of sizeable proportions isn't going to cost that much hp if the carb is not recalibrated. **Fig. 5.5** shows what happened on a highly tuned 1430cc engine which was run with and without an air filter, the particular air filter concerned being a large K & N SD-56 1360. As you can see from this 'before' and 'after' power curve, the power only changed to the tune of a couple of hp right up at the top end. In almost all instances, a little recalibration of the carb will restore power to the original level, but here's a thought to leave you with, even if it didn't the engine would lose more hp than this after one race without an air filter. On the other hand it's not likely to lose much more than this amount of power after six races with the air filter . . .

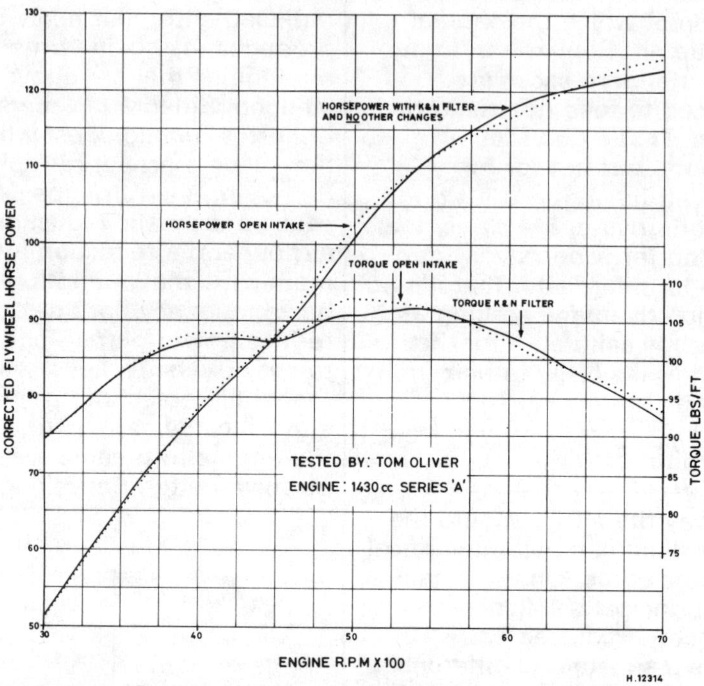

Fig. 5.5 Power output comparison between open intake and K & N filter

The open intake used 63mm long ram pipes. When the K & N filter was fitted no other changes at all were made to the engine.

6 Ram Charging

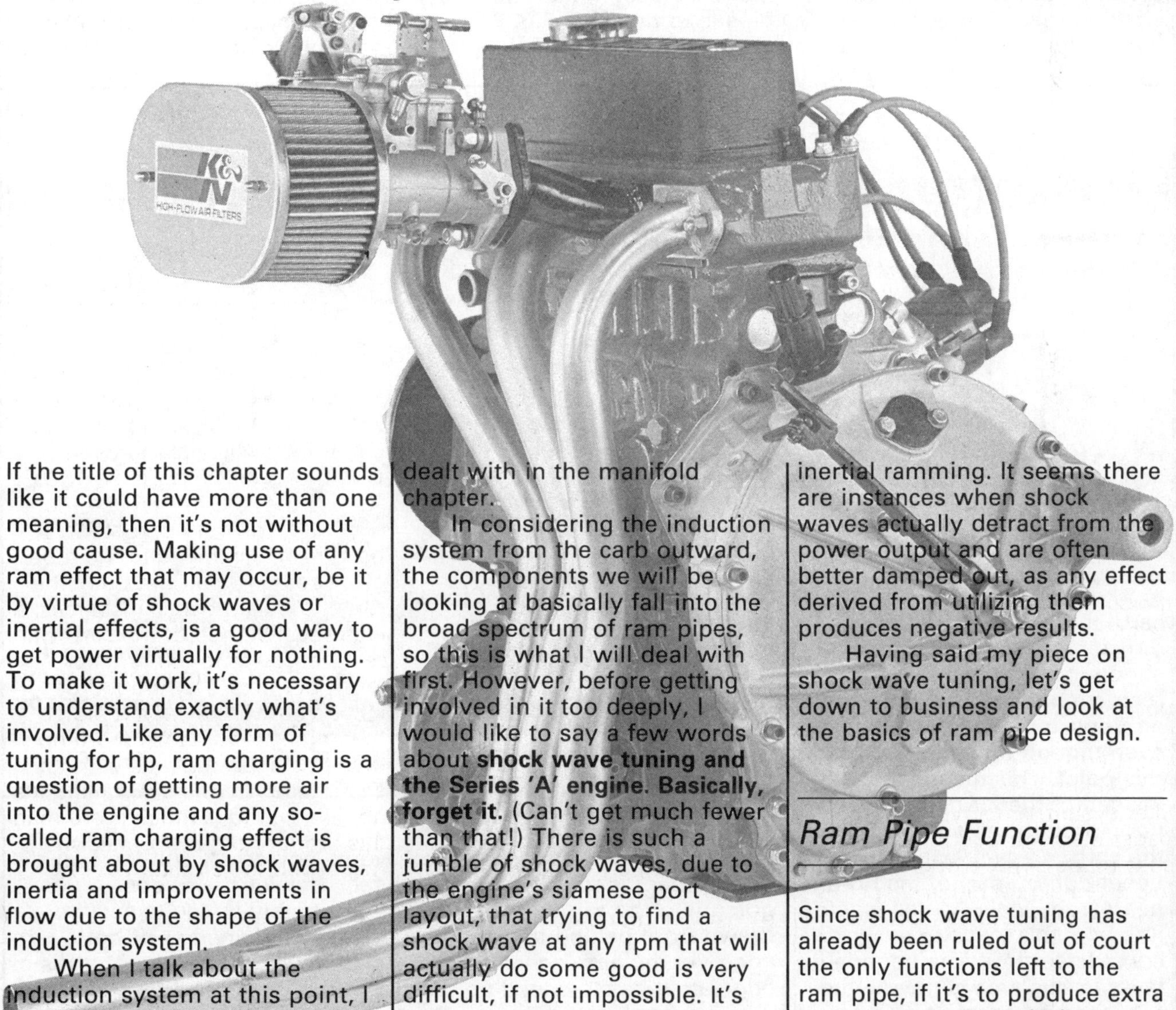

If the title of this chapter sounds like it could have more than one meaning, then it's not without good cause. Making use of any ram effect that may occur, be it by virtue of shock waves or inertial effects, is a good way to get power virtually for nothing. To make it work, it's necessary to understand exactly what's involved. Like any form of tuning for hp, ram charging is a question of getting more air into the engine and any so-called ram charging effect is brought about by shock waves, inertia and improvements in flow due to the shape of the induction system.

When I talk about the induction system at this point, I am just going to deal with the induction system from the face of the carb outwards, as the effects of manifold lengths are dealt with in the manifold chapter.

In considering the induction system from the carb outward, the components we will be looking at basically fall into the broad spectrum of ram pipes, so this is what I will deal with first. However, before getting involved in it too deeply, I would like to say a few words about **shock wave tuning and the Series 'A' engine. Basically, forget it.** (Can't get much fewer than that!) There is such a jumble of shock waves, due to the engine's siamese port layout, that trying to find a shock wave at any rpm that will actually do some good is very difficult, if not impossible. It's generally agreed by most professional engine builders that tuning the induction length on a Series 'A' engine benefits inertial ramming. It seems there are instances when shock waves actually detract from the power output and are often better damped out, as any effect derived from utilizing them produces negative results.

Having said my piece on shock wave tuning, let's get down to business and look at the basics of ram pipe design.

Ram Pipe Function

Since shock wave tuning has already been ruled out of court the only functions left to the ram pipe, if it's to produce extra power, are to smooth the airflow into the carb and provide some sort of length adjustment to make the most of inertial tuning. The truth of the matter is that any piece of pipe, so long as it's the right diameter and length to boost power in the rpm range in question, will provide inertial tune to the engine. However, the benefits of inertial tuning can easily be lost if the shape of the end of the ram pipe is not conducive to good flow.

Let's consider a carb without a ram pipe as in **Fig. 6.1**. In simple terms, this is a carb with a zero length ram pipe, having a zero radius at the intersection with the face and the bore. In other words, it's a sharp corner. You will see as the air flows into the bore of the carb, that it experiences difficulty making the turn, overshoots, and causes a sort of restriction. In other words, the air flows in a venturi shape even though no venturi exists at this point. This means that instead of the air being able to flow through the entire bore of the carb, it flows without the benefit of the part of the bore nearest the wall. As you would imagine, this leads to a constricted flow. The boundary layer in this area can easily be 1/16 inch and sometimes more, depending on the gas speeds involved. This means a 1¼ inch carb may now only flow like 1⅛ inch. To eradicate this constriction effect (vena contracta) the carb needs a ram pipe. Of course the simplest ram pipe is the radius of the entry into the carb which is fine, so long as that entry can be made compatible with the air filter case and the resultant length of carb and ram pipe, in other words zero length can be long enough to do the job. In most instances, it is not, though this is more applicable to single carb applications which we will get into later.

At this point, then, I will make the assumption that a ram pipe is to be used. **Fig. 6.2** shows the basic ram pipe, namely a straight piece of tube attached to the carburettor mouth. The only problem with a basic ram pipe of this shape is that it's a disaster in terms of airflow. If you look at **Fig. 6.3** you will see typically what happens in such a ram pipe. Airflow contraction just inside the mouth of the carb is even worse than without a ram pipe: the problem is much of the air going into the ram pipe is drawn from behind the leading edge. As the air reaches this edge, so it has to speed up and turn a tight corner. Like everything, air has inertia and does not take too kindly to being forced into a very tight, fast turn. In fact, it doesn't make it round the corner in a tight fashion. Instead it swings out and causes a greater contraction in the airflow than is the case in a carb without a ram pipe. Over a typical range of gas speeds this contraction increases with increased gas speed and finally reaches the relatively stable level over which a changing gas speed has little or no effect. Knowing that this contraction causes a reduction in airflow, we must obviously concern ourselves with the shape of the end of the ram pipe. This is an important factor because there are many ram pipes on the market which look 'powerful', but don't, in fact, deliver the goods. Fortunately it's quite easy to determine whether or not a ram pipe is likely to work in terms of airflow. There is a distinct family of shapes which do work as opposed to those which don't. To put you in the picture as to what's good and what

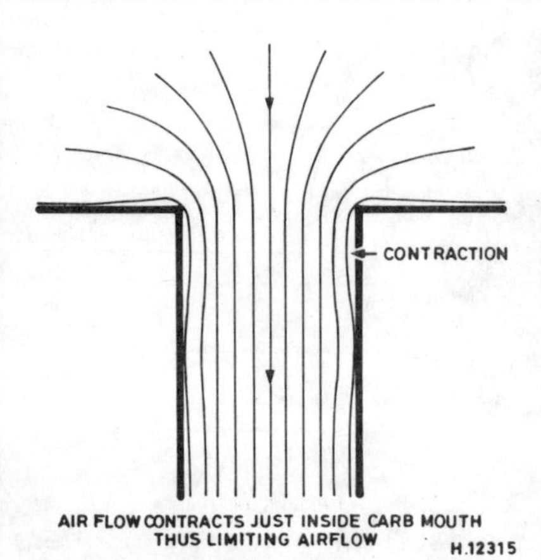

Fig. 6.1 Airflow into carburettor mouth without ram pipe

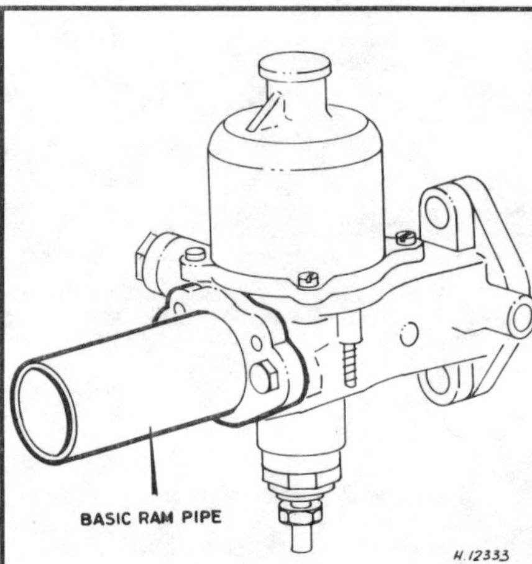

Fig. 6.2 A basic ram pipe

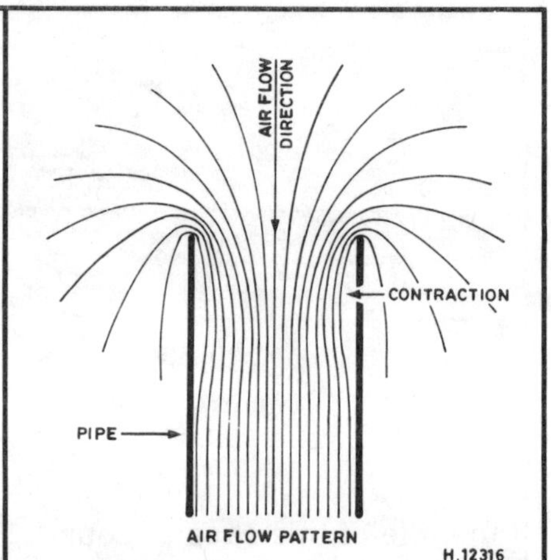

Fig. 6.3 Airflow into a basic ram pipe

A ram pipe of this form produces a greater flow contraction just inside the mouth than is experienced by a carb without a ram pipe; thus flow will be less. As is demonstrated by Fig. 6.4a generous radius is all important

isn't, refer to the flow chart **Fig. 6.4**. Except for the two basic shapes at the top of the chart, all the subsequent shapes are based on ram pipes that are, or have been, available on the market. You will see that some of these ram pipes actually produce a large negative flow change, definitely not what we are looking for. The biggest factor relating to how well a ram pipe works is the radius on the end of the ram pipe. Anything with a radius over about ⅛ inch is in with a 99% chance of working. Where a radius is not totally formed, i.e.

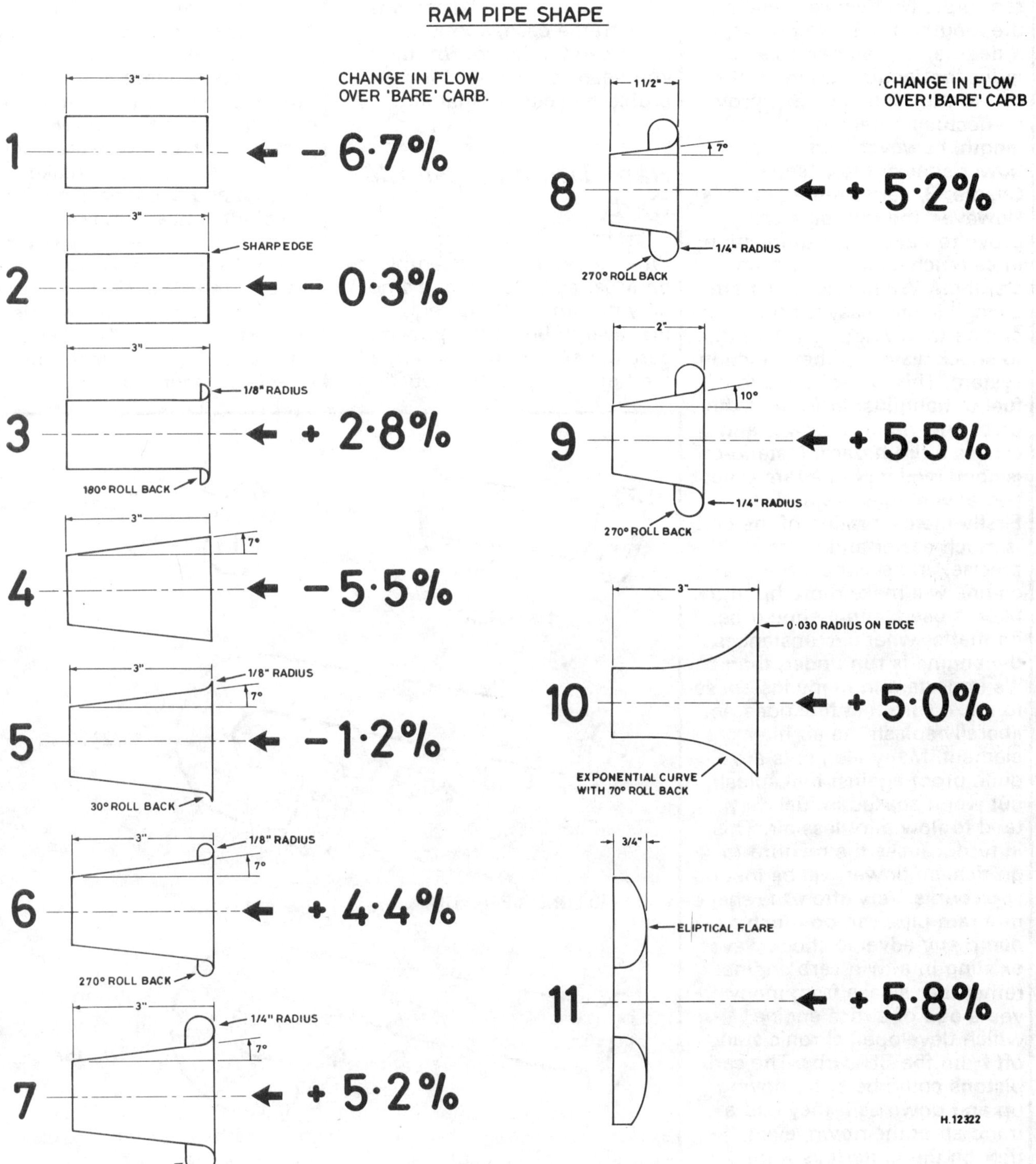

Fig. 6.4 Ram pipe cross-sections

it doesn't roll back completely on itself, then so long as it turns out at least 60°, it will work.

With the information given in **Fig. 6.4**, you should now have at least a good idea as to what actually flows air. Now let's consider the length of the ram pipe. On twin carb engines, the length of the ram pipe is, to a degree, immaterial because induction length tuning on the carburettor side, at least, proves ineffectual. Total induction length, however, can affect power considerably: see Chapter 8, Part 3 regarding this. However, the ram pipe can prove to have a useful function in as much as it can contain stand-off. When race cams are used, it's very easy for the engine to develop stand-off due to shock waves in the induction system. This causes a cloud of fuel to hang just in front of the carb. This reduces power and causes a fire hazard. If stand-off is contained within a ram pipe, two advantages are accrued. Firstly the calibration of the carb is much easier and more precise, and secondly the engine will make more hp. If a filter is used, and it should be, no matter what circumstances the engine is run under, then it's important in many instances to make sure that fuel does not liberally splash the air filter element. Many elements are quite proof against fuel splash, but when soaked in fuel they tend to flow a lot less air. This, in turn, causes the mixture to go rich, so power will be lost on two counts. Very often, the shape of a ram pipe can do much to damp any adverse shock waves existing in a twin carb engine. I remember a case from many years ago of a race engine which developed chronic stand-off from the SU carbs. The carb pistons could be seen moving up and down as if they had a mind all of their own, even though the butterflies were wide open. A pair of ram pipes having an exponential curve were fitted to the carbs. (An exponential curve has the effect of damping out shock waves.) After they were installed, the seemingly random piston motion was completely eliminated and better yet, the power came up by a whopping 34hp at the wheels! That was an extreme case, admittedly, but it can be the sort of thing you need to look out for if a problem occurs in that area.

Ram Tuning A Single Carb

When a single carb is employed on a Series 'A' engine, it's not only possible but entirely practical to boost the power in certain parts of the rev range by inertial tuning of the induction system. Any changes in length can conveniently be made from the carburettor face so it's relatively simple to 'dial in' the length required. Some years ago, engine builder and research engineer David Ray and I, did some experiments with induction lengths on a near-standard 1275 engine. The results of this have been published before, but it won't hurt to show them again to demonstrate the point. The particular engine was modified only in respect of the intake, exhaust and cylinder head. The camshaft was a standard 1275 GT item. If you look at **Fig. 6.5** you will see the ram pipes that were tested. The 2½ inch and 4 inch ram pipes are easily fitted in most situations. The next one on the list, the 17½ inch ram pipe is another matter

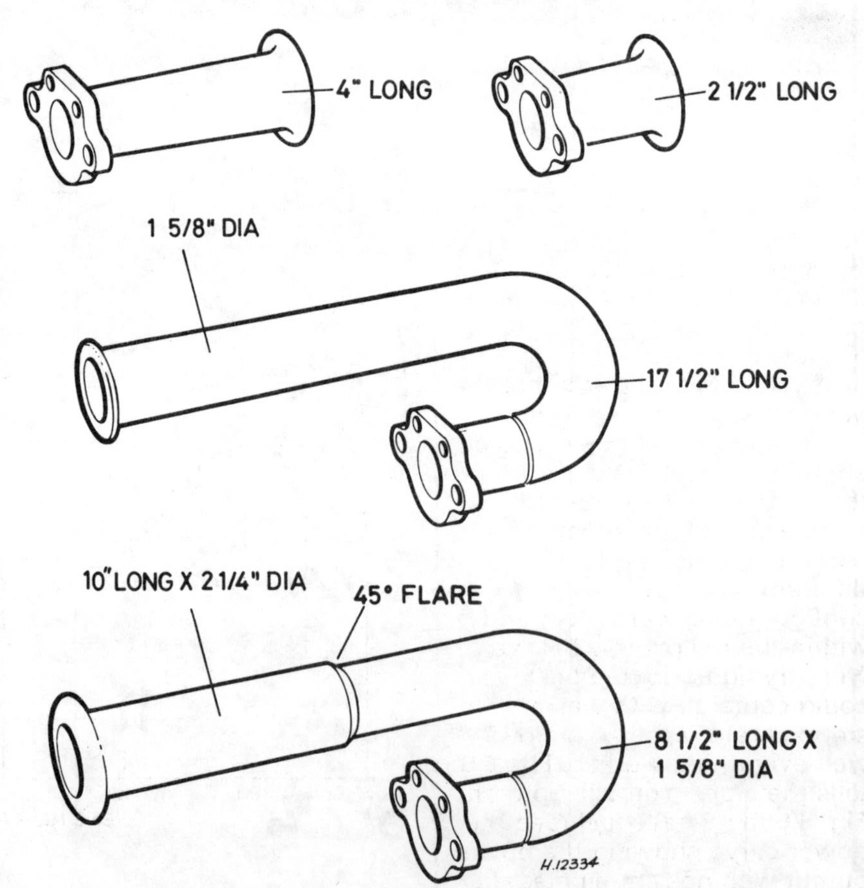

Fig. 6.5 Ram pipe designs

The only place to test and develop anything to do with engines is on the dyno. Here, a 30 inch long ram tube is tested on a 1275 motor.

altogether. Having a straight ram pipe that length would mean, on an inline engine, that it would hang out well over the wheels, and on a transverse engine, you wouldn't be far off banging your hand into it when you changed gear. Obviously it's necessary to bend such a ram pipe, and in so doing some of the effect *may* be lost, probably due to the fact that the bend cuts the flow slightly.

The fourth ram pipe was an attempt to extract the maximum torque from the engine irrespective of any other parameters. Quite a few different ram pipe configurations were tried, and within the realms of what we were trying to do, the best we could come up with was a stepped ram pipe. As to how well everything worked, **Fig. 6.6** tells the story. You will note on **Fig. 6.6** that I have included the power curve showing the power output with no ram pipe at all. From this, it's easy to establish how much power the engine can lose by not paying attention

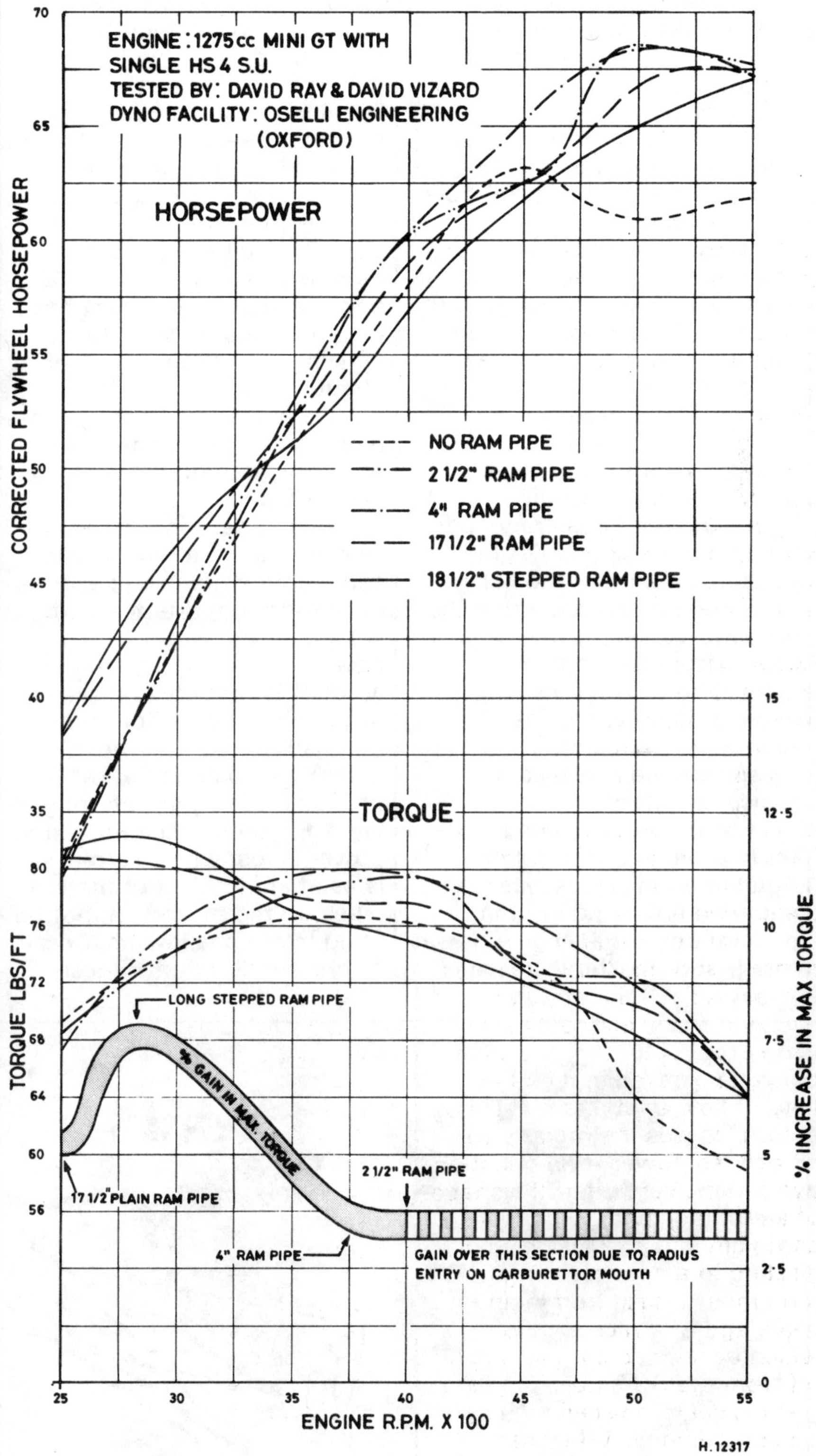

Fig. 6.6 Torque and power variations with changes in ram pipe length

to entry of air into the carburettor. The other point to note is that generally speaking, longer ram pipes not only increase the peak torque but also move it down the rpm range. The penalty paid for long ram pipes is that generally (but not always) they reduce peak power by comparison with shorter ones. Still, **Fig. 6.6** *does* tell the story.

During subsequent experiments, some time later, it was discovered whilst trying to find some quick and dirty hp, that a measure of ram effect could be achieved into a standard filter case. Many Leyland filter cases are equipped with a flap valve so that under light throttle openings, hot air is drawn into the manifold, and under full throttle conditions, the flap valve, due to depression in the case, drops and colder air is drawn from above the filter case. By removing this flap valve assembly, and installing a long pipe, both extra power and torque can be achieved. On a typical standard engine, the best pipe length seemed to be around the 12-15 inch mark. Again a good radius was needed on the end of it. In general on a 1300 engine, it was worth about a 2hp increase at the wheels over most of the rpm range.

All the tests I have covered so far have been on 1300 engines. If you intend to make up a ram pipe of the long type, patterned on those shown in **Fig. 6.5**, I would suspect that the diameter needs to be reduced if it's to be used in smaller engines. My thinking on this point is that 1100 engines will probably be okay on the dimensions shown, but 1000 engines should generally have the pipe bore reduced to about 1⁷⁄₁₆ inch to 1½ inch, and 850s about 1⅜ inch.

So much for the long ram pipes, let's get back to short ones for the moment. If the engine has been modified to any degree, especially in the camshaft area, a ram pipe of some sort is going to become even more important if carb size is limiting the engine's output. Another aspect to take into account is that on some installations space for an air filter is limited. This means a longer ram pipe is not really practical, but again referring to **Fig. 6.6**, you can see how much power is lost when no ram pipe is used at all. Most of the ram pipes on the market cannot be fitted into a shallow filter case, a common length for these being 1¾ inch. However, the Advanced Products' 'Stub Stack' is designed to fit within the confines of a shallow filter and the flow bench-developed form of this Stub Stack greatly enhances flow. This form does, in fact, make it one of the best flowing ram pipes on the market, although it hardly looks like a ram pipe. As a result of this, what it lacks in length to boost power, it often makes up for in the main, by superior flow into the carb. If your installation is such that a deeper filter can be accommodated, then a 2½ inch long ram pipe, *so long as it is of the right form* (refer to **Fig. 6.4**) will work very well in most instances. If a 2½ inch long ram pipe is not practical, then a Stub Stack from K & N (Europe) becomes the main contender once again.

Returning once more to the subject of long induction systems, this design philosophy was, at one time, taken up by **K & N (Europe)** in England. This company produced the 'Torque Ram' induction system which was designed to fulfil several requirements on the Series 'A' engine. Firstly, as the name would suggest, it was designed to boost torque, due to its long induction length. Secondly, it provided an easy way to utilize a large filter element, so that no power loss by virtue of under-sizing was experienced. Thirdly, it eliminated any undue induction noise which is often a problem with sports-type filters.

Here's an early version of Advanced Products' Torque Ram. This particular one uses a 17 inch long ram pipe and the cast body is fixed to the carb by two socket-head screws, visible on the front. Believe it or not, those two screws passing through the plenum chamber caused a massive drop in airflow potential, and so were omitted on production items.

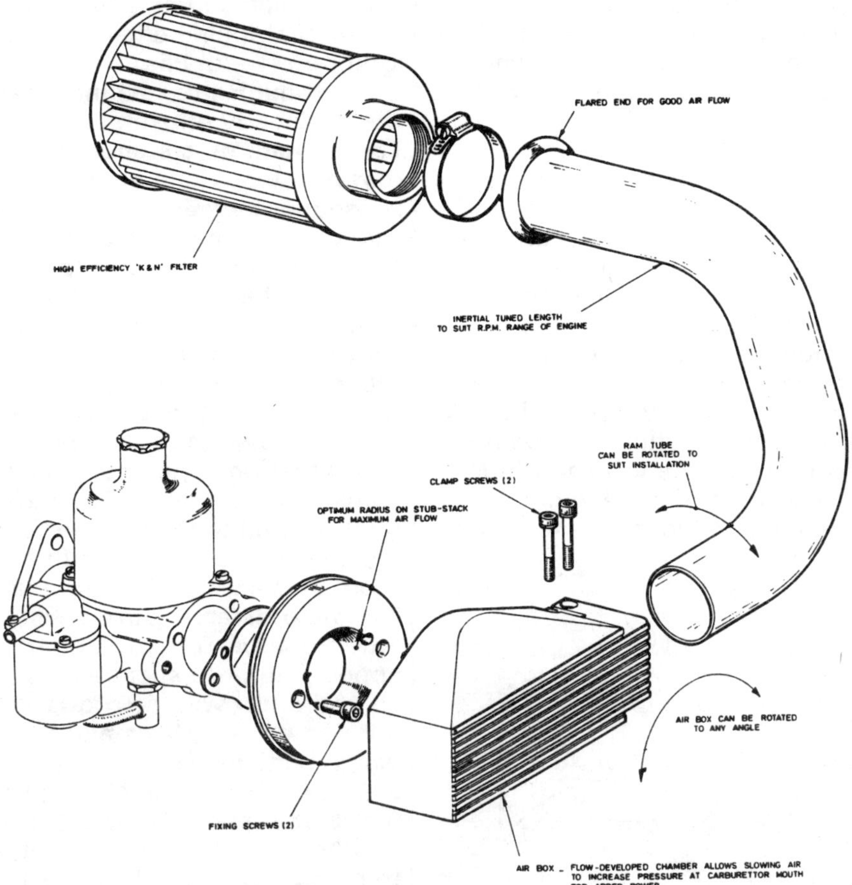

Fig. 6.7 Torque ram

As you can see from **Fig. 6.7**, the air box fixed to the carb contains a Stub Stack radiused entry which acts as a short ram pipe. The actual air box itself is designed to turn the air through a 90° angle with the minimum of flow loss. The use of this air box eliminates the problem of having to make a tight bend and suffer the consequential loss in flow due to a tight bend. The air box can be located at any angle on its Stub Stack backplate and this, in turn, means close proximity to a bulkhead need not be an obstacle as far as obtaining good airflow and the right ram length are concerned. My own dyno testing has revealed that the Torque Ram induction system is a highly functional piece of speed equipment under the right circumstances. Before embarking on the road to a long induction system it is important to understand exactly its strengths and weaknesses. For me to achieve worthwhile horsepower with the help of a sophisticated dyno is one thing: it is quite another for the enthusiast working in restricted circumstances.

The analysis of the torque ram system shows that it develops extra power by virtue of two particular aspects of its design. In general it is far more efficient in terms of airflow than a standard filtration system, through there are one or two exceptions to this. The MG Metro filter system, for instance, has more than enough flow capability to cope with the power level the engine has, and adding an air filtering system which has a lot more airflow capability does very little to improve the the power output of the engine, so little in fact, that it's best to spend money in other directions during initial stages of tuning the engine. On the other side of the coin though, we find that most standard filtration systems are far less efficient than the MG Metro one, so in almost all cases, the Torque Ram provides the power boost by virtue of its higher airflow capability. Its second aspect is that it provides a power boost by promoting inertial ramming of the engine.

Checking out the flow capability of the Torque Ram versus the standard setup is no problem. It's just a question of resorting to use of the flow bench to establish its flow capability. **Fig. 6.8** shows the results of some flow testing on a modified H.I.F.6 SU. As you can see, the system flows very much better than the standard Metro system, and in spite of the air having to come down a

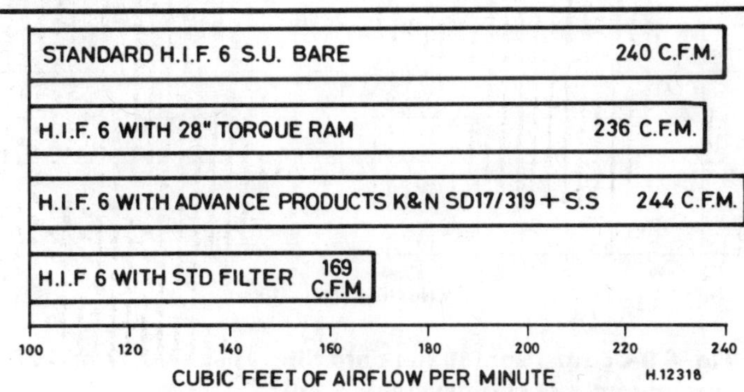

Fig. 6.8 Torque Ram flow tests on HIF6 SU

long tube and make a right-angle turn into the carb, the airflow of the Torque Ram is very nearly that of the Stub Stack setup. A point worth noting here is that the carb did, in fact, flow more air with a Stub Stack and an air filter setup on than it did without, which should be good enough justification for using these components on any engine, even if it is for a race vehicle.

As far as airflow capability goes, then, the Torque Ram does well, and **Fig. 6.9** shows the power gain on a virtually standard Metro engine when the Torque Ram is used to replace the standard filter setup. Note there is a useful amount of power gain low down. Normally, when an air filter is changed, most of the power gain is brought about in the upper rpm range simply because at low rpm the filtration system, even if it is short of maximum airflow capability, may still be able to supply the engine with the air it needs. It doesn't become hard-pushed to supply all the engine's requirements until the upper rpm range. The Torque Ram setup though, with its vast increase in airflow capability, can supply the top end of the rpm range of the engine without any loss. The low rpm range power boost seen in **Fig. 6.9** is almost certainly due to the inertial tuning effect that the long ram length has.

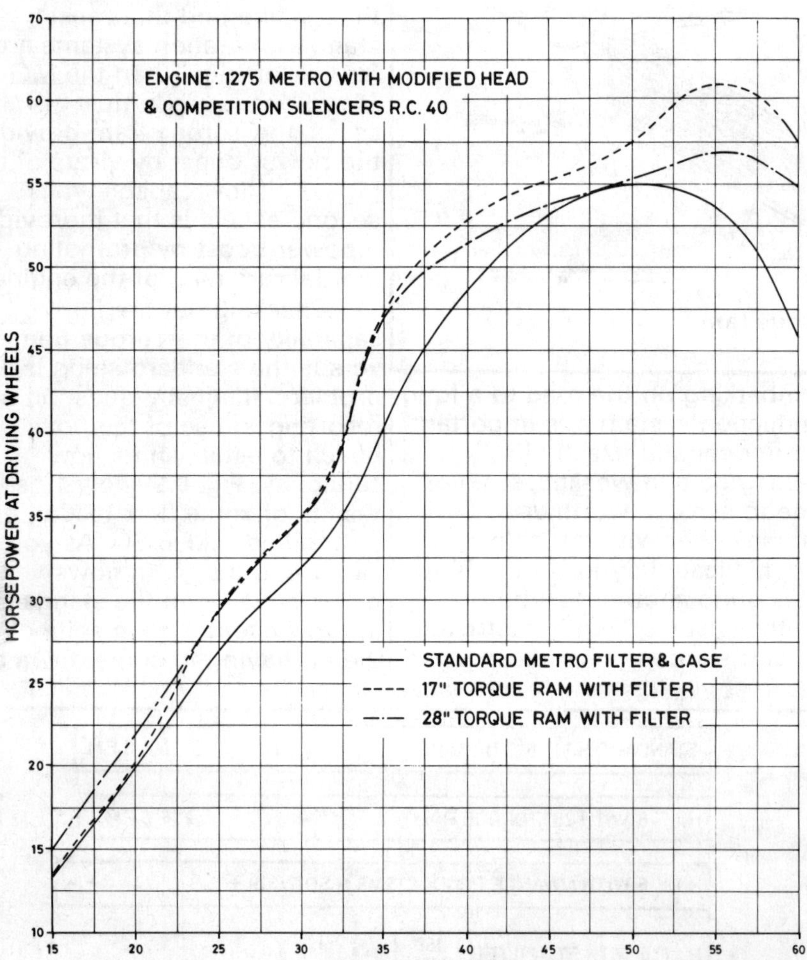

Fig. 6.9 Comparison of standard filter case and 17 and 28 inch torque ram induction systems

Although this test indicates the possibility of using a Torque Ram on a standard motor, it would appear from subsequent experience this is not its best application. The results from the Metro engine appear to be amongst the better results obtained when used on a standard engine. A good friend of mine, John Bosson, who runs rolling roads every working day of his life, has found that only minimal power gains result when the Torque Ram is fitted to the smaller engines and that those with a more efficient filter system than has the ordinary Metro, benefit little from the Torque Ram. It appears he once spent almost an entire day with a Torque Ram on an MG Metro and the best results he could obtain were exactly the same output in terms of peak power with the odd hp gain (a dubious quantity to believe in on a rolling road dyno) at low rpm. In such an instance the Torque Ram represented quite a bit of capital investment for very little in terms of extra performance. Of course the owner of the Torque Ram will never have to change an air filter again, because it's using a K & N type washable filter.

My own experience with the Torque Ram relates more to use on modified engines. I have found it particularly useful in pulling up the bottom end torque of engines which are utilizing a fairly radical cam. To give you an idea of just how effective it can be when used in this manner, it's best I quote the test results from a project engine that I did together with Nolan Pitts of **Auto Electrical (Plymouth) Ltd.** This particular power plant was to go in a project vehicle for *Auto Performance* magazine. The engine, a 1400cc unit, utilized an H.I.F.6 carb on a Torquemaster intake and a Torque Ram induction system. The cam used was Leyland's 649 profile. On Nolan's rolling road dyno, a standard 1275 'S' puts out around 60hp. The project 1400 engine gave 105 at the wheels, but in spite of using a race profile cam the engine would pull top gear at 1000rpm. Indeed you could drive round a typical roundabout in top gear and give it full throttle on the exit without the engine faltering or missing a beat. That's not the sort of bottom end power that is normally expected of an engine with a race cam. Up to around 3500rpm the engine produced little better than the power output of a standard 1275 'S'. From 3500 on up, it came up on the cam and performed like most race engines. No less spectacular in form, the modified 1275 test engine in **Fig. 6.10** showed more or less the same traits. Analysing these figures and curves, we can see that the standard air filter system fell short on hp simply because it had insufficient flow capability. Changing the standard filter to a Stub Stack setup and a K & N filter produced a hp increase almost everywhere in the rev range except down at the 2500 mark where the Stub Stack setup just marginally lost out to the standard setup. The reason for this is probably due to the standard filter and case assembly developing a measure of ram effect through the snorkel pipe on the case. At low rpm, the positive effect of ramming outweighed the negative effect of the standard filter case's inadequate flow.

When the Torque Ram induction system was installed, peak torque increased and the point at which peak torque occurred was moved down the rpm range. As you can see, the torque curve predominantly bulges in the 3000-4000rpm range as opposed to the 4000-5000rpm range when a short induction system is used.

The last test I will refer to from that graph is the power curve produced by a 40 DCOE Weber. Note that at low rpm, this induction system produced the worst power of any tested. In fact it didn't even come on a par with the SU and the standard filter setup until 3000rpm and it didn't really produce any more power than any of the other setups until about 4000rpm was reached. At the maximum power point of 7000, the Weber setup produced identical power outputs to the H.I.F.6 SU, utilizing the Stub Stack air filter system. Normally one wouldn't expect to see equality in these two systems, but the probable reason in this instance is that at 7000rpm the inertial induction length was just about right to favour an increase in power from the single SU. In other words, the SU induction system ram length was just about right for 7000rpm. On the other hand, the 40 DCOE Weber produced better power at 5000rpm due, in all probability, to slightly better

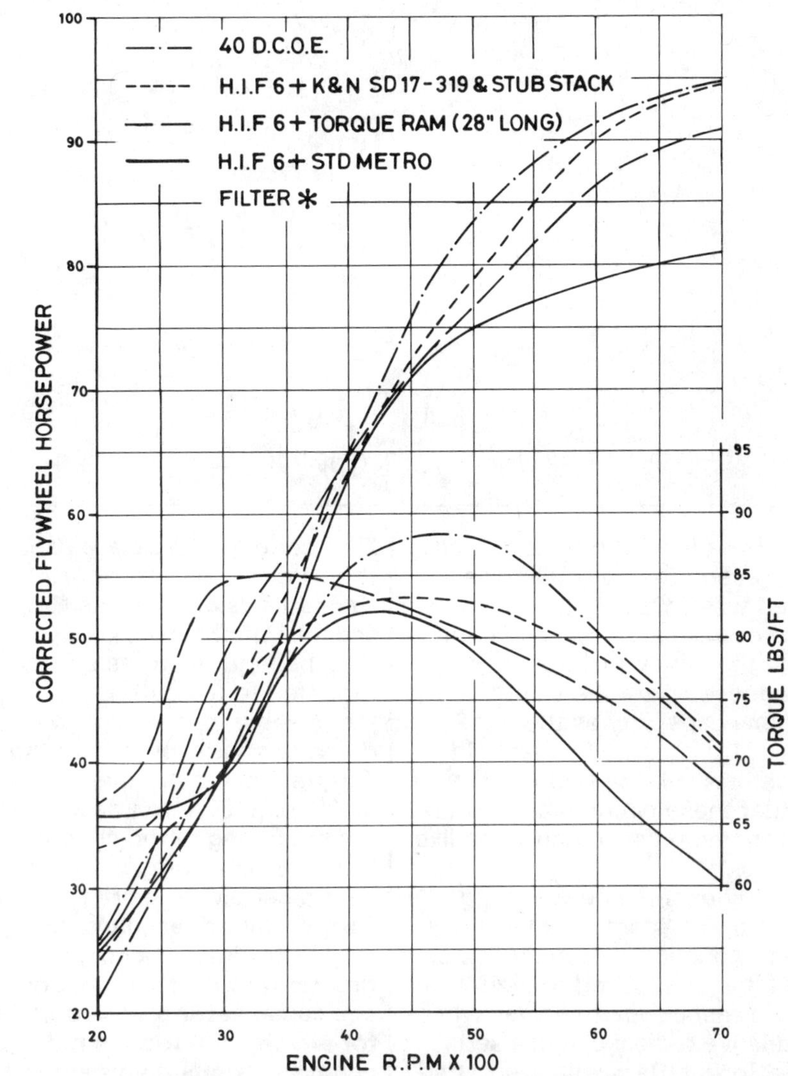

* THIS IS THE STANDARD METRO FILTER AS OPPOSED TO THE HIGH EFFICIENCY MG METRO FILTER H.12320

Fig. 6.10 Comparison of various filters and ram pipes

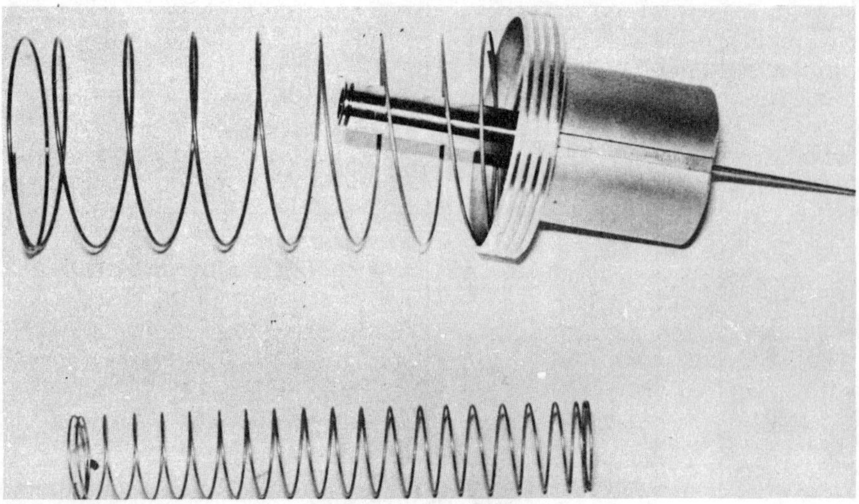

On some engines, the use of a long, single ram pipe causes so much more air to be fed into the engine that the carb opens too quickly. To avoid this, it's necessary to use two springs, such as are shown here.

mixture distribution. Incidentally, the reason the 40 DCOE was used was because the choke size of this carb, in conjunction with the 5 inch manifold used, produced about the same airflow capability as the Torquemaster intake manifold together with the H.I.F.6 carb. Under these circumstances their comparison was much more like for like.

My experience with long induction systems like the Torque Ram induction system indicates that it is best suited to high performance road engines where a relative radical cam is used. The Torque Ram will give better drivability and allow the engine to produce a civilized power curve for the road while still retaining much of the top end power associated with a high performance cam. If you buy a Torque Ram, it's well worth experimenting with different induction lengths, as there have been occasions, especially with the racier cams, where a certain length has benefited both top and bottom of the rpm range. I have also found that on occasions, bigger engines tend to favour shorter lengths. Normally, the Torque Ram comes with a 28 inch induction length. On big motors, that's 1430cc and upwards, it seems worthwhile trying ram lengths around the 17 inch mark, especially if cams of 285° period or more are used. **As the cam timing gets shorter, so the benefits of a Torque Ram seem to diminish.** It is possible that the volume in the air box mounted to the carb is responsible for the slightly different characteristics of the Torque Ram as opposed to the single pipe system I described at the beginning of the chapter. However, one thing is for sure, over a certain range of rpm, **the Torque Ram does allow the engine to draw in considerably greater quantities of air, and this can cause setting up problems, for which a solution is not always obvious if you are not experienced in this area. I understand from K & N (Europe) that this was the principal reason for the Torque Ram's discontinuation.**

The first problem likely to be encountered on fairly highly tuned engines, *i.e.* engines over about 85hp or so, is that the Torque Ram will increase volumetric efficiency to the point where the piston in the carb will be wide open long before the engine reaches peak power rpm. This causes a problem in that from this point on, calibration of the carb is lost. In other words, more air will go in but the needle has moved out of the jet to its fullest extent. To a point, more fuel will be drawn in as engine rpm increases, but the carb is now acting like a fixed jet carb. In this situation, the air flow increases at a greater rate than the fuel flow, so the engine experiences a high speed lean out. Consequently, the full hp potential is not developed. The way around this problem is to increase the spring pressure in the dashpot. On H.I.F. carbs it's often necessary to use two springs. You can either use two springs from an H.I.F. carb entwined together, or if these should prove too strong, remove the dashpot assembly from the H.I.F. carb and substitute it for an HS dashpot assembly complete with its spring, plus the H.I.F. spring. For a range of springs see the chart Fig. 6-10A. The H.I.F. carb has a ball bearing centre rod, whereas the HS carb inevitably has a plain bearing centre rod. From that point of view it's a slight step backwards, but it's swings and roundabouts. Often, using a weak spring such as a red spring, together with a standard H.I.F. spring is all that's needed, but sometimes it's necessary to use a blue spring to increase spring pressure still more. The idea is to get enough spring pressure so that the piston reaches full lift at or just before peak power rpm. On motors under the power range I'm talking about, it may only be necessary to substitute the existing spring for a stronger one. **If the carb has been modified as outlined in the carburettor chapter, the need for a stronger spring may become greater.** Modifications to the butterfly in particular affect the rate at which the piston goes up.

The next problem likely to be encountered is needle selection.

Fig. 5.10A SU Dashpot Springs

Spring Colour Code	Type	Load	Length	Part No.
Light blue	A	2.5oz	2.625	AUC 4587
Red	A	4.5oz	2.625	AUC 4387
Yellow	A	8oz	2.750	AUC 1167
Green	A	12oz	3.000	AUC 1170
Red (BBC)	B	4.5oz	1.530	AUD 4355
Yellow (BBC)	B	8oz	1.530	AUD 4398
Green (BBC)	B	12oz	1.530	JZX 1088
Red/Green (2 in carbs)	A	11.3oz	3.875	AUC 4826
Light blue/Black (2 in carbs)	A	4.5oz	3.875	AUC 2107
Light blue/Red (2 in carbs)	A	18oz	3.875	AUC 4818

Notes: *Type refers to the dashpot style. Type A is the original plain piston and Type B the later style ball-bearing chamber (BBC) the later spring is large in diameter than those for the early style dashpot. The last two springs on the list are replacements for the Red/Black AUC 5014 spring and the Blue/Green AUC 2091 spring.*

The chances of the existing needle (if it gave the correct calibration without the Torque Ram) being right are next to zero. Because the Torque Ram increases engine airflow in the lower rpm range, the needle needs to be considerably thinner in the middle range of its profile than is the case without the Torque Ram. If you are swopping from a short ram pipe to the Torque Ram, you will often find that the needle required for the Torque Ram needs to be slightly thicker towards the tip. If changing from a standard air filter to a Torque Ram, then everything from the mid-section of the needle to close to the idle position, needs to be thinner.

Fuel Flow

Be aware that if you are trying to make a lot of hp on a single SU the float bowl needle valve may become a restriction. Incidentally the HIF 6 and the HIF 44 have different needle valves in them. You will need the appropriate one. A larger needle valve (0.096 inches diameter) is available and I advise its use instead of the usually installed 0.076 inch diameter item for all high output requirements. Also I strongly recommend the spring loaded Viton tipped needle valves. They seal up better and are more consistent in operation.

Ram Pipes & Multi Venturis

I have dwelt at length on ram tuning single venturi induction systems simply because ram tuning can be very effective when applied to a single venturi system. Although it's not nearly as effective on multi-venturi systems like twin SUs, single sidedraft Webers, etc., there are still some small gains to be had. Some are easy to pick up, some are not. **Let's start off with twin SUs first, as fitted to such machines as the Sprite, Cooper 'S', etc. The induction systems on these, at least from the carb face outwards, suffer from inadequate flow capability in almost all instances.** Half the problems on Coopers and Cooper 'S's stem from the elbow that attaches to the front face of the carb. This elbow is not very conducive to flow and cuts the potential of the carb dramatically. If the elbow, together with its filter assembly, is removed and a simple air filter setup installed, the power change is not really significant, simply because although the elbow was inefficient along with the filter that fed it, not having a ram pipe on the 1¼ inch carbs is just as inefficient. **Installing some form of ram pipe within the case of a carb-mounted air filter is generally worth around 3hp at the wheels. This, of course, makes the assumption that carbs are recalibrated to suit.**

I made up these special ram pipes for the IDA Weber on my 1293 racer. They improved flexibility whilst driving around the paddock and pits, and added between 1 and 2hp to the engine's hp.

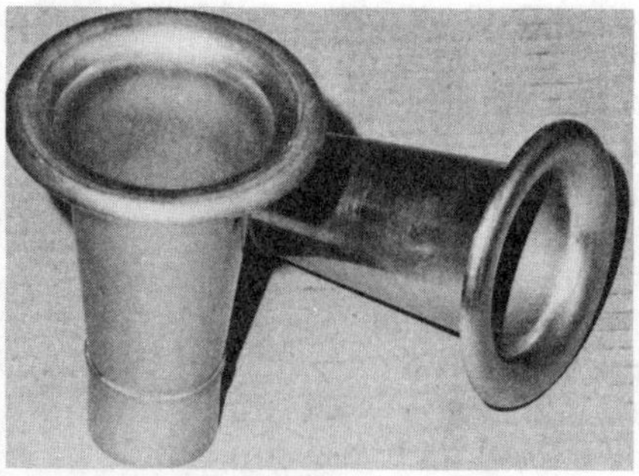

These extra length ram pipes intended for split Webers are available from Richard Longman & Co. If you want to use them on a single Weber, then it's necessary to cut a flat on the side of each trumpet.

Often what happens is that an efficient air filter, together with a ram pipe, will allow enough extra air into the engine to make the standard needle too lean. Very often, looking at the S.U. – recommended rich needle for the particular vehicle in question and installing that will remedy the problem.

If space permits, it's well worthwhile experimenting with various lengths of induction system on twin SUs, especially if you are trying to drag the last drop of power from the engine for competition purposes. When a fairly radical cam is used, stand-off can sometimes be a problem and putting on a ram pipe of reasonable length can contain the stand-off and eliminate its bad effects. Under such circumstances, it is not going to be possible to actually bolt an air filter to the face of the carb. However, it will still be entirely practical to use a clamp-on filter on the end of the ram pipe. With race cams, lengths up to 9 inches can sometimes be used effectively. However, if stand-off is not a problem (*i.e.* none is present) then such ram pipes may only produce a minimal gain in power. It definitely is something that needs to be sorted out and tuned to suit each individual engine, and when it's all dialled in right, you are talking about 2-3hp at the wheels at most. Still, often that is all it needs to tip the scales in your favour if you are racing.

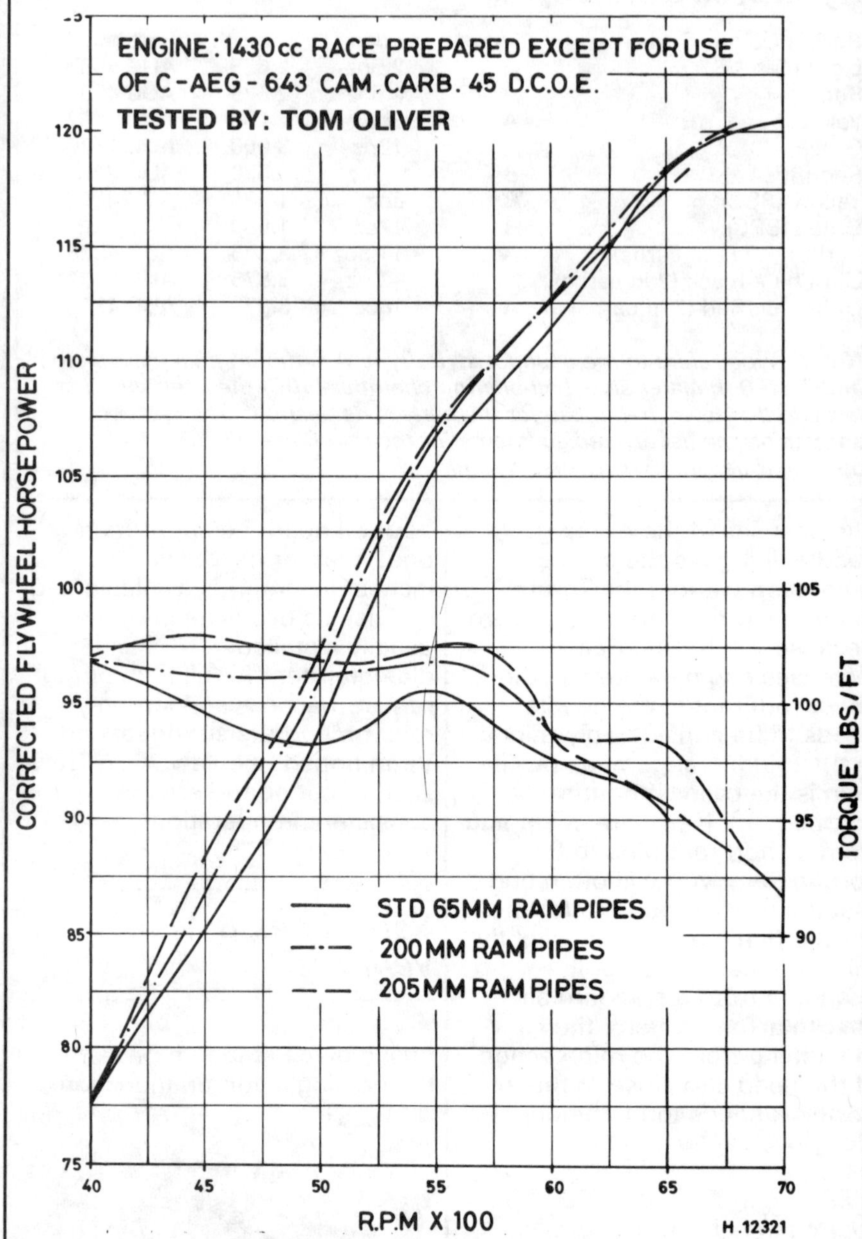

Fig. 6.11 Comparison of standard 65mm Weber ram pipes and 200 and 205mm ram pipes

Although the same basic rules apply to induction lengths when one barrel per port is used, i.e. longer gives more torque lower down, the percentage change is relatively small. A longer ram pipe works better over the greater part of the rpm range shown here, but from about 6900 rpm up the short, standard Weber ram pipe comes out on top. If this engine had been equipped with a full race cam, i.e. 300° or more instead of a 288° cam, then the short rams would have probably been as good or better over the higher rpm range involved. The manifold length, however, can have a significant influence on this.

Assuming you have an all-out race engine on sidedraft Weber carburation, just how much power can be gained or indeed lost, by ram pipe length tuning? To illustrate this question, some tests done by Tom Oliver at **Howley Racing** can be reviewed as fairly typical of changes in power due to changes in induction length. I use the term 'change in power' very deliberately since in the particular instance shown, **Fig. 6.11,** the power went up when longer, larger diameter rams were used. **The reason for the increase in power was because that's what was required for that particular specification of engine.** It's always advisable to check out the effect of such rams on your particular engine because if it should be using a different cam profile, a head modified in a different way, or whatever, a different ram length may be needed. Be sure to read about combined manifold and ram pipe lengths in the Weber manifold section of Chapter 8. Just making the assumption that long rams work because they did in the example shown in **Fig. 6.11**, could drop power from *your* engine just as easily as add it. It's also worth remembering that any time the induction length is changed by any noticeable degree, the carburation calibration may need to be changed along with it. Hopefully all the ground rules I have given you will allow you to make the most of ram pipes and ram tuning of the induction system. If it does then it will certainly be the icing on the cake, and if you do get it right, ram pipes and ram tuning in general can represent very good value for money in terms of power gained for cash spent.

Staying Cool Under Pressure

The volume of air drawn into an engine depends largely on its capacity. But an engine's potential to produce power depends upon the mass of air inhaled. In our efforts to get more power from an engine we bore and stroke and install more carburettors, flow the head and so on. These mods may help the engine to breath more deeply but how often do we see cars built to reap the benefits of optimized air intake pressure and temperature. They are a minority at the race track and almost non existent on the street. By picking up the coldest air possible and then ramming it into the induction system we can significantly improve the charge weight drawn into the engines cylinders.

Pressure Potential

Putting a hand out of the car window while travelling at even a moderate speed reveals a significant drag force acting. In engineering terms the pressure brought about by bringing a stream of air to a dead stop is known as the 'static pressure head'. It is this phenomenon that we can use to ram an induction system. Two factors affect the ram air pressure, air speed and air density. The pressure changes in proportion to air density but changes as the square of the speed. If, when using ram induction the speed of the car doubles, the ram pressure generated increases by a factor of four. That's the good news. The bad news is that it takes a lot of speed to get a worthwhile degree of ramming. The formula,

Fig. 6.12

$$\text{Ram air pressure} = \frac{Ad \times V^2}{4287}$$

Where Ad is atmospheric density. At sea level this is 0.076 lbs per cubic foot.
V = speed in MPH.

Fig. 6-12 can be used to calculate the ram pressure generated under ideal conditions by the vehicle's speed. Calculating the pressure produced by a car moving at 100mph shows that under ideal circumstances the ram affect amounts 0.177psi. That is 1.22% increase over normal atmospheric pressure. As speed increases further things begin to look better. For instance at 150mph the ramming pressure could reach 0.4psi. And at 200mph some 0.7psi could be realized. Now it won't take much to appreciate that we're talking about speeds far greater than most 'A' Series engined cars, especially Mini's, will ever reach. Although these small ramming pressures are useful to have they aren't going to be earth shattering proportions on a Mini, even if it is capable of 130mph. At best we will be looking at about 2% increase in air density. But on the track everything helps.

The Downside

The biggest problem associated with ram induction occurs when carbs, rather than fuel injection, are used. The first thing to understand is that a carburettor is nothing more than a self powered, low pressure fuel injection system. To put the 'low' into perspective the fuel metering pressure involved often amounts to no more than 1psi. To make maximum power the fuel air ratio must operate within a tolerance of about 6%. *If the mixture ratio is forced outside this narrow envelope, power losses can easily outweigh any potential gains from ram induction*. In most cases the fuel mixture problem originates at the carbs auxiliary (booster) venturi. The purposes of the auxiliary venturi is to amplify the pressure drop produced across the main venturi. This produces a stronger draw on the fuel being pulled into the engine from the main jet well. However, the system is very susceptible to outside influences. A carb intake that is directly exposed to air moving at high speed experiences considerable pressure buffeting at the auxiliary venturi. This buffeting can, at the speed of a typical race car, cause erratic pressure changes equal to *as much 30–40% of those normally experienced by the auxiliary venturi when the engine is running in still air.* This buffeting causes substantial mixture metering errors which lead to a measurable loss of the power. The cure for unwanted effects of this nature is to install a high flow air filter element into the system.

With the forward facing ram air system some thought must be given to wet weather driving. If a paper filter element is used the air flow will drop dramatically when the element gets wet. The scenario from here is that the element collapses and the engine digests the element. The fix is to use a K&N type element as this has proved capable of withstanding a heavy water splash on race boats with minimum loss of flow.

Pressure Balance

Auxiliary venturi buffeting is not the only problem that we must contend with. Ram induction increases the pressure at the entry of the carb but it may not produce a corresponding pressure rise in the carb float bowl. If this is the case it will cause the mixture to become lean for two reasons. Firstly the ram pressure reduces the auxiliary venturi gain and secondly, the extra air ingested will not draw in sufficient extra fuel to keep the mixture where it should be. The way around this problem is to connect the air space above the fuel and the float bowl/bowls to the air box being ram fed as shown in Fig. 6.13 above.

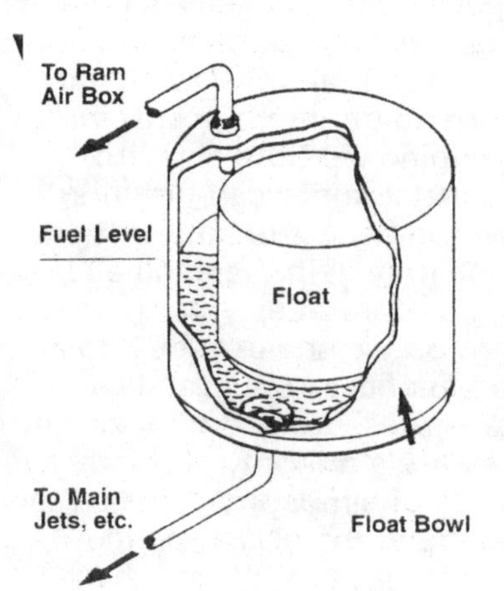

Fig. 6.13 *Here is the principle of maintaining the float bowl to booster pressure differential to compensate for any forward ramming pressure generated. Without this compensation, the mixture will lean out as speed increase.*

On DCOE Weber type carbs there are built-in holes beside the ram pipes which connect directly to the float bowls. Provided they are covered and connected to the ram air box chamber the desired effect will be achieved. If you go this route with this type of carb be sure to seal the jet covered lid on the top of the carb.

Cool Power

As the air temperature drops its density increases. The rate of change is about 1% increase in density from every 3.3 degrees C of temperature reduction. The air entering a production engine induction system can easily reach 80–85 degrees C but by drawing cold air from an outside source the air temperature at the mouth of the carb can be reduced by as much as 40 – 45 degrees C. This represent an increase in air density of some 12%. In fact this will have much the same effect as 1.77psi. of supercharge pressure *and offers the chance of a 12% increase in torque and power.* Best results are obtained if all *precautions are taken to preserve the reduced intake temperature all the way into the cylinders* and the increased air density is fully exploited by maintaining the correct fuel air ratio. In practice however, the power gain extracted from reducing the temperature of the intake by about 45 degrees C. is only 60–70% of the theoretical advantage. So where does the discrepancy occur? While you may succeed in supplying cool air to the carb mouth the air will pick up heat during its passage down the inlet track past port walls and hot intake valves. Air that is 45 degrees C cooler at the carb mouth maybe only some 30 degrees cooler by the time it enters the cylinder. This being the case we need to take steps to keep the heat out of the intake charge as it progresses through the intake system. Just as the flow bench made its mark on the performance engine in the 1970s so we find that thermal management of induction system in particular and engine heat in general, will be the most successful route toward greater power output in the 1990s and on to the new millennium.

Cool Air in Practice

If cold air is successfully picked up by a scoop your next job will be to prevent the charge being unduly heated between the carb mouth and the intake valve. The first step toward achieving this is to isolate the carb from all heat sources. This means a heat shield between the carb body and exhaust if you're using a regular five-port non-cross flow type head. Even if you're not separating the carb from the usual high under bonnet air temperature and drafting cold outside air to the carb it is a good move. The effect cool air has on the carb or injection system temperature can be further enhanced by also cooling the fuel. The easiest way to do this is to employ what is normally known as a 'cool can'. (Fig. 6-14).

This usually comprises of a container having a coiled copper fuel pipe passing through it. To cool the fuel a mixture of alcohol and ice or dry ice is put into the cool can. If going this route fails to make power then the fault lies with the *carb atomization characteristics*. Weber carbs already appear to give close to the ultimate in terms of fuel droplets size and percentage vaporiziation at normal temperatures. When the fuel is artificially cooled less of it will vaporize. This can lead to less effective combustion. Should this be the case you may need to look at using an auxiliary venturi which atomizes the fuel slightly better. However, be aware that *the five-port engines do not like to have their fuel atomized too finely*. When atomized too finely the power drops dramatically.

If you're using the cylinder heads with four intake ports then, Weber style carbs have a potential atomization problem with cooled fuel which can be exaggerated if the venturi's selected are too big for the engine, or the auxiliary venturi's are not producing an adequate signal amplification. On a fuel injected four-inlet port engine the problem of adequate fuel atomization is minimal to non-existent.

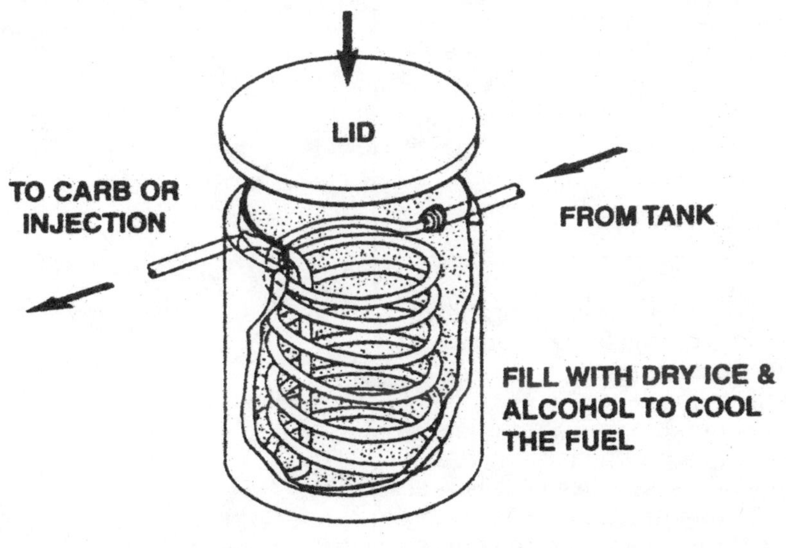

Fig. 6.14 *This is the basis of a cool can. It is most effective when steps are taken to prevent excess heat getting into the intake charge from other sources such as manifold heat.*

Having persuaded the carb to deliver a cold charge the next problem to deal with is the thermal effect the intake manifold can have on the mixture. A major problem here is that most intake manifolds are constructed from aluminium which conducts heat rapidly from the engine and *dumps it right into the intake charge*. This is not what we want. If you must use an aluminium intake manifold then do yourself a *big favour and put a thermal barrier coating on the inside of a manifold and polish the outside*. A better choice is to utilize a steel intake manifold which is polished and chromed or thermal barrier coated. The high shine does a great job of reflecting the exhaust heat resulting in a much cooler charge temperature. This move can often be more effective than you would at first think. On a hot California day we have seen chrome-plated steel intake manifolds reduce lap times by as much as 4/10ths of a second compared with an otherwise identical black steel one. With this much potential for improvement thermal management is worth consideration by any performance-minded engine builder.

If a low air intake temperature has so far been retained into the cylinder head ports all that remains is to finish the job. This will normally entail applying thermal barrier coatings to both port walls and to the combustion chamber face of the intake valves. That last point is very important as approximately 50% of the charge heat is absorbed as it passes the hot intake valve.

Pressure v Density

Forward facing air scoops may look very racy but the ramming effect does not contribute as much to the performances as the use of cooler air.

The chart Fig. 6-15 shows the effect of both ram air induction and cold air induction. Picking up cold air doesn't have quite the monster power look of a ram air system *but it is a lot more effective and works from zero mph on up!* Keep that in mind when you design your next induction system.

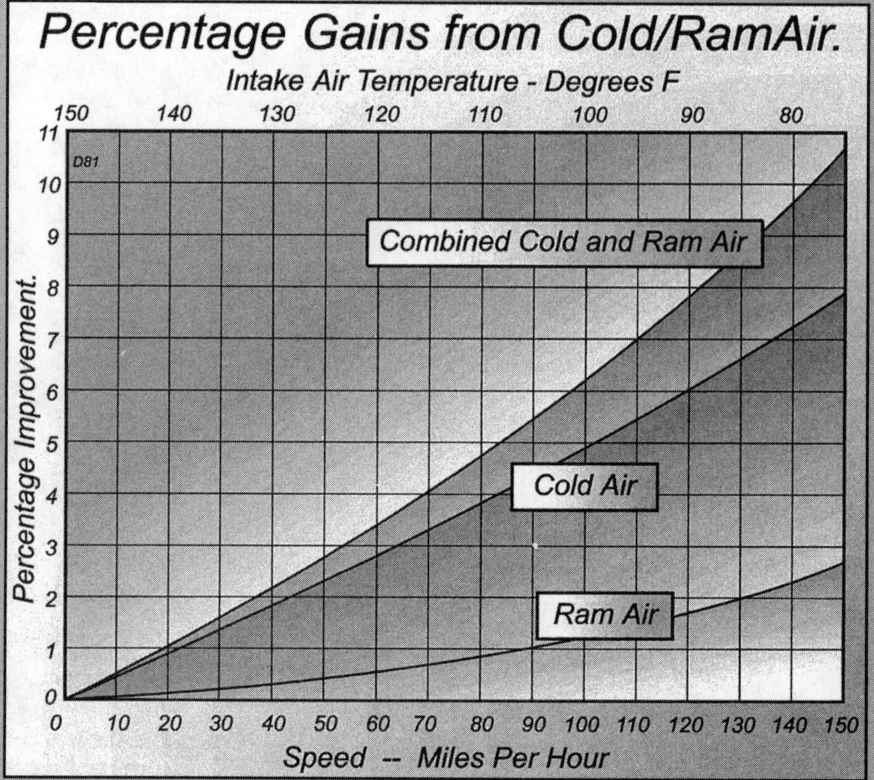

7 Carburation Part 1

One of the areas of engine tuning that is surrounded by supposedly more black magic than any other, is carburation. Namely, the selection and tuning of the carburettor for the engine.

Let's put things into perspective; firstly, the black magic doesn't exist. A good carb man is somebody who simply understands the intricate functioning of the carb, and let's make it clear, carbs need to be understood if you are going to get real performance from your Series 'A' engine. **Also, let me make it clear right now that big carbs don't necessarily mean big hp, and there never was (nor ever will be) a carb that the oil companies bought up because it gave 100mpg!** It's fairly safe to say that these days, the modern carb is a highly developed instrument. It can be set to deliver any fuel/air ratio you want, but more precisely it should be set to give the optimum fuel/air ratio that the engine wants. Often the engine tuner who doesn't know carburation also doesn't appreciate there is a difference between these two factors. **In simple terms, we can say that the carb's function is to mix fuel in precisely the right proportion with the air that is being drawn into the engine. These proportions are important, and by no means fixed.** The operating conditions of the engine determine just what fuel/air ratio it requires for best results.

In broad terms, the two main areas are full throttle and part throttle. Obviously if the engine is running at full throttle, the driver is demanding maximum power so the mixture needs to be set to be conducive towards producing maximum power. On a Series 'A' engine, a maximum power mixture is typically between 12.5/1 and 13/1. This ratio is an expression of the weight of air to the weight of fuel, so a mixture ratio of 12.5/1 involves 12.5lb of air to 1lb of fuel. When we talk about 'leaner' or 'weaker' mixtures, then the amount of air for a given amount of fuel, increases. For part-throttle operation maximum power is not the criteria. Under part throttle conditions we need to make the most use of the fuel possible so as to get the best mpg. For economy, a leaner fuel/air ratio proves to be better.

Until Leyland went really overboard trying to find ways

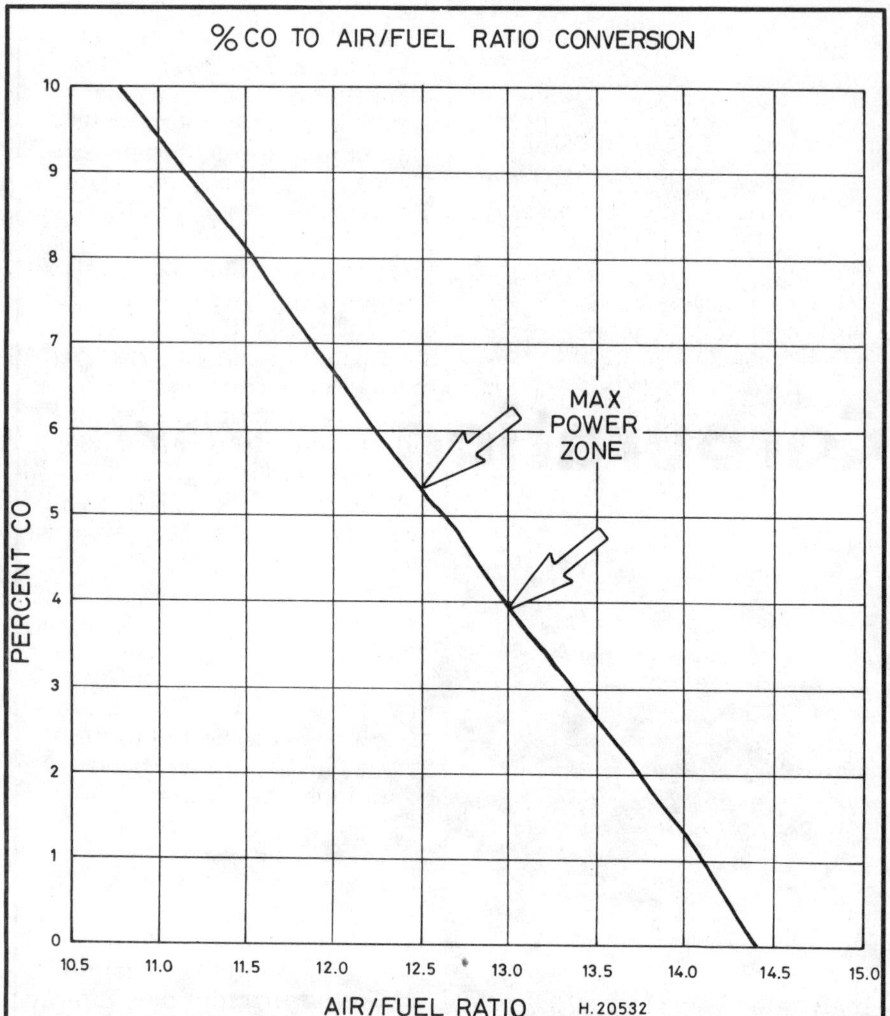

Fig. 7.0 Carbon Monoxide to air/fuel ratio conversion

Many 'tune up' or electronic diagnostic machines now have the mixture analyse calibrated in 'percentage residue carbon monoxide' (CO) in the exhaust. This has been done for emission related reasons. This conversion chart has been included as an aid to power tuning your Series 'A' engine by showing the conversion to equivalent air/fuel ratios. Best power is typically produced between 12.5/1 and 13.0/1 air/fuel ratio.

and means of improving fuel economy, most Series 'A' engines ran at part throttle, around 15/1-16/1 fuel/air ratios. The later, economy-orientated engines, tend to run much leaner than this. My own efforts at building economy engines have resulted in units which use fuel/air ratios in the region of 22/1-25/1. However, it's worthwhile at this juncture of making the point that the leaner fuel/air ratios which will run without misfire do not necessarily produce the maximum economy possible. There comes a time when leaning off the mixture any further causes a drop in economy, although the engine may still not have run into lean limit misfire.

Building an economy engine is much like building an engine for hp, in as much as everything has to be optimized and very often what you think is going to work, doesn't. This is why sophisticated development techniques will, in the end, always pay off.

There is a vast range of carbs that can be bolted, either easily or not, as the case may be, onto the Series 'A' engine. Many of these carbs are 'fringe usage' carbs, and represent a very small percentage of the range of carbs usually used on the Series 'A'. For one reason or another, I don't feel justified in dealing with these 'fringe usage' carbs, and for that reason I will stick to the mainstream of popular type carbs. These include the good old trusty SU, which in various forms is fitted as standard to most Series 'A' engines; downdraft Webers of the 28/36 progressive type; Fish carbs and sidedraft Webers and Dellortos. Each of these types of carb has its assets and deficiencies, and prior to deciding what carb you are going to use, it pays to know the strong and weak points of each type, so you can make the best choice for your particular circumstances.

SU Carbs

The beauty of SU carbs is that with the exception of some early Austin A30 & A35 engines, all Series 'A' engines have had at least one. Originally on single carb engines, the type used was the H1 or H2 1⅛ inch or 1¼ inch SU. In October 1965 the Mini automatic was introduced with a single HS4 1½ inch SU. From this point on, more and more variants of the Series 'A' engine began to use the single 1½ inch carb, and in the 1980s, this type of SU carb predominates. On the latest 1275 engines, as used in the Metro, the 1½ inch SU has given way to a 1¾ inch H.I.F.6 SU. Where twin carbs have been employed on the Series 'A' engine as standard, 1⅛ inch were used many years ago, but 1¼ inch ones have been the most common.

Many performance enthusiasts seem to fall into the trap of assuming that because a particular type of carb is fitted as standard, its performance capability is suspect. The two main advantages of the SU carb are (a) the engine comes equipped with one or more of them, and (b) it has plenty of potential for high performance, especially in its larger sizes.

For a given throat size, the SU carb has a fairly respectable airflow rate, and on top of this, it is capable, when properly calibrated, of delivering a very precise fuel/air mixture. Not only is the fuel/air mixture well metered, but also fuel atomization characteristics appear to be close to optimum for good, all-round performance in terms of power, drivability and economy. So, is the SU perfect? Hardly. 1¼ inch SUs tend to be a little on the small side, even for mildly tweaked 850s. Any SU with a side mounted float chamber can, on a car with a highly developed suspension system, run into the problem of fuel surge under cornering and braking conditions. SUs with concentric float chambers such as the H.I.F. series, effectively overcome this problem. However, they are not without problems themselves, problems incidentally, shared by some of the earlier type carbs. For instance, any carb (H.I.F. included) that has a bias needle, tends to suffer from jet wear. On carbs with a bias needle, the needle itself is not fixed into the piston in a rigid manner. It is lightly spring loaded so it contacts the side of the jet. This, among other things, overcomes the need of having to centre the jet, as was the case on earlier carbs. In curing one problem it has brought about another in the form of jet wear, which along with it, brings a mixture maintenance problem. However, once aware of this problem, it's only necessary to make a point of changing the jet at regular intervals of, say, 20,000-30,000 miles to counter it.

The latest HIF-type SU carbs are most certainly performance-with-economy devices. Note on this HIF 6 the latest cam-type throttle lever. This gives more sensitivity at low throttle openings, and faster opening in the last half of the opening phase.

If you are concerned with using the best of the SU range, then at this point in time (mid-'80s) my vote goes to the H.I.F. range of carbs. These carbs fall into the 'excellent' category. They are available in a number of sizes, of which the carb code number denotes the number of eighths above 1 inch diameter that the throttle bore is, or the size in mm. For instance, an H.I.F.4 is 4/8ths over 1 inch, (i.e. 1½ inches) and in metric it's an H I.I.F. 38. The 6 is 1¾ and the 7 is 1⅞ths. These carbs tend to have a higher flow capability than the earlier carbs of the same size, which is very much a point in their favour. Apart from this, the concentric float bowl also virtually eliminates fuel surge in hard cornering. One point that does need to be watched is the bi-metal strip fuel temperature compensating device in the float bowl. With the exhaust so close to the base of the float bowl, fuel in the float bowl can get to a relatively high temperature. Under these circumstances, the bi-metallic strip can sometimes over-compensate, causing a high-speed, high-temperature lean-out of the carb. This seems to be a problem only when the car is either run on a rolling road dyno and there is insufficient air speed to take the exhaust heat away, or a lot of town driving on a hot day has been done. The fuel in the float bowl gets hot and then a short stretch of road presents itself where the car can be accelerated to high speed quickly. Under these conditions, the temperature compensator may not have reset itself to give the correct fuel/air mixture. Of course new fuel coming into the float bowl will be cooler, and this will eventually bring the situation back to normal, but only after it has absorbed the excess heat present.

Downdraft Webers

The most popular progressive downdraft Weber is still the old 28/36 DCD. I say 'old' because the carb has been around in a relatively unchanged form, for over 20 years and it's still going strong, simply because it performs so well.

Another carb which seems to be gaining popularity is the 32/34 DMTL Weber. This is of much more modern design, and after market kits seem to have sprung from the fact that Leyland seriously tested this carb for OE use on a performance 1275 engine. There

It's on the dynamometer that a device proves its worth. The SU proved to make hp, but the gauze cover over the ram pipe sapped 3hp. Since it doesn't filter, this proved it to be a worthless addition to a performance engine.

are other, two-barrel, downdraft Webers that occasionally see use on the Series 'A' engine. Many enthusiasts simply adapt a downdraft, progressive choke, progressive opening Weber carb from one of the many types used on the European 4-cylinder Ford range of vehicles. This move can meet with varying degrees of success or failure as the case may be. It's worth remembering that the characteristics of a carb in terms of choke sizes, emulsion tubes, jets, etc. are for a Ford engine rather than a Leyland engine, and the airflow characteristics of these engines are quite a bit different. If you pick up one of these carbs cheap, then it's well worth giving it a try, but be prepared to be disappointed. It's well worthwhile considering the fact that even if you do get the calibration right, you will only be able to expect about the same sort of power output as a 28/36 Weber can give, or about the same as a 13/4 inch SU.

Another carb that I have had personal experience with on a Series 'A' engine, is a 38 DGAS Weber used as original equipment on the European 3-litre Ford V6 engine. This carb has quite a high airflow rating at around 300cfm and consequently on a tuned engine, it can give pretty good results. Since this carb has simultaneously opening butterflies, it can be used mounted crossways, that is with the barrels parallel to the crankshaft centreline, on a Mini engine, whereas it needs to be mounted at 90° to the crankshaft centreline on an inline engine. Again, because both barrels open simultaneously, fuel distribution within the manifold will be pretty even, so long as the manifold used was not originally designed for a progressive opening carb. If it was, then the manifold opening needs to be made symmetrical so as not to bias fuel towards one end of the manifold.

By comparison with the SU carb, the downdraft Weber in many instances is easier to calibrate. With the SU it's necessary to keep on changing needle profiles until you get the right one, although there is a short cut to that which I will discuss later, in the tuning section.

With a Weber carb, curing a fuel mixture situation is simply, in most instances, a question of selecting a larger or smaller jet, be it air jet or a fuel jet. The only really difficult parameter for the novice to deal with is the emulsion tubes. In a Weber carb, these can be likened to the profile on an SU needle. The emulsion tubes tend to have a moderate overriding factor which profiles the mixture curve to suit the engine's characteristics. However,

The 28/36 and the later Weber carbs have always been popular with performance enthusiasts and with good reason – it works. The particular example shown here would work a lot better without the restrictive air filter.

changes of emulsion tubes are often unnecessary, simply because once an engine's characteristics are established, the same emulsion tube will cover a wide range of engine permutations. At this point, it's simply a question of plugging in the right main jets and air correctors and possible pump and idle jets, to get respectable results.

The latest offering from Weber for the Series 'A' engine is the 32/34 DMTL. Depending upon the state of tune of the engine, it's capable of results similar to a 1½ inch to 1¾ inch SU.

Here's a Weber kit put together by A.P.T. utilizing all the best components available. The manifold, a TWM item, has high flow and power producing capabilities. The air filter is a high flow K & N item. Not the cheapest kit, but in terms of horsepower for what you spend, it's good value.

Shown here is a 28/36–32/36 Weber manifold from TWM. Flow tests produced excellent flow figures but it has a clearance problem with the centre pipe of some tubular exhaust systems.

To sum up then, we can say that the main assets of a 28/36 Weber or one of its many more modern counterparts, is fast and easy calibration by comparison with the SU, and power potential about the same as the H.I.F. 6 SU, so long as the carb is mounted on a good intake manifold.

Fish Carbs

The Fish carb, invented by one Dr Fish in the U.S.A. is produced in England by two companies, **Minnow Engineering** and **Leonard Reece & Co.** The main advantage of the Fish carb over any carb covered in this book, is the fact that it is totally adjustable without having to resort to substituting new jets, etc. In other words, to calibrate a Fish carb, all that is needed is a screwdriver, an Allen key and a ballpoint pen! There is no necessity for having a stack of main jets, emulsion tubes, chokes, etc. That's the positive side. On the negative side, it's difficult to calibrate a Fish carb in anything else, other than single carb form, simply because you don't necessarily know exactly what the setting is. Unlike a Weber carb or an SU, one can make the assumption (and, within limits, reasonably so) that if they have the same needle and the same emulsion tubes, jets, etc., then they are delivering the same fuel/air ratio. Because the Fish carb is infinitely adjustable, there is no means of telling if two carbs are, in fact, delivering the same fuel/air ratio. On the other hand, with one carb on the engine, there is no problem. The carb is adjusted to give maximum power by virtue of road testing, or by analysing the exhaust and setting the mixture to give maximum power on the chassis dyno.

Although I have used both types of carb and found them equally functional, the type I am most familiar with and have used most of is the Reece Fish carb. With its adjustable pivoting float bowl, the Reece Fish carb can be set at any angle from fully downdraft to completely horizontal. This type of carb in 1½ inch form was once very popular for 850 Mini racing. It's less popular now simply because many people prefer to use one barrel of a 45 DCOE for carburation. In terms of power, a 1¼ inch or 1½ inch Reece Fish carb can, on a typical Series 'A' engine, be expected to give 2% and 5% more hp than the equivalent size SU, but though it wins out on hp, it loses out on economy. I can be quite specific about this, having spent many, many hours doing fuel loops and economy testing with both the SU carb and the Reece Fish. The problem with the Reece Fish is that at part throttle it delivers fuel droplets which are too large, and as a result, the combustion efficiency tends to deteriorate. However, at full power, the often slightly superior airflow, depending on which SU we are comparing with, plus fuel droplets nearer

As with anything that comes from the Reece establishment, these Reece/Fish carbs are finely engineered. All surfaces are either machined or finished in a fuel-resistant, hammer-finish enamel. The carb on the left is 1¼ inch, the carb on the right is a 1½ inch.

an optimum size for best hp given by the Fish, seem to be the factors which endow it with superior power producing capabilities.

On the negative side, another factor that I have occasionally found when setting up engines on the dyno, is that the metering spread of the Fish carb from the bottom to the top of the rpm range is insufficient to set it up perfectly throughout the range. In other words, if the mixture is smack-on for power at, say, 5000rpm, it could be slightly rich or slightly lean at 2000rpm. On the other hand, if it's right at 2000rpm, it could be slightly out at 5000. However, in no case have I found that the calibration is out far enough to affect the rpm range that is normally used when maximum performance is required. In other words, if there are any metering inaccuracies, they are small enough to be insignificant over the rpm band normally used when operating the engine for maximum performance.

This difficulty over the mixture calibration may sound like a damning blow for the Fish carb, but it needs to be viewed in its true perspective. Consider if you will the quoted figures obtained when the engine is very carefully optimized for every situation on a dynamometer. Most engines in normal use will have carbs far from perfectly calibrated. Even an absolutely standard car with absolutely standard carb settings can be out by a greater percentage than the amount I am referring to now. If you don't feel inclined to make umpteen needle changes or jet changes to an SU or Weber, then the Fish carb is a really viable proposition simply because it is better to have a Fish carb set as near optimum as possible, than a Weber which is one or two jets out, or an SU that is a needle or two off the mark. **As a result of the Fish carb's infinite adjustability, many people end up with better settings than would normally be the case, simply because they can adjust them ad infinitum until they are as near optimum as humanly possible.**

Sidedraft Webers & Dellortos

Since these two types of carb bear such a remarkable resemblance, I have taken the liberty of discussing them together. However, I'll get to the point and answer the inevitable question: which of these two lookalikes produces the most hp? Again a considerable amount of dyno testing indicates that I can be relatively specific about an answer. However, I would like to add that there is probably going to be a case which will prove me wrong here and there, but in general, the following applies in the majority of cases. The sidedraft Weber produces better hp than the sidedraft Dellorto. An immediate and natural reaction to this is that the Weber must flow more air than the Dellorto, hence the power difference. In fact the reverse applies. For a given choke size, the Dellorto tends to have the odd few cfm more of airflow capability than the Weber, so

Shown here is a 40 DCOE Weber and it's difficult to come up with something to say about this fine carb that hasn't been said before.

The Dellorto carb looks very similar to the Weber and at a glance, even an experienced observer could mistake one for the other.

how come the Weber produces more hp? Again I suspect it boils down to fuel droplet size, because although the Weber produces the odd 2% or 3% more power on road engines, and as much as 5% to 6% more power on race engines, the Dellorto almost inevitably gives the best brake specific fuel consumption figures at both full throttle and part throttle. As a result of these factors, I personally use Webers for race or competition engines, and Dellortos for high powered road engines, but that's the way my priorities go. If you are building your own engine, then you must decide which of these two excellent carbs best fits your needs.

If we are going to talk about the pros and cons of Webers and Dellortos, then we find that in the main their best asset is their power producing capability. On top of this, we must also add their almost infinite adjustability due to the wide range of calibration components available for them. Although to the beginner they look like a maze of parts which seem totally unfathomable except maybe to the designer, this is not necessarily the case.

If you are going to install a Weber or Dellorto carb, then I would thoroughly recommend that you have your engine set up on a dyno, either an engine one or a rolling road one, so as to make sure that the carb's calibration is somewhere near spot-on. Failing this, in the case of the Weber I would suggest you get from your Weber agent, the very comprehensive Weber tuning manual which is 60-plus pages of valuable information for the would-be Weber carb tuner. Unfortunately at the time of writing, no such book exists for the Dellorto though the principles of tuning touched on in the Weber book will apply to Dellorto carbs.

The deficit side of sidedraft carbs is their tendency to be more expensive than any of their other counterparts in terms of initial outlay. The insurance man definitely doesn't like them and will inevitably load your insurance premium by a sizable amount. On transverse engines they tend to put a lot of induction roar straight into the passenger compartment, so you are faced with a noisier car. Also, a very important factor is to select carefully the intake manifold you intend to use. On transverse engines, very often those that work best won't clear the bulkhead without some re-sculpturing being done, and those that do clear often won't work. **Before buying an intake manifold, I would suggest you read the manifold chapter** to see just how important intake manifolds are, not only for the sidedraft Weber carb, but indeed for all the carbs we have discussed.

As far as power and economy is concerned, these sidedraft carbs will deliver the goods, but calibration is important. A sidedraft carb that is a jet size out here or a choke size out there, an air corrector off somewhere else, will produce less power and use more fuel doing it than, for instance, a perfectly set up SU. If you hope to get the best from a sophisticated sidedraft carb, then it must be properly calibrated.

Weber IDA Downdraft Carbs

The 46 and 48 IDA Weber carbs have never proved to be really popular on the Series 'A' engine. However, it is a carb I have had excellent results with, mostly in terms of hillclimb engines. The only drawback to using an IDA Weber is that it is necessary to protect the mouth of the carb from unnecessary buffeting in its exposed position: inevitably sticking out through the bonnet. If this is not done, top end performance is lost. However, with speeds up to about 70mph, the IDA Weber scores in as much as it is always drawing air at the coldest possible temperature, i.e. the ambient temperature, whereas sidedraft carbs under

Weber and Dellorto carbs are renowned for their high airflow capabilities. Topping the list for Series 'A' engines is the IDA downdraft Weber which pulled a monstrous 290 cfm+ through each barrel with a 38 mm choke installed.

Here's a 48 IDA on my own Mini. Those of you who are very observant will notice the nut on the flat-spot on the intake manifold. This controlled a mixture distribution device. There was one in each port. It was an experiment to attempt equalization of mixture to the cylinders. It didn't work.

the bonnet are drawing from the hotter air that exists in that region. However, at speeds above 70 or 80mph, the exposed carb suffers in terms of top end performance, and on engines reading identical power on the dyno, we find that with an IDA Weber out on the track, speed may suffer to the tune of 5mph. To get the best from an IDA it is necessary to build a duct above the carb which slows the air down, so the carb can draw from relatively undisturbed air. At the time I was running an IDA-Webered Mini, the rules did not allow the changes in bonnet profile, and therefore a carefully constructed duct had to be removed from my Mini racer. Three months later the rule was changed (wouldn't you know it!) Anyway, to sum up on this carb, really all I can say is that if the installation is done properly and the carb is given every opportunity to function correctly, then an engine so equipped is very responsive and will produce very good power.

Here's an IDA Weber set up being put through its paces on the DV dyno. As in every other configuration tested by the author, the IDA set up proved capable of excellent results.

Carburation - How Big?

One of the first questions that will be asked by the novice engine builder and, indeed, some that are less than novices, is 'How big a carb do I need on my engine?' Well of course this does depend on the engine, but the question itself 'how big' really ought to be re-phrased. The whole point of a carb is that it passes air into the engine and mixes it with fuel. **The size of the carb is not the criterion by which it should be judged, as this is not as relevant as its airflow potential.** Obviously, to a point at least, a big carb will flow more than a small carb, but carb efficiencies vary quite substantially, and a 1½ inch carb of one type may not flow the same as a 1½ inch carb of another. Rather than try and assess the carbs by an arbitrary measurement performed with a rule, it's best to assess the carbs in terms of their flow capability, as is done by the majority of U.S. carb manufacturers. This, incidentally, is a policy which does not seem to have picked up any followers amongst European carb manufacturers and so the average guy purchasing a carb is largely in the dark as to what that carb's airflow capabilities are. Well, I'll see that that's put right in short order, but first let me explain how carbs are calibrated for their air flow capability.

It won't come as a surprise to any of you to realize that air is an elastic medium. A cubic foot at sea-level contains a far greater amount of air than a cubic foot at 30,000 feet. This is simply because the molecules are more densely packed at sea-level due to higher atmospheric pressure than at 30,000 feet. When air is drawn into a cylinder, it flows into the cylinder because a partial vacuum is formed by the passage of the piston down the bore. Just as heat can only flow from a hot area to a colder area, so air can only flow from a high pressure area to a low pressure area. The steeper the pressure gradient, the more air flows. In other words, if we are going to have air flowing from an area of 14.7 lb per square inch (psi) into an area of 4.7psi, we can expect it to rush in in far greater quantities than if we were going to flow air from an area of 14.7psi to 13.7psi. In the first instance, there is a pressure drop of 10psi to drive that air from the high pressure area to the low pressure area. In the second there is only 1psi. Even though the air may be flowing through the same orifices, the amount of air flowed will be substantially different. It's because of this factor that it's necessary to rate the carbs at a certain pressure drop. In other words, the carb will have a given suction applied to it, and this value will be held constant for all carbs.

For calibrating high performance carbs in the States, a pressure drop equal to 1½ inches of Mercury is used. Basically that's the amount of suction necessary to draw Mercury up a column 1½ inches from the surface of the Mercury. The reason that a Mercury measurement is used is that it is accurate and absolute but, to give you a more usable reference, we can say that 1 inch of Mercury is approximately equal to half-a-pound per square inch, which in turn is equal to 13.4 inches of water. The depression of 1½ inches of Mercury used for carb calibration then, will be equal to 21.3 inches of water. In other words, if you had a 2 foot long straw and sucked water up 21.3 inches, the suction required to do so would be the same as required to draw up 1½ inches of Mercury. (Don't try this with Mercury - it's a poison!)

Having established the technique for measurement, let's look at the carbs so far discussed to see how much air they flow. The chart, **Fig. 7.1**, gives the typical flow figures. You will note when I am quoting the flow figures for the sidedraft Webers and Dellortos and the downdraft IDA Weber, I have quoted the flow for one barrel only. This may seem to be under-selling the carb because its flow capability is surely the value of both barrels? At first sight this would appear to be the case, but when a cylinder draws air, it sees the flow capability of only one barrel of carburation. Bear in mind that on a four cylinder engine, none of the induction pulses actually overlap to any great degree. As a result, although all cylinders may share a carb, they do so at a different time. Having two barrels of carburation on a Mini engine does not necessarily mean that when a cylinder draws air, it sees any more carburation. The proof of the pudding is the fact that if the intake manifold can be designed with sufficient efficiency, one barrel of a sidedraft Weber carb will produce power figures very similar to an engine using two barrels. In certain places in the rpm range, due to inertial ramming, a single barrel setup will actually show better power. However, the big problem is, that designing a manifold so that the two ports share one barrel and flows as efficiently as a set up where each port has its own barrel of carburation is difficult.

How Much Does The Engine Need?

Scanning the chart **Fig. 7.1**, you

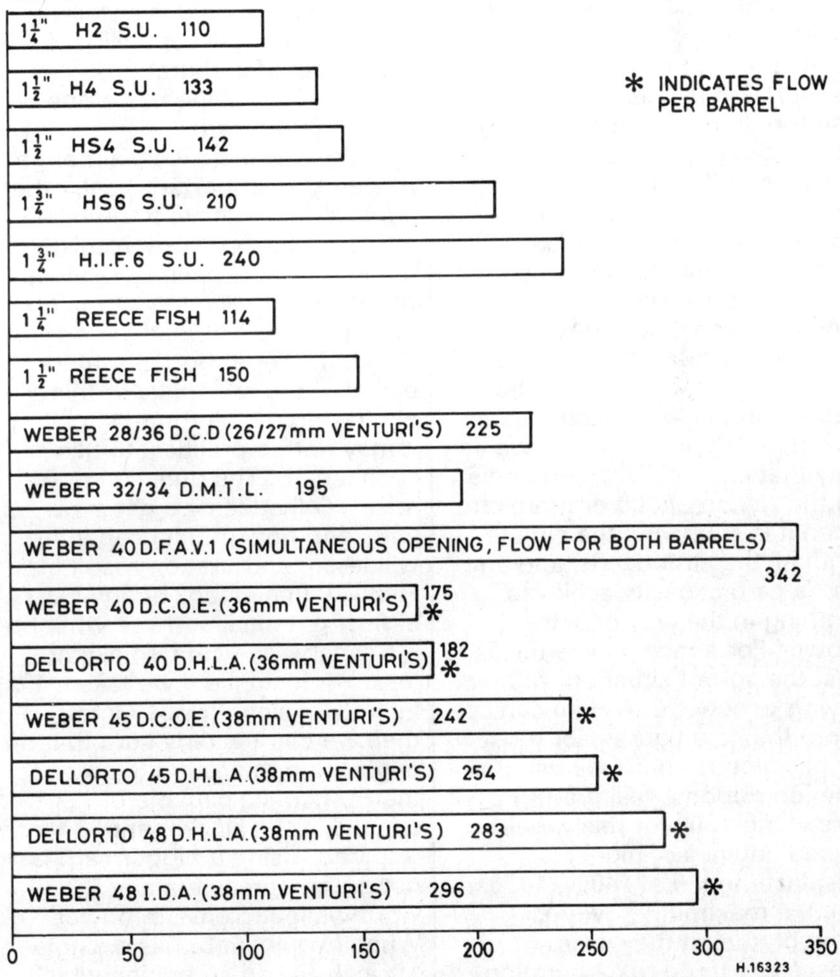

Fig. 7.1 Airflow capabilities of popular carburettors

This chart gives a clear indication of the airflow capabilities of most of the popular carburettors used on the BL Series 'A' engine. Use of the information in this chart, plus that in the text, should allow correct carburettor selection.

will have a fair notion of the capabilities of the carbs you could use, but rather than answering the originally posed question, the fact that you now know these capabilities merely raises another question, namely 'How much carburation in terms of flow capability does my engine need?' As it happens, now we have rated the carbs in cfm that question can be answered more easily.

Let's start off by assuming maximum power is required from the engine. If enough carburation is fitted the carb itself will not be the prime restriction in the chain of events that produces hp. What will inevitably limit the hp is the cylinder head. The question is knowing just how much carburation it takes to go over the top. In other words, how big is just too big? Assuming the intake manifold is up to scratch, referencing carb airflow figures to dyno tests indicates that between 2 and 2.2cfm is needed for every potential 1hp the engine is capable of developing. In other words, let's assume we have an engine which, if it had a perfect induction system, would produce 100hp. From this we can say that if the carburation flows in the region of 200 or more cfm then it is supplying the engine with all the air it needs, and any more carb capacity achieves nothing in the way of extra power. For a race engine this is just the sort of situation we are trying to achieve. As you can see from the flow figures, for most applications a 45 DCOE will do the job but on a real fire-breathing 1300, or really well-tuned engines of bigger displacement (*i.e.* 1400s, 1430s, etc.), if maximum power is sought, then it may seem necessary to go to carburation with a greater capacity such as a downdraft IDA Weber or sidedraft 48 DCOE. In practice however, this often proves not to be the case. On engines up to 115–120hp, a sidedraft 45 Weber will be found very difficult to improve on. Indeed, on the five or so occasions I have made relevant tests on my dyno the 45 DCOE has just pipped the 48 DCOE or output throughout virtually the whole RPM range. Even if you have an engine with a power potential of as much as 135hp, achieving it with normal aspiration is extremely unlikely.

If you get to this high level, then the difference in power output between a 45 Weber and 48 Weber is likely to be small.

Again, it's probably worthwhile for me to make the point that the carb capacity I am referring to is the capacity that one cylinder uses. In other words, we are only ever talking about the airflow capability of one barrel of carburation, whether the engine has a one or two barrelled carb.

In many instances, although power is an important goal, other considerations have to be taken into account and compromises made: this is especially relevant to road-going engines. It may be nice to be able to install a sidedraft Weber but, as stated earlier, there are snags to this, and taking the all round situation it may not be practical to use such a carb. The question now arises, 'What can we get away with, and how much drop in hp will lesser carburation result in?' Firstly, you can look at any carb's airflow potential and see what hp it's capable of with virtually no loss. For instance if we take a 1½ inch SU which flows 133cfm, then we can be fairly sure that on engines up to around 60hp or so, the SU is quite capable of supplying the air demand required. Using a bigger carb is not going to result in any worthwhile increase in power. What if we want to use a single 1½ inch SU on an engine which is potentially capable of 90hp? How much power is that single 1½ inch going to rob from the engine? Firstly, below the 60hp level, it won't rob significant power from the engine. In other words, as the engine works its way up the rpm range, until it passes the 60hp mark that SU carb is not costing 1 ounce of power or torque. It's only when the carb passes its power potential that it starts to rob the engine.

48 IDA down draft Weber's are getting increasingly difficult to find. Also, for the large displacement 'A' Series engine something a little larger is needed. Gene Burg Enterprises in California produces their own version of this carb with up to 58mm throttle bore. This unit takes all the usual Weber calibration components and has the same bolt pattern as a regular 48 IDA. Gene Berg Enterprises specialize in air-cooled VW engines not the 'A' Series engine so do not expect any technical information from them regarding this carb in the context we are dealing with here. Also, if you intend to use one of these excellent carbs bear in mind that you will either have to find an IDA manifold or go to a specialist manifold maker such as JanSpeed or Manifflow for a manifold.

Putting it another way, big carbs help power only from the point that a smaller carb becomes inadequate, so this means that the power increase given by a bigger carb tends to occur towards the top end of the rev range. Of course this makes the assumption that there is no shockwave or inertial ramming going on to cause a bump in the volumetric efficiency curve elsewhere in the rpm range.

At first sight the potential amount of power lost by having an undersized carb is quite substantial, but in fact this proves not to be the case. To a slight degree this situation is self-adjusting because if a carb is too small, it causes the engine to 'suck' harder on the carb. However, this is not totally compensating, so an undersized carb does cause some power loss. It's just not as much as you may think. A few examples here may serve to show just how a multitude of carbs, tested on a high powered engine, performed. **Fig. 7.2** shows the power figures obtained by a selection of carbs, and it's worth noting that there is no real difference in the power output at the wheels until over 70hp has been generated. From these figures one would suspect that 13/4 inch SU on a good intake manifold is only marginally less effective than a 48 IDA downdraft Weber. In essence you would be right. A big, well sorted SU on an effective intake manifold, does certainly produce the goods and as a carb for a high performance road engine, it is certainly one of the best ways to go.

If you are observant, you will have noted that the figures given in **Fig. 7.2** are from a rolling road dyno, and though at the end of the day a rolling road dyno should be considered THE final stage of engine tuning, rolling road figures themselves can be a little suspect. Granted, the figures concerned in that particular instance were taken one after another, all done the same day and so are representative. However, if we are really going to look at the situation hard, we need to consider more accurate figures, such as are taken off an engine dyno. The numbers in **Fig. 7.3** show the difference in power output on a highly tuned 1435 engine when the carburation is stepped up from single 1½ inch to single 1¾ inch. As you can see, hp numbers are initially quite close together when we are talking about power below about 60hp. That's because the single 1½ inch can cope just as well as the 1¾ inch. However, once the power outputs get onto the top side of 60hp, so the power curve given by the 1¾ inch begins to diverge quite rapidly from that given by the 1½ inch. At 7000rpm the power

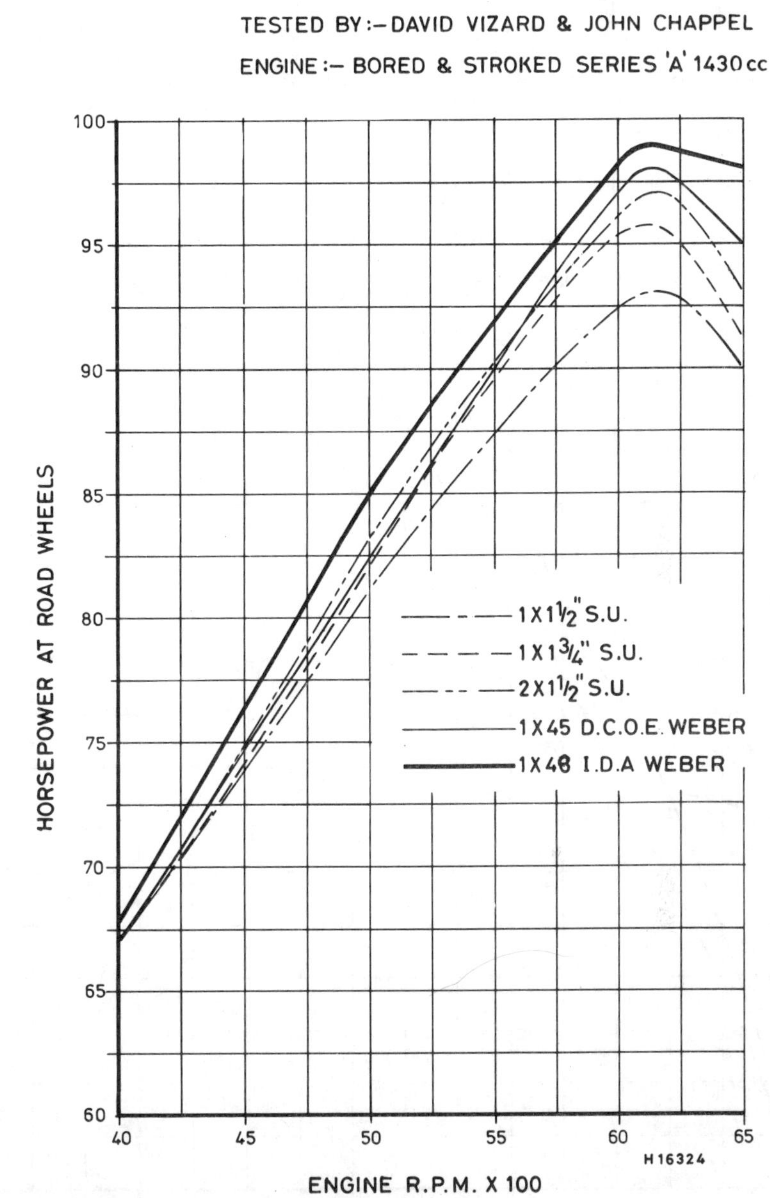

Fig. 7.2 Power changes resulting from the use of different carburettors

Carburation changes often do not result in the big power increases anticipated. This graph shows fairly typical changes. The low end increase seen with the 48 IDA setup is due to the longer intake length with the required type of manifold, not an increase in carb air flow.

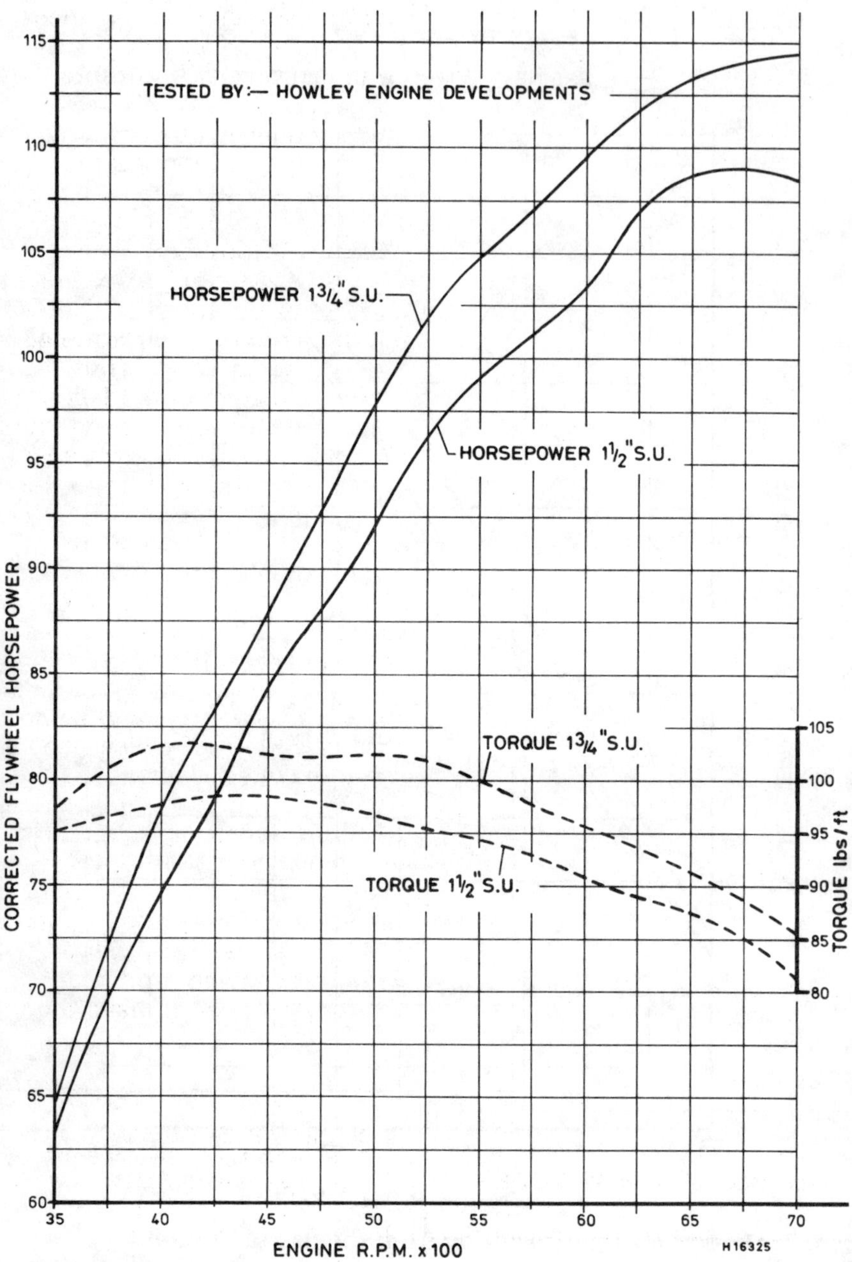

Fig. 7.3 Comparison of single 1½ inch HS4 SU and 1¾ HS6 SU on 1430cc Series 'A' engine

Experience has shown that a single 1½ inch SU can generally deliver the required airflow for engines up to about 60-65 hp. Above this power level, more carburation is needed. Replacing a 142 cfm carb with one of 200+ results in a useful gain in hp on an engine requiring a lot of air, as does this highly tuned 1430.

difference is some 6hp. Proof enough in itself that the 1¾ inch carb was needed on an engine of this power output. However, the 1¾ inch carb should, according to our rating earlier, be capable of supplying the air requirements for an engine of around 100hp, again assuming the intake manifold is of good design. So we can now pose the question, 'How much difference in power would there be going from a carb with a flow capability of a typical 1¾ inch carb, to the flow capability of a single sidedraft Weber?' With a sidedraft 45, we would be talking about flow capabilities in the region of 240-245cfm per barrel. Remembering that carburation capacity is dependant on one barrel rather than two, then we are not actually stepping up the airflow capability that much. However, the Weber manifold will inevitably have a straighter shot to the port than a typical single SU manifold, so not only is the airflow capability of the carb up a little, but also it's on a freer flowing intake manifold. Bearing these two factors in mind, let us see how much power a single 45 is worth over a single 1¾ inch.

For the answer to that, refer to **Fig. 7.4**. In essence the engine used in this is again a 1435. However it differs slightly in the cam department from the engine used in **Fig. 7.3**. As you can see, in this particular instance, the engine produced 116.8hp on a single 1¾ inch SU. When this was changed to a single sidedraft 45 on a 6 inch manifold, power increased to 120.5hp. A gain of only 3.8hp. That's probably not as much of a gain as many readers would have expected. However, under many circumstances it is pretty representative of what will be achieved, but like most things there's more to it than meets the eye. For example, the inlet manifold design is significant in actual power output; as I will show later, a good SU on a good manifold, will provide more power than a good Weber on an indifferent manifold. The particular engine used for dyno testing in Fig. 7.4 had a 288° cam in it. Engines with longer period cams, say around 300° or more, tend to show a noticeably bigger gain when a 45 Weber

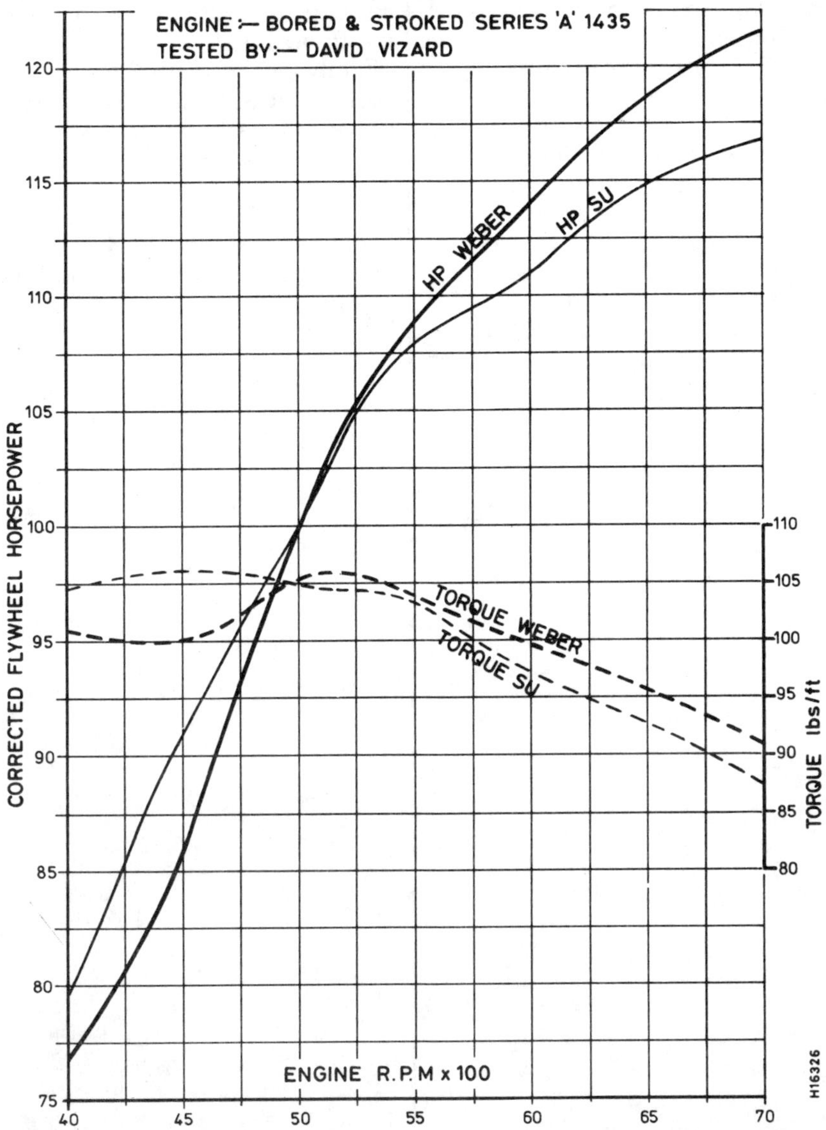

Fig. 7.4 Comparison of single 1¾ inch HS6 SU and 45 DCOE Weber on a 5 inch manifold

A single SU on a good intake manifold can perform well. In this example here, the SU was mounted on a Howley intake manifold which had been reworked to take the 1¾ inch carb. The port runners were also opened up slightly for the large capacity engine. Generally, port sizes were 1⅜ inch diameter flaring out at the plenum behind the carb. Up to 5000 rpm the SU beat the Weber 45 DCOE on torque, but from 5000 up, the superior breathing of the Weber and manifold combination pays off.

replaces a single 1¾ inch SU, and under such circumstances on the bigger motors, gains of up to 6hp are fairly common; gains greater than this are unlikely unless a particularly efficient Weber manifold is used. On smaller engines, the gain using a sidedraft Weber is typically proportionally less than is seen in the larger units.

At this point I think I should put in a commercial for the 48 IDA and 46 IDA downdraft Webers. In spite of the fact these carbs have to be mounted on manifolds which form a 90° turn to feed the mixture to the ports, the only manifold available happens to be an exceedingly good one. As a result, the excellent flow capabilities of these IDA carbs is not lost on the Series 'A' engine. However, its use entails some good features and some bad ones, both of which need to be appreciated. Firstly, on Mini and Metro engines, mounting a sidedraft Weber on a decent intake manifold normally means that the bulkhead has to be cut away. With the IDA units, this isn't necessary. They will fit without bulkhead modification, but they do need to have the

DETAILS OF AIR SCOOP AND FILTER SET UP TO MAKE THE BEST OF AN I.D.A. WEBER CARB. ON A MINI.

Fig. 7.5 Air scoop filter arrangement

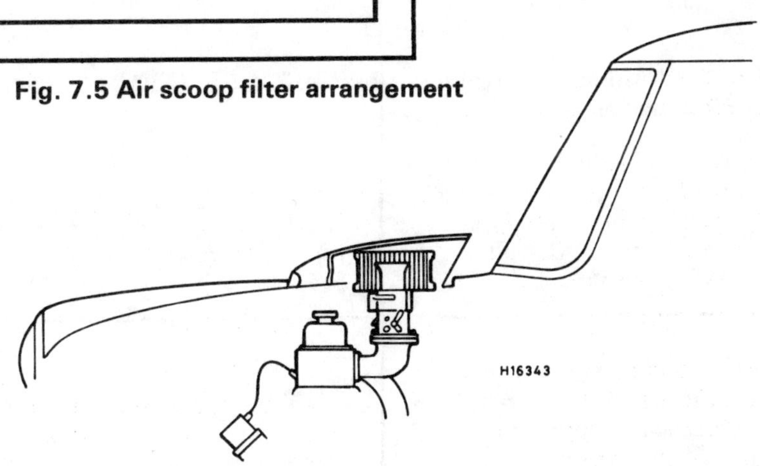

bonnet/hood modified. For many years it was common practice just to have the trumpets of the IDA Weber protruding through the bonnet/hood. Often things were just left at that. Much of the reasoning was that at least this carb could pick up cool air. However, the carb's function appears to be quite disturbed by the turbulent air passing over the top of it. My own experience has been that unless the carb is covered and air ducted to it, a great deal of top speed is lost. On the Club circuit at Silverstone which has a very long back straight, I found that running an open IDA as compared with one having air ducted to it, there was a speed difference of some 6mph at the end of the straight. It was difficult to evaluate what this was in terms of lap times, because testing was always done on different days, and the biggest factor affecting lap times was not the engine's power so much as the temperature and humidity on the day effecting just how much the tyres gripped.

To make the IDA induction system really work, the air needs to be slowed down in the vicinity of the trumpets, and I would suggest the best way of doing this is firstly to add a K & N filter to steady up the airflow over the carb, and then build a duct over that which slows up the air in the vicinity of the carb. Something along the lines of **Fig. 7.5** has worked for me.

Carburation Part 2

Carburettor Tuning

In the following part of the book I am going to look at basic carb tuning so you can get your engine set up in the right ballpark, prior to having it set up on a rolling road dyno. This is especially important as it is not practical to set up a brand new engine on a rolling road dyno. It does need some breaking-in first, and if it's a road motor, you should put at least 500 miles on it before a dyno session.

What this section is not intended to do is replace the full overhaul manual for the particular carb you have in question. What it will do is give you some pointers to get those carbs working and maybe some things you can do which are not necessarily in the manual.

So let's start the ball rolling. The first carb we'll look at, and rightly so, since most engines have them as standard, is the SU.

Tuning SU Carbs

In order for a carb to work properly, assuming it was designed correctly in the first place that is, it must be (a) unworn and (b) adjusted correctly. So the first thing to do with SUs, assuming they are not brand new, is to check them out for wear. Fortunately the SU, in spite of having some precision mating surfaces, does not suffer unduly from wear. The biggest wear problem, so long as the carb has never been maltreated or damaged in any way, is caused by the shaft and/or the jet on later bias needle designs wearing in the body. If you are buying second-hand SUs then always check these points to see if the carbs are acceptable to you. If there is any appreciable slop in the butterfly shaft, an air leak will develop at this point and could detract from a potentially good idle. Such leaks will cause erratic running at low rpm and small throttle openings, but will not make any difference to the full throttle operation of the carb, so will not affect power. Worn throttle shafts on twin carb installations can produce a real headache when trying to balance the idle airflow of each

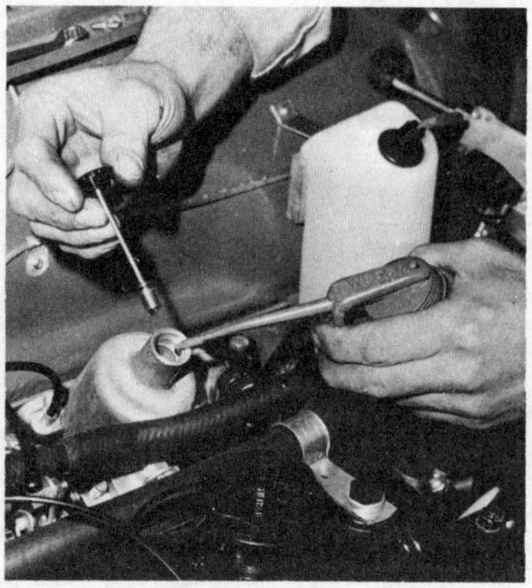

A common 'tuning' problem with SU carbs is a flat spot when the accelerator is depressed. This is inevitably due to the fact that the carb hasn't been serviced regularly and the dashpot level is low.

carb. **Again, though, the need to balance twin carb airflow at idle, does not, in any way, shape or form, produce any extra full throttle hp. So long as both throttle plates are fully open when the throttle is depressed, the airflow at idle is quite irrelevant.**

The other point you should check, and this is very important otherwise you will never get the calibration of the carb right, is the jet itself. Basically there are two types of jet: there's the old type which required centring and used a fixed needle, and there is the newer type which does not require centring, but has a biased, spring-loaded needle. Let's deal with the older type first.

Under normal conditions, the jet won't wear by any noticeable amount. Only on very old carbs will there be a small amount of wear due to fuel errosion, this is very small and can almost always be compensated for by jet adjustment. However, what can happen is that the carb may have been previously overhauled by some ham-fisted amateur, and during the overhaul, the needle may have been bent when the piston and dashpot assembly were re-installed, in which case the needle may have scuffed on the side of the jet, and worn it oval. If this is so, the engine will almost inevitably run rich because the jet is simply too big. The solution is simple: replace the jet, and remember, for SU carbs, you don't change the jets to alter the mixture. Jets come in one size per carb only, i.e. 90 thousandths for 1¼ inch and 1½ inch SUs, 100 thousandths for most 1¾ inch ones and 125 thousandths for the biggest 2 inch SUs.

If your particular carb is a later type, then it will have a biased spring needle. If so, jet wear is not a 'maybe' but a definite certainty. Since the needle is always pressing on the side of the jet, it will, during the passage of time, wear the jet. Of course this action will wear both components. However, the wear on the needle is spread over a much greater area than the wear on the jet, so needle wear is at a relatively slow rate compared with the jet. On any high performance car, I would recommend changing the jets in bias spring carbs such as the HS4 and 6, and the H.I.F. range of carbs, every 20,000 miles. As for the needle, try to leave this undisturbed so that the same part of the needle is always presented to the jet. Since the needle has been rubbing against the jet it will be smoothed off and it will wear less than turning the needle around and starting on another patch.

If it's the older type of SU you are using, then it will be necessary to check the jet centring. If the jet isn't centred properly, the needle will wear the jet when the needle is in its lowermost position. It will also wear the needle in this position, and this is the most critical part of an SU needle if good idle and low speed economy are to be had.

Another point is that if the needle is in contact with the jet, it will stick. This means that as the throttle is opened, so the mixture will become excessively rich because the piston is not lifting as it should. When the engine is given a little more throttle, if the needle is going to unstick in the jet, it will, and it will rise suddenly, possibly causing a temporary flat spot. All this leads to unnecessarily bad drivability.

The easiest way to determine if the jet is centralized properly, is to unscrew the damper rod from the top of the carb, then lift the air piston, via the mouth of the carb, right to the top of its travel. When released, it should fall freely and strike the jet bridge with what SU in their overhaul manual describe as 'a soft, metallic click'. Personally, I think a hard, plastic clunk is a better description! However, if it doesn't fall freely throughout the extent of its travel, and sticks just before the end of it, then it's a fair bet the jet is not in the right place. To centralize the jet, disconnect the choke lever from the bottom of the jet and remove the large coil spring just above the jet adjusting nut. When you replace the jet adjusting nut, screw it up as far as it will go and replace the jet. This brings the top of the jet virtually flush with the jet bridge inside the carb. Now slacken off the nut which contains the jet assembly. This is the large nut just above the jet adjusting nut. Slacken it off just enough to allow slight free movement of the jet sideways. Now press down, using a pencil or something equally soft, on the bottom of the piston rod bore. This should press the piston well and truly onto the jet bridge, thus bringing the needle into such a position that it will line up the jet. Once you have achieved this situation, tighten the jet locating nut and check the piston drops freely as described earlier. If it doesn't, repeat the centring procedure until it does.

After you've gone through the rigmarole of centralizing the jet on older carbs, you will begin to appreciate the advantage of the spring-loaded bias needle. If you have a carb fitted with this type of needle, then the only thing you need do at overhaul time is physically replace the jet.

Drop Test

Another aspect which can cause a performance problem is a sticking dashpot piston, but unless there is physical damage to either the chamber or the circumference of the large piston, then any sticking is usually cured by a thoroughly good clean. The best bet for this job is to use a proper carb cleaner, or failing that, cellulose or lacquer thinner as used in automobile painting. Once the carb components are thoroughly cleaned, the next step will be to find out if they are actually functioning properly. Assuming you have all the parts stripped down, check them all for physical damage. If they are okay, lightly oil the steel guide in the middle of the bell chamber. Using a piece of Plasticine (modelling clay) block up the transfer holes around the face of the air piston. These holes are usually about a ¼ inch to ⅜ inch away from the hole that carries the needle. In earlier carbs, they are in the side of the piston, just around from the face where the needle protrudes. Once the steel rod has been lightly oiled, install the air piston into the bell chamber and turn it upside down and time how long it takes to drop out, making sure, of course, that it doesn't hit the floor in the process!

For smaller carbs like 1¼ inch and 1½ inch ones, it should take between 3 and 5 seconds. For larger carbs, between 5 and 7 seconds before the piston fully drops out. If you have a pair of SUs, check that the drop time is about the same. If it isn't, swop one piston over to the other bell chamber and see if that evens things up. If it doesn't, take the piston from the slowest falling carb, and using a drop of metal polish, polish the outside edge of the largest diameter of the air piston and the largest inside diameter of the bell chamber and see if this does the trick. If the pistons fall within about a second of each other, then that's close enough for the job. If they don't, it could be that you have a totally mis-matched piston and bell chamber, or something is amiss sizewise. As a last resort only, you can try using some 400 grit wet and dry emery on the outside of the piston, but it is a last resort, and it's not recommended as standard procedure.

If the pistons take much more than the times quoted, go back and check for dirt somewhere in the assembly. When the drop tests are within spec, re-assembling can be done.

Calibration

There are two principal components that affect the calibration of SUs. The one people most obviously think about is the needle. However, the dashpot spring is also a calibration device. The stronger the spring is, the less the piston opens for any given amount of throttle opening and engine rpm. In other words, the spring holds the piston further down. This does not necessarily mean less fuel issues from the jet, even though the needle is further in. Because a higher depression is created in the bridge area of the carb, the engine is now drawing harder on the jet. The spring's function is to make sure the piston only reaches the fully open position when the engine reaches maximum power rpm. If the piston reaches the full open position before maximum power rpm, then the calibration function begins to be lost. What happens in this situation is, as engine rpm increases, more air is drawn in, but not necessarily much more fuel, so the engine leans out irrespective of needle profile changes. The function of the spring is to make sure the piston only opens a certain amount anywhere in the rpm range and that it reaches full open only when full power rpm is reached. Outside of that function, the spring doesn't do an awful lot. If the spring used is too heavy, the carb may not reach full open, in which case it is necessary to install a weaker spring or indulge in the frowned-on practice of cutting a few coils off the spring.

Selecting the spring for the carb is, in actual fact, a straightforward procedure. What isn't so straightforward is selecting the needle profile. **If the engine in question is modified, and I don't doubt most of you are reading this book to do just that, then the original profile will, unless you are incredibly lucky, NOT be the one for the job.** Please don't write and ask me which needle will be needed for your engine which utilizes a head done by Joe's Super Porting, a Magnahauser Ramasonic intake manifold, and a Compucams Magnabump Superdyne profile cam, because **neither I nor anyone else will be able to tell you anything that's other than an inspired guess, and just for the record, I've logged all my inspired guesses within these pages.**

In a real-world situation, the only way to establish the best needle profile for your modified engine is to have the engine set up on a rolling road dyno in conjunction with a mixture analyser. This should establish within reasonably close limits, which of the 600-odd different needles is going to do the job. A good dyno operator can tell by using his instruments, which needle is going to be right within two or three goes. This approach will find you a lot more hp than trying to guess it. If you really know what you are doing with SUs, you can go rushing up and down the road and eventually arrive at a needle that's pretty functional, but if you know that much about SUs, you don't really need me to tell you what to do!

I think it's a good idea at this point to mention that if you are calibrating one SU as opposed to two, the needle selection situation can be made a lot simpler. If the calibration is taking place on one SU only, there will be no need to replicate that needle, i.e. you do not need a pair of needles. So long as a needle profile is arrived at which produces the mixture ratio the engine needs, you are not really worried whether you know what that needle profile is, or not. In which case it's possible, to a degree, to cheat on the whole system. Here's how; firstly it's necessary to appreciate that the shape of the needle in the jet isn't important. What is important is whether or not it delivers the right flow of fuel for the prevailing engine conditions. If you can accept this conclusion then, we can say within reason that the needle doesn't necessarily need to be round. To richen the mixture up at any particular point in the rpm range, the area of the needle actually in the jet must be reduced. I think you will all agree that it's easier to take metal off a needle than it is to put it back on!, so **the technique is to make your own needle from a needle known to produce a mixture that is too weak.** In other words, the needle is too big everywhere down its profile.

With such a needle installed, set the carb up with a satisfactory idle mixture and speed. I will detail how that's done in a moment. Now have the dyno operator drive the car through its normal operating sequence from low throttle speeds, low cruise to wider throttle openings, high cruising speeds to full throttle. At each stage, check how much the carb is open and check with the mixture analyser to see how much too lean it is. If it's running really lean, don't persist in testing too long, just long enough to establish that it IS too lean. Having noted how far up the piston is at any given point, pull the piston and dashpot assembly off the carb and then using a very fine file, i.e. a needle file (co-incidentally, but aptly named in this case) file a flat on the needle. Yes, you did read that right. Until you get the hang of how much filing is necessary to have a certain effect on the mixture, don't go overboard with the file. Just file smoothly down the length of the needle on the appropriate section. How much is being taken off can be gauged quite accurately by the width of the flat you are producing. In general, the drawing **Fig. 7.6** shows which part of the needle is functioning over which part of the rpm and throttle opening range.

To be quite honest, the old mixture analysers calibrated in air/fuel ratios are better for setting up carburation than the newer type which are calibrated in the percentage of residual carbon monoxide in the exhaust. Depending on which type is being used, maximum power occurs between 12.5/1 and 13/1 on the older type gauges, and for part throttle use on a well set up engine which has plenty of spark, ratios of 17/1-18/1 perform well for economy at part throttle. If a newer type gauge is being used and full power, rich mixture occurs when the CO gauge reads between 5% and 6% then, on cruise, the engine should be set to run just down to the 0% and then a tad leaner again. The problem with the CO gauge is that it runs off the scale when the mixture is around 15.5/1-16/1 and ideally the engine needs to be slightly leaner than this. If building an economy engine, which we will deal with in another part of this book, then it's possible with the right ignition system, to be running fuel/air ratios of 20/1 or

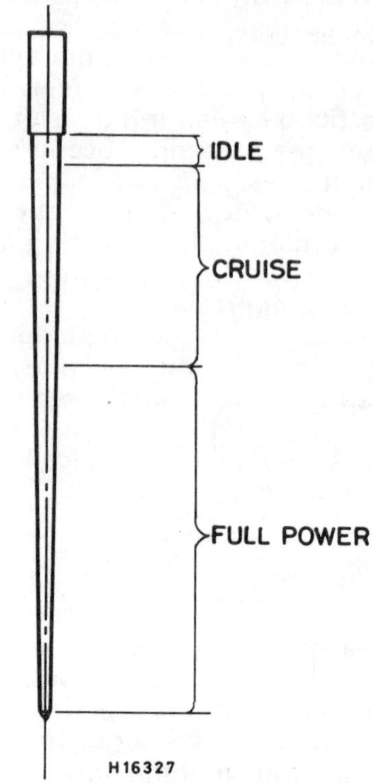

Fig. 7.6 SU needle

Before attempting to recalibrate an SU carburettor it is important to know which part of the needle affects which part of the engine operating range. This diagram gives a rough guide only. A more exact knowledge of which part of the needle is in the jet is needed to determine where the needle needs to be fatter or thinner.

more. These ratios are well off the calibration of a CO_2 gauge.

Idle Mixture Settings

One of the most difficult phases of engine operation to produce is a smooth idle. Mixture and timing error spoil results at idle to a far greater extent than they do when the throttle is open wider for cruise and full throttle operation. The reason is the mixture is the least combustable under high vacuum, slow engine speed conditions. Because of this, any erratic combustion shows up as a rough idle. Setting the idle mixture just right contributes toward a more pleasant, smoother running engine. If, at this point, the assumption is made that the needle profiles are correct for the engine, then adjusting the idle mixture airflow is done by the relevant adjusting screws as shown in **Fig. 7.7**. By the way, the schematic shown here is for an H or HS type carb. The H.I.F. carbs have the mixture idle adjusting screw in the side. If you know the carb is in perfect working order as a result of passing all the tests previously mentioned, start the engine and let it warm up at a fast idle. When it reaches running temperature you can think about setting the idle mixture and the idle airflow. On singles, the balancing operation I am about to describe will be redundant, but for twin carbs it applies, so as I go through the procedure sort out whichever case applies to your particular installation.

There are a lot of tools and gadgets on the market for balancing carbs. Some are very good; others will do the job, and some are downright useless. As far as balancing idle airflow is concerned, I would recommend you do use some instrument for measuring airflow, as opposed to the hose-down-the-mouth-of-the-carb-and-listen-to-the-hiss technique. This will work at a pinch, but it definitely is not as good as using a proper instrument. Of course, if it's a single carb you are concerned with, no airflow balancing operation is required so any device for such will be redundant. Assuming two carbs, then it will be necessary to adjust each carb so that each draws the same amount of air. With many twin carb installations, it is necessary to disconnect the linkage between the two carbs. It's usually sufficient to undo one of the clamp nuts so that the pickup arm on one of the carbs is slack. From here, adjust the airflow into each carb consistent with maintaining the required idle speed, i.e. 800-900 rpm. It will be necessary to go backwards and forwards between the two carbs until they are balanced out.

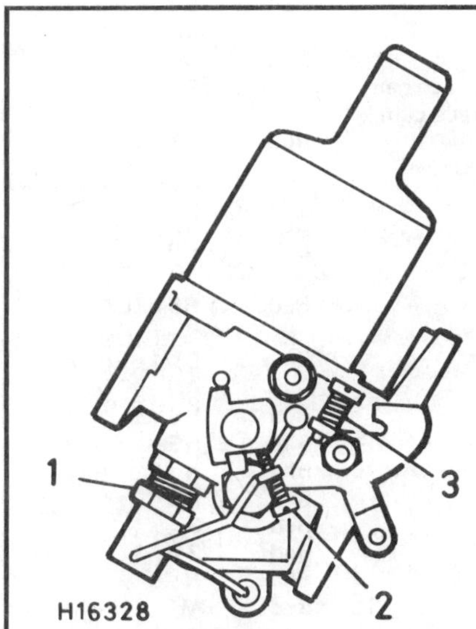

Fig. 7.7a SU adjustment screws

1 Mixture adjustment nut
2 Fast idle screw
3 Idle adjustment screw

You can balance airflow the traditional way by listening down the mouth of the carb and equalizing the noise, or ...

Balancing twin carbs is not really a big hassle. The first step is to isolate the linkages of each carb. Over the years, many types of linkages have been evolved for twin carbs, so check yours thoroughly and be sure you understand it. The relevant Haynes overhaul manual would certainly help here.

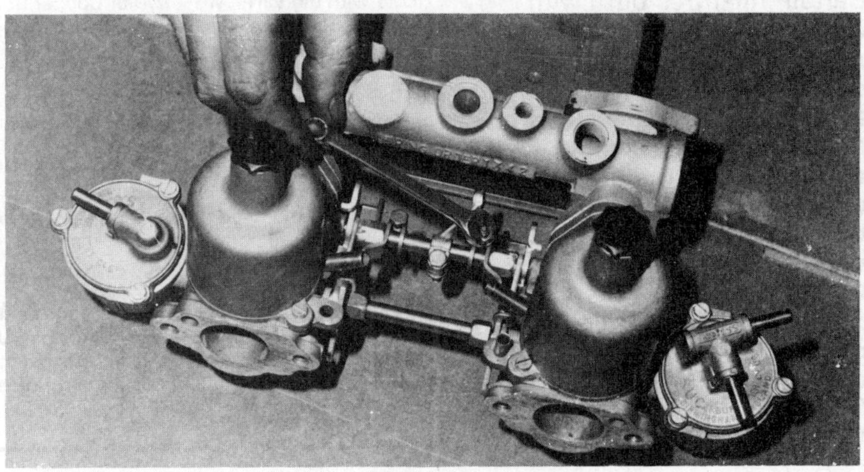

You are now at a point where you can adjust the mixture. Again, there are three ways you can do this. Firstly you can use a simple mixture analyser such as a **Gunsons Colortune**. I say 'simple' here, not as a term which is in any way detrimental to the function of this device. **The Colortune is one of those devices which is so basic it cannot actually give the wrong answer.** Essentially a Colortune is a transparent spark plug which allows you to look

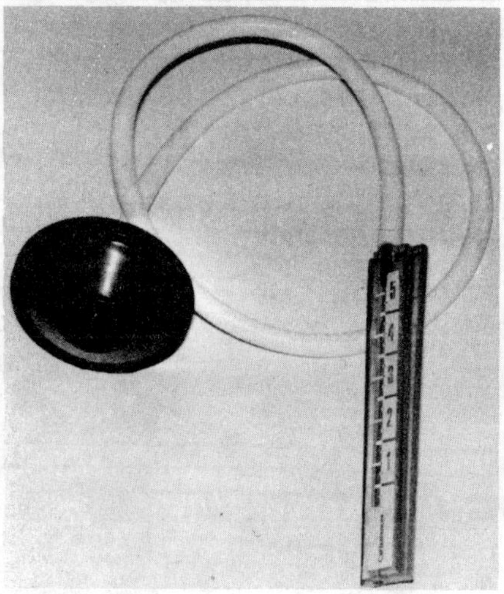

... you can use a simple device such as this Gunson carb balancer. Almost always, better results are obtained from the latter method.

into the cylinder and see the colour of the combustion flame. Certain mixtures burn with certain colours, and this will be more or less constant. This device will, at certain points, allow very accurate setting of the mixture. Sufficiently accurate, in fact, that it will pick up the mixture bias that exists between each inner and outer pair of cylinders. To use a Colortune, simply screw it into one of the pair of cylinders on one end of the engine, adjust the carb until the flame goes from a rich orange to just blue. Pull the Colortune out, install it in the equivalent cylinder of the other pair, i.e. inner or outer, and adjust the other carb to give the same flame colour. Having done this, you may find it necessary to go back and re-set the idle speed, in which case you will also have to re-balance the carbs. When all that is done, it will be necessary to set the pickup arm so that it picks up the throttle linkage of each carb at the identical moment, otherwise at anything

Fig. 7.7b SU needle/spring selection chart

Carb: 1 x $1\frac{1}{4}$ inch SU

Engine & Modification	Spring	Needle
Std. 850 Mini normal mixture	Red	EB
Std. 850 Mini rich mixture	Blue	M
Std. 850 + ram pipe	Blue	E3
Std. 850 + modified 850 head	Blue	H6 or M
Std. 850 + modified 850 head + ram pipe	Blue	E3
Std. 998 Mini normal mixture	Red	GX
Std. 998 Mini rich mixture	Red	M
Std. 998 + modified 998 head	Red	H6
Std. 998 + modified 998 head + ram pipe	Red	E3
Std. 1098 Mini normal mixture	Red	AN
Std. 1098 Mini rich mixture	Red	H6
Std. 1098 + modified 1098 head	Red	E3

Carb: 2 x $1\frac{1}{4}$ inch SU

Engine & Modification	Spring	Needle
Std. 850	Blue	EB or GG
850 + modified 850 head	Blue	M or GY
850 + modified 850 head + extractor exhaust	Blue	M
850 + modified 850 head + extractor exhaust + semi-race cam	Blue	H6
850 + modified big-valve 998 Cooper head + full-race cam + extractor exhaust	Blue	E3
Std. 997 Mini Cooper normal mixture	Red	GE
Std. 997 Mini Cooper rich mixture	Red	AH2
Std. 997 + modified 997 Cooper head	Red	AF
Std. 998 Cooper normal mixture	Red	GY
Std. 998 Cooper rich mixture		M
Std. 998 Cooper + modified 998 Cooper head	Blue	M
998 Cooper + modified 998 Cooper head + semi-race cam	Blue	E3
998 + modified big-valve 998 Cooper head, full-race cam + extractor exhaust	Blue	E3/1
1098 not originally fitted with twin carbs, i.e. 1100 saloon (sedan)	Red	EB
1098 with modified head	Red	M
1098 with modified head + 997 Cooper cam	Blue	H6
1098 with modified 1098 head + semi-race cam + extractor exhaust	Red	H6 or E3
1098 with big-valve MG 1100 or Cooper 998 head + semi-race cam + extractor exhaust	Blue	E3/1
970 S standard normal mixture	Red	AN
970 S standard rich mixture	Red	H6
970 S with modified head	Blue	H6
970 S with modified head + semi-race cam	Blue	No.5 or H4
1070 S standard normal mixture	Red	H6
1070 S standard rich mixture	Red	3
1070 S with modified head	Red	3 or 4
1070 S with modified head + semi-race cam	Blue	No.6
1275 S standard normal mixture	Red	M
1275 S standard rich mixture		AH2
1275 S with modified head	Red	AH2
1275 S with modified head + semi-race cam	Red	AF
1275 Non-S type engine with 2 x $1\frac{1}{4}$ inch SU	Red	E8
1275 Non-S with modified head + S exhaust	As for Cooper S models	
1275 Non-S with modified head + semi-race cam + S exhaust	As for Cooper S models	

other than idle, the engine will be running on one carb more than the other.

Well, that was the technique using simple equipment. Now, using more complex equipment as in your regular tune-up shop, i.e. an electronic mixture analyser. By the way, portable ones can be acquired which run off the car battery or the mains, if you choose to get that sophisticated. Using one of these devices, all you can do is to check that the final exhaust analysis is correct. To set the mixture in this instance, turn the adjusting screws to the fully up position, in other words the mixture is set dead lean. This can be done without the engine running because it would probably stop anyway. Now turn the adjusting screws down 2½ turns and start the engine. Check to see what the mixture is doing. If it is too rich, lean both carbs the same amount. If it's too lean, richen up both carbs the same amount. Try and aim for a mixture of about 14/1-14.5/1 on the old type fuel/air ratio analysers, or about 1% on a CO analyser.

The third way of doing the job is without any equipment whatsoever. I am assuming at this point the air filters are still off the carbs, in which case, screw the mixture screws up to the top of their adjustment and then back down 2½ turns. Start the engine. To one side and just below the bell chamber, you will find a small push-button which, when pushed upwards,

Here's an interesting shot taken down the mouth of an SU carb. I took this one whilst investigating atomization of fuel in various carbs, and you will note the SU is delivering a fine spray of fuel from the jet, and it does so at all throttle openings. It's interesting to compare this with some of the atomization characteristics of other carbs shown later in the chapter.

Carb: 1 x 1½ inch SU

Engine & Modification	Spring	Needle
Std. 850 Mine automatic [HS4] normal mixture	Red	AN
Std. 850 Mini automatic [HS4] rich mixture	Red	H6
Std. 850 Mini [H4 or HS4]	Blue	H6 or M
Std. 850 Mini + modified 850 head	Blue	E3
850 + 997 Cooper cam + modified 850 head	Blue	E3
850 + semi-race cam + modified 850 head + extractor exhaust	Blue	E3 or No.6
Std. 998 normal mixture	Red	Ac
Std. 998 rich mixture	Red	M1
Std. 998 + modified 998 head	Blue	H6 or E3
998 Mini + modified 998 head + 997 Cooper cam	Blue	H6 or E3
998 Mini + modified 998 head + semi-race cam + extractor exhaust	Blue	E3 or No.7
1098 + 1½ inch	Red	M1 or BQ
1098 + modified 1098 head	Red	H6 or AN
1098 + modified 1098 head + semi-race cam + extractor exhaust	Red	E3 or No.7
1275 Non-S engine normal mixture	Red	DZ
1275 Non-S engine rich mixture	Red	BQ
1275 Non-S + modified head	Red	BQ
1275 Non-S + modified head + extractor exhaust	Red	BP
1275 Non-S + modified head + extractor exhaust + semi-race cam	Red	No.7

Carb: 2 x 1½ inch SU

Engine & Modification	Spring	Needle
850 + modified big-valve head + full-race cam + extractor exhaust	Blue	No.5 or 6
998 + modified big-valve 998 head + semi-race cam + extractor exhaust	Blue	E3
998 as above with full-race cam	Blue	No.6 or RO
970 S modified head + semi-race cam	Blue	M1 or E3
970 S modified head + full-race cam + extractor exhaust	Blue	CP4
1071 S + modified head + semi-race cam	Blue	M1 or E3
1071 S + modified head + full-race cam + extractor exhaust	Blue	MME
1275S + modified head + semi-race cam	Red	No.3 or 4
1275 S + modified head + full-race cam + extractor exhaust	Blue or Red	BG

Carb: 2 x 1¾ inch SU

Engine & Modification	Spring	Needle
1275 S full-race only	Red	RR

Tested by Tom Oliver and John Chappell on the Howley Racing dyno, this 1430 engine produced around 117 hp on a single 1¾ inch SU, and better yet, it was still street-drivable.

lifts the dashpot piston a small amount. Locate this and press it to slowly raise the piston. As the piston is raised, the rpm should pick up momentarily and then drop back more or less to its original rpm. If it does this, the mixture is right. If the rpm picks up and stays high, the mixture is too rich. If the rpm drops as soon as the piston is raised and continues to drop until the engine stalls, then the mixture is too lean. Adjust each carb in turn until you get the correct response from raising the piston. If this tuning is being done with standard air filters, i.e. the vehicle is a restoration type vehicle and so you are looking for originality, then it's best done with the air filter assembly in place. If the air filters are correctly sized K & N air filters, it won't make any difference whether the air filters are in place or not, as the air filter restriction is, at this point, virtually zero. If you are tuning a single carb it's only a question of adjusting the idle speed and then lifting the dashpot piston to see if the response is right.

Float Chamber

If you find difficulty setting up the carb and achieving the sort of results expected, or the carb always tends to run too rich at idle and often smells of fuel, then it's a good idea to check that the float and float chamber assembly is working as it should. On carbs other than the H.I.F. this is easily performed as the float chamber is off to one side of the carb. All that is necessary is to undo the three securing screws on the float chamber lid, lift the lid, disconnect the fuel line and take it away to inspect it. The first thing is to wipe everything clean, hold the lid upside down, letting the weight of the floats close the needle valve, and then try blowing through the fuel input pipe. If the needle valve is doing its job, it should seal against your blowing through it. If it isn't, the fact it's leaking is a sure sign it needs replacing.

Many of the older carbs are equipped with plain, brass needle valves. Replacing these with viton-tipped ones (Flo-jet) will instantly help the situation, but if that doesn't do it, then the seat of the valve is worn as well and the whole assembly needs renewing. The other thing that needs to be checked is the float level. If you are getting this far into the carb, it makes sense to get yourself an SU carb overhaul manual, as float levels vary from one type of SU to another. Past that, there is not a lot else I can advise you to do, except in rounding off this section on tuning SU carbs, a little advice on the oil in the dashpot.

Firstly, always keep this topped up to the level required. Secondly, it's well worthwhile experimenting with viscosities in the dashpot. Thicker oil will cause more accelerator pump action, but that's only worthwhile doing if the engine needs it. Thinner oils will allow the piston to open quicker, but if it's allowed to open too fast, the mixture will go lean. Quite often it's a good idea to mix oils on twin carbs so that you can get both pistons to open at the same rate when the throttle is blipped. **Personally I like to use synthetic oil in the dashpots, simply because its viscosity stays at a more constant level over a wider range of operating temperatures.** On twins, in order to get them both to open right, I will either use watch oil to thin down the oil in one carb, or a heavyweight oil to increase the viscosity in the faster opening carb.

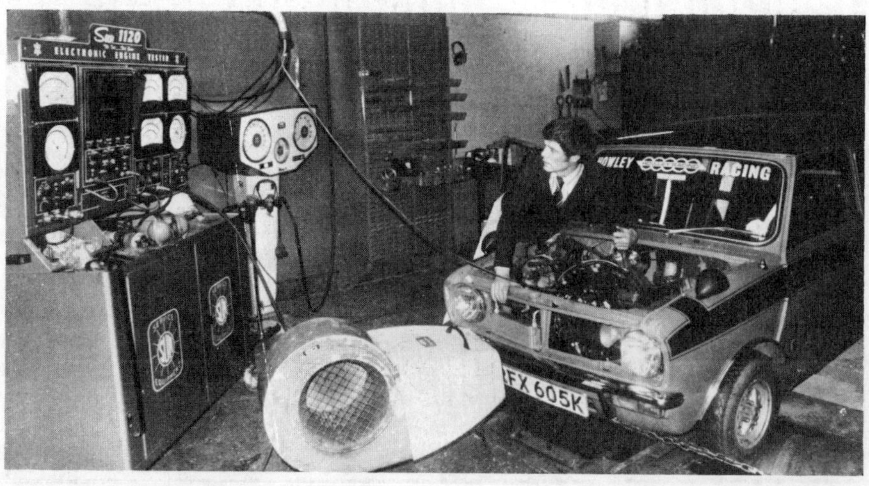

Even if you are building an engine to a relatively common modified configuration, it will be necessary, if you are to get the best from your SU carb, to have it set up on a rolling road dyno.

Tuning the 28/36 DCD Weber

Although a 34 DMTL downdraught Weber is gaining popularity, next to the 40 and 45 sidedraughts, the 28/36 is still one of the most-used Webers for the performance and tuning industry. Since this is the case, it will follow that there are plenty of you who want to know how to troubleshoot and set up this type of carb, so on with the plot.

Making the assumption that your carb is in reasonably sound condition and not falling apart with corrosion and that you are at least able to start the car, this is how to proceed. Start and warm-up the engine as a prelude to any fault-finding or tuning. Once the engine has warmed up, set the idle to something sensible, say, around 1000rpm. Having done that, we will turn our attention to the idle mixture adjusting screw. Screwing this in will lean the mixture; unscrewing it will richen it. Turn this adjusting screw inwards until the rpm start to drop off due to a weak mixture. Now slowly start screwing it back out. The revs will rise until they get to a maximum. At this point they may even start to drop if you continue to unscrew the adjuster. Stop at, or before, this point so that the engine is turning at its maximum rpm. Now turn the adjusting screw half-a-turn back in. If the right idle jet is in the primary barrel, the idle mixture screw should be somewhere around the middle of its adjustment and not at the extreme ends of being either fully in or out.

Although I've made no mention of it, if the calibration of the idle jet is right, the idle mixture when tested with either a Colortune or a mixture analyser should be around the 14.5/1 or 1%-2%, depending on the type of analyser being used and just blue on a Colortune. If the idle mixture adjusting screw is not that close to the centre of its adjustment, it could be that the idle jet is the wrong size. Selecting the right idle jet is very important on these carbs, as it not only supplies fuel to the idle mixture adjusting screw, but it also dictates the mixture strength on the progression or, as U.S. readers prefer to call it, the transition circuit. This is the phase of carb operation where the butterfly is just past the idle stage and is at a very small throttle opening, but not enough to cause the main jet to come into action. Believe it or not, most driving, especially around town, is done on the progression stage of the carb. If it is not calibrated properly, the vehicle will not deliver good mpg. If you suspect the idle jet is out, then here's how to go about determining what it should be.

Adjust the **idle speed** screw to give around 2000rpm. Now, noting where your idle mixture adjusting screw was, try turning it in one turn. If the rpm rises, the mixture is too rich, indicating that your idle jet could have been too big. If the rpm drops, the mixture was probably too weak, indicating that the idle jet may have been too small. If the rpm remain virtually unchanged, then the idle jet is probably about right, but this technique is only a guide. Proof positive has to be obtained by driving the vehicle on the road and determining that the transition circuit operates as it should. To do this, drive along in high gear, say third or top, and depress the throttle the merest amount to give the carb just a shade more than idle throttle opening. If the motor falters, try going up one size on the idle jet. If it doesn't falter, it could be either that the mixture is just right or slightly too rich. Try going down a size to see if it does falter. If it does, then you know that the right size had been previously installed. If it doesn't, you now know you have been using a jet size which was too big. With the primary barrel a little wider open, the engine will start to draw on the main jet. If the motor feels 'soggy' it could be one of three things causing it. If, as you open the throttle, the motor hesitates, then it could be that the pump jet is not large enough. If there is no sign of hesitation, and the

The perennial 28/36 Weber carb is seen here, being put through its paces on the dyno. When correctly calibrated, it's a highly functional piece.

throttle feels unresponsive after the throttle has been opened, then it could be that the mixture is too lean or too rich, or that the venturi (choke) is an incorrect size.

Taking first things first, check that the accelerator pump action is correct. On the 28/36 the pump works only on the primary barrel. It's not really necessary to have one on the secondary barrel simply because when the throttle pedal is fully floored, the primary barrel delivers a big enough shot to cover both barrels opening. Apart from that, the secondary barrel only comes into action from two-thirds of the throttle opening upwards, so dumping fuel from an accelerator pump into a closed barrel would be a rather fruitless exercise.

Metering the injection quantity of fuel is done by pump jet sizing and the calibration of the bleed-back valve in the bottom of the float chamber. The bigger the bleed-back valve, the greater the amount of fuel returned to the float chamber rather than being delivered to the engine. When opening the throttle suddenly, a very definite squirt of fuel should be seen to go down the primary barrel. If it doesn't, look for a blockage in the system. If all is clear, but there's a suspicion that the pump jet action isn't sufficient, the first thing to do is to try a larger pump jet. These are available in sizes up to about 70. (That is a jargon way of saying that the pump jet sizes go up to 0.7mm. They normally go up in steps of 0.05mm and are referred to by a whole number of hundredths rather than as a decimal.) If going up on the pump size jet doesn't do the trick, try installing a smaller bleed-back, or spill valve as it's sometimes known, in the float chamber. **Remember, when calibrating the accelerator pump, always use the smallest amount of fuel that will get the job done.**

Now for the main jet circuit. Unlike the DCOE series of carbs, the main jet is not all in one with the air corrector and emulsion tube. The main jet is, in fact, housed just level with the bottom of the float chamber on the outside of each side feeding into each barrel. A long-winded description that, but looking at the drawing **Fig. 7.8** will show the location of all the relevant jets. To actually get at the main jet, undo the hexagonal covering screw and you will find the main jet itself screwed into the end of this. Also don't assume that the main jet for the primary barrel is likely to be the same as the main jet for the secondary barrel, because each works independently, and they are usually quite a bit different in size. If you suspect the mixture delivered by the main jet is too weak, try going up one main jet size. If it's too rich, come down. If you have no idea where to start, then assuming there is a 180 air corrector in the emulsion tube, try using a main jet one-50th of the size of the choke for that barrel. For instance, if the carb has a 26mm choke, try using a 1.3mm jet, i.e. a 130.

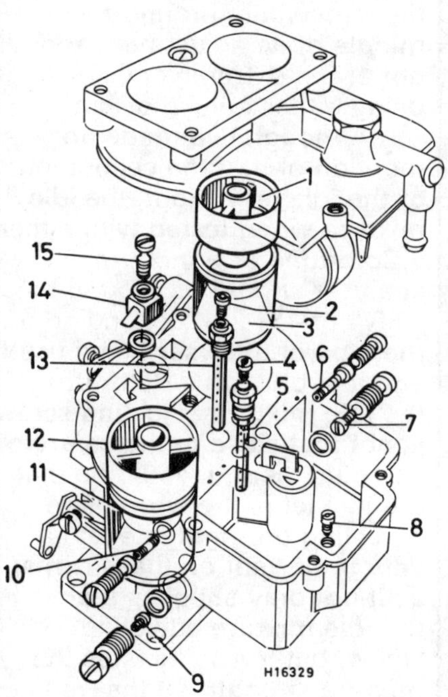

Fig. 7.8 Weber 28/36 DCD carburettor

1 Auxiliary venturi (sec)
2 Air corrector jet (pri)
3 Choke (sec)
4 Air corrector jet (sec)
5 Emulsion tube (sec)
6 Idling jet (sec)
7 Main jet (sec)
8 Pump bleed - back valve
9 Main jet (pri)
10 Idling jet (pri)
11 Choke (pri)
12 Auxiliary venturi (pri)
13 Emulsion tube (pri)
14 Pump discharge assembly
15 Pump jet

A third fault you may find that was mentioned earlier, is that the venturi or choke size is not really suitable. This is a factor which is only likely to affect performance to a limited degree in most instances. The selection of choke sizes on a 28/36 isn't as critical as it is on sidedraught carbs, but it does need to be right if you are going to get maximum performance and economy from the engine. Since these carbs are set up to be a compromise between maximum power and flexibility, we do have a reasonable amount of leeway. Now the question of selecting the choke size is quite simple. Just use the chart **Fig. 7.9a** and this will put you in the right ballpark straight away. To use this chart, all you have to do is estimate how much hp you think your motor should give, but do yourself a favour - be honest. Don't over-estimate.

When talking to people about their Series 'A' engines, I usually find that on average, the quoted hp is about 20% more than exists in reality. If you over-estimate the hp in the first instance, you are likely to lose power because of it, because your carb will end up with incorrect venturi sizes. In fact, it's best to err on the low side than the high side.

Having arrived at a reasonable guess for the hp figure of your engine, move up the graph scale until you meet the diagonal line, then turn at right angles to the left margin and read off on the vertical scale the choke size your carb needs. Always go for the smallest one on the chart that will do the job. If you have a relatively standard motor, with a relatively mild street cam in it, choose a primary choke 1mm smaller than the size indicated on the chart. This tells you the size of the secondary choke. If you have a hot cam in the motor, you may want to calibrate your carb so as to give you best flexibility at the bottom end without necessarily losing any of your top end potential, in which case choose a secondary venturi 1mm larger than indicated on the chart, but come down 2mm on the size of the primary venturi. Having a smaller primary venturi, often means that the engine will be more flexible at smaller throttle openings.

So far we have covered idle setting, accelerator pump calibration and settings for small throttle opening on the primary. Let's take the primary side to the full extent of its opening without incurring any motion of the secondary. In real terms that's about two-thirds open. Under these conditions, the mixture can be either right, lean or rich. If you sorted out the mixture at less than this throttle opening and it was satisfactory, then you are going to have to look to the air corrector to do the relative trimming of the mixture. The air corrector on these carbs is screwed into the top of the emulsion tube under the top cover of the carb. As you can see, calibrating the mixture at two-thirds throttle opening on the primary barrel is going to be difficult unless you have the car on a chassis dyno, proving once again the indispensability of this piece of equipment for the performance enthusiast (how else would you simulate 70mph driving and yet still be able to work on the car?) If the mixture looks like it's becoming too rich at this point, go up on the air corrector size. If it looks like it's going lean, come down on the air corrector size. An easy way to make sure the carb is still on the primary venturi only, and the primary is open as far as it will go without bringing the secondary in, is to wire the secondary closed. Another point to remember is that you are NOT calibrating the primary barrel for maximum power, it should be calibrated for maximum economy. Maximum power calibration should only take place when the throttle is opened to such an extent as to bring the secondary barrel well and truly into operation.

Before moving on from the primary barrel, let's consider the primary barrel emulsion tube. The effect of the emulsion tube is fairly similar to that of the sidedraught carbs and the same principles apply with its selections. Fortunately, the 28/36 is not as touchy on emulsion tubes as its big sidedraught cousins, and probably the easiest way out for selecting an emulsion tube is to ask the advice of a Weber expert such as Weber themselves, or **Derrington, Redline, Lynx**, etc., to name but a few in various parts of the world. However, having said that, you may not need to ask advice if you are utilizing the services of a good rolling road operator, he also,

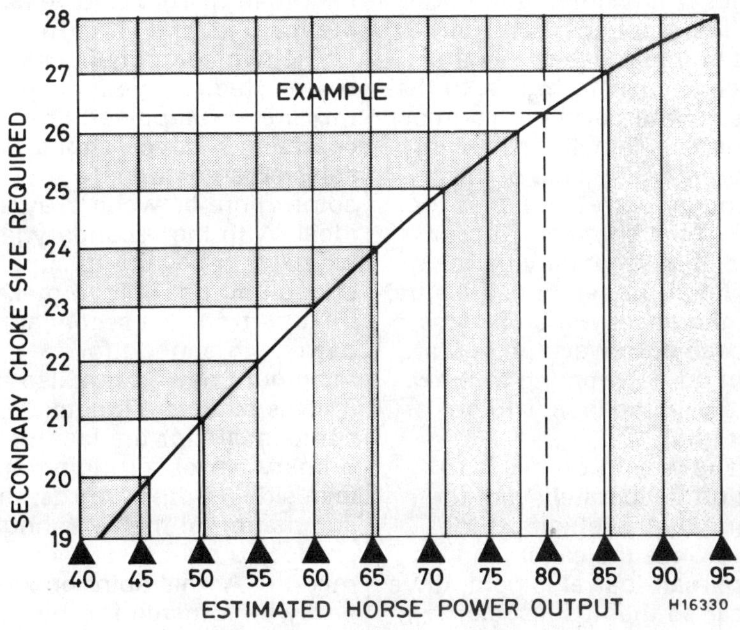

Fig. 7.9a Weber 28/36 DCD choke size selection

Estimate the horsepower you expect your engine will produce. Follow the vertical line up from this figure until you get to the diagonal line. Turn at right-angles to the left and, when you reach the left-hand margin, read off the figure.

The dotted line shows the example. Estimated output is 80 bhp and the recommended secondary choke size comes out at 26. This figure should be taken as a guide only but should be close to the actual choke size required.

from past experience, will be able to assess which emulsion tube is most likely to be needed. Even so, the chances are that if your Weber kit was bought for the Series 'A' engine, then it will probably have a suitable emulsion tube in it in many instances.

Secondary Barrel

Past two-thirds open, the secondary barrel will start to come into action. At this point there is a critical range where the secondary barrel is starting to feed air into the engine, yet its butterfly isn't far enough open for it to pull on the main jet. This is where the so-called idle jet of the secondary barrel comes in and the action of this jet at this point is very important if a smooth transition from single barrel to twin barrel operation is to be obtained. Although this jet is called an idle jet, it has nothing to do with idling except that it serves the same purpose in all other respects, apart from idle, to that of its counterpart in the primary barrel. This jet will be feeding the progression side of the secondary barrel and if it's too weak, a flat spot could occur just as the secondary barrel is coming into operation. If it's too rich, the effect will be far less noticeable. However, if you are intent on not wanting to throw fuel away, the idea is to get it just right.

The way to do this is to start off deliberately with the secondary barrel idle jet too small. Once the calibration of the primary barrel is right, drive the car so that it's just starting to open the secondary barrel. If the engine feels just a little fluffy, maybe even hesitant, with a hint of a misfire, go up onto the next size idle jet and give that a try. Continue doing this until progression onto the secondary barrel is found to be smooth. I'll make no bones about it, actually hitting this secondary phase can be quite difficult, because you haven't got to push the throttle down very much more to bring the quick–acting secondary barrel past the progression stage. Under normal conditions if a terrific flat spot is experienced at this point, it's a fair bet that the secondary idle jet is blocked. It's the sort of size that a grain of sand or something similar could shut off.

Of course the same situation can arise in the primary barrel whereby dirt in the idle jet produces a far more noticeable problem as it causes a bad idle immediately as well as a complete functional failure of the progression/transition circuit.

At this point then, you should be aware of what it takes to get the secondary barrel progression sorted out. Assuming it is, let's go on to calibration of the secondary main barrel. This involves sizing the main jet and air corrector. These two components need to be selected to give a rich mixture at full power. From the chart **Fig. 7.9a** you should have selected a venturi size somewhere between okay and ideal. With the secondary jet, we have to be sure to supply enough air not only to make the mixture from the secondary barrel rich enough for maximum power, but also that it gives sufficient fuel to compensate for the fact that the primary barrel is running on the lean side. In other words, the summation of the two should produce a full power, rich mixture. At this point one could easily be forgiven for thinking that the main jet for the secondary barrel must, of necessity, be larger than the one for the primary. This is not necessarily so; sometimes it is larger, sometimes not. Basically we must remember that the main jet has to operate in conjunction with the air corrector jet and the smaller the air corrector used, the richer the mixture delivered at the mid- and top end of the rpm range. A good place to start on jet size, as previously, is one-fiftieth of the venturi size. However on the secondary barrel, try starting with an air corrector about 0.2mm down on what you have in the primary.

Assuming you have the primary right, let's look at an example: say you have a 180 air corrector in the primary barrel, try a 160 in the secondary. Since the secondary has a venturi larger than the one in the primary, the main jet in the secondary will, at this stage at least, be larger. If, when testing such a combination, it is found that when the secondary barrel is fairly well open, the mixture is rich, try coming down on the main jet size. If, at this point it is found that with the throttle fully open, the mixture starts to lean out, then come down on the air corrector size. If it all happens round the other way, then reverse the procedure. It could be when wide open or near wide open throttle situations are used at near maximum rpm the mixture goes weak, and no matter what main jet changes are made, it stays weak. Under these conditions, it's a fair bet that the needle valve in the float chamber is too small and is not supplying enough fuel for the engine's requirements. The remedy here is simply to go larger on the needle valve. If this condition is suspected, don't be tempted to run with the engine at wide open throttle settings too long, because too lean a mixture at full throttle is a fairly effective way of blowing holes in pistons, especially if the engine is running near its detonation limit due to high compression. Remember, a rich power mixture is less likely to detonate because of the cooling brought about by the extra fuel.

One further point on full power settings: up to now you may have been doing all your calibration with the air filter off. With almost every air filter you use other than K & N, you can be fairly sure the mixture will richen up when you put the filter on. If the correct size of K & N element is being used, it will not make any measurable difference to the mixture.

Economy Tuning The 28/36 Weber

We have looked at the hp side of the 28/36 DCD, now **let's look at a few things which will effect economy, as these are vitally important when you consider the price of fuel could replace the gold standard as an economic yardstick.** The first thing is the float bowl needle valve. In the 28/36 Weber, they seem to take quite a pounding, and they don't last forever. A worn valve can cause excessive fuel consumption because it will richen up the mixture, especially in the idle and progression phase of the primary barrel. It does this simply because it increases the fuel level in the float bowl, i.e. fuel is coming in faster than it's going out and the needle valve is not shutting it off. If the needle valve is worn, calibration of the light throttle openings can be almost impossible and, in fact, if you do get it right and then find out you have a worn needle valve, it will be a case of starting the calibration all over again. Inspect the needle valve. If there's more than a minimal amount of wear, renew it. It may look an expensive item for what it is, but believe me, you will very quickly get the cost back in fuel saved. Also the float level is important. Checking the float level should be one of the first things to do after you have either okayed your existing needle valve or fitted a new one.

Setting the float level is

Fig. 7.9b 28/36 Weber jet selection chart

All use F30 emulsion tube

Std. 850 Mini	Choke	21	22
	Main jet	105	115
	Air cor.	180	190
	Pump jet	60	–
	Idle	45	45
850 + modified 850 head	Choke	22	24
	Main jet	115	125
	Air cor.	190	200
	Pump jet	70	–
	Idle	45	50
850 + modified 850 head + semi-race cam + extractor exhaust	Choke	23	25
	Main jet	120	130
	Air cor.	190	200
	Pump jet	70	–
	Idle	45	50
Std. 998 (non Cooper engine)	Choke	22	24
	Main jet	115	120
	Air cor.	180	190
	Pump jet	70	–
	Idle	45	45
998 + modified 998 head or std. 998 Cooper head	Choke	23	25
	Main jet	125	135
	Air cor.	190	190
	Pump jet	70	–
	Idle	45	45
998 + modified 998 head or std. 998 Cooper head + semi-race cam + extractor exhaust	Choke	24	25
	Main jet	130	140
	Air cor.	190	180
	Pump jet	70	–
	Idle	45	50
1098 engine (single carb type)	Choke	23	24
	Main jet	120	130
	Air cor.	190	190
	Pump jet	70	–
	Idle	50	50
1098 engine + modified head or 998 Cooper head	Choke	23	25
	Main jet	120	135
	Air cor.	190	180
	Pump jet	70	–
	Idle	50	55
1098 engine + modified head or 998 Cooper head + semi-race cam + extractor exhaust	Choke	24	26
	Main jet	130	145
	Air cor.	190	180
	Pump jet	70	–
	Idle	50	55
Non-S 1275	Choke	24	26
	Main jet	130	140
	Air cor.	190	180
	Pump jet	70	–
	Idle	50	55
1275 + modified head Cooper S exhaust + Cooper S cam	Choke	25	26
	Main jet	135	140
	Air cor.	190	170
	Pump jet	70	–
	Idle	50	50
1275 + modified head semi-race cam + extractor exhaust	Choke	26	27
	Main jet	140	150
	Air cor.	190	170
	Pump jet	70	–
	Idle	55	60

done by holding the carb cover in a vertical or near-vertical position so that the float tab just touches the end of the needle valve. Under these conditions the floats should be 5mm from the gasket face of the float cover (measured without a gasket). If the float level proves to be incorrect, the carburation will suffer. If the float level is too low, then it will be found that under severe cornering conditions the engine will suffer fuel starvation. Apart from this, it will tend to make all the settings lean if they were right with the correct float level. If the float level is too high, then it's possible to run into a flooding situation, especially if you have occasion to pull up suddenly at traffic lights, or do a violent manoeuvre of any sort such as a handbrake turn on the loose, or a twitch through a traffic island. A high float level will also invalidate otherwise correct carb settings by making them all too rich.

Lastly, a few points which I think are worth mentioning, especially if you are considering rebuilding your Weber. First of all, before you start stripping, it's always a good idea to arm yourself with a new gasket set. It costs next to nothing and it can save a lot of frayed nerve ends if you happen to break up a gasket trying to remove a certain part. Secondly, if you have just acquired a 28/36 Weber from a breaker's yard and it is badly corroded, you can find it a bit of a problem getting rid of the corrosion. In the States, excellent carb cleaners are available and so de-crudding the carb is no real problem. However, in the U.K. these carb cleaners are less than instantly available. As a last resort, and I only advise this in severe cases of corrosion, get some paint stripper such as Polystrippa, brush onto the corrosion, then rinse the whole lot off in water. After doing this, dry off the excess water and dump the whole carb in paraffin/kerosene overnight and make sure you get rid of all the water from the fuelways.

Tuning The Fish Carb

Before going into details of how to set up this carb, I think it would be a good idea quickly to run through the intricacies of its functions because it is not immediately apparent, from looking at this rather ingenious device, how each part functions, or indeed what the workings of it are as a whole. Simply because it's a good reference point, I'll start at the fuel's point of entry.

The first thing the fuel meets is the needle valve in the float bowl. Nothing unconventional here. The needle valve is operated like the needle valve in any other carb, by a float. As the fuel level rises, so the float rises and shuts off the flow of fuel when it has reached the desired position. In most carbs, the float chamber contains a float only. Here we come to the first difference between a Fish and any other carb. Within the float chamber there is a swinging arm, most commonly referred to as a pick-up arm. This arm swings through an angle of approximately 45° when the throttle is opened, because it is also fixed to the throttle shaft. The swinging arm also operates in a part of the float chamber which is partitioned off from the float area. The reason for this will become apparent in a moment. One side of the swinging arm is flush with the side of the float chamber; the other side is flush with the plate which divides the float chamber in two. Machined in the wall of the float chamber near the tip of the swinging arm is a tapered groove which is very shallow in the position occupied by the arm when the butterfly is closed, and deep in the position occupied by the arm when it's fully open. Here we have a primary stage of fuel metering. A hole in the pick-up arm, which is hollow, is opposite this groove. When the throttle is near the closed position, the amount of fuel flow area is limited by the shallow depth of the groove. When the throttle is in the wide open position, a larger throughflow area is allowed because the groove is deeper. **However, it should be emphasized that calibration by the groove is a very rough calibration, and on lower powered engines it's not likely to be a calibration factor at all.** After fuel is drawn up the swinging pick-up arm, it is transferred into a hollow butterfly shaft. However, just before it enters this shaft it comes to the second, and principal, metering point; namely a screw which varies the amount of restriction to the fuel at this point. Turning the screw in, simulates a reduction in jet size. Turning the screw out, does the reverse. Following the fuel along the throttle spindle we find that, again unlike any other carb, the fuel delivery is not out of a booster, auxilliary venturi or jet such as the SU, but it's discharged out of a series of holes in the throttle spindle. Basically there are two sets of holes, one set positioned pointing towards the mouth of the carb, the other set pointing down the barrel towards the engine. When the butterfly is at a relatively narrow angle, that is the engine is still being throttled, we find that air is drawn into the holes in the top side of the throttle shaft and mixed with fuel in the throttle spindle. This emulsion is then delivered out through the holes in the bottom of the spindle. As the throttle is opened further, so the holes in

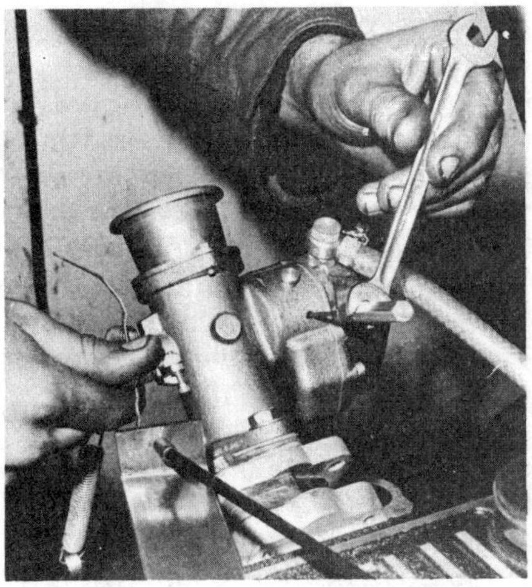

This is all it takes to alter the calibration on a Fish carb. No jets, no emulsion tubes, no needles. The Fish carb has an infinitely adjustable system. Here, the main metering system is being calibrated.

Here's the view down the throat of a Fish carb. The three holes in the butterfly shaft admit air to emulsify the fuel at small throttle openings and discharge fuel at large throttle openings.

the butterfly spindle turn to a point where they are 90° to the airflow. Under these conditions the butterfly, in conjunction with the throttle bore, creates a sort of inverse venturi shape which draws fuel out of all the holes on both sides of the butterfly. Obviously as the throttle is opened the transition from the throttle spindle acting as an emulsion tube to it acting almost solely as a main jet is gradual, but before I deal with full throttle conditions, it is necessary to take a look at light throttle calibration as this is not really controlled to any large extent by the main jet.

The butterfly on a Fish carb is not rigidly fixed to the spindle. In actuality, it is clamped in position. By virtue of a clamp screw, it can be released and turned relative to the throttle shaft. This is the key to the part-throttle calibration of this carb. Imagine the butterfly is opened such that the engine is propelling the car along at something in the mid-rpm range. Under such conditions a certain amount of fuel is being discharged from the spindle. Let us also assume at this stage that the amount of fuel coming out, is fixed. If, at this point, the mixture being delivered is weak, instead of altering the amount of fuel going into the engine like a normal carb, the Fish carb alters the amount of air going in. To do this, the butterfly adjustment facility is used. For the particular instance under discussion, the butterfly is unclamped from the throttle spindle and moved such that it closes slightly and restricts the amount of air going in. This richens the mixture bringing it nearer the correct ratio. Conversely, if the mixture being delivered is rich, then to lean it off it's a question of adjusting the butterfly so that it is further open on the spindle, thus allowing more air into the engine.

From this it can be seen that by altering the position of the butterfly, we can make

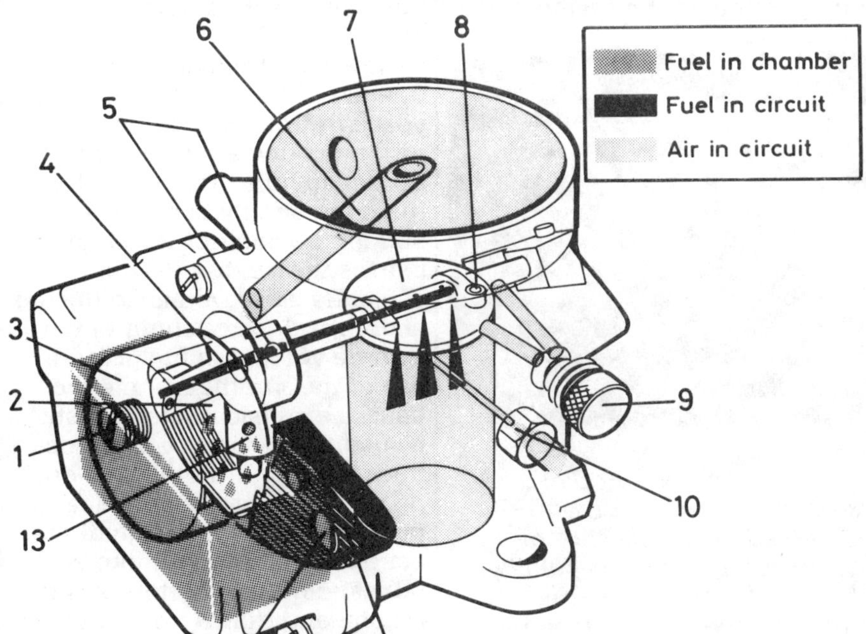

Fig. 7.10 Fish carburettor

1 *Fuel sight plug*
2 *Tapered fuel groove*
3 *Float*
4 *Fuel pick-up arm*
5 *Fuel metering adjustment*
6 *Balance pipe*
7 *Throttle butterfly*
8 *Butterfly clamp screw*
9 *Idle mixture screw*
10 *Vacuum take-off*
11 *Separation plate*
12 *Feather valve*
13 *Float needle valve*

compensations for mixture errors in the part-throttle operation phase.

Idle Circuit

If the Fish carb does have any similarity to more conventional carbs, then it's only in the idle circuit that it occurs. Even so, it can be viewed as being slightly unconventional. To visualize how the idling circuit works, we need to go back to the swinging arm in the float chamber. At idle, the butterfly in the throttle bore is closed and therefore the swinging arm will be at the shallowest part of the tapered groove in the float chamber. This means there is a minimum amount of fuel being transmitted up through the swinging arm to the hollow throttle spindle which is consistent with what we would require at idle. On the side of the carb, there is an idle mixture adjusting screw, but it doesn't adjust the fuel going into the engine, it merely adjusts the air. Removing this idle screw will reveal two holes underneath. One is positioned above the butterfly when the butterfly is in the closed position, and the other is in the position below it. Adjusting this screw in, or out, alters the amount of air that can come through the top hole and out through the bottom hole. Obviously the more air that goes through here, the weaker the mixture is, and vice versa.

Sudden throttle opening on the Fish carb causes fuel to be ejected from the throttle spindle to produce the required accelerator pump effect. Note the coarse fuel droplets being emitted. This was one of the reasons why the Fish carb proved impractical in meeting modern emission standards. Even when the throttle was ...

... fully open, fuel droplet size was still larger than that delivered by an SU carb of comparable airflow capability. As a result, fuel economy suffered and hydro-carbon emissions were high.

Acceleration Circuit

At this point we have covered all the steady-state conditions that a carb is likely to have to cater for, that is idling, part throttle and full throttle. This only leaves the transition stages, *i.e.* what is happening as the throttle is opened. On normal carbs, an accelerator pump is fitted. The function of this pump is to cover the leanness caused when a rush of air enters the engine because the throttle has been opened suddenly. If you were to just look at a Fish carb, you could be looking for hours and not see what you would expect to see in the way of an accelerator pump, but nevertheless it has one. It's just that it, like most everything else on the Fish carb, is a little unconventional. Remember I said earlier on, that the swinging pick-up arm was housed between the wall of the float chamber on one side, *i.e.* the wall with the tapered groove in it, and a plate which partitions off the swinging arm from the float. In this plate are two holes of about ⅜ inch diameter. One hole is unobstructed and it is toward the end of the float bowl which coincides with the position the swinging arm would be at at idle or low throttle openings. At the other end we have a hole which is in a similar position but it has a 'feather valve' on the inside of it. This feather valve is really a non-return valve. Imagine that the swinging arm is trapped between the wall of the chamber and the partition plate. When the throttle is suddenly opened, the fuel within the space in front of the swinging arm has nowhere to go except up the spindle and out of the main jet, simply because a sudden opening of the throttle causes the feather valve to close. Because of this combined action, sudden opening of the throttle will cause fuel to squirt out of the throttle spindle. Of course a considerable amount of fuel leaks past the sides of the swinging arm, but this is of no consequence. If none leaked past the side, the injecting action would dump about a tablespoonful of fuel into the engine and that would be too much by a pretty large margin.

When the throttle is released or comes to a standstill, the feather valve opens and allows the cavity behind the swinging arm to refill with fuel. That, my friends, is the entire accelerator pump action in a Fish carb, and if you have understood all I've said so far, then you are well on the way to understanding the basics of the Fish carb.

Setting Up

Setting up a Fish carb is straightforward. All you need to do the job is a ballpoint pen, a stop-watch, a flat piece of road which is little used, and the help of a rev counter, but the latter is not essential. This is, of course, assuming you don't have a rolling road dyno handy to do the calibration. However, let me add a point here that's very valid for the Fish carb. Because it's totally adjustable, it is a lot more amenable to calibrating by trial-and-error by driving the car than any other carb. Although it's still best to set up a Fish on a rolling road dyno, the difference between skillful setting up by driving the car after each adjustment and setting it up on a rolling road, is much smaller than any other carb, if we exclude the luck factor.

This shot shows a Minnow/Fish carb with the float bowl removed. Indicated by my pen is the start of the metering groove. In the middle of the float chamber in a vertical position is a swinging arm pickup.

Firstly, let's check that the carb is functioning properly. To do this, remove the air filter and determine that the accelerator pump action is functioning as it should. This is easily done by opening the throttle quickly whilst looking down the bore of the carb. Of course this should be done without the engine running. What you should see if all is well, is fuel squirting out of the holes in the throttle spindle. If it doesn't do this, then the chances are there is a blockage somewhere in between the point at which the fuel comes into the swinging arm and the point at which it comes out the throttle spindle, or the feather valve is not working properly. When the feather valve is pressed against the dividing plate, it should cover the ⅜ inch hole entirely. If it doesn't, position it so that it does. In its normal relaxed position, the feather valve should be about 1/64th away from the dividing plate, so that during acceleration it can quickly seat down on the dividing plate and provide the restriction that it ought.

The next thing to check is

This time the pen is indicating the flap valve that is an intrinsic part of the accelerator pump system on a Fish-type carb.

the fuel level. In the side of the float chamber you will find, on most Fish carbs, a brass screw. With the engine switched off and the vehicle positioned on level ground, undo the brass screw. The fuel level should be such that it is flush with the bottom of the hole. If it's too low, then make adjustments to the float, *i.e.* bend the float arm to bring the fuel level up. If it's too high when you undo the brass screw or sight plug, fuel will run out. Make adjustments to rectify this by bending the float arm down. However, don't be over-ambitious in your adjustments. Usually the amount of bending required is pretty minimal.

Lastly check that when full throttle is given at the pedal, the butterfly opens to the full-open position. If all these things check out okay, then you can start the calibration.

Part-Throttle

Start up the engine and set the idle via the idle speed screw on the throttle arm so that the engine turns over at around 2000rpm. Take the ballpoint pen mentioned earlier and put the tip of it in the centre hole of the throttle spindle. As a result of doing this, one of three things

Although the Minnow and Reece/Fish carbs operate on the same principal, you can see from the view down the barrel of this Minnow/Fish that it is substantially different in layout. Here, the pen indicates the hole that should be plugged to ascertain the correctness of the calibration during part-throttle operation.

will happen: the revs will go up, down or stay the same. (What else can they do?) If the revs decrease, then the mixture is rich. To compensate for this, unclamp the butterfly from the throttle spindle and open the butterfly very slightly. When I say 'very slightly', I mean VERY SLIGHTLY. 1° can make quite a bit of difference. While you are doing this you must leave the throttle spindle in its original position. Just remember, you are, in fact, turning the butterfly on the spindle.

If, when you put the pen in the centre hole, the revs increase, then you do the opposite: close the throttle very slightly. In fact you go on making such adjustments until placing the ballpoint pen in the centre hole makes no difference to the engine rpm. For the moment that's your part throttle mixture set.

Full Throttle

Now onto the full throttle mixture. To get at the main jet screw you will need to remove the little socket grub screw which plugs the access hole to the main jet screw. Once you have removed this screw you can get at the main jet adjusting screw by opening the throttle fully, again, with the engine switched off, and inserting an Allen key into the head of the jet adjusting screw. By turning the screw in, the mixture is weakened; by turning it out, the mixture is richened. The idea is to use your stopwatch and time the vehicle acceleration between, say, 40 and 60 in third, and make adjustments until the best acceleration times are produced. It's always best to start slightly on the weak side and work towards enrichening the mixture. You will find that as the mixture becomes nearer right, so the difference in recorded times will become smaller until finally there comes a stage when richening the mixture up makes no difference at all, and then after that point, the performance will start to drop off.

To get the best compromise between power and economy you will need to come back from this setting until you can detect the performance dropping off again. From here, just tweak it a shade richer and leave it at that. **Just remember, it is possible to have an excessively rich mixture without losing much performance. Though easy on piston crowns it does accelrate bore wear and will almost certainly burn holes in your wallet.** After the full throttle mixture has been set, it is a good ideal to go back and check the part throttle setting, although in most instances it is not likely to have changed to any measurable degree.

Idle Mixture

This now leaves us with only the idle mixture to do. Reset the idle speed screw on the throttle arm to give an idle speed of around 800 or 900; at most 1000rpm. Of course that figure is going to depend on the stage of tune of your engine. Obviously it should be set to whatever speed the engine runs most smoothly. Having settled on a suitable idle speed, screw the idle mixture adjustment screw right in, and then back it off about a quarter of a turn, and from there just experiment with minor adjustments a little one way or the other, until the smoothest idle is achieved. Most settings are between a quarter and half a turn out. Adjusting this screw may cause quite a bit of rpm change, so it is necessary to go back to the idle speed adjustment screw and reset it. Once the idle is set, as a final operation, re-check the part throttle mixture and, on completion of this, your Fish carb will be set up.

Carburation Part 3

Sidedraft Carbs

Because of their marked similarity, I will deal with Webers and Dellortos sort of comparably. I say 'sort of' because in essence I will deal with Webers in great detail. Afterwards, rather than flog through similar facts and figures again, I'll look at the Dellorto and see how it differs, bringing out any pertinent

No matter what angle you view it from, a Weber carburated Series 'A' engine gives the impression of power. I built this particular engine as a street burner. Its 1400 displacement produced 119 hp.

points along the way.

Generally speaking you can assume, that where no differences are mentioned, what applies to a Weber also applies to a Dellorto.

Because these carbs represent the pinnacle of power development for the Series 'A' engine, I will go into them in pretty heavy detail, and the best way to explain my reason for so doing is to use an analogy. Basically we can say the difference between a sidedraft Weber or Dellorto and, say a lawn mower carb, amounts to the same level of difference as that found between a simple snapshot camera and a sophisticated camera like a Minolta, Nikkon or Pentax. With a snapshot camera, an inexperienced user only has one or two choices of settings to make, and therefore has only one or two settings to foul up on. As a result, the chances of getting a reasonable picture are quite fair. Give the inexperienced photographer a highly technical camera and he or she will have a combination of several hundred different settings to choose from. This also means a combination of several hundred different settings to adjust incorrectly: and the chances are they *will* be incorrectly adjusted, reducing the possibility of getting a good picture to almost nil. Nevertheless, for an experienced photographer, the technically sophisticated camera is in a class of its own by comparison with a snapshot camera, and so are the results achieved.

This camera analogy holds good for the type of carbs we are now going to look at. Both Webers and Dellortos are very sophisticated instruments, with hundreds of different jets, emulsion tubes, air correctors, venturis, idle jets, pump jets, etc., available with which to calibrate them. All these parts can produce millions of combinations. Within reason there is almost always one to suit your engine, but the point is, without proper calibration to suit the characteristics of your particular engine, a sophisticated carb is worse than useless. Indeed, since each barrel can be calibrated as a separate entity, it's quite possible to double the calibration error associated with an engine which has only a one barrel carb on a four cylinder engine. This means that incorrect calibration can, and more often than not does, produce greater losses in power and fuel mileage than the incorrect calibration of a single barrel carb on a Series 'A'. If you are going to install Webers or Dellortos on your car, and you want to calibrate them yourself, then it would pay you to learn about the instrument concerned first. The manufacturers want their customers to know about the carbs so they can make the best use of them. After all, dissatisfied customers, especially when they are spending as much money as these types of carb cost, reflect on their product, and how many dissatisfied customers actually blame their own incompetence for lack of performance or economy? Human nature being what it is, few are prepared to be that honest with themselves.

To help owners understand their carbs, Weber produces a very informative booklet and this can be obtained from the many Weber agents or direct from main suppliers such as **Weber Carburettors U.K.**, **Redline** or **B.A.P. Geon** in America, or **Lynx** in Australia. For Dellorto carbs, similar though less informative information can also be obtained. Again, Lynx in Australia can help out here, whereas **Contact Developments** in U.K. can supply information, and in the U.S.A. **Claude's Buggies** is the sole agent for this type of carb.

Apart from publications issued by the relevant manufacturers, there are several books on the market on Weber carbs at least, so there is no excuse for not getting acquainted with this type of unit. Of course you may not feel inclined to calibrate your own carb, especially if you go by ads often seen in magazines where carbs are claimed to be pre-set and ready jetted for the job.

Shown here, removed from the carb, are the principal calibration parts of a sidedraft Weber, these being the auxiliary venturi, the main venturi or choke, the idle jet and its holder and the main jet emulsion tube and air corrector.

From this Dellorto, I have pulled out the same 'calibratable' parts as those shown in the nearby photo of the Weber. Notice the marked degree of similarity.

Well, unfortunately that's not quite true. In many instances the carbs are closely jetted for the job. The problem is that with modified engines there are so many combinations of variable factors that can affect the required calibration of the carb because, in reality, no two engines are exactly alike. Just because you have a modified head on your 1400 Mini, does not mean that the carb settings for your engine are going to suit another 1400 engine with somebody else's modified head on it. The two head grinders that prepared the heads in question, probably did a completely different job, so the carb settings required will not be the same.

Of course most of the main agents for Weber and Dellorto carbs, have on the premises experts on the particular carb in question. **These experts can often jet a carb close for many applications, by drawing on past experience, but it is very rare that they will be spot on, for the simple reason that NOBODY knows what a particular engine will need to make it spot on.** The only accurate way to ensure near-100% carb calibration is to employ the services of, yes, you've guessed, a rolling road dyno, and it's best to choose one whose operator is well acquainted with the ins and outs of Weber or Dellorto carbs. Remember, there are quite a few rolling roads around which are used only for simple service tune-ups and their operators are not in the business of trying to dial in carburation to the ultimate. They simply set up the carbs to the manufacturer's spec and use the dyno to check output. You do not need to go to a place like this. Be sure your choice of establishment is one in business for the purposes you require.

Many people are reluctant to use a rolling road dyno, simply because they have already spent a lot of money on their conversion and they simply don't feel justified in going out and spending another handful of notes to have the necessary calibrations done. If

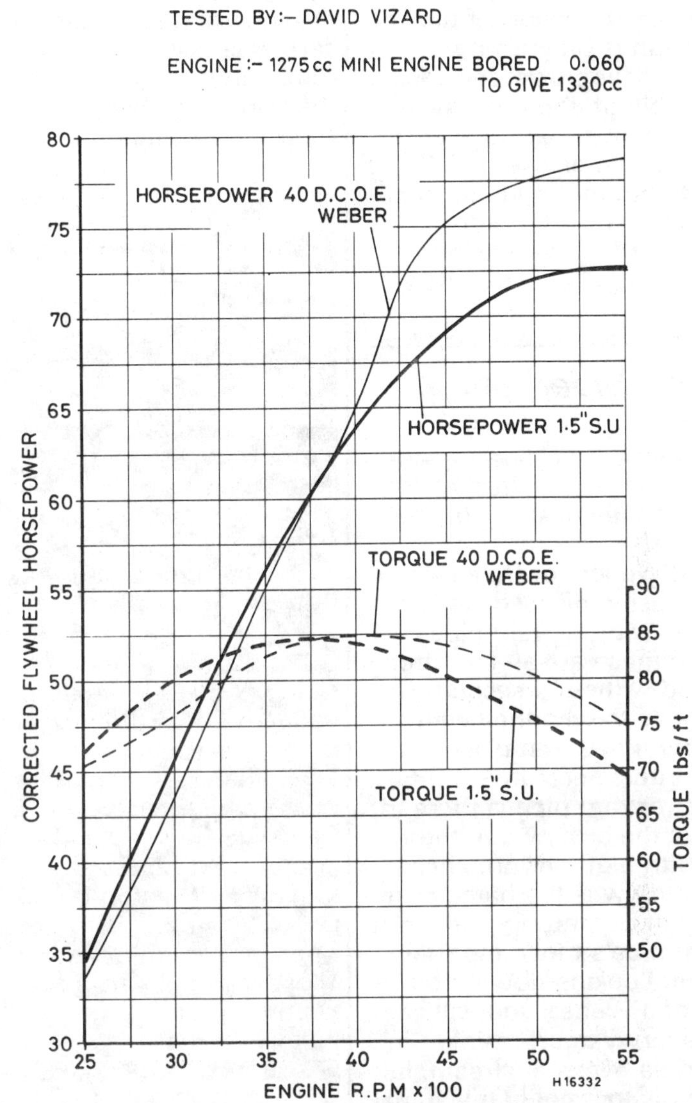

Fig. 7.11 Comparison of 1½ inch SU on Oselli Torque Master (as cast) and 45 DCOE Weber on 5 inch manifold

This 45 DCOE Weber is equipped with what, at one time, were known as Group 2 ram pipes. Note that they are cocked up at an angle. This apparent mis-alignment does not seem to cause any reduction in airflow.

you fall into this category, ask yourself the question: if the conversion is only giving a fraction of the power increase it's capable of, then has the money spent so far been cost effective? The answer is NO. It's wasted until the modifications have been fully utilized by virtue of proper calibration.

Making Sidedrafts Work

As I have previously said, because they are such complex devices, Webers or Dellortos can be easily misused. Let's deal first with the ram pipes, because these are all too often removed without a second thought as to why they were put there in the first place . . .

Not unexpectedly, an intake trumpet or ram pipe is there to improve the airflow into the carb. With high performance carbs, airflow is the name of the game, but optimizing this is not the ram pipe's most important function. Looking down the throat of a Weber, you will see an auxiliary venturi residing in the middle of the airstream. Its function is to amplify the signal generated by the main venturi. It does so by virtue of speeding up air flowing through it. Air speed is increased because a greater vacuum is created at the end of the auxiliary venturi as it resides at the point of greatest depression in the main venturi. If the airflow approaching the outward end of the auxilliary venturi is not reasonably well orientated, a substantial amount of vacuum 'signal' is lost. The signal is the means by which the carb meters the fuel. The signal should increase at some proportional rate to the weight of air going through the carb. By removing the ram pipe, the flow at the mouth of the carb can be considerably disorientated and as a result, the metering capability of the carb is changed. In most cases removal of the ram pipe causes the metering ability to be reduced, resulting in the same effect as if a carb of too great a size had been selected for the engine. The outcome can be a measurable reduction in drivability and low end torque.

A common mistake made by first-time users of Webers and Dellortos is bolting the carbs up solid to the intake manifold. This can, especially with Webers, cause a float bowl vibration problem which leads to frothing of the fuel. To prevent this, it is necessary to use rubber O-rings between carb and intake manifold, and the securing nuts should only be tightened sufficiently to effect a seal. There should be a degree of up and down movement left. This Maggard kit contains all the necessary parts for the correct installation of a Weber carb onto its manifold.

Another side effect is that usually fuel droplet sizes are increased, principally because a larger main jet and a smaller air corrector are necessary when ram pipes are not used. Carbs used without ram pipes also tend to use more fuel from the higher cruising speeds on up. Having the ram pipes installed and producing a more uniformly orientated airflow improves many aspects. Usually there are only two reasons for removing ram pipes from carbs in the first place, and these are (a) to clear the bulkhead or inner valance because the intake manifold used is too long, or (b) so as to allow the fitment of an air filter case which is too shallow to be used with ram pipes. Basically either problem is really a question of bad planning. In almost every instance there is a manifold short enough to accommodate a Weber or Dellorto complete with ram pipes. **Although short manifolds aren't necessarily the best, they are often a better solution to a clearance problem than a larger manifold and ram pipe removal. Short filters are available, and though these may not house the standard ram pipes, short ram pipes are also available from those people who produce the short filters.**

Okay, enough on that. Let's look at the next mistake I often see, which is the case of the wire mesh grid over the end of the ram pipe. There are two things these wire mesh grids do. **Firstly they reduce hp** because in spite of their very open appearance, they do impede airflow a lot more than you would expect. A wire mesh gauze over the opening of the ram pipe will, in most cases, effectively dial out about 3hp on a 100hp engine. That's a lot of

Any engine development session requires frequent changes to carb calibration, especially when different items of speed equipment are required. Here, Nolan Pitts and Alan Jeffery are re-calibrating a Weber carb in a sequence of tests evaluating a Janspeed intake manifold. Note the modified Longman intake stacks used on the carb.

hp loss when you consider that a correctly sized K & N filter will not reduce the power output one iota. **The second aspect of these wire mesh elements is their total lack of filtering capabilities. Sure, they may stop small rocks going in, but the particles that actually cause wear, damage valve seats and score bores can pass through such elements with no difficulty whatsoever.** The last thing I am going to comment on is the use of ladies' nylon stockings over the mouths of Webers. If you have to resort to such practices because the correct filter is not to hand, the chances are your engine will make more power on a single SU with a decent air filter!. Believe it or not, those stockings can cost 6 or 7hp without any great difficulty at all. Worst yet, those nylons don't filter either, so a power loss has been incurred for no good reason. . .

Sizing

The most common pitfall where sidedraft carbs are involved, a pitfall incidentally, that is not confined to inexperienced users, is the choice of carb size in the first instance, and choke (venturi) size in the second. In almost every case where a choice is available, the one settled on is too large. It would be as well to bear in mind that a pair of 50mm sidedraft instruments on an 8-port 1000cc Mini engine is not going to give any more performance than a pair of 40s. In fact, in all probability the reverse is likely to be true. **The trick with carb sizing and venturi selection, is basically one of choosing a size which provides the engine with the airflow it needs without going over the top. Such a situation produces the best ratio of airflow to main jet signal strength for the most accurate calibration possible,** and that usually results in the best all-round performance.

If you look at the chart, **Fig. 7.12**, you will see the airflow versus the venturi size of a 40 and 45 DCOE. Each is equipped with the most commonly used auxiliary venturi. You will note that where the small chokes are used, the 40 Weber flows considerably more than the 45 for the specific depression chosen for the test. For instance, a 40 DCOE equipped with a 30mm choke flows almost 160 cfm. per barrel, whereas a 45mm carb equipped with 30mm chokes flows only 135 cfm. This should illustrate the error of going overboard on carb size in the first place. If your engine finally requires choke sizes in a 45mm carb that are down below the 32 - 33mm size, it is a reasonable indication that you selected a carb which was too big to start with. **It is not until venturi sizes above 33mm are reached, that the 45 pays off in terms of extra**

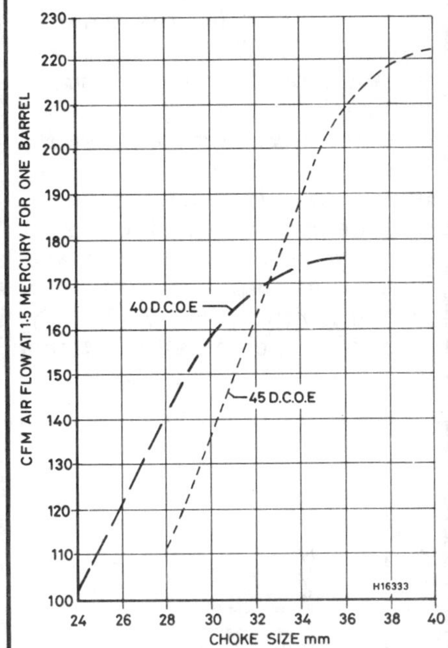

Fig. 7.12 Airflow comparison of 40 and 45 DCOE Weber carburettors

In this comparison the 40 DCOE was fitted with a 45 mm long trumpet and 4.5 auxiliary venturi. The 45 DCOE was fitted with a 56 mm long bent (10°) trumpet and 3.5 auxiliary venturi.

usable airflow.

These flow figures relate to a bare carb. When the carb is mounted on a manifold, and it, in turn, is on the engine, then changes in venturi size have far less of an effect on the overall flow. This is because the valve and the cylinder head represent a greater restriction to flow and, as a result, they, not the venturi, limit the airflow. **Fig. 7.13** illustrates such an effect. Here a cylinder head was set up to draw air through the intake port of the manifold at a rate approximately proportional to a Series 'A' engine producing around 85-90hp. You will note that variations in venturi size have far less of an effect than those in **Fig. 7.12.** Also worth noting, is the fact that the difference between a 40 and a 45 in terms of airflow in an 'on engine' situation is much closer. **Basically what it boils down to is this: the venturi size should not be too large as any gains in airflow are more than counter-acted by the reduction in signal at the auxiliary venturi.** The main jet must have a good signal so as to feed the engine the right amount of fuel, properly atomized. The question is, how to select the basic venturi size. For one choke per cylinder **Fig. 7.14,** which incidentally is from a Weber tuning manual, should help out. As you will see, the chart gives a good guide to the venturi size needed in the carb. This chart can also be used with a fair degree of accuracy for Dellortos. If the venturi size indicated by the chart is around 34mm diameter, then you are on the cross-over point between 40s and 45s.

At this point I would say, if the engine is a fairly well modified 1100, or a lesser modified 1275, then go for the 45.

Again, using hp as a basis, the chart **Fig. 7.15** can be used to select venturi size. Don't over-estimate the point at which peak power will occur as you

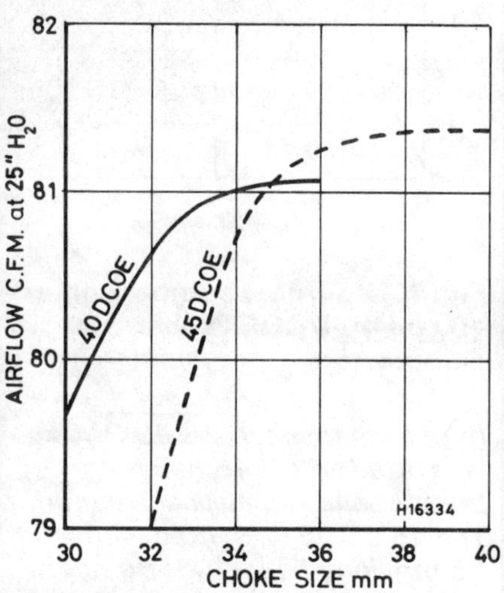

Fig. 7.13 Airflow comparison of 40 and 45 DCOE Weber carburettors

In both tests the carburettors were mounted on a 6-inch manifold and then installed on to a modified Leyland head.

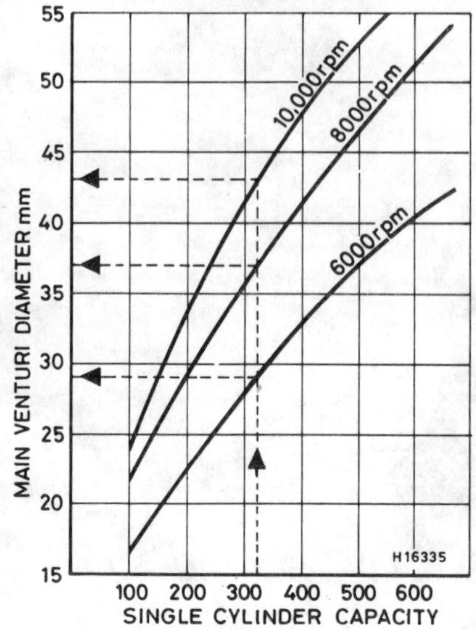

Fig. 7.14 Main venturi diameter selection

This chart is for four-stroke engines fitted with one down- or side-draft carburettor per cylinder. The three curves refer to maximum output speeds.

There are three things worthy of note in this shot that contribute towards making an engine a better, more practical performer. The first is the Montagu/Timavo throttle linkage. The second is the installation of the Maggard spacers and O-rings between the carb and manifold, and the third is the highly efficient Maniflow intake manifold, which you can read about in the following chapter.

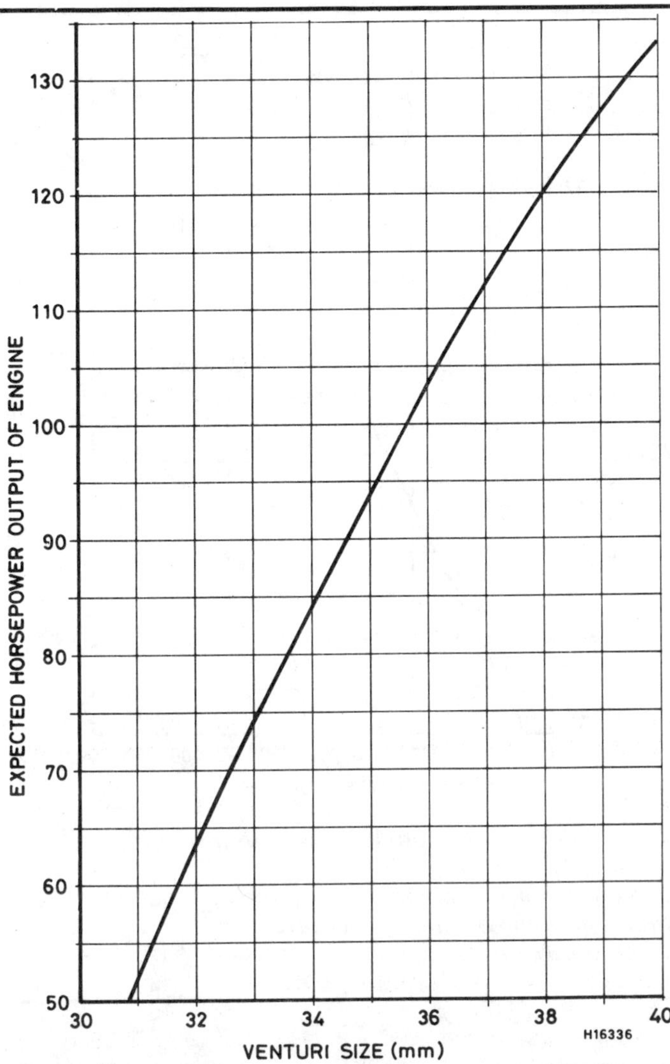

Fig. 7.15 Venturi selection graph for siamese port engines

To use this chart, first estimate the power level expected of the engine. It's best to be a little on the conservative side, because the engine will lose more overall performance by going oversize on venturis than under. Generally, a couple of mm down from optimum venturi size has little or no effect on peak torque, and only reduces top end horsepower by 2 or 3 bhp at most. On the other hand, a few mm over-size reduces both maximum torque and horsepower.

When selecting the venturi, the size indicated by this chart should be modified if it is greater than 5mm less than the rated carb size. Thus a 45 carb should have a venturi not exceeding 40mm. This 'minus 5mm' rule should be considered a minimum; often a reduction of a 6 or 7mm contributes to greater power output.

will only end up fooling yourself into believing your engine needs bigger venturis or indeed a bigger carb than is really the case. Over-estimate and you will fall right into the trap that I am attempting to steer you clear of.

I would like to make a point here, just in case there are enthusiasts amongst you who are using this section of the book as a guide for engines other than the Series 'A'; because the Series 'A' engine has only two intake ports the chart for choke (venturi) size selection applies only to the Series 'A', not to any other engines.

On Series 'A' engines, a 40mm instrument will give good results up to power outputs around 70-75hp, almost irrespective of engine size. However, there is a proviso to add here. If you are seeking absolute maximum hp from a small displacement engine, say an 850 or 1000 for competition use, the 45 seems to have the edge from about 70hp upwards. However, let me add that this is by no means a foolproof guide. There are instances on 850s where the use of a 40 has produced better hp, and these are not also-ran 850s I am talking about but front-running, highly competitive 850s. Since any engine's function is a question of getting the right combination of parts, it's a good idea, at least on 850s, to try both 40s and 45s. It must, of course, be done on the dyno so the best possible hp can be wrung out of each combination.

When the engine in question is over 80hp, then virtually irrespective of capacity, it should be equipped with a 45mm carb, and that is about as cast iron a rule as I can offer in the carb sizing area for these engines.

Jetting

Having established the venturi size within reason, the next calibration area to sort out on either a Weber or Dellorto will be the jetting components. Of all these components, probably the biggest headache for most is selecting an emulsion tube. Basically, an emulsion tube is a little like an SU needle in disguise. The letters and numbers stamped on them, be they Weber or Dellorto, have little to do with the characteristics of the emulsion tube. As it happens, for Webers of the 40mm variety, F.16s

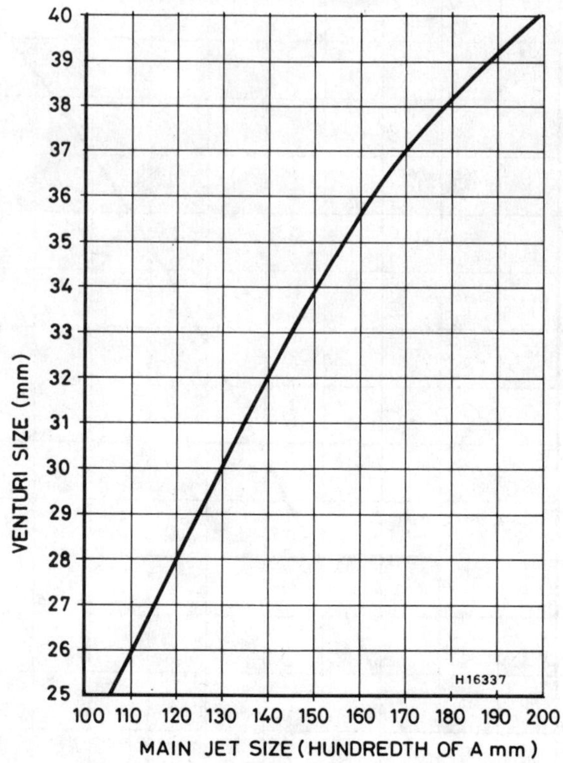

Fig. 7.16 Main jet size v. venturi size

Use this chart as a starting point during initial calibration of side-draft carburettors. The chart assumes the use of a 180 air corrector.

almost always work in anything except highly modified 850s, where F.2s or F.9s will often do the job. When we are talking about 45s, then F.2s tend to be the most effective although, occasionally, F.9s are required.

For a 40mm Dellorto, you can figure that almost all will run well on a number 5 emulsion tube. When a 45 is used, most engines up to 1100cc seem to run better on a number 6 emulsion tube, whereas those of 1275 and upwards seem, in the main, to be at home with a number 5 emulsion tube. Still, the jetting chart at the end of this section should give you a good starting guide as to what's needed with these types of carbs.

Moving onto air correctors, a good starting range for this jet is in the 160-180 range. Remember, with air correctors, the smaller the air corrector is, the richer the mixture becomes at higher rpm and vice versa. It's by juggling around with the air corrector and main jet, that the correct mixture can be achieved throughout the rpm range under full throttle conditions. The question is, if you were starting with an air corrector jet of 160 - 180, what size of main jet should you use? The chart, **Fig. 7.16** should help you out on that one. The graph has been plotted out for the average jet size needed for one venturi per two cylinders, as in Series 'A' engines. Don't apply this chart to anything else. This graph assumes there is a 180 air corrector installed. After this, the main jet selection is a function of carb, main venturi or choke size. The jet size arrived at by this chart is nominal: a great many vehicles will fall into the range of two or so jet sizes up or down on this. A lesser number will fall into the group of two to four jet sizes up or down on size, and a very small number may be out a little more than that. So basically what you have here is the information to get an average setting for your average modified vehicle, so please bear in mind that this is only a starting point.

The main jets and air correctors may be the key factor as far as maximum power is concerned, but whilst driving at part throttle, the most important jetting is, in fact, the idle jet. Just as the idle jet on the 28/36 Weber controls idle and progression/ transition, so it does on both the Dellorto and the Weber sidedraft carbs. Basically it's a question of selecting an idle jet which will cope with both the idle and progression requirements. Again, there is quite a range of jets that will do the job, and probably the best reference you have as to what's needed will be contained in the jetting chart mentioned earlier.

Accelerator Pump Jet

When we start looking at accelerator pump jets in earnest, the situation gets relatively involved as it is not just a jet which decides the total calibration property of the system on either of the carbs being dealt with. There are also other factors to take into account such as the accelerator pump bleed-back valve, and the accelerator pump stroke. First of all let's deal with the accelerator pump bleed-back valve. This can be either a closed one or one calibrated to allow a certain

Here's one of the most bullet-proof and versatile throttle linkages I have seen in a long time. Produced by Janspeed to their normal high standards, this particular linkage can be used with either single or dual sidedraft Webers, although the latter application is of little importance to a Series 'A' owner.

amount of fuel back into the float chamber when the accelerator is depressed. The function of this valve is two-fold. First of all it allows fuel to replenish the cylinder of the pump system, and secondly it allows enough back into the float chamber so that excess fuel is not drawn into the engine. Some engines need more than others. With the Series 'A' engine, it will inevitably be found that a very restrictive or even closed bleed-back valve is needed, as one pump jet has to supply two cylinders rather than the more normal case with other types of engine where a pump jet per cylinder is used. The calibration number of the bleed-back valve is stamped on the side so it's easy to see what it is. The bleed-back valve tends to give some control over not only the amount of fuel going into the engine but also the rate at which it's injected. However, its operation has to be considered in conjunction with the accelerator pump jet itself. A big jet with a closed bleed-back valve will give a heavy shot of fuel over a moderate period. Dropping the size of the pump jet will reduce the amount of fuel injected, but also if the bleed-back valve is a restricted one, it will extend the period for which it's injected. On a Weber at least, for a Series 'A' engine, a 50 accelerator pump jet will often suffice. If the engine is highly tuned, or of big displacement, then this may need to be enlarged to 55.

Now let's deal specifically with Dellortos for a moment, as the Dellorto accelerator pump system is noticeably different to the Weber's. The accelerator pump unit itself is underneath the main body of the carb and it is of a diaphragm type. The amount of fuel that can be injected is adjusted by means of turning the nut up and down a threaded rod on the accelerator pump mechanism. Usually this is set by the supplier of the carb, and it rarely seems to need adjustment. The only time you are likely to need to change it is if you have just borrowed a pair of carbs off a twin carb installation and you are using one of them on your Series 'A' engine. Carbs used in pairs are inevitably set to inject less fuel per barrel than a carb which has to feed two cylinders per barrel. Other than the odd occasion when this adjustment needs to be done, the only calibration you really need concern yourself with is sizing of the pump jet. A 40 pump jet seems to be a good starting point, especially for smaller engines, but if the engine experiences a flat spot when the carb is opened suddenly, then a bigger pump jet is going to be called for. In some engines I have found that pump jets need to be as high as 55, and on rare occasions, I have found it necessary to size the pump jets out to 60.

Idle/Progression Circuits

Up till now, I've dealt only with the power production side of the calibration of side draft carburettors. However, with the exception of a race vehicle, much of the driving of any side draft carburated Series 'A' engine will be done on the idle and progression circuits. The fact that you fitted or are contemplating fitting an expensive side draft carburettor to your engine indicates that the engine may be substantially modified in other areas. The fact of the matter is, some modifications, in particular a big cam, can detract from the low speed driveability of the car. This being the case, it is necessary to optimize carburettor calibration. This will ensure as near as possible the best of whatever is available in terms of a smooth idle and good driveability. First let's deal with function of the progression holes as this will give you a better insight to what is going on. The progression holes in the throttle body of either a Weber or Dellorto are those holes positioned just after the idle screw holes. The butterfly moves by the progression holes as it starts to open. These holes serve a dual function. As the butterfly passes the holes the ones on the manifold side of the butterfly supply a mixture of air and fuel. Some of the air is drawn through the progression holes that still reside on the air filter side of the butterfly. As the butterflies are opened further, so more vacuum is transmitted to the progression holes furthest from the intake manifold end of the carb. This causes them to stop supplying air and start supplying fuel. The fuel for the progression circuit is delivered to that circuit via

Fig. 7.17 Weber and Dellorto jet selection chart

Note: When split Webers are used, the bleed-back valve must allow a greater amount to bleed-back, because there are two pumps feeding the accelerator pump system. A closed or very nearly closed valve is typical for a single Weber; use a 50 or more bleed-back valve with split units.

Carb: 40 DCOE Weber

850 + modified 850 head + semi-race cam + extractor exhaust	Choke	31
	Main jet	115
	Air cor.	150
	Em. tube	F2 or F9
	Pump jet	35
	Idle	45F9
	Aux. vent	4.5
850 + modified 850 head with big valves or 998 head with near-standard valve size + full race cam + extractor exhaust	Choke	33
	Main jet	135
	Air cor.	155
	Em. tube	F16
	Pump jet	40
	Idle	50F2
	Aux. vent	4.5
998 + modified head + semi-race cam + extractor exhaust	Choke	33
	Main jet	140
	Air cor.	180
	Em. tube	F16
	Pump jet	40
	Idle	50F2
	Aux. vent	4.5
1098 + modified head + semi-race cam + extractor exhaust	Choke	34
	Main jet	145
	Air cor.	180
	Em. tube	F16
	Pump jet	45
	Idle	50F2
	Aux. vent	4.5
1275 + modified head	Choke	35
	Main jet	145
	Air cor.	180
	Em. tube	F16
	Pump jet	45
	Idle	50F2

Carb: 45 DCOE Weber

Full-race 850 short stroke (destroked) S unit)	Choke	36
	Main jet	185
	Air cor.	175
	Em. tube	F2 or F9
	Pump jet	35-40
	Idle	50F2
	Aux. vent	3.5
	Restrict ball valve in float bowl	
Full-race 850 normal stroke	Choke	36
	Main jet	175
	Air cor.	175
	Em. tube	F2 or F9
	Pump jet	40
	Idle	56F2
	Aux. vent	3.5
998 + modified head + semi-race cam + extractor exhaust	Choke	34
	Main jet	150
	Air cor.	180
	Em. tube	F2
	Pump jet	40
	Idle	50F2
	Aux. vent	4.5
998 Mk. V Sprite head, sprint cam, etc. i.e., full-race specification	Choke	38 m/m
	Main jet	195 m/m
	Air cor.	175
	Em. tube	F2
	Pump jet	45
	Idle	50F11 or 50F2
	Aux. vent	3.5
1000 S using S head type valves and equipped to full-race specification	Choke	38
	Main jet	180
	Air cor.	180
	Em. tube	F2
	Pump jet	45
	Idle	50F2
	Aux. vent	4.5
1071 + modified head + semi-race cam + extractor exhaust	Choke	35
	Main jet	155
	Air cor.	175
	Em. tube	F2
	Pump jet	40
	Idle	50F2
	Aux. vent	4.5
1071 S with S type head, equipped to full specification	Choke	38
	Main jet	185
	Air cor.	175
	Em. tube	F2
	Pump jet	50
	Idle	50F11
	Aux. vent	3.5
1275 standard	Choke	35
	Main jet	150
	Air cor.	180
	Em. tube	F2
	Pump jet	40
	Idle	50F2
	Aux. vent	4.5
1275 + modified head + sem-race cam + extractor exhaust	Choke	38
	Main jet	165-170
	Air cor.	180
	Em. tube	F2
	Pump jet	45
	Idle	50F2
	Aux. vent	4.5

the idle jet. In essence, the idle jet does less to control the idle mixture than it does to set the mixture on the progression circuit. In other words, you can imagine that the idle jet is really a miniature main jet system designed to cater for the smooth driving through the progression range. It is dependent solely on the calibration of the idle jet and the little emulsion tube it's used in conjunction with. For the Weber, the emulsion tube is, in essence, built in. For instance, for a 50F9 idle jet, the 50 designates the size of the fuel orifice in the end of the idle jet, and the F9 designates the air hole pattern in the rest of the assembly. The actual idle mixture can be controlled over an extremely wide range by the idle mixture adjustment screw. So when selecting an idle jet, you will be looking for one which satisfies the progression circuit with the actual idle mixture being set by virtue of the idle adjustment screws. Exactly the same rules apply to setting the idle/progression circuit as they do on the main jet circuit, at least as far as driveability through the progression range is concerned. As a starting point for your idle jet selection, go for a jet size that gives the correct mixture when the idle adjustment screws are 2½ turns out. Next,

Fig. 7.17 (cont'd)

1275 full-race	Choke	40
	Main jet	200
	Air cor.	175
	Em. tube	F2 or F9
	Pump jet	55-60
	Idle	50F11
	Aux. vent	3.5

Carb: Dellorto 40, all use emulsion tube 5

850 + modified 850 head + semi-race cam + extractor exhaust	Choke	30-31
	Main jet	135-145
	Air cor.	200
	Pump jet	40
	Idle	55
	Aux. vent	4.5
850 + modified 850 head with big valves or 998 head with near-standard-size valves + full-race cam + extractor exhaust	Choke	31-32
	Main jet	140-150
	Air cor.	200
	Pump jet	40
	Idle	55
	Aux. vent	4.5
998 + modified head + semi-race cam + extractor exhaust	Choke	32
	Main jet	145
	Air cor.	200
	Pump jet	40
	Idle	55
	Aux. vent	4.5
1098 + modified head + semi-race cam + extractor exhaust	Choke	32
	Main jet	145
	Air cor.	200
	Pump jet	40
	Idle	55-98
	Aux. vent	4.5
1275 + modified head	Choke	32
	Main jet	140
	Air cor.	200
	Pump jet	40
	Idle	55

Carb: 45 D.H.L.A. Dellorto
All use emulsion tube 6 except wnere noted

Full-race 850 short stroke (destroked) S unit)	Choke	38
	Main jet	180
	Air cor.	150-170
	Pump jet	40
	Idle	60
	Aux. vent	4.0-4.5
	Restrict ball valve in float bowl	
Full-race 850 normal stroke	Choke	37
	Main jet	160-180
	Air cor.	160-170
	Pump jet	40
	Idle	60
	Aux. vent	4.0-4.5
998 + modified head + semi-race cam + extractor exhaust	Choke	35
	Main jet	140-160
	Air cor.	170
	Pump jet	40
	Idle	58-60
	Aux. vent	4.0-4.5
998 Mk. C Sprite head, sprint cam	Choke	38
	Main jet	150-185
	Air cor.	170
	Pump jet	40
	Idle	58-60
	Aux. vent	4.0-4.5
1000 S using S-head-type valves	Choke	39
	Main jet	180-200
	Air cor.	170
	Pump jet	40
	Idle	60
	Aux. vent	4.5
1071 + modified head + semi-race cam + extractor exhaust	Choke	36
	Main jet	155-165
	Air cor.	170
	Em. tube	5
	Pump jet	40
	Idle	58
	Aux. vent	4.5
1071 S with S-type head	Choke	38
	Main jet	180-185
	Air cor.	170
	Pump jet	40
	Idle	58-60
	Aux. vent	4.0-4.5
1275 standard	Choke	34
	Main jet	135-140
	Air cor.	200
	Em. tube	5
	Pump jet	40
	Idle	55
	Aux. vent	4.5
1275 + modified head + semi-race cam + extractor exhaust	Choke	38
	Main jet	170
	Air cor.	170
	Em. tube	5
	Pump jet	40
	Idle	58-60
	Aux. vent	4.5
1275	Choke	40
	Main jet	195-200
	Air cor.	170
	Pump jet	40
	Idle	60
	Aux. vent	4.0-4.5

try driving the vehicle at very small throttle openings, about 1/16th inch of butterfly movement past idle, and see what the mixture is like. If it's too rich, the answer will almost inevitably be the need for a smaller idle jet, and a larger one if the mixture is too lean. However, you always need to consider idle jet size in conjunction with the air bleed hole being used for the idle circuit. If the progression appears to be fine just off idle, but then leans out, it may be necessary to use the same fuel orifice and a smaller air bleed, or vice versa if the mixture richens up. Like calibration in the main circuit system, calibration in the progression circuit is best done on a chassis dynomometer where small changes in throttle opening can be made consistently and the effect on mixture analysed accurately. With the idle and progression circuits correctly set up, a Series 'A' engine can run and drive a lot smoother than you would expect even though it may have a pretty monstrous cam.

Modifying SUs

The SU carb has been with us for a long time, in fact, since the turn of the century, or at least thereabouts, and there was a

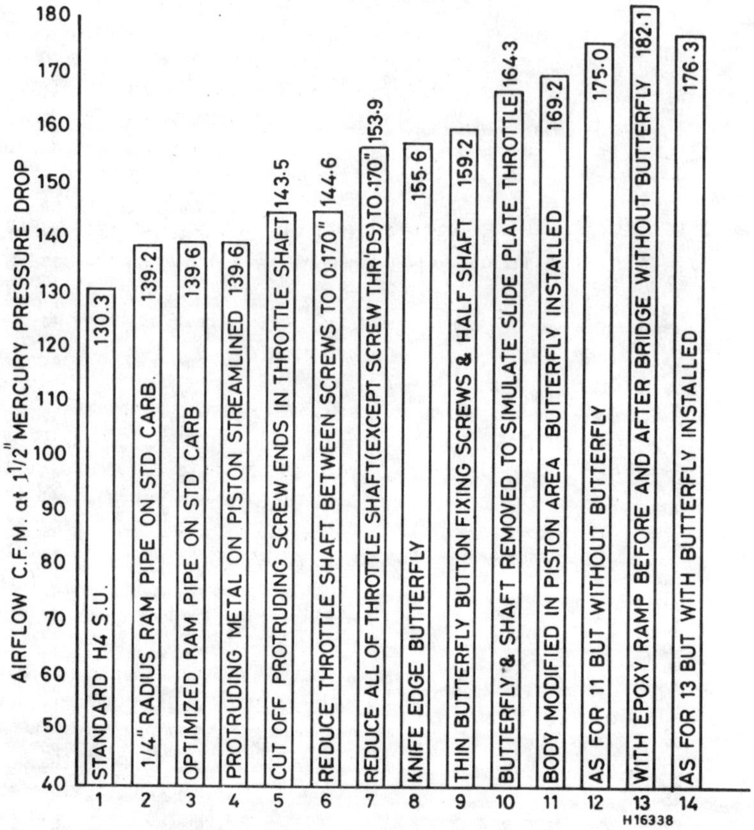

Fig. 7.18 Airflow improvements on a 1½ inch H4 SU

The 'guinea pig' carburettor used was a 1½ inch H4 SU, but the same technique can be applied to other carbs. For instance, it is possible to uprate an HIF6 from 240 cfm to over 300 cfm.

time many years ago, when it was a cheap carb as well as being one which was very effective. These days inflation, and maybe a bit of Government bureaucracy has pushed the price of the SU up until it isn't too far off the price of a twin choke, sidedraft carb. Since a sidedraft Weber or Dellorto isn't exactly cheap, you can see the implication. In fact, these days, if it's got 'carburettor' stamped on it, the chances are, it's quite expensive, and essentially what I am leading up to is that the cheapest carb you will ever own is the one already on your engine. Since most of these are SUs, it's worthwhile looking at the carb to see if there is anything that can be done to make it more suitable for whatever application is in mind. An example here should make the point. Suppose you have an 850 Mini, an early one. You have plenty of time on your hands but no money, you have succeeded in modifying the head, separating the intake from the exhaust manifold, slipped in a re-profiled cam and maybe a few other incidental but relatively inexpensive mods. You are now confronted with the fact that the single 1¼ inch carb is just a shade on the small side for your engine. What options are open to you? Basically two present themselves. You can attempt to locate a cheap, single 1½ inch SU from the wrecking yard and rebuild it, or you can, for even less money, modify the carb you have. My intention now is to show you how to take up successfully the second option, namely the upgrading of whatever SU carb you have.

In essence, this upgrading will just about make a smaller SU flow and perform like its next larger size. For instance, applying the basic modifications to a 1¼ inch unit, will make it flow about as well, and in some cases slightly better than, a single 1½ inch. On a 1½ inch, it will increase the flow to that approaching the 1¾ inch, and so on. In the example I am going to quote, all the flow

For performance applications, an SU without a ram stack of some sort, is as bad as a car without wheels. The 2½ inch ram stack here, proved to be a very cheap and effective aid to performance. However, ...

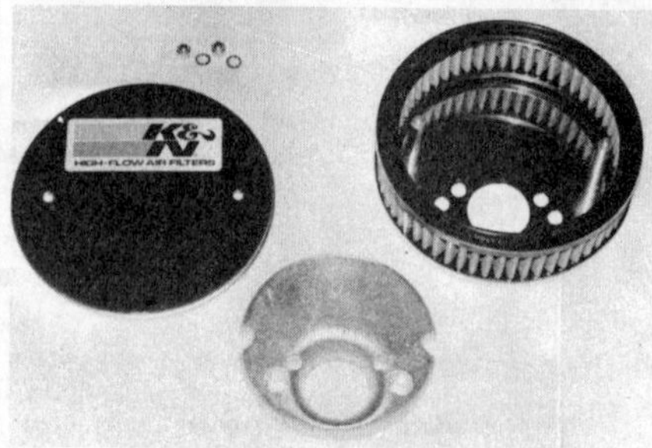

... clearance for the installation can be a problem, but something like an Advanced Products/K & N setup with the stub stacks can be installed simply as it was specifically designed for such installations.

figures will be measured at 1½ inch of Mercury depression, which is consistent with the measurement of airflow in high performance carbs. It might well be worth mentioning for American readers who are used to looking at cfm carb flow ratings, that two barrel carbs are flowed at 3 inch of Mercury, and only four barrel, high performance carbs are flowed at 1½ inch of Mercury.

If you are going to compare a two barrel American carb with the flow figures we are talking of here, then it will be necessary to divide the flow ratings by 1.41 in order to be able to make any comparison. In other words, a 350, two barrel Holley carb will be, at 1½ inch, 350 divided by 1.41 = 248.2.

Having got all the theory over with, let's take a look at a standard SU. The guinea-pig I'll use here is an old H4 SU which is a representative carb, and the results from these tests can be applied to HS4s, or indeed, any other SU carb including the H.I.F. series.

If you look at the chart, **Fig. 7.18**, you will see that a standard H4 flowed 130 cubic feet per minute. This represents a bare carb flowing into an open volume. There was no streamlining on the intake side or the exit side of the carb, and indeed through the whole test, there was no streamlining on the exit side whatsoever, this being standard practice when rating carb airflow.

The first and most obvious modification to obtain more carburettor airflow is to streamline the entry into the carb mouth. I know we have dealt with ram pipes in the previous chapter, but since this is all part of this test, I'll just cover it briefly again here because it's applicable. The first move in this series of demonstration tests was to make a clay bellmouth at the carb entry to demonstrate the

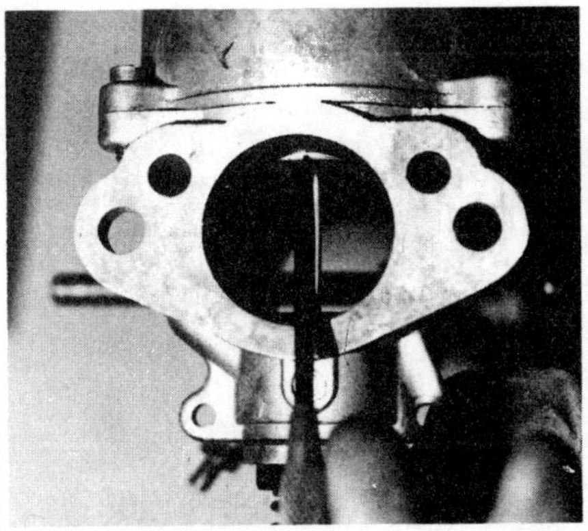

On many SU carbs when the piston is at the top of its stroke, some of the piston still intrudes into the airstream. As a matter of course, I remove the part indicated by the screwdriver by chamfering it off, as seen ...

sort of effect it would have. The increase in airflow was substantial: a 9 cubic feet increase resulted, as you can see from the second column of our bar graph, **Fig. 7.18.** On the face of it, one would think that the larger the radius going into the bellmouth, the better the flow would be. However, testing various radii at the bellmouth showed that after the radius had reached about ¼ inch, no

... in this shot here. The arrow A points to the leading edge of the piston which is chamfered off at an angle of about 30°. The trailing edge, arrow B, is chamfered off at about 15° to match it to the bore on the exit side.

additional gains were seen; therefore a ¼ inch radius is all that's needed. If the radius was reduced below ¼ inch, the effectiveness of the radius dropped substantially. Look at the third bar of the graph, and you will see the results of building a ram pipe and moulding various shapes to optimize the airflow into the carb. You will notice the best that could be achieved only produced a ½ cfm increase over a simple bellmouth. This is a very relevant point here, as the Advanced Products/K & N Stub Stacks for SUs which have the benefit of an intense flow test programme in their design, are in fact short ram pipes that fit within the filter case and at least as far as flow goes, appear to be near impossible to improve upon. The tests done here tend to show why.

From the foregoing we can conclude that the mouth of the carb is critical in as much as it must have some sort of lead-in, but it does not necessarily have to have a long ram pipe for good airflow. The ram pipe length modifies the inertial effect of ramming the induction system, and this, of course, won't show up on the flow bench. The indication so far is

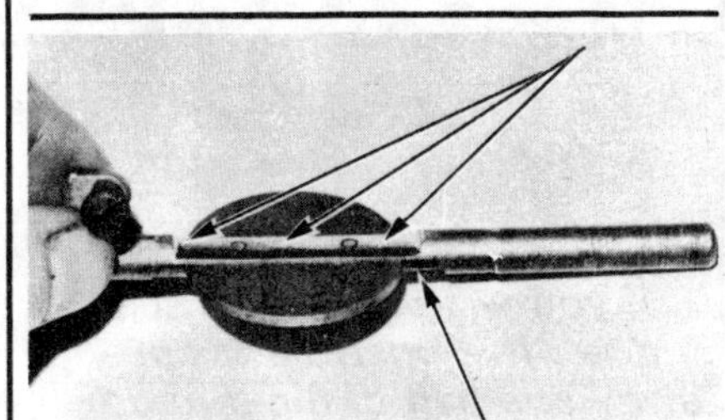

Flow bench studies have shown that one of the biggest impediments to airflow in an SU carb or indeed any carb, is the butterfly and shaft itself. A simple modification is to slim down the shaft as shown by the three arrows on the top. The underside is done in a similar manner. However, ...

that if we use air filters on an SU carb, airflow figures very close, or even virtually identical, to the open intake airflow can be achieved if an internal ram pipe having a radius of ¼ inch or more is used at the mouth of the carb. Again, though, let me reiterate, this does not mean that ram pipes with their extra length are not worth having. Apart from inertial ramming, very often a ram pipe on a carb will contain fuel stand-off which may occur when a fairly radical cam is used. A short radius into the mouth of the carb will not perform this function, but there again, standoff can be reduced by other techniques, either in the exhaust or induction systems. For instance, although a short ram pipe may not cure standoff, an air filter case (often used with a short ram pipe), will contain the standoff in much the same way as a longer ram pipe.

Well, I've got through three measurements so far and I've not made any mods, so here comes the first. Take a look down the throat of an SU carb and lift the piston to its full lift. You will notice that more often than not, the piston does not lift completely out of the bore of the carb. It usually stops short so as to leave an edge just hanging into the bore. In the past I have seen many well-known engine tuners modify this edge in an effort to get more air into the engine. In fact, if my memory serves me correctly, it was the done thing as a matter of straightforward practice on many of the engines prepared by the now no longer existing, but in its day highly successful, Downton Engineering. The edge concerned can sometimes hang into the airstream as much as ⅛ inch, and other times it's a much lesser amount. Test 4 involved filing off the sharp edge just where it protrudes into the bore of the carb. Subsequent flow tests show that the small amount that protruded into the bore was not affecting the flow one bit. In spite of this, I would advise taking off this edge because it seems to improve throttle response a little. The way to determine just how much to take off is to push the carb dashpot up to its full lift and then, with a scriber, scribe the amount it's protruding into the bore. Then, using a file, angle it off to that line.

... if you wish to go one step further, it is possible to cut away one-half of the throttle shaft completely, then file down the other half (that's the half with the threads in it) to give a streamlined approach. Then countersink the butterfly as necessary and use countersink screws to hold the butterfly in place on the remaining half of the shaft. In this shot here, the arrows indicate where the top side of the throttle shaft has been removed completely.

The next modification that's worthwhile doing is so simple and so effective that I find it difficult to believe it's not more commonly done. Take a look at the screws which hold the throttle butterfly to the throttle shaft. These are retained by virtue of the fact that their ends are split and spread. The test, bar 5 of **Fig. 7.18**, shows what happens when the ends of these screws are filed off so they are not protruding into the airstream. In practice, once these ends have been filed off it will be necessary to retain the screws by some other means such as a drop of Loctite on the thread. Remember, if one screw comes out, it's going to get eaten by the engine, and that could bend a valve, ding a piston, or whatever. Filing off the two screws resulted in an increase of some 4 cubic feet of flow. **This suggests that the butterfly shaft and anything in that region is a major obstruction to flow and a lot of subsequent flow testing has shown this to be the case.** In the following tests I demonstrate just how much of a restriction the butterfly and shaft assembly is. Bar 6 shows the effect of filing the throttle shaft evenly on both sides between the fixing screws to reduce the thickness to 170 thousandths. Again, the flow increased, this time by no less than 11 cubic feet. If you really want to go the whole hog in this direction, then here's how to do it.

Firstly, cut away the throttle shaft on the one side, so it's no longer a slotted shaft, but a shaft with a flat on it. Replace the countersunk screws with some dome-headed button screws, preferably Allen socket button-headed screws. The remaining semi-circular shaft should be cut down between the threaded holes to about 75 thousandths thick and have its form streamlined. Next, if you have access to the equipment, mount the butterfly on some double-sided sticky tape and very carefully fly cut it to half its width. Then knife-edge the remaining edges. If you can't fly cut it to half its width, then simply do the knife-edging operation with it at standard width. If you take the simple way out and just knife-edge the butterfly without reducing its width, the flow goes up by almost 2 cfm to that shown in Bar 8. If you do manage to get to grips with the mill, and face

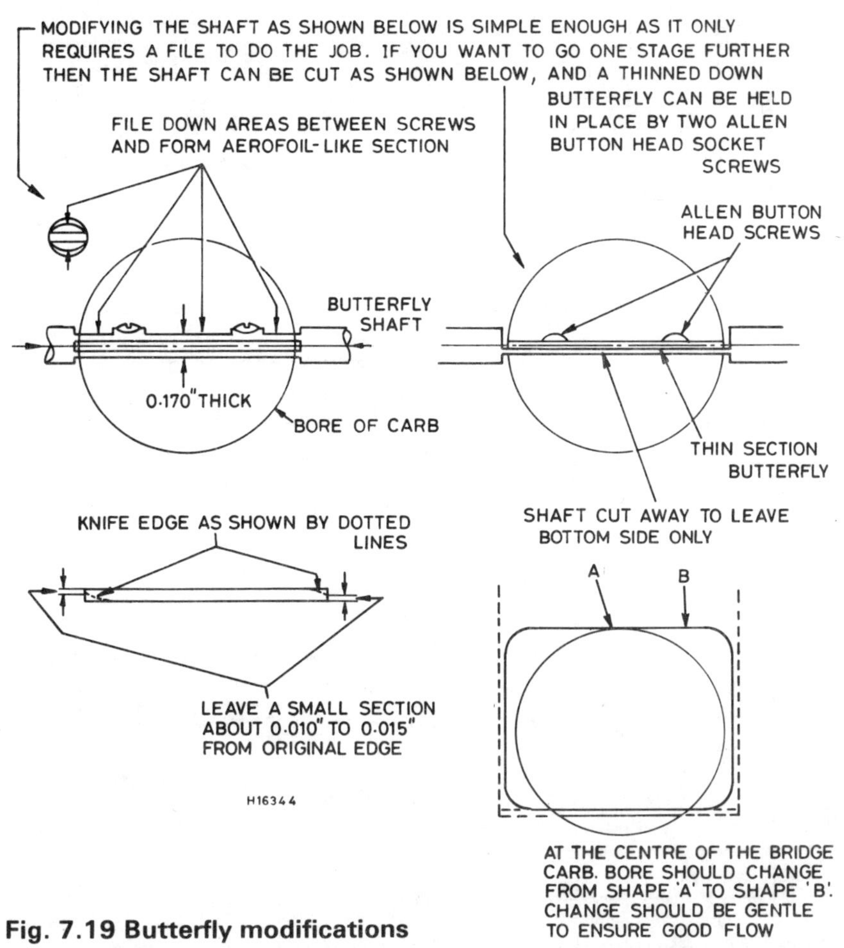

Fig. 7.19 Butterfly modifications

This shot shows what I mean by squaring-off the throttle bore in the vicinity of the piston. You see how the bore goes from the round shape at the mouth of the carb to a square shape where the piston intrudes into the bore of the carb. The square shape then blends back into a round shape further downstream. This contributes greatly to the airflow capability of an SU carb.

If you are trying to remove the dashpot assembly from an HIF SU, don't forget to take off the circlip from the top of the steel sleeve. If you don't, your efforts at dismantling will be in vain.

off the butterfly to half width, and you use the button-headed screws, the flow climbs to 159.2 cfm. Of course all the modifications in this area tend to beg the question, 'How much would the flow be if a slide-plate throttle was used, i.e. no butterfly?' Well I have seen examples of slide plate throttles used. They have never gained popularity because they are complex to make, but it would be interesting to see what would happen if the butterfly and shaft assembly were removed. This would allow us to determine how close to optimum the modified shaft was. Removing the throttle plate and shaft brought the flow up to 164.3 cfm as shown in column 10.

It has been common practice, for SCCA race-prepared engines in the States, to bore carbs out so as to get the maximum airflow. This boring operation simply does away with the bridge area completely, and of course the jet has to be modified to cope with this metal removal in two ways. Firstly the venturi effect is vastly reduced, and secondly the jet position is altered to the point where the fuel level in the 'before' and 'after' situation can be substantially different. My contention is that this route is not necessarily the best way to go, as the engine only runs properly at full throttle, and such an operation can really only be done on a race engine. **Boring out a carb for a road engine would be a disaster.** However, the point I want to make here is that I am not sure that boring is necessarily the best way to go, even for a race engine. Personally I have had great success with squaring off the back of the carb just after the bridge. This produces a remarkable increase in airflow, but does not adversely affect the way the carb works. In order to get an idea of what is needed, take a look down the throat of an SU carb with the piston in the raised position. Basically what you see is the round bore of the piston intruding into the round bore of the throttle body. Where the two intersect, there are some corners which stick out into the airstream. As the air passes over the bridge, so it encounters these edges. It's almost like having to start again and go down the mouth of a carb that has no ram pipe. If the throttle body is flared out and squared off at the point where these two bores meet, the movement of the air is considerably eased. **Without touching the bridge area at all, it's possible to achieve a substantial flow increase.** Column 11 of the bar graph, **Fig. 7.18** shows just how much the flow improved. This particular test was done with the butterfly in place. As you can see, the improvements in this area produced a flow figure higher than would have been the case if a slide plate throttle had been used, but no modifications done in the bridge area. Still, these mods don't preclude the use of a slide plate throttle. Should you wish to go that far to improve your existing SU, the airflow will rise to 175 cfm.

Up to this point, everything that has been done to the carb will allow the engine, if it needs

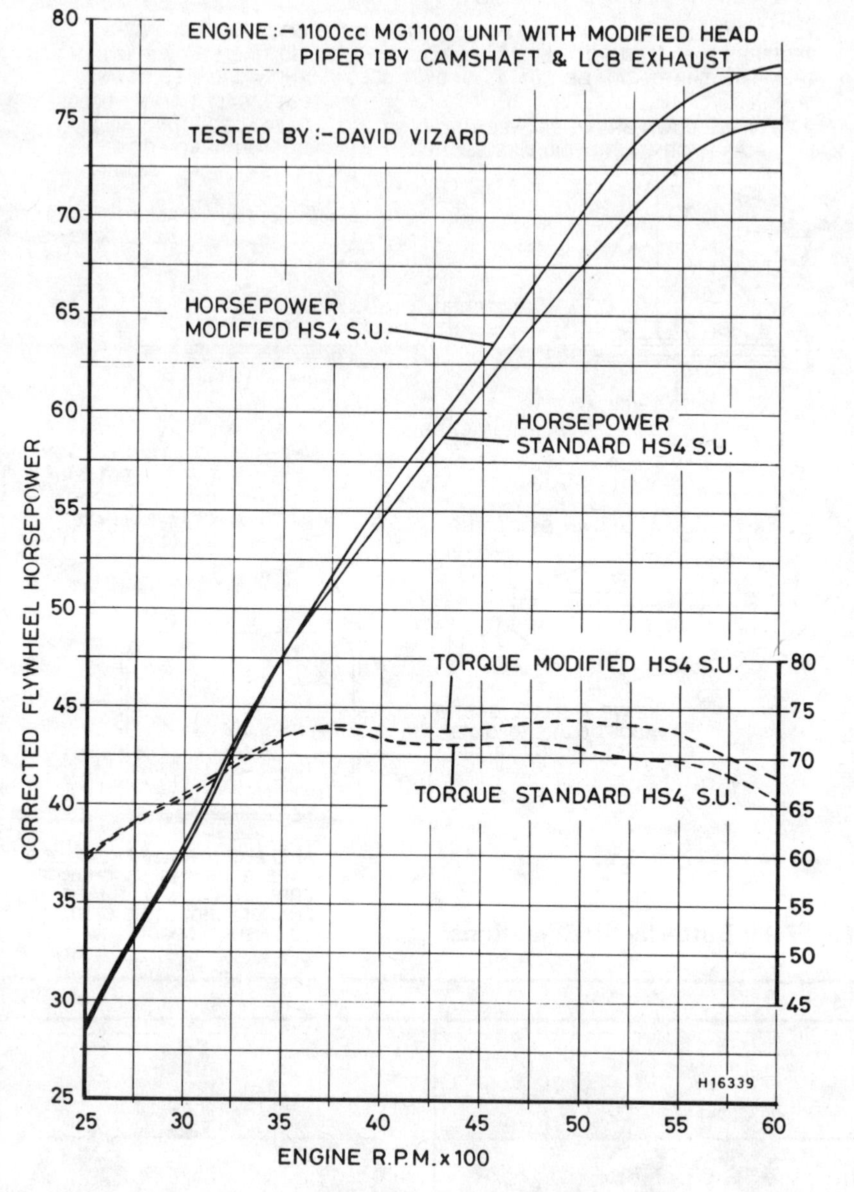

Fig. 7.20 Comparison of standard and modified $1^1/_2$ inch HS4 SU carburettor

the extra carb capacity, to produce increased hp without necessarily affecting driveability. When substantial increases in airflow are made, the carb will almost certainly require a different needle profile to produce the correct mixture, so **if you do modify your carb, don't assume the existing needle profile is anything like correct.** Even if the carb is only modified to the point we have so far discussed, it is producing an increase in airflow of some 30%, so it's bound to want some corresponding fuel flow changes.

If the carb is to be used specifically for racing, then there are a few more modifications that can be done to find more airflow and hopefully more hp. However, I do not advise taking these steps on a road engine, as in order to get the throttle response correct without any undue flat spots, it's necessary to run a much richer mixture than would normally be

This is how to make the most of a pair of SUs. These $1^1/_2$ SUs are airflowed which increased the flow capability from just under 140 to a little over 175 cfm. The route to this increased flow starts with the use of a pair of APT stub stacks. The innards of the carb are then modified as explained nearby. To further aid power a heat shield is installed between the carbs and the exhaust manifold.

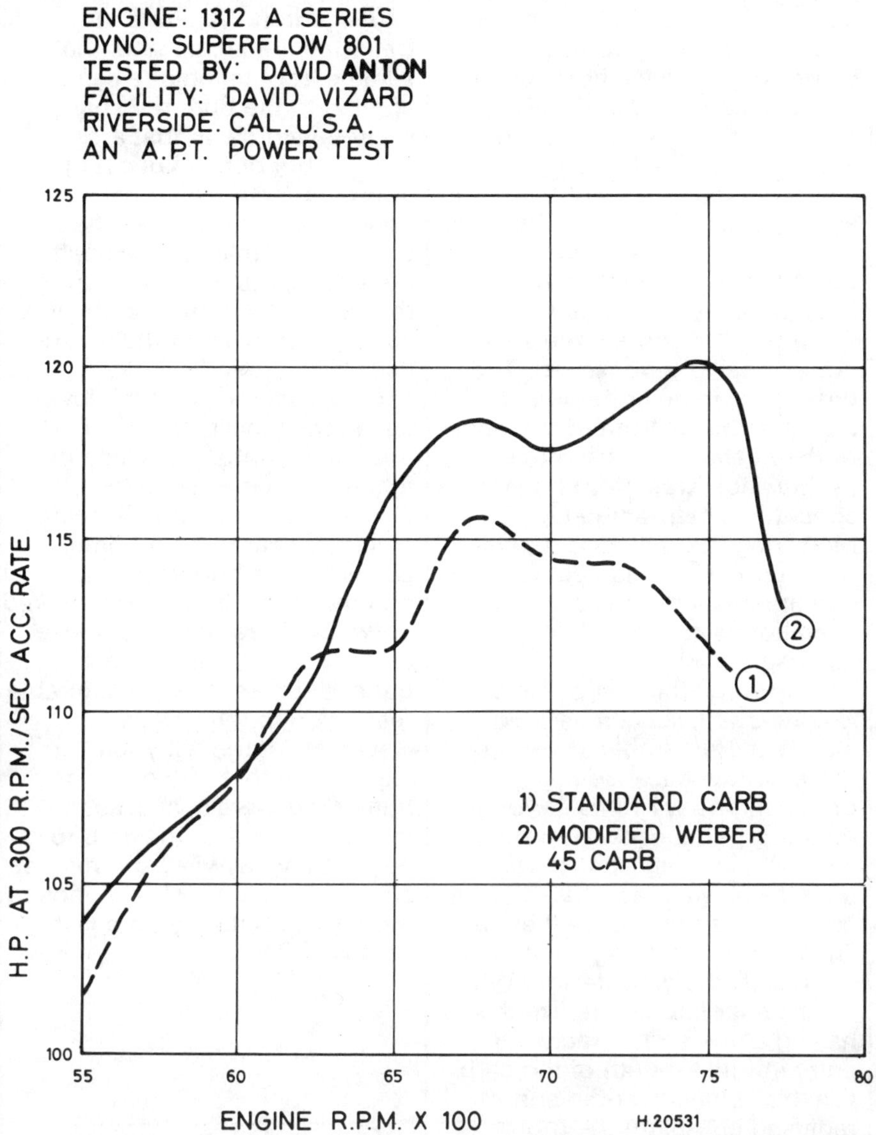

Fig. 7.20a Carb modification test *An engine can be starving for air because of insuficcient carburettor airflow, but fitting a larger carb doesn't always work. Our efforts at A.P.T. have shown that a 48 DCOE rarely produces positive results. So to get the most from a 45 DCOE it was 'airflowed'. This involved mods which made the carburettor more efficient NOT bigger. The higher gas speeds, plus the increase in airflow, seemed to be what the engine liked for power output. As would be expected, the biggest gains occur up at the top end of the rpm range.*

the case. The basic reason for this is that the next modification seeks to streamline the bridge area of the carb. This move causes the discharge co-efficient (that's the flow efficiency of the piston and bridge area) to increase, and in so doing it tends to back off the effectiveness of the pump jet action on an SU. No doubt somebody is going to say 'What pump jet?' An SU doesn't have one.' Not as such it doesn't, but the pump jet action is achieved very subtly by having the air valve, that is the space which allows the air to go into the engine past the air piston, to be less efficient than the fuel valve. The fuel valve in this instance, is the needle in the jet. When the throttle is opened suddenly, the oil in the damper assembly stops the piston rising as fast as it would otherwise normally do. In so doing, the air over the bridge area is speeded up, but the amount of air passing into the engine is not increased proportionally. Because the efficiency of the air passage into the engine is lower than the efficiency with which the fuel is discharged, there is a momentary increase in fuel/air mixture. This produces exactly the same effect as having a pump jet, hence the term 'inbuilt SU pump jet action'.

By streamlining the bridge area, we are going to invalidate much of that action. The only way it can be made up for is to run the carb much richer in the first place. On a race engine, this won't really be important. It will be possible to run the mixture richer without any loss of hp. In fact, at fuel/air ratios of 12.5-13/1 where a Series 'A' produces its best power, flat spots should not be encountered. At leaner mixtures for road use, the flat spot may well be experienced, hence the reason for not recommending this mod for road use.

Okay, having got that caution

over with, let's get down to the nitty gritty. Basically this modification involves pouring liquid epoxy resin into the area before and after the bridge to form a gentle slope rather than the abrupt one it has at present. The easiest way to do this is to tilt the carb at an angle and then pour the epoxy resin in so that it just comes up to the bridge, and then leave overnight to set. Do one side before the bridge in one 24-hour session, and the othe side, on the trailing edge in another 24-hour session.

It's obviously important to make sure that the epoxy stays firmly attached to the body of the carb. Having a lump of epoxy passing through a Series 'A' engine has yet to prove beneficial in any way. An easy way to make sure the epoxy does stay keyed, is to drill some small holes at an angle to each other, into the metal in the area where the epoxy will reside. Streamlining the bridge in this fashion will boost the airflow from 175 cfm to a little over 182 when a slide plate throttle is used, or conversely, from 169.2 to 176.3 if a conventional throttle butterfly is used and modified as detailed earlier.

This sort of flow work then, has produced quite a bit more airflow potential from a standard carb. To give you a comparison of our guinea-pig carb and, say, a 40 DCOE, the modifications so far described will produce flow figures about the same as a 40 DCOE with a 34 – 35mm venturi in it. However, the carb I started with in this particular example was relatively smallfry, and of older type. Using an H.I.F.6 or an H.I.F.7 it's possible to pass the 300 cubic foot per minute barrier with ease, and flow figures like that are pretty big. That's more, in fact, than one barrel of a 48 IDA Weber.

As for power increases, well **Fig. 7.20** shows the power increase on a modified 1100 cc engine when a flowed carb has been used.

If SUs can be modified to flow more air, how about other carbs? Well I'd like to pass on some sound advice here: firstly let's consider the 28/36 Weber carb and carbs of that ilk. Mods in the area of the butterfly and shafts can be done, but there are a few points of caution. Thinning down the shaft presents no problems in terms of carb function after it has been done, so long as, of course, the shaft is not excessively weakened. **The butterfly can be knife-edged also, but the section of the edge of the butterfly that sweeps past progression/transition holes should not be modified.** Modifying this will tend to alter the progression characteristics and the chances are, any alterations here would be adverse.

As far as the rest of the carb is concerned, leave it as standard. Remember, if you need more airflow because the power of the engine warrants it, you just go for larger chokes. However, when choke sizes start to reach the maximum sizes available, we find it is not the chokes that are limiting the breathing so much as the rest of the carb design. What will help air into a carb like this is having some sort of radiused entry into the mouth of the carb like this is having some sort of radiused entry into the mouth of the carb like an Advanced Products/K & N Stub Stack. On higher powered engines, a good radiused intake stack can add the odd hp or two to the engine's output.

As far as the downdraft IDAs and sidedraft Webers and Dellortos are concerned, the same sort of technique can be applied to the butterfly shafts to improve airflow in that area slightly, but don't expect big improvements. The most noticeable thing you will find is that by streamlining the shaft and butterflies, the carb pulls up onto the main jet sooner and the low speed tractability of the engine is usually improved slightly. This, of course, assumes that the low speed tractabililty hasn't been destroyed in the first place by fitting a very radical cam.

At first sight, the Fish carb would appear to be a prime candidate for butterfly modifications. If you have looked down the bore of one of these carbs, you will see that the butterfly presents a considerable restriction to the airflow. However, let me tell you from experience that re-working the butterfly generally causes the metering curve of the carb to go completely haywire, and my best advice to anyone who intends to modify the butterfly on a Fish carb is, DON'T.

Carburation Part 4

Carb Comparisons

One of the biggest fears of most engine tuners, especially those who don't regularly have recourse to a dyno, is that they are going to under-carb the engine. **Sure, putting on too small a carb is not really the way to go, but I hope to demonstrate with figures from various tests that it is not necessarily as bad as is often thought.** However, when it really boils down to it, there's more to selecting a carb than just the bare bones of how much power it produces, and that in itself is dependant on more than just how much airflow it's capable of.

Although I don't have hard and fast data, evidence does point to the fact that the Series 'A' engine does not like fuel too finely atomized if outright power is the ultimate goal. On the other hand if economy is a major aspect, then the fuel should be as finely atomized as possible. The type of fuel droplets the carb delivers seems to have as much influence on the total power output and economy the engine is capable of, as the carb's airflow capability.

For that reason I think it's best to show the power output of a range of engines of different hp levels, with different carburettors. The first engine, **Fig. 7.21**, is a comparison of a Weber, Dellorto, Fish, SU and 28/36 Weber, all on a relatively low-powered engine, *i.e.* around the 70-75hp range. As you can see from the graph, the spread is only a little over 5hp between a standard 1½ inch SU at one end of the scale and a 40 DCOE Weber at the other end. More to the point, however, is that the hp at 4000rpm is more or less the same for all the carbs, thus indicating that the 130 cfm odd that a 1½ inch SU has is adequate to supply the needs of a 61hp engine, and that putting on carburation of some 150 cfm per barrel, as in the case of the 40 Weber, does little or nothing for a power increase at that level. It's only above 4000rpm that the carb of least flow capacity, starts to lose out.

Let's take another case, on a higher powered engine, but this time we will also throw in fuel-injection. The particular injection system concerned is the Piper RDA system and the

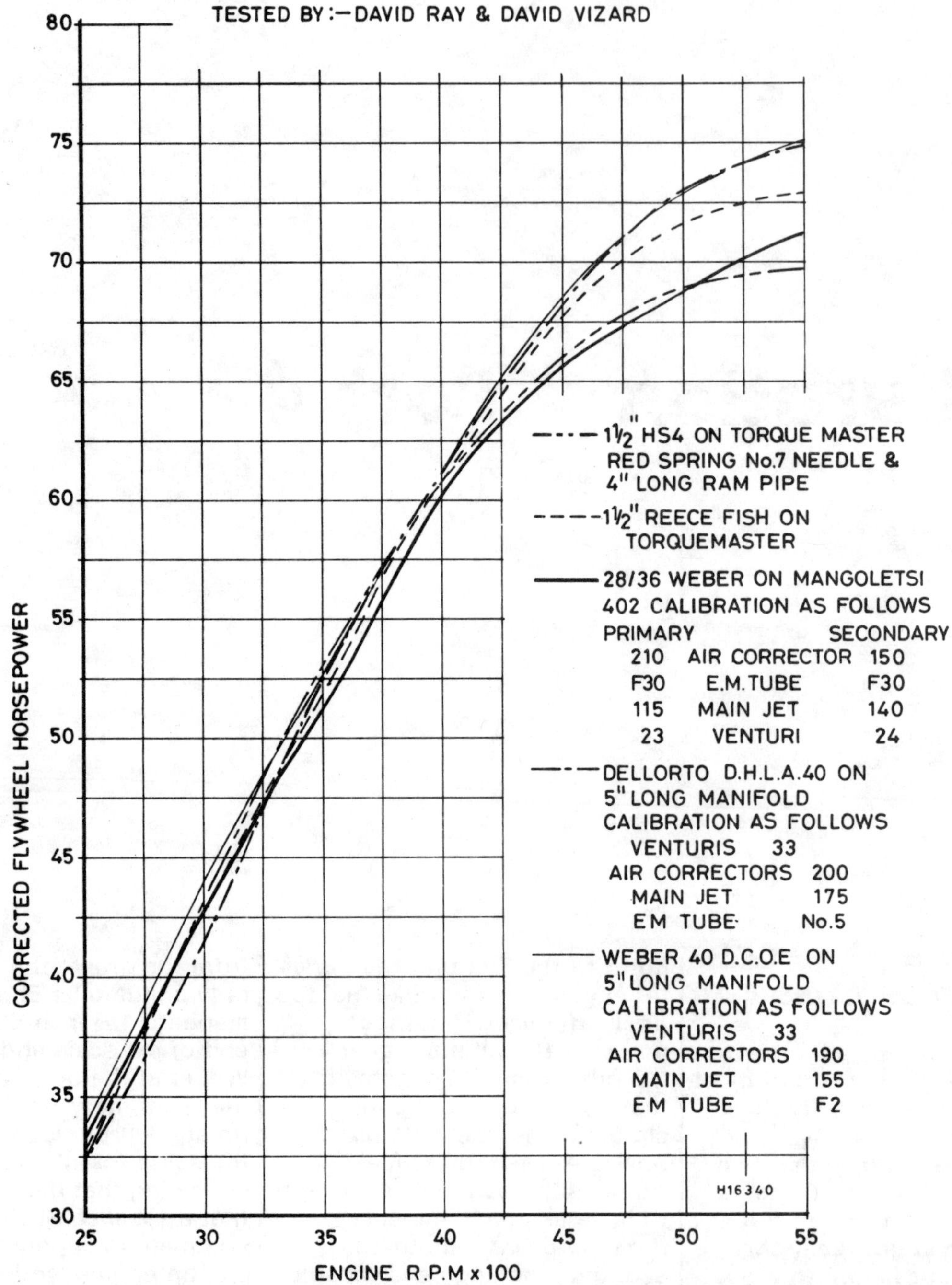

Fig. 7.21 Comparison of popular carburettors on street modified 1275GT Mini engine

test engine is a modified 1275. The carbs tested were a single 1¾ inch SU, a 45 DCOE, 45 DHLA and the injection system. The results, **Fig. 7.22**, adequately demonstrate two things - the difference between single barrel induction in the form of an adequately sized SU, and an induction system employing 2 barrels, as in the case of the rest of the units tested is not as great as many people would often suppose. In fact, the Dellorto was slightly down on hp by comparison with a Weber, even though it flowed the same amount of air, and was indeed tested on the same manifold. **The injection, however, the highest flowing unit used, produced the least hp**. This is not to say there is anything wrong with the system but since it injected fuel at high pressure, atomization is incredibly good by comparison

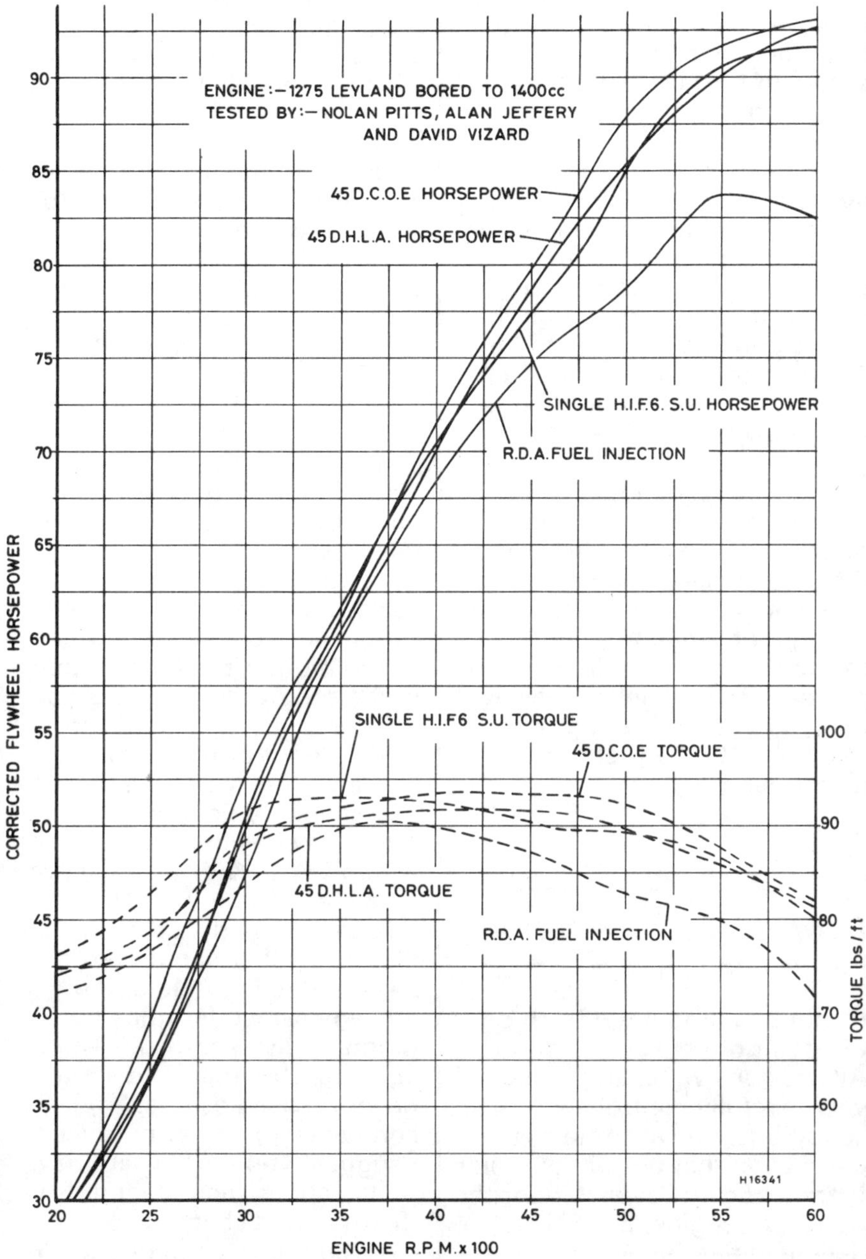

Fig. 7.22 Power output comparisons of 45 DCOE Weber, 45 DHLA Dellorto, HIF6 SU and Piper RDA fuel injection

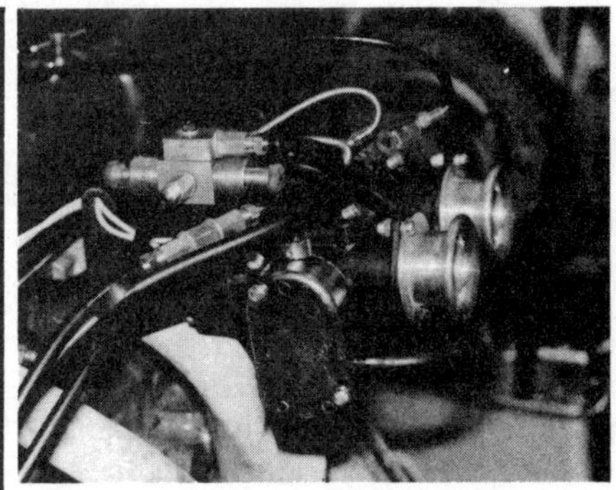

The Piper RDA fuel injection system, when tested on a Series 'A' engine, proved to be very effective from the point of view of smooth running, good throttle response, very good brake specific fuel consumption figures and excellent drivability. What it did lack was the ability to produce hp principally, I believe, due to atomization which was too fine for the requirements of a Series 'A' engine.

with the carburated fuel atomization. It's evidence like this which leads me to believe that too fine a fuel atomization on the Series 'A' causes a reduction in power. This could be due to the fact that the very fine droplets of fuel vapourize before reaching the cylinder and thus displace air. It's worth noting that a droplet of fuel takes up about one-hundredth the volume when liquid than when vapourized. If fuel takes up space in the manifold, then that space is not available for air. On the other hand, the fuel-injection would probably be the best way to go for a super economy engine. However, I hasten to add that I have not done any accurate fuel economy measurements to determine whether or not this is so, but an educated guess indicates that it should be. Really, the high point of this test, is that it shows beyond a shadow of doubt, that an H.I.F.6 on a good intake manifold makes an incredibly good setup, as it produces a power curve very similar to the best of the bunch.

Of course so far I haven't really demonstrated the practicality of twin SUs. However, **Fig. 7.2** should set that right. These are some figures taken at Howley Racing some years ago, and rather than the engine dyno, the rolling road dyno was used, simply because the suitable test engine was already installed in the car.

The main point to note about these tests is that the graph starts at 4000rpm. The reason being all the carbs on this particular engine gave identical power curves from 2500 to 4000rpm, and since they didn't diverge much above that, the graph has been chopped to show the differences more clearly. What you should note here is that the tests with the twin SUs were done on intake manifolds which had been internally reworked and were better than the standard Leyland manifold. Even so, the power production of twin SUs over a big single, is hardly worth all the hassle of having to have that extra carb.

The best carb tested was the 48 IDA unit on a Derrington intake manifold, the patterns for which have now been sold to T.W.M. in the USA. As I have often found before, this carb gives very good reponsiveness but it can sometimes be problematical to set up. It is important to remember that it needs big pump jets and relatively large idle and main jets to allow it to run smoothly through the transition and on the main circuit.

Moving on to a real top-of-the-line engine now, let's go back a few pages and have a look at the effect of the 1½ inch SU versus the 1¾ inch SU that we dealt with earlier. Bear in mind that tests were conducted on a very high powered engine, and will stretch the capabilities of a 1½ inch and possibly a 1¾ inch to their limits. Any power differences will, therefore, show clearly. The power differences on a smaller engine will, as one would expect, be less.

As can be seen from the graph, at 3000rpm the difference in power was only a mere 0.8hp, but by the time the engine is turning at 7000rpm, the power differential has risen to 6. The next logical step from here would be to see how a 1¾ inch SU fares against a sidedraft Weber on an engine of this power level. **Fig. 7.4** shows the difference between 200 cfm of SU carburation in the form of an HS 6, and a 45 DCOE having an airflow capacity of around 210 cfm per barrel. Again, no staggering change with a difference of 4hp. However, to be fair, there are occasions when the Weber seems to turn up just a little more in the way of extra power than this example. On engines of 115hp or so, the gap between SU and Weber is about 6hp. At these high power levels it would appear that some engines can be a little more touchy on their carburation than others and it just seems to be a question of what combination of parts are used in putting the engine together in the first place.

Having waded through what must appear to be dozens of power curves and the like, the only conclusion I can come to is hopefully I have passed on enough information to put carburation capacity into perspective, **as well as making many of you realize that the good old SU is a pretty stout performer**.

Economy Carburation

I'll not beat about the bush on the point of carbs for economy. After extensive testing under very carefully controlled conditions, I have come to the conclusion that out of the major brands of carbs available for the Series 'A' engine, the SU is definitely number 1. Its economy figures are closely followed by the 28/36 Weber. Now I have to point out that these tests were done on a Mini engine, and as such, the 28/36 Weber tests were done with the carb barrels parallel to the engine's centreline. If you look through the manifold section, you will see there is a problem with the 28/36 manifold when the barrels are lined up this way in as much as at part throttle there is an uneven flow distribution between each siamese port of the cylinder head. This is due to the fact that the primary barrel is nearer one end than the other. Okay, many of the manifolds make compensation for this by putting in a deflector underneath the primary barrel, but even this doesn't seem to compensate totally.

At this point I am going to express an opinion. If a carb such as the 28/36 Weber is mounted with its barrels 90° to the axis of the Mini engine, I have little doubt it will equal the SU figures as far as economy is concerned. I haven't tested the 32/34 DMTL carb for economy, but I suspect it will do well in this area.

Even though they are performance items, sidedraft carbs can produce quite good economy figures. If calibration of the progression circuits and the pump jets is toward the lean side without actually getting into a misfire situation, then the fuel economy figures, when driven at similar speeds, can be within 5%-10% of that achieved by a carefully set up SU. However, experience has shown that if a Weber is put onto a swan-necked manifold, the mileage can suffer quite dramatically as a result. There again, such a manifold should never be used as it isn't even conducive to power production. I suggest you read what I have to say about such manifolds in the manifold chapter.

The art of setting up a carb for economy really hinges on being able to calibrate it as lean as possible without actually running into a lean misfire situation. Sometimes it's not really possible to hear when an engine is starting to run unevenly, so a degree of experimentation is always required, especially if you are having to set up your motor and test the mileage on the road as opposed to carefully controlled conditions in the dyno cell. Just how lean the motor will run depends on a number of factors, of which one of the most important is the intensity of the ignition system and, to a lot lesser degree, the charge temperature. My own dyno tests indicate that too much heat of the intake charge, though it may cause good vapourization in the fuel, does not help steady state fuel economy. Exhaust heat on the intake manifold is too much heat. On the other hand, a water-heated intake manifold appears to be just right. A cold running manifold, *i.e.* one with no water, is too cold under most conditions.

Another relevant point, especially if you are sticking with the SU to get your fuel economy, and there's no reason why you shouldn't (after all it is the best carb to use), is the question of intake air temperature. Many vehicles have an adjustable snorkel which can either pick up cold or hot air. Experience in the area of hot and cold air induction indicates that from a cold start, and for the first, maybe, 7 miles a hot air intake pays off in terms of fuel economy. However, at distances over 7 miles and with the engine already at running temperature, the hot air intake detracts slightly from the mpg.

Let's return to the subject of spark intensity for a moment. The more energy that's put into the spark and the bigger and hotter the spark produced, the more possibility of running lean mixtures without encountering a misfire. Under ideal situations, it would be nice to have an engine which was not sensitive to the mixture ratio at all, and whatever amount of fuel was put into the cylinder, it could still be fired. Under these circumstances it would be possible to do away with the air metering function of the carb. It would simply be a case of metering fuel like a diesel. For those of you not aware of a diesel's function, a diesel is always at wide open throttle as far as air is concerned. The engine's power output is controlled solely by the amount of fuel injected. In other words, at idle the fuel/air ratio is literally hundreds to one, but at full throttle it's probably down to 20/1 or so. The reason a diesel engine gets good fuel economy is that it operates at the highest effective CR possible at all times. When an engine is throttled, the amount of air drawn into the cylinder is vastly reduced, so although the engine may have a theoretical CR of say, 10/1, in practice, because the cylinder is not filled, it may only be 1½ to 2/1 at most at low throttle openings. As a result, the Otto-cycle efficiency or thermal efficiency of the engine, is reduced quite considerably. When lean mixtures are used, it's necessary to open the throttle more to get the sort of power required to cruise along the road. Opening the throttle more, of course, causes more air to pass into the engine, and for a given power output, only a slightly lesser amount of fuel is likely to be put into the engine, but because the engine is now operating at a higher effective CR, it will deliver the same power as before, hence a saving is made in fuel. The big problem with spark ignition engines is the fact there is a limit on just how lean the fuel/air ratio can be before a misfire sets in, and as soon as a misfire does set in, of course the fuel economy will go to pot anyway.

If you read through the ignition chapter you will discover that there is a big difference between the lean misfire ignition capabilities of various types of ignition systems. As a result, it's not possible to say that for any one vehicle it should have a fuel/air ratio of such-and-such for maximum economy. This brings me round to the whole point of this discussion, namely a device for adjusting the lean limit of the carb whilst having no effect on the full power rich mixture. With some SUs, namely the HIF 38S (metric HIF 4) fitted to some of the 1-litre Metro engines, we find that the carb is equipped with what SU call 'an additional weakening device'. This device causes the part throttle mixture to be leaned out to a greater extent than is normally the case, but it does not affect the mixture at full throttle. The amount of fuel delivered by a main jet is governed, not only by the area of the jet presented to the airstream, but also the pressure differential between the end of the jet, (the bridge area) and the air pressure above the fuel in the float bowl. Normally the fuel in the float bowl is subjected to standard air pressure, whatever that may be. However, if that air pressure is lowered during part throttle operation, then the mixture will be leaned out, because the pressure differential between the float bowl and the bridge area (which acts as a venturi) will be reduced. It's upon this principal that the additional weakening device functions.

Take a look at **Fig. 7.23**. To see how the system works, we'll just follow things through the normal cycle of events between idle and full throttle. At idle the butterfly is on the downstream side of the drilling (5), so this drilling is not communicating with engine vacuum. Rather it is being subjected to near-normal air pressure so little or no reduction in air pressure takes place over the fuel in the float bowl, hence the mixture delivered is as per normal settings. When the throttle is opened slightly, as would be the case under cruising conditions, the drilling (5) will eventually end up just downstream of the butterfly, so any engine vacuum created will cause air to flow through the

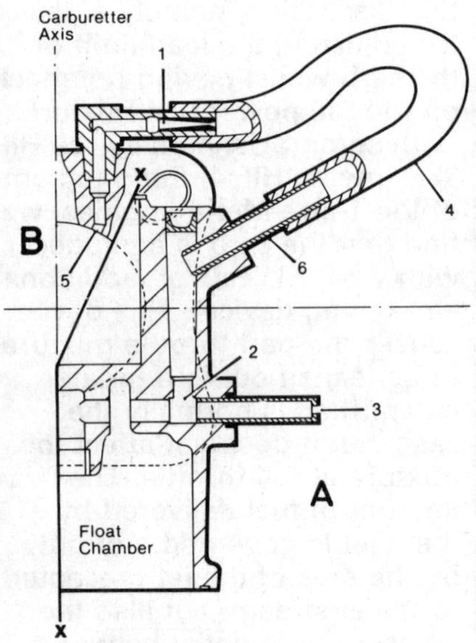

Fig. 7.23 Additional weakening device for HIF SU carburettor

Section through HIF 38 carburettor body

An engine normally requires a richer mixture when running at full throttle than under cruising conditions. The SU carburettor automatically achieves this on the majority of installations due to the pulsating nature of the air flow at full throttle as compared with the steady flow when cruising with the throttle partly shut. This effect, known as 'mixture ratio spread', is also affected by the design of the air intake and induction passages.
Certain engines however demand a greater 'mixture ratio spread' than can be met by the above factors. In such cases the additional weakening device is used.

The rate of fuel discharge from the jet is governed by the difference in air pressure between that existing over the fuel in the float chamber and that over the jet.
The weakening device is a fitting attached to the carburettor designed to reduce air pressure (create a depression) in the float chamber when the throttle is partly shut, thereby reducing the rate of fuel discharge from the jet.
The fitting consists of a venturi (1) which is connected at one end to a drilling in the carburettor body close to the throttle disc edge (5); the other end, which contains a calibrated air bleed (3), is connected to the air intake passage by a flexible tube (4).
A drilling (2) communicates with the float chamber.

Idling
With the throttle in the normal idling position, the drilling (5) in the body emerges on the carburettor side of the disc and is only subjected to the slight depression exerted in that condition. This will cause a flow of air through the venturi (1) but the effect of this on float chamber air pressure is negligible.

Full throttle
As with the idling position, the depression produced is slight and any effect on air pressure in the float chamber is compensated for in the design of the jet needle.

Cruising
When the throttle is partly open, the drilling (5) is on the engine side of the disc and the high manifold depression causes air to be drawn through the venturi (1). The use of a venturi (instead of a plain orifice) ensures that the air velocity through it will reach a maximum value which remains constant once the predetermined depression figure has been attained.
The air bleed (3) admits air into the system and the resultant float chamber depression produces the required reduction in fuel discharge.
This arrangement allows the maximum weakening effect to be produced when the throttle disc is closed a small amount from the full open position (when only a slight increase in manifold depression is obtained) and ensures that further closing of the throttle does not increase the weakening effect to the point at which misfiring may occur.

General
The size of the venturi is standard, the only variable on different vehicles is the size of the air bleed.
The flexible tube (4) connecting the air bleed union to the air intake passage has a substantial effect on mixture stength. As the air velocity through the air intake increases, a depression is communicated to the float chamber, the effect of which is compensated for in the jet needle design, thus the removal of the connecting tube would cause some alteration to the mixture strength.
The additional weakening device is not usually applicable to types of carburettor on which it was not fitted as original equipment, since the resultant weakening effect would be excessive.

venturi (1). This venturi is connected by a pipe (4) to a drilling in the body of the carb which communicates to the float chamber. Also communicating to the float chamber is a bleed orifice (3). If the bleed orifice is large enough, then air being drawn in through the drilling (5) will be completely supplied by that drilling (3) and so the air pressure in the float chamber will not change to any appreciable level. On the other hand, by reducing the size of the drilling, it's possible to calibrate just how much extra depression can be created in the float bowl. Also, the shaping and sizing of the initial orifice, that's the venturi (1) in the drawing, will also have an effect on just how much depression can be created. Under normal conditions, the depression at (1) will quickly fall to a certain level, beyond which it will not go due to the fact that the venturi will have reached the limit of how much air it can pass. As you can see, at part throttle this device will cause fuel mixture to lean out. If the throttle is now opened to the wide open position, all the manifold vacuum will go, so there will be virtually no vacuum communicated at the port (5) and the weakening effect of this system will be eliminated, therefore the mixture will be restored to the full power mixture required at full throttle opening.

From that description you should be able to see that the system is quite straightforward and simple, and, more to the point, effective. However, for anyone building their own economy engine, it lacks one major factor of importance, namely adjustability. It is this aspect that I'll now delve into. What will be covered is how to make your own 'additional weakening device' which is suitable for use on any carb, not just an SU.

The problem with building a replica of the SU system, working on the same principal, is that we won't know what the calibration sizes are for the air bleed jets, for two reasons. Firstly, we won't know how weak our particular engine will want to run at part throttle, and secondly, outside of SU I doubt that many of us have the calibration equipment necessary to size the orifices, even if we knew what was needed. The foregoing is, in essence, the total argument for building an adjustable system. As it happens, this is relatively straightforward. The schematic **Fig. 7.24**, shows a system that can be applied to an SU, Weber or indeed, almost any carb. It consists of a few lengths of plastic hose, plus two adjustable air valves of the type commonly used in airation systems for fish aquariums. When buying these air valves, be sure they are resistant to fuel, and that at least one of them is able to pass a reasonable amount of air. You should be able to blow through it relatively easily when it's fully open.

Starting at the carb, the major points of the system are as follows: firstly you will need to drill a hole 1½mm from the edge of the butterfly in the relevant side of the carb. Remember, when the butterfly opens, the hole must be in such a position that the butterfly will sweep over it, effectively moving it from upstream of the butterfly to downstream so that it communicates vacuum to the system. Using a suitable adaptor, connect this hole via the hose, to the first tap in the system. The output of tap 1 goes to a T-piece, one end of which goes into the connection going to the float bowl, and the other to atmosphere (preferably the inside if the air filter case). Now here's how it works: by putting an adjustable valve between the bleed in the carb body and the float chamber, adjustments can be made to the rate at which the air is drawn out of the float chamber to fill the partial vacuum downstream of that tap. However, to maintain a certain pressure drop over the fuel, it will be necessary to allow air back into the system to replenish, at a controlled rate, that which has been removed. If too big a vent is used, the air will flow back into the float bowl as fast as it is being evacuated, cancelling out the effect we are trying to achieve. On the other hand if it's too small, then the vacuum created in the float bowl will be sufficient to stop the fuel going into the engine with obvious results. This problem is circumvented by having the second tap on the vent side to adjust the amount of air flowing back into the float bowl.

By means of these two taps, it is possible to control the way in which the mixture becomes weakened by the device over a fairly wide range of operating conditions. Because no vacuum exists in the intake manifold under full throttle conditions, the effect of this device, like the SU one, is cut out when maximum power is required, so the mixture returns to a full power rich mixture.

However, if your engine is a relatively powerful one, using a relatively restricted carburation setup, the engine may pull fuel from the float chamber fast enough to overtake the air supply back into the float chamber via the tap to atmosphere. This is tap (2) in the system. If this is the case, you will have to compensate for this effect by one of two means, either increase the richness of the mixture by having a slightly slimmer needle at the top end, or sacrifice some of the potential weakening effect by opening the 'to atmosphere' vacuum tap slightly wider.

The situation I have just described, however, is not likely to present a problem in most cases. For the purposes of setting up this device, it's handy to mount the 'to atmosphere' tap inside the car via a fairly long length of hose. Again the other end of it must go back into the air filter case. Once the system is set up, it is best to store the whole thing in the engine compartment because it's not a good idea to have anything connected to the fuel system leading into the car; it's a potential fire hazard even

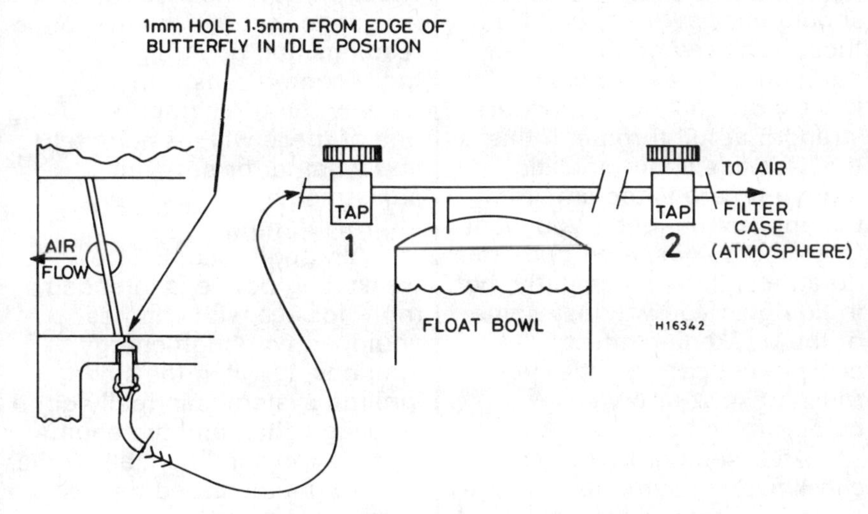

Fig. 7.24 Part throttle weakening device

though, in this case, it's a pretty remote hazard only.

One point I should mention here, is that it's necessary to make sure that the float bowl is sealed to atmosphere in all other respects. **In other words the number 2 tap from the filter case to the float chamber should be the only source feeding air into the float chamber.**

Setting up the device is relatively simple, though a little time consuming. The first move is to establish just how much the 'to atmosphere' tap needs to be open. To do this, first close the vacuum tap connected directly into the carb, so that no vacuum is communicated to the float bowl. Next, set the 'to atmosphere' tap fairly well open so that the float bowl is adequately vented. Now take the car out and check the third gear 30-70 or 30-65, whatever you feel comfortable with, acceleration time, flat out. To do this, it's best to start below 30 and hit the stopwatch as the speedo passes 30. Hit it again when it reaches the upper speed, but always drive on past that speed by a few mph. When you are sure times are consistent, close the tap down slightly and do the test again. Continue repeating the test and closing the tap down until you find the point at which the performance starts to deteriorate. In some instances you may find that the tap can go all the way closed without affecting the performance, in which case open the tap about two full turns. If you find that performance starts to deteriorate, open the tap slightly to restore the engine's performance.

Now turn your attention to the vacuum tap (1). Open this very slightly and check that the engine picks up properly. It should do. Take the car out on the road and drive at your most used cruising speeds, and these can range, incidentally, from your in-town cruising speeds of around 30mph up to motorway cruising speeds of 70mph. If you have a vacuum gauge, note the vacuum at each speed you test at. A point here when using vacuum gauges on the road, is that unless the road is dead flat, those vacuum gauge readings won't repeat because it takes only a minimal gradient to make a substantial difference to the vacuum reading. What you are looking for at this point is a lean surge in the driving characteristics of the car. As the engine approaches the lean limit of combustion, a lean surge will develop. If you have never experienced the effects of lean surge before, you will know when it starts because of the slight hesitant way in which the car drives. If the car drives perfectly okay with just a small amount of vacuum tap opening, then open the vacuum tap a further small increment. Take the car out and try it again. Keep on opening the vacuum tap until such time as you feel the car run into a lean surge condition. When this happens, close the vacuum tap down slightly so that the lean surge just disappears. Now check your maximum performance, *i.e.* acceleration from 30mph to whatever speed you chose as a top limit. The acceleration time should not have changed from those achieved originally. If it has changed, it's because there is a slight reduction in mixture strength at full throttle. If this is the case, it's a fair indication that your carb is too small for the application, but if you want to correct things, then open the 'to atmosphere' tap slightly, but in doing so you will lose some of the weakening effect achieved at part throttle: you must strike your own compromise.

After setting the device up, check fuel consumption. Over a period of time, under consistent conditions, just to make sure things are optimized, try turning the vacuum tap slightly more open or closed to see if either way improves the economy. As a matter of interest, the best economy is usually achieved just before the engine runs into lean limit surge conditions. Incidentally, due to mixture adjustment, best economy is not achieved when the manifold vacuum is at its highest. In fact, if we start off with a mixture way too rich and plot manifold vacuum versus economy at constant speeds, it will be found that the vacuum reading will go up as the mixture is leaned out towards the best power mixture and then, as it passes the best power mixture on the lean side, it will start to come back down to a lower vacuum and best economy. The leaner the engine is able to run at part throttle, the lower the vacuum will be. On the other hand, if you throw ignition timing into the equation as well, then you will find the nearer to optimum the ignition timing is for a given mixture, the more the vacuum reading will increase.

Another factor that must be taken into account is that the leaner the mixture becomes, the slower it burns, so the ignition advance may need to be altered to suit. Obviously a quick way of determining exactly what's needed for best mpg is to ignore vacuum gauge readings and go straight to a more complex instrument, namely the trip computer. Using one of these will instantly tell you whether or not your adjustments are positively improving mpg.

Although this mixture weakening device is intended more for use with engines equipped with high energy ignitions, it will, if the standard ignition system is in really good working order, and the engine is in tip-top condition, allow the user to get increased mpg simply because the mixture can be optimized under cruise conditions

Pressure Regulator Valves

Most European carbs tend to operate at relatively low fuel pressures, 2-3 psi being usual. I mention this because many U.S. carbs operate at two or more times this pressure. The principal reason for these higher fuel pressures is that on higher powered cars, the acceleration can be great enough to counteract the pump pressure, simply because it's trying to force a column of fuel against the force produced by the acceleration of the vehicle. With smaller vehicles the acceleration is a little more limited and so fuel pump pressures need not necessarily be as high. In fact, European carbs have never been made to handle high fuel pressures. On a modified vehicle you may have the situation whereby a fuel pump with a higher pressure is needed simply because the high initial acceleration can cause a weakening of the mixture. On the other hand, if this pressure is maintained at the carb during other aspects of the engine's operating sequence, it can be too high and the mixture might become overly rich due to increasing fuel level, or even the float's inability to fully shut off the fuel. For this reason, it's a good idea to mount a pressure regulating valve in the fuel line, just ahead of the carb. By doing this we can feed the pressure regulator with fuel at a relatively high pressure. The short run of fuel line between the regulator and carb will have fuel at a more normal pressure, *i.e.* 1-2 psi. However, because this line is short, there will be little or no actual fuel pressure drop, thus getting over the problem of fuel 'back-up' in the supply line during hard acceleration.

Another factor that the pressure regulator allows is the setting of the minimum fuel pressure possible. From the point of view of economy, it's a good idea to have as low a fuel pressure as possible at the carb. By so doing, the float, which is easily influenced by bumps, does not necessarily have to suffer fuel injected into the float bowl at times when the fuel level is already adequate. My own testing of pressure regulator valves has shown that on the dyno at least, they show no gain in mpg, BUT on the dyno an engine is not subjected to bumps, so one really wouldn't expect a pressure regulator to show any extra economy. On the other hand, out on the open road, indications are that a small amount of fuel can be saved, especially on bumpy terrain.

Fuel Pumps

All the carburation in the world isn't going to do any good unless the carb has an adequate supply of fuel, and doing this is the function of the fuel pump. The standard mechanical pump used on many late Series 'A' engines does a pretty good job of supplying fuel to the engine, and can generally cope with engines of around 90-95 hp. However, an engine-mounted pump is essentially a suction pump, because fuel tanks are usually in the rear of the car. For a really high powered vehicle, one that can accelerate very rapidly, it is best to have a pump which is situated close to or even inside the tank. Of course tank immersed pumps are something of a problem for most amateurs to build if the vehicle hasn't already got one, so the next best thing is to mount a powerful electric pump close to the tank. The pump, of course, must deliver adequate fuel to the carb. If you reckon on one Imperial gallon per hour for every 10 hp the engine puts out, or one U.S. gallon per hour for every 8 hp the engine puts out, then the fuel supply you arrange should be more than adequate. This does, of course, assume that it's petrol/gasoline that the engine is running on, not alcohol, nitro-methane or any other exotic fuel.

8 Intake Manifolds Part 1

Let's start this discussion on intake manifolds on a note of caution. It is difficult to separate any part of the induction system and talk about it on its own, especially the intake manifold, as it is affected upstream by the airflow of the carb and filtration system, and downstream by the cylinder head. Additionally, the intake manifold characteristics will modify the equation. The overall view must take into account the airflow capability of the 'system' *and* how the manifold affects the quality of the mixture that arrives in the cylinder. Since an understanding of the whole will result more easily by my doing so, I will deal first with airflow capability and then mixture quality, keeping the two subjects as separate as possible.

Just how much the system's airflow capabilities contribute to the results of fitting a higher performance intake manifold to an engine is often determined by the demands placed on the manifold. this is important to understand, so let me expand in case it isn't clear. It's fairly obvious that a msmall engine is going to draw less air than a large engine. Also, an engine which peaks at low rpm, due to the cam and cylinder head design, will draw less air than an engine which has a racey cam and a head that breathes well. A manifold which may work well on an 850 may be less than a adequate on a fire breathing 1300. On the other hand, a high-flowing big port intake manifold on a fire breathing 1300 engine may be way off the ultimate for an other wise standard 850. Some manifolds are more sensitive to being matched to the engine than others, and the reasons for this will hopefully be made clear as progress is made through this chapter.

At this point, a look at the different types of manifold likely to be encountered is necessary. **The main sub-division for all intake systems hinges on whether or not both intake ports draw from a common source or an independant one.** For instance, a common source manifold would be one utilizing a single SU or 28/36 Weber. Although the 28/36 Weber has two barrels, they both dump into a common plenum chamber with both ports drawing from this. By contrast, an independant system is one whereby each port draws from an essentially independant source. A good example of this is a single 45 Weber. The left-hand port draws from the left-hand barrel, and the right-hand port from the right-hand barrel. Of course making sweeping statements is all very well, because there is always an exception to the rule. For instance, twin SUs have a balance pipe between the ports, and depending upon the type of manifold, it is possible for one port to draw from the diagonally opposite carb. However, that is something of a special case with a special manifold, and we'll deal with that separately.

Design Criteria

So as to clear up a few long outstanding myths and general misconceptions about Series 'A' induction systems, let's determine what the basic requirements are for a high performance manifold. **Of course, the number 1 asset an intake manifold must have is, as ever, high airflow capability**. Of equal importance is the fact that it must also promote good fuel distribution between the cylinders, and lastly it should have, within reason, as high a port velocity as possible, consistent with maintaining a good flow.

The first requirement, airflow, is an asset that the manifold must have. Any component which is supposed to produce additional power must, in some way or other, aid total engine airflow. If it doesn't, then unless it corrects some other malady, it won't really produce any extra hp.

If a little thought is given to the exercise, it will be quite obvious that high airflow can be achieved with large ports. However, this conflicts with a third requirement, namely that the port velocity be as high as possible. A high port velocity helps do two things, firstly it promotes better fuel suspension, and secondly it aids inertial ramming, a subject which was covered extensively in a previous chapter.

As far as fuel distribution goes, the requirement for being effective in this area is also as obvious as having good airflow. However, there is only just so much the intake manifold can do, because the Series 'A' engine, with its siamese intake port design, inherently builds in a mixture distribution problem which though often exaggerated, is difficult to overcome. Part of this problem is due to the firing order of the engine, and the pairing of the cylinders. If we stop and take a look at just one intake port, we find that two intake pulses occur, one after the other, and then there is a long idle period where the port does nothing, whilst the other two cylinders draw from the opposite port. Although the mechanism by which fuel maldistribution occurs, may be more complex than the basic description. I will outline here, it will give at least a rough idea of the events taking place.

Let's start off the sequence of events assuming that an idle period in the port has just taken place. The first cylinder of the end pair starts its induction stroke. For the sake of the analysis, we will assume that the air in the port just prior to the start of the induction stroke, is stationary, so also will be the fuel. As the air in the port starts to accelerate, due to the induction stroke, so the fuel, being heavier, will tend to get left behind to a slight degree, so the first part of the charge entering the cylinder can be slightly deficient of fuel. As the induction stroke continues, so the fuel builds up speed to match the air. At this point, the air entering the cylinder reaches a more normal fuel/air ratio. As this cylinder approaches the end of its induction stroke, so the air and fuel mixture in the port will begin to slow. However, the air/fuel mixture further up the port won't slow up as much as the air/fuel mixture immediately adjacent to the valve. This is due to the slight compression effect that is occurring at the high gas speeds in the port. Even when the intake valve is fully closed, up by the carb, there will still be air moving into the port. However, before the intake valve on the first cylinder to draw, is actually closed, the intake valve on the second cylinder of the pair will have opened and started to draw air. It will not be drawing from a source that is anything like as close to stationary as the first cylinder. Consequently, the air and fuel are travelling at more of a uniform speed, so there will be no leanness in the initial part of the charge passing into the cylinder. Consequently, throughout the second cylinder's induction stroke a more uniform fuel/air mix will be delivered. As the second cylinder gets near to the end of its stroke, so the air being drawn in will slow up, but the fuel will not slow quite as quickly, so during the last part of the stroke a slightly rich phase will occur. Consequently, the second cylinder to draw will run slightly rich by about the same proportion the first runs lean.

Many experiments have been done with the intake manifold in an attempt to cure this fuel distribution problem. At best, they have only been partially successful, but at least that's better than nothing. For what it is worth, the scatter pattern cams discussed in the cam section have done more to fix the fuel distribution problem than has ever been achieved with manifold design.

It's inevitable that when we start talking about induction systems, the discussion will get around to the question of just how much better twin carbs are than a single. The truth of the matter is, they aren't necessarily better at all, and the usual assumption that two carbs must be better than one does not really apply unless the conditions under which the two carbs are used are clearly stated. **In fact, I have to say that, for high performance use for virtually all applications, except out and out race, a single carb setup has the potential to produce a better power curve than twin carbs or a single twin choke such as a Weber or Dellorto. At first this**

may seem to be a statement that contradicts the fact that the making of extra hp requires increased airflow. In fact no such contradiction is occurring. Let's consider the engine as a suction pump which has four suction strokes per sequence of events. In other words, for each cylinder to go through the Otto four-cycle sequence, the engine will turn 720° and in that time, four induction pulses will have taken place. The key to understanding what is happening is to appreciate that the significant part of the induction pulses do not take place at the same time. Except for the small amount of sharing during a period where both valves are open in any one port, each induction stroke occurs quite independently of any other induction stroke. Now bear this point in mind because it's important if you are not to miss the next point. Let us assume now, that we have the choice of two types of carburation. It can be either a single carb, says an H.I.F.6, with 240 cfm of airflow capability, or two 1½ inch SUs, each having 120 cfm of airflow capability. On the face of it the carburation capacity is identical in each case, both being capable of 240 cfm of airflow.

Now let's view the situation from one cylinder. On a single carb setup, when one cylinder draws it does so through a 240 cfm carb. THE single carb. On a twin carb setup, when one cylinder draws, it does not draw on 240 cfm of airflow capability, it draws on the 120 cfm of one carb. **In other words, although the engine may be equipped with two carbs, any one cylinder will only use half of that capability to supply it with air. Unless there is carb sharing amongst the ports, twin carbs offer little or no advantage over a single carb.** Of course this makes the assumption that the losses in the intake manifold which obviously have to follow a more tortuous route on a single carb setup, are no more than the losses on a twin carb setup.

Another factor in favour of the single carb setup is that a fair degree of inertial ramming can be used to boost the power. On a twin carb setup, this inertial ramming is largely lost. So in theory, the single carb, IF it is on a good manifold, has much to commend it by comparison with a conventional twin carb, or single two-barrel carb setup, and in this particular instance the theory is well borne out in practice. Dyno testing shows that if the induction system flow capacity, when checked through one port of a single carb setup, is comparable to that when checked through one port of a twin carb setup, then the power potential of the single carb setup, tends to be superior to the twin. The experiments leading to these conclusions were carried out under controlled conditions. Unfortunately few people have dynos, flow benches, etc., to test manifolds to establish the value. It is this very factor which lends another point in favour of the single carb setup, especially if we are talking about single SUs versus twin. **At the time of writing, there were only three good twin SU manifolds, these being the Janspeed item, for the metro, the Steve Harris/ Manflow tubular steel item and the Vizard Schattily VSI discussed in Part 3 of this chapter.** Almost all others are of very doubtful benefit. If we add to this the fact that twin carbs usually result in a higher insurance premium, we can see that the single carb, although it doesn't appear as powerful, is in real-world terms, more of an asset. An asset on the pocket, economy and performance and, for that very reason, we will now delve into single carb manifolds in depth.

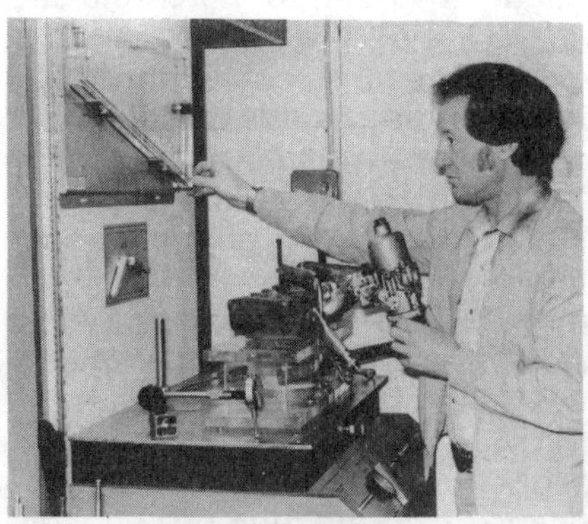

Here I am, looking more studious than usual and checking the flow potential of an MG Metro induction system on a Superflow SF 300 flow bench.

Single SU Intake Manifold

There are numerous single intake SU manifolds available to the performance engine builder. Several designs are produced by Leyland as well as a substantial number by independant speed equipment manufacturers. We can generally lump the factory single intake manifolds into four groups, the first being the single 1¼ inch intake manifold which is cast in one piece integral with the exhaust. This is generally used on manual transmission 850 engines in Minis or single carb 948 inline engines, and many of the early 1100 engines and 1000 Mini engines, again of the early type. This manifold was very similar to the single 1½ inch SU manifold introduced on Mini automatics and later used on all transverse engines. These single 1½ inch SU manifolds came to dominance from the early 1970s onwards. Again the intake is integral with the exhaust.

Although the 1½ inch SU manifold gave slightly more power than the 1¼ inch, and the operative words are

'slightly more power' the basic design of the manifold was, to an extent, self-defeating. With this type of intake, heat from the exhaust is conducted through to the intake in large quantities so as to vapourize the fuel. The idea behind this was to cut the time the choke was required, so as to improve mileage. Incidentally, this heating of the fuel to vapourize it also minimizes the cylinder-to-cylinder fuel distribution problem, discussed earlier. The real snag to heating the intake arises due to the dramatic air density drop produced. The obvious result is a cylinder which draws in a less dense charge. **Because it contains less oxygen, less fuel is burned and this in turn leads to less hp by virtue of a substantial reduction in torque.** Remember, if we increase or decrease the hp at any particular rpm, it's because the torque has either increased or decreased.

Eliminating the hot spot in the intake system allows the engine to produce quite a bit more power from the standard cast iron manifold. However, elimination is not as easy as it would at first appear. It is difficult to actually separate the intake from the exhaust without breaking through into one or other of the passageways. A better route to follow is to start off with two castings and saw the intake off one and the exhaust off the other so that you are left with a complete casting for each manifold. It's then a question of mounting these on the cylinder head and making sure they clear each other at the point at which they used to join. The graph, **Fig. 8.1**, shows the difference in power output when the hot spot is removed. At this point there has been no change in the airflow in terms of cubic feet per minute through the manifold, the only change is the density of the air arriving at the cylinders.

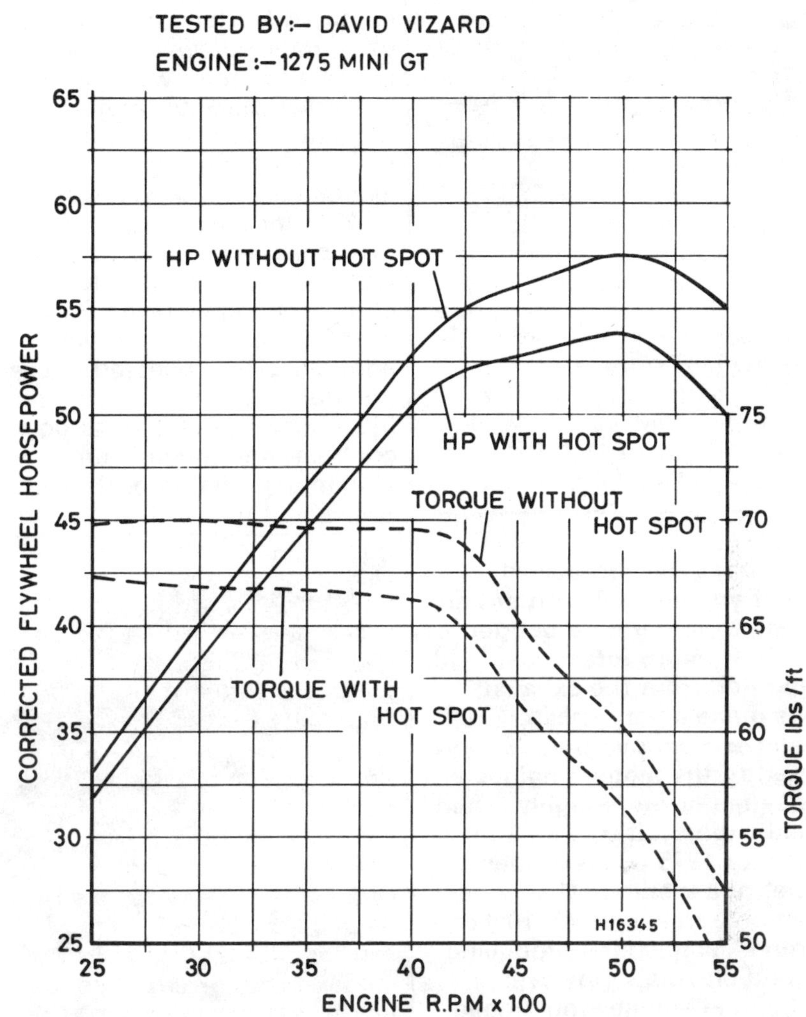

Fig. 8.1 Comparison of standard $1\frac{1}{2}$ in SU cast iron intake manifold with and without hotspot

If you are picking and choosing from standard manifolds, then the manifold on the 1300 Metro, which takes an H.I.F.6 SU, has better airflow capability than those fitted to the previous 1275 engines or indeed, any of the other engines in the range. With its H.I.F.6 carb, the cast iron Metro manifold flows as well, if not better than some after-market intake manifolds with either 1½ inch or 1¾ inch carb. However, it still suffers from the fact that it has a hot spot. Removing that hot spot by utilizing the technique just described can be expected to produce at least a similar gain in power and torque to that shown in **Fig. 8.1**.

In terms of outright airflow capability, the Metro manifold is pretty good. Also, it's very compact by comparison with most of the after-market intake manifolds. However, it is not without a snag or two . . . The high airflow on the Metro intake manifold has been achieved largely by the use of big ports: this means that the port velocity is down and this in turn results in slightly less low and mid-range torque than some of the more efficient, but smaller ported after-market intake manifolds. Such items may flow about the same but do so with

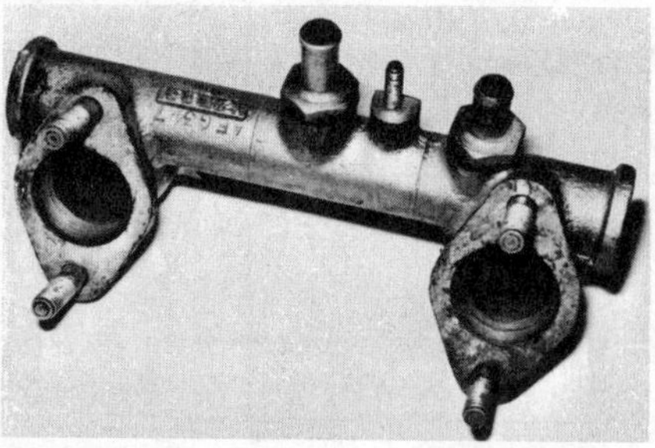

Although it may seem difficult to design a badly flowing twin carb intake manifold for the Series 'A' engine, it has been done, and here it is, the standard Leyland item.

a much higher port velocity. **Still, if you are trying to put an engine together on a budget, a cast iron Metro intake manifold, separated from the exhaust, does make a very cheap, effective manifold, but it's more suited to the bigger engines in the range or more highly tuned small engines.** It will work on a standard 1275 GT as it does on a Metro. It would probably be okay on a mildly tuned 1098cc engine also, *i.e.* one producing around 60-70 hp. However, on an 850 the engine would need to be modified quite extensively and be capable of turning up to about 7000 rpm before the manifold started coming into its own. Even so, it's still not capable of the power that the best after-market manifolds are.

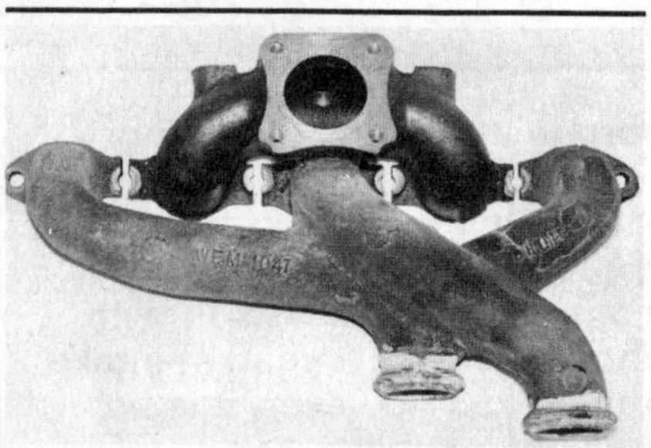

By using an HIF 6 carb and a re-designed intake manifold, this setup for the standard Metro definitely moved up a notch or two on the performance scale. This setup flows well but suffers from excessive heat input into the airstream due to the hot spot.

When Leyland introduced the MG Metro in normally aspirated form, they were looking for more easy hp from the Series 'A' engine, and they went the route that performance enthusiasts have taken for the past 20 years, namely they adopted a water-heated intake manifold. The manifold on the MG Metro is, in terms of port configuration at least, an aluminium version of the original cast iron Metro manifold. The fact the manifold is water-heated, does cut the power, but by a very small amount in comparison with the older exhaust-heated manifold. Depending on the water jacket temperature, the drop in power may amount to anything from zero to about 1½ to 2 hp. Again, the MG Metro manifold does suffer from excessively large ports, but it should be pointed out that it's difficult to get a manifold to flow this well within the confines that this particular unit dictates. No doubt if the original designers had been allowed, say, another inch of space from the manifold face to the cylinder head face, they would have managed to produce a more efficient item.

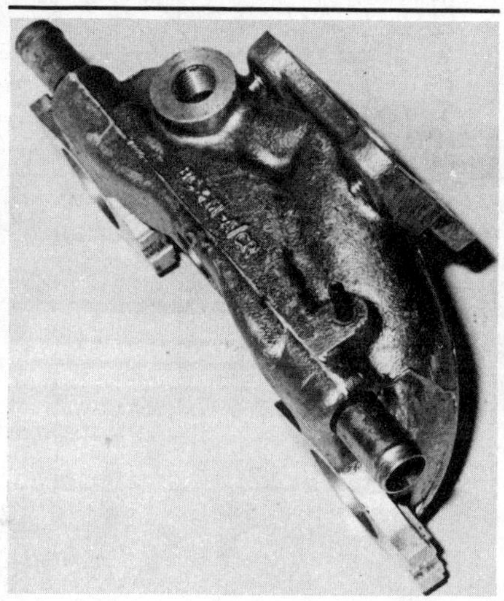

This is the MG Metro intake manifold and it is, without a doubt, one of the best intake manifolds to have been produced by Leyland.

Although the MG manifold is not perfect, it does, without a doubt, work well, especially on the bigger engines in the Series 'A' range, and for overbored 1275s it is very much a practical choice, especially when you consider that it's between ½ and 2/3rds the cost of the manifolds produced by the speed equipment industry.

After Market Manifolds

Maybe it's in us all to be naïve to a greater or lesser degree, but it's often assumed by many enthusiasts, especially, and understandably so by those newer to the game, that after market proprietary performance parts must be better than the original, factory part (otherwise why would they be made?) Unfortunately, this could not be further from the truth. The sad

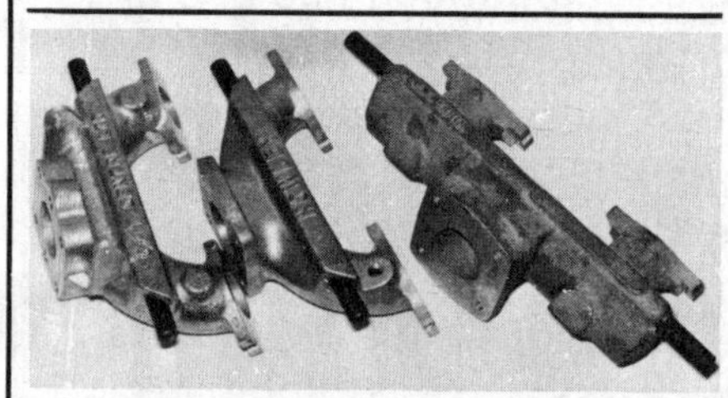

Here we see what almost amounts to the good, bad and the ugly, except that we have two good, and one ugly. The two manifolds on the left flowed well and would produce good hp figures. The log manifold on the right flowed less than the standard intake manifold and so a performance item it was not!

thing is that there are many intake manifolds on the market which only produce more power by virtue of the fact they do not have an exhaust heated hot spot. In other words, their only virtue is that they are running cooler than the typical standard factory manifold. **In terms of airflow, they can be considerably worse, and if dynoed against a separated factory cast iron manifold, will produce less power.**

The only thing that has kept some of these manifolds on the production line is the fact that few people realize there is so much power to be gained by eliminating the hot spot. Apart from that, the average enthusiast doesn't have a dyno to check these things out, so it does make it harder to find out what does and doesn't work. In this particular part of the chapter, I intend looking at the majority of manifolds that were available to the general public as of the early 1980s. The analysis of these manifolds should allow you to determine which manifold is likely to work on your engine and which will not. It should also put you in a situation of being able to avoid buying intrinsically bad designs.

At this stage I would like to demonstrate the point of having good airflow potential and high port velocity. To do so I shall lean heavily on the data taken from the flow bench and dyno. All the manifolds tested fell into three groups: long conventional, medium conventional or log. **Fig. 8.2** is a superimposed drawing of each type and clearly shows the port routing differences that exist between these designs. Although there are many different brands of the medium conventional type of manifold, the overall length and port routing between them only differs marginally. The biggest variance is in the port diameter.

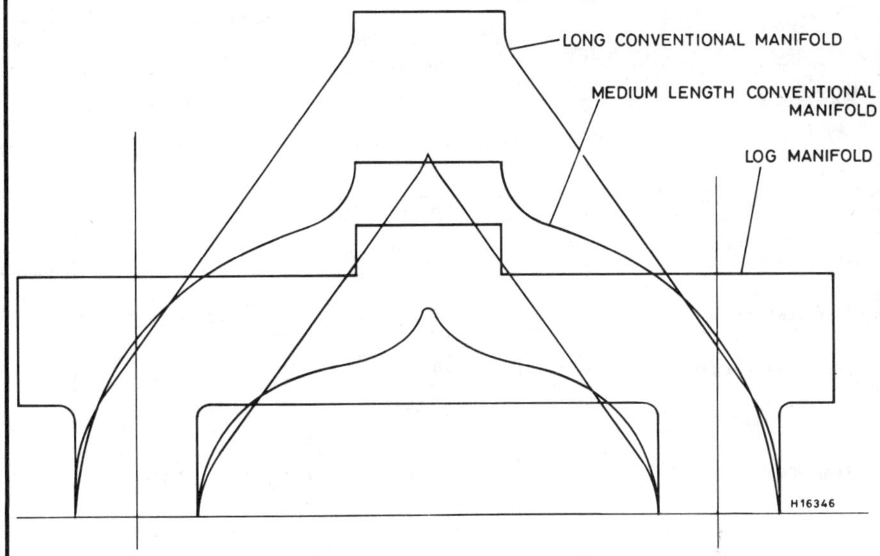

Fig. 8.2 Manifold shape and flow efficiency

Manifold shape has a distinct effect on flow efficiency. In almost all cases, the straighter the 'shot' to the port from the carburettor, the better the efficiency.

Now for some serious flow testing. When investigating an induction system, we need to consider not only the efficiency of the component in question, but also how effective it may be when used in conjunction with associated components such as carbs, etc. For this reason, flow testing was done in several stages. At this point it's worthwhile referring to the charts that comprise **Fig. 8.3.** Section A shows the manifolds in question, together with the sizes of carb they typically take. I say 'typically take' because it's worthwhile making a point here. If a manifold is made, say, for a 1¾ inch SU and you only have a 1½ inch SU, then so long as it actually fits, there is no real disadvantage to utilizing such a manifold with a smaller carb. It is not necessary to have a total match between the carb and intake manifold. So long as the carb is smaller than the hole in the intake manifold there is little flow loss. In other words, if you are hell-bent on using a small carb because you happen to have one, the fact that the manifold was for a bigger carb is of only minor consequence. **The flow bench shows the step formed this way (from small to big) only makes minimal reductions in airflow, most of which rarely show up on the dyno as a loss of hp.**

In Section B we have the airflow of the bare manifold. This is the airflow capability of one port of the manifold tested without cylinder head or the presence of the rest of the induction system. To eliminate the edge effect at the entry into the manifold, a radiused entry piece was used, having approximately ⅜ inch radius. This was done because it completely eliminates the effect of having a sharp edge or a slightly chamfered edge at this point: a single factor that can greatly affect the flow. For instance, having a manifold freshly machined and flowing it to establish its flow rating, and then taking a needle file and putting a 10 thousandths radius on that sharper edge, can make as much as 10 cfm difference to the flow. This is why it's

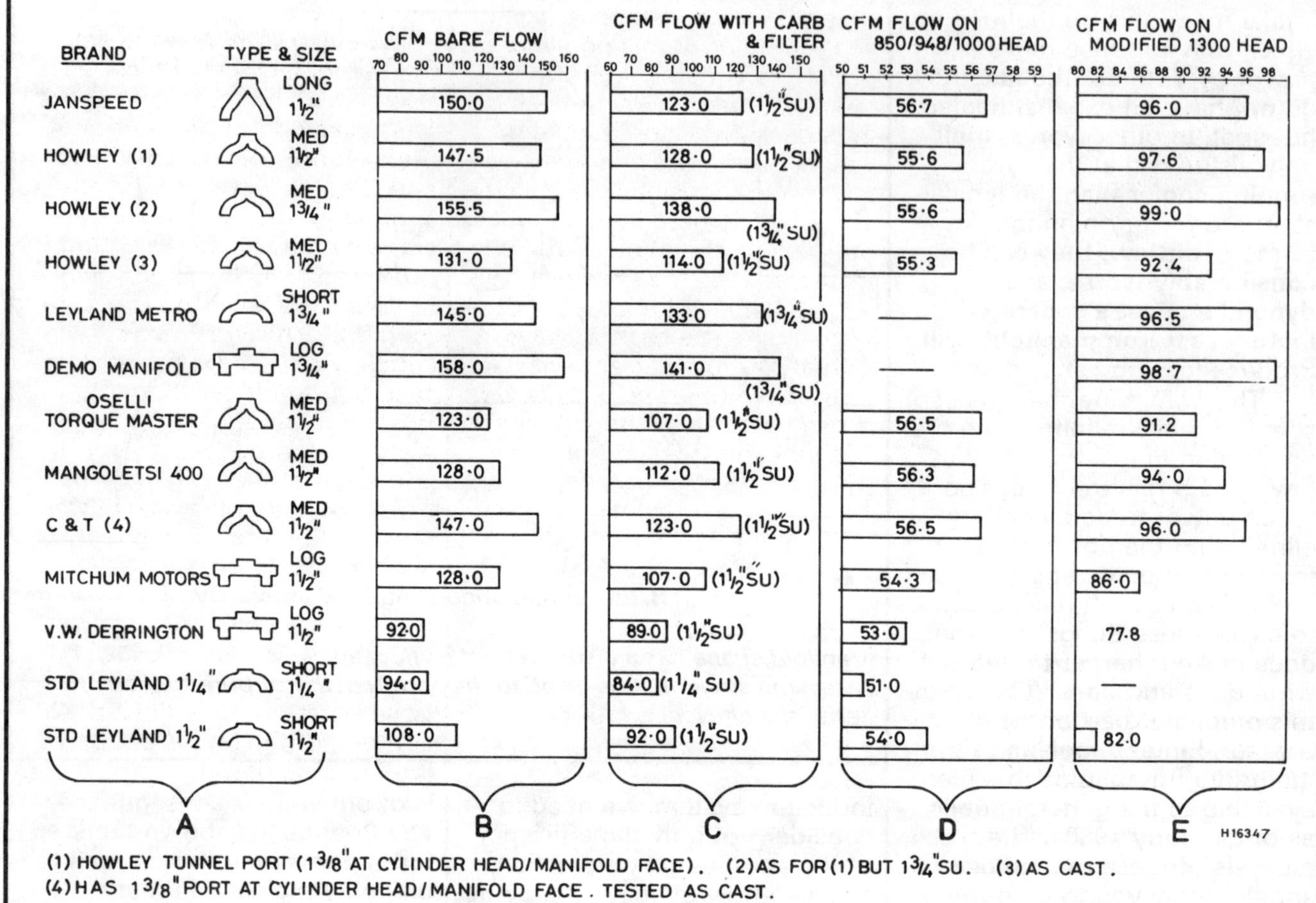

Fig. 8.3 Intake manifold flow comparisons, using a single SU

important to get rid of the edge effect with a radius. Once the radius is over about ¼ inch, any increase has virtually zero affect on increase in flow.

Well, so much for that digression. If you look at the effective flow of each of the manifolds, you will clearly see which has the most flow potential. However, let's not run away with the idea that it's quite as clear cut as all that. As previously explained, it's necessary to look at port velocity as well as the flow capability to get an assessment of a manifold's potential. If the manifold runners were a constant size all the way through, it would be easy enough to quote an overall velocity figure for it. However, this is not necessarily the case, so an assessment of port size has to be judged on what the average port size appears to be. This really is a subjective assessment rather than objective. It's because there is a variation in port diameter as the port runner traverses its length through the manifold, that makes this a difficult factor to assess, but having looked at enough manifolds in my time, a subjective assessment will suffice at this stage.

Let's consider the manifold at the top of the list, the **Janspeed** tubular steel intake manifold. This flowed 150 cfm and it did so by having efficient ports: the port size was not particularly large, so a high port velocity was produced.

Next on the list is the **Howley** manifold for a 1½ inch SU. This flowed 147.5 cfm, close to the results produced by the Janspeed manifold. However, except for the area of the port up against the cylinder head manifold face, this manifold had noticeably larger ports than the Janspeed one, but still not overly large. Though this manifold was shorter, and its turns more tortuous, it did flow almost as well as the Janspeed manifold.

The next Howley manifold was for a 1¾ inch carb. This had the same 1⅜ inch diameter at the manifold face as the first Howley manifold, but the plenum volume is larger, as is the opening in the manifold face as it is for use with a 1¾ inch SU.

Next on the list is an as-cast Howley manifold. Generally this

Shown here, we have just about the best power producing, single 1½ inch intake manifold for the smaller Series 'A' engine. This item, made by Janspeed, was originally designed for 850 Mini racing.

manifold is about 1/16 inch smaller in bore size all the way through, so it lost around 10% of potential flow area, and actual flow was down by about 12.5% so the cast finish which, incidentally, had a few flow disturbing ridges in the side, was only costing some 2.5% flow over the ported, bigger bore Howley manifolds. However we'll study some significance in this in a moment.

The next manifold on the list is the Metro intake manifold and these figures apply either to the cast iron manifold or the water-heated alloy MG Metro manifold. The first thing one is likely to notice about this manifold is it's much shorter length by comparison with most speed equipment manifolds. It also uses a 1¾ inch SU rather than a 1½ inch SU. As can be seen from the flow numbers, it delivers a very respectable 145 cfm. However, due to the way the machining is done at the manifold face, I suspect that number could vary by plus or minus 5 cfm from one manifold to another. Much of this high flow was achieved by virtue of largish ports throughout the manifold, so port velocities were only moderate.

Next on the list is a demonstration manifold. You'll see the point of this one later on in this chapter, when we come to do some dyno testing. This demonstration manifold was a log manifold of which the relevant dimensions were as in **Fig. 8.4**. The bare flow on this manifold is the highest of those tested at 158 cfm. However, this flow is achieved more by virtue of having large passageways rather than by efficient ones. As a result, the port velocity is way down by comparison with some of the lower flowing but smaller port manifolds.

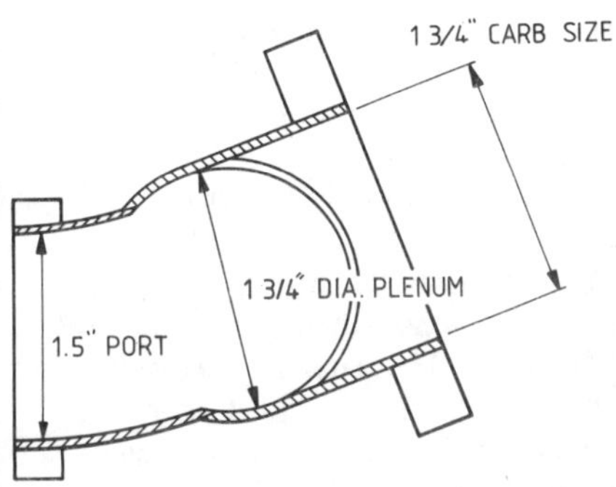

Fig. 8.4 Section through 'demo' manifold

Next down the list is the **Oselli** Torquemaster. As tested, this manifold was in the as-cast condition. As such, its bare flow was 123 cfm. However, in the manufacturer's instructions, it is advised that if this manifold is used on a 1275 engine, the port should be opened up to 1.375 inch diameter at the manifold face and this should be tapered back to the original casting about 1 inch in from the manifold face. Of course if you are buying the manifold from Oselli and you know you are going to put it onto a big motor, then it will be no problem for them to do this work for a small additional charge over the basic manifold price. Anyway, as tested, this manifold was in a configuration for a small bore, small port engine, such as an 850 or 1000.

Moving on down the list, the **Mangoletsi** 400 which, as tested, had a light clean up on the intake end of the port and the exit into the cylinder head end, was basically a relatively small bore manifold about the same as the Howley one, and it produced 128 cfm of airflow.

Next we have the **C & T** manifold. This was most certainly a big bore manifold as cast, but no bigger than the Howley manifold for the single 1½ inch. It produced a very good, out-of-the-box 147 cfm. Again, equating the size of the ports to the flow shows that the port velocity was potentially quite high.

The **Mitcham Motors** 1½ inch manifold is the first of our *proprietary* log manifolds. It had large ports but only flowed 128 cfm due to the inefficiencies of the log-type design.

The **V.W. Derrington** manifold is also a log manifold, but it had small ports and flowed only 92 cfm, the lowest flowing manifold of all the ones tested, including the standard ones.

We move on now to the standard manifolds: the 1¼ inch SU manifold fitted to 850s, 948 inline engines, 998s and many 1100 engines. This manifold flowed 94 cfm, but again, looking through lots of manifolds, it can be seen that the alignment of the casting and the surface finish inside varied quite a bit, so that 94 could easily be plus or minus 5 cfm.

Last on the list is the standard manifold for 1275 engines, as well as those engines utilizing the single 1½ inch SU, *i.e.* Mini automatic, later Mini 1000, etc. This particular manifold flowed 114 cfm, but again a small variation between individual manifolds should be expected. Both the standard cast iron intake manifolds had relatively small ports, which , in the main, contributed to their low flow capability, as did the rather abrupt bends in the port routing.

Moving onto Section C of **Fig. 8.3**, we have flow bench results after adding a carb and high flow K & N filter to the manifold. Essentially the numbers speak for themselves, but there are a few areas that at first sight appear to be anomalies and so could well do with expanding upon. Firstly, the Janspeed manifold. When fitted with a 1½ inch SU, it flowed 123 cfm, but the Howley manifold, which flowed less on its own, in combination with its SU, flowed more than the Janspeed item. Basically, the cause is a flow pattern interference. With an HS4 carb on the Janspeed manifold, the butterfly, in the fully open position, is placed well into the plenum area of the manifold. That is the small chamber in which the two runners join. Essentially the plenum area of the Janspeed manifold is too small but this can easily be rectified by installing a spacer between the carb and manifold. If a 1 inch spacer is added to the Janspeed setup, then the flow typically increases to around 128–129 cfm. The addition of a 1 inch space proved equally effective on a variety of sizes of series 'A' engines The only drawback is that in some installations, space may be a problem when an extra inch is added.

Moving onto the Howley intake manifold designed for a 1¾ inch carb, we find that when fitted with a single 1½ inch, it flowed an identical amount to the manifold matched to 1½ inch. This is in spite of the fact that the bare manifold flowed some 8 cfm more without the carb. On the other hand, when a 1¾ inch HIF6 carb was added to this manifold, the flow figures went to 138 cfm.

The next manifold of importance on our chart is the demo manifold. With an HIF6 on, this manifold flowed 141 cfm and was notable in as much as it was the highest flowing manifold in this configuration.

When carbs are added to intake manifolds, we find that the higher flowing manifolds, in the main, lose around 25 cfm of flow. When we get down to very low flow manifolds, like the one flowing 92 cfm bare, we find that adding a carb only drops it another 3 cfm. With a high flowing manifold, the carburation provides quite a restriction. With a manifold which is restrictive in its own right, adding the extra restrictiveness of the carb does little to affect the overall flow capacity.

The lowest flowing combination on the list is, of course, the standard cast iron manifold which utilizes the 1¼ inch SU. The primary reason for this is due to the fact that the 1¼ inch SU flows substantially less air than a 1½ inch.

The real truth as to how effective a manifold is, or isn't, is shown up in sections D and E. Dealing with section D first, we see the results of flow tests of the intake system from section C mounted on a virtually standard 850/1000 Mini cylinder head. Again, I will deal only with what may appear to be an anomaly. Firstly, although the Janspeed manifold wasn't the highest flowing setup when tried without the cylinder head, with the cylinder head, it does prove to be (by a small margin) the best at 56.7 cfm, thus establishing that the way the air was routed into the port, *i.e.* taking a long curve to direct the air into the port, is beneficial. If you run down all the rest of the numbers, you'll find that the results virtually speak for themselves. In many instances, manifolds which did not particularly shine when flowed on their own, proved to be among the most effective when they are mounted on a cylinder head. Also, another point to note, is that the spread between the best and worst was narrowed once mounted on the head. **Again, this is an indication that the main problem with flowing air through an induction system does not lie in the induction system itself but in the cylinder head, and this is one of the reasons why the cylinder head chapter in this book is very extensive.** Anyway, I digress, for the next series of tests, induction systems were mounted on a modified 1300 head. The particular head was a **Swaymar** 'Rally Comp' head which has proved effective both on the flow bench and on the dyno. With this dramatic change in head flow capability, the order of overall flow capability changes. The Janspeed manifold which was clearly in the number 1 spot on the 850 Mini cylinder head has dropped back a number of places. Why? Principally because the ports are too small for a big engine because such an engine places a greater air demand on the manifold. In actual fact it is probably better to look at the situation in terms of power output rather than engine size. The Janspeed manifold which, incidentally, is definitely the best on 850s and 1000s in terms of power output, loses out to some of the other higher flowing setups when installed on engines with the potential to produce more than about 75 hp.

Running down the list, we find that the Howley tunnel port manifold clearly has the edge over the rest, with the exception of an Oselli Torquemaster, modified as per Oselli's instructions.

Although it wasn't done for these tests, past experience shows that the Oselli Torquemaster, opened up, flows very comparable figures. When used with an HIF6, the 1¾ inch

Howley manifold showed 99 cfm and wrested first spot on our chart. The second place went to the demo manifold which flowed 98.7 on this cylinder head. Again, it should be pointed out that it's doing so with low port velocity as opposed to the much higher port velocities of the Howley manifold.

Running down the list, we find that the log manifold, other than the demo one, did not fare very well, turning in 86 cfm and 77.8 cfm respectively. The lesser of these two actually produced less flow than the standard 1½ inch cast iron manifold, and both of them producing less than the standard 1¾ inch Metro manifold.

At this point we now have an extensive number of flow tests under our belt, and I suppose it's valid to raise the question as to whether we should draw any conclusions. Well, the answer to that is 'yes' if we were all driving flow benches. At this point, it's really necessary to see how the flow bench results equate in terms of the power output the engine can produce, and essentially this is what we'll delve into next.

Intake Manifolds Part 2

Dyno Time

We have already looked at the effect that removing the intake manifold hot spot has on hp. Now is an appropriate time to see what happens when we substitute a higher performance intake manifold for a standard one. The results of doing just that are shown in **Fig. 8.5**. Here, an almost standard 1275 engine is used as a test base. The intake manifold, an Oselli Torquemaster, was opened up at the manifold face, to suit the cylinder head as per instructions, and, after a run through with coarse emery mop to clean out the ports, it flowed 145 cfm.

As you can see, this manifold is worth about 5 hp over the standard one, but if you relate back to **Fig. 8.1**, you will note that separating the standard intake manifold from the standard exhaust, resulted in an increase of around 2½ hp. From this we can reasonably safely deduce that at this particular power level, half the power increase is due to the fact that the after- market intake manifolds run much cooler.

At this point it might well be worth mentioning another aspect of this carburation installation, since we are looking at carburation in cfm and relating it back to the dyno. It seems worthwhile raising the question of how much power this engine would have made, if it had been equipped with an extremely high flowing induction system (one that had far more flow capability than the engine could reasonably use)? Assuming no untoward problems occurred with mixture distribution, atomization, etc., how much hp would the engine have made? Well, experience rather than back-to-back testing indicates that at this power level there would be very little to gain by adding say, a 45 Weber or Dellorto carb in place of the single SU. The reason being, the SU is capable, at this power level, of supplying the engine with the air it needs with very little loss due to flow restrictions. **The question of 'adequate carburation' for an engine's needs is important, so it's analysed in some depth in a summary towards the end of the chapter.**

Let's move onto the freest flowing manifolds under discussion to see what happens

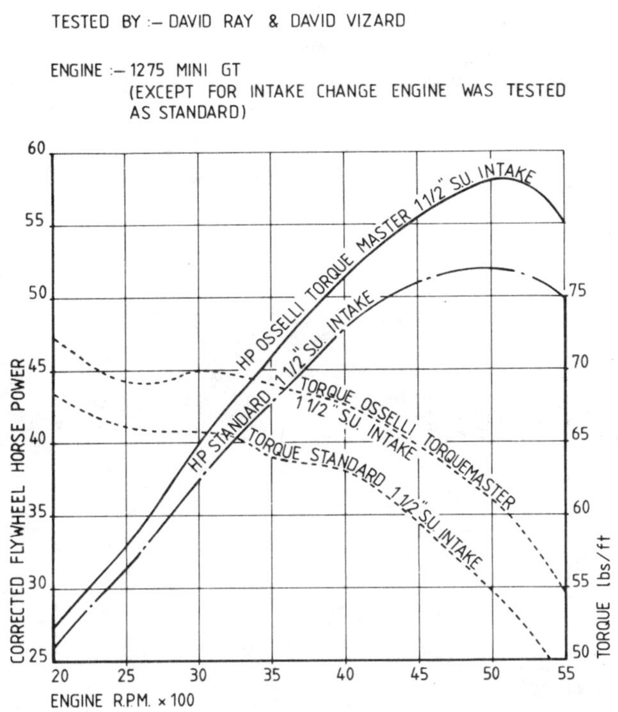

Fig. 8.5 Comparison of Oselli Torque Master water-heated intake manifold and standard factory exhaust-heated intake manifold

A standard exhaust manifold was used for both tests.

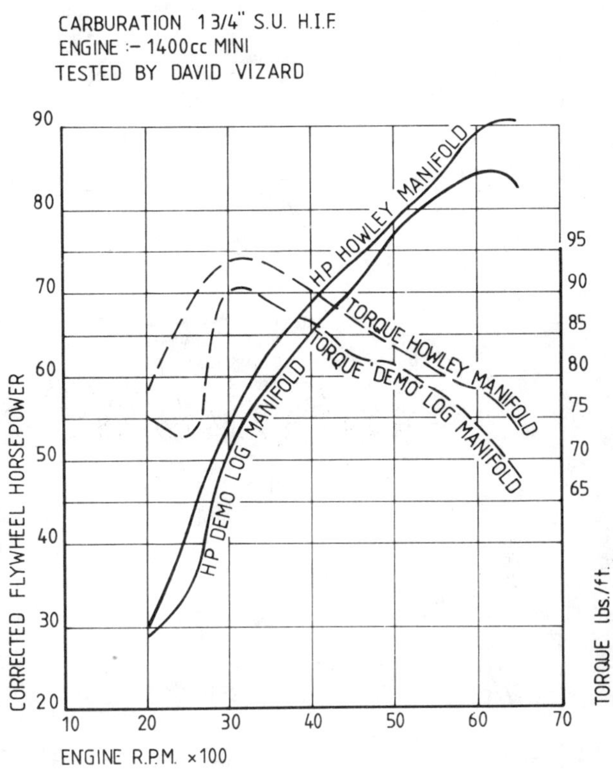

Fig. 8.6a Comparison of demo log manifold and Howley medium length conventional manifold

The important factor to note from the figures presented by these graphs, is that there is more to manifold design than just simple airflow capability. The two manifolds flowed similar quantities of air but the port velocity in the Howley manifold was substantially higher. Extra power has probably resulted due to better inertial ramming of the cylinders or better fuel suspension or a combination of both. (See Fig. 8.3 for flow figures of these two manifolds).

when they are used on a real fire-breathing Series 'A' engine. **At this stage I think it's well worthwhile to demonstrate the importance of manifold efficiency rather than just total flow, and to do this I intend to compare the power curves of the two highest flowing manifolds on the list.** One is the 'demo' manifold, which is, in fact, the highest flowing one tested of all those in **Fig. 8.3**, and the second one is the Howley manifold, utilizing the 1¾ inch HIF6 SU. The graph, **Fig. 8.6a**, gives you the lowdown on exactly what happened. **Note that the Howley manifold, despite flowing slightly less, produces more torque and more hp than the 'demo' log manifold.** Since operating temperatures of both manifolds are the same, it can only be assumed the difference lies in the fact that the Howley manifold has a much higher port velocity, thus producing better fuel suspension or better inertial ramming, or a combination of both.

To Date We Have Learned ...

After perusing the results of the manifold testing, the most obvious conclusion we can arrive at is that both airflow and port velocity are important. Therefore it's necessary to select our manifolds carefully, not only in relation to the bare flow and velocity performance but also how well they perform on the intended engine.

The first rule of thumb is to choose the manifold with the highest flow consistent with good port velocity. The next to choose a manifold that has the

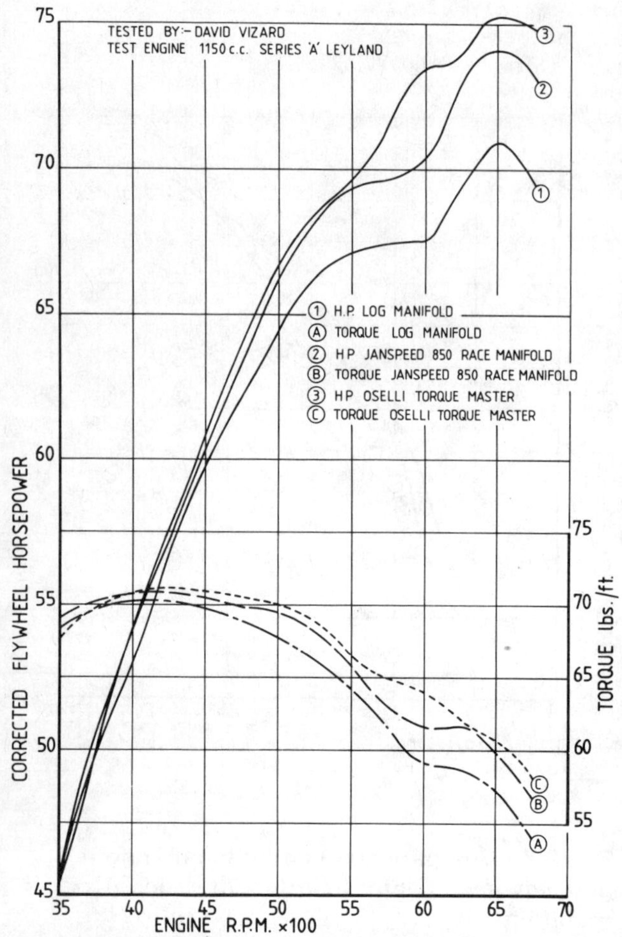

Fig. 8.6b Comparison of Janspeed 850 race manifold, Oselli Torque Master and Mitchum Motors log manifold, each fitted with a 1½ in HS4 SU

The curves here demonstrate that for a reasonably well-modified 1150 cc motor, the larger port diameter of the Torque Master paid off. The ports of the Torque Master were rough polished, and the port size on average was around 1¼ in diameter. The inadequate flow of the log manifold placed it last in line as a power producer. On a smaller motor (see Fig. 8.6c), the effect of a smaller manifold port can be seen.

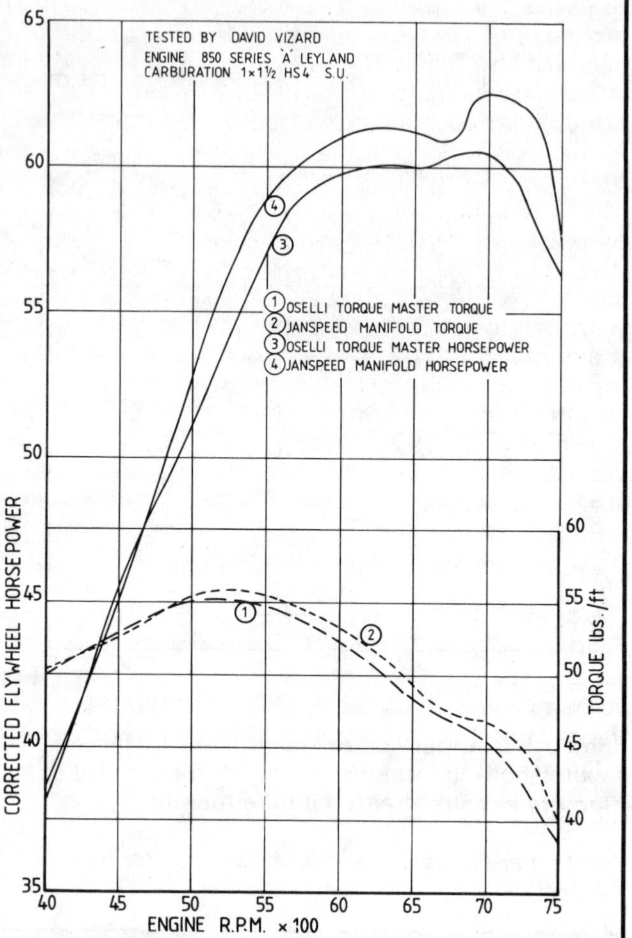

Fig. 8.6c Comparison of Janspeed race manifold and Oselli Torque Master

This test was conducted on an 850 engine in a fairly high state of tune. The power and torque improvements shown here are more than likely due to better inertial filling of the cylinder. However, some of the improvements may be due to slightly better mixture distribution, as the Janspeed manifold approached the cylinder head port at less of an angle.

right characteristics in relation to the intended engine. For near-standard engines in the 850, 1000 and maybe 1100 range, we need to think in terms of high efficiency small bore manifolds, otherwise the engine will lose potential inertial ramming. From this we can conclude that generally speaking manifolds with 1³⁄16 inch to 1¼ inch port size at and near to the cylinder head face, will get the job done. At the other end of the scale, we find that engines in the 1400cc+ range equipped with big valve heads, high performance cams, etc. are likely to need a manifold with all the flow capability possible, again consistent with good port velocity. This generally means ports 1⅜ inch diameter, and if you are prepared to spend the time working on the manifold, maybe a bit larger yet. However, much over 1⅜ inch diameter and you will find that either the carb or the port in the cylinder head itself will become the prime restriction. **Manifolds to avoid are log manifolds of any sort, as these simply do not produce the results.**

As far as factory manifolds are concerned, the only design really worth using is the Metro manifold. An ordinary cast iron Metro one, separated from its exhaust, will perform just as well as the MG Metro aluminium item. However, it's worthwhile to remember that these manifolds were intended for use on 1275cc engines. By comparison with many of the aftermarket manifolds, their port velocity is a little low on the smaller displacement engines, and they will hardly come into their own unless the engine concerned is tuned to a relatively high output, say around 60 hp. This means, if it's an 850, it's pretty much a screamer and even if it's 1000, then it's still got to be tuned to a fairly advanced spec to produce a genuine roadable 60 hp. Of course with 1100 engines, that 60 hp is much more achievable by virtue of the extra capacity.

Getting The Most From A Single SU Intake

With possibly the exception of the Janspeed tubular steel manifold, all the manifolds we have so far examined have the potential to be improved to a greater or lesser degree. From that statement, you mustn't conclude that the Janspeed manifold is perfect, because it isn't. What it is, is a purpose-designed intake manifold which, when used for its designed purpose, is excellent. As a power-producer on small displacement engines, (850s, 948s, 1000s) it is virtually unbeatable but due to lack of any heating, it will ice up in the winter. On big engines it's starting to lose out on flow capacity and it really doesn't have enough material thickness, due to its form of construction, to be modified. Therefore as far as this particular item is concerned, we can say use it as intended, or turn your attention to alternative manifolds of cast construction. Virtually all the other manifolds in contention are water-heated. In warm weather, there is no need for the water-heating. If the water flow to the manifold is shut off or disconnected, you can expect your engine to produce another 1 hp, maybe 2 if it's a really highly tuned, big displacement engine around the 100 hp mark. It's principally on such engines that the problem of providing enough air for it to produce its true potential is found. It is on big, tuned engines that the need to optimize the airflow of the intake manifold becomes of paramount importance. Fortunately there are a few modifications possible which effectively improve the manifold's efficiency.

Although many of the intake manifolds on the market follow the same sort of port routing they all, to a degree, have been designed around the simplest form: namely that the ports are round where possible and the only real complex area is where the two ports join the common plenum behind the carb. If you are prepared to spend some time with a cutter (not a grinder) then it is possible to make some small improvements on many of the intake manifolds available. The easiest to work on are the factory-produced HIF 1¾ inch manifolds (standard Metro or MG Metro) which are either cast iron, and can be ground with a grindstone, or aluminium. Basically these manifolds benefit from a port tidying up job. They seem very susceptible to varying production tolerances, and sometimes the machining forms lips within the port itself, especially on the short side radius adjoining the cylinder head manifold face. Blending the manifold port at this point will often help to the tune of 3 or 4 cfm airflow for the bare manifold.

Another problem area is the short side radius turn just behind the carb flange. This situation can also be improved by reshaping. If just the

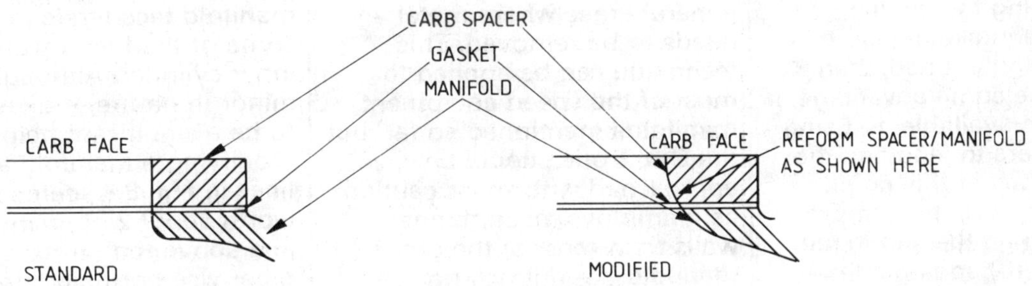

Fig. 8.7 Modification to a standard BL 1275 Metro intake manifold

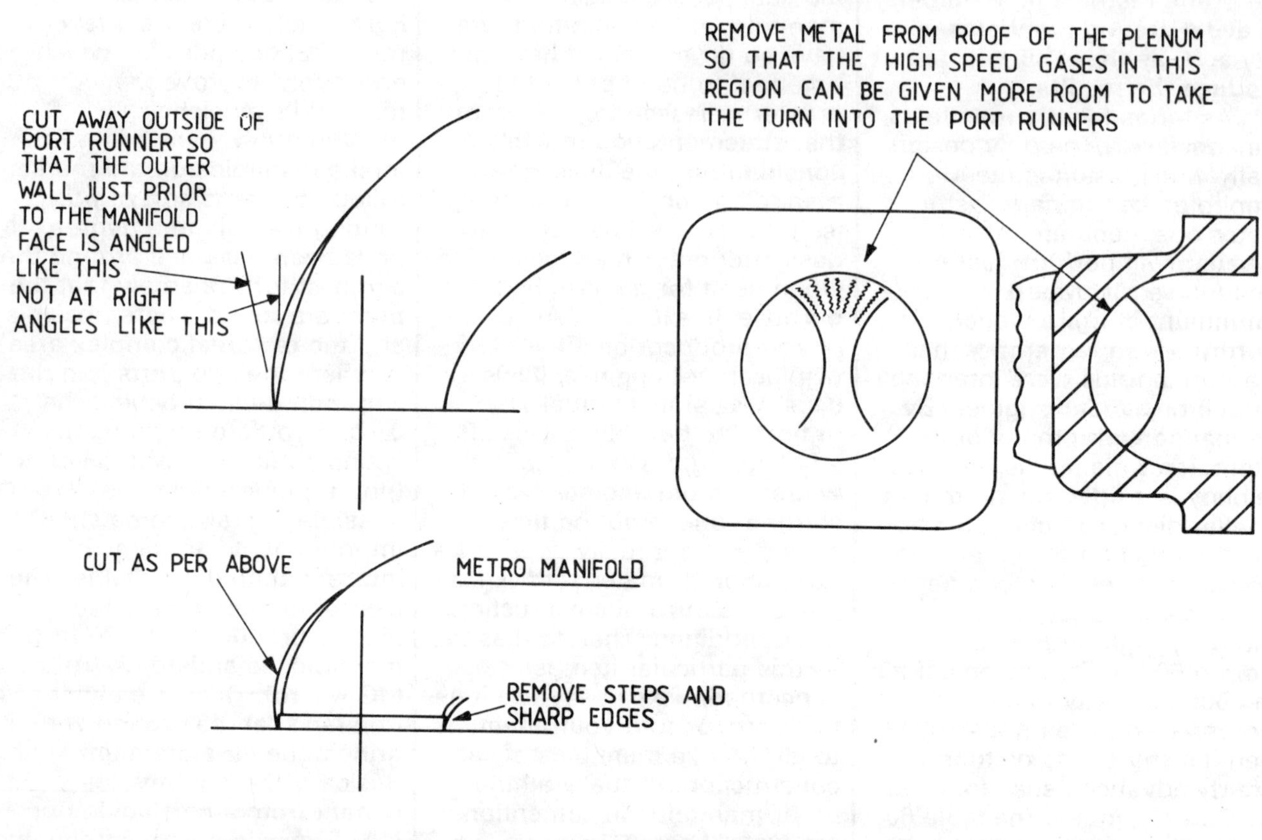

Fig. 8.8 Methods of improving cast single SU intake manifolds

manifold itself is considered, then there's insufficient metal to increase the radius at this point to any marked degree. However, the addition of a carb spacer, substantially changes the situation. By utilizing a standard SU spacer, either an aluminium one or one of the heat resistant phenolic spacers, the reshaping of the short side radius can be carried on right up to the outside face of the carb spacer: **Fig. 8.7** shows the technique.

Assuming the engine utilizing the intake manifold is relatively highly tuned, then there can be some advantage, if the space is available, to using 2 or 3 spacers to lengthen the induction tract at this point, because (a) it will help improve the shaping on the short side radius, and (b) adding a little extra volume to this manifold has been shown to improve the top end hp on engines capable of 70-80 hp or more.

Many of the after-market manifolds such as the Howley, Oselli, Mangoletsi and to a lesser extent because of its thinner castings, the C & T manifold, can be improved. All these manifolds have the same general form, hence their very similar performance and power results. However, some minor improvements can be made in these manifolds, both from the point of view of fuel distribution and airflow. **Fig. 8.8** shows the general areas where metal needs to be removed. This technique can be applied to most of the speed equipment manifolds mentioned so far, but the C & T one should be approached with some caution as it employs much thinner walls than most of the other manifolds, and it will not withstand too much material being removed before the possibility of cracking occurs.

Although the bare flow of the intake manifold does not necessarily go up that much when the modifications outlined in **Fig. 8.8** are performed, it does improve the total flow when used in conjunction with the cylinder head and carb. Generally such mods are worth an extra 2½ to 4 cfm when used in conjunction with an HIF6 or bigger carb and a high flowing head, as in our example **Fig. 8.3.** The undercut on the turn just before the intake manifold face tends to bias some of the fuel towards the inner cylinder. Although only minor in nature, it does appear to be a significant help. Modifying a manifold along the lines in **Fig. 8.8** seems to be worth about 2 hp increase over and above that given by an otherwise normally prepared speed equipment manifold when tested on a near-standard 1275.

Progressive Weber Manifolds

With its relative ease of calibration, good airflow properties and accurate metering potential, the Weber progressive choke carb offers much in terms of performance potential, but like any high performance carb, its potential will only be realized if used on a good intake manifold. In much the same way as the SU manifolds were dealt with, we will now look at a cross-section of Weber manifolds.

A glance back through the carb chapter, will refresh your memory that there are a number of progressive, downdraft Weber carbs that can be successfully utilized on the Series 'A' engine. However, the one that most of the tests shown here will be based on is the 28/36 Weber, simply because it shares, in common with the racing and high performance sidedraft carbs, the facility of being able to change the main venturi to optimize it in relation to the size of the particular engine it's intended for.

Depending upon their application, the manifolds themselves can vary in design. Some manifold manufacturers consider the alignment of the carb axis in relation to the direction of vehicle motion important, and we must remember when dealing with the Series 'A' engine, that it can be a transverse or inline installation. Because the float chamber is off to one side of the induction barrels on a progressive Weber carb, it is possible under severe acceleration conditions, to bias the fuel flow through the main jets. **For this reason the carb is usually mounted on the intake manifold so the float chamber points forward.** Under accelerating conditions, the weight of fuel acting on the main jets, will increase rather than decrease. This will cause a slight enrichment of mixture rather than a less desirable lean out. In truth, the importance of this factor largely depends upon how powerful the engine may be and how light the car is, as under more normal acceleration rates the changing fuel level has largely zero effect on performance. Only in terms of very high performance cars does float bowl orientation become a problem. Anyway, having got that out of the way, let's look at the intake manifolds themselves.

Essentially the list comprises a number of log manifolds, with various port runner lengths from the log to the cylinder head, and one long conventional manifold designed by Derrington an now produced by T.W.M. in the USA. At this point it may well be worth making a note of the fact there are some factory-produced cast iron Webber manifold which were used in the first couple of years of Metro racing. These manifolds were not sold as original equipment on any Leyland car, only to completitors in the Metro Challange Series. Although they were used on competition cars, they are not, strictly speaking, high performance items. Rumour has it that these manifold were left over from a still-bore Leyland project. They were intended for a production line engine that never came to fruition . . . **Their basic problem is that they are exhausted-heated in the same fashion that standard factory SU manifolds are. Consequently a lot of power is lost due to induction heating.** Also, the high intake temperatures involved sets a lower detonation limit on the CR. This in itself will markedly reduce the power potential of the engine.

Another factor detracting from the factory's Weber manifold is the relatively indifferent airflow characteristics it has. So basically if you are offered one of these manifolds, unless it is dirt cheap and you are replacing an otherwise standard SU manifold, don't bother with it, as you will gain only a fraction of the power increase that can be obtained with a better, after-market one.

A list of the type of manifolds tested for progressive choke Webers is shown in part A of **Fig. 8.9**. In essence, **Fig. 8.9** is the Weber version of **Fig. 8.3**. The same test heads were used, so all the flow figures between the Weber and the SUs are directly comparable. In part A of **Fig. 8.9**, we have the first on the list, namely the V.W. Derrington manifold. This is a non-log type manifold, in fact it's best referred to as a conventional manifold which is intended to run on inline engines. However, its use has reportedly been successful on transverse engines, although you need to watch for potential space problems on some installations.

Next on the list is the Mangoletsi 406 manifold, again intended for inline engines. This is a log-type manifold as are all the rest on the list.

Third on the list, and intended for transverse engines, is the Chris Montague Timavo-type manifold. Although this is a log type manifold, some careful thought seems to have gone into its design. The junction where the port runners lead off the log part of the manifold appears to be shaped to promote good airflow. Another point worth noting on the Montague Timavo manifold is the deflector placed under the primary barrel. This has a notable degree of importance when a progressive choke carb is mounted with its barrels in line with the cylinders of the engine, as is common on a transverse installation. Under such conditions we find when

The flow bench showed the only log manifold to flow with any degree of efficiency was this Montagu/Timavo setup. Excellent results are produced on all Series 'A' engines, but watch out for a bonnet clearance problem on 1275 Minis. There seems to be about minus ⅛ inch clearance.

An air filter, carb and manifold kit from Weber UK came into my possession for testing just as I was completing this book. Provisional testing has shown very promising results.

the primary barrel only is open, the incoming mixture is dumped nearer one cylinder head port than the other. This results in one pair of cylinders running slightly richer than the other. The Timavo manifold, together with some of the others, has a deflector formed in the manifold which tends to direct fuel and air toward the pair of cylinders which would otherwise be starved.

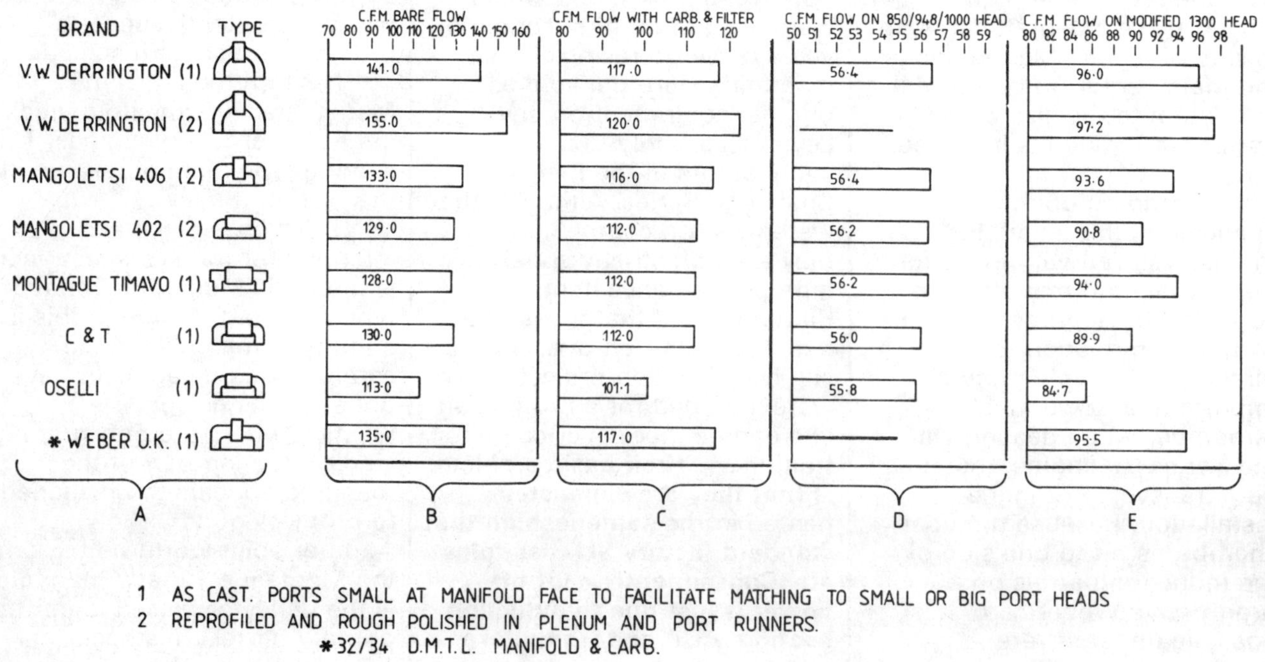

Fig. 8.9 28/36 Weber and 32/34 DMTL intake manifold flow comparisons.

Mangoletsi, C & T and Oselli manifolds that comprise the remainder of the list, are relatively straightforward, conventional log-type manifolds, although they are without the buffer chamber at each end of the log, which is often a feature of the log-type manifold.

In section B of **Fig. 8.9**, are the flow figures produced by the bare manifolds. **These numbers clearly show that the Derrington manifold is head and shoulders above the others.**

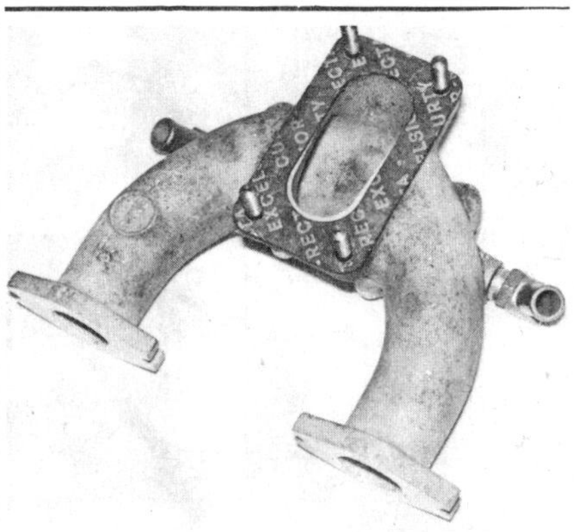

The most effective Weber manifold I have tried is this high flowing Derrington-designed, T.W.M. produced item.

However, this number is artificially low for the simple reason that the Derrington manifold has the ports tapered down at the cylinder head to manifold face so that it can not only be used with small port engines, but enlarged as necessary for use with big port engines. With the manifold port enlarged to 1⁵⁄₁₆ inch diameter at the manifold face so as to suit a typical street modified, big valve head, this manifold flowed over 155 cfm bare. That's right up there with the best of any single carb manifolds.

The next manifold on the list, the Mangoletsi 406, was tested with the port runners matched to suit the big port castings, *i.e.* 998 Cooper, 1100 Sprite/Midget or 1275 engines. This produced fairly good flow figures at 133 cfm for the bare flow, but it was approximately at the limit of its flow capability due mainly to the severe bends at each end.

The first of the transverse manifolds, the Montague Timavo manifold, flowed 128 cfm. However this, like the Derrington one, was reduced at the manifold face for use on small port heads with the optional capability of being enlarged to suit large port heads. This manifold was not tested with enlarged ports. However, experience has indicated that it is capable of roughly another 10 cfm when matched to the large port cylinder heads typically used on big bore engines. Although it doesn't have the total power capability of the Derrington manifold, which must be ranked as THE top progressive Weber intake manifold currently available, the Montague Timavo manifold is probably the best intake manifold for transverse engines if you are hell-bent on having the carb placed around the so-called 'right way'.

As far as the rest of the manifolds are concerned the numbers speak for themselves. A minor point but worth mentioning is the fact that two manifolds with similar port runners, *i.e.* the Mangoletsi 402 and 406, differ only in as much as the carb mounting is changed by 90°. Following flow performances right through the charts will reveal that the manifolds which position carb barrels 90° to the axis of the engine tend to have better flow performances in all the tests. Flow bench results seem to be supported by dyno figures, as carbs mounted with the barrels at 90° do tend to show a slight improvement in both torque and hp over their counterparts, the transverse engines. This could be due to both superior flow and slightly better fuel distribution.

Section C of **Fig. 8.9**, shows the flow figures produced with a 28/36 Weber, having a 25mm venturi in the primary barrel and a 27mm one in the secondary. Basically these would be about the venturi sizes you would use for an engine of around 65-75 hp. As you can see, the Derrington manifold still comes out on top with 120 cfm for its big port conversion and 117 for the small. In spite of being tested only in 'small port' configuration, the Montague Timavo manifold acquits itself well at 112 cfm. Of the two Mangoletsi manifolds which differ only in carb positioning, the one which positions the carb with its barrels at right-angles to the engine, proves to be superior by a 4 cfm margin over its transverse counterpart.

Section D of **Fig. 8.9** is the flow produced when the intake system is mounted on an 850/1000 Mini cylinder head. If a comparison is made between these numbers and the numbers in section D of **Fig. 8.3**, it will be seen that the Weber induction systems acquit themselves very well. The highest flowing Weber systems are almost equal to their SU counterparts, but the lowest flowing Weber installations tend to do better than their low flow SU counterparts when mounted on a cylinder head. This point is worth noting. **If you look at the porting on a typical Weber manifold, you will see that in the main, it enters the head in a fairly straight fashion, whereas virtually all the SU manifolds have a pronounced turn just prior to entering the cylinder head port.** Flow happens mostly on the outside of a turn. This being the case, if a turn exists in the manifold just prior to the port in the cylinder head, then the intake port in head is not totally utilized. In other words the flow has not straightened out to fill and utilize the whole port in an even fashion. Because of the turn in the manifold just prior to entering the head the outside of the port is utilized more than the inside.

With the Weber manifold, the turn is made a long way from the cylinder head and as a result, more of the port in the cylinder head is utilized. This factor shows up to an even greater extent when you look at Section E of **Fig. 8.9**. Here it can be seen that with only 120 cfm of carb and manifold flow capability, 97.2 cfm was actually achieved at the cylinder head. The extra flow being accounted for by the straighter run leading to the port in the cylinder head.

The time has come to bring the subject of mixture quality back into the picture. Usually for best results an engine will require finely atormized fuel in complete suspension in the air stream. As noted in Chapter 7, however, the Series 'A' engine seems not to like finely atormized fuel. The quality of fuel most suited to this engine seems to be measurably affected by the way the manifold presents itself to the cylinder head port, the best results being obtained from a straight or even slightly outwardly splayed presentation.

With few exceptions, the typical Weber induction system produced a lower percentage of flow drop when installed on the head than the equivalent SU system.

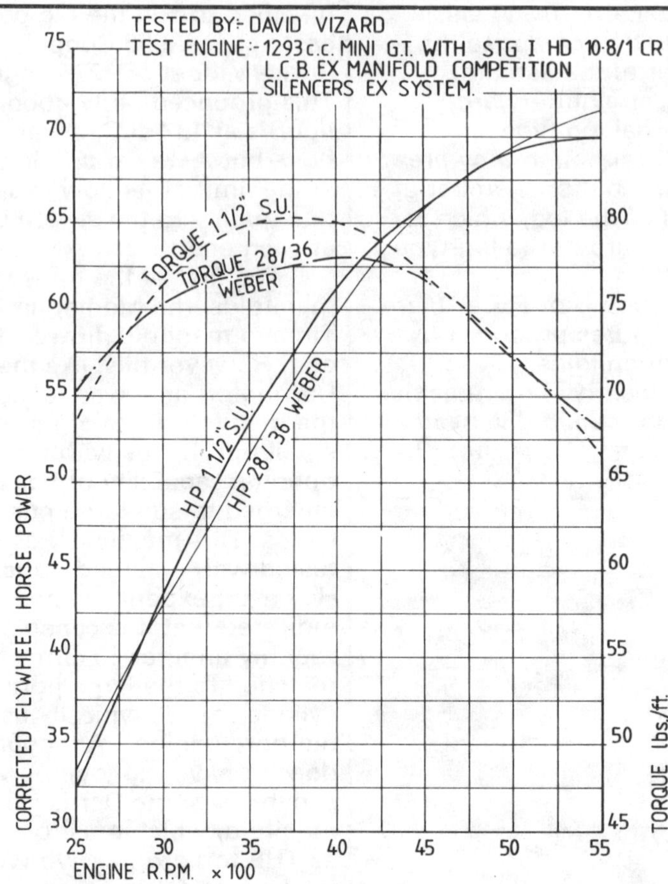

Fig. 8.10 Comparison of 1½ in SU and 28/36 Weber

SU induction system, flow @ 25 in = 90.6 cfm on head.
28/36 induction system, flow @ 25 in = 90.8 cfm on head.
The SU has a 4 in ram pipe.

A comparison of SU and 28/36 Weber carburation reveals distinct characteristics of the two. Firstly, peak power differed little between the two. Such differences that do exist in the whole power curve can, to a large degree, be attributed to the effects of inertial ramming. The 'bump' in the torque curve is mainly due to a 4 in long ram pipe on the SU carb. However, below 2600 and above 5000, the 4 inch ram length detracted from the potential power. As a result, the 28/36 Weber was superior below and above these two rpm limits, whilst between them the SU was distinctly superior.

Inevitably the question will be asked, 'Which produces the most power, a single Weber or a single SU?' Let's face it, there are a lot of factors to take into account. What size of SU? What manifold is it on? Which Weber manifold? What choke sizes are in the Weber? The only way we can separate the effects of the carb atomization, fuel distribution, etc., from the airflow capability, is to take two matched intake manifolds and test them back-to-back on the same engine. The two intake manifolds concerned are the Oselli Torquemaster with the small-sized ports and a single 1½ inch SU. On the test 1300 head as in **Fig. 8.9**, this flowed 90.6 cfm. Of the Weber induction systems, a Mangoletsi 402 manifold was used. With 25/27 venturis in the Weber, this system flowed 90.8 cfm on the same test head as the Oselli system. When dyno tested on an engine, these two induction systems produced the power figures as in **Fig. 8.10**. As can be seen, there isn't an awful lot of difference in the performance potential of these two installations. In fact, the figures were so close, it was necessary to run the tests on each carb three times and average out the results, to get meaningful figures.

Reece Fish Intake Manifold

In the chapter on carburation I looked extensively at the pros and cons of the Fish type carb. In the U.K. examples are manufactured by both Reece and Minnow. The carbs themselves then, are fairly well covered, but what does concern us here, is the Reece Fish intake manifold for the Series 'A'. In the past, this manifold has been popular amongst 850 Mini racers as has the carb. To see just where it stood in relation to SU manifolds, I tested it both on the flow bench and on the dyno to see if its form was any more conducive to hp than any of the good SU manifolds. I make the mention of form here, because unlike the SU manifolds, the Fish manifold has no hint of a divider in the plenum. Flow bench testing,

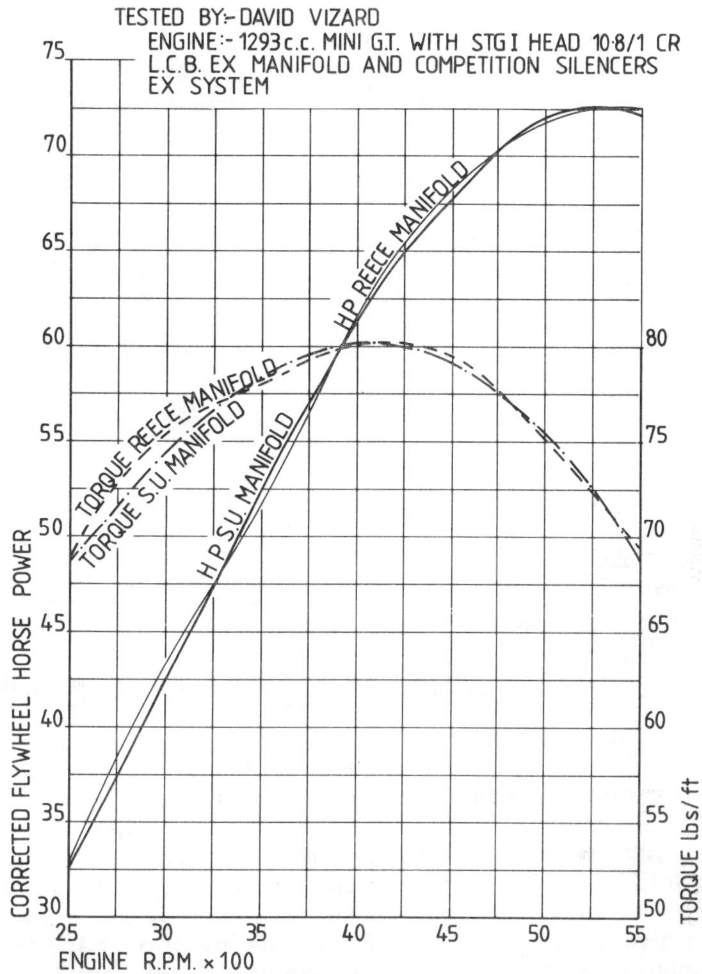

Fig. 8.11 Comparison of a Reece Fish carburettor fitted to Oselli Torque Master and Reece Fish manifolds

The Oselli Torque Master is a top performing intake manifold. From the curves shown it can be seen that the performance of the Reece Fish item is almost indistinguishable.

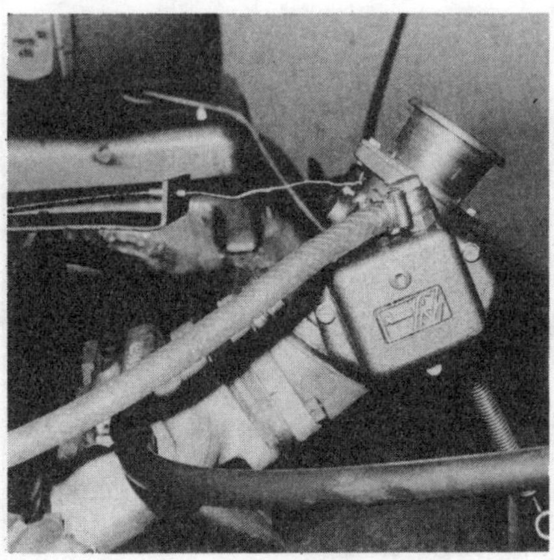

In an effort to find out which manifold produced the best power with a single 1½ inch Reece Fish, an Oselli Torquemaster manifold was tested against ...

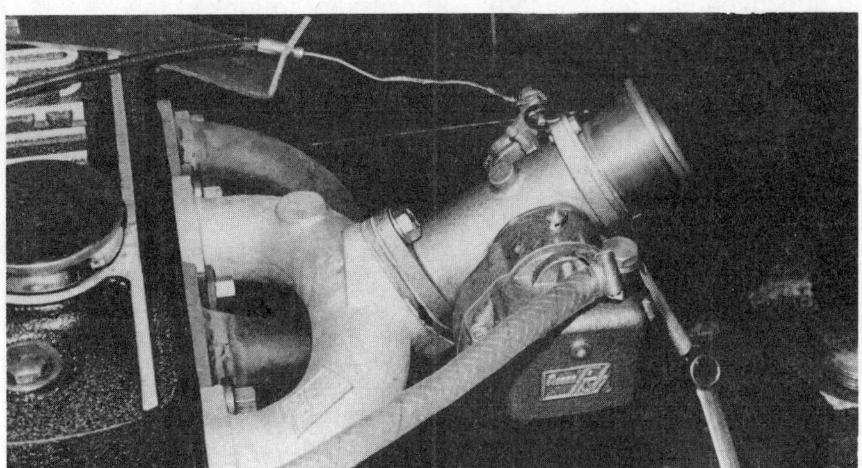

... the Reece Fish intake manifold, and the accompanying chart shows the results.

drawing through one port, as was done with all the other manifolds we have tested so far, does not necessarily prove whether that divider is an asset or not.

The flow bench testing indicated that on a high flowing 1300 head, the Reece Fish manifold flowed 93 cfm and that figure is directly comparable with the figures for the Oselli Torquemaster manifold shown earlier. These manifolds were tested back-to-back with a Reece Fish carb mounted on it. It flowed 92.7 cfm, and the dyno test, **Fig. 8.11**, shows the difference in power and torque produced by the Reece and its SU counterpart. Again you can see that the figures are extremely close and for them to be meaningful it was necessary to average out the results over a number of runs. Even so, these numbers can be taken as a good indication of the real-world situation, as the Fish manifold consistently showed slightly superior results to the SU manifold it was tested against. Admittedly the difference in power is small, but it would seem fairly safe to attribute the difference in power to the airflow difference in the two manifolds, rather than any form difference that exists within the runners of each manifold.

My own conclusion here is that the Reece Fish manifold is an effective item, but the carb may be better utilized on a manifold which flows more air still.

One Barrel Weber Induction

A popular move in 850 Mini racing is to utilize one barrel of a 45 DCOE Weber, as the regulations call for one barrel of carburation: all of which must go through a single 1½ inch diameter hole. Most racers utilizing the one barrel Weber system are also utilizing a venturi of 1½ inch, *i.e.* 38mm. Such an installation will give more airflow than either a single SU, which typically flows around 135 cfm, or a 1½ inch Reece Fish which tends to be around 150 cfm. By comparison, the single barrel of a 45 sidedraft Weber with a 38mm choke flows typically around 240 cfm. Basically that's enough airflow to satisfy the demands of an engine of over 100 hp, far more, in fact, than can be expected from an 850 race engine.

Having established that the sidedraft Weber concept offers the best in terms of carb airflow, I can come to the point of the discussion, namely that from here on, it is very necessary to look at the induction system so as not to spoil the potential of the carb.

Installations similar to this are currently the hot favourite for 850 Mini racing. The intake manifold here, a Maniflow item, has since been superseded by one which produces more power by virtue of a larger plenum area behind the spacer. The later Maniflow intake manifold showed noticeably superior results.

Double Barrel Induction

By far the most impressive induction system in terms of 'power aura' is the twin choke, sidedraft setup, either a Weber or a Dellorto. As stated before in the carburation chapter, for

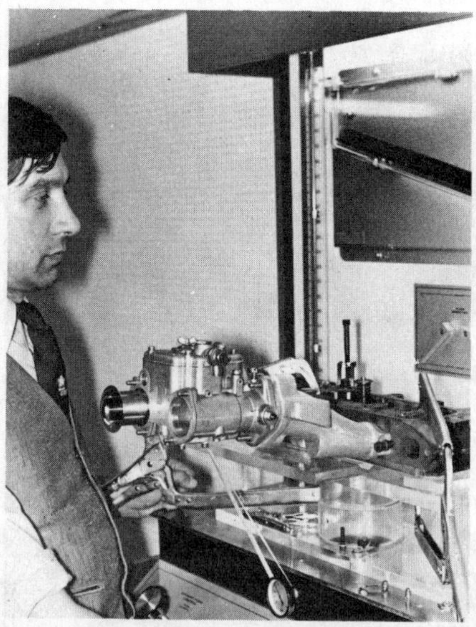

Martin Goodall of Weber Concessionaires Ltd is seen here using a Superflow SF 300 flow bench to check the compatibility of a 40 DCOE Weber with a 5 inch long intake manifold. The results showed it to be a good working combination.

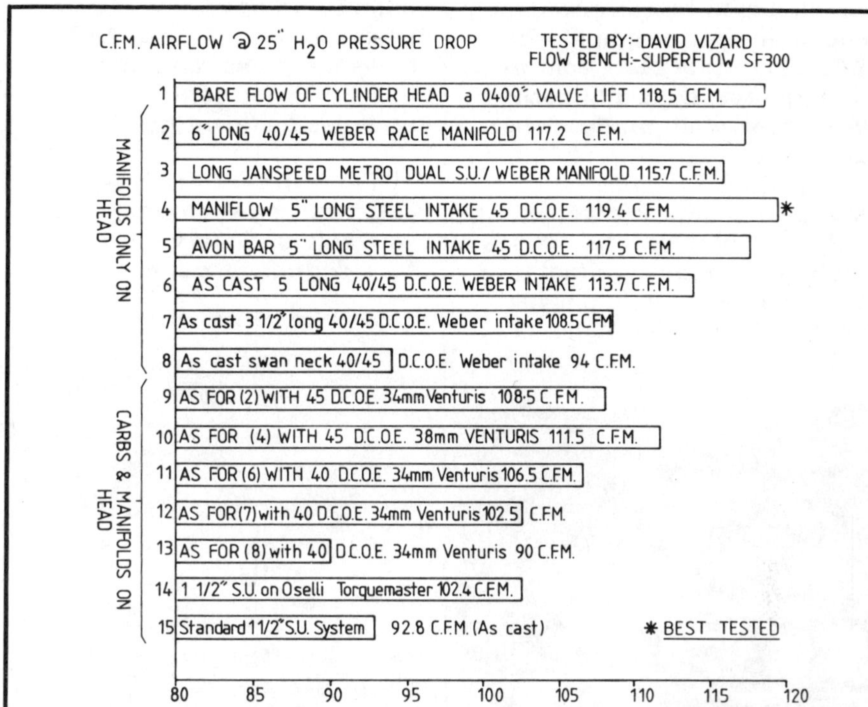

Fig. 8.12 Comparative flow tests of side draft Weber manifolds and SU manifolds

Unless otherwise stated, the manifolds were flowed with the ports matched to the head and all rough finishes smoothed out. The 5 inch long 'as cast' manifold did not trail far behind the hand-finished items, the difference shown here being typical of what a smoothed manifold port is worth in terms of flow. As the manifold get shorter, so it becomes increasingly difficult to get good flow because of the dog-leg routing necessary. From these figures it can also be seen that the swan-necked manifold is disastrous in terms of flow efficiency. Indeed, **it is so bad** *that* an entire *single 1½ in SU induction system utilizing a K & N filter and an Oselli Torquemaster intake outflowed just the bare swan-necked Weber manifold. Manifolds of note here are the Janspeed Metro unit which, installed in a Metro without body alterations, can utilize either twin SUs or a single DCOE Weber. The Maniflo and Avonbar steel tube manifolds are also strong performers.*
Note: *the Maniflow manifold, when installed on the head, actually produces more flow than the 'no manifold' situation. This is rare, but does happen from time to time with a well-developed manifold. This particular item seems to be a hot favourite amongst racers of 1000 cc engines.*

smooth road use where power output is of secondary importance I favour a Dellorto carb, whereas for road or racing where maximum power is paramount I favour a Weber. At this juncture I am really only interested in discussing intake manifolds, and their effect on power. Almost all the manifold testing done in this section, was done together with Weber carbs. However, the same basic ground rules apply, whether you are using a Weber or a Dellorto, so don't be put off by the fact that Weber carbs will be all you will hear talked about in this section.

I'll start the ball rolling by saying that just because you bolted on a sidedraft Weber carb, does not necessarily mean your engine is going to produce more hp. **These days, high performance carbs are manufactured with considerable expertise, and they are fine instruments which function very well. Unfortunately, the same degree of expertise is not seen in the intake manifold.** There are a substantial number of good intake manifolds on the speed equipment market. However, there are also some commonly available items which are only a little better than junk. In this section I hope to steer you in the right direction when it comes to buying a sidedraft intake manifold.

Without putting any emphasis on priorities, let's establish some basic rules concerning sidedraft carburettors. The first rule is pretty simple and, if it comes to that, pretty obvious. Namely, that **up to a point**, the straighter the route between the carb and the cylinder head the manifold runner makes, the better the intake manifold will be. However, there can be over-riders on this simple concept, as we shall see later. This is not to say that an intake manifold which has a considerably curved passage cannot be made to function. It takes a lot more thought, careful design, and flow bench testing to get the air to flow round corners easily than it does to go in a straight line. Unfortunately few intake manifolds currently available are designed with the aid of a flow bench. **There are manifolds which have curves in them which flow well and produce good results. There are others which do not.** To give you an idea of just how much the intake manifold can affect flow and power, we will start off with a test on four intake manifolds.

The first is our reference manifold. It utilizes a standard 1½ inch SU on a port matched Oselli Torquemaster. The flow capability of this manifold, together with the others with which it is to be compared, are shown in the bar graph, **Fig. 8.12. As can be seen from these flow figures, the swan necked Weber manifold produces less airflow as a bare manifold than the Oselli Torquemaster manifold produces complete with carb and air filter.** This is a fairly good indication that the swan-necked manifold is bad news. On the other hand, the engines in a relatively low state of tune. The graph, **Fig. 8.13**, shows the power and torque curves with the same four manifolds as flow tested in **Fig. 8.12.**

From the results in Fig. 8.13, it can be seen clearly that the whole performance concept of a sidedraft Weber can be completely destroyed by the use of a manifold having inadequate flow. Of course the idea of the swan-necked manifold is to allow the installation of a Weber in an engine compartment that's a tight fit for space, like a Mini's. However, it's worthwhile pointing out here that it's only other two manifolds with their more direct routing, produce good flow figures. The effect an inadequate manifold has on hp is quite startling, even on

Here is the infamous swan-necked manifold. This design, and many like it, represent the height of inefficiency in terms of airflow and ...

... the dyno bears out the validity of the flow bench results. These manifolds don't produce the hp and they are very heavy on fuel. Slightly longer manifolds ...

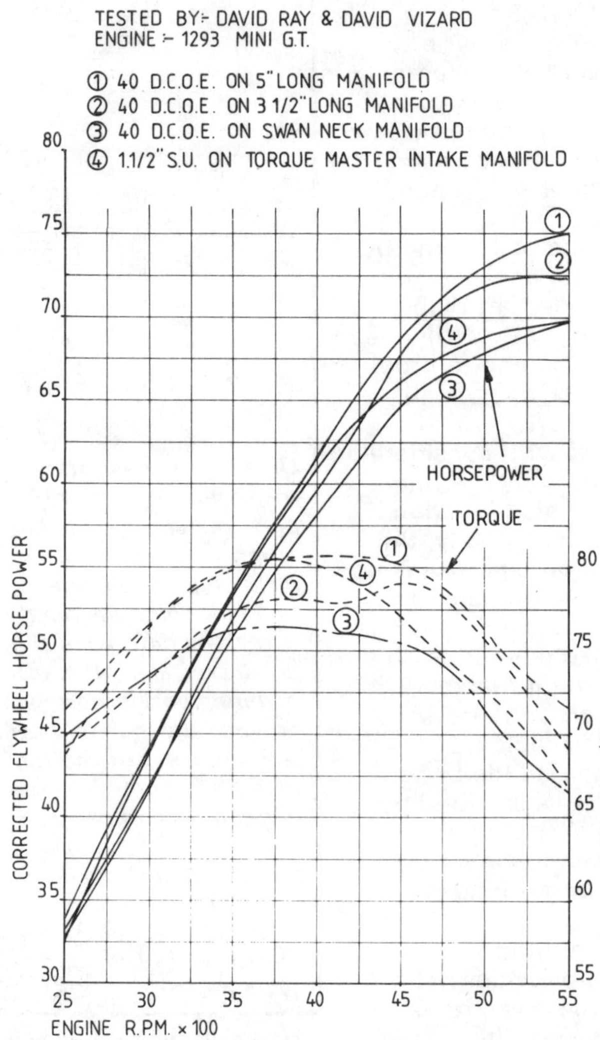

Fig. 8.13 Comparative dynomometer test of three types of side draft Weber manifold and a reference single $1\frac{1}{2}$ in SU induction system

This graph shows the results of dyno testing the manifolds and carbs as per columns 11, 12, 13 and 14 of Fig. 8.12. The dyno figures bear out the fact that the best flowing manifolds produce the best power. Note, however, the strong torque curve around 3000 - 4000 rpm given by the SU. Once again, this is more than likely the effect of inertial ramming, a phenomenon largely denied the twin choke Weber installation.

... such as this one shown here, produce far superior results and ...

... slightly better results yet can be obtained with even longer manifolds. The particular example shown here is a 5 inch long manifold. Increasing the length any more than this does not really result in any more hp due to airflow. The extra length may or may not help from the point of view of length tuning, but that factor is more dependent on matching intake length with camshaft characteristics. Also the mixture quality entering the cylinder head ports may have decided a influence on power. This is discussed at greater length in Chapter 9.

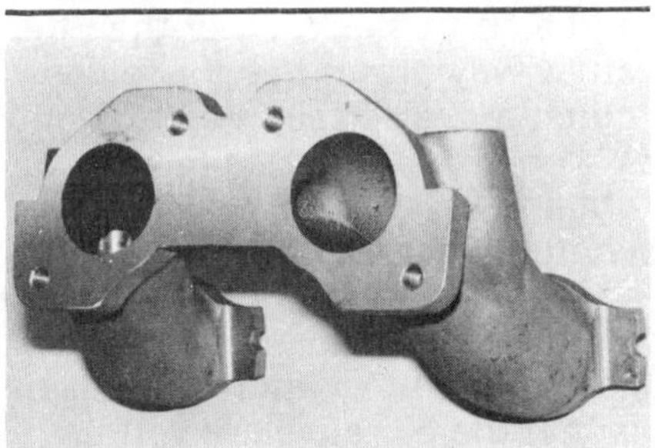

Here is a manifold which to the uninitiated observer, looks like the dreaded swan-necked item. However, it would be a great mistake to confuse this with the highly inefficient swan-necked manifolds shown a few pages earlier. This particular item from Janspeed Engineering has proved to be a highly efficient intake manifold, designed specifically for use on the Metro. It is very unusual in that it can mount both ...

ever any use to fit speed equipment if it actually gives more speed. An SU manifold will fit into that same space and it will give more hp, at a lot less cost, than a sidedraft Weber mounted on a swan-necked manifold.

At this point, to avoid confusion, it may be as well to categorize swan-necked manifolds. There are manifolds which bear a definite resemblance to the manifold tested, except the swan-neck is not quite so pronounced. Some of these manifolds actually produce extremely good results. A prime example is the intake manifold commonly used in Mini 1275GT racing. If installation space is a problem, then this is the manifold to use. The object lesson here, then, is to choose carefully the intake manifold you intend to use.

... a single side draft Weber as shown here, together with the variable ratio linkage or ...

... a pair of 1½ inch SUs. It's easy to think of this intake manifold as a Weber intake manifold only. However, when it is used in its SU form as shown here, it is amongst the most efficient SU manifold commercially available for the Series 'A' engine. This, together with the filter box for use with a remote mounted filter, produces a very quiet and effective induction system for the Metro, and possibly for Minis, if few clearance problems exist.

Intake Manifolds Part 3

Induction System Length

I realize we have dealt with inertial and ram tuning elsewhere in this book, but it's a subject which has a relevance here as it concerns induction system length on Weber carburated engines. As I've said before, one induction barrel per port on a 5-port engine, is not overly sensitive to overall induction length. This does not mean to say that it is totally insensitive; nothing could be further from the truth. The best I can say concerning Series 'A' induction lengths is that it is sensitive as to where the length is put. That may sound a bit ambiguous, but the truth of the matter is, it depends on whether the length of the induction system resides mainly in the manifold or mainly in the ram pipe. In other words, you could have two systems of the same overall length, one with a short ram pipe and a long intake and the other with a long ram pipe and a short intake and find they produce significantly different results. Much of what I'm about to say has come to light since I did the first edition of this book, and here a conglomeration of tests will prove the point I'm trying to make. Originally my conception of the intake system was such that I felt that the overall length was the criterion of main importance and that if the ram pipe length was adjusted, this would have the same effect as lengthening or shortening the intake. Whilst working with Maniflow in an effort to achieve a better intake manifold, I found that my own tests proved the fallacy in this thinking. Dozens of tests with many different manifolds on my own dyno have now revealed that getting the best from an intake manifold on a Series 'A' engine is anything but simple. I feel that the pressure change within the ram pipe, especially a long one, affects what happens at the auxiliary venturi but these pressure fluctuations may not be communicated effectively to the float bowl. This may be why we often have trouble getting long ram pipes to work as effectively as long manifolds. In essence they create their own little environment between the auxiliary venturi and open air. Since fuel injection is not affected by these relatively small pressure fluctuations it would be easier to get an idea of what might be happening here. Unless you have access to a dyno it is for this reason that I strongly advise the use of a long intake and a short stack rather than the other way around as this gives much more certain results. With a Weber good results have been obtained with a 5.0 auxiliary venturi. Perhaps this is not so sensitive to some of the pressure fluctuations that I suspect occur at this point. An interesting aside is that the 5.0 auxiliary venturi also appears to allow about a 1% increase in engine air flow as measured on the dyno. Still, the theme we're dealing with is overall length, so let's start the ball rolling by considering the effect of adjusting the overall length of the induction system by virtue of making changes to the ram pipe length. For this we'll use some tests done by Tom Oliver when he worked at Howley Racing. These show a typical case of how the power varies with a change in induction length; but it's worth pointing out that these figures don't necessarily indicate exactly what you can use on your engine. The particular test engine was a 1435cc engine with a C-AEG-643 cam installed. No doubt, if the capacity had been different, and a different cam or exhaust system, or even cylinder head had been used, the optimized ram lengths could have altered. These figures are only practical as a guide, and basically all they prove is that the last bit of power can be extracted by getting the ram pipe length just right for the existing combination. However, if guessed, it's just as easy to get it wrong, so it really is a question of putting the engine on a dyno and physically sorting out the most appropriate length.

Let's turn our attention now to the effect of intake manifold length. Here you need to refer to **Fig. 8.14.** In simple terms, what you see here is the effect of running with three different intake manifold lengths. For this test I had Dave Dorrington at Maniflow make up three different intake manifolds. The extra length was added into the port at the cylinder head flange face. The form from the carburettor end of the manifold remained largely unchanged until it encountered the extra length. On the flow bench all of these manifolds flowed comparable figures, the short manifold being the best flowing and the long manifold being the worst. However, there was only a 2% difference in the open flow capability of all the manifolds. When installed on the cylinder head with a carburettor on, this flow difference dropped to something in the order of ½%. Next, changing an intake manifold will obviously change the overall induction length. This was catered for **as far as possible** by changing ram pipe lengths to several different off-

The particular engine used here was used to test the effect of varying induction lengths when a sidedraft carb was used. To simulate the change in manifold length, different length ram pipes were used. One of the checks was to see if any power was lost using the Group 2 ram pipes, which are turned up at an angle, in comparison with the straight ones. Power produce was identical.

the-shelf lengths available. However, the tests in **Fig. 8-14a** did not hold the total induction length constant. What was done here was that the manifold would be changed and then the optimum carburettor ram pipe length established amongst those that were available. These were 16, 26 and 60mm long. With the 3.5 inch long intake manifold, the 60 mm ram pipe gave best results. With the 5.5 and 6.5 inch length intake, the 16 mm gave best results. So a comparison between the 3.5 intake and the 6.5 intake shows the difference in intake manifold when the overall length is held constant to within ¼ inch. As you can see, the longer intake manifold (known as a 'VP Special') paid off. However, it is necessary to stress that a great deal of attention with this longer intake manifold was paid in the way the direction of the manifold took as it joined the cylinder head face. What was attempted here was to get as straight a run as possible prior to going into the cylinder head. Based on the evidence of these tests, the only valid conclusion I can come to is that manifold length in itself is not a super critical factor. What appears to be significantly important is whatever effect the shape of the intake manifold and its length of straight run before the cylinder head port has on the mixture quality itself. Whether it allows a more homogenous mixture to be formed or whether it allows the fuel to coagulate into larger droplets is not apparent. The only apparent factor is the longer manifold makes more horsepower.

This particular 1000cc Longman-modified unit is the result of many patient hours on the dyno. Tuned induction lengths, improved airflow and the containment of stand-off have all been achieved with these special long rams, rather than by increasing the manifold length.

Additional Power Producers

You could easily be forgiven for thinking that now we've dealt with intake manifold length in it's entirety. This is not the case. Firstly, let's consider the variety of intake manifolds that can be had. In essence, the induction manifolds are available either as a casting in aluminium or, as in the case with the Maniflow manifolds, a fabricated steel item. Although aluminum would seem like a good material because it is light, for an intake manifold it must be one of the worst. This is because it conducts heat at a very high rate, a factor we do not want on a race engine. Of significant importance is the fact that on my engines I stick exclusively with Maniflow manifolds which I have chrome plated. The importance in the chrome plating lies not in achieving any special finish on the inside of the manifold, but achieving a highly polished surface on the outside. This acts as a heat reflector and has been shown to be worth as much as 2hp on the dyno due to the fact it keeps the intake charge temperature cooler. If you want to go one step further, the internal surface of the intake manifold can be coated with a ceramic thermal barrier such as supplied by Techline in the USA or Nimbus in England. If a chrome finish does not appeal to you these two companies can supply a less shiny heat rejecting ceramic finish. Both types of finish can be applied in a home workshop with a small spray gun. Another helpful fact is intake manifold diameter. Maniflow can produce the manifolds we've discussed here with either a 1.4-inch or a 1.5-inch diameter at the cylinder head face. In terms of peak power, the effect of changing from one manifold to another is not totally clear. However, very often the larger an intake manifold especially, on well-developed, larger displacement engines, allows the engine to hang on longer after peak power. This means the engine can be buzzed higher in the gears before a gear change and out on the track this results in better overall acceleration. If you have time to spend on the dyno, it is well worth checking out to see which manifold suits your combination. As you would expect, the port size is not only dependent on the cam being used, but also on the size of the engine. Bigger engines almost always need a bigger intake manifold. Yet with some combinations the smaller intake manifold produce measurably more torque around the 5000rpm mark. My advice concerning the diameter of intake manifold used is simple. If in doubt, use the 1.4-inch diameter. If you know the combination you are working with and you know it needs the 1.5-inch diameter, then go for that.

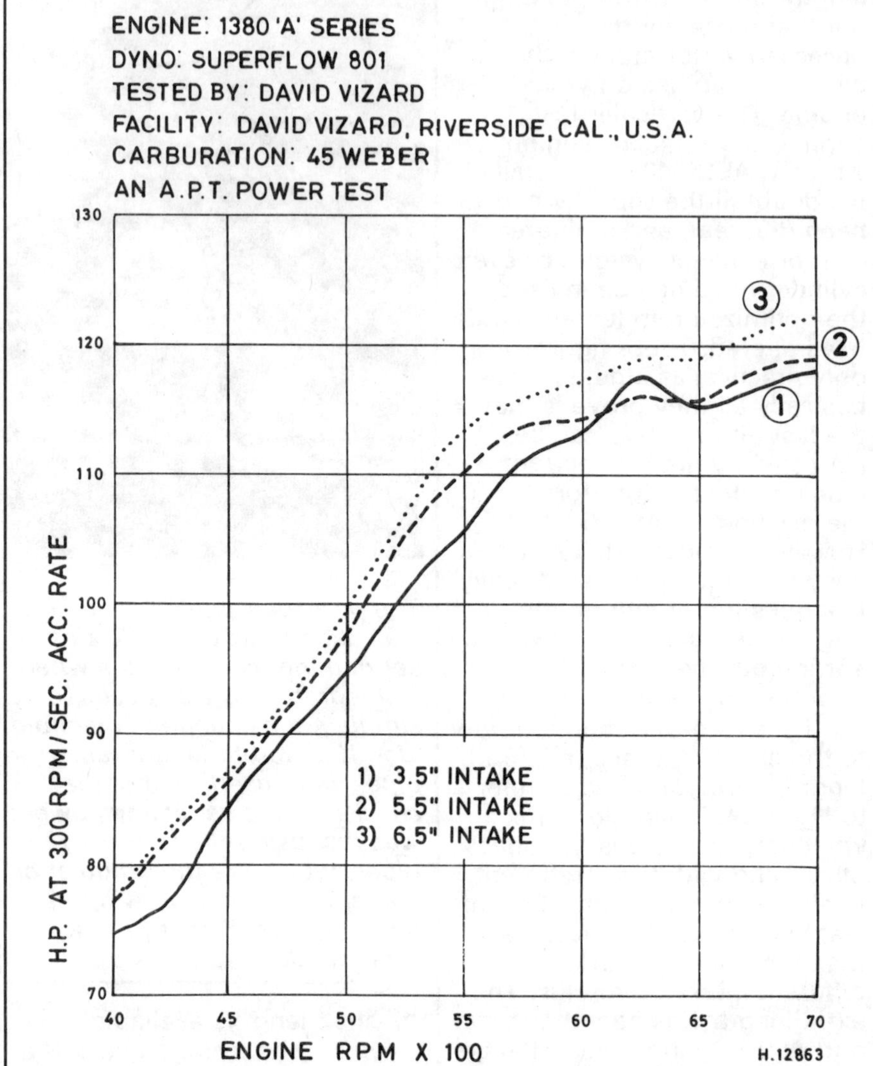

Fig. 8.14 Intake manifold length
From the curve shown here we see the intake manifold length and form can significantly affect the power output of a Series 'A' engine. Whereas overall induction length has less of an effect on power output. This is a strong indication that it is manifold form which affects the quality and distribution of the mixture within the port, the longer manifold producing a more favourable situation.

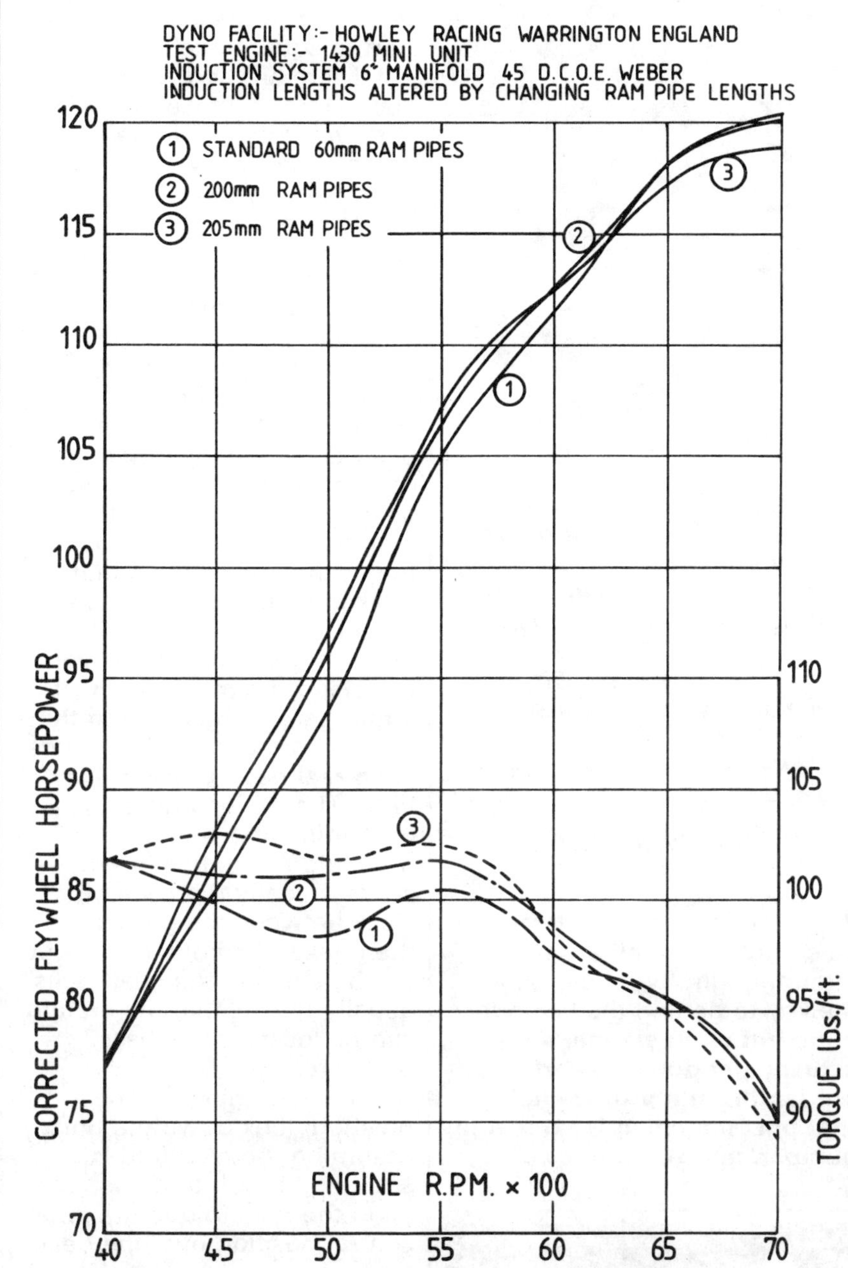

Fig. 8.14a Comparison of induction lengths on 45 DÇOE Weber caburettor

These figures, produced by Howley Racing, show the relative insensitivity of the Series 'A' engine to ram length when one barrel of carburation per port is used. Although beneficial changes in the power and torque curve can be made, the magnitude of the changes is far less than that produced by induction length tuning on a single barrel carb setup.

As is usually the case, the longer induction lengths favour the lower rpm range and the shorter lengths the upper range. This particular engine was equipped with a relatively radical camshaft. Most induction length tuning is as much 'cam' sensitive as anything, so a cam change means a redetermination of the best length.

Here, being contemplated by Don Loughlin of Aldon Automotive, is a split Weber, 1293 engine that I had a hand in building many years ago. The car appeared to be very competitive until it did a 32-foot high excursion which involved a triple somersault.

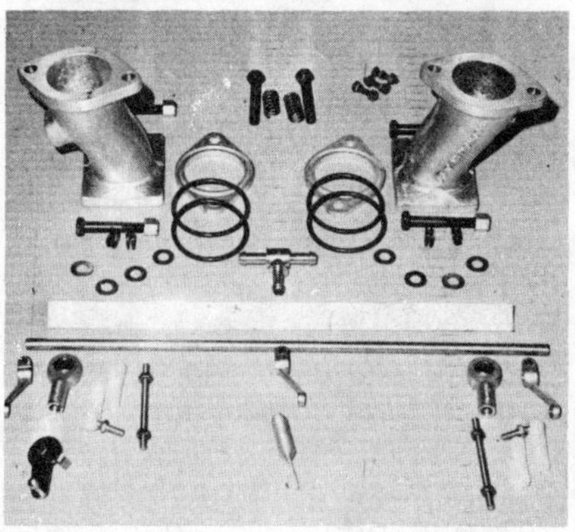

If you want to build yourself a split Weber setup, here's the kit of parts you require. This kit is available from Richard Longman & Co.

Split Twins

A move that seemed to gain popularity a few years ago was to use one barrel of each of a pair of Webers to feed the two intake ports on the Series 'A' engine. This allowed the use of the most direct intake route possible, *i.e.* carb aligned 'straight shot' to the port. Such an induction system obviously offers the best chance of getting the highest power from the engine. Occasionally it did show superiority over its underdeveloped single carb counterpart. With the introduction of the long manifold used in **Fig, 8.14,** shown in the nearby photo, the power advantage has swung to favour the single rather than split set-up.

Apart from this, building a set of splits represents quite a bit of extra cost over and above the price of a single Weber. Add to this a lot of work, and at the end of the day splits may be worth little in terms of cost effectiveness.

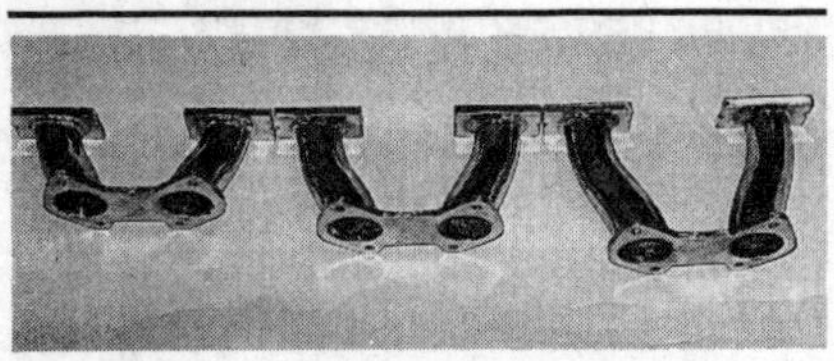

Here are the three intake manifolds that produced the test results seen in Fig. 8.14 . The short intake manifold is the off-the-shelf Maniflow item that I've favoured latterly. The only thing done to it here is the bright chrome plating to reflect and cut heat conduction from the exhaust manifold to the charge passing through the intake manifold. No additional polishing has been done on the bore (it doesn't need it). On the flow bench this manifold flowed, to all intents and purposes, 100% efficient at the flow levels we normally test at. The second and third manifolds were ones made by Maniflow to my specifications. As the test figures show, longer proved better, and there's no indication to the contrary, that 7 inches is optimal. Could be that a greater length yet is required, but. . .

Here's an engine that I built in 1982 which had a state-of-art induction system. On the dyno at least, nothing I tried showed any significant improvement over this setup. By 1987 I could better this by 5hp and 5ft lb minimum, with the intake shown below.

Manifold Matching

Port matching is considered to be something of an art and is often thought by many to be the key to getting the last ounce of hp out of an engine. **If we actually get down to sorting out fact from myth, we find that port matching is no real hp bumper, and sometimes a perfectly matched port may even cost power.** However, let's start at square one . . .

. . . .the long manifold, when installed on an engine (especially a transverse engine) may cause additional installational problems.

There are two ways of matching the intake manifold port with the one in the head. There's the hard way and the easy way.

The hard way is to bolt the manifold to the head with a gasket in place, drill through the manifold and gasket into the cylinder head and fit dowels on which to align the manifold perfectly each time its assembled. Once the head has been doweled, the gasket is used as a template to match the head to the gasket and the manifold to the gasket. Using this technique ensures that the intake manifold will always align close enough for all practical purposes. **However, matching the manifold in such a manner is a real pain in the neck to do, especially as there is no real need to do it this way!** The easier way is to determine just where the manifold aligns best on the studs. Usually there's a bit of slack between the grooves in the intake manifold and the fit on the studs in the head. This usually means that the manifold can be located in a 'best position'.

To determine the 'best position' look down the port, assuming the manifold is straight enough to do so, and align the manifold in the most central position over the head port and note which way it has to be pushed to get that alignment, whether it's up, down, across or whatever the case may be. Next determine approximately how much slack there is due to the loose fit on the studs, and then grind a chamfer on the edge of the intake port as shown in **Fig. 8.15. This will have the effect of making the intake port in the head bigger at the entry point, than the exit port in the manifold. A step in this direction causes virtually no**

Even in the D.V. workshop, things used to be done the hard way. Here, my good friend Colin Ashdown-Pogmore is seen matching up a manifold by doweling it to the head. The flow bench has shown us that this technique is redundant.

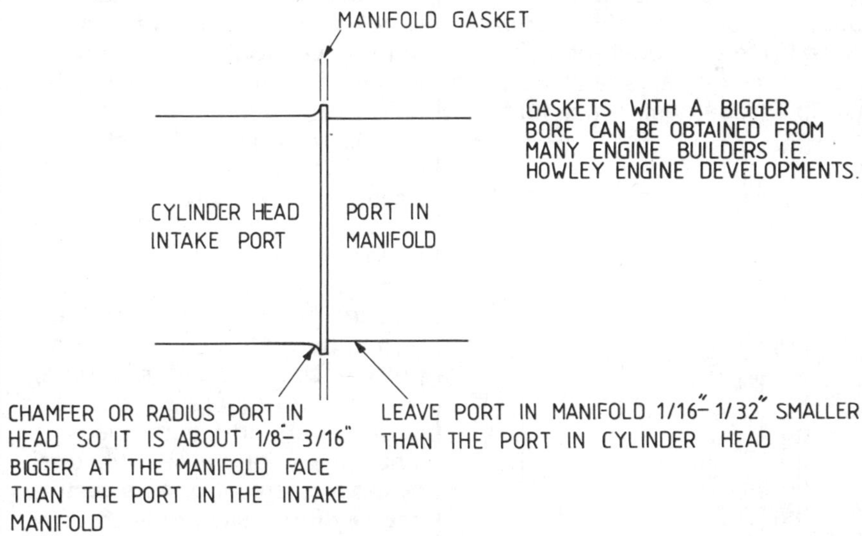

Fig. 8.15 Port matching...the easy way!

loss in flow. In other words it will flow just as well as a perfectly matched port. However, it can have one side benefit, namely that if any fuel puddling occurs, there is a tendency for that fuel to be re-introduced into the airstream as the air goes off the step from the intake manifold into the head port. If the manifold is a tight fit on the studs, and on occasions they are, then try filing the grooves in the manifold so as to align the ports as best you possibly can, before chamfering the edge of the intake port on the head. Of course if the type of intake manifold you are using has locating rings because (a) it's a stock one or (b) it's the type that had them as an after-market unit, then the matching up to the head is no problem.

Twin SUs

Many people, whose Series 'A' engines have twin SUs as standard equipment, mistakenly believe that there is little they can do to the carburation to improve it. In reality this is not the case. **Twin SUs in essence, are an asset, but the standard Leyland intake manifold is a very inefficient design, so much so that most of the asset of twin carbs is lost.** Proof of this can be had by considering the fact that a good, after-market, single carb intake manifold can, with the same carburation area, produce more power than the standard factory manifolds. Only when the size of the carbs gets very large, *i.e.* 13/4 inch or 2 inch, do we find the advantage going to twins. This comes about for two reasons. Firstly a lot of the inertial ramming is lost when twin carbs are used. This tends to rob the mid-range torque and power. Also, since the factory intake manifold is such a bad design to start with, the power potential of the carbs is largely invalidated.

Manifold Gaskets
If your engine is stripped on a regular basis because it's a competition machine or you do a lot of experimenting, you will probably get fed up with having to repeatedly scrape off the manifold gasket. If you wish to avoid this, use graphited gaskets. Not only do they not stick but they can be re-used several times.

So just how bad is the standard factory manifold? For that, we can again turn to the flow bench to see just how it fares in terms of airflow potential. **Fig. 8.16** is the graph you should be looking at. It's worthwhile noting that these figures cannot be directly related to the earlier flow figures given because they are on a different head. They were, in fact, done on one of my race heads, and the baseline flow figures are quite a bit higher than anything used as a reference elsewhere in the book. However, they will serve to put you in the picture as to what works and what doesn't.

Column 1 shows the open intake flow of the cylinder head. Column 2 shows what happens to that flow when the standard intake manifold alone is bolted on. As you can see, the flow drops appreciably. With the balance pipe blocked off so that no air is drawn across from the diagonally opposite carb, the flow drops a shade more, and when the carbs are fitted to the intake system the flow goes down another small amount. Essentially we can see from these figures that there was a flow drop of 17+cfm experienced by the cylinder head when the intake manifold was bolted on, but only 4cfm when the carbs were bolted on. Just for reference, we can see that an Oselli Torquemaster manifold flowed almost 117cfm, thereby losing only 15.5cu. ft of air per minute. Obviously these flow figures do not bode well for the twin carb manifold.

At this point we mustn't fall into the trap of thinking that just because a twin carb manifold employs two carbs half the work is done by each because only two cylinders have to draw through each carb. It must be remembered that all induction pulses occur at a different time, and therefore there are no overlapping induction pulses since at any one time, only one cylinder is drawing. **As a result, each cylinder sees only the cfm of the carb on the end of its particular port.** It's for this reason that although the twin carbs have more total airflow capability, no extra power is realized because that total airflow capability is not available to any one cylinder. Probably by now you are a little disappointed with your twin carb manifold, well fear not. It's not necessarily the end of the world. Fortunately this manifold can be improved quite markedly with just a few minute's work. Take a look at the cross-section of the intake manifold **Fig. 8.17**. At the top of the port you will see the junction between the balance pipe and the main part of the port. It is the edge nearest the manifold face which is the real bugaboo in the system. The greatest proportion of the air flowing in the manifold does so along the top side of the manifold. The sharp edge formed at the back of the balance pipe causes a considerable loss of flow efficiency, so a simple remedy is to put a generous radius at this point. **This simple move does more to improve the flow capability of the intake manifold than all the polishing and hacking and grinding in the main part of the passageway.** Sure, smoothing it out and blending in all the lumps and bumps does improve flow efficiency, but the major obstacle is that corner, so be sure to apply a generous radius at that point.

Once the preceding modifications are done, a twin carb setup will flow as well, and sometimes slightly better than a

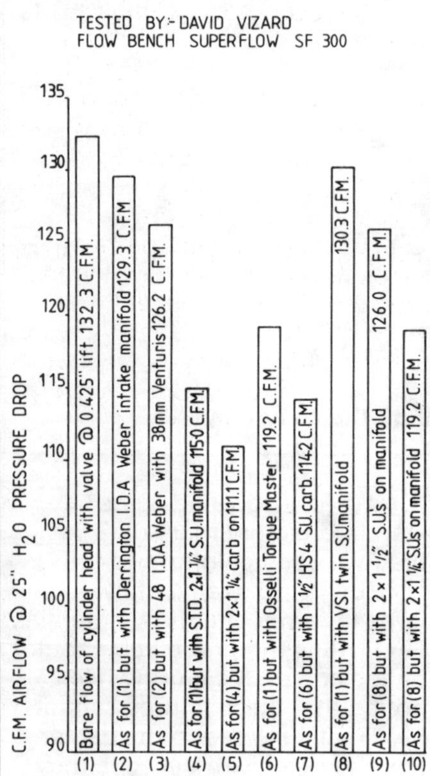

Fig. 8.16. Comparison of IDA Weber and various twin carburettor systems.

From this bar graph, you can see just how inadequate the standard Leyland 2 x 1¼ in SU installation is. In spite of having two carbs, its flow capability is bettered by a single 1½ in SU on a port-matched Oselli Torquemaster. It is probably the discovery of the fact that two 1¼ in SUs are not that good, that led to Leyland employing a single 1¾ in SU on a much more efficient manifold for later engines, including high performance examples such as the MG Metro. The VS1 twin SU manifold, designed jointly by the author and Carl Schattilly of Competition Grinding, produces flow figures on a par with the best Weber manifolds. Dyno tests show that the hp figures follow suit over most of the rpm range.

I was looking for the very best in SU induction systems (the Janspeed one at the time was not in production) which meant designing one from scratch. The design work started at the drawing board and was quickly transferred to the flow bench. Here, I worked closely with airflow specialist Carl Schattilly of Tucson, Arizona. After months of testing with a wooden mock-up . . .

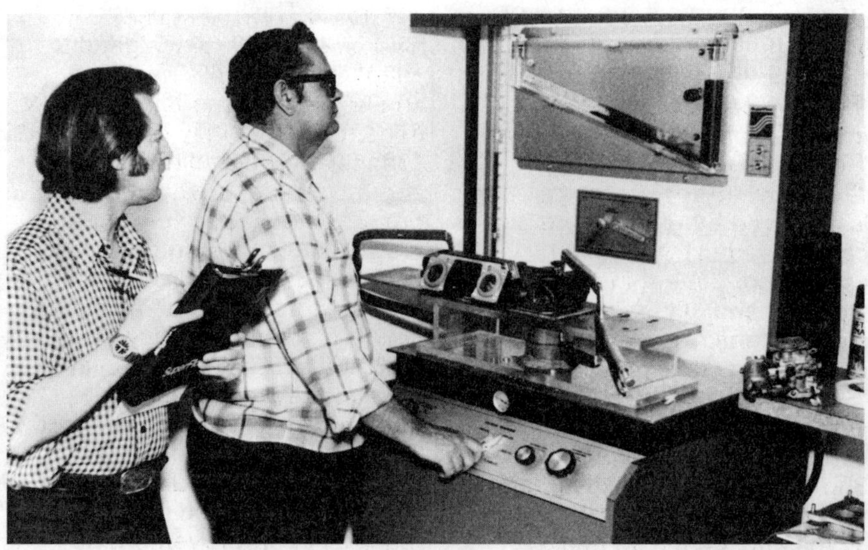

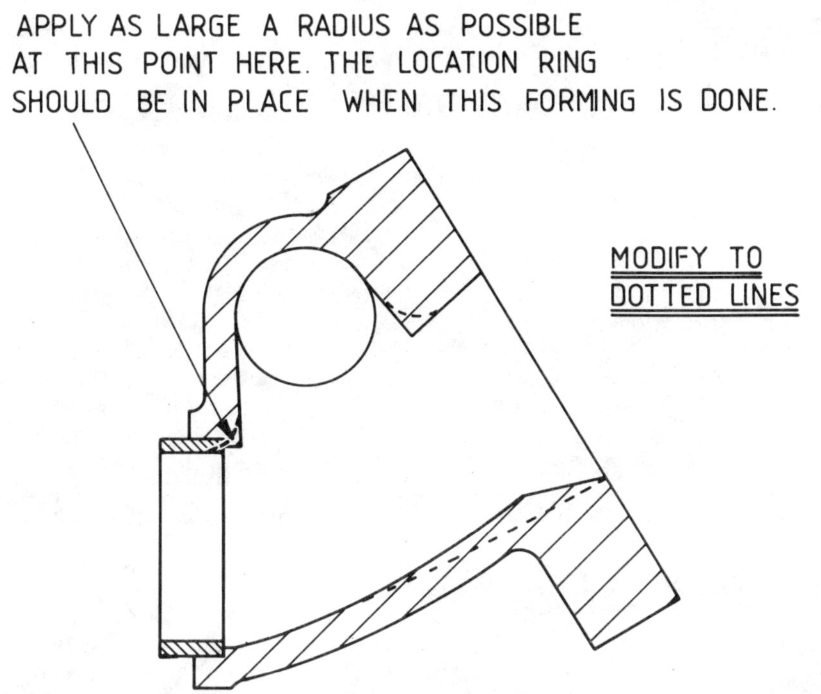

Fig. 8.17 Modifications to standard Leyland twin carburettor intake manifold

single 1½ inch or 1¾ inch intake manifold. Any standard intake manifold employing an H2 SU which they virtually all do, can also benefit from being modified to accept the larger H4 1½ inch SU. All that's necessary for the adaption is to place a gasket on the intake manifold at the carb end, and using a rotary cutter, cut the manifold out to the larger size, tapering it down to the port size at the other end. The business of radiusing at the point the balance tube joins is just as important, if not more so when bigger carbs are involved.

As to how much extra power is achieved by utilizing the bigger carbs is largely dependant on the capacity of the engine involved. On 850s and 1000s in a moderate state of tune, the power increase is insignificant. Highly tuned 1000 'S's begin to make use of the extra breathing of the bigger carbs, and generally show 2 or 3hp more when in typical racing trim. On bigger engines such as 1275s and 1400s, an intake manifold modified to take 1½ inch will generally show 2 to 4hp increase. This increase usually occurs from 4000rpm upwards.

The VS1 Manifold

A few years ago it occurred to me that there had never been a true race manifold designed to adapt SUs to the Series 'A' engine, so an American friend of mine, Carl Schattilly of **Competition Grinding,** and I set about developing what we thought would be the ultimate induction for SUs on the Series 'A'. The VS1 manifold was the result. As you can see from the flow figures in the graph, **Fig. 8.16,** it's capable of holding its own with even the best manifolds available, of any sort. Since it was capable of running at efficiencies of 99+% you can see that it would be hard to make much of an improvement upon it.

The concept of this manifold was a little different to that of an ordinary SU manifold. In essence, it can best be described as a tunnel port manifold. One of the basic reasons it flows well is that the balance pipe no longer really functions as a balance pipe, but as an additional port, so that air can be drawn from not only the carb directly aligned with the port that's drawing, but also the one diagonally opposite. This means that any one cylinder can see not one carb to draw through, but two. Well, it sounds simple enough in concept, but there is unfortunately much more to it than that. It's fairly obvious that it's easy to get the flow from two carbs into one port to function, but what is not an obvious point of realization is that the air that is crossflowing from the diagonally opposite carb must, sooner or later, reverse direction to feed the nearer cylinders. If the cross port or tunnel is made too small, the inertia of the contained air will quickly prevent it from being easily drawn from the diagonally opposite carb into the port that is making the demand. Because of this, the volume of the manifold is quite critical. As a result, it has been found that on bigger engines, a bigger cross-sectional area of cross or tunnel port is needed. Primarily this is to keep the gas speeds in it from being excessively high. Tuning the 'tunnel' was achieved by building the manifold up in a number of pieces. Basically there was the engine section and the section upon which the carb is bolted. These two halves could either be fitted directly together, or a spacer could be interposed between the two to increase the plenum volume. The manifold could be tuned on the dyno by adding spacers into the system until maximum power was seen. The bigger or the higher revving the engine was, the greater number of spacers it required.

Dyno testing showed that when sufficient plenum volume

... we progressed to testing a metal prototype, seen here with the carb in position. Flow tests with the metal prototype duplicated those of our wooden mock-up. Such factors as ...

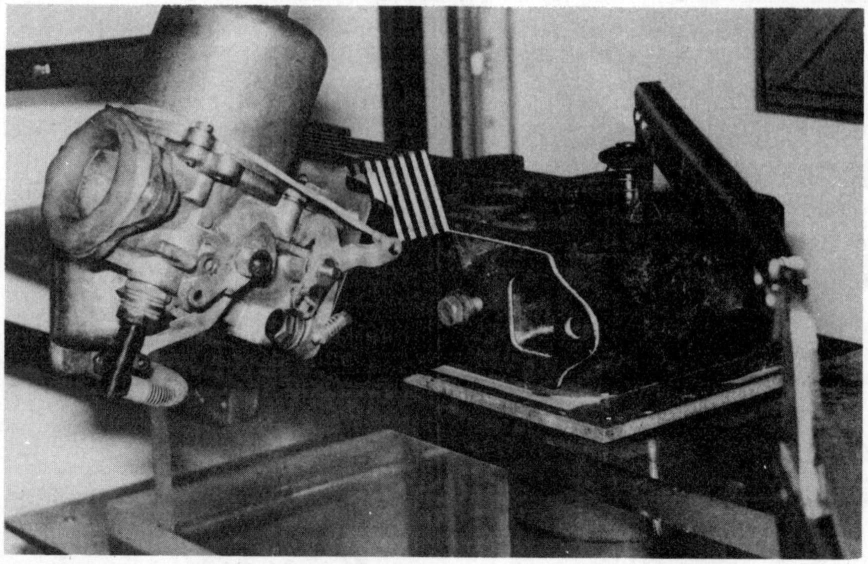

was adjusted into the system this manifold, with a pair of 1½ inch carbs, would match the power of a sidedraft Weber on a typical 5 inch manifold, even on engines in the 100+hp range. The only drawback to the intake manifold is, of course, that it's expensive to produce by comparison with manifolds having a single casting.

No doubt if you have looked at the photographs of the VS1 intake manifold, you will see that it has rather complex internals by comparison with most of the intake manifolds we have looked at so far. For instance, it has a vane just at the edge of the tunnel. Essentially this is used to counteract air rotation down the port. During testing initial designs, it was found that there was so much 'cross draw' from the diagonally opposite carb, air would swing into the port and develop a high-speed rotation on it's way down the intake port. This would have the effect of centrifuging the fuel onto the port walls. Placing the fin in the intake port largely eliminated this phenomenon.

Astute readers may well ask, why go to all that trouble to design an intake manifold to utilize twin SUs that only produce the power of a single 45 sidedraft Weber? This is, indeed, a good point to raise, but anyone who has ever driven a Weber carburated Mini will know that the hp produced by Weber is also accompanied by lots of induction noise. Also, adapting a Weber to a long intake manifold on a Series 'A' engine installed in a Mini, inevitably means cutting away the bulkhead. The VS1 manifold can produce virtually the same power as a 5 inch long Weber intake manifold, and since there is plenty of room between the carbs and the bulkhead, adequate space is left for large air filters which would have the effect of deadening the already quieter running SUs to a level very close to factory original. This means a useful increase in power, without a big noise penalty.

... airflow rotation were investigated. I believe that rotating the air in the port causes it to centrifuge the fuel out and the fin seen here in the photo, was added to prevent air rotation in the port. A piece of thread going into the port was used to check the steadiness of the flow right down to where the port splits up to feed the two valves.

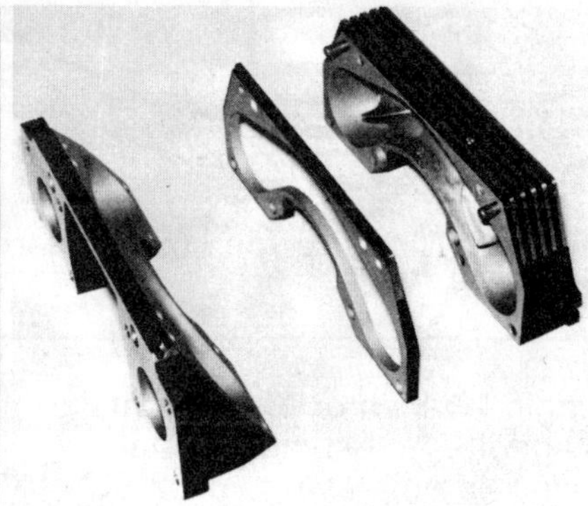

Here's a shot of a dismantled VSI intake manifold. The addition of more spacer plates allows the volume of the manifold to be tuned to suit different capacity vehicles.

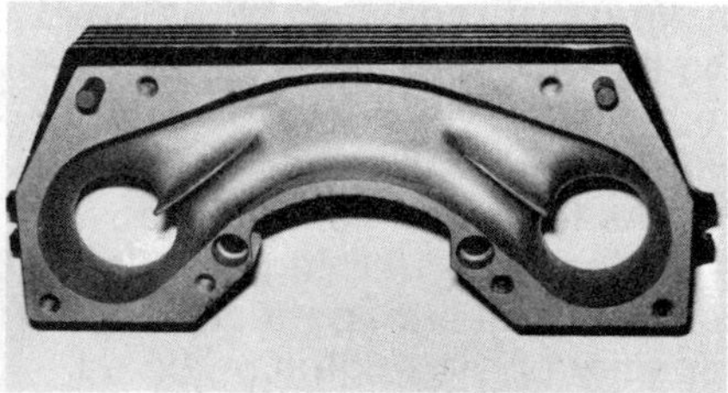

Here's a view of the internals of the intake manifold showing the positioning of the fins in the linking tunnel port. The flow bench tests showed that the location of these proved quite critical.

IDA Weber Manifolds

From the carb chapter you will probably have gathered that the 46 and 48 IDA Weber carbs are amongst my favourite carburation instruments for the Series 'A' engine. Of course they will only function well if they are on a good intake manifold. **As it happens, there is only one intake manifold manufactured for the Series 'A' engine to take an IDA carb and, fortunately, it is good.** On the face of it, it would appear that a manifold with a right-angled bend immediately under the carb, could not possibly flow as well as a sidedraft manifold which has a straight shot to the head. As it happens, this is one of those rare instances where the turning of the air in the manifold has been catered for remarkably well, and as the flow figures in **Fig. 8.16** show, the IDA manifold ranks joint best with the best sidedraft manifolds.

The only company producing this manifold is T.W.M. in the USA. It will be necessary to match the 'as supplied' manifold to the port because the port diameter towards the cylinder head is necked down to suit those cylinder heads utilizing a relatively small port. Port diameters should be opened up to a minimum 15/16 inch, for best results on almost any engine from 1000 'S' to a 1600 stroker, and a well sorted combination can satisfy 1.5 inch port diameters.

Since the super trick manifold shown in the adjacent illustration never made it into production this item from Maniflow is your next best choice. APT uses this manifold extensively and it always delivers good results.

This is what the fully assembled manifold looked like. Dyno tests showed it to be very efficient, but its complexity, and the resulting high cost, prevented it from ever reaching production.

9 Cylinder Heads Part 1

It Pays To Know What's Available

Since the Series 'A' engine was first introduced as the diminutive 803cc power plant for the Morris Minor and Austin A30, it has undergone, amongst other things, numerous cylinder head changes. For the most part, these different cylinder heads were each designed to fulfil a specific function and, as such, it is not difficult to recognize the difference between potential high performance heads and those which are not.

To make life easier, it's best to split these heads into two categories and then analyse the pros and cons of various heads within each group. The differentiation of these categories is one that is used frequently throughout the book, namely small and large bore engines. Engines defined as large bore, are all those using the bore of the 1275 unit, *i.e.* 2.780 inch. This range of engines includes 970s, 1071s and 1275 engines of all types. All other engines are small bore engines, whether they be the very small bore of the 803 or the 850, or the slightly larger bore of the 998 or 1100.

Although there are some major differences in design, heads from one group of engines can be fitted to the other. Apart from the very obviously different chamber form, heads on the big bore engine have wider spaced valve centres. This in itself allows larger valves to be installed in the heads and large valves are always an asset if you are looking for high rpm/hp. Indeed, it is possible to fit heads from large bore engines onto small bore engines, a move frequently made by engine builders specializing in 850 Mini race engines. However, the exhaust valve overhangs the block and it's necessary to cut away the block to give clearance to the valve.

How about changes in the reverse direction? Yes, no problem here. However, putting the head from the small bore engine on a big bore motor will, in all probability, cause a fairly severe reduction in hp, simply because most small bore heads have small valves and in standard form, have restricted breathing capabilities.

Fig. 9.1 Cylinder head specification guide chart

Group	Engine type	In. Valve Diameter	Ex. Valve Diameter	Chamber Volume	Casting Numbers	Notes
A	803 Austin/Morris, 850 Mini, 948 Morris Minor, A35 Austin, A40 Austin, Sprite MkI	$1\frac{1}{16}$ in	1 in	24.5 cc	2A 628, 12A 1456, 2A 629	Round port on intake.
B	997 Cooper Mini, Austin/Morris 1098, MkI MG Midget MkII Sprite	$1\frac{5}{32}$ in	1 in	26.1 cc	12G 202	Square intake ports.
C	998 Cooper Mini, 1098 MG 1100, MkII MG Midget, MkIII Sprite	$1\frac{7}{32}$ in	1 in	28.3 cc	12G 206 12G 295	Large square intake ports with location rings.
D	1275 Austin/Morris, 1300 Austin America* 1275 Sprite & MG Midget, 1275 Mini GT	$1\frac{5}{16}$ in	$1\frac{5}{32}$ in	21.4 cc	12G 940	Good ports in standard condition.
E	MG 1300, 1300GT & Last 'S's	1.4 in	$1\frac{5}{32}$ in	21.4 cc	12G 940	–
F	All 'S' engines other than in group E	1.4 in	$1\frac{5}{32}$ in	21.4 cc	12A 185, AFG 163	12A 185 was early casting coded 'yellow'. AFG 163 is later colour coded brown or blue.
G	Turbo MG	$1\frac{5}{16}$ in	$1\frac{5}{32}$ in	21.4 cc	12G940	Sodium cooled exhaust valve and improved cooling but thinner deck.

**Includes air–injection 'smog' head under other casting number*

Incidentally, always use the gasket which corresponds to the cylinder head in use, not the block.

If you are looking for inexpensive hp from your engine then it's important to be aware of the heads available, especially if the engine concerned is a small bore one. There are swaps that can be made that will result in an easy route to extra power, but the reverse can also apply.

Take a look at the accompanying cylinder head specification guide chart, **Fig. 9.1**. Within that chart I have grouped engines into seven different groups from A-G. The first three of those groups are heads relating to small bore engines; the last four to big bore engines. Generally speaking, each group of engines has a specific type of cylinder head or pattern of head. Although there may have been casting number variations, the design within each group has, over the years, remained substantially the same. Also, the way they are listed in the specification chart gives a guide as to their performance value; cylinder head A having the least performance value while heads at the other end of the chart tend to have the most. I say 'tend' because at this point in time we are looking at the power potential of bolting an alternative head straight on. I bring this point up because the potential for a reliable high performance head is better from a 12G-940-style casting (especially the air injection 'smog' castings) than for any of the 'S' castings including the last of those produced.

Starting off with the heads in Group A, which include such machines as Austin A30s, on up through the performance range, all the way to the 948 Mark 1 Sprite, we find that this basic head configuration was used on all single carb engines and twin carb Mk.1 Sprites. Although functional in many respects, *i.e.* the head produces good economy figures, it is certainly not a big power producer. As far as performance is concerned, the head has inadequate flow due to (a) small valves, (b) having those small valves shrouded, and (c) not particularly good porting. It is for this reason that modification of this particular head, even by someone with very limited knowledge of what is required of a high performance head, generally results in a very noticeable power improvement. **Because this type of head is bad in standard form as a power producer it's hardly surprising that any reasonably well thought-out modification produces good results.**

In fact, many head grinders cut their teeth on this type of head and because acceptable results are easily produced many became firmly convinced they were the 'bees knees' at modifying heads. This very factor could have been influential in bringing about the very slow introduction of flow benches to improve the quality, and speed up the rate, of cylinder head development in

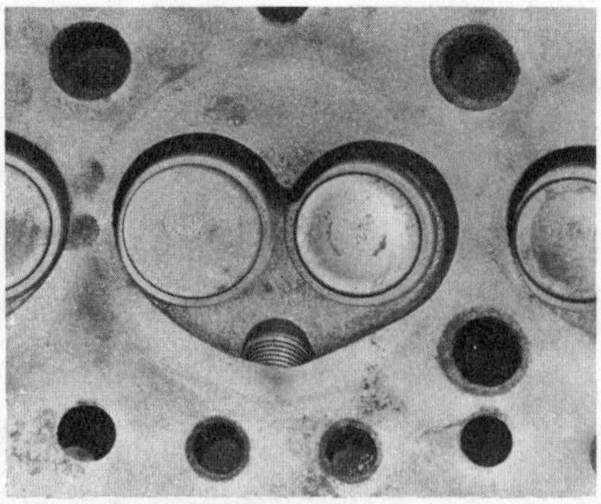

Shown here is the original Wes lake design for the Series 'A'. It has efficient combustion but it lacks good breathing capability due to valve shrouding and other, more obvious, factors such as small valves.

the U.K. Anyway, I digress. Although there is power potential to be extracted from heads in Group A, they are not necessarily the ones that you should choose if you are looking for bolt-on power from a standard head. The heads in Group B have larger valves and intake ports, and generally flow about 5% better than the heads in Group A. However, they also have a larger chamber volume at 26.1cc nominal. If you replace the cylinder head from Group A with a cylinder head from Group B having the same combustion chamber volume, the engine will generally be found to give 2 to 3 hp more. Apart from the casting number, identification of these heads is simple enough. Take a look down the intake port and you will see that it's of square cross-section, as opposed to a round port of some 15/16 inch diameter (a little less on the very early 803 heads and some of the current production heads.)

Although they can give a little boost in power when replacing a lesser head, heads in Group B cannot be considered real stormers; but the heads in Group C are. These heads were used on the 998 Cooper and the 1098cc engines powering MG saloons, Midgets and Sprites. This type of head has a much larger inlet port and bigger valves than the previous two types of head, plus a combustion chamber which is much less shrouded and superior intake and exhaust ports. It also has the largest chamber volume of any Series 'A' head at 28.3cc. If this type of head is used to replace any of the previous types of head so far discussed, then a very worthwhile power increase can be expected, simply because it flows so much better as **Fig. 9.2** shows. Assuming the head is skimmed to bring the compression ratio at least back to that of the original head it replaces, then a power increase of around 6 hp can be expected on a typical 1000cc engine.

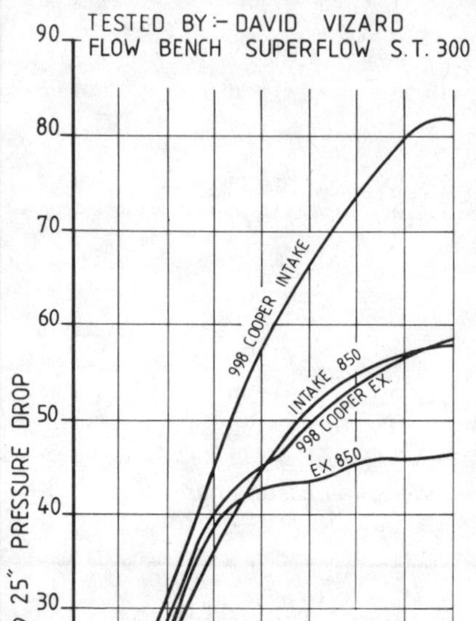

Fig. 9.2 Comparison of 850 Mini cylinder head (2A 628 type casting) and 998 Mini Cooper cylinder head (12G 295 type casting)

The superiority of the 'open chamber' 12G 295 head is clearly shown here by flow bench tests. Even in standard form, it is capable of outflowing most modified heads from the lower performance categories such as those fitted to 850 Minis, 948 engines etc. It is a good head to use as a basis on a high performance engine, but less improvements in flow can be made because the factory have already done the lion's share of the work.

Ignoring the very rare 12G-206 head, the 12G-295 head itself has not been in production for many years, and although they are still obtainable, they are becoming rarer as time goes by. As a head for use on small bore engines for anything but competition purposes, this head has a lot to commend it, especially if you are considering modifying it.

Moving onto the group for big bore engines, we find that all these heads bear a distinct resemblance to the 998 Cooper head. The principle difference between these and small bore heads is that they have bigger valves, the valve centres are further apart and the cylinder centres are much nearer to being evenly spaced. The first use of these heads was, of course, on Cooper 'S's. The Cooper 'S' head had valves as big as could be accommodated without their touching. These were 1.4 inch for the intake and 1.218 inch for the exhaust. Good for breathing and power. The trouble was, these heads tended to crack. In most instances the cracking occurred at the narrowest point between the valves. Although casting changes were made on the later 'S', the cracking problem was never really resolved until the reduction in the exhaust valve size was made. The last 'S' cylinder heads used a 1.4 inch intake and a 1.156 inch exhaust. The type of head casting used

Sometimes the factory will change a particular part but not the part number associated with it. Here is an early pre-MG turbo 12G 940 cylinder head. Notice that the valve cover face of the cylinder head is one flat, machined surface and is only sculpted within the confines of the valve cover. This is the way to tell the difference between the early non-turbo head and ...

on the final production run of Cooper 'S's was the common 1300 12G-940 casting. When used on the more mundane vehicles, this cylinder head utilized a 1.3 inch intake valve instead of the 'S' 1.4 inch valve. (The 1.4 inch intake valve size was also shared by the MG 1300, which also shares the 'high performance' category with the 'S'.)

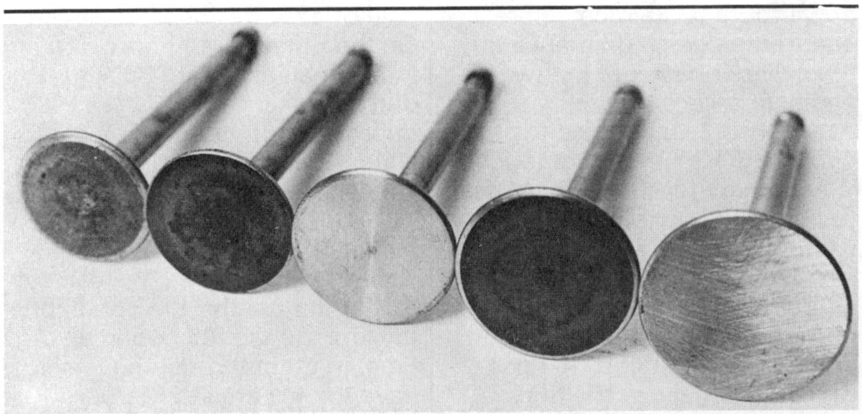

The range of inlet valve sizes for the Series 'A' engine is shown here. The valve on the right, for a Cooper 'S' has just about double the area of its counterpart at the other end of the line.

Looking at valves just for a moment, we find that although many cylinder heads used a similar size of valve, there was a tendency, where heads were used on high performance engines, to make the valves of a superior material. For instance, all 'S' engines used Nimonic valves. This material is well suited for competition, but the inlet valve used in the MG 1300 and 1275 GT as well as the last of 'S's was of different material. Apart from this, all heads based on the 940 casting as opposed to 'S' castings had a different spring deck height, and as a consequence, used a different length of valve to compensate. If you are swopping 'S' valves into a 940 casting, be careful. These valves are 50 thousandths shorter and springs

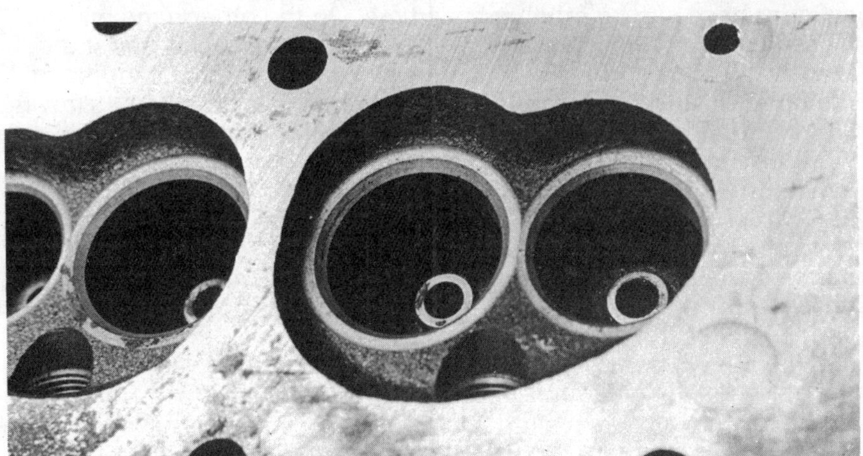

The current combustion chamber shape for 1275 engines, unlike its early counterpart, shows almost no unnecessary shrouding of either intake or exhaust valve.

. . . many later 12G 940 heads including the turbo head. Here you can see that the top face is sculpted around the thermostat cover, the water heater take-off at the other end of the head, and around the cylinder head fixing bolts that are outside the valve cover area.

used with them will experience coil binding at lesser lift figures.

Modifying Cylinder Heads

The only real difference between a Cosworth BDA and a Series 'A' engine is the cylinder head. The Cosworth BDA engine will produce in excess of 160hp per litre, whereas the Series 'A' engine is hard pushed to make 100 per litre. This difference in power output is almost entirely due to the design difference in the cylinder heads of the two engines.

To enable an engine to produce a high power output, it is necessary for the cylinder head to flow large quantities of air without necessarily having overly large ports. **Assuming the engine's bottom end is capable of high rpm, and a Series 'A' engine's is, then getting hp depends largely, in the first instance, on how effectively the cylinder head can be modified. Of secondmost importance is how well the cam can be matched to suit the dynamic characteristics of the head. Here I will discuss the cylinder head, leaving the cams for treatment in Chapter 11.**

Let's put a few things into perspective here; if we look at relative flow efficiencies at various points in the induction and exhaust systems, we find that it is relatively easy to produce an induction system which ranges from 90% to 98% efficient. An exhaust manifold can fall into a similar category, but to actually come up with an intake and exhaust port, or more precisely, an intake or exhaust valve, that flows as efficiently is another thing altogether. The flow efficiency of a standard valve arrangement in a Series 'A' engine cylinder head, is typically only 40% to 45%. If we raise that figure to 55%, an overall flow gain of 22% has been achieved. On the other hand, improving an intake manifold from 88% to 98% only represents an 11.3% improvement, but looking at efficiencies isn't the only criterion it's necessary to consider.

Looking at an entire induction or exhaust system, excluding the valves, we can say that under full throttle conditions these ports are fully open all the time. In other words, the gases have the full area of the port available to them 100% of the time. On the other hand, valves open and close, and whilst a valve is only 50 thousandths off its seat, it is incapable of allowing high airflow. **From the foregoing you have probably already deduced the fact that it is the valves that are the worst impediment to gas flow in the engine.** This means if the engine is to produce any worthwhile power, we need to really work overtime on the cylinder head to make it flow as effectively as possible, and in this section of the book, that's exactly what will be dealt with

Valve Ratios

If maximum power is the principal goal, then the flow capability of both the intake and exhaust must be improved as much as possible. In most Series 'A' cylinder heads there is enough metal around each of the valves to be able to increase their size quite substantially. The question which now arises is, which valve should we favour most? Should we improve the intake port and cram in as big an inlet valve as possible, letting the exhaust take care of itself? After all, the exhaust gas is being pushed out. Conversely, should we increase the size of the exhaust valve and improve the exhaust port as much as possible on the assumption that the gas volume it has to handle is much higher than the intake, simply because it's so much hotter and has therefore expanded? These are very valid questions and need serious consideration, though not necessarily in terms of valve size as is often done, but more in terms of port flow capability.

The first point that needs to be clearly established is that the biggest problem with a normally aspirated engine is getting the air into the cylinder in the first place. The only pressure available to fill the cylinders is that of the atmosphere around us. On the other hand, on the exhaust stroke, the piston can forcibly expel most of the exhaust. Over the years, experiments by many engine research establishments, not just with Series 'A' engines but with all types of engines, has shown that generally speaking the best usage of the available valve space is to distribute the flow potential such that the exhaust valve flows around 70% to 75% of the intake.

However what can be easily overlooked here is the additional factors which may influence this number and may not have been adequately catered for in the auto industry tests so often quoted. Most testing of this nature was, by necessity, done on dyno mule motors. Such motors must be repeatable so the engines concerned cannot be as highly tuned as a race or even high performance road engine. One aspect about a typical research mule motor is that it would have a compression ratio suitable for a road engine rather than a race engine. As it happens the compression ratio influences the optimum intake to exhaust valve ratio by quite a margin. *Understanding this can give the informed engine builder a significant advantage over an engine builder that does not. What you will learn in the next few pages will be a great help regardless of the type of engine involved.*

First let me say that there has been, especially in the popular automotive press in the US, a considerable amount of words written about this subject by writers who apparently have no idea of the controlling factors involved. After reading pages on the subject the reader ends up as confused as the writer. I do not pretend that I have an easy explanation that is 100% accurate but I will give you a very simple explanation that allows the understanding of the principles involved. Knowing these principles will allow a more educated selection of the valve size/cam spec/and CR combination. If you give that a little thought you will realize that having *a working understanding of the interaction of these elements is a major step toward developing an engine combination with more power than your competition*.

Although it is strongly influenced by the CR it is best to forget, for a moment, the compression ratio and deal only with the expansion ratio. This rarely mentioned factor was defined in an earlier chapter but it won't hurt to reiterate that it is nothing more than the other side of the coin to the CR. The expansion ratio is what happens on the power stroke and, the way we are dealing with it here, is numerically, exactly the same as the CR. To see the significance of the expansion ratio take a look at the drawing Fig. 9-2A. Illustrated here are the pressure decays curves from 1,000psi for two cylinders, one with a 2:1 expansion ratio and the other 15:1. If each starts with the same pressure we find that the pressure drops more slowly as the piston goes down the bore on the 2:1 cylinder than the 15:1. When the piston reaches the bottom of the stroke the pressure remaining in the 2:1 cylinder is 37.8% of the starting pressure and the 15:1 is 2.26% of the starting pressure. The significance of this is that there is a lot of potential energy left in the 2:1 cylinder

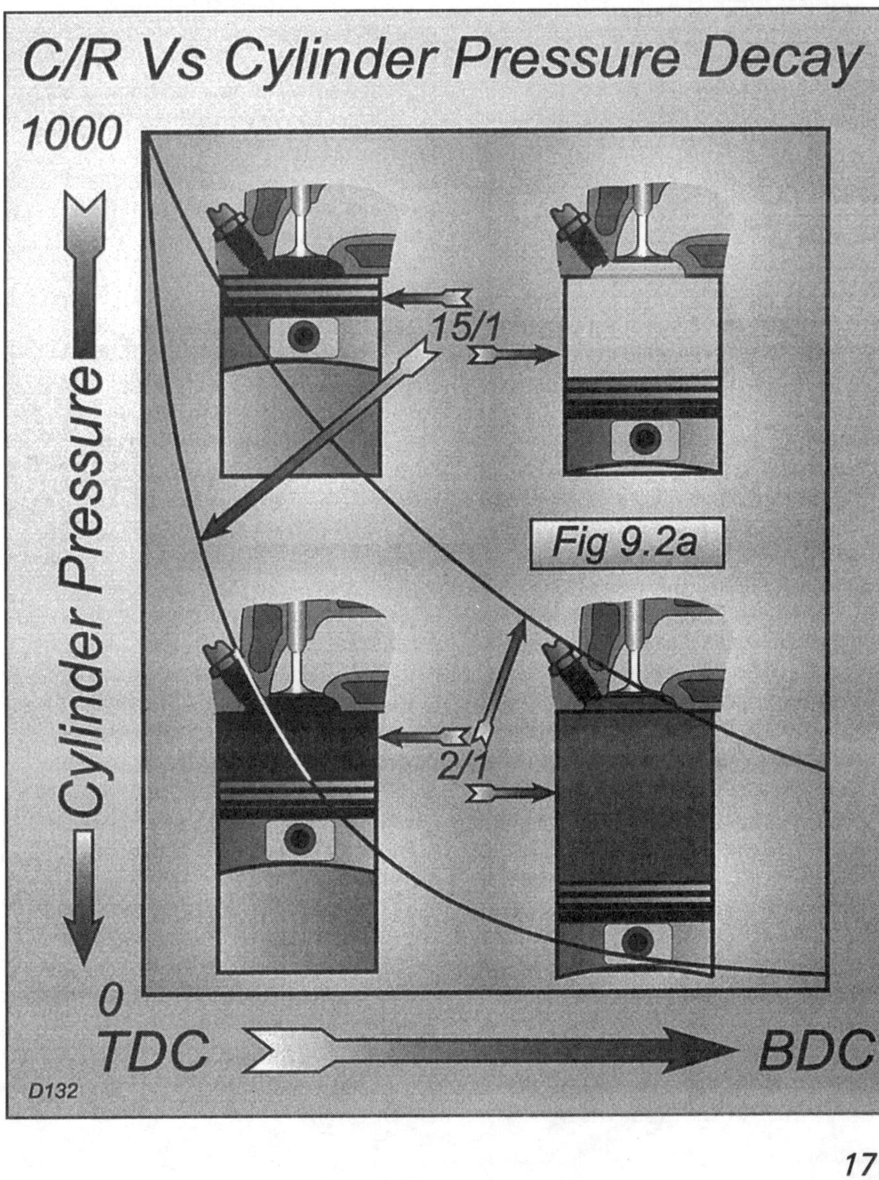

even when the piston has travelled all the way to the bottom of its stroke. In such a cylinder opening the exhaust valve much before the bottom of the stroke is going to dump pressure that could be used to turn the crank. Now let's consider the 15:1 cylinder. Its pressure has dropped to a fraction of the starting pressure so opening the exhaust valve a little earlier won't dump anything like as much usable gas pressure as the 2:1 cylinder. The implication here is that a low compression cylinder needs to have the exhaust valve opening delayed until the piston is much nearer bottom dead centre than a high compression cylinder. By delaying the exhaust valve opening we are in essence giving the exhaust a shorter time to blow down the cylinder before the piston starts its upward journey on the exhaust stroke. This in turn means that if exhaust pumping losses are to be minimized the exhaust valve needs to flow correspondingly more, and that usually means having a bigger exhaust valve than is the case for the high expansion ratio cylinder.

Let's continue with an analysis of the 15:1 cylinder here and see how this fares. Because the pressure drops faster and finishes at a lower pressure there will be much less of a penalty if the exhaust valve is opened earlier. If this is the case we can open the exhaust valve longer and achieve the desired outflow of exhaust with a smaller exhaust valve. This in turn allows a bigger inlet to be used. From this you should get the picture but to put it into words, we can say that the higher the CR becomes (and as a result the expansion ratio) the less exhaust flow is needed in relation to intake because the valve can be opened earlier and held open longer. High compression engines are almost always high rpm units and as the rpm the engine is called upon to turn at increases, so the difficulty of filling the cylinder increases. The ability to use a larger intake valve at the expense of the exhaust valve becomes very much a positive asset toward increased volumetric efficiency at high rpms.

In our hypothetical example comparing two widely differing expansion ratios we used 1,000psi as the starting pressure in both example cylinders. In reality the combustion pressure in a 15:1 cylinder will be significantly higher than a 2:1 cylinder. A working rule of thumb is that the peak combustion pressure reached is about a hundred times the compression ratio. This means, in round numbers, that a 10:1 ratio motor will see peak cylinder pressures in the order of 1,000psi. We are now talking about the compression ratio rather than the expansion ratio. The compression put into an engine, especially an 'A' Series is very important and should be done with due consideration for the cam to be used as well as the fuel octane. For more information on the compression ratio and its effect on fuel requirements and cam timing refer to the end of part 2 of this chapter. However, because the Series 'A' has siamesed intake ports, it cannot really make full use of intake pressure wave tuning. In fact to be honest, I really don't think it can make ANY use of pressure wave tuning because such a jumble of waves exists. Because of this factor, it seems beneficial to favour the intake valve a little more on an engine utilizing a siamesed intake port. On engines of 13:1 or less I size the exhaust so that it flows about 65% to 70% of the intake. Looking at it the other way round, the intake needs to flow between 43% and 54% more than the exhaust.

Let's continue for a while on the business of valve size/flow distribution, so that the point is fully driven home. When building a high performance engine, it's fairly obvious that valves as large as possible will be required. The sizes of these will be limited only by the space available for them to pass down the bore and the margin needed to avoid the head cracking between the valve seats. In other words, if maximum power is sought, valves need to be as big as possible short of causing a mechanical problem. Under these conditions, the valves for a five port Series 'A' engine should be sized such that the exhaust valves' flow capability meets the 65% to 70% criteria mentioned earlier. In other words, the intake valve should flow between 43% and 54% more. If a cylinder head is modified such that the intake flows 60% more than the exhaust, this does not mean that the intake flow should be reduced to put matters right. Basically what it means is that the exhaust valve size should be increased slightly at the expense of some intake valve size. If all the available space in the combustion chamber is not taken up by valves, then it's obvious that the engine is not being utilized solely for the purpose of producing maximum power. Under these conditions the ratio of flow of one port to another becomes less critical.

For cylinder heads being modified with a strong bias towards economy, a different situation exists. So much so that this subject will be dealt with separately later in this chapter.

Porting Basics

To be an expert cylinder head modifier, one really needs the combined skills of a sculptor and an engineer. Also, to really modify cylinder heads right up to the level of the current state of art, and past it if you are trying to break new ground, requires certain basic equipment. Apart from the tools *i.e.* grinders, cutters, mills, presses, etc., one really needs equipment to test the effectiveness of any modifications.

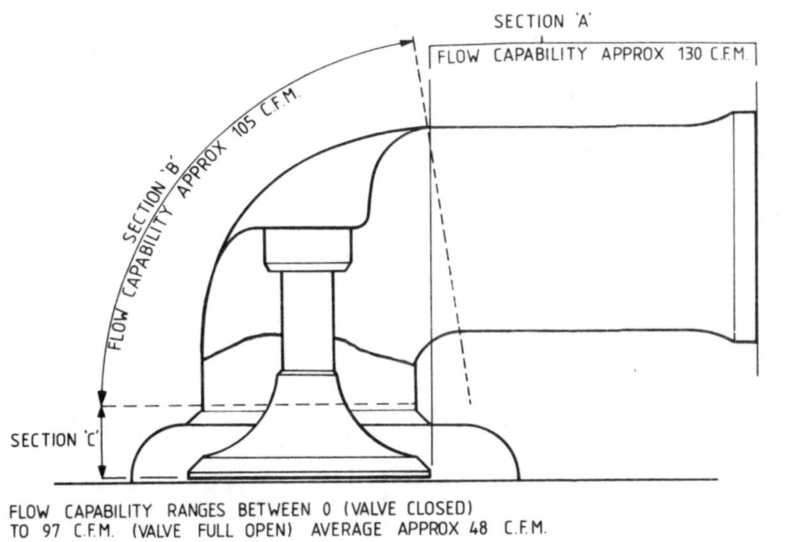

Fig. 9.3 Flow capabilities of various parts of a Series 'A' intake port
This illustration shows a 12G 940 casting as used on big bore type engines.

First and foremost of such equipment is a flow bench to see that the airflow has, in fact, been improved and secondly a dynamometer to determine that the airflow increase has resulted in more power. As would be expected, due to both expense and training, most enthusiasts do not have formal engineering training in automotive engine design, nor do they have dynamometers and flow benches. It is for these reasons that I will detail the cylinder head modifications very thoroughly. This should obviate the need for you to own any expensive test equipment and/or go on a five year engineering course! However, if you are going to produce your own cylinder heads, you will still need the sculptural skills which, incidentally, aren't that difficult to acquire. Practice on a few scrap or unwanted head castings will usually give you sufficient proficiency to produce a good working cylinder head.

Detailed elsewhere in this chapter is the sort of equipment you will need to do the donkey work to grind your own cylinder heads.

In putting this chapter together, another aspect I have considered is that many readers may not want to grind cylinder heads, but this doesn't mean the following is wasted. Those not wanting to do the work of modifying their own heads should still read on as it will supply the knowledge of what is required of a cylinder head. This should allow the purchase of a good, rather than bad, cylinder head. If ignorance prevails in this area and an indifferently ported cylinder head is purchased, the engine will certainly not produce of its best.

If you ask a professional tuner to port a head for you based on a drawing given in this book do ensure he follows the profiles, etc to the letter. If a variation or error occurs, no matter what the explanation, it would be quite reasonable for you to expect full recompense in the form of a replacement head or the price of the same item.

Port Priorities

Before getting into the fine detail necessary to produce a super Series 'A' head, let's look at a few factors to put things into perspective so that you don't waste time modifying parts of the head that achieve nothing and then leave other parts that DO need to be modified.

One of the favourite areas for amateur porters to modify is the siamese section between the pushrods. **The average uninitiated head modifier will rework the siamese section with a grinder, enlarging and smoothing this area to the utmost, believing that this will produce good airflow. Unfortunately it doesn't.** To see why, let's consider the flow capability of various parts of the Series 'A' intake port. **Figs. 9.20 & 21** (pages 203-204) is a typical standard port for the 12G-940 casting. You will see it's divided up into three sections: section A, that is the section from the manifold face right up until the port starts to make a turn. This section includes the siamese port. Section B is the section that attempts to turn the air through 90° to direct it to the back of the valve. Section C is the valve and valve seat from its closed to fully open position.

If the cylinder head was cut and each section flow tested independently, section A, at 25 inch of water pressure drop, would flow about 130 cubic feet of air per minute (cfm). This is because it is a relatively straight, streamlined part of the port, and as such, is not bad!

Section B, with its bend, plus the valve guide boss and the stem of the guide, is somewhat less streamlined, but nevertheless quite open in its form. This part of the port will flow about 105cfm.

Section C will now be considered. Here, in effect, is an aperture of changing shape. When the valve is closed there is nowhere for the air to flow. As the valve is lifted off its seat, so the available flow area increases from zero to a maximum value usually (but not always) at full valve lift. Unfortunately the passage around the valve is not only a tortuous passage, but also it ranges from zero area when the valve is closed, up to approximately the area of the valve minus the area of the stem when the valve is at full lift. The last factor does depend on the amount of valve lift. As such, this part of the port will flow somewhere between zero cfm to about 100cfm but because the valve goes through an opening and closing phase, it only averages around 50cfm. **Looking at the entire port from the manifold face to the chamber side of the valve, we find, depending on the amount of valve lifted, that the port will flow somewhere between zero cfm and 100cfm and will average around 50cfm.**

At this point it's worth posing a question: how much more air will pass through the port if section A is improved from 130cfm to 150cfm? Answer: almost none whatsoever simply because this area is not the flow bottleneck. The worst part of the port is right in the vicinity of the valve seat. Improvements to section A only show an overall port, improvement when section C has been modified in such a way as to improve its own flow capability dramatically. Hopefully this little analysis of the port will serve to show that the most important part of an induction system, not just the cylinder head, but the induction system as a whole, is the area ½ inch before the valve to about ¼ inch after the valve. Although gases are going out the exhaust port in the opposite direction, the same sort of criteria applies. What happens to the gases in the immediate vicinity of the valve eventually determines the overall effectiveness of the port.

Polishing of ports is a favourite pastime, but polishing ports does little or nothing to improve flow. A smooth, cast finish will normally flow within a hair's breadth of the figures produced by a highly polished port. In practice it is sometimes found that highly polished ports cause problems in terms of fuel puddling and therefore may not, in some instances, produce the results being sought. What is important for an effective cylinder head is shape. **A ground finish the right shape will produce hp; a polished finish the wrong shape will not.**

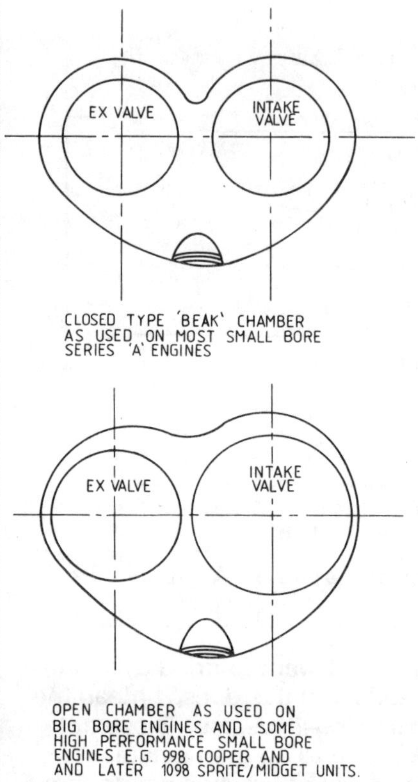

Fig. 9.4 Combustion chambers of 'beak' and 'non-beak' type

Another porting trap not to fall into is that big is beautiful. Remember, big holes may flow a lot of air, but port velocities will be reduced and as a result, any inertial ramming of the cylinders will be reduced. The right size port is one which is not too big and not too small.

Simple Porting Modifications

Just how much effort you are prepared to put into modifying your cylinder head depends upon time and inclination, as much as whether or not the necessary equipment is available. In an effort to provide something for everybody, even those who have the barest essentials to work with, I'll start off with some very simple but effective modifications to the Series 'A' cylinder head.

At this point, let's put the cylinder heads into two groups as defined by combustion chamber. One group has the closed combustion chamber; the other, the open combustion chamber. The closed type has the pronounced beak as shown in **Fig. 9.4.** This type of cylinder head is used on 850, 998 and 1100 engines with the single carb specification. All these engines are small bore units, *i.e.* non-Cooper 'S' size bore.

There is a variety of casting numbers for this general type of head, but at this stage they can be lumped together and treated as one.

As far as open chamber cylinder heads are concerned, there are only two types fitted to small bore engines. One of them is pretty rare and goes under the part number 12G-206. The other more common one being the 12G-295 casting, although now even this is becoming comparatively rare. These heads were used on 998 Coopers and the MG110 and 1098 Sprites and Midgets from about 1964 on.

The other type of open chamber head casting can be termed the 'S' type, 'big bore' casting. All these head castings were fitted to big bore engines, and came with a variety of intake and exhaust valve sizes ranging from 1.312 inch to 1.406 inch for the intake and 1.156 inch to 1.215 inch for the exhaust. The small bore, open chamber head bears a remarkable family resemblance to the big bore open chamber heads, and although used on a substantially different bore size, at this point in time the two can be considered jointly. Although

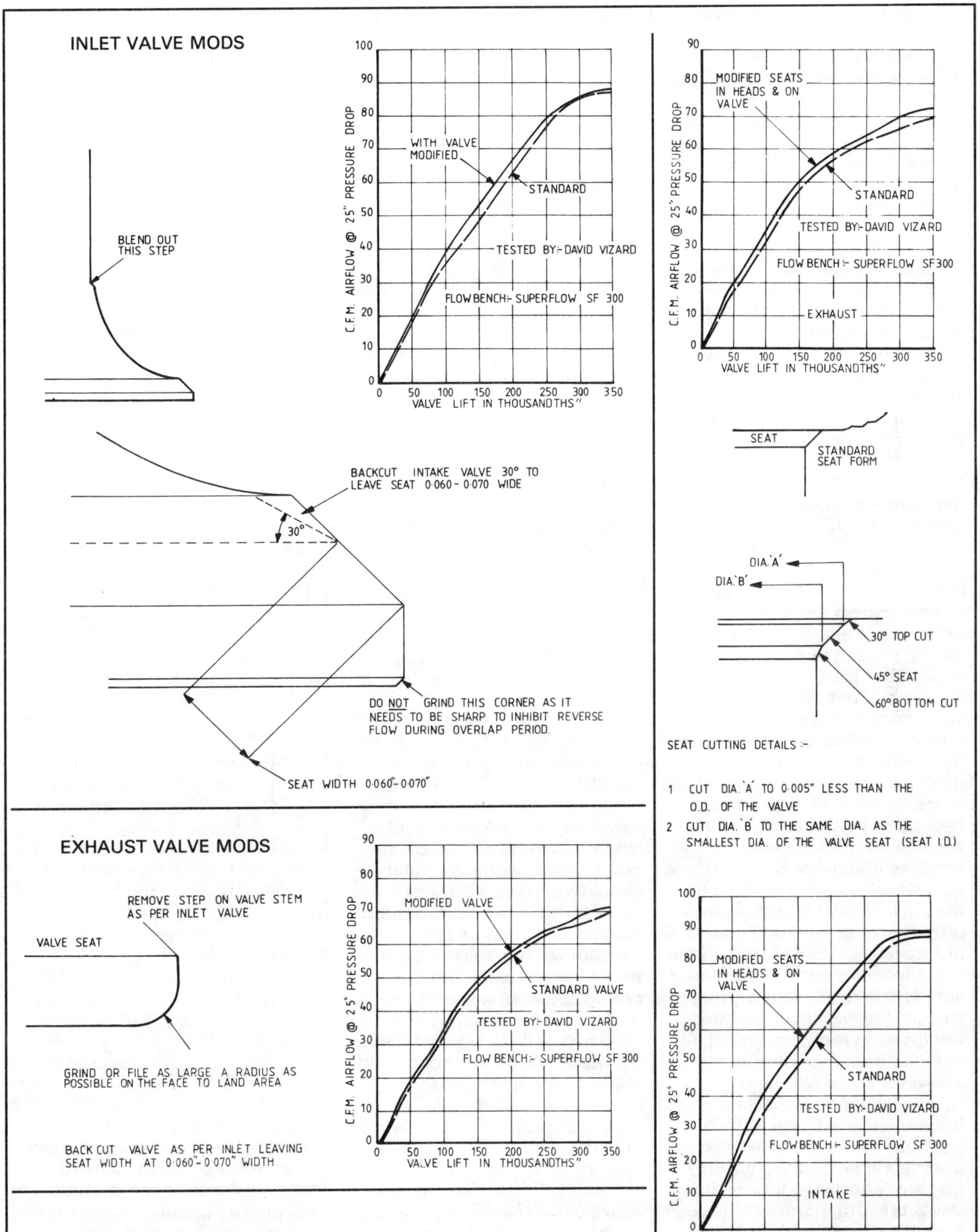

Fig. 9.5 Valve seat and face modifications

the numbers as far as cfm is concerned may go up or down depending on the valve size a good middle of the road head to use as a demonstration to show how much simple modifications can improve the airflow is the 12G-940 head casting. This is the most common head casting currently available on the big bore engines.

In keeping with the promised simplicity, the first modifications to be dealt with relate to the valve and valve seat. **Fig. 9.5** shows these modifications applied to both the intake and exhaust along with the airflow increases given on an otherwise standard cylinder head. In simple terms, these mods entail a 30° back cut on the valve, and 'topping' and 'tailing' the seat in the cylinder head. They are simple enough in themselves, but can easily fail to work unless you are aware of a couple of important facts. At first sight, thinning down the valve seat so that there is a larger aperture for the air to pass through seems like a good idea and one that should be continued until the valve seat is an absolute minimum width consistent with being able to support itself mechanically. This is a typical line of thinking for anyone who has never airflow tested a cylinder head. However, it's forgivable because it appears so logical that the thinner the seat is, the better the flow. The truth of the matter is that, in the real world it does not work out this way: if the seat width is made too narrow, the airflow, especially at low valve lift, is actually reduced. If the valve seat is too wide and has too sharp an edge on its approach it can reduce the airflow throughout the lift range. However, if the sharp edge is taken off the inner edge of both the valve and the seat, we find that at low lift, the air is directed towards the seat through the gap between the cylinder head seat and the valve seat, and directed out into the chamber much more efficiently. The net result is more low lift airflow. A very important factor, since the valve spends more time at low lift than it does at high lift.

The argument concerning valve seat widths applies to both the intake and exhaust, but more so on the intake.

However, the exhaust valve seat shouldn't be thinned down indiscriminately for another reason concerning reliability. A substantial amount of heat from the valve head is dissipated through the seat into the cylinder head. If the valve seats are made too narrow, the exhaust valve will have a tendency to overheat. This can cause two things (a) the exhaust valve will burn out easier, and (b) the engine will detonate easier. The last factor is definitely not so good, considering that worldwide fuel octane values are dropping and a high compression is needed for a Series 'A' engine to produce of its best.

As far as valve shapes are concerned, here again we find there is a subtle difference in the requirement between the intake and exhaust. On the exhaust valve, the radius blending the valve face into the valve land, needs to be large so that the flow of gases around the edge of the valve and into the port can be aided as best possible. **On the other hand, the intake valve should not have a land to chamber face radius. It should have as sharp an edge as possible without being so sharp that it causes a detonation problem.** Because the valve gets cooled by the incoming charge on every cycle, it need only have the merest hint of a radius at this point, so to all intents and purposes it is a sharp edge. The reason a sharp edge is needed here is to reduce the flow of residual gases from the combustion chamber past the intake valve during the overlap period. In other words, the intake valve is required to allow effective flow into the cylinder,

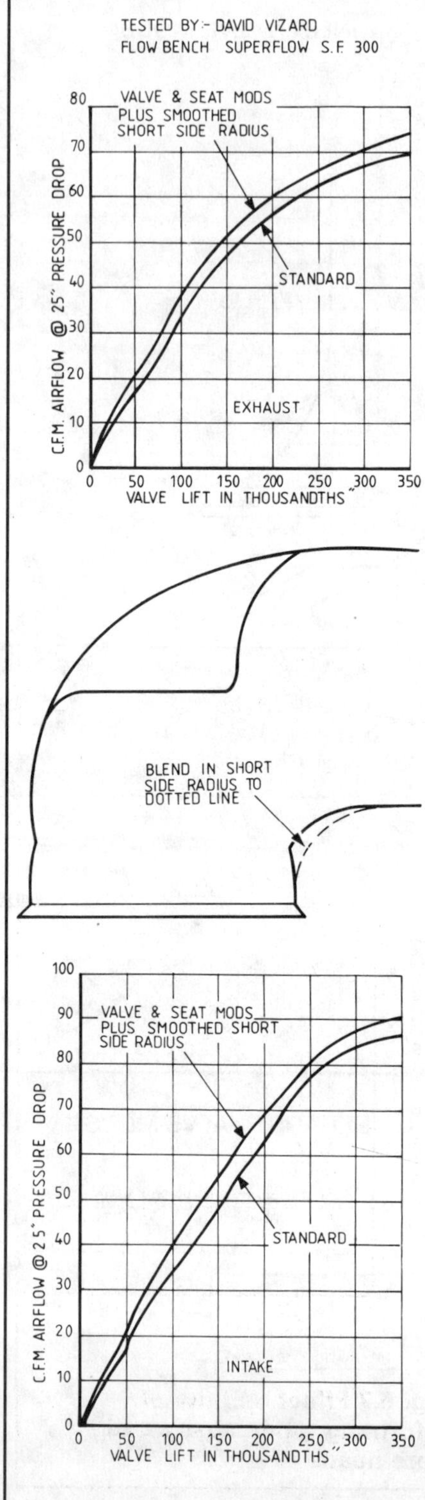

Fig. 9.6 Effect of blending short side radius of 12G 940 head

but not out into the intake port.

If you feel so inclined, whilst you are cleaning up valves (if you are using second-hand valves,

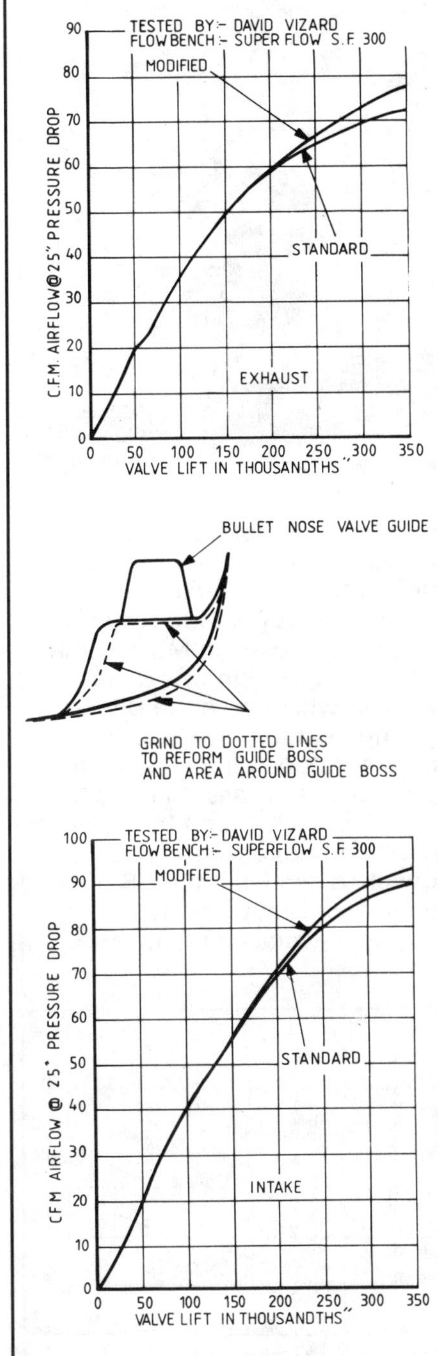

Fig. 9.7 Effect on flow of reforming guide boss on big bore heads

that is) it's a good idea to spend a little time with a grindstone to remove the forging step that is often left between the radius forming the back of the valve and the stem. This does add the odd couple of cfm to the airflow at

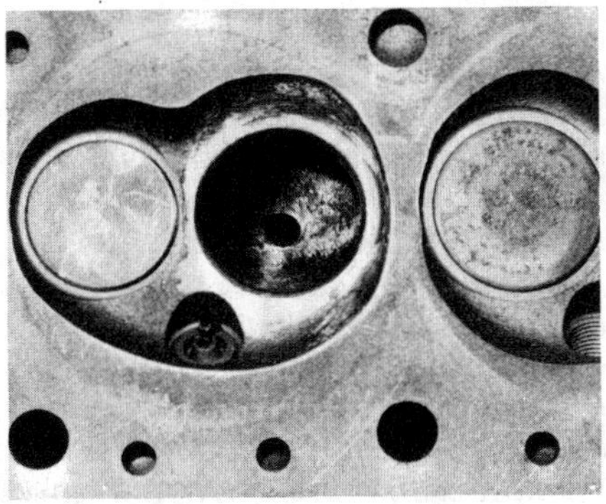

The port throat, or pocket as it's known in U.S.A., has been smoothed out and the valve guide boss cut to the shape shown in the accompanying drawings. After this work had been done, the new bullet-nose guide was installed. This work produced a respectable increase in airflow and subsequent polishing of it achieved no further gain in airflow.

the mid- and upper valve lift points. However, it's worthwhile remembering that flow at high valve lift is only important if the cam used happens to lift the valve to the point flow improvements occur. For instance, it is pointless doing a modification to the valve, port or whatever that improves the flow at 450 thousandths lift, if the engine uses a cam giving only 400 thousandths lift.

The next mods to do are going to entail some carving away of obtrusive cast iron down the throats of the port. If you refer to the cross-section of the intake port, **Fig. 9.6**, you will see on the short side radius, just a little way from the valve seat, a sharp edge left from the valve throat machining operation. It is difficult enough in practice to get air round a tight radius, and putting a sharp edge into the equation further complicates the matter. Fortunately, the problem is simply remedied. Using a grindstone of approximately ½ inch to ¾ inch diameter, this edge which exists in both intake and exhaust, can be simply removed, and hey presto, the airflow in the mid- and upper lift ranges, that is from typically 150 thousandths upwards, is improved. **Fig. 9.6** shows what happens in both intake and exhaust when this simple modification is done, together with the other mods previously described.

Now for the valve guide boss and adjacent areas. In most open chamber heads, the

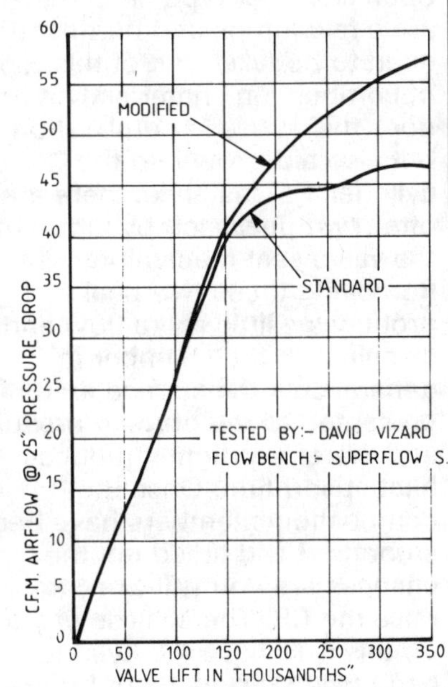

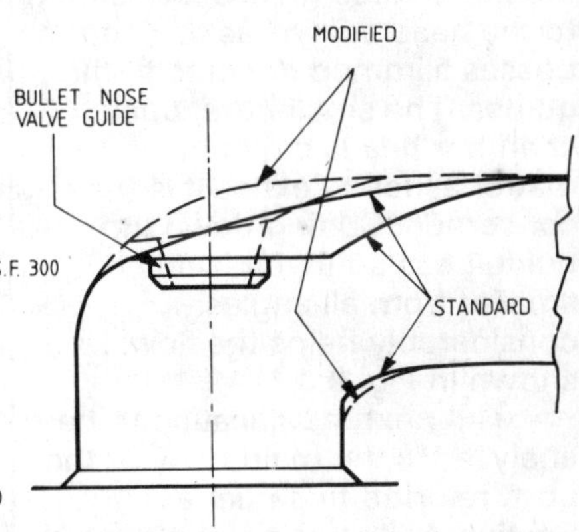

Fig. 9.8 Effect of guide boss removal on flow in exhaust port of heads fitted to small bore engines

guide boss is, to a degree, intrusive and has an adverse affect on flow in the mid-to-upper lift ranges. On the small bore open chamber cylinder heads, the guide boss can be very intrusive, affecting the flow even at quite low valve lift figures. Because it's so large, it considerably cuts the cross-sectional area through which the gases can pass. There are times when the cross-sectional area in a port can be reduced, as its effect on flow is minimal. On the other hand, the opposite situation can easily exist, this being one such occasion. In this instance, the reduction in cross-sectional area occurs right on the bend, this being the very point at which the cross-sectional area, if anything, needs to increase, so as to slow the gases down to allow them to take the turn into the straight part of the port. On big bore heads, there isn't a lot of metal removal, comparatively speaking, that has to be done to blend the guide boss in closer to the valve guide. **Fig. 9.7** shows approximately the trend on the intake of all open chamber heads when the guide boss is slimmed down to the guide. This job is most easily done with the valve guides removed from the head, and whilst they are out, it's a simple operation to bullet-nose the guides. These flow figures represent what happens when the bullet-nose guides are refitted to the head as well as having the bosses slimmed down to fit the guides. The small bore, open chamber heads are another matter as far as exhaust is concerned. Cutting down the guide boss so that it is much smaller from all angles considerably helps the flow, as is shown in **Fig. 9.8.**

The next modification to be analysed is the main body of the port, referred to earlier as Section A. At this point it's really only worthwhile removing the large lumps and bumps, and just generally smoothing out. It is not a good idea to consciously

Here's the intake port at the manifold face on a 12G 940 casting. The counterbore is the position in which the intake manifold locating ring sits to locate a twin carb intake manifold. It makes little or no difference to the flow whether or not the ring remains in place, as long as the port is modified to give the minimal obstruction to flow as the cross-sectional area changes.

enlarge the port to any degree. The smoothing operation will get the job done, and it will improve flow in the region of a couple of cfm in the high valve lift ranges.

Combustion Chambers

The last area you need to look at is the combustion chamber. On the open chamber castings, life is pretty straightforward but on the closed chamber heads, things are a little more complicated. Let's deal with the open chamber type first. There are a few lumps and bumps that need to be taken out of this type of combustion chamber. Not only does this improve combustion but also aids flow into the cylinder. For instance, there are often two peaks left by virtue of the valve seat machining. Removing these two peaks produces a little more flow and cleaning up the chamber in general cuts the surface area presented to the burning mixture, and therefore slightly reduces heat absorption. Once the combustion chambers have been smoothed and tidied up, the chances are you will want to raise the CR. The easiest way to do this is to have the cylinder head face skimmed, but before doing so you should be aware of the limitations involved. I would suggest a wise move would be to read the sections of this book concerning CRs, fuel octane and the factors that make an engine detonate. You should also be aware that some of the later engines, as used in Metros, may already have high CRs to start with, in which case the CR can't go much in an upward direction before detonation becomes a problem; don't assume you can just skim a chunk off the cylinder head and hope for the best. Check to see what the CR of your engine currently has by physically measuring it, as

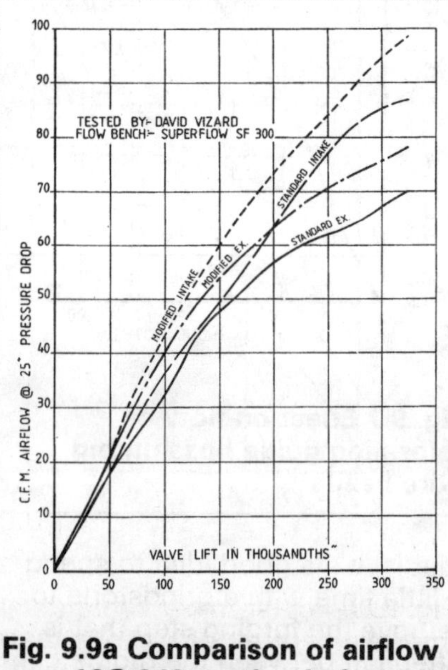

Fig. 9.9a Comparison of airflow on a 12G 940 casting before and after simple porting modifications

The first simple mod to the combustion chamber of an open chamber type head is to remove the lumps and bumps that are in the chamber area adjacent to the cylinder wall. After that, some simple cleaning up around the valve seat is all that's necessary to produce some worthwhile results in terms of extra airflow.

detailed elsewhere in this book, and then determine how much has to come off the head. Let's assume at this point that you can mill the cylinder head to raise the compression. This actually achieves two things; firstly it increases the thermal efficiency of the engine, and secondly, in most instances, it has the effect of slightly increasing the airflow of the cylinder head. I should point out that this phenomena is, in my experience, peculiar to Series 'A' cylinder heads as in virtually all other cylinder heads I have ever worked with, skimming the head has not changed the flow characteristics any measurable amount. However, skimming a Series 'A' cylinder head, often tacks on a couple of cfm.

If you have carried out all the basic mods that have so far been described, you should with a typical 12G-940 head casting, sporting 1.156 inch valves and 1.312-intake, see a flow improvement on the intake and exhaust about as depicted in **Fig. 9.9.** For what it's worth this is typically as good as a full-blown, stage 1 cylinder head from most of the proprietary cylinder head producers.

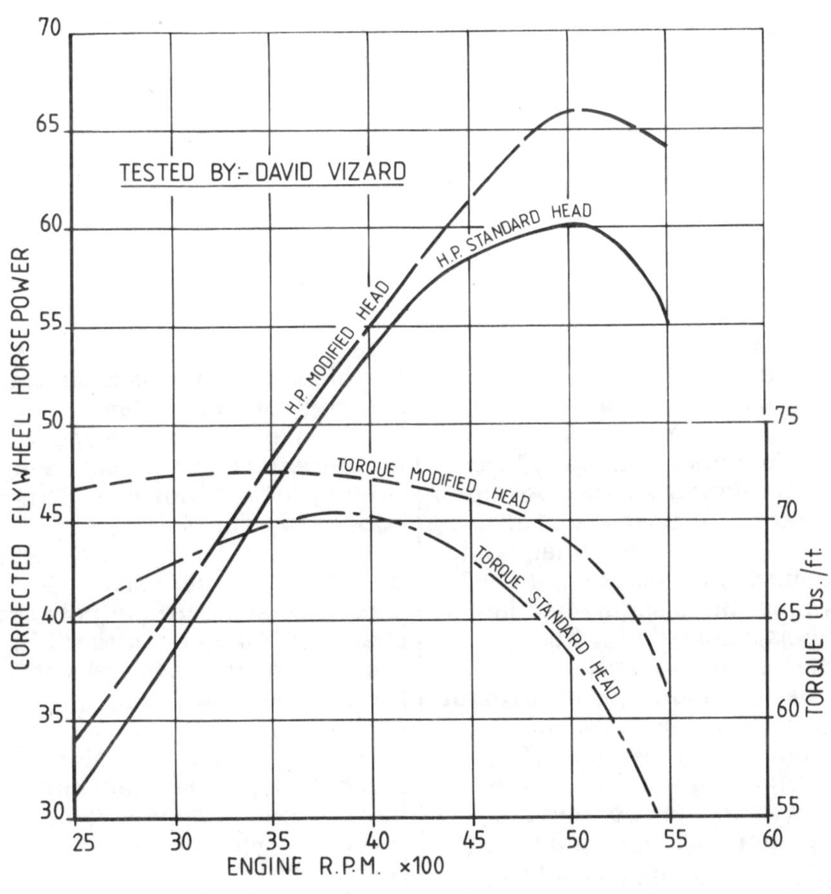

Fig. 9.9b Power improvements achieved from basic flow improvement modifications tested in Fig. 9.9a
As a result of the modifications the compression ratio changed from 10.2:1 to 10. 8:1. The test engine was a near-standard 1970 vintage 1275 GT unit.

Cylinder Heads Part 2

Closed Chamber Heads

Modifying a closed chamber head for high performance is far more complex than any of the open chamber heads, and this stems from the fact that there are two conflicting requirements within the combustion chamber design. Take a look at any of the photos or drawings depicting the closed chamber form. You will see the wall of the combustion chamber fits very closely around the intake valve, thus shrouding that part of the valve. This shrouding prevents this part of the valve passing any significant amount of air out into the chamber. There is, as it happens, a simple cure for this problem. Since the combustion chamber wall is very thick, it is possible to cut a substantial amount of the chamber wall away to relieve the shrouding. Before looking into the effectiveness of this mod, let's consider what would happen if, for instance, an 850 cylinder head simply had larger valves installed say, from a basic 1100 cylinder head. **If a larger valve is installed without any changes in combustion chamber form, what we find is that the edge of the valve is even nearer the combustion chamber wall, and as a result, even more valve shrouding occurs.** When the valve initially lifts off the seat, the larger valve has a slight advantage, but as valve lift increases, the shrouding factor increases at greater rate than the increase in valve area. **The result is that the big valve flows less air than the original small valve.**

I make this point because on numerous occasions I have seen cylinder heads modified, supposedly for high performance, by the installation of a slightly larger valve in an otherwise standard but polished chamber. **Take my word for it, these cylinder heads will not produce anything near optimum results; you will merely be paying for a shiny finish.** Incidentally, that shiny finish is one thing you can't see when the cylinder head is bolted on.

Fig. 9.10 shows exactly what happens to the airflow when the 1 3/32 inch diameter 850 Mini valve is replaced by the 1 5/32 inch 1100 valve. Note that during initial lift from the seat, the flow is marginally better, but once the valve reaches around 170 thousandths lift, the situation reverses - the smaller valve flows better. This flow characteristic is due solely to combustion chamber shrouding, as will now be demonstrated. Refer to **Fig. 9.11**. Here you will see a typical 'closed' combustion chamber. To highlight the reduced flow potential caused by shrouding, the intake valve was progressively unshrouded by successive cuts on the offending part of the chamber wall. A point worth mentioning here is that it's only necessary to angle the combustion chamber wall back: it does not have to be cut vertically down to the base of the chamber, the reason being that when the valve is a short way off the seat, the flow restriction is not due to the closeness of the combustion chamber wall to the edge of the valve, but due to the proximity of the valve seat with the seat in the cylinder head. In other words, it's the physical position of the valve in its lift curve that is limiting flow, not the wall of the chamber. As the lift reaches greater values, so the wall of the chamber must be further away so the area around the valve is at least a reasonable proportion of the curtain area due to the valve lift, **Fig. 9.12**.

Now it's obvious there are certain areas that can't be ground away, namely those areas upon which the gasket must lie, so in **Fig. 9.11** the limits of the gasket are shown. Using the centre of the valve as a reference point, the chamber

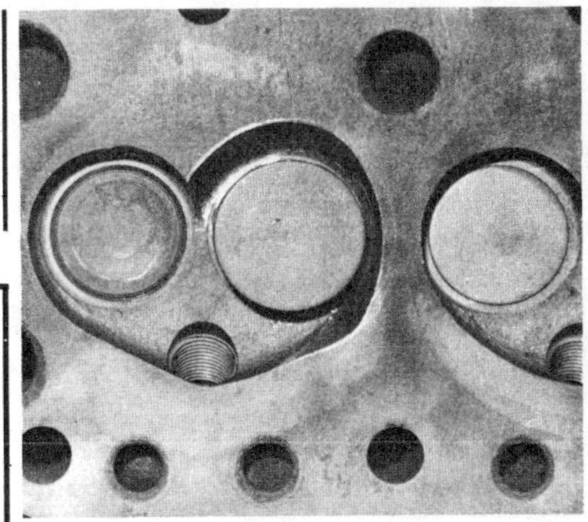

On small chamber heads, a little re-shaping of the combustion chamber can make a substantial difference in airflow. Compare this shape with the shape of the standard chamber shown a few pages back.

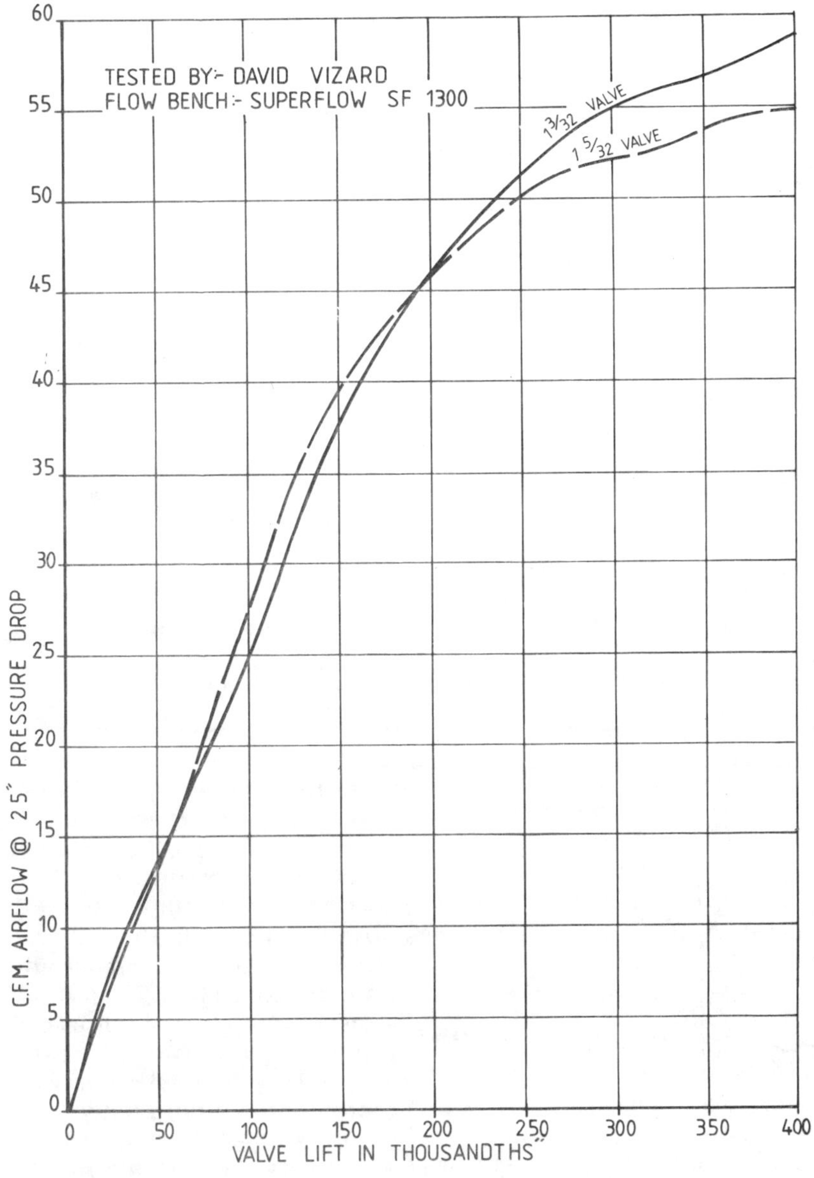

Fig. 9.10 Comparison of standard (1 3/32 in) valve and big (1 5/32 in) valve in a standard 850 Mini combustion chamber, demonstrating the effects of valve shrouding

The graph demonstrates clearly that modifying a head without reference to a flow bench can easily result in less *flow. In an unmodified 850 combustion chamber the larger, 1100, valve showed slight improvements between 75 and 200 thousandths lift, but above this progressively worse results were recorded.*

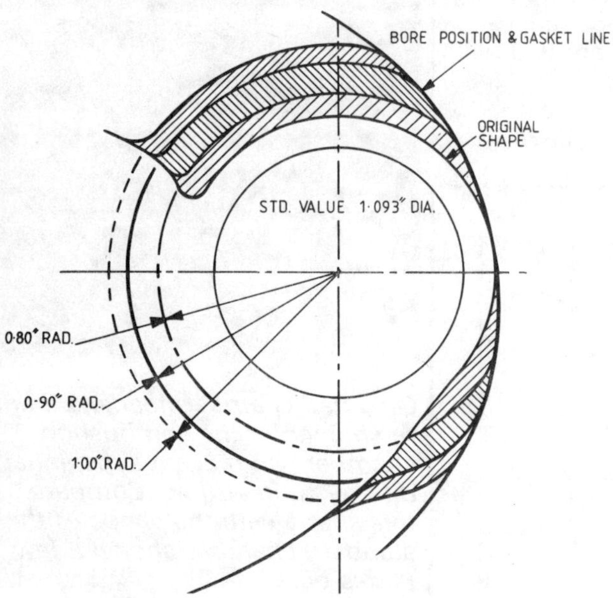

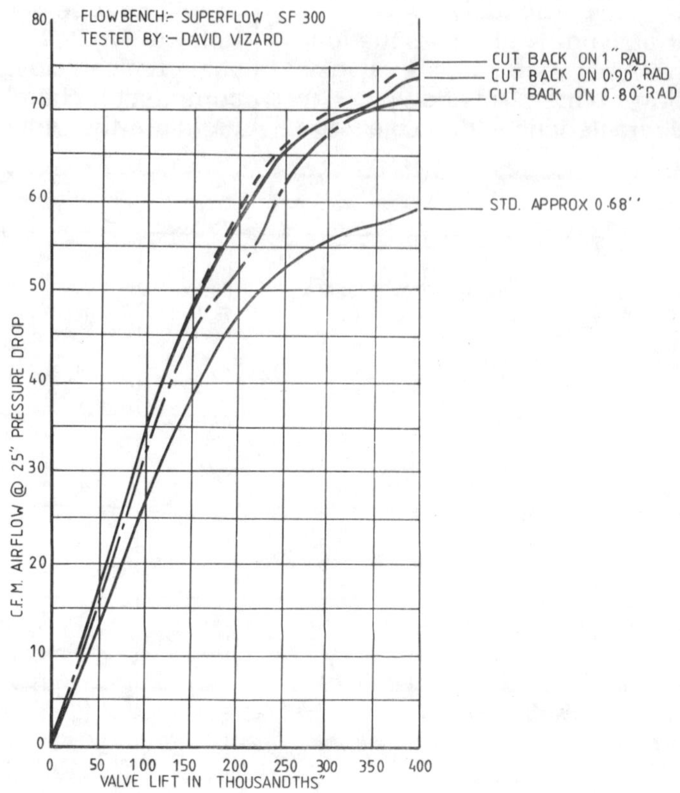

From the graph a law of diminishing returns clearly applies to deshrouding combustion chambers. The first cut-back at 0.80 inch radius gave a substantial improvement, but subsequent cuts made only relatively small improvements at the expense of a substantial compression ratio reduction.

Fig. 9.11 Effect on airflow when a closed chamber head is unshrouded to varying degrees

The cylinder head type used is that for 850, 948, 998 engines with the smallest valve size.

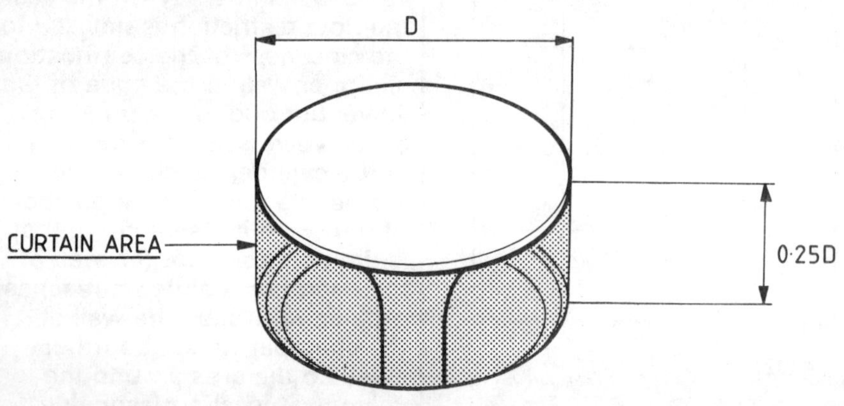

Fig. 9.12 What is the curtain area?

The curtain area is the circumference of the valve times the lift. Obviously the curtain area increases with increased lift. When the lift is equal to quarter the valve diameter (0.25 x D), the curtain area is equal to the valve area. If flow occurred at 100% efficiency further increases in valve lift would be pointless. In practice, however, 100% efficiency is not obtainable and as a result the valve's flow capability continues increasing to 0.35 x D, and sometimes more.

was cut away except in those areas which encroached on the gasket seating surface. The results are shown on the accompanying graph. You will note the first amount that was cut away had the greatest effect on increasing flow. Each subsequent amount cut away produced proportionally less increase in flow. Obviously there comes a time when cutting a chamber wall away any further produces no increase in flow because the shrouding has been removed as completely as possible. As it happens, there is physically enough metal thickness in the combustion chamber of a closed chamber head to virtually eliminate shrouding, but this entails removing a substantial volume of metal from the chamber, and here's

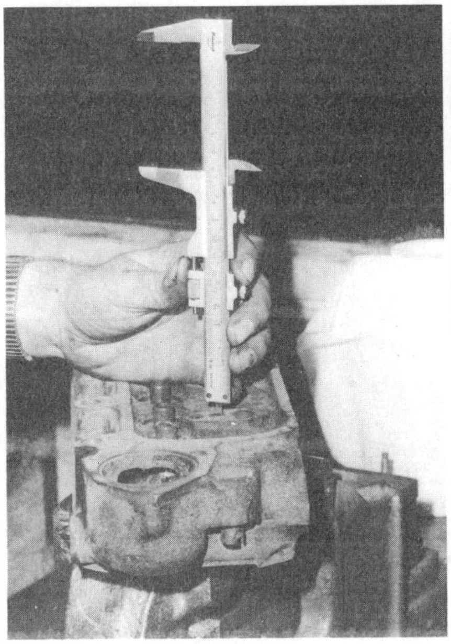

If you are modifiying a cylinder head of the closed chamber variety for use on a small displacement engine, first check to see that it has sufficient material down through the oil hole to allow reasonable machining of the combustion chamber face. If it hasn't, it may mean some quite complex brazing and grinding modifications so that the head can be skimmed sufficiently. Often, if maximum CR is not the aim, it is easier to sort through some cylinder heads down at the breakers yard and find one that is more suitable for the purpose.

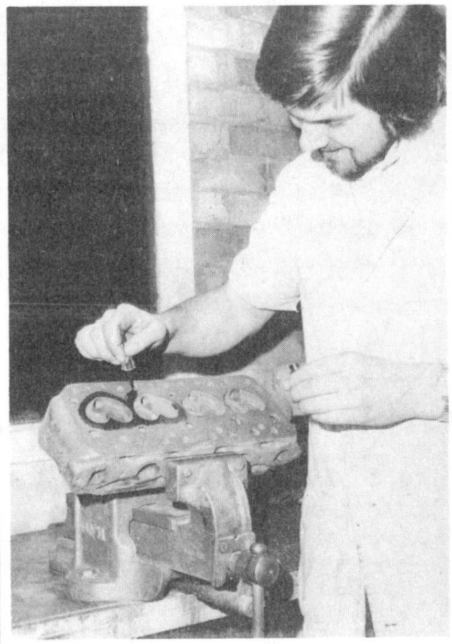

Once you have a cylinder head that has passed initial inspection as being suitable, blue the cylinder head face so you can scribe out the lines the head is to be cut to.

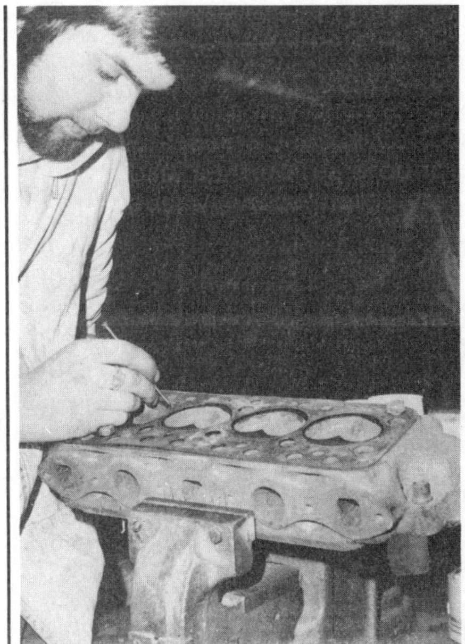

So you don't undercut the gasket area, position a used gasket on the cylinder head and scribe around the inside edge of the fire ring to locate the position of the gasket relative to the combustion chamber. From this point on, consider a no-man's land the area outside the ring you have just described. Grind past this line and the head won't hold a gasket.

where a snag arises.

Firstly, it's worthwhile stating that Series 'A' engines do like high CRs due to their relatively poor breathing efficiency. Basically it's a good adage to remember that if you can't fill the cylinder very well, it's best to squeeze the heck out of what is in there, if you are looking for all the power possible. In this situation, that very much holds true. The problem we run into is that there is barely enough metal on the cylinder head to skim it sufficiently to bring the CR back up to the level needed to give maximum power. If you bear in mind that even with a standard cam and a high enough fuel octane, an 850 engine likes to have a CR around 12.5 for maximum power, then you can begin to appreciate the problem. Even with flat topped pistons, it's necessary to skim the cylinder head by as much as 3/16 inch, and generally speaking, most heads just won't stand this much being taken off. At this point a compromise of some importance is necessary. Because more material has to be taken from a head used on a small engine, this compromise is very significant for 850 cc engines. Obviously the bigger the displacement (swept volume) of the cylinder, the easier it will be to achieve the CR required. A 12.5:1 for a 998 engine is relatively easy, though it can still entail a few snags. Because of an 1100 engine's even bigger displacement, there is an adequate amount of metal to produce the CR required. Remember, the bigger the engine is, the greater the combustion chamber volume will be for a given CR, and therefore the less the head needs to be skimmed. The trick to getting the best is to deshroud the chambers as much as possible consistent with achieving the desired CR.

For the moment, let's get back to 850s as they present the most difficult case. The best plan here is to cut the chamber around the intake valve to the 0.080 inch radius line as shown in **Fig. 9.11**, and simply clean up the rest of the chamber. The exhaust valve, incidentally, is not shrouded to any such extent as would cause a substantial power loss, so the area around the exhaust valve can largely be

If you intend installing bigger valves in a cylinder head, a piloting single point tool mounted in a drill is a quick and easy way of cutting the first 3/16 inch of the port. This machining of the initial part of the port considerably aids the accuracy with which you can cut the port in the vicinity of the valve seat.

Unless you want to play havoc with your lungs, have a vacuum cleaner handy to suck up the grinding grit as you produce it.

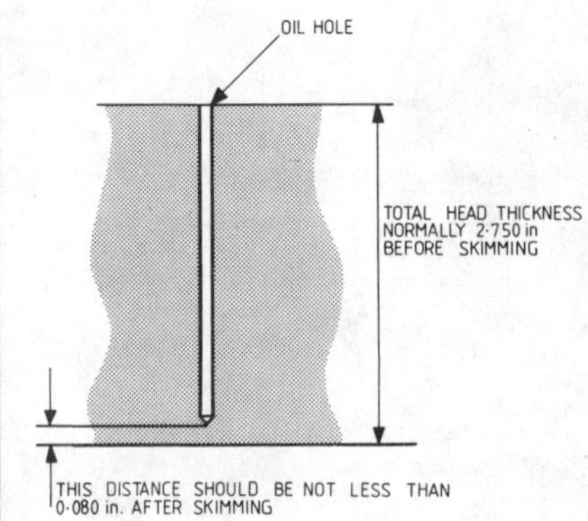

Fig. 9.13 Cylinder head oil passage depth

left as it stands. For normal road use, it's difficult to justify the recommendation of anything more than about a 10:1 CR, especially when bearing in mind the fact that fuel octane numbers now are typically in the region of 96 to 97 RON (Research Octane Numbers). As unleaded fuels become more prevalent, such as is now the case in the US, a 9:1 CR will be a more reasonable figure, taking into account the reduction in octane ratings (typically 94 to 95 RON). Usually the limiting factor on how much a closed chamber head can be skimmed is dictated by the depth of the oil hole drilling. This hole transfers oil from the block up to the rocker pillar. For a 'simple skim' measure the depth of the oil hole as shown in **Fig. 9.13** and then face the cylinder head so as to leave not less than 60 thousandths of material thickness. If the cylinder head is to be used on a competitive engine and a higher CR is needed, then some rather more complex procedures have to be followed to reduce the chamber volume to get a sufficiently high CR. These we will deal with later.

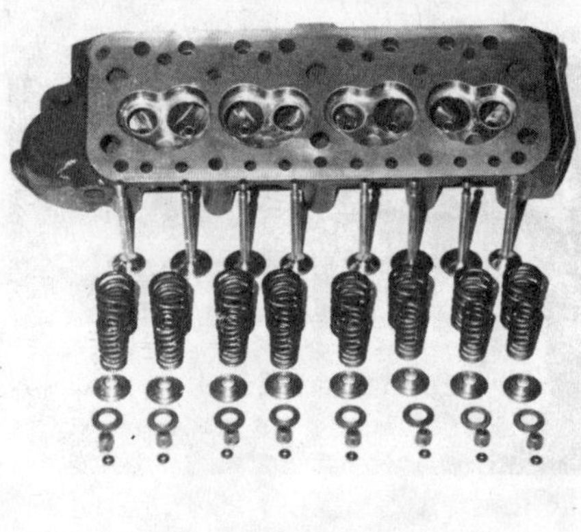

Here's a head I did in the early '70s. The flow bench results with this head show that it flowed very well, and at the time I wondered what else could be done to improve the cylinder head's performance potential. In the ten years that elapsed ...

Getting Down To Some Serious Head Modifying

So far, what we have looked at in terms of head modifying has been relatively basic stuff. If the object of the exercise is to extract every last ounce of hp from the engine, it is necessary to explore the flow potential of the head to a much greater degree. Since it is by far the most popular basic type of head, we will start off with the big bore, open chamber head. The smallest valve version which was available (12G-940), had 1.312 inch intake and 1.156 inch exhaust. Early 'S' engines, of course, had larger valves than this, namely 1.4 inch on the intake and 1.215 inch on the exhaust. Unfortunately, many of these early 'S' heads were prone to either overheating or cracking or both. Either way, they are not the most reliable heads to work with. Currently, the best easily available head to use on a big bore engine is the 12G-940 head casting. I say

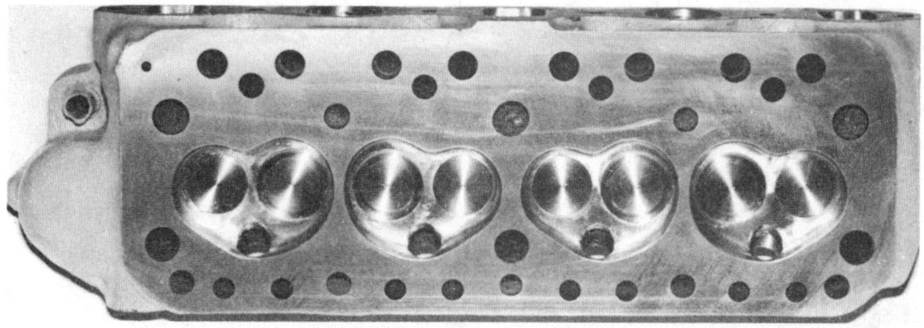

... heads like these have evolved, utilizing intense flow bench development programmes which have brought about better port and valve shapes. This particular head utilized Paul Ivey's Rimflo valves.

'easily available' because the rarer MG Turbo Metro head may well have more potential because of its better waterway routing around the chambers, ports and more especially, between the valves. On the debit side, the 'turbo' type casting which eventually will become common across the board on big bore engines, has less 'deck' thickness so cannot be skimmed much. However, for the present, the 'turbo' head casting is another ball game, so let's concentrate on the non-'turbo' 12G-940 casting.

Firstly, if this casting is to be used on an 'S' engine, or indeed, on any highly tuned engine, it should have the two extra holes drilled in it. This is easily done if you have an old 'S' head on top of the 12G-940 head, spotting through with a snug-fitting drill, and then drilling the holes out. The metal to do this is present in all.

To get hardware data for the books and articles I produce I spend most of my time either here, in my machine shop or ...

Testing – Getting Realistic Results

Although I am a technical writer I spend very little time actually writing. About 80% of my time is being spent designing, developing and testing the hardware you read about. This being the case, I thought it might be a good idea to fill you in on basic details of not only how I test, but what I use for testing. After all, this book contains a lot of information in the way of graphs and charted tests results. Firstly, air flow figures. After making several of my own flow benches, the first in the early sixties, I graduated to a relatively sophisticated self-built set up. When finances allowed, I purchased some commercially available flow test equipment. The particular flow bench I now use is a Quadrant Scientific unit. It can not only flow test cylinder heads but can also perform a host of other cylinder head related measurements. In addition to this it can also test cylinder heads, but can also accurately test air filters up to flow values somewhere in the region of 3000cfm. The flow bench's support computer is programmed to do the very substantial amount of number crunching for the flow benches many inputs. With all the little bells, whistles and gizmo's on this flow bench, it is

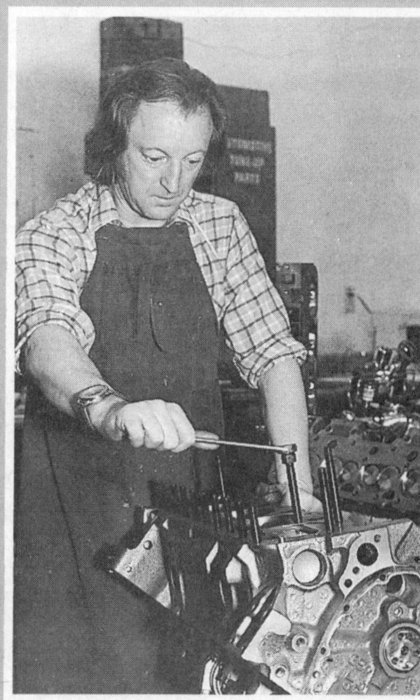

... building engines. As seen here, they are not all Mini engines. Some are big V-8s producing over 1,000hp.

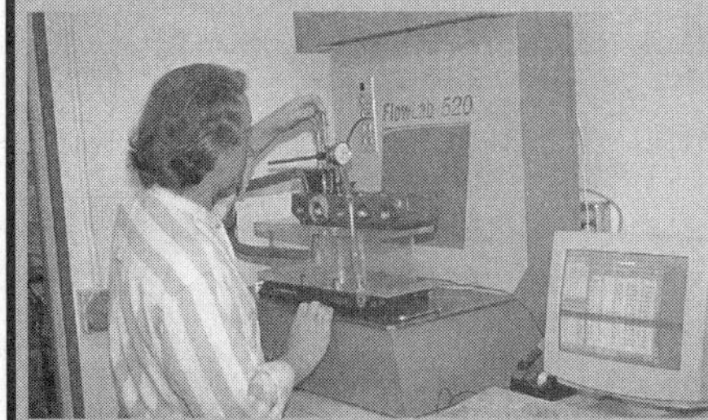

The process of developing almost any engine in the DV workshop starts off with an analysis of the cylinder head. For convenience . . .

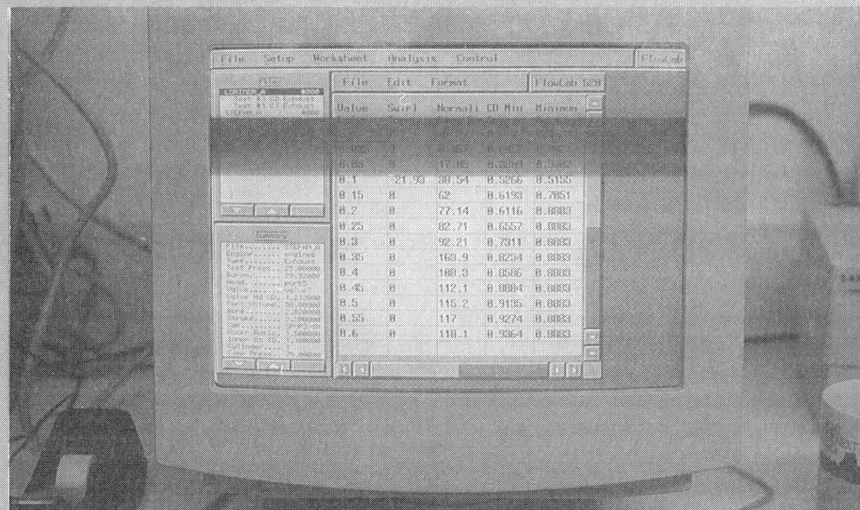

. . . the computer shows some of the results directly on the screen as the tests are being conducted. Far more results are condensed and put through a number crunching programme which prints out all the results in tabular form on a full page.

Here I am at the dyno console running some tests on Series 'A' engines with the late Bob Munt of Kent Cams looking on.

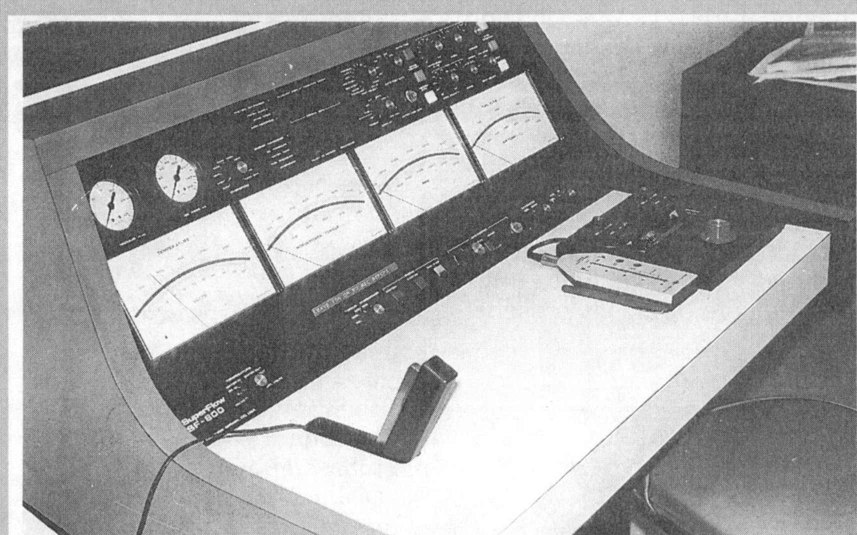

Here's a close-up of the dyno console. In addition to all the normal Superflow functions there are some features that have been added. The principal of these, seen on the flat surface of a console to the right, are the engine management controls. These control both mixture, ignition timing and the advance curve. The little box just to the left of that is the sound measuring equipment read-out.

possible to determine a wide variety of cylinder head/induction characteristics that are not normally investigated if just simple flow testing alone is done. For instance, cylinder heads aren't just tested for flow capability. Apart from the basic flow tests, such things as the discharge coefficient (port flow efficiency) is compared with a port of 100% efficiency both by the SAE method and one of our own methods which give nearer a true picture of what's going on. Ports are also probed for velocity and energy density. On top of dry flow testing, wet flow tests are often conducted. This adds yet another dimension to what is achieved with flow testing to the point where if I were to deal with Series 'A' cylinder heads as per the workshop practice, it would fill a book.

Outside of two or three big companies (and they are not in the performance business per se) these are factors very rarely investigated in the cylinder head industry. Interrelated with flow and head design is the swirl, produced by the cylinder head, so this is also measured. As you can see, the number of factors are mounting up and even in small print, a cylinder head test

This is the business end of the dyno. In this particular shot a 1400 Mini engine is being put through its paces. All the testing that is done is performed to very vigorous standards and nothing less than three runs repeating to within 0.5 per cent are considered valid.

Sometimes engines are on the dyno a considerable length of time, maybe as much as a month. This is because testing may be covering a vast range of a particular component. To make sure the basic engine doesn't change, tests are done every few hours to check the leakdown of the cylinders and the compression pressures to see that they are as per the original baseline. Also torque tests are done to establish that the engine hasn't freed up anymore than it was when the engine was originally baselined.

on my flow bench produces an entire page of numbers. The sheer weight of numbers is the principal reason I only show flow figures in this book. If all the other data was included, most of which is only any good to an experienced cylinder head engineer, it would fill this book with numbers. At this point it's probably sufficient to say that the basic technology and modifications detailed within these pages are amongst the most comprehensive on the subject to be found . . . anywhere. Most of the information is first hand and I would go so far as to say 95% of professionals dealing with the Series 'A' do not have this information, but look to tests as performed here to garner it. Of course, putting a head through its paces on the flow bench is only one step down the road to power. Just because it flows well, doesn't mean it's guaranteed to produce power, although with the means at my disposal, I'm pretty sure of the results prior to testing on the dyno. Nonetheless, whatever positive results the flow bench indicates, dyno tests are used to verify. The dyno I use is a much-modified 800 series Superflow. It functions in conjunction with a computer and the measurements that the dyno picks up from its various, and numerous, sensors are normally fed back into the computer memory. Not only are the obvious tests for horsepower and torque made, but also such factors as air consumption, fuel consumption, exhaust temperatures, air intake temperature, air intake volume, fuel temperature, etc. As a result, the computer punches out a print-out of engine related data. So much so that I can even determine (within limits) whether a poor engine performance can be attributed to an intake lobe or an exhaust lobe not working due to its design. The computer does a great deal of number crunching to come up with calculated values based on the information supplied by the various dyno sensors mounted on the engine. These sensors, together with input data, produce figures for such things as Volumetric Efficiency, Brake Specific Fuel Consumption, Brake Specific Air Consumption, and so on. On top of this, a high speed exhaust sampling system has been installed which allows a very fast read-out of the mixture to be taken immediately behind the exhaust valves. Additionally, noise is measured to see just how well silencers/mufflers are working.

With its ability to be programmed, the dyno can not only do steady state power tests – which incidentally, are rarely done – but accelerated power

The dyno, via the computer, can be programmed to run a series of tests, store the results on disk and . . .

test. These are performed with the engine in a more realistic accelerating mode rather than being held at a steady speed. These tests are quicker to perform and they also allow the effect of lightweight engine internals to be assessed in terms of available power on the track. The various acceleration rates can be selected by the operator and, if required, the dyno can be programmed via the computer to do a simulated run through the gears. These figures, or any suitable portion of them, can be used together with relevance vehicle data to 'build' a computer simulated 'vehicle'. The computer can then run this 'vehicle' to produce an accurate performance curve of the vehicle in question. The programme takes into account gearing, tyre grip, wind resistance, centre of gravity, etc.

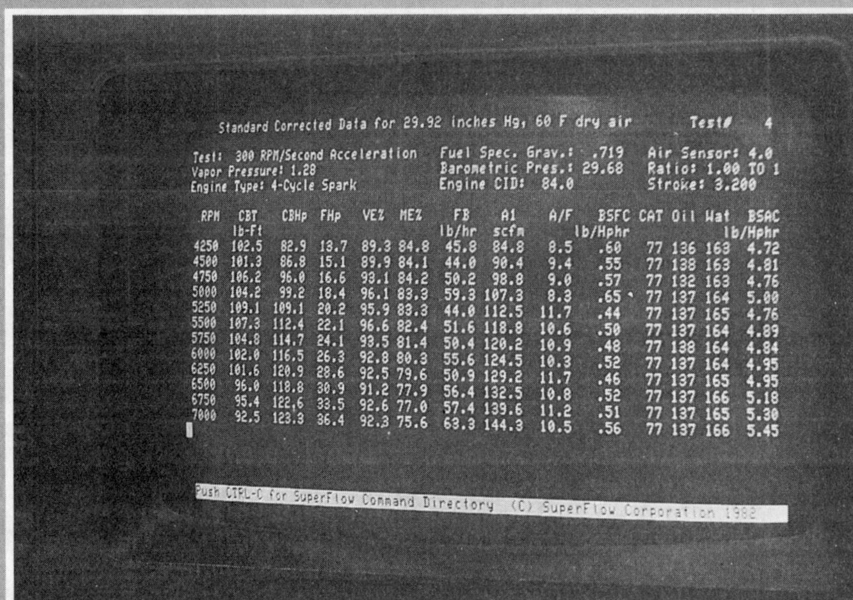

. . . display them on the screen as required. With so much data available from the dyno, tabulating results by hand would be much too time consuming. So when required . . .

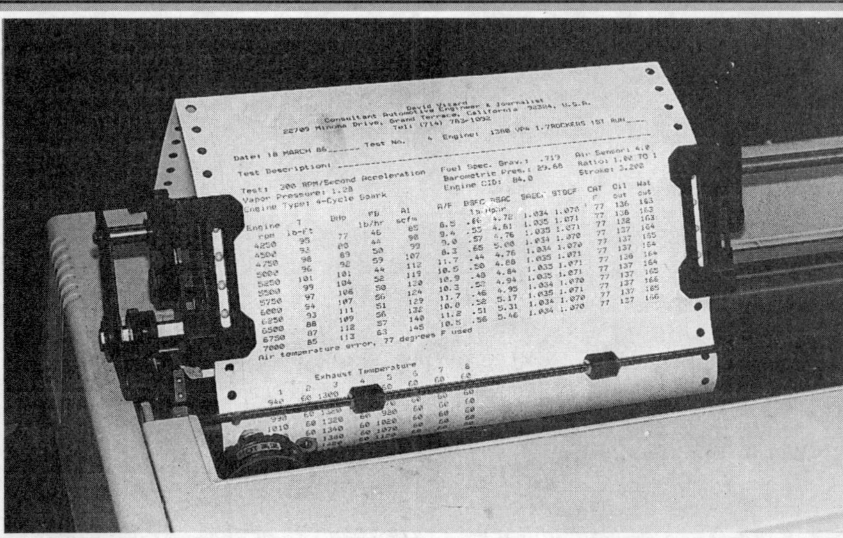

. . . they can be printed out directly on the printer. Either in tabular form like this or in graphic form via a pen plotter. The graphic capability allows any one factor of about 40 different functions to be plotted, one against the other.

Results are very realistic. Typical 0–60 and S/S 1/4 times can be computed usually to better than 2/10 of a second accuracy. But of greater significance is the fact that repeatability is virtually 100%, so the effect of small changes in the power curve can be seen whereas in the real

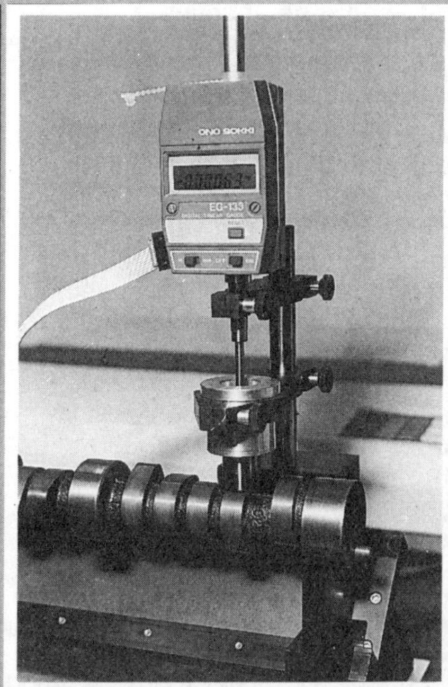

What you are looking at here is a comparator capable of resolution of 40 millionths of an inch. This is part of the system used for analysing cam profiles. The output from this electronic comparator, plus an angle comparator, are fed into the computer for analysis. This . . .

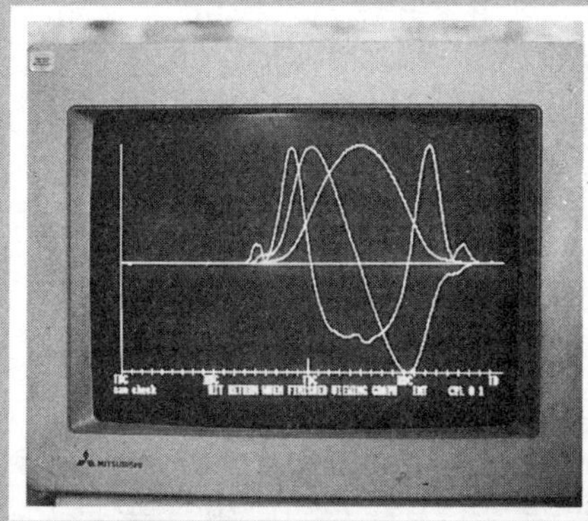

. . . can be displayed on the screen as we see here or graphically plotted out.

world, such small changes would be overwhelmed by prevailing inconsistencies in weather, track conditions, driver, etc. The dyno and the flow bench constitute the biggest piece of test equipment (as opposed to manufacturing equipment) in my workshop, which is also where David Anton runs A.P.T. There are lots of other pieces of test equipment which do their small, but important, part to help build better high performance engines. For instance, dyno testing is all about optimizing whatever combination is being tested and the biggest facet of this is ignition timing and mixture calibration. The dyno ignition system is made up of about six different units plus three different coils. Switching from any system to another usually takes less than five seconds. Both the ignition curve and timing are adjustable from the dyno console. If I need to know what one degree more or less advance does, it's only a question of going click, click, click, pull the throttle and re-run the test. Carburation is often handled the same way and mixture adjustment up to about five jet sizes is handled electronically from the dyno console for some types of carburettors. The next most adjusted engine parameter (at least in my shop) is the camshaft. This is something that is not 'controlled' via the dyno, but it is worth mentioning that most engines run on my dyno have been converted so that cam timing is externally and speedily adjustable, It's amazing what having the cam timing 'right on' will do for power and it's not always where the 'official' cam specs call for it to be. Still on the subject of cams, the next most important piece of equipment on the list is a computerized cam checker which not only checks timing figures, but also allows the determination of velocity, acceleration, jerk, etc. From the data generated, one can establish if the cam is well-designed or indeed, well-produced.

Not all the equipment used is tied to the workshop and here's a prime example of something mobile. Sooner or later engines must be tested in race cars and one of the vehicles I use as a test bed has been instrumented with this 'on board' vehicle dynamics computer. With various sensors all over the car, this allows the acquisition of a lot of data not only about the chassis, but also the engine. Having fed in a rolling resistance curve to the computer, it can return via . . .

. . . a mini printer an installed engine power curve.

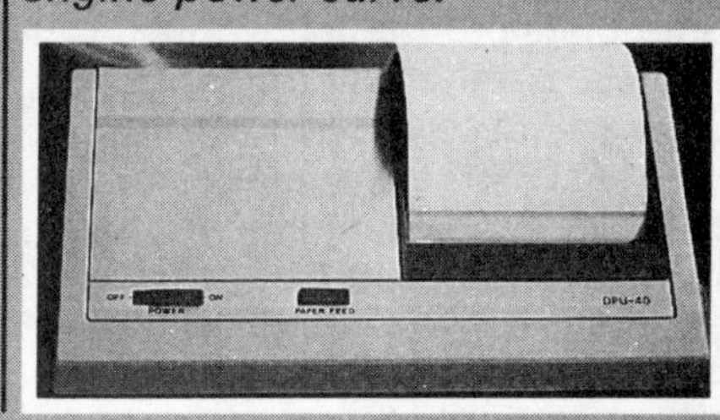

With a lot of more minor, but equally important, test equipment I go to work to find out what makes engines tick. I have well-known companies come to me to see if what I've got in my workshop can be used to help them prove or improve their products. During the process of attempting to make their speed equipment better, I learn more about what I like to do best and that is to make engines go. So next time you wonder where all the information comes from in this book, just look at the little credits on the graphs, charts and power curves and you'll see that 98% of it came into being as a result of my efforts with whatever equipment was at hand.

Intake Port

To get a clear understanding of what needs to be done, you will need to read the following text and refer to the drawings which comprise **Fig. 9.14**.

Virtually all big bore heads have a manifold location ring which goes into the entry of each port. There is often a debate as to whether that location ring should be kept or not. If it does remain in place, then obviously it reduces the size of the intake port at the manifold face. **However, you may have figured out by now that the size of the port at the siamese section is not a critical factor in producing effective overall flow figures from these heads, though there is an interrelated port size power factor – this is dealt with later.** It makes virtually no differences to the following figures whether the port is opened out to the internal diameter of the location ring, or the location ring is removed and it is opened out to the external diameter of the ring.

The question whether or not the ring should be retained, really relates to what manifold is being used. If a standard Leyland twin SO type manifold is being retained, you may as well keep the location rings. However, most after-market intake manifolds do not have provision for that ring, in which case there is no disadvantage involved in doing away with it. Whether the port, at this point, is cut to the inside diameter of the ring or the outside diameter of the ring, is of little consequence: what does matter is the way the port wall is shaped around the pushrod holes just a short way in from the manifold face. **Some clumsy porting in the region of the pushrod holes can effectively reduce the flow, and remember, shape is all-important.**

At this point, it's just as well to get into a serious discussion on the size of the siamese section of the intake port. Hopefully the significance of this section of the port, and its influence on the flow, can be put into some perspective. First, before putting grindstone to port, it is necessary to consider what valve size will be used, and how much the cam is going to lift that valve. In simple terms will the head be a 'big valver' for a race engine, or will the valves be standard size for a budget road engine, or something between these extremes? If the head is to be used on a high performance road unit, retaining 1.31 inch intake valves, and the valve lift does not exceed about 380 thousandths, then the siamese section of the port need not be enlarged any further than whatever is required to clean it up. In other words, the 20 or 30 thousandths layer of metal removed from all over that area of the port is all that's necessary. **If the port is made larger than this, it will not help the flow but it will reduce the port velocity, and this, in turn, can reduce the**

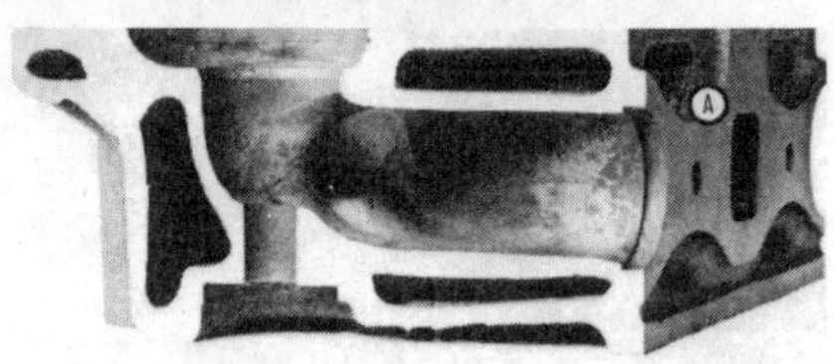

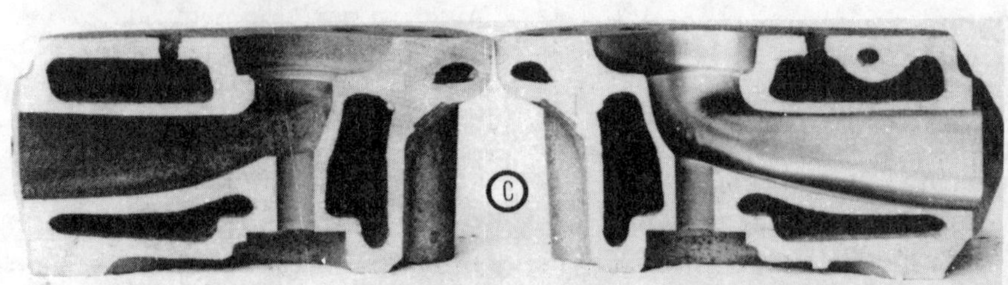

Porting open chamber 1000/1100 heads. *In these photos are the port configurations of the 12G 295 casting. This casting is, of course, similar to the 12G 206 casting. It differs only in that the 12G 206 casting has not got a location ring on the intake port. However, in all other respects they may be considered the same casting. Photo* **A** *shows the standard intake port. In* **B** *we see the form the port takes when modified to accept 1⁵⁄₁₆ inch intake valve. In photo* **C** *a comparison between the standard exhaust port and a modified exhaust port is shown. The exhaust port utilizes a 1¹⁄₁₆ inch exhaust valve. (Photos courtesy of Swaymar Race Engines)*

Fig. 9.14a BIG intake ports. Just how much are they worth? *This table demonstrates their worth as regards flow; to get the whole picture regarding their effects on power read the relevant text. To prove a point here's a comparison between a well modified port using a 1.40 inch valve but retaining a virtually standard port size at the siamesed section (cleaned up only) and the same port with the siamesed section opened up to 1.350 high x 1.225 wide (just breaks into pushrod holes).*

Valve lift Thousandths in	Standard port CFM @ 25 in	Big port CFM @ 25 in
50	22.5	22.5
70	31.4	31.2
140	62.4	61.9
210	87.5	88.0
280	101.0	102.0
350	109.0	114.0
420	112.0	124.0
Average Effective Flow	72.3	74.1

Notice the flow only increases right at high valve lifts. If the cam in your race engine lifted to 420 thousands (most lift only to 390!) the average gain would only be 1.8CFM. Modifying the port from standard to that of column 1 results in an average gain of 14.8CFM that's 822% more than enlarging the port was worth!

FOR THE SAKE OF CLARITY STD VALVE SEAT WIDTHS ARE SHOWN ON THIS DRAWING ONLY.

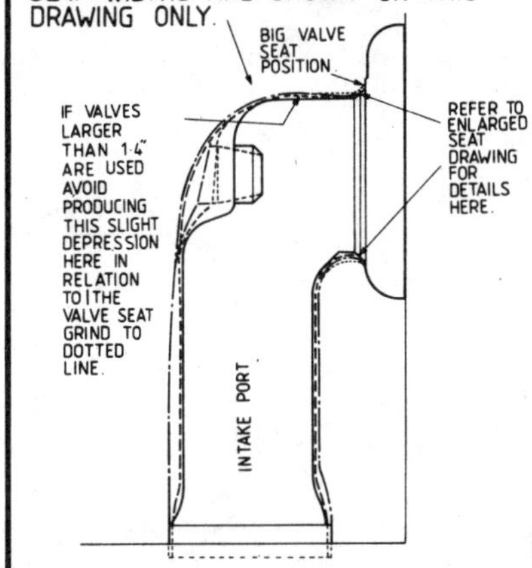

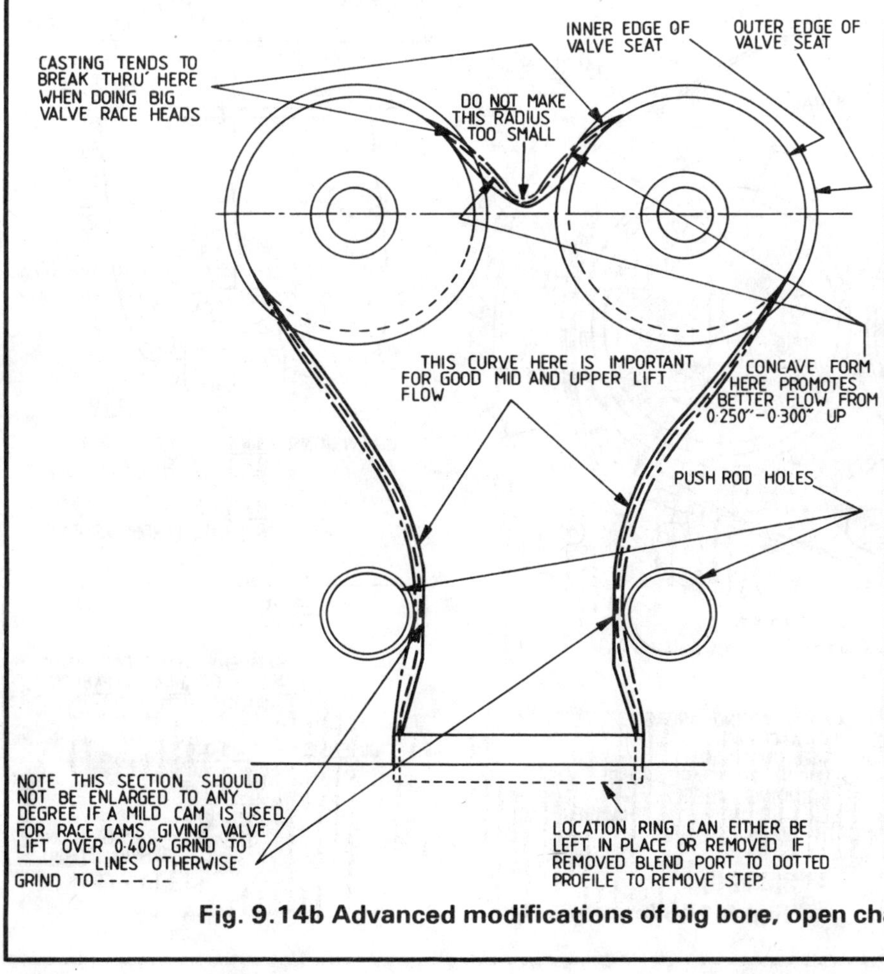

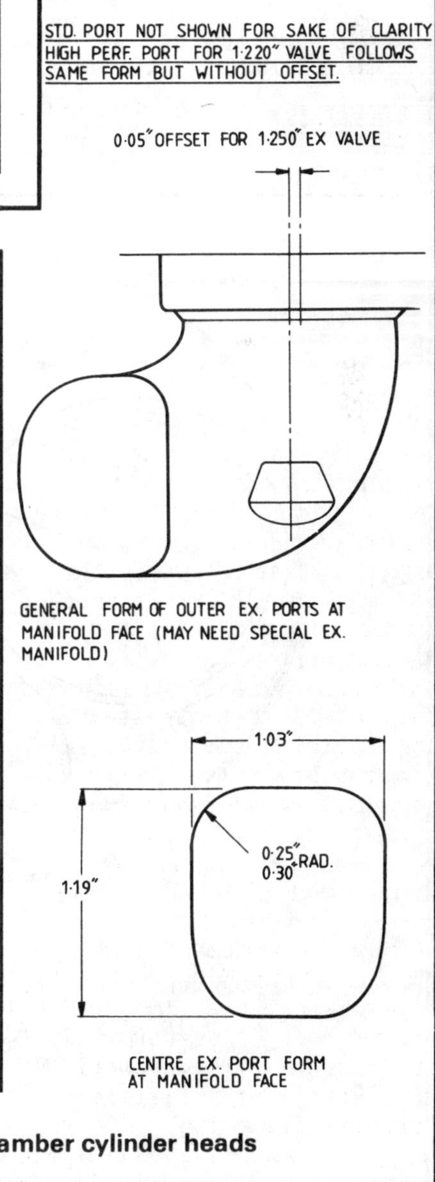

Fig. 9.14b Advanced modifications of big bore, open chamber cylinder heads

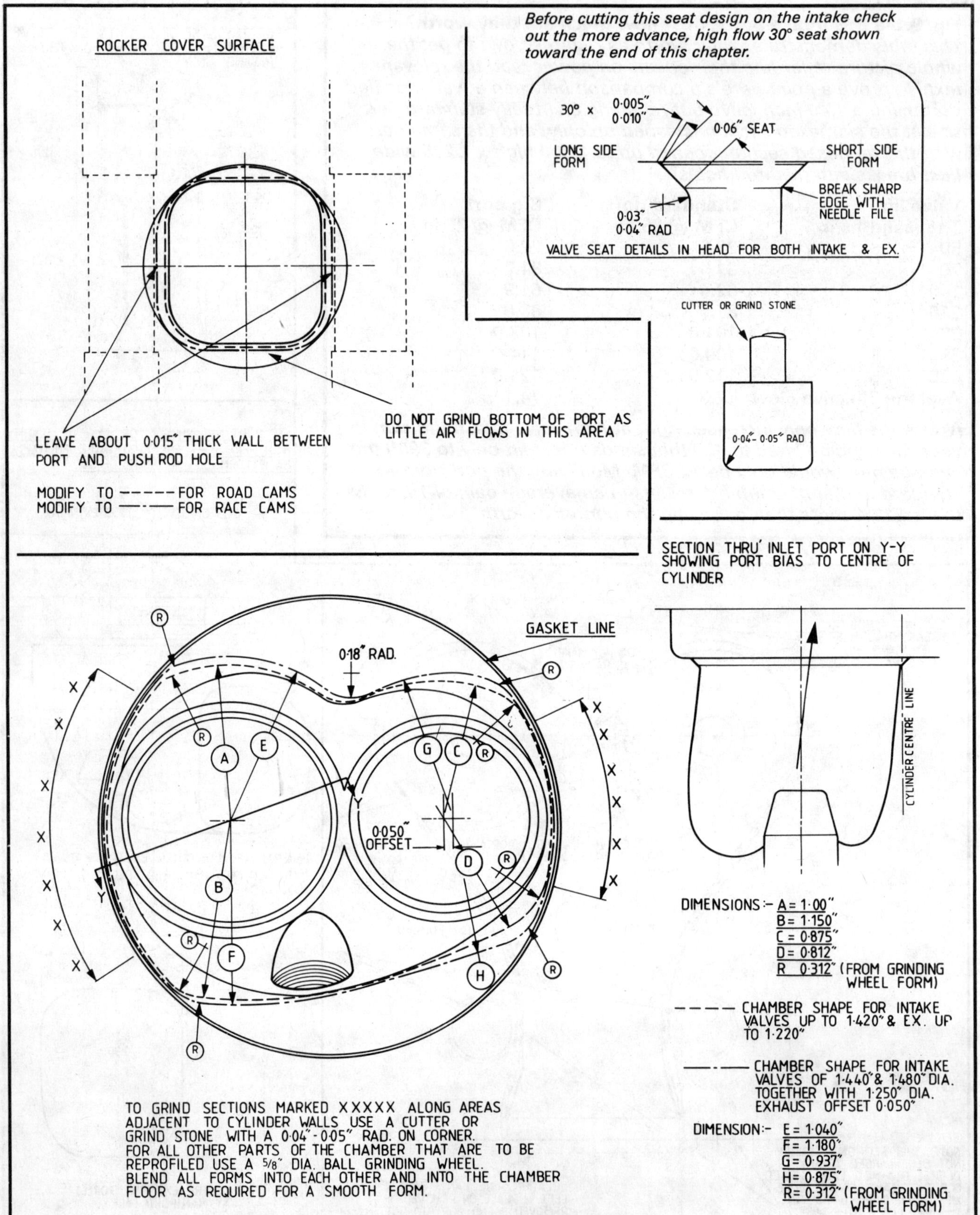

Fig. 9.14b (continued)

low end torque output of the engine. If the valve size used is the 1.4 inch 'S' size, then the siamese section can stand enlarging a little more. Again, it's not necessary to go overboard unless a high lift cam is used.

It became the fashion during the '70s and early '80s to sleeve the pushrod holes so that the siamese section of the port could be made even larger. With the valve efficiencies that are typically obtained, even on the best heads, this move is unnecessary. In fact, it may even be detrimental to hp. There is enough available area between the pushrods, both in width and vertical height, to cut a port that will satisfy the demands of all but the most efficient valves. On a well-prepared head additional power can be gained by making the ports just a little bigger than dictated by flow bench results. This applies not so much to peak power but in terms is how well the engine 'hangs on' after peak power is reached. This applies equally whether standard or high-lift cams are used, but to reap any benefit a cam duration of at least 285° is required. Considering the siamese section of the port in fine detail, it is useful to realize the airflow through the siamese section is not by any means uniform. In round terms, some 60% of the air flows in the top 40% of the port. In other words, the lower half of the port, that is the side that eventually leads up to the short side radius, is not as active as the top side of the port. If the objective is to cut a port for maximum effect, the floor of the port is the last place that should be reworked. It is better to widen the port to within about 30 thousandths of the pushrod hole, clean up the floor, but move the roof up.

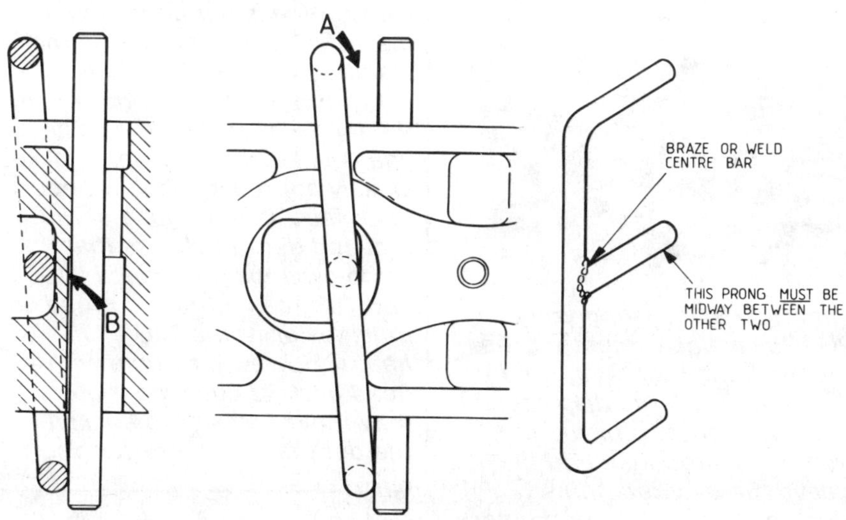

Fig. 9.15 Intake port grinding guide tool
This tool helps to prevent breaking into the push rod hole when grinding the siamesed section of the intake port. Make up the 'E' bar of something reasonably substantial, e.g. 1/4 in bar, to about the proportions shown here. Make sure that all the prongs are on the same plane or the object of the tool will be defeated. To use the tool press a 5/16 in (or 8mm) bar up against the wall of the push rod hole as shown above. Put the 'E' tool into the port and tip it as shown in the right hand drawing. Dimension 'A' can then be measured. It will be twice the thickness of dimension 'B' the port wall thickness.

By the way, many people find it difficult to determine just how close they are to grinding into the pushrod hole. The drawing **Fig. 9.15** shows a tool, (devised by Hugh Venables, an Australian acquaintance of mine) which obviates this problem. As you can see, it allows accurate gauging of port width so as to avoid breaking into the pushrod holes. Once past the pushrod holes, the air moves into one of the more critical areas if the port is to function well, namely the outside curve which joins the siamese section to the actual underside of the valve. The radius on the outside of the port is an important factor toward producing good flow. The radius here needs to be made as generous as possible. Again, it's worth referring to and studying the drawings, detailing various parts of the port. This is a particular section of the port which is difficult to see, so it's necessary to work by feel. A nice, large, smooth radius is the objective.

At this point in the guided tour of the intake port, we are approaching the valve guide boss and the guide itself. Many enthusiastic first-time head modifiers are often tempted, and indeed many succumb to the temptation, to grind the guide and valve guide boss flush with the roof of a port in the belief that this will help flow enormously. Unfortunately it does nothing for the airflow until relatively high valve lifts are reached, *i.e.* measurable at 300–350 thousandths but not usually significant until around 600 thousandths. What is guaranteed by this move (unless steps are taken in terms of guide material to reduce the effect) is a serious shortening of guide life to the point where the valve doesn't seat properly, oil leakage occurs and an appreciable amount of hp is lost as a result. To be effective, the valve guide needs to be its full length, and the guide boss itself needs to be slimmed down

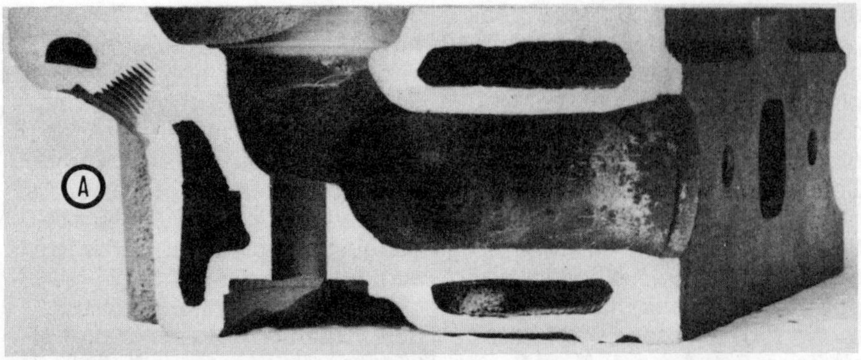

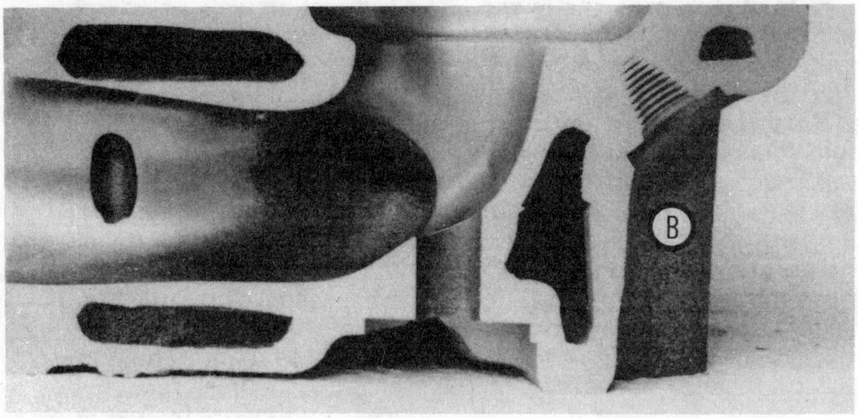

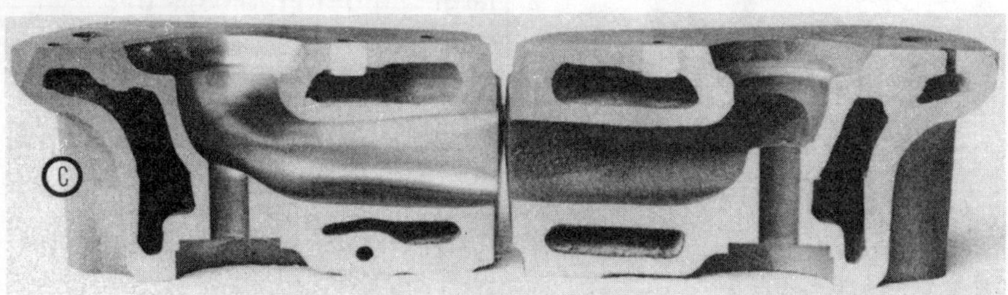

Porting big bore heads. *In this sectioned 12G 940 casting you can see the family resemblance to the 12G 295 casting. Except for the fact the port is bigger, the form is very much the same. In photo* **B**, *the port is modified for use in big valve form in a large displacement Mini engine, for utilizing a cam close to 500 thousandths lift. In this particular instance, the port broke into the pushrod hole and the next operation would have been to sleeve the pushrod holes. Widening the port to this extent is not a necessary operation unless very high lift cams together with 1.48 inch intake valves are being used. In photo* **C**, *a comparison of the standard and modified exhaust ports is being made. Note here how the exhaust port is restricted by the valve guide boss to a greater extent than the 12G 295 head. (Photos courtesy of Swaymar Race Engines)*

A cross-section through the turbo head reveals that in many places the casting thickness between ports and water jacket is thinner than its pre-turbo counterpart. This was done to improve cooling. The general port configuration is much the same, but note the wall thickness on the long side radius of the intake port prior to the valve guide is thinner than on the 12G 940 casting. The same situation also exists on the exhaust port. Also in photo **B**, *in the top LH corner of the casting, you will see that the water jacket closely approaches the cylinder head face. It is because of the reduced deck thickness that turbo castings produced in 1983 and 1984 cannot be skimmed as much as previous 12G 940 heads. However, trouble with head gaskets on turbo motors may move Leyland to thicken up the deck face of these heads.*

around the valve guide. The guide should be bullet-nosed from the point it enters the port and its edges should be rounded off. By the way, it's not a good idea to bullet-nose guides so that a sharp edge is formed at the bore of the guide. Basically, the metal at this edge needs to be around 30 thousandths thick and a radius applied to the outside edge.

Valves have become more important in power production as fuel characteristics have changed recently in conjunction with the reduction of lead. *See my further comments on this topic in Part 4 of this chapter.*

At the point where the guide enters the port, several things need to be done. Firstly, since the guide acts as a disturbance to airflow, it's a good idea to widen the port at this point. Obviously there's a limit to how much it can be widened, but whilst you are reworking the port, consciously widen it without forming an abrupt change of direction as the port approaches the back of the valve seat. Another area to allow the air to take the turn to the back of the valve better, is to grind either side of the valve guide to deepen the port. Any metal removed here needs to be gently blended to the outside edges.

Let's turn our attention now to the area which divides the port into the two sections, each feeding the back of its respective valve. Here we come to the first really problematical area. A popular move, especially on heads that have not had the benefit of flow bench development, is to make the port divider sharp-edged. Obviously, the thinking here is that the air will be 'knifed' into its relative passage better than if the divider had a rounded, less streamlined edge. Unfortunately the reverse is true, simply because, for most of the induction cycle of any pair of cylinders air only flows into one valve at a time. As a result the knife-edge actually obstructs air being drawn from the area diagonally opposite. **Fig. 9.16** shows the route the air takes into the port. As you can see, a certain amount of air is drawn from behind the edge of the divider. If a sharp edge existed, air would not make it round the corner as easily. The rounded edge allows air to rejoin the 'busy' route of the port, feeding the open valve more easily.

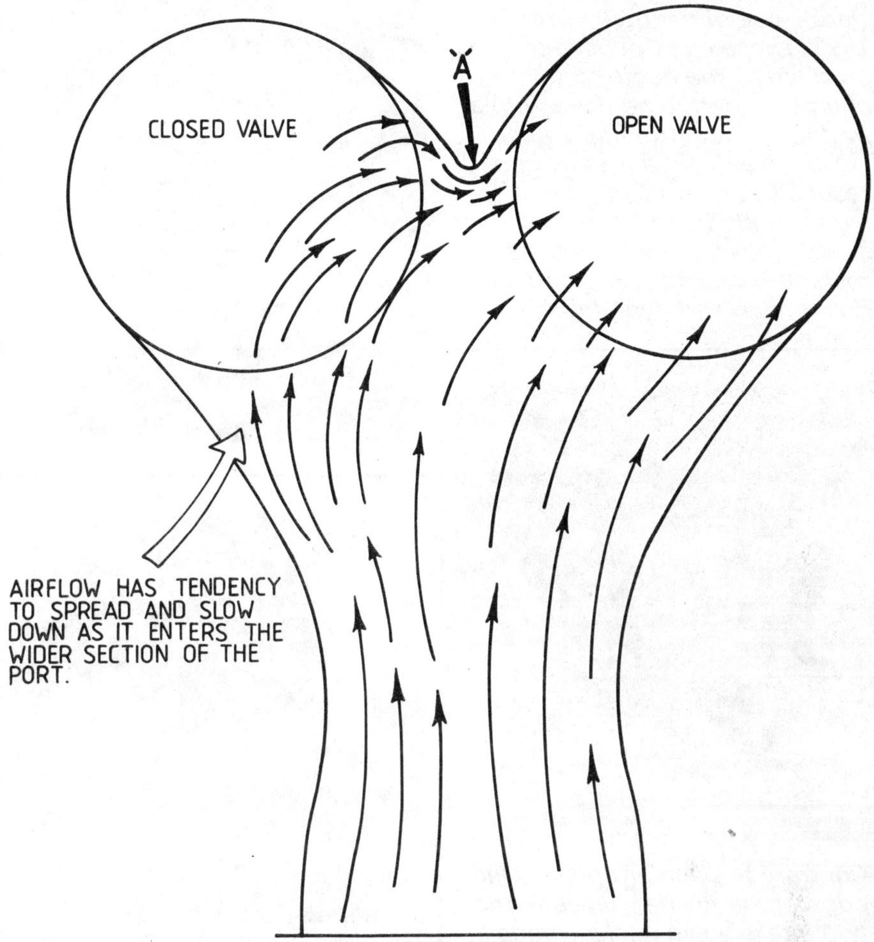

Fig. 9.16 Port divider 'beak' radius
This illustration shows why a sharp edge on the 'beak' in the port does not improve airflow. It is wrong to imagine that the beak acts as a splitter to divide airflow neatly into two streams running to both valve pockets, because (for most of the time) only one valve is open and drawing air and no splitting action is occurring. A sharp edge at 'A' would obstruct the flow through the open valve of any air that had tended to flow initially into the pocket of the idle valve.

The next problematical section of the port concerns the area on each side of the divider between, and just adjacent to, the valves. Basically, this part of the port wall needs to be concave. In standard form, it is markedly convex in shape. Unfortunately, the casting thickness in this area is not very great. It's possible to get the form shown in the drawings if relatively small valves are used, *i.e.* 1.312 inch, although in many cases that form can still be achieved with a 1.4 inch 'S' size valve. **However when large valves in the region of 1.460 inch, maybe 1.480 inch are used, then it is almost a surefire bet that applying this form will cause the port passage to break into the waterway. You can hold back on the grinding so that this doesn't happen, but, if this is done, much of the effectiveness of the large valve is lost.** Indeed, in some cases, all the assets of the large valve are lost because it simply does not flow any more air than a smaller valve. This has more than a passing tendency to make the installation exercise pointless . . .

To really make a big valve work, it is necessary to grind the concave form in question. If the port wall breaks through, it will be necessary to chisel it out until a thick edge on the break is formed, and then have the hole welded. Sure, it sounds like a pain in the neck to do, but it's the only real way I know of to make

Here's one of my trick heads of 1973. These were difficult to produce as the combustion chambers had to be done with a profile-formed grinding wheel. Heads like this did produce good results but I am now getting significantly better results from simpler, less critical mods, as a result of thousands of hours of flow bench and dyno testing.

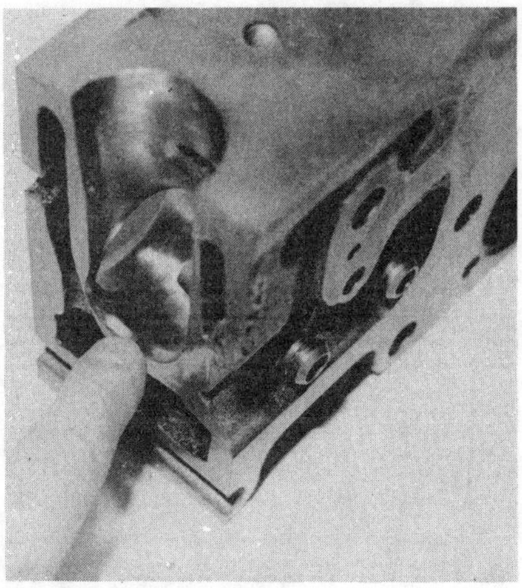

Contrary to popular opinion, the edge of the divider between the two intake valve pockets does not need to be sharp. It should have a radius of about 3/16 inch on it for best results.

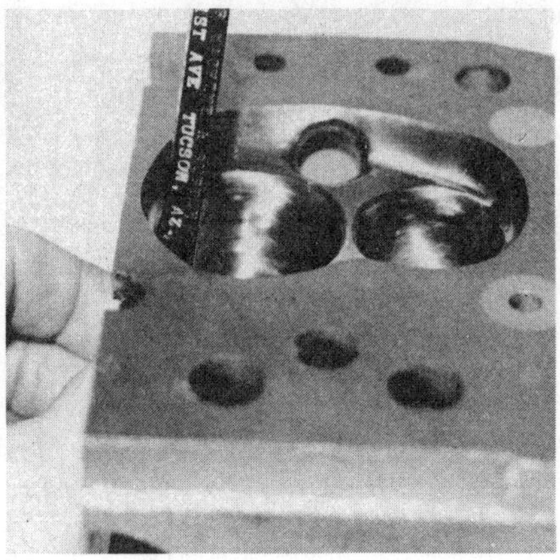

The slope on this pencil, which is laid against the port wall, illustrates the bias of the port toward the centre of the cylinder that is discussed in the text.

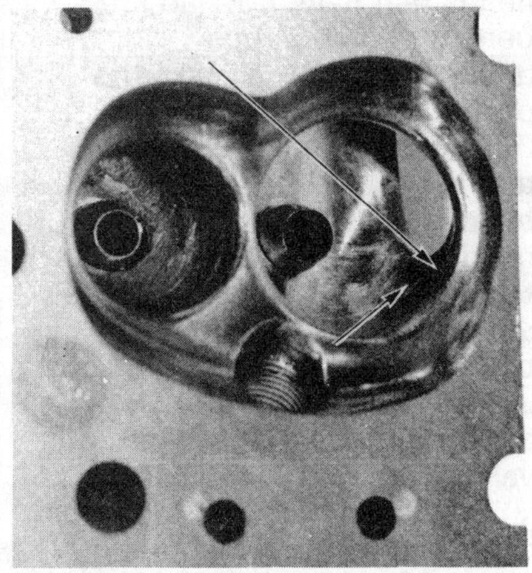

For good, high lift, flow, the area between the two arrows needs to be concave rather than the convex form that is so often produced.

the super big valves effective. If this is not done, the largest valve that can be made to work is about 1.43 inch, and even then, certain provisos must be considered which I'll discuss when I get around to valves and valve shapes.

Many hours on the flow bench have indicated in no uncertain terms that the area immediately adjacent to the valve seat can be very influential on

When putting in extra large intake valves, the head casting may break through into the water jacket in the area indicated here by my pen.

overall port performance. At first sight, a good form just prior to the valve seat would be a radius of reasonable proportions blending into the rest of the port in a smooth manner.

Although this looks like a good idea, it does make one assumption: namely that the airflow all around the valve is of basically the same pattern. In practice nothing could be further from reality. The flow patterns on the long side of the port are totally different in form to the airflow on the short side. To this must also be added the fact that close proximity to an opening and closing valve, means the port has a varying shape. If this leads you to the conclusion there is more to developing a high flowing valve seat than meets the eye, you'd be right. At low valve lift, air is discharged into the cylinder in a fairly even manner from the entire periphery of the valve. As the valve lift increases, the air finds increasing difficulty making the turn on the short side radius. However, it is forced, by virtue of the shape of the port, to negotiate the turn on the long side radius. As a result, a greater percentage of the air transits the port through the valve on the long side radius than the short side. At mid- and high lifts, the shape of the valve seat adjacent to the long side radius needs to be distinctly different from that on the short side. At such valve lifts, there is little airflow of any consequence around the short side. **In fact, the short side of the valve can almost be written off in terms of airflow at high valve lifts, so what must be done is make the best use of it when it is most likely to flow, namely at low valve lifts.**

Two things to note from this picture. Firstly the chamber shape is slightly different to the one used in the drawings and is experimental as of 1988. It has, under certain circumstances, already produced positive results. What is not experimental is the Rimflo valve. Numerous tests have shown these always give positive results.

Here's a cylinder head cut as per the drawings a few pages back. This particular example, crafted by Bill Quine of Manx Racing, was used on a road going 1400 Mini engine which crossed the 140hp barrier.

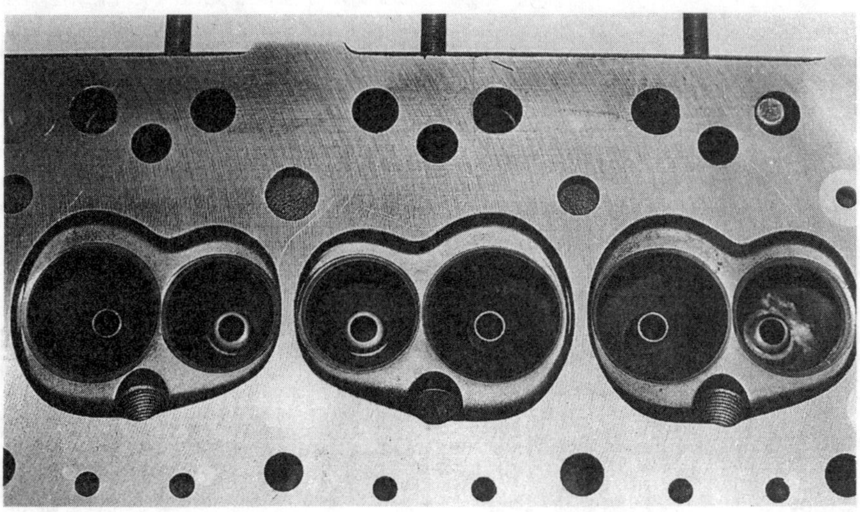

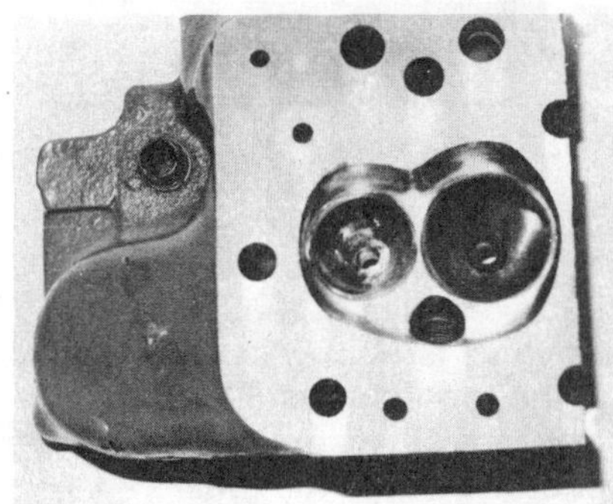

Here is a combustion shape and the throat of two ports which represent the state of the art in 1983. Note that the chamber walls are not cut back as far as they were on my heads of some ten years earlier. This is because more of the flow takes place in a direction toward the centre of the cylinder from the intake, or away from the centre of the cylinder for the exhaust than on previous heads.

Anti Pollution Heads

Many Series 'A' engine fitted to vehicles exported to countries where pollution controls are far more stringent than they are in Europe, have heads which have air-injection on the exhaust ports for 'after burning' any unburnt hydro carbons in the exhaust. During 1985/86 work done by my A.P.T. partner, David Anton, has shown that air injection 'smog' heads were better for all-out racing modifications. Thicker castings in general, but around the exhaust port in particular, have all allowed the sculpting of more favourable port forms. It is no problem to plug the air holes up and then work with the casting in the normal way. If the air injection has to be left or you don't feel inclined to plug up the holes, it's no big deal. Shaping the ports as outlined in the relevant section concerning your particular head and ignoring the presence of the holes, will result in a port which flows very close to what it would have done had the holes been plugged.

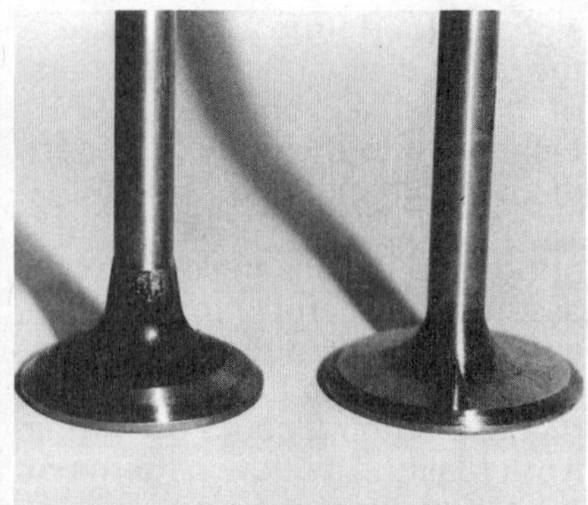

It's easy to take valves for granted, but they can make a great deal of difference to the performance potential of the cylinder head. On the left is a standard exhaust valve. The thick, under head profile is used for two reasons, firstly cheapness of manufacture and secondly, the valve needs to have sufficient mass to be able to conduct heat into the stem because the material has limitations on the temperature it can take. On the right we have a valve made of 21 4N stainless. It can stand very high temperatures and so the valve's form can be designed to produce good gasflow and light weight rather than satisfy thermal considerations.

Pushrod Clearance

If you intend to use a pushrod larger than ¼ inch diameter (and there are good reasons why you should) the head and, in some cases the block, will need cutting for clearance. While being built it is not a bad idea to drill the block pushrod holes to 7/16 inch diameter. This is best done freehand as the holes are at a 2½ degree angle from the vertical. As for the head the pushrod holes need to be ground or milled elongated for extra clearance. This move has the added benefit that it also allows the oil to drain back faster.

In other words, the seat adjacent to the short side radius should be cut to suit the characteristics of the port when the valve is at low lift. However, on the long side radius, the converse case applies. The valve seat, or at least the area immediately adjacent to it, needs to be cut to suit the airflow characteristics at high valve lift. Because at low valve lift, the air is travelling at a relatively low speed on the short side radius the port form can join the valve seat with almost no radius whatsoever. Because low flow speeds are involved, a measure of streamlining can be usefully traded off for opening area. The shape of the seat on the short side is shown in the set of drawings (**Fig. 9.14**) detailing the intake port. On the long side, the air is going to be travelling at high speed as it approaches the seat, so it needs a radius to aid its passage around the corner into the chamber. On the long side then, a generous radius needs to be applied just prior to the seat, again as detailed in the relevant drawing.

Seen here is an early experimental flow bench-developed, flat backed anti-reversion valve. The thin stem and the flat profile on the back of the valve, produced excellent flow figures in the forward direction whilst the . . .

. . . recessed front face of the valve reduced back flowing during the overlap period.

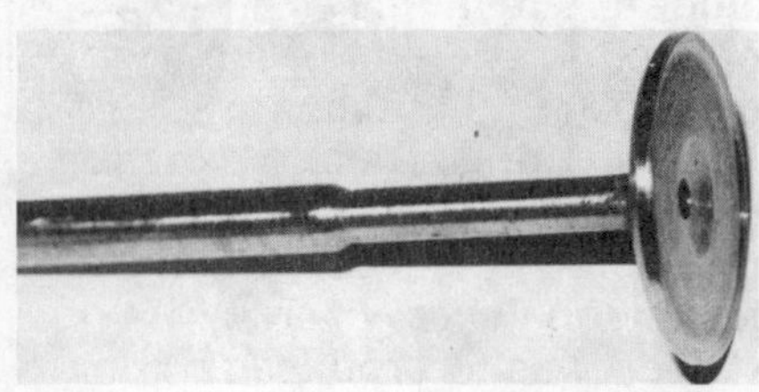

The intake seats themselves need to be exactly at 1/16 inch wide. Any narrower than this and they will lose flow both at

the low and high ends. Any wider will cost high lift flow, although wider may well marginally improve low lift flow.

It may sound like I am going overboard on valve seat detail, but basically this area should be given number 1 priority. Any slipshod workmanship here can cost a lot in the way of flow, and ultimately hp. At this point consideration needs to be given to the fact that the valve seat comprises of two parts, one in the cylinder head and one on the valve itself. So far I haven't delved too deeply into valve characteristics, but since the subject has reached this point, now is as good a time as any to deal with the subject.

Valve Shapes For Airflow

Whilst working on a development engine, I decided to do an investigation as to what was the biggest useful intake valve that could be installed into a Mini cylinder head. This should not be confused with the biggest valve that can physically be installed, but what the largest EFFECTIVE valve is. The two can be quite different. Basically I started this project as a result of a conversation I had with Dave Martin of Swaymar Race Engines. His own flow testing on Series 'A' heads had indicated that it was not practical to get a valve larger than about 1.42 inch to 1.43 inch to be effective. **In other words, over this size, an increase in valve size did nothing to enhance flow.** All a 1.5 inch valve would do (and it's possible to squeeze one in this size) is reduce the strength of the head, especially in that thin area between the intake and exhaust valve, and increase the amount of reciprocating mass the motor has to cope with. What it DID NOT do was to produce any

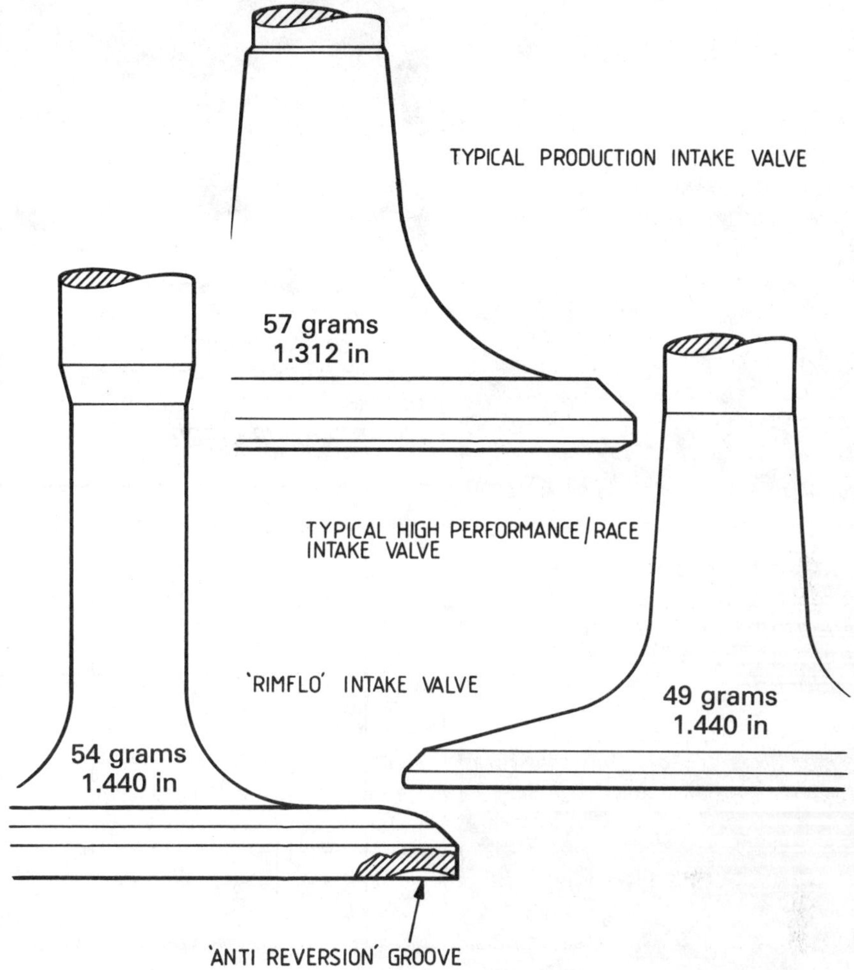

Fig. 9.17 Comparison of standard, conventional race and Rimflo valves.

An analysis of intake valve shapes shows that until the introduction of Race Engine Components' 'Rimflo' valve, valve design had more or less stagnated. Use of conventional valves over 1.42–1.43in had shown no real gains in the Series 'A' heads, simply because the shapes used were not totally compatible with the port flow patterns. Air in the intake port tends to flow ***across*** *the back of the valve from side to side, rather than down around its circumference. This is why a tulip valve produces bad results in a Series 'A' engine. The 'Rimflo' valve's flat back and waisted stem provide less flow obstruction simply because there is less valve bulk in the way of the airflow. This accounts for the substantial increase in high lift flow. On the other hand, the radius just prior to the valve seat gives a great improvement in low lift flow. These improvements can be as much as 16%–17% over the best conventional race valves. Another unique feature of the 'Rimflo' valve is the anti-reversion groove. Unlike the other two valves shown in the drawing, the 'Rimflo' valve flows worse backwards than forwards.Most intake valves flow better backwards (as exhaust) than forwards. This tends to exaggerate any camminess the engine may have, due to the use of a big cam. The 'Rimflo' valve tends to reduce camminess, thus improving low end power as well as top end power.*

Tests of standard valves versus Rimflo anti-reversionary valves were done on Nolan Pitts' dynamometer at an interval of 30 minutes. The differences produced . . .

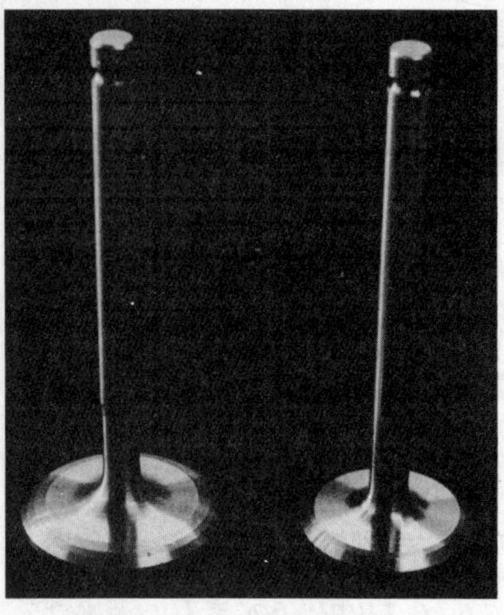

. . . by the higher flowing forms on the back of the valve and . . .

increase in hp. At this period in time, it appeared there were only two general shapes of valve in use, these being the standard form of valve and the generally accepted high performance/race valve with the small radius on the stem to the back of the valve and the back having a 10° to 14° angle, depending on who manufactured the valve.

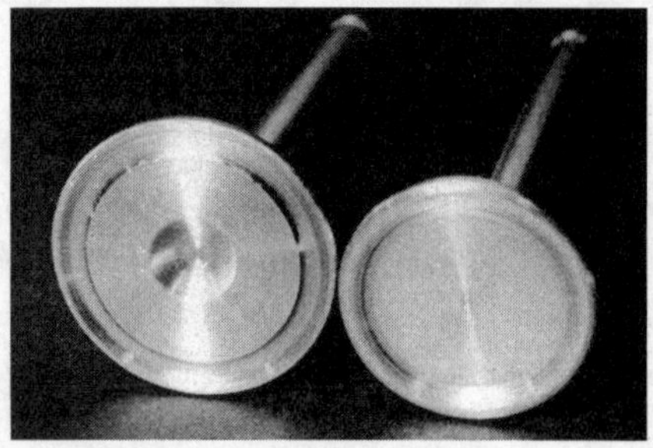

. . . by the anti-reversionary properties produced by the shape on the front of the intake valve, resulted in the power increases shown in the accompanying graph.

As a result of the foregoing, I decided to embark on an intensive campaign of valve shape development and worked for some months on various valve shapes in conjunction with various subtle port changes. **The result of the development project was the flat-back valve. This showed a marked improvement in flow, and also proved to be worthwhile in diameters up to 1.480 inch.** In other words, the flat-backed valve shows a flow improvement over either the standard or the angled-back race valve, and

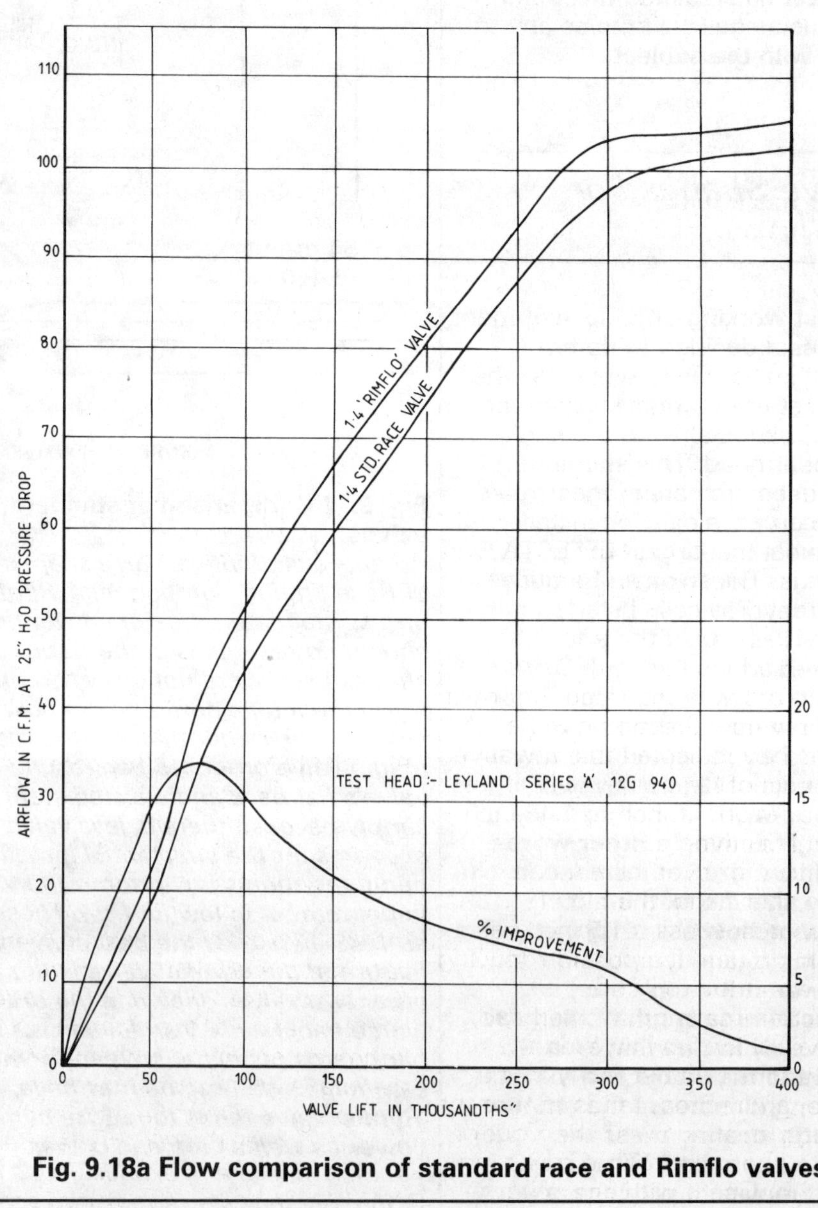

Fig. 9.18a Flow comparison of standard race and Rimflo valves

continues to show flow improvements in sizes up to some 50 thousandths larger diameter than either of the other two valve types. The form of this valve can be seen in **Fig. 9.17.**

Since the development work on this, **Specialized Valves** have continued work along these lines and come up with the 'Rimflo' valve. This valve has the added and significant advantage of not only improving forward flow, but has a tendency to inhibit reverse flow. **It's an unfortunate fact of life that most intake valves in Series 'A' heads tend to flow better as exhaust valves than intake valves.** In other words, if the intake port is tested as an exhaust port, it would flow more air at a given lift in the reverse direction than it does in the normal intake forward direction. **This has a tendency to exaggerate the camminess of the engine.** In other words, if a 300° cam having a lot of overlap and valve lift at TDC is used, the residual gases are confronted with an efficient port from which to exit the cylinder, namely the intake port. Of course, as rpm rises, the situation has a tendency to cancel itself out because the exhaust velocity out of the exhaust port increases and the gas doesn't have time to stop and reverse flow.

However, if the head has valves which tend to flow well in the correct direction but poorly in the reverse direction, then a tendency to reduce the negative aspects of high overlap is brought about. For instance, rough running at low rpm is reduced or even eliminated. On the other hand, the high forward flow means less to inhibit the high overlap cam from producing power at the top end.

Summarising, we can say a valve which flows well in the right direction but badly in the wrong direction, tends to take the rough and lumpy running out of the bottom end of the rpm range, and replace it with smoother running as well as adding top

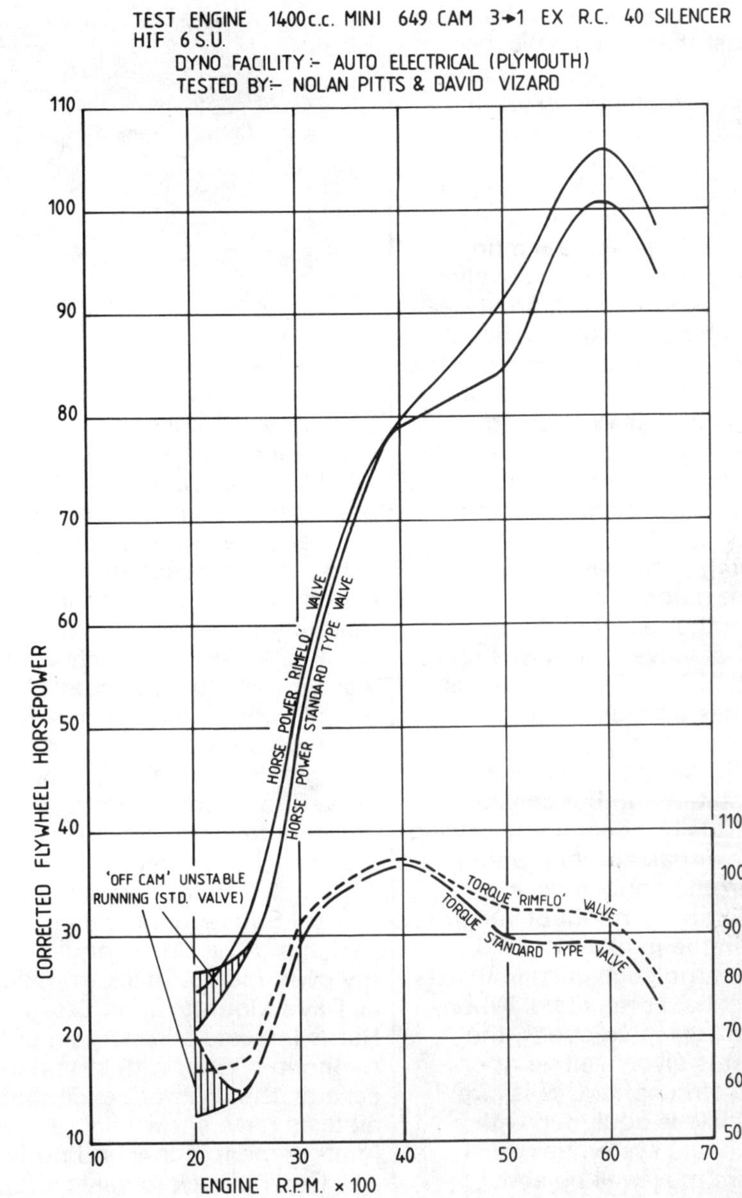

Fig. 9.18b Power output comparison of standard race and Rimflo valves

Here's one of the more successful tests using 'Rimflo' valves in a bigger Series 'A' engine. At low rpm the 'Rimflo' intake valve's tendency to reduce backflow improves the low speed running of the engine. With conventional valves the engine 'hunted' at the 2000rpm mark; a typical characteristic of an engine which is about to fall off the cam completely. With the 'Rimflo' valves, the engine ran perfectly satisfactorily at 2000rpm. The reduced backflow obviously cut exhaust contamination of the incoming charge and the result is a marked improvement in the low end torque. As rpm rises, so contamination of the intake with exhaust is reduced so there is no problem for either type of valve. Over the mid-range there is little difference in the output, but past 4000rpm the effect of the superior flow of both the 'Rimflo' intake and exhaust leads to the production of significantly more horsepower. A max power increase of 5hp is shown.

end power. The 'Rimflo' valve design from Specialized Valves, does just that. On a typical 1000cc race engine, the motor will sometimes pull up 'on the cam' 200 or 300rpm sooner. However, a Rimflo valve is not solely designed for use in race engines. If you look at the flow difference between a Rimflo valve and a conventional valve, you will see there is a great deal of extra flow potential, even at low lifts (**Fig. 9.18**). This extra low lift flow has the advantage of making engines employing streetable cams produce more top end power simply because the extra flow occurs within the lift envelope used and enhances breathing potential throughout the rpm range.

Another asset of the Rimflo design of valve, and this is really important for a road engine, is that it has a tendency to produce better economy figures. At first sight this seems to be a rather bland statement, but consider this: virtually all economy driving is done at part throttle. Under part throttle conditions, a considerable amount of vacuum exists in the intake manifold. However, no such vacuum exists in the exhaust manifold. When the exhaust valve opens, the exhaust is always either at or around atmospheric pressure. If the vehicle is equipped with a bad exhaust system, exhaust pressure may well be above atmospheric pressure. During the relatively small amount of overlap produced by a road cam, we find the exhaust, being at a higher pressure than the intake, easily backflows into the intake port, contaminating the fresh charge. The Rimflo valve reduces reverse flow under part throttle conditions, and allows a purer, more ignitable charge to be drawn in. At this point, although the engine can run cleaner and smoother, no real economy gains are likely to have been made. However, what can now be done is that the part throttle mixture can usefully be leaned out a little, so as to promote better economy. Because of contamination, the leaner mixture would not previously have been of ignitable quality. Granted, it will need some fine tuning to extract the possible increase in economy, but the potential is there if you choose to explore it.

Shown here is the cylinder head for the standard valve versus the Rimflo valve test, just before it went back on the engine with the Rimflo valves in. The valves may have looked a bit strange but they did produce some very positive results in this particular test.

Fig. 9.18b represents what was virtually the first controlled test of a Rimflo-style valve in comparison with a standard racing valve. The Rimflo valve proved to be much more effective than anticipated, and, at the time, I was having nightmares that the test might be a fluke. Subsequent tests on perhaps half-a-dozen engines of my own and feedback from such as Dave Mountain and Steve Harris indicate a likely gain of 2 to 4hp in 105hp, with a worse-case of 1hp. Without exception **all** tests have shown an improvement in low end output.

Getting back to valve sizes for a moment, super large valves up to 1.48 inch on the intake are okay for hp, but they can in the long term, be unreliable for a road engine. My personal preference for such an application, considering not only possible cracking between valve seats, but also the probability of grinding into the water jacket, would be to use a 1.4 inch Rimflo valve. If the prospect of having to weld the head in the divider section doesn't overly bother you, then sure, go to a 1.440 inch Rimflo valve. I would only advise the 1.480 inch size valve for all-out race engines and even then, because of the inlet valve size the head should be ground by someone with a good deal of experience of flow bench testing in this area. In the States my own company, A.P.T., can undertake this work. In the UK, I recommend Mike Parry (Race Techniques), Bill Quinne (Manx Racing) or Steve Harris (Steve Harris Racing).

Since the Rimflo valve does not appear to be particularly streamlined in form by comparison with the more common tulip valve, it may come as a surprise to many of you that it, in fact, does flow better, and you may wonder why? The reason is simple enough once you realize that the airflow in the port just behind the valve is anything but what it at first may appear to be.

As I pointed out earlier, most of the air comes out from the long side radius. It is not moving vertically downwards into the cylinder as it goes by the valve. The actual direction of motion just prior to the valve is about 45° to the valve stem, so flow is across the back of the valve, not down onto it. **Fig. 9.19** should give you an idea of the airflow

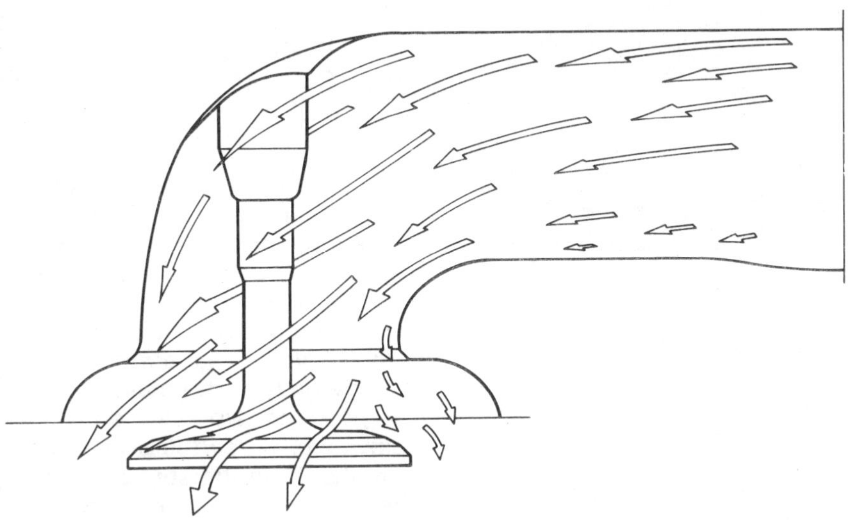

Fig. 9.19 Flow pattern at high lift with Rimflo valve
The illustration shows why a 'Rimflo' intake works. Much of the airflow is across the back of the valve, so the waisted stem and flat back present less obstruction to the flow than does a conventional valve. Most of the air is flowing along the top of the port and out of the long side. Only a small percentage flows out of the short side.

direction involved. As you can see, the airflow scoots across the back of the valve and out the far side of the port. The Rimflo valve's thin stem and flat back presents less of an obstruction to mid- and high-lift ranges than does the conventional angled-back, thick stem valve.

However, at low lifts, a different situation exists, and you will note that the progressive form transition from the valve back to the valve seat gives, at low valve lift, a near ideal approach to the seat. Remember, at low valve lifts, air is going to flow fairly uniformly from all around the periphery of the valve. A good approach to the seat enhances the low lift flow. There is enough thickness of valve close to the seat to allow a good approach form to be developed, hence the much higher airflow at low valve lifts. No doubt from this explanation you are beginning to see that the valve form cannot be just a shape plucked out of thin air. The shape of the back of the valve has to be a compromise simply because the port form varies considerably throughout the valve lift cycle. The Rimflo valve represents an excellent compromise in all areas. It is the result of hundreds of hours of flow testing, whereas the other two valves in **Fig. 9.17** are not.

The Importance of Valve Seat Design

In the development of any production cylinder head for high performance use it is very important to identify where the major flow restrictions occur. Unless just short of a totally disastrous design then in any induction or exhaust system the principle restriction occurs at the valve. The first step toward appreciating this is to take a typical cylinder head as seen in Fig. 9.20 and visualize what each major portion of it would flow. The example cylinder head shown here would, when sectioned as shown in Fig. 9.21 produce flow figures typically as indicated.

From this, we can see how the section that flowed the least is the portion that contains the valve. Analyzing the lift curve of a

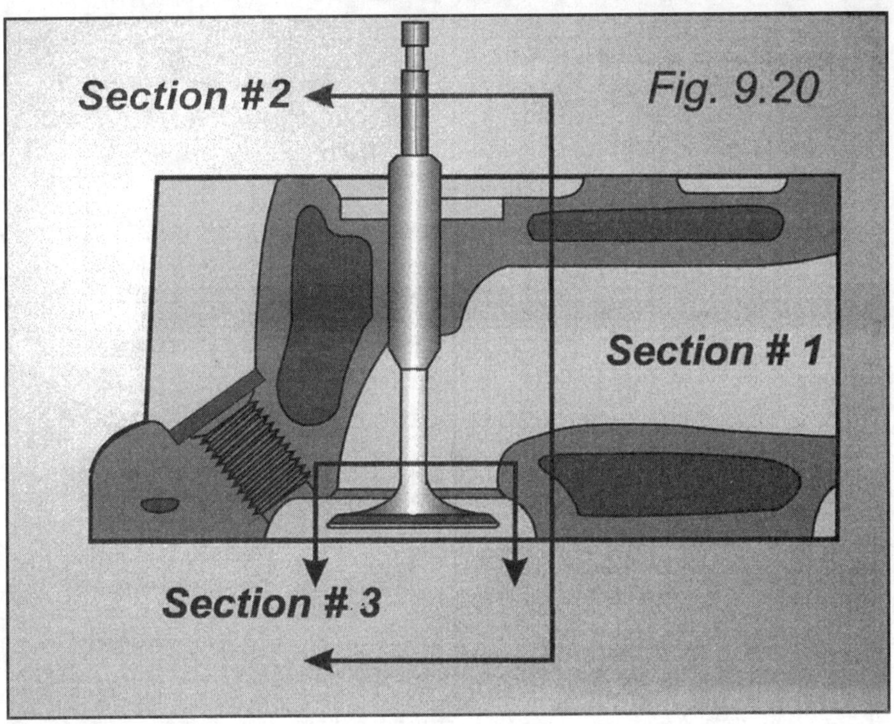

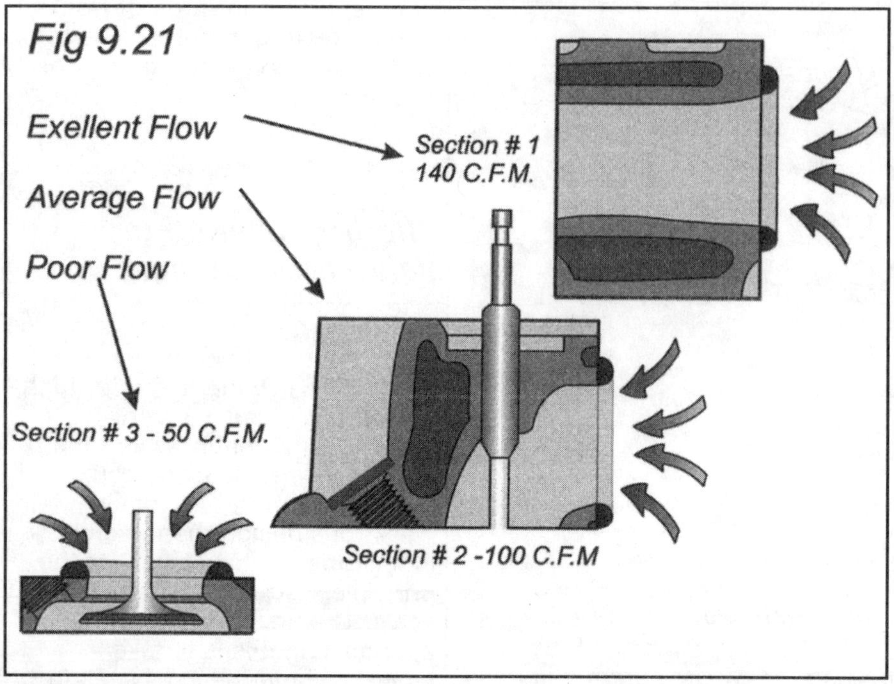

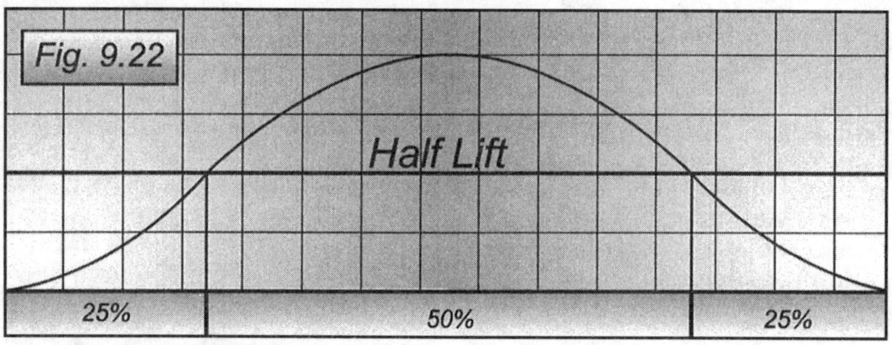

The Relative Importance of Ports and Seats

By investigating port velocity we can establish at what point the seats influence is diminished to the point of irrelevance. At the first approximation we can say that when the velocity of the air past the valve seat equals the port velocity in the main part of the port, the two share equal relevance. When the port velocity is higher than the seat velocity the port is the main priority. When the seat velocity is higher than the port velocity the seat form is the main priority. The graph, Fig. 9.23, compares the mean seat and port velocity for a standard 1275 'A' Series head. Note port velocity does not reach that of the seat until about 300 thousandths lift. This gives us direct comparison of the relative importance of seat and port. Below the crossover point the seat form is the main priority. Above the crossover point the port is our main priority. This puts a far greater emphasis on the normally accepted importance of the seat typical cam (Fig. 9.22) is also revealing. It shows that the valve spends as much time below half lift as it does above. In an earlier part of the book it was shown how valve seat form can dramatically affect low, and to a lesser extend, mid rangeflow. Just as important to a head development programme is the need to appreciate at what point the valve seat is of no further significant influence on the airflow.

This will be our next topic of discussion. It is a topic that is important to the output of the 'A' Series engine especially a large 'A' Series engine. If you are intent on extracting the most from your engine you will do well to absorb this information.

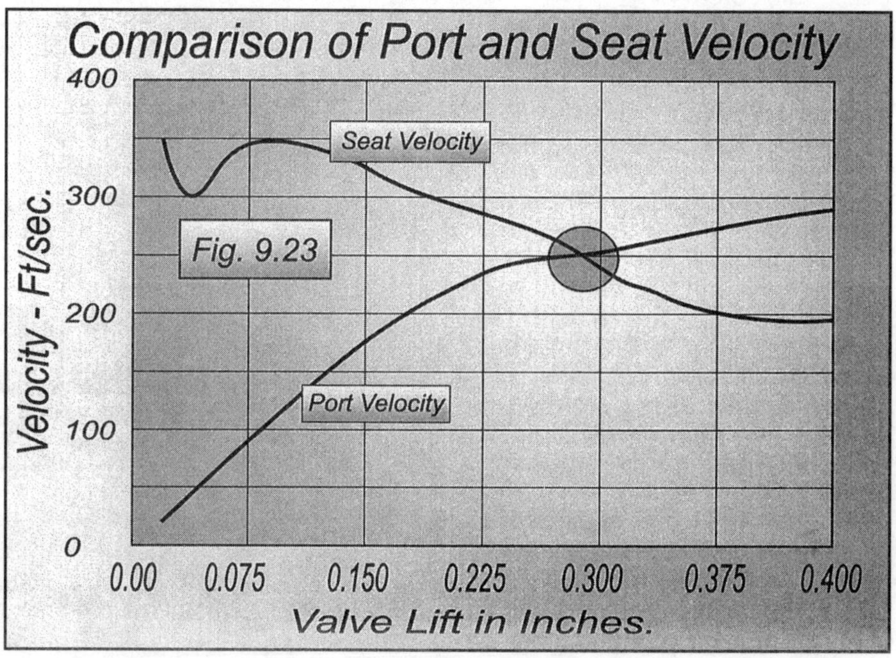

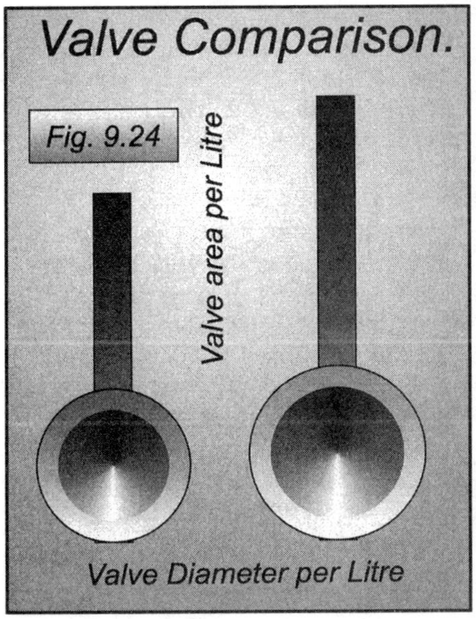

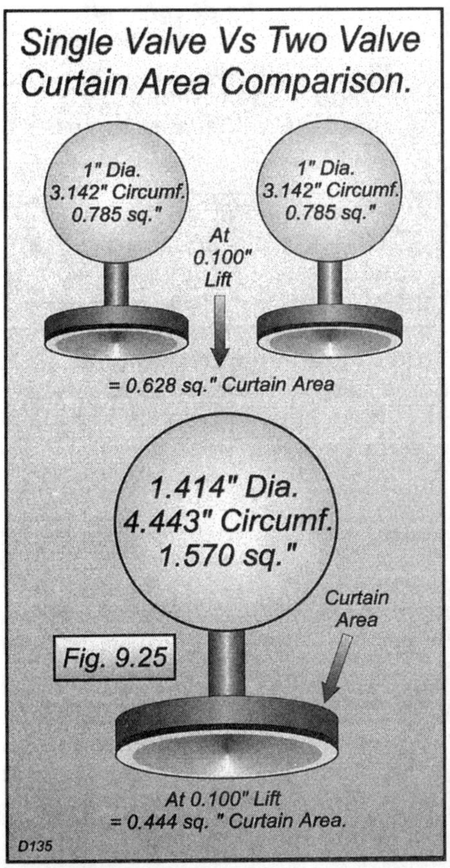

considering many street cams lift the valves about 320 thousandths.

Now let's take a look and see how under valved the typical two valves per cylinder 'A' Series engine compares with a typical four-valve engine of the same bore size. Chart Fig. 9.24 shows how a 1.44 intake valve and 1400 Mini compares in terms of the valve area per litre with its typical four-valve counterpart. From this we can see that the two-valve engine falls well short of its four-valve counterpart but that's not the only aspect of a single intake valve that proves a disadvantage. A game played extensively in the camshaft industry is to see how fast the valves can be opened. Looking at it another way we can say that it's all about presenting the most breathing area to the cylinder as quickly as possible. However, it would be wrong to think that this is only a function of the camshaft. Check out the numbers in Fig. 9.25. This drawing compares two intake valves of the same area with a single intake valve and looks at what happens to the breathing or curtain area presented to the cylinder as the valves are lifted from the seat. With the diameters used in the example two intake valves lifted 100 thousandths of an inch presented to the cylinder with a curtain amounting to some 0.628 sq inches. Lifting a single intake valve of identical area 100 thousandths off the seat only presents the cylinder with 0.444 sq inches of curtain area. From this we can see that the intake valve's circumference is, at low lift, of *more importance than total valve area*. The greater the total intake valve circumference is the faster breathing area presented to the cylinder. In other words it is one big step nearer instantaneous valve opening. This is one of the reasons why four-valve engines don't need as long a cam duration as two-valve engines because the valve opening area happens 41% faster. This is a huge advantage for a four-valve per cylinder engine to have and a very difficult deficit for a two-valve per cylinder engine to make up. However there are ways we can cut down that difference.

Cutting the Margin

The graph, **Fig. 9.26**, shows the difference in developed curtain area between two intake valves and one having one intake valve

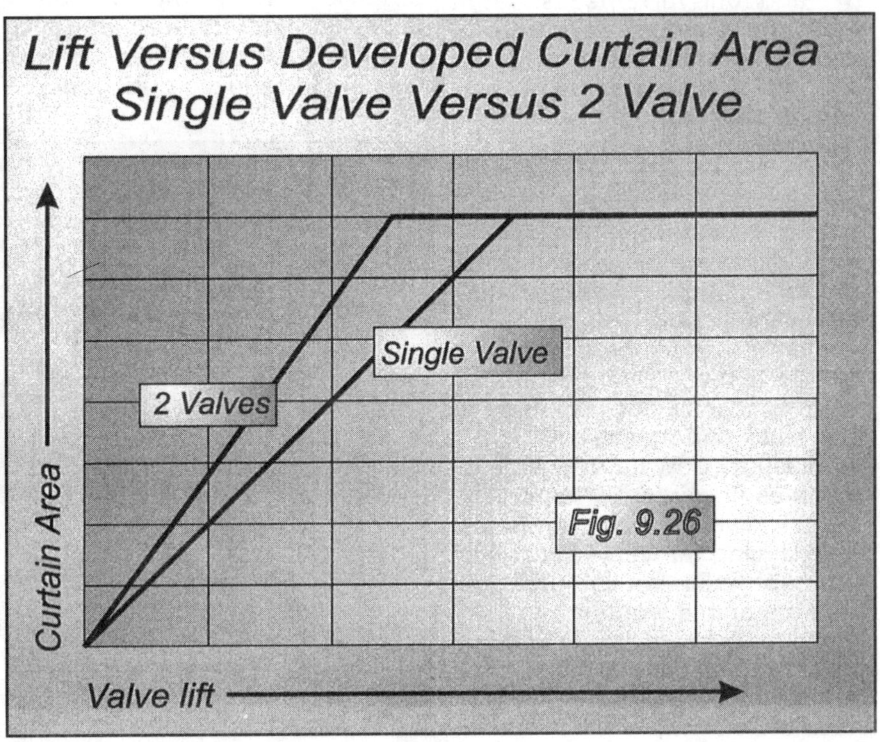

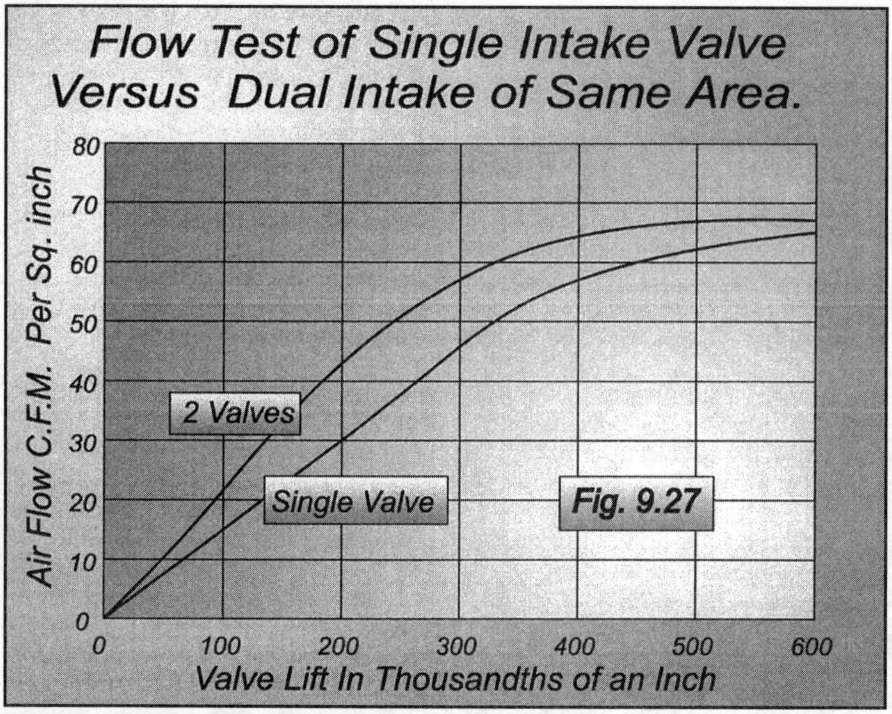

Fig. 9.27

of the same area. As you can see, ultimately they reach the same level but the two intake valves get there much sooner. When this is put into practice on a real cylinder head we find that the flow curves follow a similar pattern. The flow bench figures in **Fig. 9.27** were generated on two and four-valve heads that fit the same engine. The reason the single intake valve head failed to reach the same peak flow as the two-intake valve head was that it was as a production design with far less port down draft to meet clearance constraints on a production vehicle. Needless to say the race design four valves per cylinder head was not hindered by such constraint. If we are to make a single intake valve function better we need to find ways and means to increase the low lift flow as much as possible. Fortunately, there is a way we can do that. By reducing the valve seat angle from 45° to 30° we can pick up a considerable advantage. Historically however, 30° valve seats have usually failed to deliver any more power – but I have the fix!

30 Degree Seat Evaluation

Seat angles prove far more significant to engine performance than they first appear. Changing the seat angle from 45° to 30° can produce dramatic improvement in low lift flow. The following chart, **Fig. 9.28**, analyses the true flow area produced during the initial departure from the seat. The drawing on the right hand side of the graph clearly shows why a 30° seat has the advantage. If an intake valve having a 45° seat is lifted 50 thousandths then the gap between the seat on the valve and the seat on the head is only 35 thousandths. If we perform the same exercise with a 30° seat the gap between the two seats is 43 thousandths. The difference between the two as they transition through the initial part of the lift curve is shown in the graph to the left. *The advantage delivered by the 30° seat represents a 21% increase in breathing area over that given by a 45° seat during the first 100 to 150 thousandths of valve lift*. If we applied this to a single intake valve and compare it to the breathing area presented by a conventional 45° seat dual intake valve set up *the advantage of the two intake valves is cut from 41% down to only 20%.*

I'm sure this sounds like a pretty good deal. However, 30° seats have been around a long time and the fact that they present more initial breathing area has long been understood by those knowledgeable in the field of cylinder head development. So why hasn't the

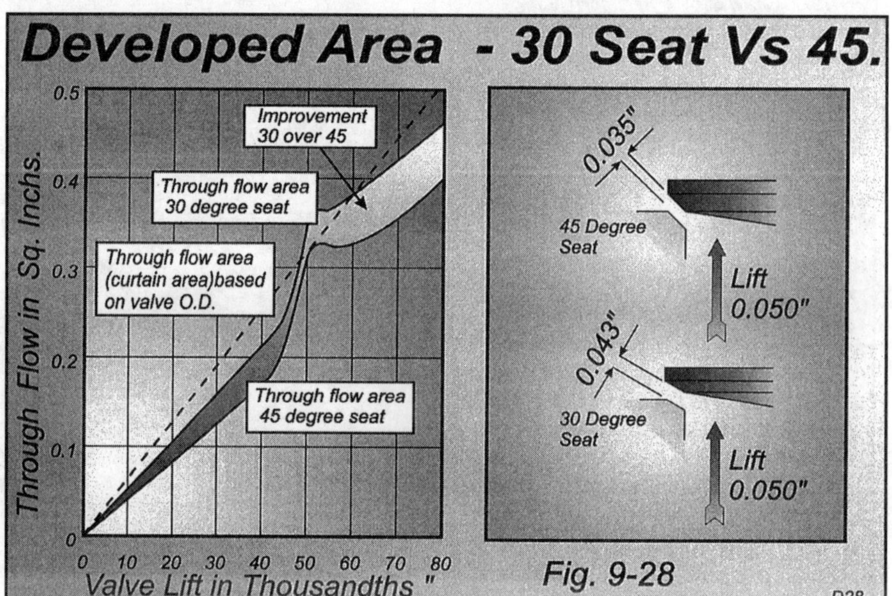

Fig. 9-28

30° seat found popularity? To understand why a 30° seat has historically been unsuccessful we need first to look into valve seat to port blending geometry. This we will deal with now.

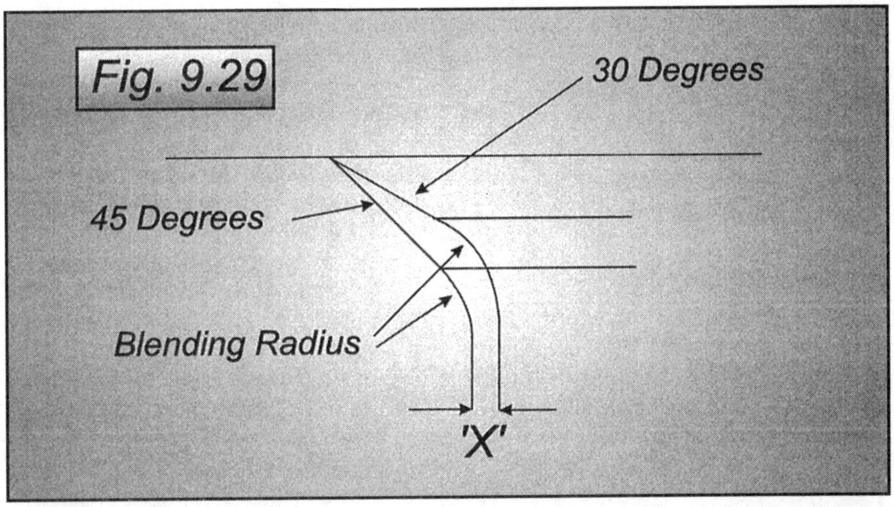

Seat Angles, the Pros and Cons

Although the 30° seat has an advantage when the valve and seat is in close proximity the price paid can be constricted flow at high valve lifts. For the most part it is those high lift flow figures that the 'A' Series engine needs to make horsepower. Assuming that the valve seat is blended into the port with a typical blending radius we find that the constriction from the 30° seat comes about because a finished throat diameter just below the valve seat is smaller, as shown in **Fig. 9.29**. For an 'A' Series head dimension 'X' is typically 25 to 30 thousandths which makes the throat diameter of a 30° seat prepared in a conventional fashion some 60 thousandths smaller. *This represents a loss of valuable high lift breathing area on an engine which is already in short supply of such*. However all is not lost. The day is saved by a little known phenomenon concerning the way air in small conduits *ie* the size of typical cylinder head ports, negotiates minor turns up to 15°. In **Fig 9.30** we see two methods being used to redirect air through a 15° angle. The top one employs a conventional radius bend as could be expected in a typical ported head. However, we find that for angles 15° or less there is no apparent loss of flow if the turn is made sharply as shown in the lower drawing. In effect this means that instead of having to have a radius between two surfaces at a 15° angle *we can just as effectively have a sharp edge*. This can be applied directly to a valve seat. The drawing **Fig. 9.31** shows how this is done on a valve seat. Note how, compared with conventional blending, the throat diameter of the 30° seat is restored almost to that produced by a 45° seat. In addition to this

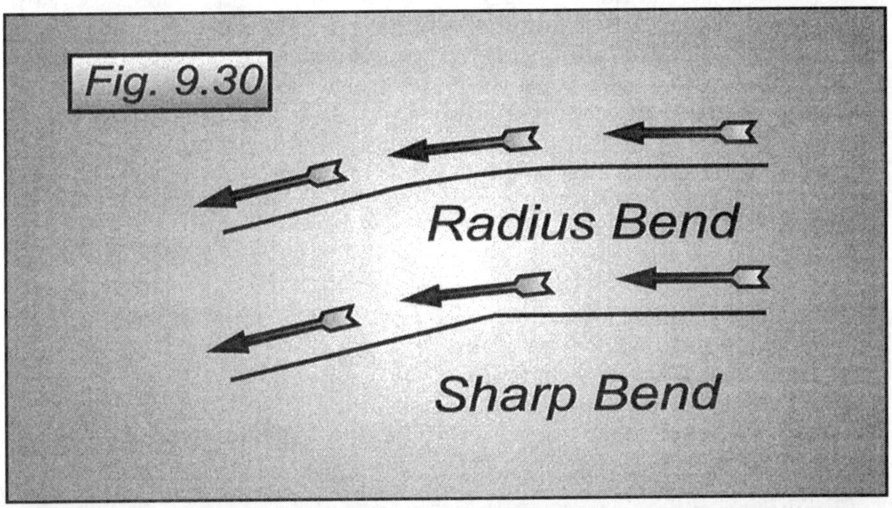

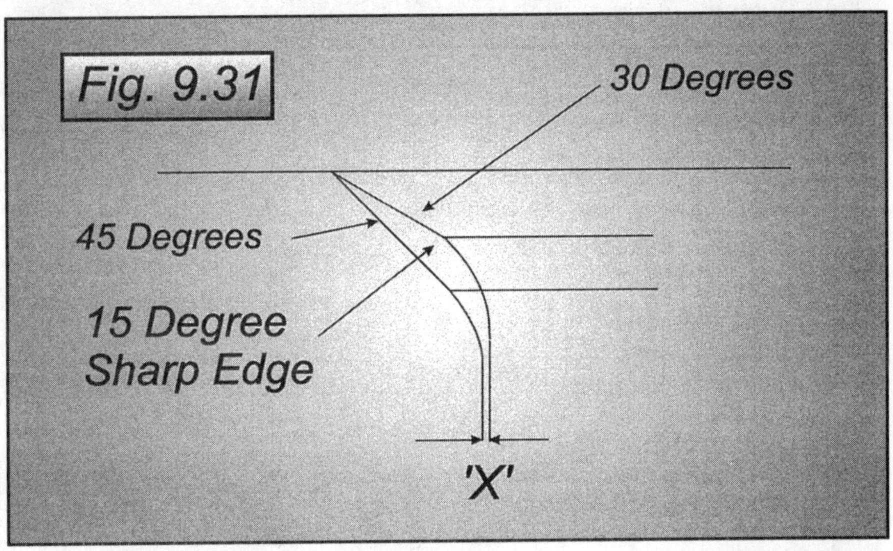

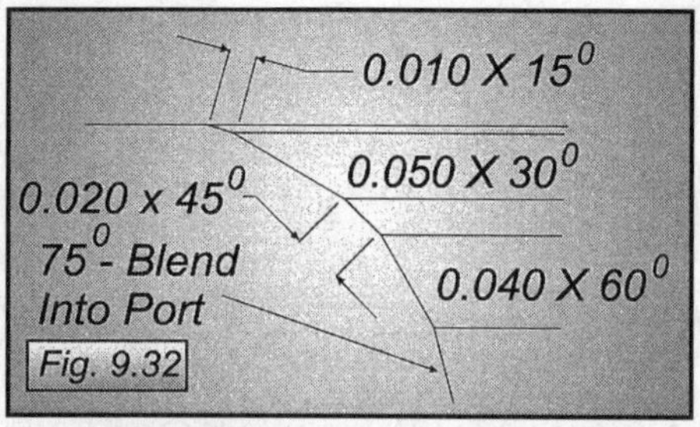

Fig. 9.32

one more factor comes into play *which virtually restores parity* in terms of efficient throat diameter for the 30° seat. A question which often crops up is 'how wide should a valve seat be for best flow results?' My own extensive testing in this area has strongly indicated that for a conventional 45° seat a 60 thousandths wide seat provides about the best overall results. If the seat is made narrower the low and mid lift flows suffers because of a less efficiency form for the air to pass around. If the seat is made wider the flow in the low and mid range is good but the high lift flow is constricted. When a 30° seat angle is used the width that gives optimum overall results appears to be around 50 thousandths instead of the wider 60 thousandths of the 45° seat. This being the case we find that on a Mini cylinder head the throat diameter, which is normally the principle constraint, a 30° seat has toward producing less top end flow is restored to virtually that of a 45° seat.

From the forgoing you can see that the seat dimensions and form are far more critical than may first appear.

In all drawings so far shown the seat is blended into the port with the traditional radius but we've already established that a 15° angle will produce virtually the same results as a radius. Applying what we have learned to what appears close to an optimal 30° seat produces the design shown in **Fig. 9.32**. In terms of flow a typical Mini head generally delivers improvements in the order of those shown in **Fig.9.33.**

Turning Flow into Power

The only justifiable reason for a 30° seat is not whether it produces more flow but whether it produces more power. For this to be the case it proves necessary to overcome yet another often unrecognized 30° seat stumbling block. With a 45° seat we find that the wedging action, which considerably aids sealing, due to the 45° angle is considerably higher than a 30° seat. Indeed, the 30° seat can fall short to such an extent in this area that seat leakage problems offset any potential gains. The problem with seat leakage cannot, as it might at first be thought, be cured by simply making everything more accurate. The real root of the problem is one of thermal gradients. If you look at the temperature gradient across the intake valve we find that, because the exhaust valve is so much hotter, that the expansion of that side of the intake valve seat is much higher. The end result is that the seat distorts out of round when viewed from above and develops a wave when viewed from the side. **(Fig. 9.34)**

The result is that the valve seat can distort by in excess of 4 thousandths of an inch. The only cure for this is to make the seat area of the valve head sufficiently flexible so it can accommodate this dimensional 'error' which a 45° seat does without any real difficulty simply because greater 'wedging' action is involved. To make the valve seat face conform we simply make the front face of the valve into a 'rimflow' form. This is achieved by cutting a valve seat 'confirmation groove' in the front of the valve. The typical dimensions for this are as shown in **Fig.9.35.** This you will need to take care of yourself as, at the time of writing, no commercially available 30° seat valves with confirmation grooves are being made. A good valve to convert to this form is the valve part No. V188 from A.P.T.

Having gone through all the antics of getting the valve to flow more and provide an effective

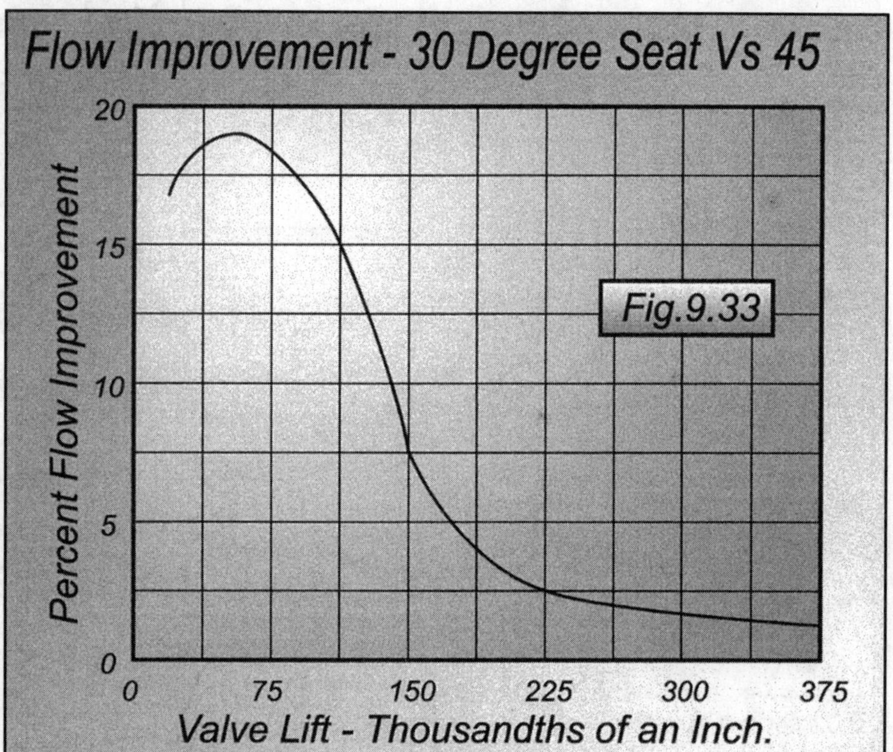

Fig.9.33

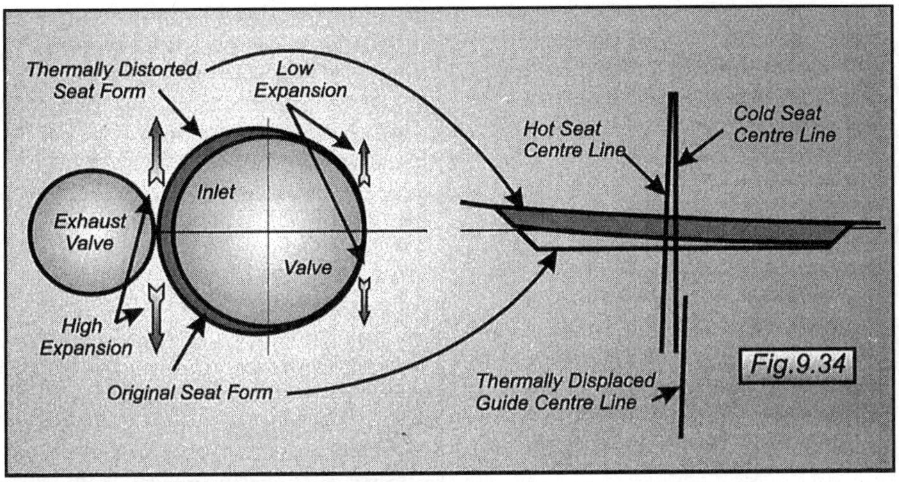

Fig.9.34

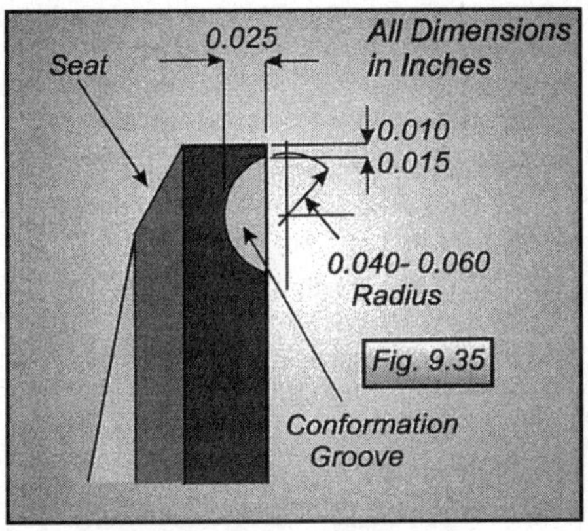

Fig. 9.35

seal what's it worth in terms of power? If you have a short cam engine a valve set up like this can be worth as much as 4 horsepower on a typical 1275 or bigger engine. In addition to this a significantly extended rpm range can be realized. This is one of the benefits of the increase in low lift flow. Although peak flow has a greater influence on maximum horsepower low lift flow is the prime ingredient toward extending the power band width. At a thousand rpm past peak power a 30° seat can be worth as much as 6 or 8 horsepower on a typical 1275. This means the engine can be *usefully* turned more rpm before making a gear change and this shows up as substantially improved time to speed or time to distance performance.

A great deal of airflow work has been done in my shop in regard to valve seat design. Because results can be influenced by very small dimensional changes I use this Serdi precision seat and guide machine to cut the test seats.

If the engine having the 30° seat applied to it already has the cam timing well optimized for the cylinder head flow characteristics of a 45° seat then you will find that the increase low lift flow of the 30° seat can produce the same effect as there being too much overlap. This means it will be necessary to re-evaluate what the optimum cam timing may be. Typically when making the change from a 45 to a 30° seat on a 1275 to 1400 cc engine you will need to use a lobe centerline angle about 1° less than whatever was optimum with a 45° seat. Conversely, if you are just going to change the opening point of the intake valve it should be delayed by about 4°. However, we won't dwell unduly on cam event timing at this point. Suffice to say I detail an exclusive Vizard technique for evaluating the cam event timing required *for your motor* in great depth in Chapter 11 which is where valve trains and associated cam stuff is dealt with.

Combustion Chambers

The standard combustion chamber shape used for open chamber heads is pretty well developed, but nonetheless there are areas in which improvements can be made. If you recall that earlier on it was stated that skimming the head increases the flow, it will be seen that there is some leeway here for developing the chamber shape by making it shallower. Of course, this can only be done if the CR doesn't exceed the value required. **However, if a choice is possible, then for a given CR, it is better to skim the head and have a dished piston, than have a flat topped piston and less skimming off the head.**

Basically, there is plenty of material to allow skimming on all pre-'turbo' 12G-940 cylinder heads, so problems in achieving the required CR are unlikely. In some situations, a reverse problem to that found with 850s can be experienced. If a big overbored motor is being built, the combustion chamber cannot be made large enough to get the CR down to a sane level. However, I will deal with that as a separate issue in a while. For now, let's just look at the chamber mods aimed at promoting gas flow.

The first mods to consider are those which reduce valve shrouding to a minimum. A step towards achieving this is to locate a cylinder head gasket on the face of the cylinder head and scribe around each bore. This defines the grinding at the position limits where the valve is closest to the bore. As a result you will be able to see how much material can safely be removed to achieve the maximum unshrouding benefit. Again, **Fig. 9.14** shows how the chambers should be cut for various sized valves in the area immediately around the intake valve. On big bore motors there is also an indication that if some of

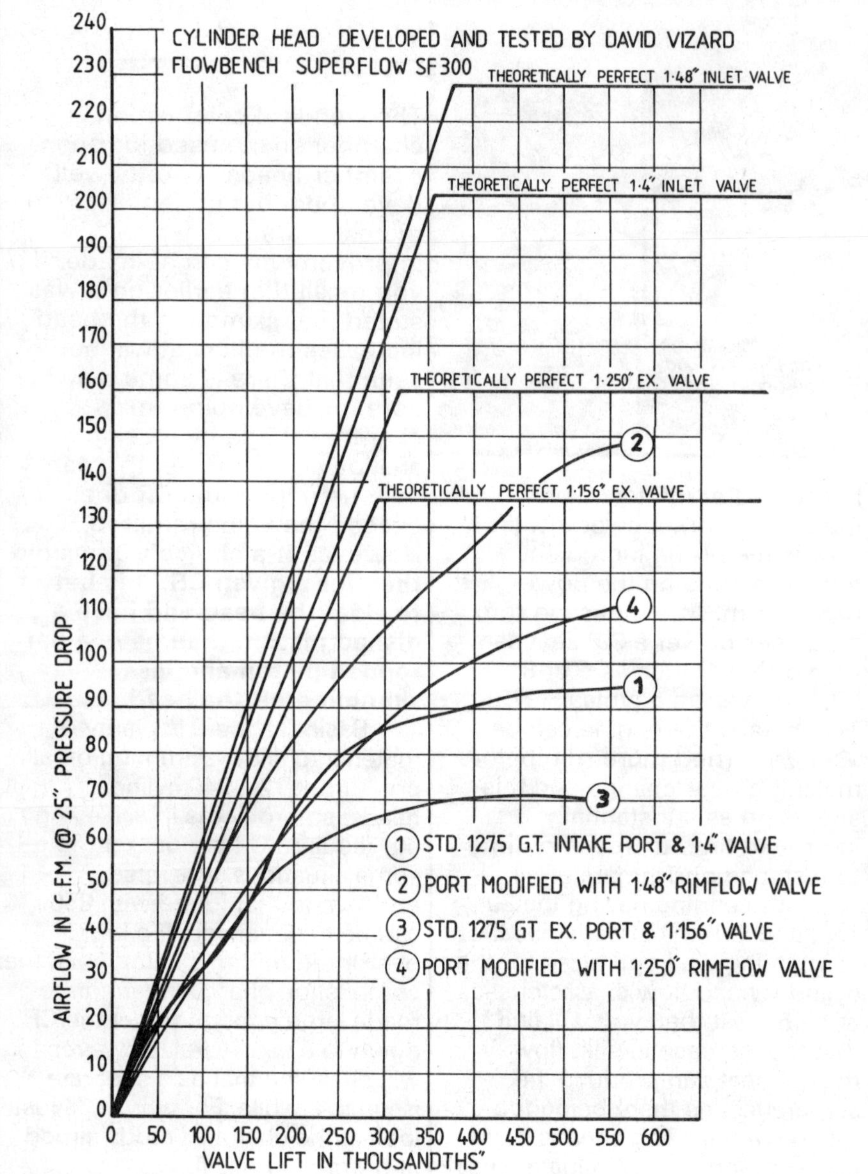

Fig. 9.36 Comparison of flow of standard 1275GT head (12G 940 casting) and head modified as detailed in Fig. 9.14
What can be expected in terms of airflow increase after a lot of hard work has been put into the ports? This graph is very informative. For instance, at 0.500in lift, intake flow is up from 94 to 148, an improvement of more than 55%. At the same lift, the exhaust flow has improved 35%. Although the exhaust increase is not as great, the flow ratio between the two valves is about right, though tending toward too great an exhaust flow. Big improvements have been made, but to keep things in perspective, the flow that should be produced by theoretically perfect valves has also been put on this graph. As you can see, even a well-modified Mini head is a long way from perfection.

the squish area in the region of the spark plug is reduced, then a small gain in power is achieved. However, if these heads are to go on small displacement engines, *i.e.* racing 850s, it's necessary to limit the amount of metal removed from the combustion chamber to a minimum, as a lot of metal needs to be skimmed from the cylinder head in order to get a respectable CR on these small engines. **Fig. 9.14** shows the metal that needs to be removed in the region of the spark plug. Don't remove the metal from the plug side of the head if the head is to be used on a racing 850.

At this point it's well worth analysing what we have managed to achieve in the way of induction flow, assuming the most sophisticated modifications have been done. Namely the biggest effective valve, the most modified port, etc. As you can see from this graph, **Fig. 9.36,** the flow increases with increasing lift at a pretty rapid rate, all the way up to a little over 500 thousandths lift. This in itself indicates that we should be using cam profiles which give at least 500 lift and preferably around 525 to 540, as it never hurts to 'overshoot' on valve lift, so long as the engine doesn't run into valve train problems as a result. It also may be well worthwhile pointing out here that the sort of flow figures produced by the ports (intake and exhaust) in **Fig. 9.36** were the best that my A.P.T. partner, Dave Anton, and I could manage in 1986. They are also substantially better than any proprietary head that I have flow tested. To date, the best flow figure at 560 thousandths lift is 132cfm as opposed to the 149 given by the one in the graph. I have flow tested heads with 1.5 inch valves that have only flowed in the region of 115cfm at this lift. Of course, having coaxed the intake to flow, it's necessary to make the exhaust work proportionately as well, so that's that we'll deal with next.

Exhaust

If the passage of the gases is followed through the engine, then investigation of the exhaust system starts right in the combustion chamber. The first aspect worthy of investigation is the potential shrouding of the exhaust valve.

If the most common standard size valve is used, *i.e.* 1.156 inch diameter, then the amount of unnecessary shrouding is almost nil. The total extent of the work around the exhaust valve is to move the combustion chamber wall out so it coincides with the bore over the area shown in **Fig. 9.14**. Not until larger valves are used is it necessary to cut back a little on the combustion chamber walls. If the intake port is modified along the lines previously given, and a big intake valve used, *i.e.* 1.44 inch to 1.48 inch, then I've found it necessary to go for every trick in *this* book to get the exhaust flow up to a compatible level, bearing in mind the ratio between intake and exhaust flow mentioned earlier in part 1 of this chapter.

At this point, a potential problem can arise. If a valve much larger than the original Cooper 'S' size is used (1.215 inch), then it will interfere with the big inlet valve. Unfortunately, at 1.215 inch, the flow potential of the exhaust valve is still short of what is needed unless the high-tech loaf-shape ports developed at A.P.T. are used.

If the cylinder head is going on to an overbored engine, *i.e.* 73.5mm or 74mm to make 1400cc from a 1275, then it's entirely practical to do a little re-positioning of the exhaust valve so as to achieve a situation more likely to produce the desired flow. Basically, this involves offsetting the exhaust valve 50 thousandths further away from the intake. Doing so will allow the use of a 1.25 inch exhaust valve, together with a 1.48 inch intake valve. This makes a very good working combination for a short stroke big bore engine. If a very high compression ratio (14:1+) is to be used then the offsetting of the exhaust valve can be beneficial in another way. By offsetting the exhaust valve say 30 thousandths toward the cylinder wall and

To get super big exhaust valves in on overbored big bore engines, it's necessary to offset the exhaust valve guide. Obviously this entails the use of a special guide, seen here, or an offset bush as explained in the text.

Another way of offsetting the guide is to bore the guide hole out larger as seen here . . .

offsetting the intake half this amount in the same direction we can now accommodate a larger intake valve. Going this route with a little more offset has allowed me to use a 1.53 intake with a 1.218 exhaust. This produces a better flowing intake but the ratio of intake flow to exhaust flow *is only optimal when a very high compression ratio and extended exhaust valve timing is used.* In case you skipped it the rational for this was detailed in Chapter 2.

When an offset exhaust guide is used, the radius on the edge indicated by my finger, can be made larger and the amount of port bias reduced. As a result, port efficiency increases quite dramatically.

. . . then press in a sleeve and the appropriate guide as seen on the end guide of this cylinder head.

If a 1.440 inch intake valve is used, then it's just about possible to get by with a 1.215 inch exhaust valve, although even at that point, if there's room down the bore for a 1.250 inch offset valve, it's the way to go. If a 1.4 inch intake valve is used, and this is what I would recommend for 1275 engines utilizing a standard bore size, then the 1.215 inch exhaust valve copes adequately. The relevant part of drawing **Fig. 9.14** shows the shaping around the exhaust valve for offset and non-offset exhaust valves.

Cylinder Heads Part 3

Valve Shapes

Just like the intake, the shape of the exhaust valve can be quite critical in terms of how it affects flow. However, the exhaust valve also has to cope with a much more hostile situation than the intake valve. Heat has to be conducted away from the head of the valve, otherwise excessive valve head temperature can lead to the early onset of detonation. Not so good for an engine which needs all the CR it can muster.

One of the areas that does prove to be quite critical in terms of flow is the land of the valve. Most proprietary exhaust valves have a land too thin to be able to make the best use of any form applied to the edge. Specialized Valves 'Rimflo' valves, however, have a relatively thick land, with a generous radius between land and chamber face. This is very conducive toward promoting high flow at low valve lift. For the exhaust valve this can be very important for a number of reasons. Firstly, if any shockwave tuning of the exhaust is to be brought about, then it's intensity is increased if the shockwave front is steepened; this occurs when the low lift flow is improved. The greater the flow a valve produces at low lift, the more it simulates the effect of rapid valve opening. This high flow at lowlift also allows the cylinder pressure to drop much faster when the valve opens so reducing the hp lost to pumping on the exhaust stroke. It can also allow a delayed exhaust valve opening which in some instances can increase power. The effect of a large radius on the edge of the valve does not really diminish until mid-lift so it is beneficial over a wide range. However, from the mid-lift point on the under head shape of the valve becomes critical, and here again, the Rimflo valve has been shaped to give good results.

Flow bench testing indicates that on average, a Rimflo valve is worth about 5% more flow due to its shape than a typical high performance valve, and as much as 10% in some places in the lift range. Relating to standard parts, it has been found to be worth an average of about 15% more than the standard factory off-the-shelf-type valve. If a valve other than a Rimflo item is being used for reasons of economy (at this point I am considering the fact that the valves you currently own are going to be the cheapest) then, it's worthwhile noting that there are a few mods to do to the exhaust valve to get the best from it. Basically, these mods are as detailed earlier (**Fig. 9.5**), and should be applied to any exhaust valve that is not so formed.

Moving onto the seat in the head, I have found that the 'asymmetric' seat previously described for the intake, also works well for the exhaust. The port needs to come straight off the seat on the short side, but on the long side, the seat needs a reasonably generous radius, something in the region of 60 thousandths to blend it into the rest of the port.

At this point it's time to consider that there are two distinctly different exhaust ports to work with. The two centre cylinders are dumping into a siamese exhaust port, whereas the outer cylinders dump into individual ports: each end being a mirror image of the one at the opposite end.

It may be worthwhile explaining exactly what port bias is, as related to a vertical valve pushrod engine in general, and a Series 'A' exhaust port in particular. Take a look at **Fig. 9.37**. Shown here is a cutaway of an exhaust port in relation to a cylinder. Because the valve is close to the cylinder wall on one side, most of the exhaust passing out through the port comes from the centre of the cylinder. This means the exhaust is not going out of the cylinder vertically but is moving across at an angle. One of the secrets of good airflow work is to try not to force the air into a route it doesn't want to take, but to try and find out which route it does want to take and then let it. This is the sole justification for

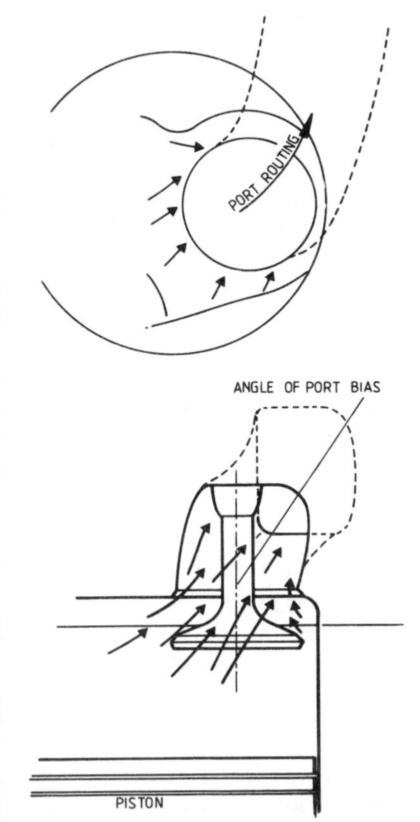

Fig. 9.37 Exhaust port bias

The exhaust flows predominantly from the centre of the cylinder so better flow is achieved if the port is biased, or leaned at an angle, to favour the natural flow. The same goes for the port routing. The end cylinders with one valve per port are more sensitive to correct bias and routing.

biasing or leaning a port to one side. However, this biasing can be overdone. If you look at the centre two exhaust ports, you will see how they naturally tend to lean away from the centre of the cylinder to a certain degree. If you check the end ones you will see that the routing is such that they are **very** biased away from the centre of the cylinders. As it happens, the biasing on the two end cylinders seems a little too much, whereas the centre two cylinders appear just about right, or at least close enough to what's required that some simple grinding will almost always fix the situation. **For the end cylinders in particular, cutting down the amount of bias improves the port efficiency and the easiest way to do this is to install a**

So that I have a record of port shapes, I often take molds like this of ports that are cut in my shop. This particular one is an end exhaust port the standard one is on the right and the modified one on the left. It shows very clearly the bias that I'm talking about in the adjacent text.

larger valve offset away from the intake valve, as discussed earlier.

Principally because of shrouding but also because of other relevant but lesser factors, installing bigger valves rarely pays off in proportion to the increase in valve area. In other words, a 10% increase in valve area rarely gives a 10% increase in flow, 5% being more the norm. Often, as is the case with the intake valves, increasing valve area may do nothing to improve flow, and in some instances can reduce flow. **However, in the case of the offset exhaust valve we find, if starting with a 1.156 inch valve, that valve area increases 16.9% when diameter is increased to 1.250 inch, but typically the flow capability of the port goes up by 18% or 19%.** Of course that figure does in part depend on how the port is cut after the offset valve is installed, but it's generally of the right order.

Valve Seats

Before going on to look at the rest of the port, it's relevant at this point to consider valve seats, in particular with offset valves, the valve seat has been found to be quite critical. As with the inlet, the exhaust valve seat affects the amount of flow, especially at low valve lift. **The wider seats favour better low lift flow than the narrow seats.** On the other hand as valve lifts get higher so the port, because of its small area, tends to favour the narrower valve seats, but there's a distinct limit on just how narrow the seats can be made, even if airflow alone is the only consideration. For road engines, it's best to have a relatively wide valve seat, simply from the point of view of reliability. This is good because with cams lifting up to 400 thousandths, the seat width is best left on the wide side. 60 to 65 thousandths seat width is a good figure to aim for. When race cams are used, valve lifts significantly over 400 thousandths are achieved, then the valve seat width should be reduced for maximum flow. However, the trend these days is to use high lift road cams, so even if the valve lift is above 400 on a road engine, I would say stick with 60 to 65 thousandths, but if seat life is not critical because it's a race engine and the lift is well over 400, then drop the seat width down to 50 thousandths.

Whichever width of seat is used, it is important that the valve seat on the valve itself has the same width and matches the seat in the head. If the seat width is too large, back-cut the valve 30° to reduce the width to match that in the head.

Like the intake port, the area immediately below the valve seat should be cut to favour the type of flow pattern that exists in the vicinity of the port. On the long side of the port, the seat needs a radius to blend it into the rest of the port. Typically this radius needs to be in the order of 20 to 30 thousandths, whereas on the short side, the port can come virtually straight off the seat with nothing more than the use of a needle file to break the sharp corners, and a 5 to 10 thousandths radius at most is all that's needed.

End Exhaust Ports

Since we have two styles of exhaust port on the five port Series 'A' engine, it's necessary to deal with them separately. Since each end port is simply a mirror image of the other, the same rules will apply, but in the reverse form. As with the intake port, it's necessary to find as much flow as possible by virtue of efficiency increases rather than area increases. High efficiency ports mean that port areas can stay relatively small, thus promoting high exhaust gas velocities. Higher exhaust gas velocities tend to inhibit exhaust reverse flow, and this in turn leads to an engine which will produce more low and mid-range torque even when race cams are used. To get the exhaust port to flow, it's generally necessary to cut the exhaust port in 'busy' areas rather than in stagnant areas, hence ports modified without the aid of a flow bench typically portray nothing other than an enlargement all the way through the port.

Take a look at **Fig. 9.38**. This should give you some idea where the mass of air is flowing in the standard port. It's obvious that cutting the port in an area where it's not flowing will do little or nothing to improve the flow. Of course, **Fig. 9.38** only shows the pattern as it exits the manifold face. Just as important is the flow direction out of the cylinder, and the gas route all along the exhaust port. The shape of the port starts to become critical in the mid- and upper valve lift ranges, and so it's the flow pattern in these lift ranges that must be considered most when developing the port.

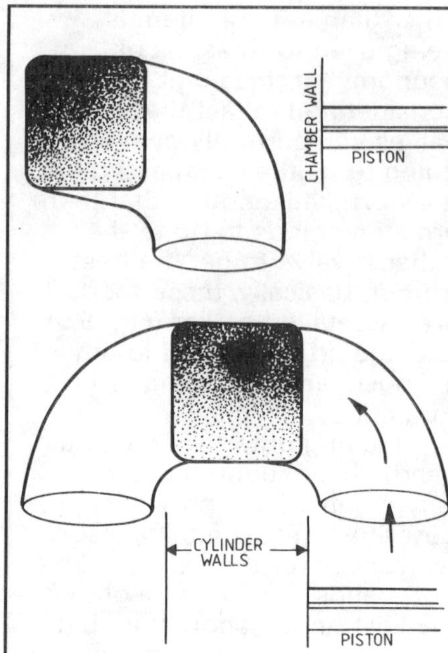

Fig. 9.38 How do the exhaust gases flow?

The flow rate within the exhaust port is by no means uniform. the dark areas show the high flow areas whilst the lighter region indicates flow at lower speeds. On the centre port the pattern will reverse as each valve opens in turn.

At higher valve lift values most of the flow takes place along the roof of the port, that is the long side radius as it comes out of the cylinder. Then, as it takes the turn directing it towards the manifold face, the flow tends to move across the port and concentrate itself mostly along the upper, outer wall of the port. On approaching the manifold face, the stream of flowing gases start on a slightly downhill path so that at the manifold face the flow pattern shown in **Fig. 9.38**

The intake and end exhaust port shown here represent the latest out of the DV/A.P.T. workshop as of 1999. Difficult to see is the form on the intake port. If you look carefully you'll see that the intake port roof has been raised to just a little above the locating ring bore that's machined into the head as standard. This is an attempt to get the port area required consistent with the minimum amount off the port floor. Refining the exhaust port shape to the form seen here has allowed us to achieve even more flow than the ports shown in the cylinder head drawings some pages back. This is how they are now looking and the figures produced are as shown on the computer screen on page 188.

is produced. Essentially this means that in order to make the most of the port, it's necessary to grind only in the areas where the gases are flowing.

The first move is to virtually eliminate the valve guide boss except for a streamlined wedge downstream of the guide itself. Next, bullet-nose the valve guide. It's a good idea to widen the port slightly where it passes the valve guide to compensate for the fact that the valve guide is in the gas stream. From this point, work on the outside port wall and take off as much metal as possible without actually breaking through into the water jacket. Roughly 3/32 inch can be ground from this wall without any real fear of breaking through. The same applies to the roof of the port past the valve guide. However, metal removal here needs to be done in such a way that more of a bias or slope towards the outer wall of the port (the one nearest the end of the cylinder head) is produced. By the time the port gets to the manifold face it should, ideally, be roughly loaf- or D-shaped as shown in the adjacent photo. Of course using this shape can present a problem in as much as your exhaust manifold may not match it. If you want to get the port right, then it's not too much of a problem to have an exhaust manifold made up with a round exhaust pipe at the manifold face. Many people, used to more mundane or conventional techniques, will say that the exhaust manifold won't match the exhaust ports. **Well the news is that, matching exhaust ports to exhaust manifolds is old hat and nothing more than outdated technology.** Sweep it under the carpet!

Matching up exhaust manifolds to the cylinder head port, in many instances, is not the way to develop maximum power. Often, having a step at the port/manifold interface works better. It can improve flow in the forward direction while impeding flow in the reverse direction . . .

As far as the manifold gasket is concerned, there is no problem here. Most composition big port gaskets (as opposed to standard metal shim bonded gaskets) can easily be cut to match up with the exhaust manifold. Such gaskets are generally available through Series 'A' specialists and engine builders. Most well-known exhaust manifold manufacturing companies such as Janspeed, Maniflow, Manx Racing, etc., can produce exhaust manifolds with round exhaust pipes at the manifold face.

Of course, there is the other possibility that you already have an exhaust manifold you wish to use, and the chances are it has rectangular ports, in which case you will not be too happy about having to grind your

Here's a head that was cut to the drawings you see within these pages. This particular head was done by Bob at Oselli Engineering and he was given strict instructions to forget everything he had ever learned on cylinder heads and to cut this one exactly to the drawing. The result was a head that flowed better than any proprietary head I had then tested. This means that it's amongst the best available even now. If you have never cut a head before, you can, using the drawings in this book, cut one for yourself which will give you A1 results. Note that this cylinder head uses Rimflo valves.

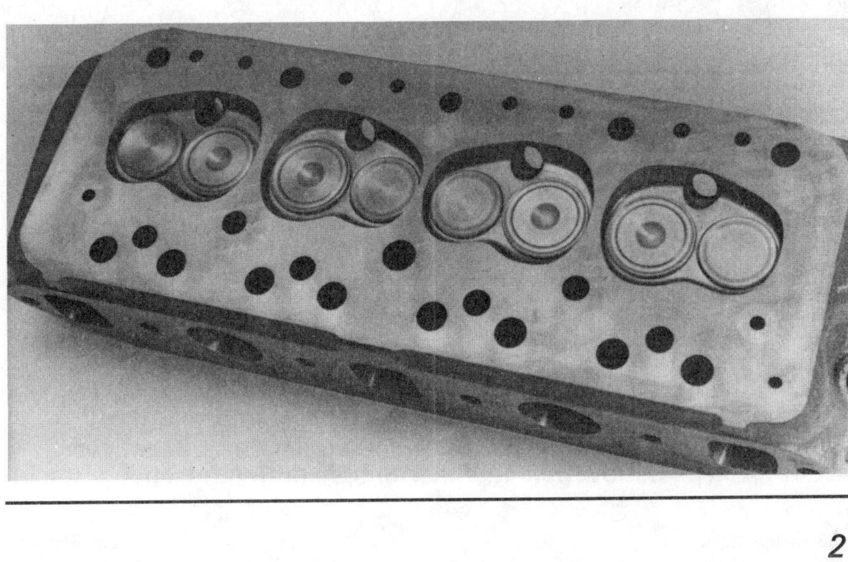

exhaust port to a D-shape. Well, this is not necessarily the end of the world. The way around this problem, although there will be a small sacrifice in flow, is to grind the exhaust port to the profile just discussed, right up to about an inch from the manifold face, and from that point, blend it back into a rectangular profile at the manifold face. Whichever port shape and valve size you decide to use, the exhaust flow curves in **Fig. 9.36** should give you comparative information on the assets of each in relation to the standard port.

Modifying Small Bore, Open Chamber Heads

Performance modifications to small bore, open chamber heads generally follow the same lines as those for big bore, open chamber heads. If the cylinder head is for use on an all-out small bore race engine, then it's best to use a 12G-940 casting as a starter. However, if you are looking for more performance from your existing open chamber small bore head, be it a 12G-295 head or the very much rarer 12G-206 head, then the porting procedure simply follows what was done on the big bore heads with the exception of the mods to the combustion chamber, as shown in **Fig. 9.39**.

If you are modifying a cylinder head that has seen some use, you may well find that the valve seats have sunk a little. The easiest way around this problem is to go up slightly on valve sizes. Tranco produce valves at 1¼ inch for the intake (that's 1⁄32 inch up in diameter) and 1 1⁄16 inch for the exhaust. Similar sizes are also available from Specialized Valves in 'Rimflo' pattern.

If it is necessary to use larger valves in an attempt to extract the most from this casting, then probably the best route is to use the smallest of the 1275 size valves, namely 1.156 inch exhaust and 1.312 inch for intake. Again, if you have a serviceable set of valves to hand and the budget is limited, you can utilize reshaped standard valves. The sizes we are talking about would be from a 1275 GT or the like. These valves are about 50 thousandths longer than the valves generally used in small bore motors, or Cooper 'S' valves.

This extra length is no real problem. It can be compensated for in the valve train adjustment one way or another. It is important to remember though,

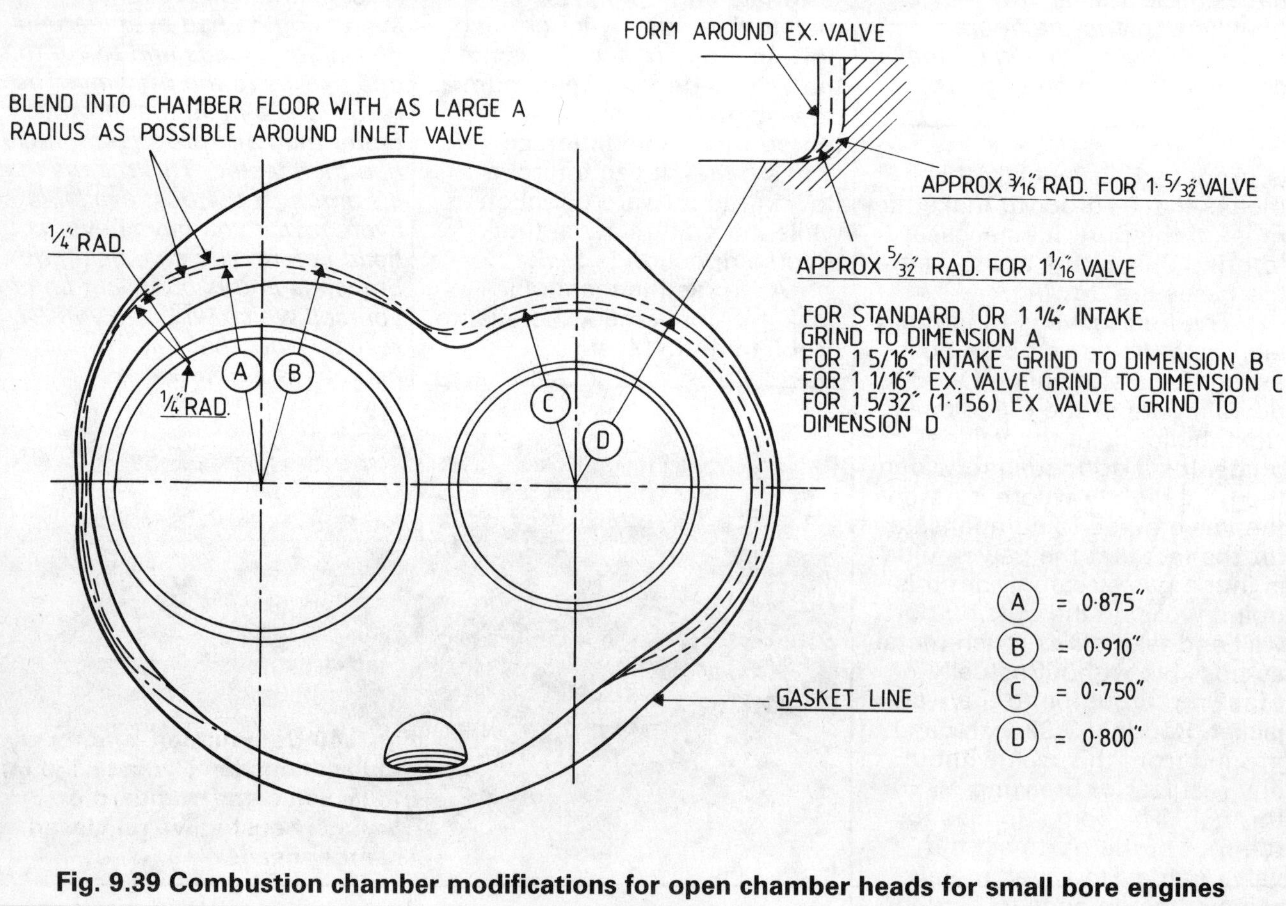

Fig. 9.39 Combustion chamber modifications for open chamber heads for small bore engines

that the fitted spring height will be greater by close on 1/16 inch. If the valve springs were already marginal on spring pressure for the rpm and cam combination being used, it may be well worth remembering that you can see up to 50 thousandths thickness of shim before you get back to the original fitted height. If this larger size of valve is used, the combustion chamber should be modified as illustrated in the relevant part of **Fig. 9.39**. Again, the rest of the porting procedures for the big bore head castings apply, so no more need be said in that quarter.

Modifying Closed Chamber Heads

We have already dealt with the problematical side of modifying the closed chamber heads on 850 engines, so we will not delve into that any deeper. Suffice it to say, the 850 head is modifiable up to a point, after which it is simpler to get the head casting from another vehicle and use that. If you take a look at the ports on 850s, very early 803s (if there are still any left out there) or 948 engines, you will see they have a round intake port. If standard size valves are being used, the easiest mod for this first part of the port is to simply run a 15/16 inch drill through the port.

If small bore, closed chamber cylinder heads are being modified for use on the larger displacement small bore engines, i.e. 1100s, then it's possible to de-shroud the valves to a much greater extent, as getting the CR to a suitable level is not so much of a problem.

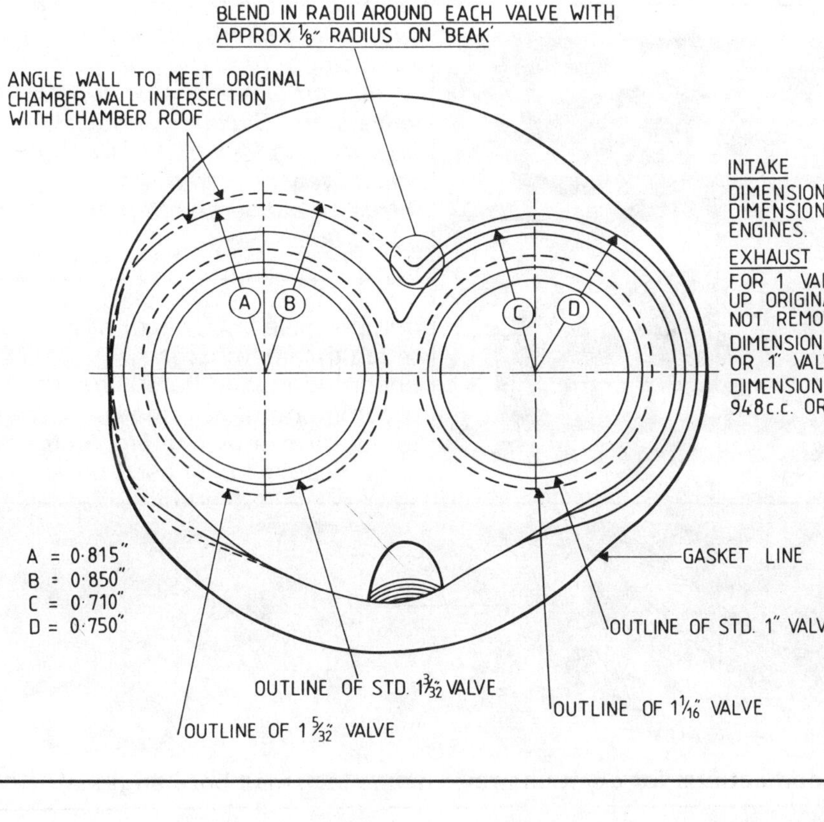

Fig. 9.40 Combustion chamber modifications for 1 5/32 in (1.156 in) intake valve and standard or 1 1/16 in exhaust valve on closed chamber heads

(Many of the early ports are close to this size already). That's a relatively safe mod. In fact on early heads, that's prior to about 1975 or so, it's a fairly safe bet that you can run a 1 inch drill down through that port. **A 1 inch diameter port or its equivalent in slightly squared-off form is quite adequate for the standard 1.093 inch diameter valve size.** From here on in, the general rules for porting the Series 'A' head that have so far been described, can be successfully applied.

Shaping of the combustion chamber has been well documented earlier on in this chapter, as have techniques for grinding the exhaust port. However, let's consider the best route to take if you intend to install larger valves in the head. The next size up for the intake is the $1\frac{5}{32}$ inch size used in 997 Coopers and 1098 transverse and inline engines. Because of the increased valve size, we need to consider two things, firstly the port size at the siamese section may need to be a little larger in area, and secondly the combustion chamber may need to be cut back further so that the valve is not shrouded. Again, it's necessary to strike a compromise in terms of how much the chamber is unshrouded, and here I would recommend the chamber is cut to the form shown in **Fig. 9.40**, if the head is being used on an

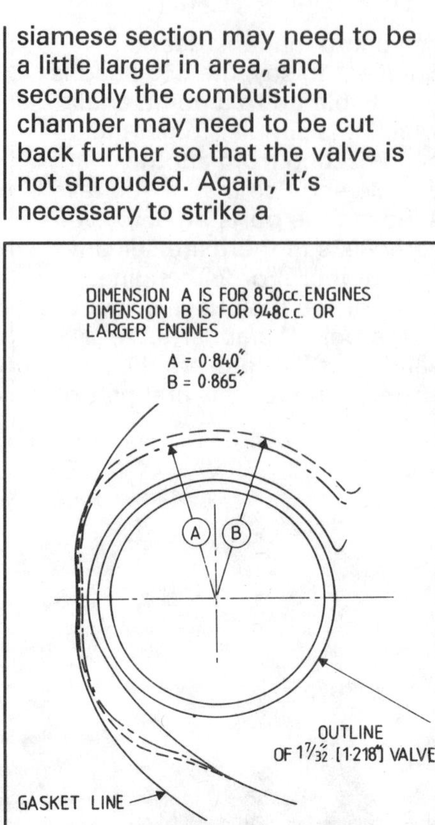

Fig. 9.41 Combustion chamber modifications for $1\frac{7}{32}$ in (1.218 in) intake valve in closed chamber heads

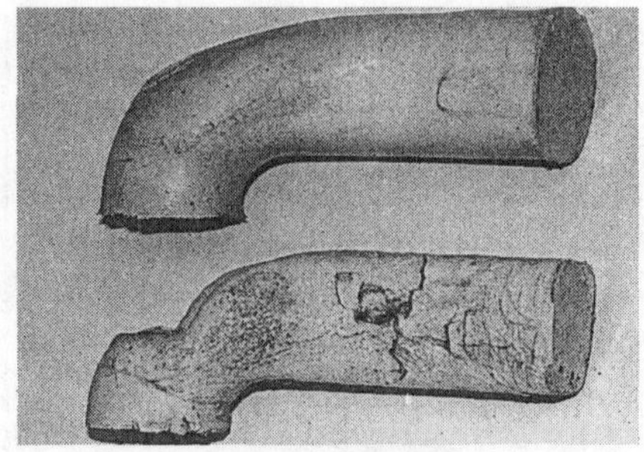

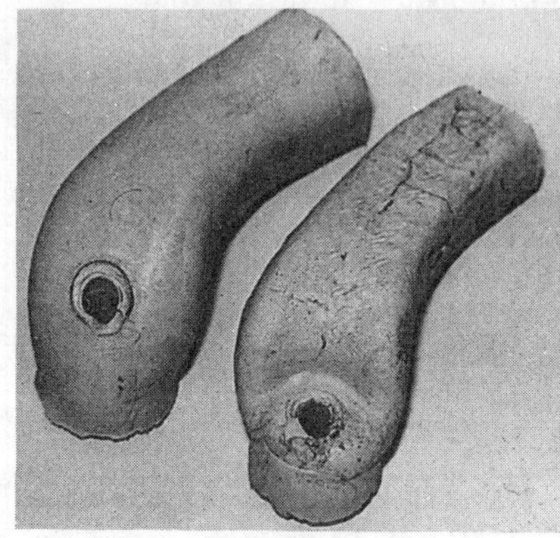

Very often to get a better idea of the shape of the port involved it is better to look at a male version of the port as seen here moulded in silicone rubber. In these two shots of both the standard and modified port it is easy to see the two features talked about in the text and the captions of nearby photos. Firstly, note how the port gets wider and expands around the valve stem. Secondly, notice how the modification of the port has largely removed the intrusion caused by the valve guide boss.

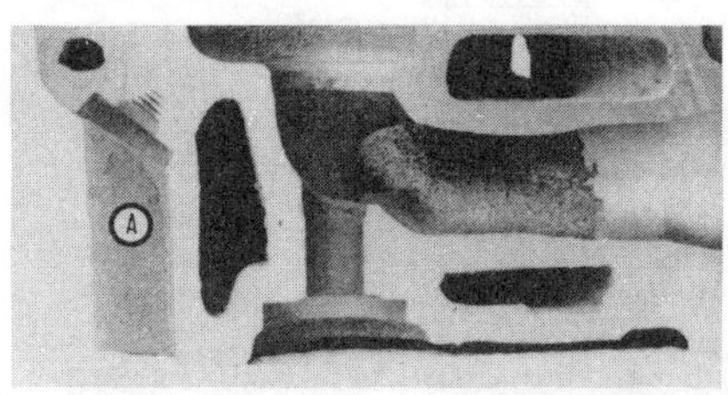

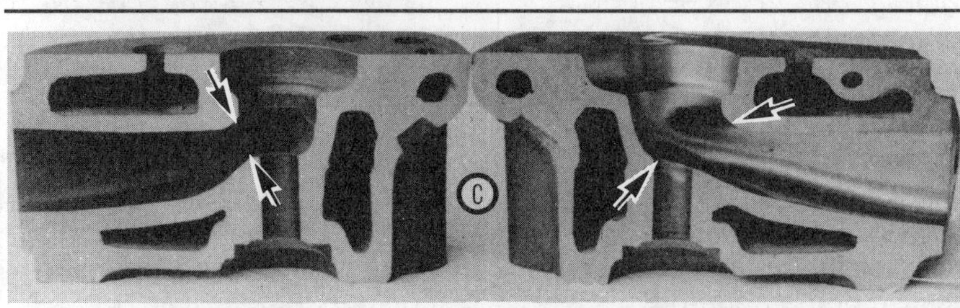

Porting closed chamber 850/1000 heads. *Photo* **A** *shows the cross-section through a standard 850/1000 intake port.* **B** *shows what the same cross-section looks like when it has been re-worked to a flow bench developed form. In photo* **C**, *we have examples of a standard exhaust port and a re-worked exhaust port. The cross-section through the standard port adequately demonstrates the restriction that occurs between the sharp edge of the short side radius and the valve guide boss. The ported example on the right flows substantially more than the standard one. Although a slightly larger intake valve is used, the heads for the 997 Cooper and the small chamber 1100 heads follow the same form as the head shown here. (Photos courtesy of Swaymar Race Engines) 9-22, 9-23, 9-24*

850. On the other hand, if it is being used on a 998 engine or an 1100, there is more latitude because the CR can be raised to a higher level more easily, in which case the combustion chamber can be cut to the second line of that drawing. As for the port itself, drilling to 1 inch diameter is not far off what's required, but it can stand being a little larger.

If you are starting with a casting other than a round port 850 Mini-style item, such as an 1100 or 997 Cooper casting, then these have square ports and the port size is just about adequate for a 1 5/32 inch valve. So, other than cleaning the port up, it's not really necessary to go overboard on enlarging it. If you are considering installing a 1 5/32 inch valve into the small valve 850/948 casting, then the port really needs to be enlarged quite a bit. Probably the best technique I have found to date is to drill the port 1 inch and then work on the top part of the port and grind about a further 1/16 inch from the roof, thus forming a slightly elongated oval. Once the port is past the pushrod holes, start flaring it so that it joins up in a large smooth radius with the outer walls of each side of the port.

The furthest I would recommend going with this type of casting in terms of valve size is the 1 7/32 inch valve from the 998 Cooper, 1098 Sprite, etc., or possibly the 1 1/4 inch valve. Again we are into a combustion chamber volume situation so I don't recommend this head to be used on an 850 unless the compression ratio required (due to the use of low fuel octane) is no more than about 9:1. The benefit of a big valve can be seen at low valve lifts but because we can't cut away the chamber shrouding enough, the big valve cannot produce the high lift flow of which it is potentially capable because of the trade-off between CR and good breathing. Anyway the drawing **Fig. 9.41** shows how the chamber should be cut around the intake for use on 850, 998 and 1100 engines, when intake valves up to 1 1/4 inch are used.

As far as the exhaust side of the small bore heads is concerned, the 1 1/16 inch exhaust valve is as good a valve as any to use. Cut the chamber around the valve as detailed in **Fig. 9.40**, it's then far enough away not to cause any significant shrouding. Again, the porting procedure for the exhaust has been covered previously, so it should be clear what needs to be done.

Raising The Compression Ratio

At this point, I don't intend to talk about the amount the CR needs be raised, but rather the physical ways of actually doing it. On any of the big bore castings, with the exception of the 'turbo' casting, getting adequate CR is rarely a problem. Assuming the pistons used don't have an outrageous dish, it's simply a matter of skimming the head until the desired ratio is achieved. Of course there are situations, alluded to earlier in the chapter, when the CR given by a head, even at standard thickness, is too high. For instance where a 1275 engine has been substantially increased in capacity. Coming up with the right ratio on a 1400 or a 1430 is rarely a problem, but when displacements over 1500 cc are involved, then some thought has to be given as to how an acceptable ratio can be achieved. On 1600 engines it is necessary to cut the chamber deeper into the head simply to get enough volume in the chamber to end up with acceptable CR, that is around 12:1. (it is also usually necessary to lower the pistons in the bore so that some of the combustion chamber volume resides in the top of the block. Some specialist head grinders have achieved as much as 31-32cc chamber volume, in which case the pistons can be left flush with the block. So long as acceptable airflow is maintained, when this is done the end results are better than when the pistons are lowered.)

Cutting the chamber deeper does cause breathing problems. Basically deepening the chamber destroys a vital portion of the port, *i.e.* the area near to the original valve seat. Because the port will now not have the last fraction of a straight run to the back of the valve seat, the port efficiency is reduced. This applies both to the intake and exhaust, but more so to the intake. **The irony of the situation is that big engines need all the breathing capacity the cylinder head can produce. The fact it's necessary to bring the CR within sensible bounds by chamber deepening techniques means some of the**

Getting a suitably low compression on big displacement Series 'A' engines, especially with modern fuels, can be a problem. What you see here is a chamber which I've milled out the minimum amount out consistent with getting the compression ratio down to where I want it and retaining good air flow. In this, as in most instances, the pistons required to be modified also.

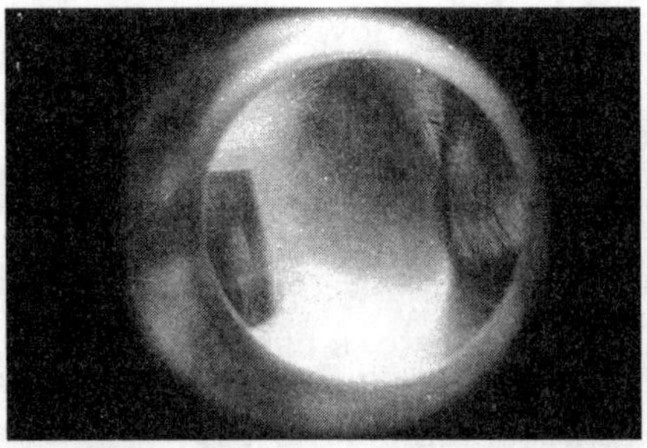

Here's a view down the intake port of one of Dave Mountain's race-winning 850 Mini-X heads. Notice the port isn't big and it isn't polished, nor does it have a sharp edge at the split between the two ports.

possible advantages of stretching the engine are offset by the fact that the breathing capacity goes down in so doing.

Let's take another look at ways and means of raising the CR on small bore heads when they are to be used on 850s. Basically the problem is that by the time the correct chamber volume has been reached for a suitably high compression for a competition engine, the head has been skimmed so much that it has broken into the oil gallery that feeds oil from the block through the head to the valve train. There are two ways that this problem can be overcome. Firstly, deliberately cut into this oil gallery where it traverses the face of the cylinder head, grind the cross-drilling out until it is a much deeper ditch, *i.e.* about 1/10 inch deeper than it is standard. Then take some appropriately sized copper tube, bend it, fit it into the deepened slot and then braze over the tube. What we have now done is effectively to move the oil gallery deeper into the casting.

Another technique, one shown to me by Dave Mountain of Mountune Race Engines, is to cut into the oil gallery going across the face of the cylinder head, for the purpose of completely brazing it shut. Once the head has been skimmed to give the desired chamber volume, a groove is cut, from the hole which picks up oil from the block, across to the number 1 rocker stud as illustrated nearby. Oil is then fed up around the outside of the stud and back from the stud, along a groove in the rocker pillar face, to the point where the oil was normally transferred from the head to the rocker pillar.

Assuming the oil gallery is not now the limitation, the question arises as to just how much can be skimmed from a cylinder head without causing a reliability problem due to cracking? Basically if you look through the waterways and assess the thickness of the casting, it's possible to skim the head so that the deck of the casting is reduced to around 80 thousandths thick. In the main, going for a deck thickness around this figure does not seem to give undue reliability problems. This isn't to say you will never have a cylinder head crack; what it does mean is that for competition purposes at least, reliability is at an acceptable level in view of the fact that hp is the prime objective.

The method of raising the ratio to its maximum by this technique can be applied to both the closed and open chamber small bore heads. However, one thing you should appreciate is that the open chamber small bore heads do have the largest volume combustion chamber of all Series 'A' heads at 28.3cc. When you consider that on a racing 850, the chamber volume has to drop to around 15cc, then it will be apparent that the head has to be skimmed to get rid of some 13cc of chamber volume as a minimum: not an easy task! On the other hand, a 12G-940 casting starts off life at 21.4cc nominal, so

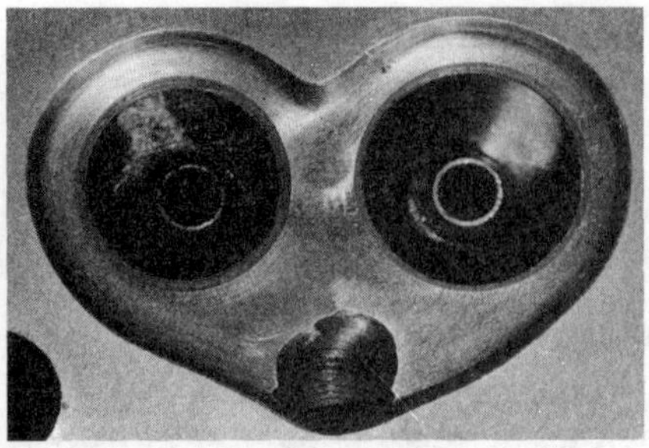

In an effort to get the CR up to a suitably high level, the amount of de-shrouding that has been done on this cylinder head has been limited. Note also that the valve seats do not fall into the super-thin category. Also the spark plug cutout has been biased towards the intake.

Raising the compression ratio by milling the head is one of the most cost effective mods to perform on an 'A' series motor. An advantage for a street motor is that it increases torque throughout the rpm range and improves mileage. Keep the intake charge temperature down and fuel octane will not be anything like as much of a problem.

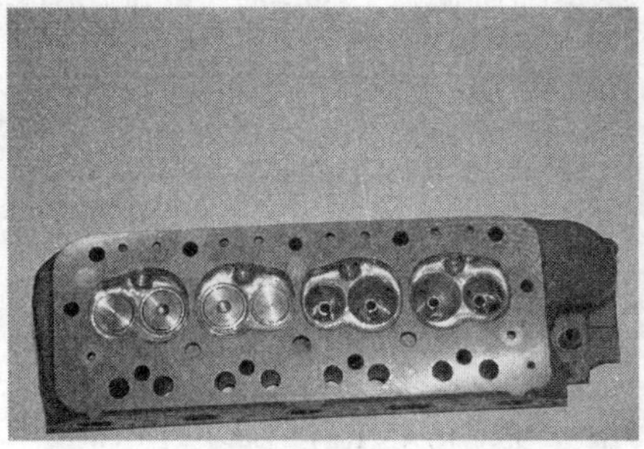

Although a regular Bridgeport style mill will do the job of milling the head, a specialised machine such as this with a very slow feed for a fine finish is best for facing cylinder heads.

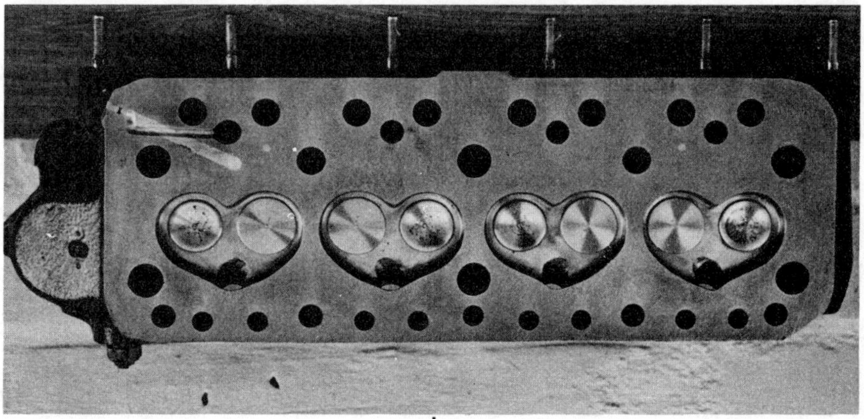

*Here's a finished **Mountune** cylinder head, prior to the time that Dave Mountain changed to Rimflo valves. The point to note here is that the oil transfer hole from the lower right-hand corner has been brazed in and a groove has been cut in the cylinder head to pass the oil up to the rocker gear, via the stud hole to the top face of the cylinder head. From here . . .*

. . . it passes via another ground channel back to the original exit point so the oil can flow up through the hole in the rocker pillar, through to the rocker shaft.

it takes less than half the amount of chamber volume reduction to achieve a similar CR. These are all points to bear in mind when selecting a cylinder head for high performance applications, but for the moment let's look at other small bore heads.

The ordinary 850 Mini head starts off life at 24.5cc, so on the face of it, it looks easier to achieve the desired compression ratio. However, the area of the combustion chamber in a head of this type is less, so it takes more skimming to remove a given volume. The net result is, it's hardly any easier to achieve high compression with a closed chamber head than it is the open chamber head on small bore, small capacity engines. The situation concerning the amount to be skimmed from the head is considerably eased on bigger engines. For instance, if an equivalent CR is required on a 998 engine to that just considered for an 850, it will be found that the combustion chamber volume needs to be around 18cc as opposed to 15. That's certainly a figure which is not out of reach in many instances. When the engine concerned is 1100cc capacity, chamber volumes in the region of 21cc are required and represent no problem at all.

Some final points concerning CRs, or at least the techniques to achieve them, firstly never forget that CR achieved by skimming the cylinder head is better on Series 'A' engines than CRs achieved by using pistons with less dish in them, or even raised crowns. Obviously if you can't get the CR required by skimming the head, then you must resort to raised crown pistons.

The last point concerning machining cylinder heads is that after the head has been skimmed, the sharp edge between the combustion chamber and the head face must be filed to avoid having an edge which may become overheated and lead to premature detonation.

Compression Ratios – How Much Or How Little?

After all the cutting and grinding of the cylinder head is done, a major decision has to be made: how much should the CR be raised or, to be more precise, to what should it be set? As in the case of a supercharged engine, for instance, the CR may actually have to come down. The compression ratio used is far more important than most tuners realize. For a street motor, especially when expected to pull from low rpm, getting the compression to just below the detonation limit *is a key factor toward* making good torque. In simple terms three factors go toward generating high torque outputs. These are volumetric efficiency (VE), compression ratio and combustion efficiency. At the low rpm *an appropriately cammed* street motor turns at increasing the VE is difficult because it is engine speed and cam limited (ie small overlap and duration). At very low rpm there is no ramming or exhaust scavenging of the cylinder of any consequence. If gains in what can best be described as low end and mid range punch are to be made *we will find most of it by attention to the CR and combustion efficiency*. As long as the basic characteristics of the combustion chamber have not been altered then the 'A' Series traditional high combustion efficiency will still be there. *This leaves the compression ratio as the major source of increased output*. Raising it to as high a level as possible is the only route

Determining Engine Compression Ratio

There is probably one of two things you will want to know about your engine's CR. What it is currently, or how much has to be machined off the cylinder head to set it to what you want it to be? Well the first thing needed here is a formula which defines CR and the following does just that. CR equals V + C ÷ by C. In this formula 'V' is the swept volume of one cylinder and, 'C' is the TOTAL combustion chamber volume.

So to work out the CR your engine currently has, you need to know the swept volume of one cylinder and the volume of the combustion chamber space in its entirety. The formula for the swept volume of one cylinder is $B^2 \times 0.785 \times S \div 1000$. In this formula, 'B' is the bore in mm, 'S' is the stroke length in mm.

Don't fall into the trap of thinking that an overbore of 20 thousandths doesn't have any significant effect on CR. It can, especially in the higher ratios, so you do need to know the REAL WORLD capacity of the cylinders of your engine. If you don't feel inclined to work it out, then look up the chart in the chapter on big engines. This gives the capacity for most stroke and bore combinations. To get the cylinder volume just divide by four.

Once you have determined what the swept volume of the cylinder is, now you need to know exactly what the total capacity of the combustion chamber is. I emphasize 'total' because part of the combustion chamber is also in the piston if a dished piston is used. Some of it is in the land just above the top ring, some of it is contained in the cylinder head gasket and, if the piston does not come to the top of the bore, more combustion chamber is contained in the resulting space. All these small volumes need to be accounted for. If they are ignored, your CR calculations will be worthless . . .

Take a look at the chart, **Fig 9.42**, and you will see the typical volumes contained in head gaskets, ring lands and the amount contained above the piston between the piston and deck of the block when the piston is at TDC. For instance, using the chart, if the piston is 10 thousandths of an inch down the bore at TDC on an 850, the volume contained above the piston excluding any dish in it, will be 10 × 0.079 = 0.79cc. If you are intending building a real flyer, then all the CR possible will be needed. Under these conditions, the pistons may be set to be actually above the top of the block at TDC. We'll say the pistons come out of the block by 4 thousandths, in which case it will be necessary to check out how much volume that 4 thousandths takes up. From the chart we can see it will be 0.32cc and that will be subtracted from the total combustion chamber volume because the piston is taking up that space.

Also, it's worth noting that the raised crown on the 998 Cooper piston takes up 4.5cc, so this, because it intrudes into the combustion chamber volume because the piston is taking up that space.

At this point let us take a typical 1275 engine to work out an example. Let's say burette measurements establish the combustion chamber in the head is 21.4cc (a rarely occurring, exactly standard head!). Also, from measurement the piston is short of the top of the block by some 50 thousandths of an inch and that the piston has a 6cc dish in it. Looking at the chart, it can be seen that if the piston stops 50 thousandths short of the top of the block, the volume contained above the piston and below the deck will be 50 × 0.105 = 5.25cc. This, plus the 6cc in the dish of the piston, gives 11.25cc. Also, we must account for the ring land. This contains, according to the chart 0.78cc and from the same chart we see that if the engine had a copper asbestos competition head gasket, there would be 4.2cc in that, so the total chamber volume is now:

Ring land volume	0.78cc
Piston dish	6.0cc
Volume above piston and below block deck	5.25cc
Volume in cylinder head gasket	4.2cc
Volume in combustion chamber	21.4cc
	37.63cc

Assuming an absolutely standard 1275 motor, the capacity of one cylinder will be 1275 ÷ 4 = 318.75cc. The CR = V + C ÷ C so, fitting in the number, we have 318.75 + 37.63 ÷ 37.63 = 9.47.

Okay, so much for actually calculating the existing CR. How about determining what chamber volume is needed to produce a required CR other than standard? For calculating this, a slightly different formula is needed. The total combustion chamber volume required equals the swept volume of cylinder ÷ CR – 1. Recalculating the example just given, except for a 10:1 CR, we utilize this formula: V ÷ CR required – 1 = total chamber volume required. The cylinder volume V, as was previously calculated is 318.75cc. This has to be divided by the CR required minus 1. We require 10:1, so 10 – 1 = 9, and 318.75 ÷ 9 = 35.41.

This is the TOTAL chamber volume required. It is now necessary to find out how much of the chamber volume must be in the head to give this total chamber volume, remembering that there are some parts of the combustion chamber volume in other places. Totalling up the other areas of combustion chamber volume, we have 0.78 in the ring land, 6.0 in the piston dish, 5.25 in the space above the top of the piston at TDC and below the deck of the block, and 4.2 in the head gasket. All this totals 16.23cc. Subtracting this figure from the total chamber volume required will leave the chamber volume that must be created in the head to give the 10:1 CR required. Doing so (35.4 – 16.23) results in 19.18cc, so to raise the CR from 9.47 to 10:1 requires the cylinder head volume to be dropped from 21.4cc to 19.18cc. The next step is to determine how much has to come off the head to leave 19.18cc.

The technique here is to stand the assembled cylinder head, complete with spark plugs, on an even surface. Set the head face horizontal with the aid of a spirit level. Using a burette, empty 19.2cc of paraffin (kerosene) or water mixed with alcohol into the chamber. Next, using a Vernier scale, carefully check the depth between the surface of the fluid and the cylinder head face. This itself is a fairl tricky operation because the fluid's surface tension tends to cause the meniscus to attach itself to the end of the Vernier calipers, so a small amount of judgment is required. The dimension you come up with is the amount the head has to be skimmed.

Determining the exact amount that has to come off the cylinder head is, in fact, a lot more difficult than checking that the right amount has been removed afterwards. To do this check, it's a simple matter to put the sheet of glass/perspex over the chamber and fill the chamber (through a hole) to see if it is down to the required volume. At the same time, it's also a good idea to check chambers are near enough equal in volume. For a race engine the chambers should be within about 0.2 of a cc, but for a road engine, less critical chamber volumes within about 0.4 of a cc will be fine. If you have been honest about marking out the combustion chambers and grinding each true to form, you will probably find that the combustion chambers will fall well within 0.4 of a cc tolerance for road use. For race engines it may be necessary to dress a chamber to remove a bit more metal, but usually by the time you get to your third or fourth head, even that is rarely necessary.

Fig. 9.42 Capacity in ccs above piston due to piston-to-deck block clearance

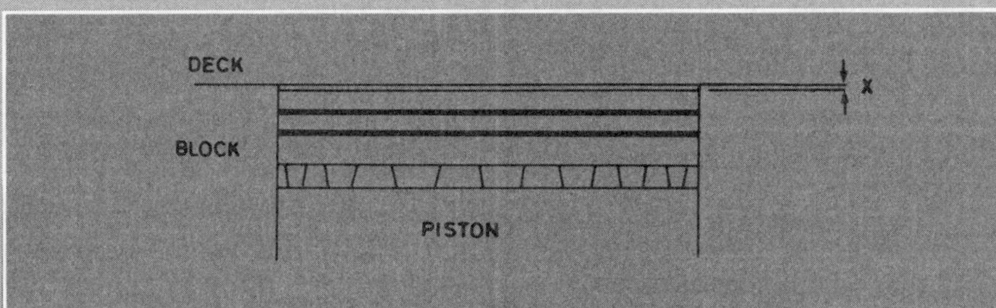

To use this chart, firstly check the bore size of your engine, then measure the distance between the block deck and the piston at Top Dead Centre. Next look up the ccs contained above the piston, and below the block deck, i.e. 1275 engine ÷ 0.020 bore with piston 8 thousandths down bore. From Chart 3 we get 0.81cc contained above piston. **Note:** *This amount does* NOT *include any dish in the piston. This must be accounted for separately.*

If the piston to deck clearance is greater than 10 thousandths of an inch (0.010 in) which will be the case in most instances, it will be necessary to compute the clearance volume by adding some of the chart figures together, i.e. 16 thousandths, piston-to-deck clearance (0.016 in) can be found by adding the volume for 10 thousandths and 6 thousandths together. From Chart 3 for a 1275 with a +0.040 overbore, this would be 1.02 ÷ 0.61 = 1.63cc.

On certain race engines, the piston may be set to come out of the top of the block at T.D.C. This produces a Negative *clearance volume and the figures must be* Subtracted *from the combustion chamber volume. Use the chart as per normal to find out how much must be subtracted. i.e. 850 with +0.040 rebore pistons 0.006 in (6 thousandths) out of the top of the block, from Chart 1 clearance volume = 0.49cc.*

1. Engine Capacity: **850/948**, gasket volume 1.4cc, ring land volume (typical) 0.6cc

	Piston to block deck clearance in thousandths-of-an-inch				
Bore	**2**	**4**	**6**	**8**	**10**
Standard	0.16	0.32	0.48	0.64	0.79
+ 0.020	0.16	0.32	0.48	0.64	0.80
+ 0.030	0.16	0.32	0.49	0.65	0.81
+ 0.040	0.16	0.33	0.49	0.65	0.82
+ 0.060	0.17	0.33	0.50	0.66	0.83

2. Engine Capacity: **998/1100**

	Piston to block deck clearance in thousandths-of-an-inch				
Bore	**2**	**4**	**6**	**8**	**10**
Standard	0.17	0.33	0.50	0.67	0.83
+ 0.020	0.17	0.34	0.51	0.68	0.84
+ 0.030	0.17	0.34	0.51	0.68	0.85
+ 0.040	0.17	0.34	0.51	0.69	0.86
+ 0.060	0.17	0.35	0.52	0.70	0.87

3. Engine Capacity: **970/1071/1275**

	Piston to block deck clearance in thousandths-of-an-inch				
Bore	**2**	**4**	**6**	**8**	**10**
Standard	0.20	0.40	0.60	0.80	0.99
+ 0.020	0.20	0.40	0.61	0.81	1.01
+ 0.030	0.20	0.41	0.61	0.81	1.02
+ 0.040	0.20	0.41	0.61	0.82	1.02
+ 0.060	0.21	0.42	0.62	0.83	1.04
+ 0.114 (73.5mm bore)	0.22	0.43	0.64	0.86	1.08
+ 0.133 (74.0mm bore)	0.22	0.44	0.65	0.87	1.09

open to us if low end that feels like added engine size is to be achieved. It is also worth mentioning here that increasing the compression ratio is one of the few mods that produces more mileage as well as more output.

For a high rpm engine, that is one with the long duration cam, a different set of conditions apply but they are equally demanding of a high compression ratio. At high rpm exhaust scavenging of the combustion chamber occurs. This scavenging works better when the combustion chambers are small which of course will be the case if the CR is high. Also, long duration cams don't close the intake valves until the piston has, on the compression stroke, moved up the bore a considerable distance. For example with a typical 300 degree cam the piston is over a third of the way up the bore before the valve closes. This late closure causes the dynamic compression ratio to be considerably lower than the measured or static compression ratio. One of the biggest mistakes made by both pro and amateur engine builders is failing to match the required compression ratio to the cam duration. There is a way to calculate a close match of the compression ratio required to the cam characteristics. This we will deal with toward the end of part 2 of the Cams, Cam Timing & Valve Trains chapter.

Until now I've talked about compression ratios in a simple way. If we're to make the most of the power from a Series 'A' engine we need to deal with the compression ratio in detail, especially in how it relates to the Series 'A' engine in particular, rather than engines in general. Firstly, let's look at the maximum compression ratio usable. We will assume for the moment that compression ratios for good power are only just barely obtainable due to the octane of fuel available. Such a viewpoint is in fact true in most parts of the world, except possibly the USA, where racing fuels up at as high as 119 octane are available. This is a special case, so let's first deal with the original premise. *If we're to extract maximum power from a Series 'A' engine, the compression ratio will have to be set so that it is just short of detonation.* In the past many books and articles have made a big deal about spending a lot of time balancing up combustion chambers such that the compression ratio is even all across the board amongst the four cylinders. Such a notion is only valid if all four cylinders have the same characteristics. In other words, we would be dealing with one four-cylinder engine. The Series 'A' engine is not one four-cylinder engine, in most circumstances it is best likened to four one-cylinder engines that happen to be joined up. That is, four one-cylinder engines with slightly varying characteristics. This would mean in essence, each cylinder should have its own highest useful compression ratio limitation. Engines that I've built to the sort of specifications that I normally use tend to show that No 2 cylinder (No 3 on Metro-style turbo motors) detonates prior to any other cylinder. Chances are, this situation exists with other Series 'A' engines whether they are built to the sort of specs I build them to or not. From this point of view we can see that No 2 cylinder should run at a slightly lower compression ratio than cylinders 1, 3 and 4. Some folks may suggest lowering the compression ratio on all the other cylinders. There again, why should we? We will get more total power by dealing with each cylinder individually and optimizing its output, irrespective of what the others put out than we will reducing the best cylinders down to the worst. As for the amount the compression ratio should be reduced on No 2 cylinder, the answer is not altogether clear. At present it would appear that between 0.25 and 0.5 (usually nearer 0.5) gets the job done. Do not reduce the compression ratio by cutting down piston-to-squish area clearance. In other words, don't skim the top of the pistons because doing this reduces a cylinder's tendency to avoid detonation in the first place. To get that lower compression in No 2 cylinder you need to look at making bigger the chamber in the cylinder head or the dish in the piston, or better yet, a combination of both.

As for selecting the actual compression ratio the engine should run at, here again we come across some factors which have changed the situation dramatically since I completed the first edition of this book in 1984. To explain what I mean, let us for the moment at least, assume that fuel octane is not a problem. If we are to extract the maximum power from an engine, our first priority is to get the volumetric efficiency of the engine as high as possible. As the volumetric efficiency approaches 100% we find that the engine's output in relation to its compression ratio becomes critical. This is simply because if the compression ratio of an engine is raised too much and that engine is capable of more than 100% volumetric efficiency, the high compression ratio may cancel out the engine's capability of going over 100%. The reason for this is that the extra 'super charge' has to reside somewhere within the working cylinder of the engine. The additional gases coming in find the combustion chamber the most convenient place to occupy. This being the case, if the combustion chamber space is reduced below a certain amount, *i.e.* if the compression ratio raised, there will be less volume available for the 'super charge'; result, the 'super charge' effect and consequently, the volumetric efficiency, is reduced right along with the torque and horsepower at that rpm. At first this may seem a backward step

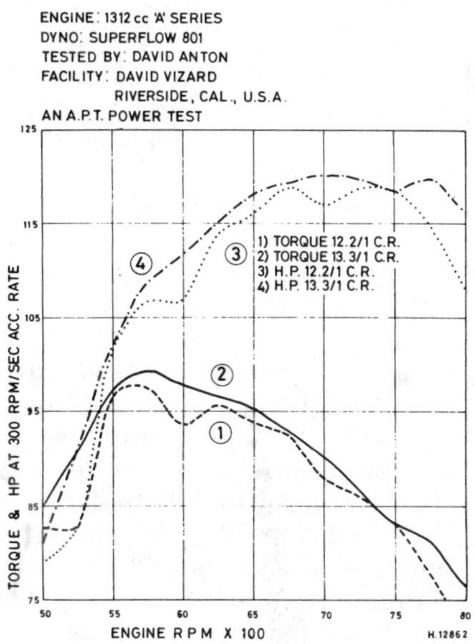

Fig. 9.42a Effect of compression ratio on power output

The effect of raising the compression varied a little from engine to engine, but generally on a high VE Series 'A' engine, the results shown by this graph are pretty typical. If the engine was at the lower VE to start with, the gain due to compression would have been proportionally greater. Unfortunately, it would have been a bigger proportion of a much smaller number to start with, thus proving that VE should not be sacrificed for compression unless it's a small VE decrease for a large compression increase.

but the fact that the engine can now breathe better at a particular rpm also means it can now be usefully turned to higher rpm. To do this it probably will need a cam with a little more duration which in turn will require more compression to be used. As can be imagined the link between VE, CR and cam duration prove to be a key factor toward developing maximum horsepower. Probably the best way to explain this power critical interaction is to consider two distinctly different cases. The first is an engine using a *five-port* cylinder head having conventional intake and exhaust valve sizes (and therefore intake to exhaust flow ratios) and a cam duration up to 300°. The 300° figure is chosen here because this is about the limiting duration that can be used before unacceptable inter cylinder robbing (detailed in the cam chapter) takes place. The second case is five-port motors with scattered cams, that is *the appropriate* different timing on each cylinder, or engines having an intake port per cylinder. With the second case engine very high compression ratios should be used together with the high ratio of intake flow to exhaust. Along with this the exhaust duration needs to be substantially longer than the intake. Let us deal with the first case scenario, that is five porters with cams up to 300°.

Up until the mid '80s, achieving anything like 100% volumetric efficiency from a Series 'A' engine was a difficult task. However, consultancy work that David Anton and I did at A.P.T. did for some prominent UK performance companies over a few years advanced the state of art to the point where volumetric efficiency figures for the Series 'A' rose quite substantially. In fact, some of our R & D engines have shown a little over 100% VE figures, 96–98% being quite common. Most of these improved VE figures, along with a corresponding output increase, have been brought about by virtue of intake, head, camshaft and valve train modifications. Up until we did this work, the apparent norm for a race engine in terms of VE was about 88%, though sometimes an engine or two would show 90%. Because of these historically low VEs, the Series 'A' engine would always produce more horsepower the more compression you put into it. Compression ratios up to 15:1 were not unheard of. Our findings to date are that if compression ratios go significantly over about 13.5, peak power starts to drop off and the engine actually loses out all the way around. I use the figure of 13.5 as a maximum here, but there is a considerable buffer below that ratio. If you look at the graph, **Fig. 9.42a,** you will see what happens when the compression ratio is raised from 12.2 to 13.3. The graph starts at 12.2 because we had previously investigated ratios up to 12.2 and found that *peak* power tended to level off there. Going from 12.2 to 13.3 shows only a minimal increase in peak power (1.5hp) but it does show that the engine hangs on longer and gives a little more torque. So far though, all of this hinges on the fact that the engine fuel must have sufficient octane. If it doesn't have sufficient octane to run a 13.5:1 compression ratio, then the figure used needs to be moderated accordingly. The fuel situation in the UK seems to have reached a disastrous all time low as far as octane values are concerned. It would appear that in 1989 at least anything much over 11:1 is getting a little dodgy in terms of detonation unless an effective octane booster is used. At the time of writing octane boosters under R.A.C.-sanctioned racing events are largely banned. I personally think this shows little forethought under the circumstances. I would like to suggest a remedy, namely that all racing be restricted to a 9:1 compression ratio, or that octane booster or race fuels be allowed, or that the R.A.C. pick up the tab for all the burnt out pistons.

Now let us consider the second situation, that is, a single intake port per cylinder type head or five porter with a long duration scattered cam. This combination of parts, especially the single port per cylinder, can be made to make some very impressive power. It is just a question of knowing how. In an earlier chapter we discussed how the compression ratio and hence the expansion ratio affected the optimum ratio of intake valve size to exhaust. Now we can see this put into practice. To reiterate the situation the bottom line is that the higher the compression ratio (and hence the expansion ratio) the greater the proportion of work done in the beginning of the power stroke and less at the end. This means opening the exhaust earlier is less of a penalty. As a consequence we can opt to use a bigger intake valve at the expense of the exhaust valve. Since the bigger intake will allow breathing at higher rpm we can, to a limited extent use a longer cam. The scenario looks something like this – a 1.53-inch intake flowing about a 155 – 160 cfm at 550 – 600 thousandths lift. A 1.218-inch exhaust valve flowing about 100 to 105 cfm. A compression ratio between 15 and 17 to 1 and a cam with about 320° on the intake and 335° on the exhaust. Getting such a high CR on a motor less than 1300 cc is difficult because piston cut outs limit just how small the total chamber volume can be, but on bigger engines it is achievable. The big stumbling block in any combination is the cylinder head but if the compression and the cam are up to the mark for an 8,500rpm engine there is enough air for a *good* 7 or 8 port, push rod head to make about 180hp! At the time of a writing it is not being done with a mini engine but push rod two valve engines with similar head flow per litre of engine displacement are making close to *150hp per litre*. These seemingly outstanding outputs all hinge about one aspect – the compression ratio! Neither the cam, head, induction or anything else will make this work *without the high compression* in the first instance and the head/cam combination in the second.

Fuel Octane

The most obvious factor affecting the highest useful compression ratio an engine can use is the octane rating of the fuel available. A less obvious factor is the camshaft. A cam having a long duration intake can tolerate a higher compression ratio than one having a short duration. Why? Because the static is reduced to a lower dynamic compression ratio which is brought about by the late closing of the intake valve. In addition to this the point in the rpm range at which detonation occurs is influenced not only by compression but by the amount of time the unburned gases are exposed to the advancing flame front. An engine will detonate much more easily at low rpm than high. At high rpm the unburned gases don't have very much time to pick up radiated heat from the advancing flame front. As a result these gases don't detonate as easily. When a long period cam is used the engine is never operating with high dynamic cylinder pressures at low rpm. As a consequence of these two factors the engine, when running between 2 and 4,000rpm tends not to detonate. This allows the engine to use a higher compression ratio simply because the long duration cam has cut the compression pressure seen at low rpm. It isn't until the action of exhaust scavenging and cylinder ramming starts to take place that we get the cylinders filled. At this point the compression pressures prior to ignition start attaining a high level. However, by this time the rpm is high enough to get the combustion process over with quickly thus reducing the tendency for the end gases to detonate. So how much compression can be used – check out the chart **Fig. 9.27**. If the maximum octane of the fuel available is only 75, and this is so in some parts of the world, the engine cannot utilize anything like the CR that it can on 105 octane fuel, which again, is just as available in other parts of the world, although not necessarily at the service station pump.

To start with, let's define what octane rating is, otherwise a point of confusion can exist. Firstly, there are two octane rating methods, these being the Research Octane Number (RON) and the Motor Octane Number (MON). These ratings are measured on a special octane rating engine known as a CFR engine. The main asset of the engine is that its CR can be adjusted. To find out what the octane number of a sample fuel is, the engine is run on that fuel and its point of detonation is compared with a mixture of N-Heptane and Iso-Octane. Iso-Octane is anti-detonant and has a figure of 100 assigned to it. N-Heptane is very detonation prone and has a figure of 0 octane rating assigned to it. A fuel with, say, 95 octane rating is one that when compared with a mixture of Iso-Octane and Heptane, detonates at the same point as a mixture of 95% Iso-Octane and 5% N-Heptane. An 80 octane fuel would be one that detonates at the same point as 80% Iso-Octane and 20% N-Heptane. With figures over 100 octane, it is necessary to make comparisons with Iso-octane with an octane booster added to it, typically tetra-ethyl lead, as this is a powerful detonation suppressant.

At this point it's well worthwhile to define the difference between the Research Octane Number of the fuel and the Motor Octane Number, as this has an important bearing on how our Series 'A' engine responds to CR increase, and whether or not it will detonate. The Research Octane Number is measured in the CFR engine by running it at 600rpm with a water jacket temperature of 212°F and an air intake temperature of 150°F. For the Motor Octane Number, the CFR engine is run at 900rpm, the water jacket temperature is unchanged at 212°F but the air intake temperature is raised to 300°F. **The Motor Octane Number rather than the Research Octane Number more closely simulates**

what happens in a typical automobile engine, and the Research Octane Number is quite far removed from reality in all except the coolest running race engines. When a typical fuel is tested by Research and Motor octane methods, it results in two different octane number ratings. Typically a 97 Research Octane Number fuel, when tested for Motor Octane Number, would be around 89 octane. The difference between these two figures is known as the fuel's 'sensitivity'. The main factor producing the difference between these two numbers is the *intake temperature* used. The rpm of the CFR engine generally makes only a small difference.

At the fuel pumps in most European countries, the Research Octane Number is shown. Fuel pump octane ratings in the U.S.A. use a slightly different method of rating. This is the R plus M divided by 2 system. That is the Research Octane Number plus the Motor Octane Number, divided by 2. In other words, it's the average of RON and MON.

Let's deal with this subject of intake charge temperatures in relation to the fuel's apparent octane rating. Anyone who has ever torn down a Series 'A' engine will realise that nearly all single carb units have an exhaust heated intake manifold. **This is a delightfully simple device for reducing the apparent octane capability of a fuel by a very substantial amount.**

On a Series 'A' engine, I would hazard an intelligent guess that a unit using the exhaust heated intake manifold would require about 8 octane numbers more in the fuel to run the same CR as an engine with an unheated intake manifold. **So, from this point of view, for a high performance engine, the exhaust heated intake manifolds are out.** They limit the amount of compression that can be used much sooner than a water-heated or an unheated intake manifold. The same goes for water jacket temperatures. Getting the engine to run at reasonably cool levels means less octane value is needed in the fuel to run a given CR.

Whilst still on the subject of fuel octanes, you may ask, 'Is it possible to boost them?' Yes, there are octane boosters on the market where fuels of a sufficiently high octane are not generally available. Though expensive, a preferable alternative, is to use specially brewed race fuel which, incidentally, only seems to be commonly available in U.S.A. Fuels such as H & H Blue apparently have octane ratings well past the 110 RON figure, and I have heard figures as high as 119 RON quoted. **However, there is no point in using fuel with any more octane value than is necessary and I have yet to find the need for fuel octane levels more than around 105 to 106 RON.**

Personally, I have never built an engine with more than 14.4:1 compression, and this ran perfectly satisfactorily on fuel with around 101 to 102 RON. However, some of the top U.S. engine builders such as Bob Griffiths at **Bhp Developments** have experimented with CRs up to around 15:1 and, using super-high octane fuels, detonation does not occur even at this level. Anyway, I digress. Boosting the octane level of a fuel is relatively simple in most parts of the western world, as octane boosters are readily available. In the U.K. **Aldon Automotive** produce an octane booster which is especially effective as it has two good properties. Firstly, when used in the prescribed dose, it can add 4 octane numbers on almost any fuel and, on many fuels, doubling the dose will give almost another 4, so it's capable of taking 96 RON fuel and boosting it to around 103 to 104 RON. Apart from this, it's relatively insensitive to intake temperature. **This makes Aldon's octane booster especially useful on turboed or supercharged engines, as the intake temperature of such power plants can get very high, and any supercharged engine generally needs all the octane it can get.** For those Mini enthusiasts in the States I have tested 104+ extensively and found that it can consistently raise the performance levels of premium pump petrol, achieving results to equal 105 octane racing fuel. Although extremely effective 104+ does have one undesirable side effect in that it does not burn too cleanly. This means more frequent plug decarbonization and replacement. There is however a simple remedy which kills two birds with one stone. Adding about one-quarter the prescribed amount of the proprietary product TK-7, will achieve clean combustion and offset the usual valve seat recession associated with the removal of tetra-ethyl lead from the fuel.

Associated Detonation Factors

The principle factors affecting the highest useful compression ratio that can be used are intake temperature, intake closure point, mixture ratio and spark timing. There are other factors, but they are, to coin an engineering term, second order effects so we need not concern ourselves unduly with them.

Just to drive the point home let's go through the cam/compression routine again so you don't end up with too little compression for a long period cam. It will be found that the longer the cam's opening period, *i.e.* the nearer it becomes to a race cam, the more compression the engine needs. The reason is that many race cams, especially if they are short of lift, tend to have a lower trapping efficiency than

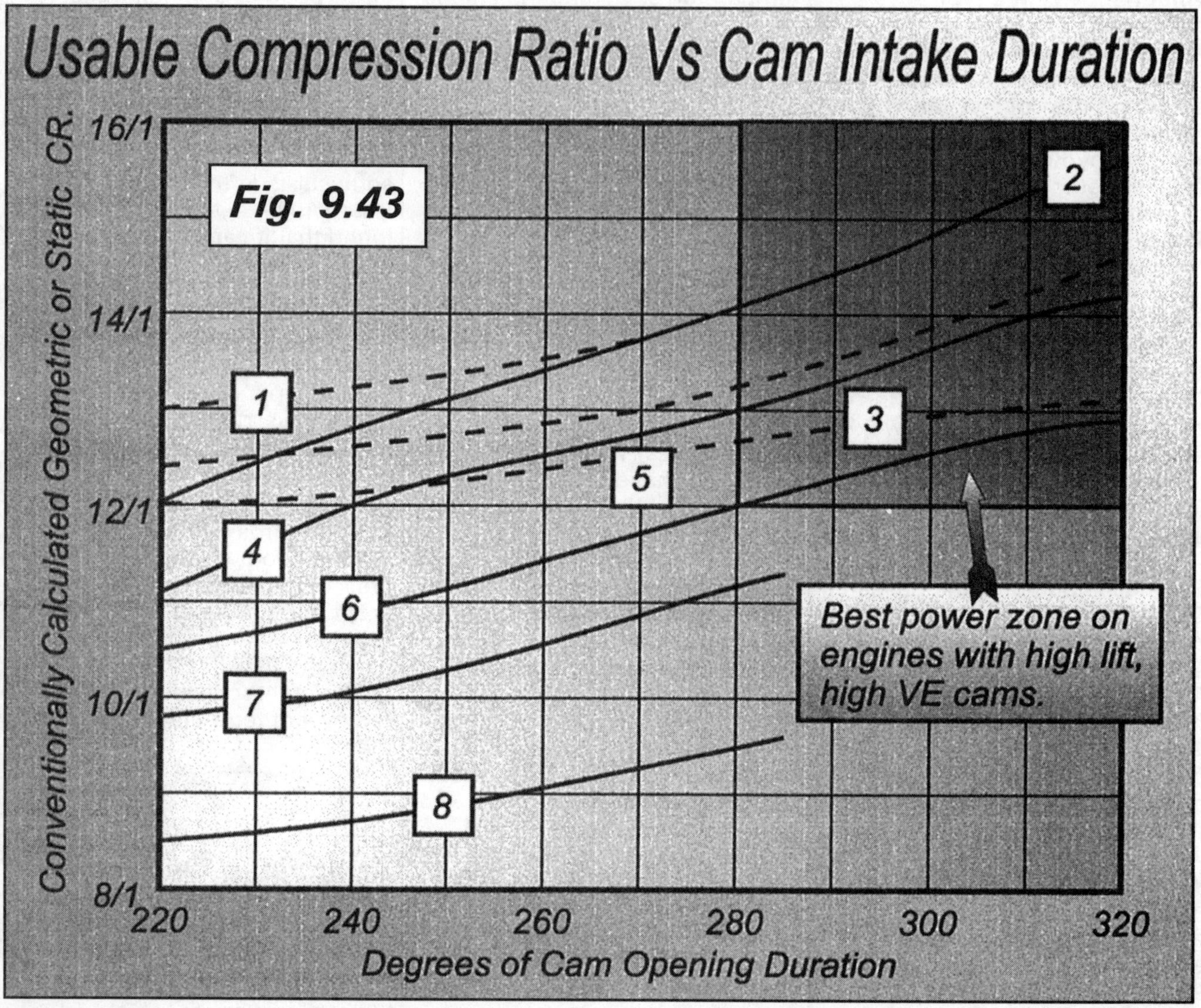

Note: 'Limited valve size' refers to those engines which are required, due to class regulations, to run with standard size valves. This applies in particular to small bore, small valve engines rather than big bore engines with 'S' size valves.

1 – 105 Octane – limited valve sizes.
2 – 105 Octane – unlimited valve sizes.
3 – 100 Octane – limited valve sizes.
4 – 100 Octane – unlimited valve sizes.
5 – 96 Octane – limited valve sizes.
6 – 96 Octane – unlimited valve sizes.
1 – 6 unheated manifold.
7 – 96 Octane with heated manifold or 93 Octane with unheated manifold.
8 – 93 Octane with heated manifold or 90 Octane with unheated manifold.

road cams. Imagine a camshaft for a street engine. This opens and closes the valves, not that much off TDC and BDC. This means that at low rpm the cylinder, assuming no undue obstruction in the induction system, is fairly well filled by atmospheric pressure prior to closure of the valve. This results in almost a whole cylinder of mixture getting compressed into the combustion chamber at TDC. On the other hand, if a race cam is used, the valve may not close until the piston is one quarter, and in extreme cases, as much as one-third, of the way up the bore. Under these circumstances, even if the cylinder was full at BDC, quite a lot of this charge could be pushed back out the intake valve prior to this closure as the piston comes up the bore on the compression stroke. Of course, it won't actually start compressing any mixture until the intake valve closes. When it does close, that cylinder has already dumped probably one-quarter of the charge. The result is that only three-quarters of the cylinder's volume of mixture is compressed into the combustion chamber space, thus resulting in a lower effective CR. This is the case at least at low engine speed, where, incidentally the Series 'A' engine is most likely to detonate.

In many instances, long period cams, though they have low trapping efficiencies at low rpm, can have a much higher trapping efficiency at high rpm, especially if the engine is shock wave tuned on the induction system. Under these circumstances, an engine may run at volumetric efficiencies close to, or even above, 100%. Unfortunately, the Series 'A' engine, due to its five port layout, doesn't easily achieve VEs near 100% unless the very latest cam, valve train and head techniques are used. We usually find that, even in race trim, most modified Series 'A' engines only achieve 85–88% efficiency.

As a result of the reduction trapping efficiency given by a long period cam at low engine speed, the engine can usefully use a higher measured CR.

As a result of the reduced trapping efficiency given by a long period cam, the engine needs more measured CR. In other words, the static compression on the engine needs to be set higher, in the knowledge that under running conditions the *effective* CR at lower engine speeds, where detonation usually takes place, will be quite a bit lower.

If by virtue of rules covering a particular form of motorsport, the volumetric efficiency of the engine is restricted, then this has a similar effect, at least at high rpm, as using a long period cam, because once again the cylinders are not properly filled. A good example of this is 850 Minicross engines. Because standard valve sizes must be used, as well as a standard cam, the engine can spin at high enough rpm to experience very limited cylinder filling. This means that the engine can use a lot higher CR for top end power. This can mean that ignition timing may be over advanced on standard advance curve at the low end: it may therefore be necessary to build a distributor with less initial advance, just to keep the engine free of low speed detonation. The upper end of the rpm range may need a more normal advance setting to produce the best power from a poorly fed but high compression cylinder.

The mixture ratio at which the engine is run can also have an effect on just how high a CR can be used before detonation occurs or power drops off. If the mixture is rich, the excess fuel tends to suppress detonation, although there is a limit to how much it can do so. In fact, if the mixture is too rich to suppress detonation in an engine with a CR which is too high, the net result will be less power than with a lower compression and a mixture slightly leaner. Generally speaking the Series 'A' engine likes to have mixture ratios between 12.5 and 13:1. On later emission type mixture analysers, that's generally around 5% to 6% on the CO scale.

The last point worth noting is that the higher the CR goes, the easier it will be to run into detonation due to having excessive ignition advance. **As the compression goes up, so the total ignition advance required often comes down, simply because high CRs tend to promote faster mixture burning.** However, there comes a time when if too high a CR is used in relation to fuel octane, the ignition timing will have to be artificially retarded to avoid detonation, even though it may be short of the ignition timing required for maximum power. Under part throttle conditions, combustion occurs more slowly, so a lot more ignition timing is needed.

On road engines, where economy is as important as all-out power, it is possible to go a little overboard on compression so as to achieve good economy and then stay out of detonation under full power conditions by using slightly retarded ignition timing. Because lower cylinder pressures are involved under part throttle conditions, the ignition timing can be brought up to 'mean best torque' timing (MBT timing) long before detonation sets in. This is the technique used by Leyland to obtain the high fuel efficiency figures for Series 'A' engines from about 1980 on. If economy is important, then high CRs are also important, because high CRs can add a considerable amount of mpg to the engine's capabilities in part throttle conditions.

Cylinder Heads Part 4

Valve Guides & Compression

An aspect which is taking on decidedly greater importance these days is the valve guides used in the Series 'A', or for that matter any high performance engine. Fuel octanes are dropping, which inevitably means that compression ratios must come down. This in turn tends to take the fine edge off the power producing capabilities of long period cams. As a result of this, newer generations of cams are coming out with higher lift and shorter periods than those they seek to replace. These cams can function much better with a low compression ratio than the longer period cams. So far, so good. Unfortunately, several factors creep in to mar the picture. High valve lift causes substantially increased side loads on the valve guides, causing them to wear much faster. A worn valve guide in itself can cost power because it prevents the valve seat from sealing up as it should and it may even cause the valve head to break off due to fatigue. Another factor comes into play which definitely falls into the subtle rather than obvious category. Worn valve guides allow oil to pass relatively freely, aided by the intake vacuum, down the intake guide to contaminate the fuel/air charge. **This oil causes a marked reduction in the octane value of the fuel**. The net result is that the engine will detonate easier, and this is the very thing we are trying to avoid. This problem was just one that David Anton and I at A.P.T. had to face long before our counterparts in the UK. The fuel situation in USA has been abysmal for high performance engines since the early seventies. One of the areas we delved into was valve guides for Series 'A', especially since we were developing very high lift valve trains (up to 0.600 inch) with short timing and rapid valve opening rates. With later generation cams/valve train combinations that we were either testing or developing for various companies in England, for the guides to have even a half-way respectable life, it was necessary to come up with a guide material and design that would survive the rigours involved. As a result, we

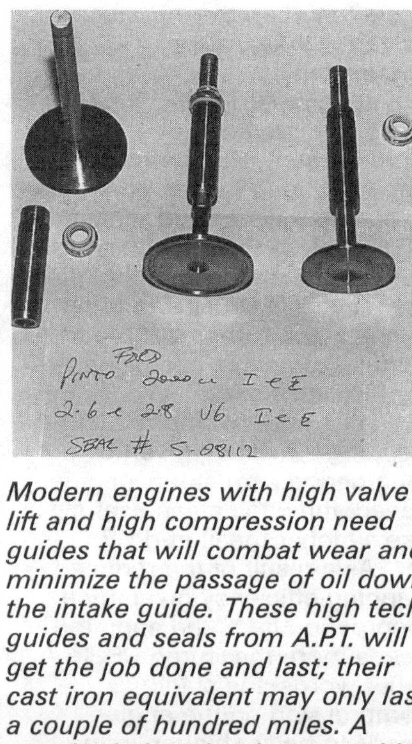

Modern engines with high valve lift and high compression need guides that will combat wear and minimize the passage of oil down the intake guide. These high tech guides and seals from A.P.T. will get the job done and last; their cast iron equivalent may only last a couple of hundred miles. A standard phosphor bronze guide may only last a couple of thousand miles.

developed guides of a manganese aluminium/silicone bronze material that required very little in the way of oiling. The material spec allowed close fits, and the shortest guide possible consistent with good guide life. We drew them up in all the various standard styles, plus ones which were machined to take a high control teflon oil seal. These guides are compatible either with a conventional valve steel, stainless or chrome stems (cast iron guides are not compatible with stainless). So far, over the years we've been using them (from 1987) they have proved very successful. Another point to consider with valve guides, and this is a secondary factor in relation to compression, is a hot spot in the cylinder can cause detonation to occur at lower compression ratios than would otherwise be the case. Apart from the spark plug, the heat range of which can be altered as necessary, the hottest thing in the chamber is the head of the exhaust valve. The exhaust valve dumps heat both through the seat and through the valve guide stem. Bronze guides with a close fit can transmit significantly more heat to the cylinder head casting than can a cast iron guide because the silicone bronze guide can be run with closer tolerances and this has a tremendous effect on heat conduction from the valve. Sometimes reducing the clearance by 0.001 inch can increase the heat conductivity by 100% and that's a big margin. Of course, you could go to a sodium-cooled valve as fitted to the turbo MG Series 'A', but that is a restrictive valve and much more costly. By far, the best bet for most practical applications is to use close fitting bronze guides (0.0015 inches in and 0.0020 inches ex.) and conduct as much heat as possible out from the valve stem via the conductive guide. By the way, *having oil on the stem of the exhaust valve is important and I do not recommend the use of an oil seal on the exhaust valve stem.* Oil consumption due to not having a seal on the exhaust valve stem is negligible. Very little oil passes down the exhaust valve stem because there tends to be pressure from the exhaust port outwards. Any oil that passes down the stem is due to capillary action and mechanical transference due to the valve motion in and out of the oil mist existing under the valve cover. Obtaining these guides in the UK is simple enough as they are distributed by K & N Europe, the same company that distributes K & N filters. This means that the hundreds of K & N air filter distributors across the country can also supply these guides. The price is not that much different from a cast iron guide from an ARG (Leyland) parts dealer. In comparison with most other bronze guides they are far cheaper, so you are doing yourself a favour by making what little effort is required to obtain them.

Well, the foregoing might be of interest, but I am sure many of you are asking, where does this leave us in terms of the CR we should select for the engine we are building? Well, here I can say that experience is really the only worthwhile teacher, so I have produced a graph **Fig. 9.43** based on what I have been able to find out about the Series 'A' engine's likes and dislikes in terms of CR. This graph assumes that the ignition timing needed to produce mean best torque at full throttle conditions can still be achieved. I add this point because on some of Leyland's economy engines this cannot be achieved as ignition timing is artificially retarded under full throttle conditions to keep the engine out of detonation, and the high CRs utilized purely and simply to promote good fuel economy. On this graph, you will see that I have included the effects of such things as long period cams and intake temperatures. **The graph Fig. 9.43, needs to be used with a certain degree of caution as octane figures can vary considerably from stated pump values.** However, on the positive side, it is generally a fairly sound guide as to what can be used and what cannot.

Heads For Economy

Although the subject of economy has had more than a passing mention so far, it's about time it was dealt with to the exclusion of all else. If economy is your number one priority rather than total power output, then cylinder head modifications need to be conducted in a slightly different fashion. Extensive dyno testing in this area has indicated that three factors affect economy: the roughness of the intake port, the CR and the flow capability of the exhaust port.

Dealing with the intake port first, I think it best to relate some experiences concerning how intake port finish can affect economy. A few years ago I was busily engaged on an economy project with Oselli Engineering, and in an attempt to improve economy, I simply installed a Stage 1 modified head onto an engine. However, the original head, though standard in every other respect, gave a high compression which the Stage 1 head did not increase. Therefore, any power increase could be ascribed to increased airflow capability from the Stage 1 head. The modified cylinder head certainly improved hp but, surprisingly, when the part throttle fuel consumption tests were done the Stage 1 head actually used slightly more fuel than the standard head!

When modifying a cylinder head to get the maximum economy from it, you might consider insulating the chamber by coating with ceramic insulation the valves and any area of the cylinder head exposed to the flame front. One company that specializes in this treatment is Heany Industries in the USA

The particular head concerned was a straight off-the-shelf unit with all the ports finely polished. **Many companies in the cylinder head business find it difficult to sell cylinder heads unless they have a fine, polished finish. This is simply because the customer demands highly polished ports in the mistaken belief that polish makes hp.** Those polished ports, as stated before, contribute nothing. However, at this point I concluded that the polished ports may be adversely affecting economy, so another cylinder head was produced with exactly the same intake port shape, but nothing more than a coarse ground finish was used. Subsequent dyno tests resulted in identical hp to the previous Stage 1 head, but fuel consumption figures were marginally better than the previous Stage 1 head although still slightly short of that produced by the standard head at part throttle and slightly better at full throttle. Of course the full throttle economy figures are important, but by far the most important are the part throttle economy figures because during economy driving conditions part throttle is the most used mode.

As a result of the second test, further heads, with less and less work done to the intake, and more and more to the exhaust, were manufactured. Eventually it was found that the best economy figures were achieved when the intake port was left virtually standard. The intake valve itself was back-cut and the face of the valve machined to leave a sharp edge so as to reduce the effect of back flow. The valve diameter utilized was 1.312 inch.

On the exhaust side, the valve diameter was increased from 1.156 inch to 1.215 inch, that's the standard 'S' size valve. Here, the exhaust valve and exhaust port were tailored for high flow capability. In fact, put in simple terms, the exhaust port was given race engine treatment.

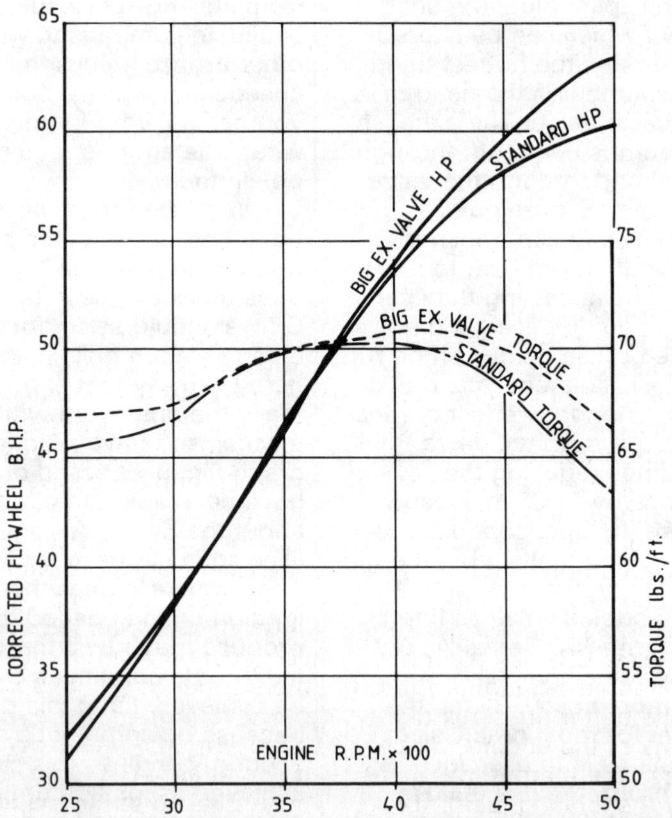

Fig. 9.44 Comparison of big exhaust valve and standard valve
Using a 1.218in exhaust valve and a modified port but with no other changes, the power and torque were increased as shown. Part throttle fuel consumption at steady speeds over the rpm range 1800–4270 (corresponding to 30–70mph for a 1275 GT Mini) showed a 2% fuel economy improvement.

Under these conditions, this cylinder head produced a 2% improvement in the steady speed fuel consumption between 30 and 70 mph. That, incidentally, was the entire range it was tested over. Under full throttle conditions, it produced a slight improvement in economy throughout the rev range, but quite a bit of extra power, relatively speaking, up at the top end of the rpm range. The power figures in **Fig. 9.44** show the results of utilizing one of these cylinder heads.

At the time of writing, my suggested formula for success in terms of economy from the head is to utilize an absolutely standard intake port with a standard size intake valve, preferably of the Rimflo design. This should be used together with a larger Rimflo exhaust valve in a fully reworked exhaust port. The combustion chamber itself needs to be polished, as this will achieve two things. Firstly, it will cut down the surface area presented to the flame and therefore less heat will be conducted to the water jacket; secondly, the polished surface will also cause some of the heat to be reflected back into the chamber rather than conducted away. Remember, heat loss is also economy loss. We are burning the fuel to expand the air in the cylinder and the higher temperatures that are achieved for a given amount of fuel, the more economy the engine will achieve.

Now for CRs. It's a simple fact of life that the higher the CR goes, the better the potential economy of the engine will be. Of course there are upper limits. Usually the limiting factor for a CR is how much the engine will withstand under full throttle conditions. However, under part throttle conditions, CRs of 14 and maybe even 15:1 can be utilized to achieve better economy. Remember, when the engine is throttled, the *effective* CR is substantially less than static or measured CR. It's because of this factor that we find that increasing the CR from, say, 8:1 to 11:1 can produce substantially better economy figures under light throttle driving conditions. Going up three ratios from 8:1 to 11:1 can, under ideal conditions, achieve as much as 20% improvement in fuel economy. However, before setting your CR to some ridiculous level, bear in mind the other factors that have to be taken into account. Re-read what I have to say on CRs earlier in this chapter so that you are sure you understand what's needed.

Exotic Heads

So far, we've looked at some sophisticated, though not exotic, modifications to Series 'A' cylinder heads. In an effort to get horsepower, many people will go to extreme lengths to try to improve on what already exists. Some modifications are doomed to failure before they ever get off the drawing board. Some modifications look like they have some potential and we'll look at a few of them here. Firstly, let's deal with 8-port cylinder heads. Eight-port cylinder heads have been around for a long time, one of the first successful ones being the downdraft Weslake iron head produced way back in the mid to late sixties. In essence, this cylinder head had the same combustion chamber shape as the Cooper S, but it sported eight ports, four on each side of the head, with the intake coming down at about 30° above the horizontal so that it made a much better approach to the back of the valve.

These heads have some good potential if you can find one. I had the opportunity to flow test a well worn one so the valve seats were far from ideal but it still returned some very respectable flow figures as seen in the chart **Fig. 9.45**. If you can find one that is not the worse for wear or cracked then it is well worth getting. Remember though that the different port layout will require a different cam. Way back in the days when they were popular among those people who could get them, they tended to produce, as near as I can find out, about 10hp more than a typical 5-port set-up. That's a 5-port set-up as of the mid-sixties; things have changed a lot since then. It could be that if you built a fuel-injected 8-port set-up exactly as used in the mid-sixties; a current 5-port set-up with latest technology would probably blow it away without too much trouble. On the other hand, if modern technology was applied to an engine which utilized one of these 8 port heads, we may see a completely different picture.

To produce my own incline valve heads, a special fixture, as seen here, was made up. This inclined the head to the right angle and located everything in the right position so that the whole job could be turned into a simple boring operation on the milling machine.

So You Want To Do Your Own Heads?

Without a doubt, one of the most expensive parts of building a high performance engine, is the cylinder head, at least it is if you get one that has been properly done. **There are, of course, dozens of cheap heads on the market, but remember you only get what you pay for.** Those cheap heads may give you very minimal power increases. I have flow tested quite a few proprietary heads that run from the cheap to the middle end of the price bracket, and have found that very few give more than 5% – 7% increase in flow, and some actually gave less flow, especially in the lower lift ranges. If you want a good head for the Series 'A' engine, that is one that will produce hp, torque and what ever else is necessary, expect to pay in the region of three to four times as much money as you would pay for, say, an exhaust manifold or something like double that of a Weber carburettor. Of course, if you are prepared to invest that money, plus a little more, you could set yourself up to grind your own cylinder heads. If you grind them exactly as detailed in this book, you will get good results. However, I would be the last to say that these heads are the ultimate. At the time of writing they were amongst the best figures I had seen of any cylinder heads tested on my flow bench, but diligent work could improve on these. If you look at the graphs showing some of the flow figures I have obtained from my own cylinder heads, and compare this with theoretically perfect cylinder heads, you will see that there is still scope for improvement, but be warned, such improvements will be hard to find.

Let's look at the first problem, that is grinding your own cylinder heads. Sure, it can be done with an electric drill and a flexible shaft, but electric drills which turn 2000, 3000 or even 4000 rpm are pitifully slow for a grinding wheel. Of course you can buy cutters **but the cutters most commonly sold in do-it-yourself shops are high speed steel. These cutters are not up to cutting cast iron in a drill, or, for that matter, by any other method of drive.** If you try cutting cast iron with them, they will last about 5 minutes at most before they are too blunt to do anything but burnish the surface. This means you have one of two alternatives: either a machine fast enough to drive a grinding wheel at the correct speed and use inexpensive grinding wheels to do the job. Alternatively, you could stick with a comparatively slow electric drill and use a tungsten carbide cutter which, although it is still operating at the wrong speed, will cut and is unlikely to go blunt until several dozen cylinder heads have been done. The only snag is, tungsten carbide cutters range from four

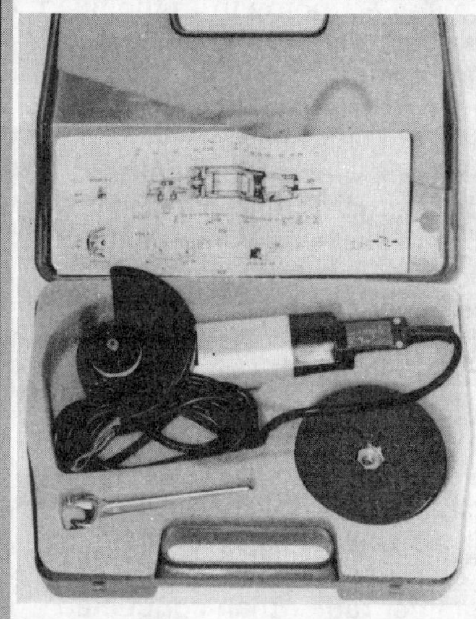

This useful little gadget from Suhner, can be used as a body grinder or . . .

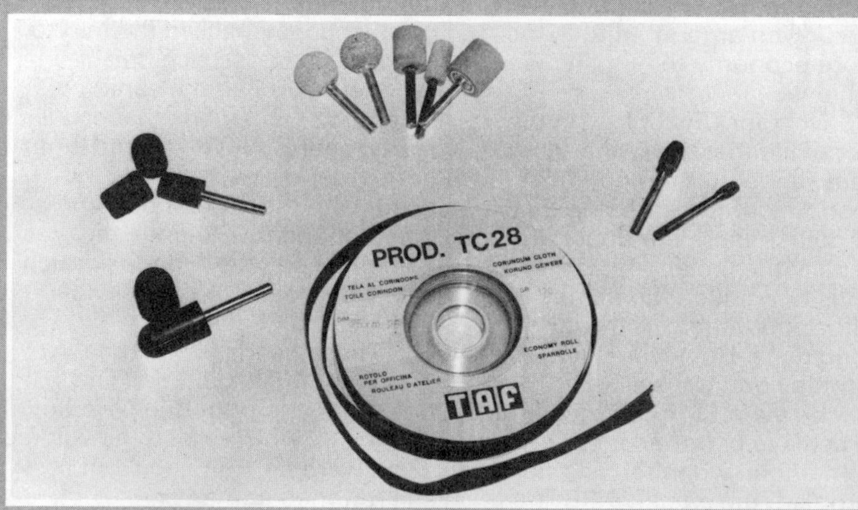

Shown here are some of the typical tools needed for cutting a cylinder head. Most of the materials are self-evident from the photo, but the two cutters in the upper RH side of the picture are well worth discussing. These are tungsten carbide cutters. They are expensive – five or six times the price of high speed steel cutters – but worth the extra money. High speed steel cutters simply wont work for long on cast iron cylinder heads. On the other hand, tungsten carbide ones really do cut the metal away and they will last five to ten cylinder heads

to ten times the price of a high speed steel cutter, but they do get the job done, so that makes them cheap by comparison.

Let's talk about investing some money in cylinder head equipment. If you are thinking of doing two, or maybe three cylinder heads to race specification, then the chances are you will just about break even if you buy proper equipment of a professional standard to do cylinder heads. In many cylinder head grinding shops, air grinders are common equipment. In the States the **Dotco** grinder is one I can recommend, and there are plenty of brands in U.K. and Australia which will work about as well. Basically you need to go for a unit that will run a grinding wheel at least 15,000rpm but preferably around 20,000. The only snag with air grinders is (a) they blow grit all over the place unless the head is in a cabinet, or a large vacuum cleaner is used (b) an air grinder needs a good supply of compressed air from a big air compressor. Small air compressors simply won't keep up with the demand. So this means that for the guy doing cylinder heads in his garage, a high speed electric grinder with a flexishaft drive becomes a much more promising bet. In fact, a good electric flexible shaft machine is as good, if not better than an air grinder. A brand of electric grinder that I can recommend is the **Suhner-**type grinder. Apart from big, professional grinding machines that are used for jobs other than grinding cylinder heads, but admirably suited to cylinder heads, they make a small, compact 22,000rpm flexishaft grinder. The beauty of this equipment is that the flexishafts have a very long life. Many of the flexible shaft machines I have tried have a very short flex shaft life. These shafts aren't cheap to replace. You only need to run through one or two shafts before the bill starts to look a little outrageous.

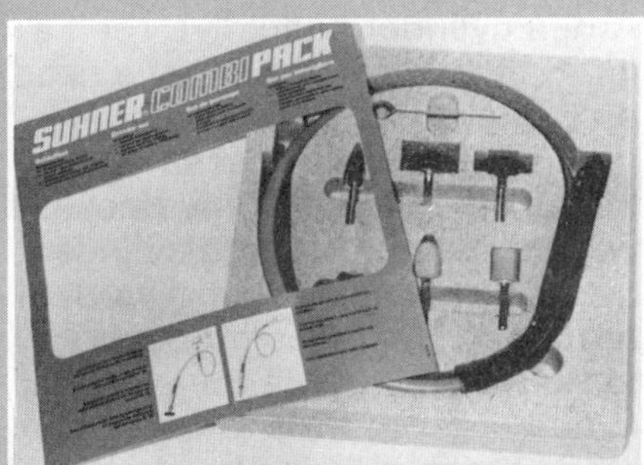

. . . as a high speed cylinder head grinder when coupled up to drive this flexible shaft.

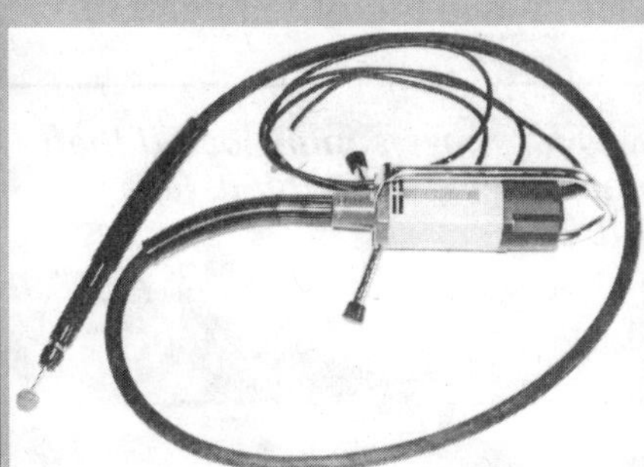

If you are really looking for the best in versatile cylinder head grinding equipment for do-it-yourself or professional purposes, this compact Suhner grinder has the capability of running between 4000 and 22000rpm with a speed controller, and it's virtually unstallable because of the high output, compact motor.

The Suhner machine shown in the nearby photo is going to cost two to three times the price of a typical full race head, but if you do a couple of heads for your friends, you will recoup much of the cost.

If you want to go in for cylinder head modifying in a professional or even semi-professional manner, then ideally you need a flow bench. Of course it's possible to build your own. Constructing a flow bench which would tell you whether or not an improvement has been made is quite simple, but if you want to compare figures with those obtained by other manufacturers, to determine whether or not your heads are better, then you are going to need a calibrated flow bench.

The least expensive on the market is the Quadrant Scientific 170. This was specifically designed to flow cylinder heads about the size of 'A' Series heads so it is admirably suited to the job. The bench I currently use is a Quadrant Scientific FL520. It has enough flow capability to deal even with the 2.5-inch intake valves of 500 inch race V8s. It has a number of facilities that I find very useful. The one that speeds tests the most is the self setting pressure – it follows the valve opening so a test is very rapidly done. Also the software that comes with the bench is extremely comprehensive and saves an enormous amount of time on number crunching. Another facility that is proving very useful is the computer port velocity gradient mapping. Although not completed yet it is expected by the year 2000 to have this mapping facility to draw 3D prospective views of the air velocity in the test port. With improved flow figures becoming increasingly more difficult to find velocity mapping is proving to be a significant advantage to the cylinder head developer.

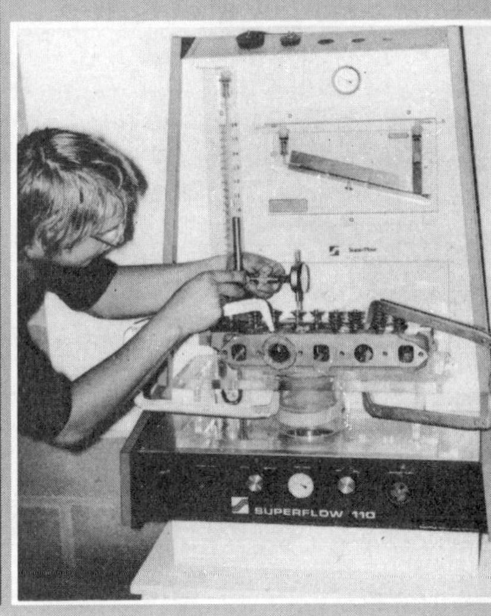

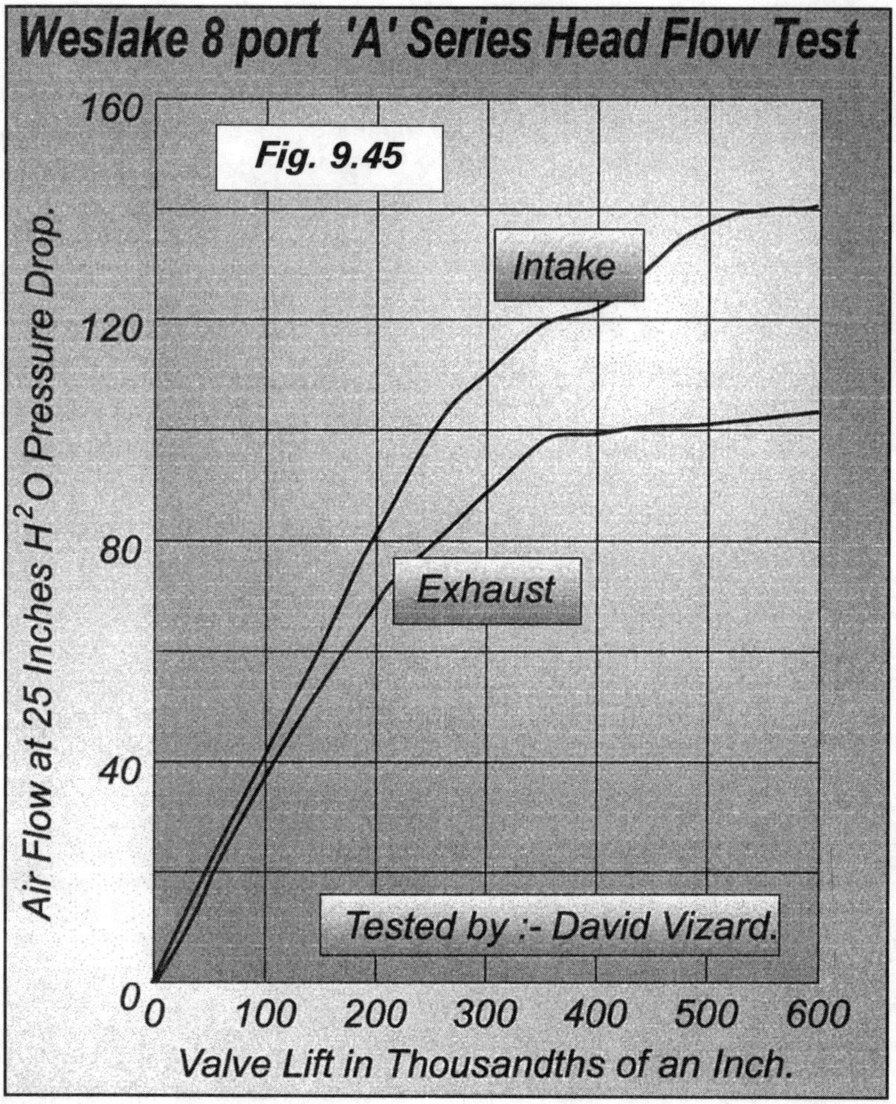

Undoubtedly, the best known purveyor of 8-port cylinder heads is Jim Whitehouse of Arden Engineering. Jim has made hundreds of 8-port cylinder heads and indeed, it was his head which was adopted by Leyland as the official 8-port factory head. It can be used either with carburettors or fuel-injection. If you bought a head from Jim, you could either have something which was pretty much off the shelf or if he was building your engine for you, the cylinder head would be tailored to whatever specification he thought the engine would need. Consequently, there are lots of minor, and a small number of major, variations on the basic Arden cylinder head. I've seen a lot of these heads, but have used only one. At the time they were popular they always seemed to be priced well out of my pocket. I was still using cast iron cylinder heads from the wrecking yard. In retrospect, that was probably fortunate as it made me work even harder to try and find what made these cylinder heads go, as opposed to simply shelling out the cash necessary for something like an Arden 8-port head. My limited 8-port flow testing to date has shown the Arden head to

This is the overhead cam, eight port Howley cylinder head, the production rights for which have been acquired by **Mini Spares**.

Although its overhead cam design was far more sophisticated than the conventional 5-port, Series 'A' head, it was, if anything, less complex; the hallmark of good design.

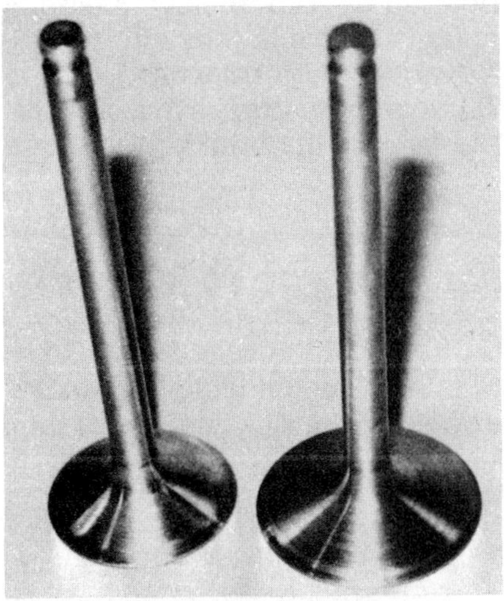

The valves used in the Howley SOHC head were the same size as those used in a typical 'S' engine. Both intake and exhaust were made of 21 4N stainless steel.

have less flow than my own '5-porters'. The flow figures themselves don't look especially good when compared with the modified 5-port head. But flow is not the only criterion. With the 8 port set-up, shockwave tuning can be employed, especially if a lot of work is done in matching the cam characteristics and induction and exhaust lengths to the rpm band that the engine runs in.

So that easy cam timing changes could be made on the Howley SOHC head, this Vernier cam sprocket was employed. It allowed timing changes to be made in very small increments so that the optimum shaped power curve could be developed from any single cam profile.

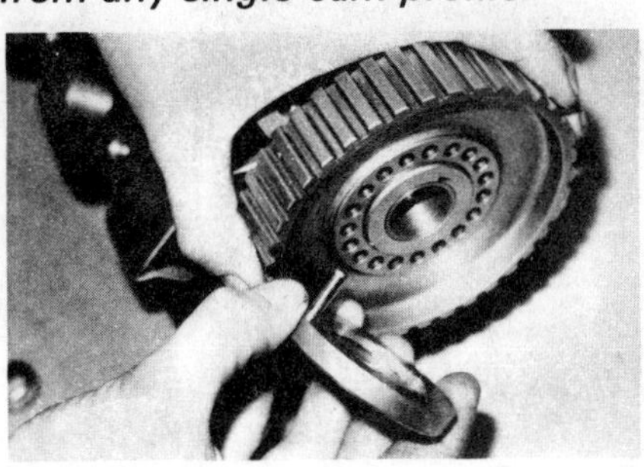

Surely the ultimate development in cylinder head technology is this 16-valve set-up for the Series 'A' engine created by Weslake engineer Gary Oldfield.

At the time of writing, there is another 8 port cylinder head that is supposedly about to appear on the market.

Unfortunately it has been 'about to appear' on the market for about two years now. As yet, I've not been able to get my hands on one to test. No doubt a test example will turn up the day after this manuscript goes to press. The particular history of this head, currently known as the Jomo head, is somewhat spotted. The particular castings and the project itself have been around for some years and have been passed from one specialist to another. Each has appeared to lose interest in it at some point or other usually because of more pressing work in other areas. Under the Jomo banner, several examples have been made and are out being tested and so far results are reportedly good. As usual, I will reserve judgment until I have proved that for myself.

The Howley-developed single overhead cam set-up that you see photos of dotted around the nearby pages seems to have fallen on hard times. With other things to do at the time like preparing some of the competition department's race winning engines, Howley Racing had much to take their minds off the single overhead cam project. As a result, it languished for awhile until the whole project was bought out by Mini Spares. After what appears to be an initial hustle to get things going, Mini Spares also found they were too busy to do anything with it. So, right now the position on the single overhead cam set-up for the Series 'A' engine is stagnant. It appears it will stay that way until someone decides they want to front the money to actually get this thing into production. That could be a pretty expensive job.

Four Valve Heads

As far as I know, the first four valve head ever made for the Series 'A' engine was built from a solid lump of aluminium by Colin Thomas of the Birmingham based company Motospeed. Having proved the basic concept worked as the prototype on a 1400 producing 155hp, several more examples were built to use for Rallycross events. This was the late seventies to early eighties era. Although I wrote to Colin for details on this cylinder head and maybe a sample port for some airflow testing, I heard nothing. Since the initial flurry of publicity in the UK performance magazines, I have seen or heard nothing of this four valve head.

The most recent endeavours on the four valve scene (1988) has been undertaken by one Gary Oldfield. Gary, late of Mountune, Piper, and Weslake started K.A.D. (Kent Auto Developments) in his spare time. He, along with some willing helpers have designed and built a four valve cylinder head, not from solid as is often the case with prototype heads, but from a casting. The first prototype casting, which if the engine was in a Mini would have the exhaust around the front of the engine and carburettors at the back, on a 1500cc unit produced in excess

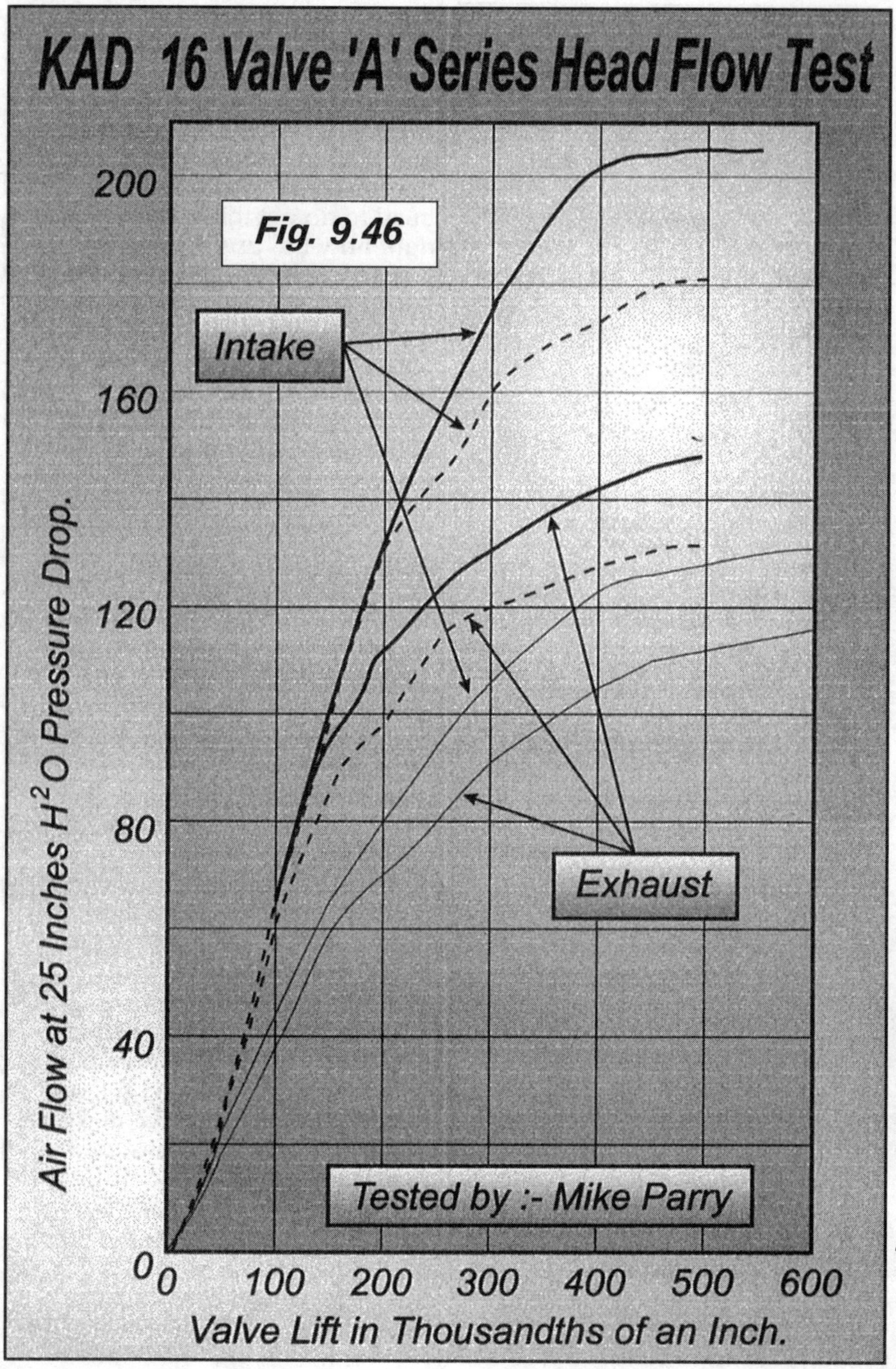

of 180hp. This was a fully specced engine with fuel injection and race cams. However, tests were halted because it was felt the installation with the intake and exhaust that way around would be more of a hassle than if it were reversed. The whole project was restarted and new patterns made up so as per normal Mini practice the exhaust would be on the back of the engine and the carburettors at the front. Two versions of this head are currently being built by K.A.D. One, for road use, is intended simply to bolt straight on to an otherwise standard Metro (and most others) short block, while the other is a race version. The race head is equipped with bigger valves and is meant to install on blocks having a minimum of 73.5mm bore. In the U.S. the cylinder head is available through Advanced Performance Technology (address available at the back of the book).

Jack Knight 16 Valver

In effect Gary Oldfield and KAD arrived at the right time with their

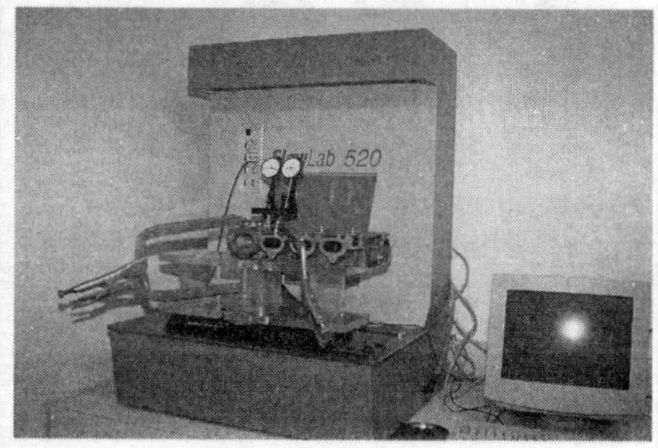

The Jack Knight head proved an easy casting to improve from very good to even better. Almost all the moves were obvious.

The junction with the two intake ports has, as can be seen here, blunt leading edges. By knife edging the adjacent walls . . .

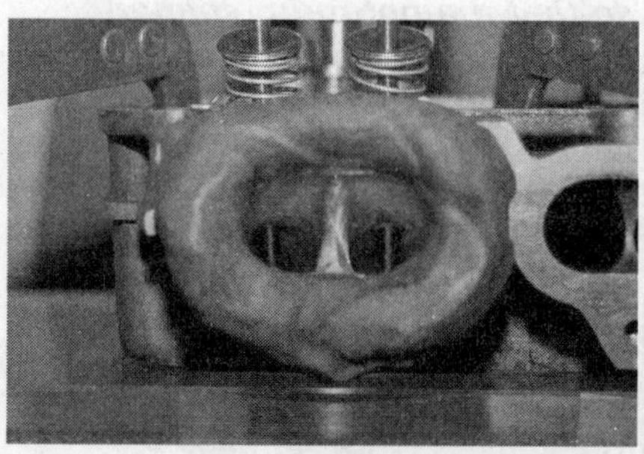

. . . in the manner shown here the flow was measurably improved.

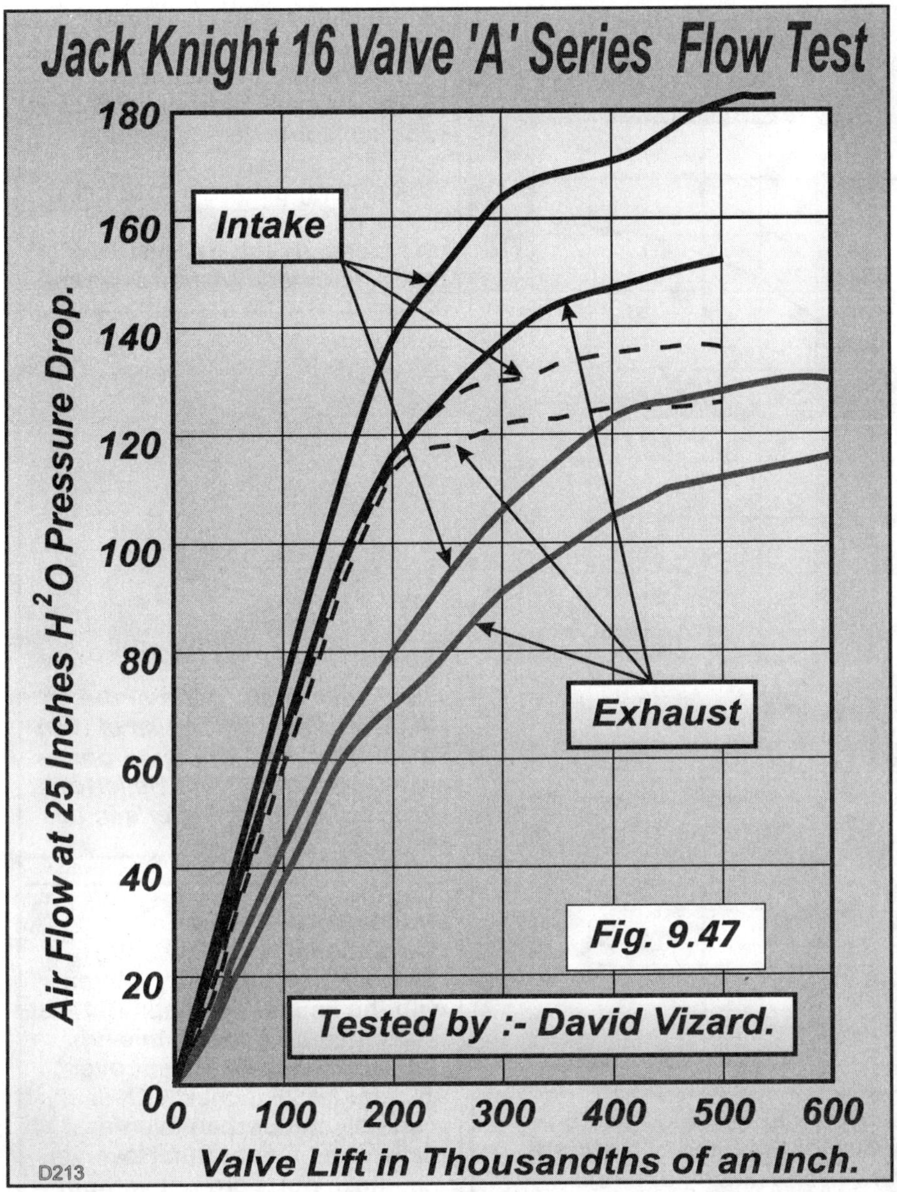

Fig. 9.47

16 valve head. Because the market was now seen to be ready this prompted others to look at producing a similar head. However regardless of when it may happen such an undertaking requires a great deal of perseverance and design/production capability. One such company that could meet these requirements was Jack Knight. Long known for their transmission parts this company embarked on the production of a four valves per cylinder head for the 'A' Series engine. To date this is the head that I have had experience with. It is in appearance very similar to the KAD head but this is to be expected. The only reason for going to a four valves per cylinder layout is more power. As it happens the basic design parameters were well sorted on such heads long before they were ever seen on 'A' Series engines. This is the primary reason for almost all of four valves per cylinder heads looking for all the world like scaled up or down Cosworth BDA heads. As expected, air flow with four valves per cylinder is markedly superior, especially at the lower lifts, than a two-2 valve head. As a consequence these heads are capable of allowing 180 plus horsepower from 1300 cc engines. Although still early days the graph, **Fig. 9.47**, shows what I had achieved as of 1997 for air flow through these heads.

Seven-Port Heads

The use of 8-port heads such as those produced by Weslake & Arden engineering (which eventually became the 'factory' 8-port head) always carried a penalty for anyone working on even a relatively good budget. When the head was changed it also meant a cam and exhaust system change as well. Since these parts were special their cost was proportionally as high as the head in the first place. In Denmark during the '60s a 7-port cross flow head was designed and produced in limited numbers for Minis competing in Denmark's national touring car championship. These heads proved to be successful in as much as they won several championships. However their continued production fell by the wayside, that is until Georg Mikkelsen of AKM Racing resurrected the design and put it back into limited production. The 7-port head configuration has the advantage of having the same valve sequence as a conventional five porter and therefore can successfully use much the same in the way of camshafts. Also the three exhaust port layout being the same as that of a five porter means that a conventional LCB exhaust system can be used. At first sight the use of three exhausts instead of four may seem like a strike against the 7-port configuration. In practice this appears to most certainly not be the case. Since the centre ports exhaust at 360° intervals there is no sharing of exhaust time however the exhaust pipe appears to be more active in

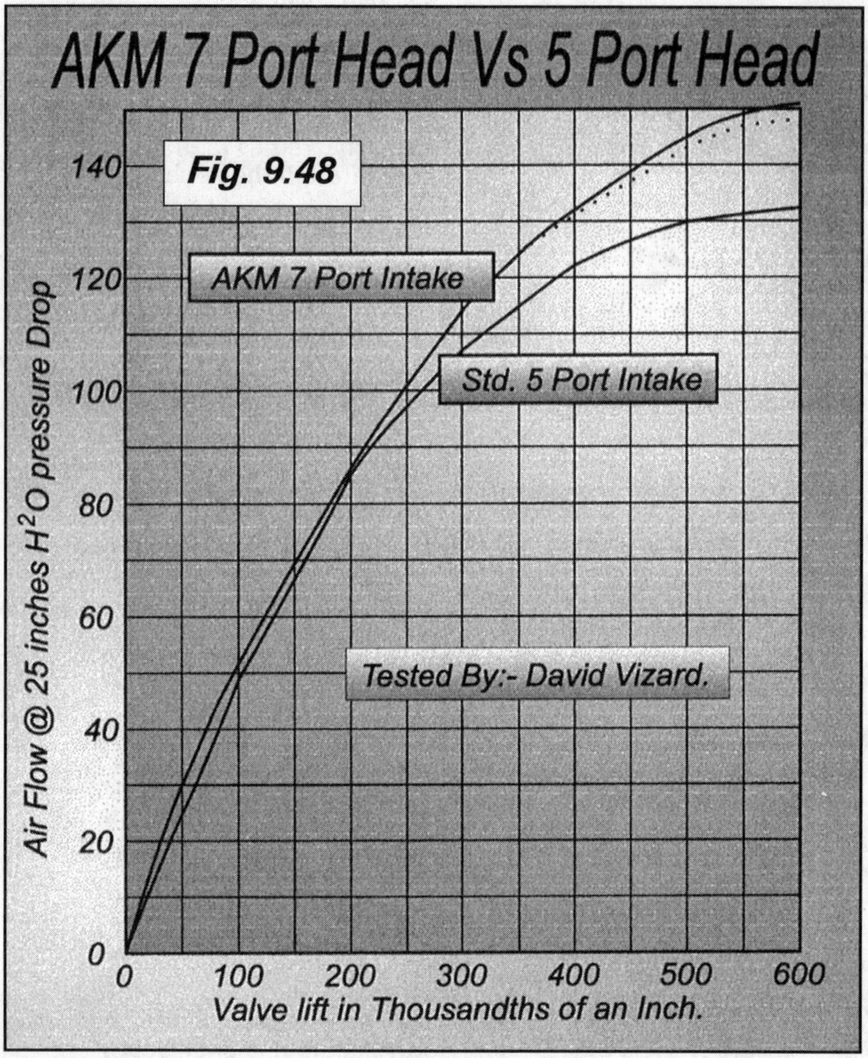

Shown here are the flow figures for a typical well modified 5-port head Vs the AKM 7-port head. Because of the pitch of the carbs in relation to the intake valves there are two distinctly different intake ports. Above about 350 thousandths lift the outer ports have a slight advantage. Exhaust ports are not shown here because they are intrinsically the same on both types of head so flow potential is fundamentally the same.

race fuel and a suitably high compression ratio can be used.

Doing the ports on the 'as produced' casting from AKD can produce some very satisfying results as shown in **Fig. 9.48**. This casting utilized a 1.48 intake and a 1.218 exhaust. Not unexpectedly the exhaust flow is really no different from what one

The power intake ports of the AKM head allows the use of twin side draft carbs. The large ports may favour the use of Dellorto carbs rather than Weber's in this instance.

would expect from a conventional head. The intake however is another story. Even with the limited development that I have had time for on this type of head the intake flow at over 150 cfm (at 25 inches H_2O) is markedly better than even a really good five-porter. However the intake ports are not without

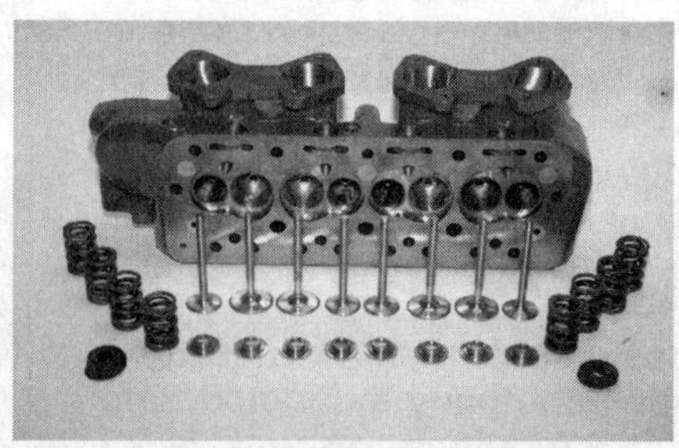

Here are the components that make up a complete AKM head. The valves, springs and retainers can be the same as those for a normal five-port head. In the case of this head the intakes used were 1.5 inch diameter.

terms of pressure wave tuning and results to date indicate that the shared centre port of a 7-port head actually seems to work slightly better than having a separate exhaust port per cylinder.

I personally have yet to build a complete motor using one of these heads but I have done quite a lot of air flow on the heads and have helped spec out some motors using this head. For the record Glyn Swift of SwiftTune spent a little time with the iron AKM 7-port head and found that with comparatively little effort he saw 160 horsepower from a 1400. However Glyn shares my opinion that such a head layout has far more potential than even these favourable numbers suggest. My own flow testing and computer modelling has indicated that as much as 180 horsepower is on if

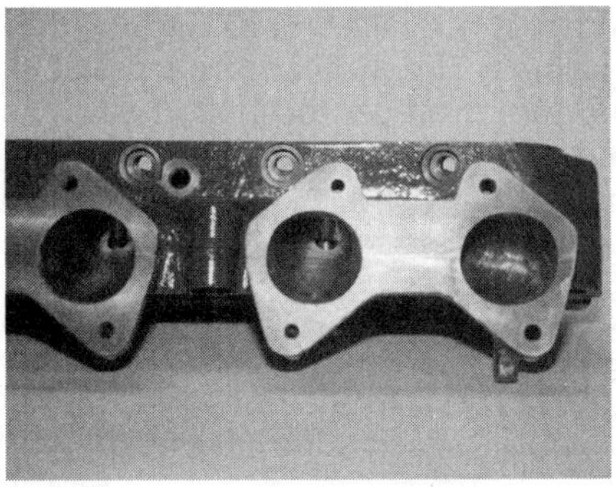

Because the pitch of the carb barrels does not match that of the intake valves the AKM head uses two distinctly different intake ports ports as is seen here. This means developing two intake port shapes for best flow and swirl rather than one.

their own set of problems. Because of length constraints and the need to match up to 45mm side draft carbs the ports are a little too big with a little too steep a taper angle on them. Because air speed will be slower in the main body of the port the likelyhood of fuel drop out will be increased. This means these heads are going to respond favourably *to finer fuel atomization than a five-porter*. My recommendation here is that if the mid range torque suffers or is no better than your 5 port set-up then change the carbs to Dellorto's as these atomize finer than Weber's in most instances.

If you are prepared to do some shifting around of the valve

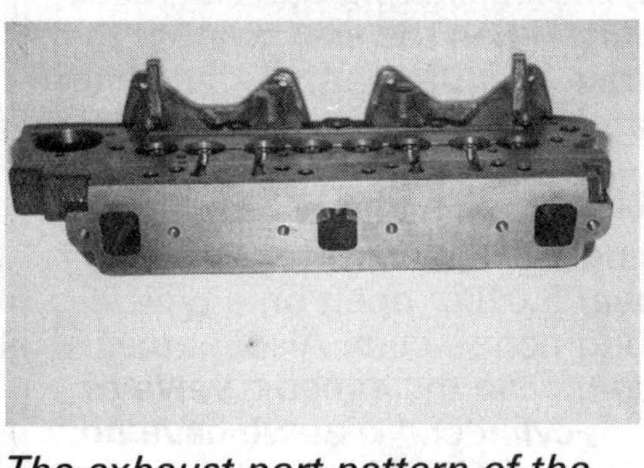

The exhaust port pattern of the AKM head is as per a five-port head but with the intakes around the other side that head looks distinctly bare from this view.

positions of the AKM seven-port head then some very useful intake flow increases can be had. By offsetting the intake valve 0.020 inches toward the exhaust and moving the exhaust 0.060 inches closer to the cylinder wall a 1.53-inch intake with a 1.218-inch exhaust valve can be accommodated. With a good 30 degree intake seat the low lift flow of this head can start to rival that of a basic four valves per cylinder type head and top end flow is up by about 5% over that achievable with the 1.48 valve.

Because of the potential of making a worthwhile power increase for little more than the cost of a head and an extra carb Glyn Swift decided to produce his own seven-port head but with a number of design differences that should make it more amenable to making power. The most significant of these is the fact that it will be poured in aluminium instead of cast iron. Secondly it will be cast by a process that will produce port shapes much nearer those that you would have if it had been fully (and intelligently) hand ported. With out-of-the-box air flow figures much higher than would normally be seen without the high cost of hand porting this head holds promise for a very good power increases in relation cost.

Exotic Five Port Mods

Probably the most common of modifications that can be termed exotic is the use of angled valves in a 5-port cylinder head. As far as I know, this type of modification was popularized by Richard Longman. What it entailed doing was angling both the intake and exhaust valve by 3° the idea being to get a better approach and departure from the back of the valves. It was expected that this would increase flow capability. Richard's initial

The guide holes were enlarged to take an over-sized guide rather than a sleeve and standard guides. This route was chosen because of superior heat conductivity.

testing on the angle valve head (and we're going back a few years now) showed that the concept worked. Yet at that time, Richard had not the flow test facilities to test his standard vertical valve 5-porters, or the inclined valve 5-porters. The only results he could go on were those from all the dyno testing which, ultimately, is the true test of whether a modification works or not. Whilst Richard was working industriously on these angle valve heads, I was industriously working on vertical valve cylinder heads because that's what 98% of my readers wanted to know how to do. As a result, many years went by before the D.V. workshop got around to producing an inclined valve cylinder head. By the time that opportunity came, my own flow test facilities had reached the stage of being nothing less than exotic. It was with this equipment at hand that I attacked

One of the last operations to be performed from this side of the head was to re-machine the spring platform so that they would in fact be 90 degrees to the spring axis.

this situation.

Surprisingly enough, I found that trying to make the inclined valve cylinder head flow anymore air than my then 'state-of-art' vertical valve cylinder head proved to be a difficult job. In fact, inclining the valves by 3° produced less flow in many parts of the lift curve, even after a couple of hundred hours work, than my best vertical valve heads. There were some advantages though to the inclined valve. Firstly, flow from about 150 to 300 thousandths was, if anything, marginally stronger than the vertical valve. Secondly, the process of inclining the valves produces a slightly wedge-shaped chamber, the wedge being thickest where the spark plug is. This appears to give a superior combustion chamber shape. If the basic course of events at Longman's shop followed a similar pattern to mine, it would appear that the angle valve cylinder head isn't worth any extra flow to speak of, but what it does do is produce a better combustion chamber shape. From this point of view, it seems to have a slightly superior power potential. Yet this leads one to the conclusion that it would be easier to angle mill the cylinder head to produce the slightly wedge shape chamber and leave everything else as it was, except maybe re-establish the manifold face to a square pattern with a deck block.

Another move that's been relatively popular, though many of the people who have done it have thought they were the first, is to 'angle port' the cylinder head. To do this, the original intake is discarded and a hole is bored down at a much steeper angle and a piece of suitably sized tube let in. The idea is to create a port having a down draft angle of about 25–30°. I personally have not tried one of these and most of the cylinder heads I've seen in various workshops throughout the UK have been done without recourse to a flow bench. Since most flow gains to be had are in the vicinity of the valve, so too are most of the porting mistakes made. I've often wondered just how much benefit the down draft port would have. However, one person who has had recourse to a flow bench is Dave Mountain and he, on one occasion, produced what he called a quick lash-up of a cylinder head where he simply bored down through the top of one of these cylinder heads, araldited some pipes in, ported the rest of the cylinder head to a known valve and stuck it onto a dyno mule motor. Because of its thrown-together concept, the cylinder head didn't run too long before it leaked water. However, Dave did say that the cylinder head did produce the best power figure he had ever seen from a Series 'A' engine. A statement like this coming from Dave Mountain has a lot of significance because Dave's dyno sees not only his motors, but all sorts of Series 'A' engines. The indications are that a down draft-ported-head has potential that needs exploring.

Drawing the Right Conclusions

At this point I need to sum up on what's been said on Series 'A' cylinder heads. What I'm going to deal with is the effects of porting and how to best utilize efficient flowing ports on a 5-port engine. At first it would appear obvious, if you've made a port flow better, that the horsepower must go up. Assuming all other criteria such as port energy, port velocity, turbulence, fuel mixing, etc. have in no way been adversely affected, this at first sight at least, seems a logical conclusion. In essence, there are situations (mostly with very long period race cams) where porting can be improved upon, yet engine output remains unchanged or even gets worse. To see how this can be so, let's look more clearly at some of the peculiarities of a 5-port engine. The port make-up consists of two intake ports and three exhaust ports, which **when a long period race style cam is used** (rather than to a shorter road-type cam), can all start to interact in negative ways. For instance, let's look at the situation concerning a pair of cylinders sharing one siamesed port. When the first cylinder to draw is at the bottom of its intake stroke, the intake valve of the second cylinder to draw (which will always be an end one with a single exhaust port) can be well over 0.0100" open on a typical long period cam. Also, at that point, can the exhaust valve of that cylinder. What we have now is a situation where if the end exhaust port and pipe combination is working well and it is scavenging right through the system, it can draw in a charge from the intake port, but in so doing it will, to an extent, rob the last part of the filling of the adjacent cylinder. In other words, the scavenging effect of the No 1 or No 4 exhaust port can scavenge a charge back out of the last part of the filling sequence of its adjacent cylinder (No 2 or No 3). Often, if a very strong scavenging effect is reduced the power can be improved. However, this is not the proper way of solving the problem. Backtracking for a moment we can see that using exhaust ports or intake ports that flow worse than we're able to get them is not really the way to go. Our primary objective is, of course, horsepower and this is achieved with good flowing ports. If a good flowing port causes a problem like this, the proper way to deal with it is to find a way to cure it without necessarily reducing the flow of the port. This can be done by timing the valve opening sequences so that they are more conducive to producing the best results from each individual cylinder; this aspect we will deal with in Chapter 11.

10 Exhaust Systems Part 1

Having plodded through to this point in the book, you will have noted that the theme is the quest for extra power, and for those who want it, extra economy. When studying various aspects relating to engine output, it is important not to equate more hp with more efficiency and resulting better economy. It doesn't always follow that way. Certain modifications, especially in the cam timing department, will produce more hp, but at the expense of part throttle fuel efficiency. However, there are certain modifications which do produce better power *and* improved economy, and the exhaust system is one area where this can be so.

Each time fuel prices increase, the number of so-called economy devices on the market seems to quadruple. Most of these 'economy devices' only provide economy for the vendors, by providing them with enough income to be able to afford the fuel that the device supposedly saves its purchaser. **However, there are many economy devices on the market which actually DO work, but they are not always recognized for what they are. Exhaust systems most certainly fall into this category.**

Before delving into just how much additional power is available through tuned exhaust systems, it would be wise to go through some basic theory and clarify why a good exhaust system will improve hp and economy.

Pumping Losses

Let us consider for a moment, the power and exhaust strokes of a four-stroke (four cycle) engine. The power developed by an engine depends on the average pressure developed in the cylinder during the power stroke. The higher this pressure is, the more hp the engine will develop. **The power stroke is the only stroke that produces energy; the remaining three strokes, i.e. intake, compression**

and exhaust, all absorb energy. The more they absorb the less there is left from the power stroke to actually drive the car along. The idea then, is to get the idle strokes to absorb the minimum amount of energy possible.

If you look at the three idle strokes, there are certain limitations with regard to minimising their power absorption. Consider the intake stroke first: a high flowing intake system will suffer minimal pressure drop across the valve, so the amount of vacuum existing in the cylinder during the intake stroke, will be minimal. Remember here that **vacuum on the top of the piston is like having pressure underneath the piston and pressure underneath the piston while it's moving down, represents a negative power stroke.** To cut down pumping losses, as they are called, on the intake stroke it's necessary to have an efficient high flow intake system. However, minimising such losses on the intake only works at full throttle, so no part throttle gains are made from this aspect.

Looking at the compression stroke in the same way, we run into a dilemma of sorts. The higher the CR is, the greater the energy required to compress the charge. On the other hand, the higher the compression, the higher the pressure produced on the power stroke. Fortunately the second aspect more than counteracts the first. If it didn't there would be no point in having high CRs, so as far as reducing losses on the compression stroke are concerned, there are really no further avenues open to us where overall improvements can be made. As the CR goes higher, the trade-off is always in our favour, *i.e.* more hp.

When considering the exhaust stroke, we find that it is an area where the biggest potential for improvement exists. Substantial improvements in efficiency are achieved by reducing back pressure to a minimum. **Back pressure is the product of an exhaust system which does not flow efficiently.** Compounding the problem is the fact that back pressure does not go up in proportion to the engine's gas throughflow capability which, of course, gets higher as the hp goes up. *Instead, back pressure rises more in proportion to the square of the engine's gas throughflow capability.* To gain an idea of just how quickly the back pressure can increase in a system, take a look at **Fig. 10.1**. The figures on this graph show what happens in a standard 1275 Mini GT exhaust system. You will note that as the rpm increase so the line indicating the back pressure gets steeper and steeper. At the top end of the rpm range, there is enough back pressure in the exhaust system to be robbing the engine of half-a-dozen hp. That's bad enough for a standard engine. If the engine had been modified to produce more hp, the back pressure produced by the standard system would go up much higher, and the percentage of power loss would be greater.

The particular exhaust system shown here is a middle-of-the-road type exhaust system as far as efficiency is concerned. There are many exhaust systems that produce higher back pressure figures than this. The

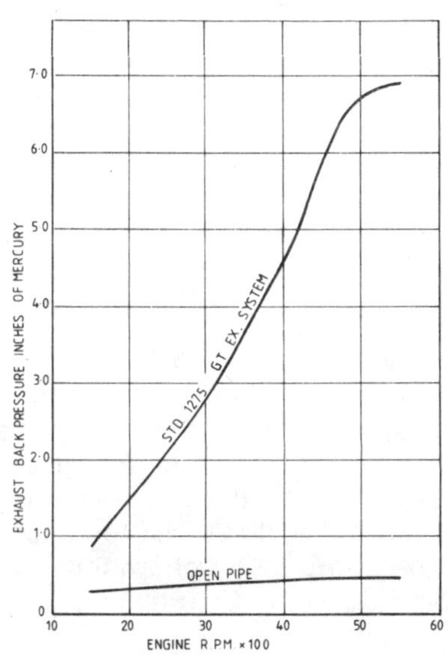

Fig. 10.1 Comparison of back pressure in a standard 1275 GT Mini exhaust system and an open pipe
The back pressure curve produced by an exhaust system is related to the engine's power potential. The greater the horsepower, the greater the back pressure and, in turn, the greater the power robbed from the engine. Once the engine has reached peak power, back pressure increases very little. The example shown here power-peaked at 5000rpm.

Exhaust back pressure and power losses

The amount of engine output is reduced by pumping losses *due solely to back pressure generated by insufficient exhaust system flow* can be easily calculated. The formula presented does not include losses due to reduced combustion chamber scavenging as occurs on a race engine. The bottom line is that the loss your engine will experience will be at least what the nearby formula predicts plus a certain amount more dependant on the amount of overlap the cam has. A word to the wise here – back pressure is death to a race style cam!

Power reduction due solely to back pressure induced pumping losses

$$HP\ Lost = \frac{BP \times Stroke \times Piston\ Area \times RPM}{792{,}000}$$

BP = back presure in PSI, Units in inches. Use piston area of all 4 cylinders.

power lost due to exhaust back pressure, *i.e.* pumping loss, is relatively easily calculated.

A typical exhaust system which has 5psi back pressure can, at the top end of the rpm range of an average 1275 engine, cause a power reduction of around 3–4hp. About 1 hp of this is due to the pumping loss and the rest is due to reduced breathing efficiency. This is brought about because any residual pressure in the combustion chamber means the piston must go part way down the bore before the cylinder pressure drops below that of the intake port and fresh charge starts to flow in. The lower the CR the more critical this is. With a 1275 engine having about an 8.5/1 CR the loss of intake charge with a 5 psi back pressure is equal to the loss of 50 to 60 cc. This adds about another 3 – 4 hp loss over that produced by the pumping loss.

Another point to note is that power developed on the power stroke will, due to back pressure, be lost on the exhaust stroke. **This means that fuel has been burnt merely to develop the power to expel the exhaust.** From this you can see that **exhaust back pressure spells a reduction in economy.** Of course, at part throttle the exhaust system doesn't have as much exhaust to handle and so the back pressure is less. On the other hand, the power for part throttle cruising is less, and so the percentage of inefficiency is not too dissimilar at part throttle and at full throttle. However, the curve of fuel efficiency versus flow capability of the exhaust system is quite complex, and as the engine's output gets lower, so the exhaust system is less critical in producing economy figures. For small engines of the type we are dealing with here, minimal exhaust back pressure is critical for economy, and there are worthwhile gains to be had by utilizing an efficient exhaust system, especially in the silencer/muffler area.

More Power Losses

Let's return to 'power only' considerations for the moment and consider in more detail losses due to reduced chamber scavenging and charge dilution. Right after the exhaust stroke comes the intake stroke. If the exhaust system has a substantial amount of back pressure, it will cause the TDC pressure in the combustion chamber to be higher than at atmospheric pressure. This means that **as the intake valve opens, so exhaust gas is more likely to go out through the intake valve, rather than a fresh charge entering the cylinder.** When the piston has travelled far enough down the bore to drop the cylinder pressure low enough for the intake charge to flow in, the first portion of the charge will be spent exhaust gases. Not until all the exhaust blown into the intake port has been drawn in, does the cylinder start to pull in a combustible charge. The piston could be a substantial way down the bore before starting to draw a fresh charge of fuel and air. From this you can see that back pressure has the effect of reducing the working capacity of the engine. In other words, instead of drawing in, say, 250cc of fresh charge into one cylinder of our 1000cc Series 'A' engine, the engine may only draw in 200cc of fresh charge, plus 50cc of old exhaust. As a result of the charge contamination, the engine loses quite a substantial amount of power. Just how much depends upon the cam timing, the CR and, of course, the back pressure the exhaust is encountering on the exhaust stroke.

Effects of Cam Timing

Let's assume an engine which has a cam with 'classical' timing in as much as the intake valve opens at TDC and closes at BDC. The exhaust valve opens at BDC and closes at TDC. In other words there is no overlap. Under these conditions it will be seen that there is the least chance of exhaust going out of the intake valve and contaminating the fresh charge. Having a cam of this timing does not guarantee that no contamination of the fresh charge will take place simply because the TDC pressure in the combustion chamber will be the same as existed in the exhaust system just prior to the exhaust valve closing. All that can be said with a cam of this timing is there would be the minimum amount of charge contamination. However, you can imagine that if the cam timing is extended, there will be a greater opportunity for the exhaust to go out the intake valve. The bigger the cam gets, the greater effect exhaust back pressure has on contaminating the new charge. **This means that on an engine with a longer period cam, the need for an efficient exhaust system becomes greater.**

So far we have talked about the negative effects of back pressure and how our efforts are best directed towards minimizing it but let's take things one step further. What if the TDC pressure in the combustion chamber can be reduced to something less than atmospheric (*i.e.* partial vacuum) during the valve overlap period. Consider the fact that the motion of the piston at TDC is very slow, and it spends a lot of time, relatively speaking that is, moving very little distance. If, either because of inertial extraction or negative shock wave generation, the TDC pressure in the combustion chamber during the overlap period can be reduced to less than atmospheric the intake charge will actually start flowing in before the piston starts the induction stroke for real. Now this has a very important implication. Let's assume we have an engine with nominally around 11/1 CR, and that the total combustion chamber volume is 25cc. Let's also assume that due to a very efficient intake system, the piston, when it travels down the bore, will pull in a full 250cc. If the induction stroke can be started so as to pull in a charge which will eventually occupy the combustion chamber area, the cylinder could, if sufficient depression existed, pull in an additional 25cc. In other words, a 250cc cylinder could pull in 275cc of fresh charge to produce what would normally be termed as a supercharged effect to the tune of 10%. Such an engine would have a volumetric efficiency of 110%, *not an impractical figure for a highly developed pushrod engine*. Normally aspirated Formula 1 engines regularly exceed 125% volumetric efficiency. However a Series 'A' engine, though it may be highly developed, is certainly not sophisticated, and indeed most tuners are hard pushed to exceed some 90% volumetric efficiency, but this doesn't matter. The goal being sought is just the same. In other words, the exhaust system needs to extract the residual charge of burnt gas contained in the combustion chamber to as great a degree as possible so as to start sooner induction of the new charge. Again **it can be seen that the more overlap there is on the cam, the more likely this can happen if the exhaust system is tuned to the engine. One of the criteria that is desirable for such an exhaust system is, of course, zero back pressure.**

Traditionally we tend to view exhaust systems as two distinct parts, namely the exhaust manifold (header) and the silencing (muffling) system itself. From the practical point of view, this isn't a bad idea, as these two entities are often sold separately and most do-it-yourself engine modifiers and tuners will buy one or both, depending on the size of their budget. If the budget is limited, which should be bought first? That question can only be answered accurately after analysing separately the efficiencies of each component for each engine considered and each stage of tune that may be done. This is not really practical so it's necessary instead to lay down some basic ground rules as a guide. **By and large, as a first step, more is to be gained by changing to an efficient silencing system, than is gained by an exhaust manifold change.** This is a general rule to which there are exceptions. For instance, as far as a production unit goes, the original (but not current, as of 1980 on) BL produced Cooper 'S' silencer was quite efficient, there being only the odd 1 or 2hp to be had by changing it to a super-high efficient system. On the other hand, there are 3 or 4hp to be had by changing the exhaust manifold. Even so, it's worth remembering that changing an exhaust manifold is a lengthy job but changing the silencer is quick so really it's a matter of time and effort involved as well as cost. In spite of the fact that changing the silencer is the first thing to do in terms of modifying the exhaust system, I am going to deal with it second, simply because the effect the silencer has on the system depends on the type of exhaust manifold installed. Apart from that, we need to equate the exhaust manifold with the state of modification of the engine and there is a little more theory to cover before I get down to actually demonstrating the gains that can be made.

Sizing

The proportioning of exhaust systems is a highly developed science, and one that is substantially misunderstood. Although it will be necessary to consider some theory behind exhaust pipe design, it will only be done when and where it's directly applicable.

First, let's look at exhaust manifold pipe diameter. In reducing the pumping losses during the exhaust stroke, it's obvious that a big exhaust pipe diameter will flow better than a small one. At first sight it would seem the obvious choice is an exhaust pipe as big as practically possible. **Unfortunately, in practice the 'big' rule (if it ever existed in the first place) does not work. Large diameter exhaust pipes do not necessarily produce positive results** because this simple view of the exhaust system does not take into account all the exhaust system 'phenomena' which occur. The reason why big exhaust pipes don't necessarily work is related to 'gas speed'. The common theory is that the gas speed within the pipe must be sufficiently high to provide some extraction effect. In other words high speed gas travelling out through the exhaust port should, during the overlap period, pull the rest of the exhaust gases with it. This theory, though it seems to work when applied in practice, is still a rather simple and less than accurate view of what actually happens. I believe that the gas speed itself is not of prime importance. What seems of equal importance, is whether or not the exhaust has time to reverse flow within the system. It's easy to make the assumption that the exhaust travels only in one direction in an exhaust system. At low rpm or light throttle openings nothing could be further from the truth, especially if a big cam is being used. By using an exhaust system with small pipe diameters, the exhaust gas velocity is increased. As a result of this increase in speed, it takes longer for the exhaust to slow down and reverse flow at the end of the exhaust stroke. If an engine has a long period cam, it can be made to come 'up on the cam' at lower rpm by the

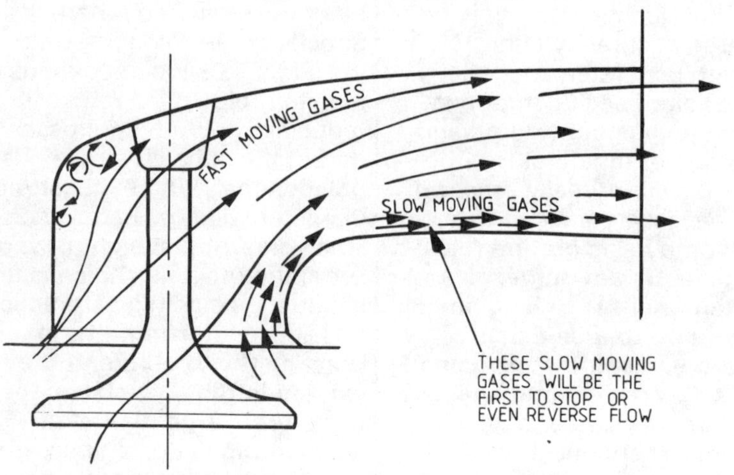

Fig. 10.2 Gas flow relative speed in an exhaust port

fitting of a smaller bore exhaust manifold. **Analysing exhaust phenomena a little further, we find the speed of the exhaust gases is not uniform across the port. Most of the exhaust flows along the top of the port** as shown in **Fig. 10.2.** Because the exhaust gas at the bottom of the port is travelling so much slower, it will stop, then reverse-flow up to higher revs than will the faster moving exhaust gas at the top of the port. In fact, the situation often occurs where the exhaust is moving in the correct direction on one side of the port, but has suffered a flow reversal back into the cylinder on the other. These 'back-flowing' exhaust gases contaminate the new intake charge during the overlap period. As mentioned small diameter exhaust pipes can delay flow reversal down to a much lower point in the rpm range. At the same time, however, they can restrict flow at high rpm, thus increasing the back pressure felt by the piston on the exhaust stroke. This will cost the engine top end hp. If the exhaust system is to be tuned by pipe diameter selection alone, and at this stage we are assuming the lengths are right, then we find that for a given rpm and engine capacity, **there is a certain pipe diameter best suited for producing maximum hp. Such a pipe diameter may not produce maximum torque and it almost certainly won't be the best one for bringing the engine 'on the cam' at the lowest rpm possible.** What it will be is a pipe diameter which gives just the right inertial effect for the best scavenging of the combustion chamber at and around peak power rpm.

If reverse flow could be reduced by some means other than reducing pipe diameter, would it have the same effect? The answer to this is a conservative 'yes'. Dyno tests have indicated that if the exhaust port could be made to flow well in the right direction and badly in the reverse direction, the engine will have a tendency to come up on the cam sooner. In fact the camminess of an engine is almost entirely dictated by the passage of back-flowing exhaust into the intake system. **If back-flowing could be effectively cancelled out, then even the most radical race cams could run at relatively low rpm without any drivability problems.** In fact theory suggests such cams would be just as docile as typical road cams. The crux of the matter is that the gas flow must be maintained in the correct direction throughout the entire engine. The principal engine feature that tends to prevent attainment of this goal is the fact the exhaust gases want to flow out of any valve that may be open.

The type of thinking which connects drivability and flow reversal is relatively new. For this reason you will see for some time to come, exhaust systems which do not take into account this factor other than by pipe sizing. However, there are ways of reducing exhaust flow reversals. In addition to helping to reduce an engine's 'camminess' it can sometimes add a little more to top end hp.

The simplest way to achieve good forward but bad reverse flow, is the mis-matched exhaust pipe. **For years, the matched exhaust pipe has been presented as THE route to maximum flow capability.** Forget it; such thinking is out of date. Flow bench tests show that a stepped port/pipe combination equals or betters the flow capabilities of a matched one, as long as the step is the right way and the mis-match between exhaust pipe and exhaust port is optimized. The Series 'A' cylinder is not as fussy in this area as some cylinder heads, and fitting the square exhaust port to a round exhaust pipe allows flow in the forward direction to be as unimpeded as it would be if the sections were matched. On the other hand, when the exhaust tries to reverse flow, it meets the step between the pipe and the exhaust port. Result: less reverse flow than would otherwise be the case.

To optimize this type of system it's a good idea to position the exhaust manifold so that the exhaust port is towards the top of the bore of the exhaust pipe. A central location is fine for the middle port, but the outer ports tend to be better if exhaust pipes are towards the top and biased outwards. **Fig. 10.3** indicates roughly what is needed.

There are alternative ways of

solving this problem and Janspeed's C.V. (Controlled Vortex) exhaust system seems to do more or less the same thing. **Fig. 10.4** illustrates how this system works. Basically the system seeks to control the slow moving vortex of exhaust on the bottom of the port by allowing easy exit but not such an easy re-entry at the lower gas speeds prevailing during the overlap period.

The discussion of ways of reducing back flow may have given the impression that the exhaust pipe diameter no longer matters. It does. It's merely that using some means to control

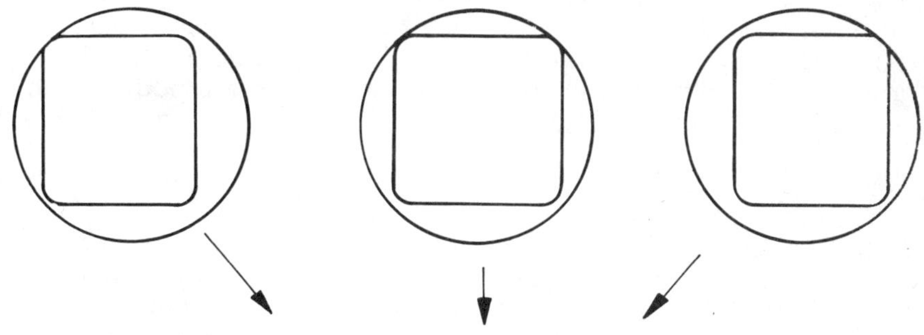

Fig. 10.3 Exhaust port mismatch

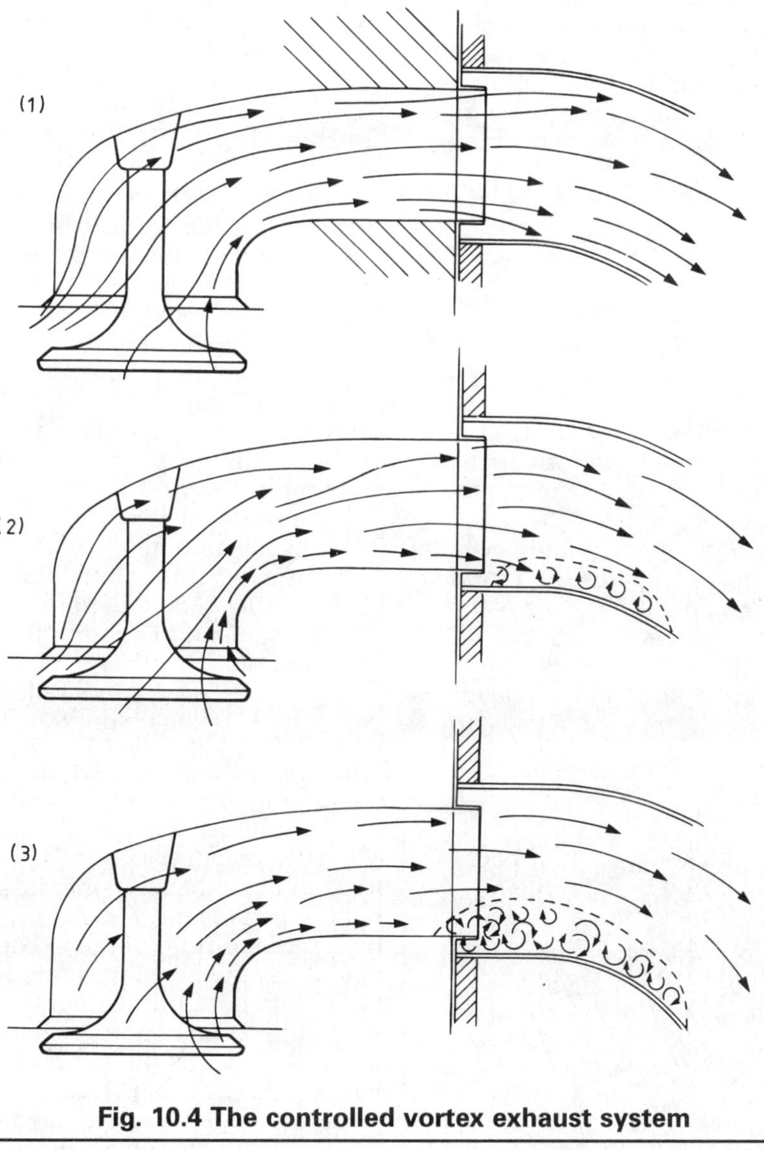

At high valve lift (1) when piston speed is high (mid-stroke) or when rpm are sufficiently high, exhaust gases will all be moving in the correct direction, i.e. out of the cylinder. Even though the direction is uniformly correct, the gas speeds vary considerably from top to bottom, the fastest gases being those at the top. As the valve starts to close (2) so the gas speeds in the port begin to drop, due in part to the slower movement of the piston up the bore. The slower moving gases along the bottom of the port which have a tumbling non-orientated flow, will try to reverse flow back into the cylinder if the pressure gradient (i.e. from high to low) is right. In a conventional exhaust port, the ability of the exhaust gases to backflow into the chamber is virtually as great, if not in some circumstances greater than, it is to flow out. The CV type of exhaust with its stub extending into an enlarged exhaust pipe provides an easy flow out and in many cases, such as the Series 'A' actually enhances outward flow, but greatly reduces flow in the reverse direction. Once the exhaust passes the stub and enlarged section, the exhaust pipe itself should reduce in size to about that of a conventional exhaust pipe, though in some cases, such as big displacement engines, up to 1/8in larger can be beneficial.

Fig. 10.4 The controlled vortex exhaust system

reverse flow allows the pipe diameter to be selected for maximum hp, and whatever means is selected for cutting down the reverse flow will go towards improving the engine's low and mid-speed range torque. Apart from this it will also, as stated before, pull the motor up on the cam sooner.

Another way of looking at the situation is that a flow selective system such as a stepped pipe or CV system will allow the use of a more radical cam profile in an effort to obtain more top end power.

Pipe Lengths

There are three main reasons why pipe lengths need to be considered. Let's take the extreme case first, as is typified by most standard exhaust manifolds. If the exhaust pipe lengths are too short, it's possible for the last phase of the exhausting of one cylinder to interfere with another cylinder just starting its exhaust event. Depending upon the shape of the manifold, the new slug of gas coming out of the one exhaust port, could actually run into the tail end of a slug of gas which may well be going in the opposite direction. Lengthening the exhaust pipe to the first junction and shaping the pipes right at the junction will, to a large extent, eliminate this, even if the back pressure isn't changed appreciably. Assuming minimal back pressure, adjustment of exhaust pipe lengths can enhance the extraction of residuals from the combustion chamber during the overlap period. This is done in two ways; inertial tuning of the exhaust system and pressure wave tuning. Inertial tuning is the easier to understand and has been touched upon briefly earlier on in this chapter. To reiterate, if exhaust is moving at high speed through a free flow, unrestrictive exhaust pipe, the contained slug of gas will have a considerable amount of momentum. If the exhaust is not dumped into the atmosphere too soon, that slug of gas will tend to cause a *slight* depression to exist behind it. If the depression occurs at the valve during the overlap period at TDC, it will start the intake charge flowing into the cylinder sooner, possibly even before the piston has started down the bore. If the pipe is too short, the slug of gas will expend its energy into open atmosphere and little or no inertial extraction will take place. Any inertial extraction achieved depends largely on the speed of the exhaust in the system. If anything impedes the exhaust flow it will cut gas speed and inertial extraction will suffer as a result. The conclusion one must reach here is that apart from correct sizing, a good free-flowing exhaust by virtue of an efficient shape, is important.

Pressure Wave Tuning

Here we get to the heart of a functional high performance exhaust system. Since the last update of this book in 1988 I have been able to study cylinder and port pressures on a running race engine. It has, to say the least, been an eye-opener and has significantly altered the degree of importance I have placed on the exhaust system. The most significant factor that stands out from these tests is that I had underestimated both the magnitude and the rpm range over which it was possible to make an exhaust system work. This is significant to a five port 'A' Series engine because it has both positive and negative implications. The first real eye-opener was that *the suction caused by an effective exhaust system scavenging the combustion chamber was significantly greater than the suction caused by the piston going down the bore on the intake stroke*. The second was that if the pipe's length, and just as importantly the diameter, were even just close to optimal the benefit of the system could cover in excess of 4000rpm rather than the 4-500 I had previously supposed. Now if that piques your interest let's get down to some basic theory and see how to best apply what has been learned to an 'A' Series engine.

The principle of pressure wave tuning can best be understood by considering just one cylinder and one exhaust pipe. When the exhaust valve opens towards the end of the power stroke, depending on the engine's specification, the gas pressure at the time of valve opening can be anything from 50 to about 70psi. The release of these high pressure gases produces a positive pressure wave, or as it's often called, a 'shock wave'. This wave passes down the exhaust pipe at the local speed of sound. I say 'local' because in high temperature gases, the speed of sound through the gas is substantially higher than in gases at normal atmospheric temperatures. At the temperatures existing in a typical exhaust system, the pressure wave travels through the gases at some 1300–1500 feet per second. Upon reaching the end of the exhaust pipe, it undergoes a 'reflection'. This is not a reflection in its normal sense, but the process exactly simulates a real reflection, so it's valid to call it that. In the process of being reflected back down the pipe, the pressure wave is changed from positive to negative. What now exists is a low pressure wave substantially below atmospheric pressure travelling back toward the exhaust valve. If, upon reaching it, the exhaust valve is (a) still open and (b) in the overlap period, the negative pressure will cause not only the combustion chamber residuals to be drawn out but also cause a start to the intake process even while the piston is still on its way to TDC. It can readily be seen that correct timing of the pressure wave's return is dependant on three factors, namely engine rpm and exhaust pipe length and cam event timing. The time in which it takes for the shock wave to travel down the exhaust pipe and back is going to be almost a fixed quantity because its speed is dependant on temperature and pressure only. The engine rpm is something which continually varies as the car is driven up and down through the gears. The exhaust pipe length then, is the only 'adjustable' avenue open to us. By choosing the appropriate exhaust pipe length, it is possible to pressure wave tune the engine in the middle of the rpm range most useful to us. The question is, what length is best? That depends upon a number of factors such as the cam timing, the rpm range and the type of intake and exhaust used, etc. Because pressure wave tuning requires a 'right' length, many people seem to become over concerned with precise pipe length equalization, a worry which is, in general, misplaced. Ask yourself why a certain pipe length is needed in the first place. The length being sought is one that causes the exhaust system to resonate at a certain frequency to increase hp over a corresponding rpm band. The rpm band within which a particular length of a single otherwise unconnected pipe works is narrow.

However by joining pipes from a number of cylinders together to generate second and even third sources of reflections the effect of a tuned exhaust can be had *over in excess of 4000 rpm*. If this band is not wide enough then the effect can be widened a little more by having slightly different primary pipe lengths. This may be something to consider if you have the luxury of an 8-port head but consider this – for the more usual three exhaust ports the 'A' Series has the popular LCB manifold which does just that!

With the wide gear ratio spacing that most production vehicles have in relation to the power band that the exhaust works over we still find that a little more is needed to extend power both up and down the rpm range. These areas of the rpm range are best filled in by whatever techniques are available in other tuning areas such as the intake system and the cylinder head. Just to remind you of the possibilities here we have induction length tuning, long is good for the street, and high flow at low valve lift head mods to consider. If the engine is mated to a close ratio gearbox then the working range of the exhaust more than covers what is needed.

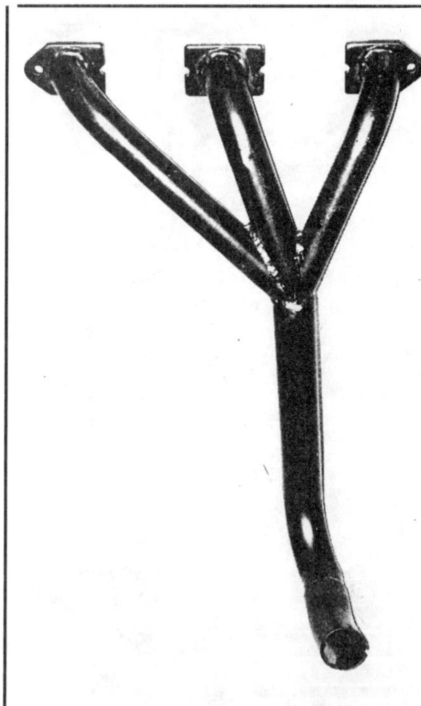

Although it doesn't look impressive, the Cooper 3-into-1 exhaust system, or its aftermarket replacement as shown here, does produce good results.

In Practice . . .

Okay, so much for theory, now let's see how these exhaust system techniques apply to the Series 'A' engine, as far pipe diameters, lengths and configurations go. Basically a number of exhaust manifold pipe diameters are available and, not surprisingly, **over the years, these pipe diameters have been fairly well optimized for various power specification of Series 'A' engines.** Since the pipe sizes used for a Series 'A' exhaust manifold aren't all the same, various systems are commonly known as 'small', 'medium' and 'large'. As far as pipe lengths are concerned, there are two distinct styles of high performance exhaust manifolds, namely the LCB system and the 3-into-1 system. The LCB, known as such because of the Long Centre Branch, has the two outer cylinders connected in the same sort of style that a normal four cylinder engine having four ports would have if it had a 4-into-2-into-1 system. In other words it

I have spent a lot of time on the dyno testing exhaust systems and each test session reinforces my opinion that there is always more to exhaust systems than meets the eye.

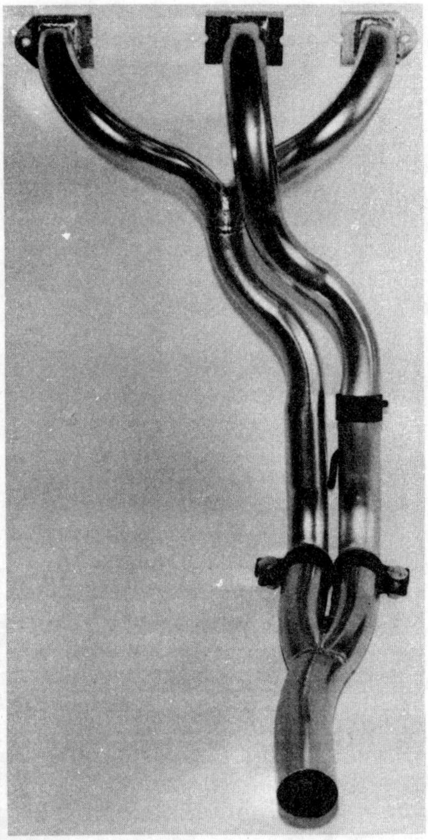

Shown here is what can best be described as the 'classic' LCB system. This system has been popular for over a quarter of a century. The example shown here, made by Maniflow, is a big bore item for engines of 100hp+.

has a short primary running into a secondary pipe. With this system the tuned length on the primary pipes is double the length of each primary pipe. In other words, the system it works on is that the shock wave produced at one cylinder is used to extract exhaust from the other and is sometimes known as an interference working system.

The other system commonly used is the long, race-style, 3-into-1 system. This is the Series 'A''s equivalent to a 4-into-1 system commonly used on many four cylinder, four exhaust port race engines. The Series 'A' engine of course only having 3 exhaust pipes because the two centre cylinders share the same exhaust port, a phenomena not common to many other four cylinder engines.

Exhaust Manifold Power Gains

The most common exhaust manifold used on the Series 'A' engine is, of course, the log-like 3-into-1. This is the manifold usually found attached to the intake manifold. Usually, when an intake manifold change is made, an exhaust manifold change happens right along with it. Because of this, there seems to have been very few tests involving an exhaust manifold change only. As a result, few people seem to know just how much extra power a relatively standard motor will produce when the exhaust manifold is changed. To put that right, a near standard test engine was set up with the intake manifold separated from the exhaust so exhaust manifold changes could be made without interfering with

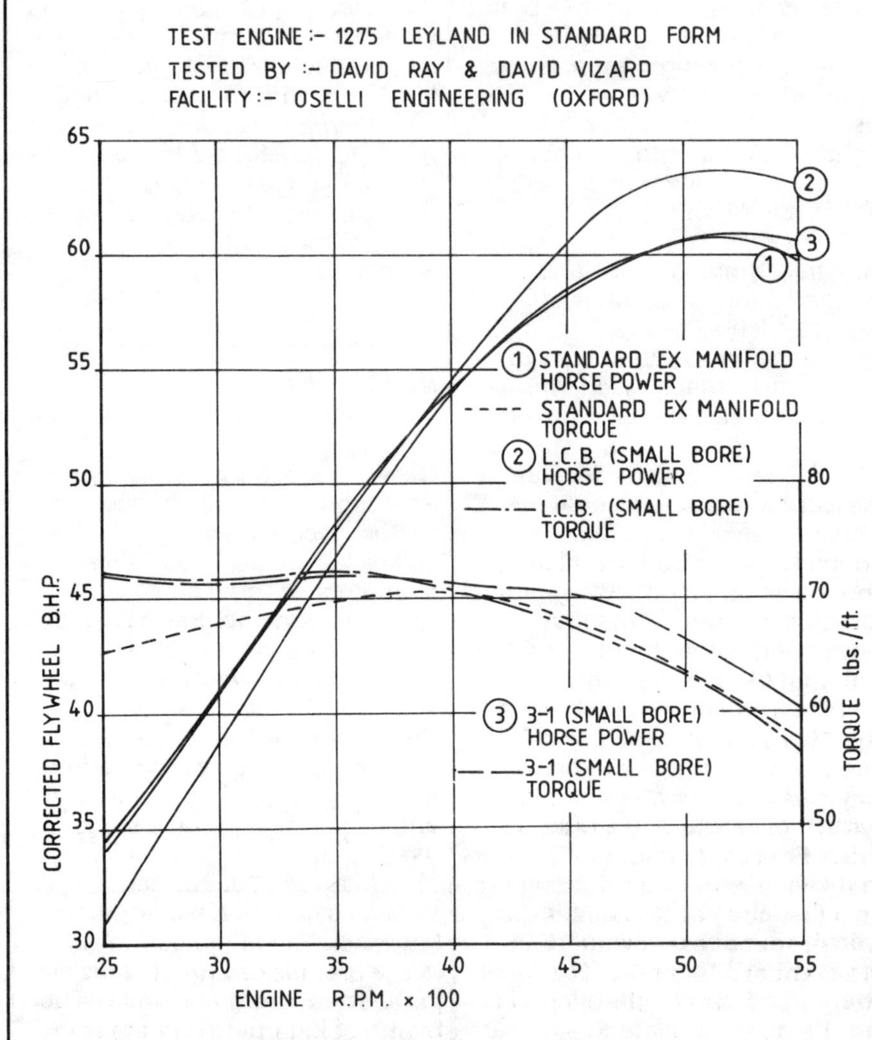

Fig. 10.5 Comparison of manifolds

Used in this test were a standard cast iron very short 3-1 manifold, a small bore long centre branch (LC8) manifold, and a small bore 3-1 manifold

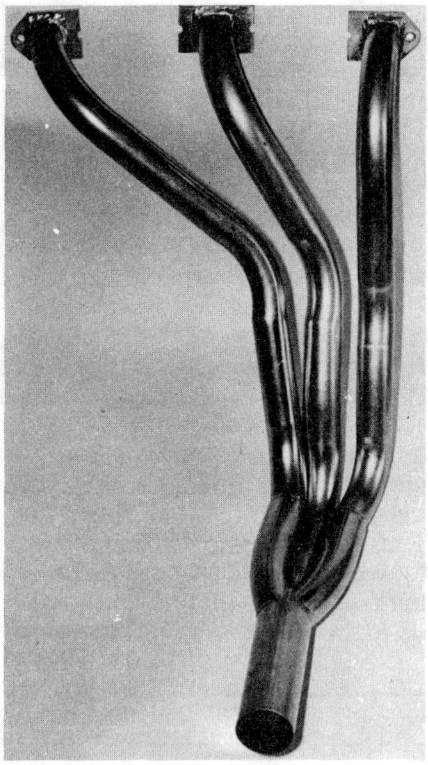

This exhaust system is a race 3-into-1 system, intended for engines with very long period race cams. This type of system was pioneered by Richard Longman in the early '70s.

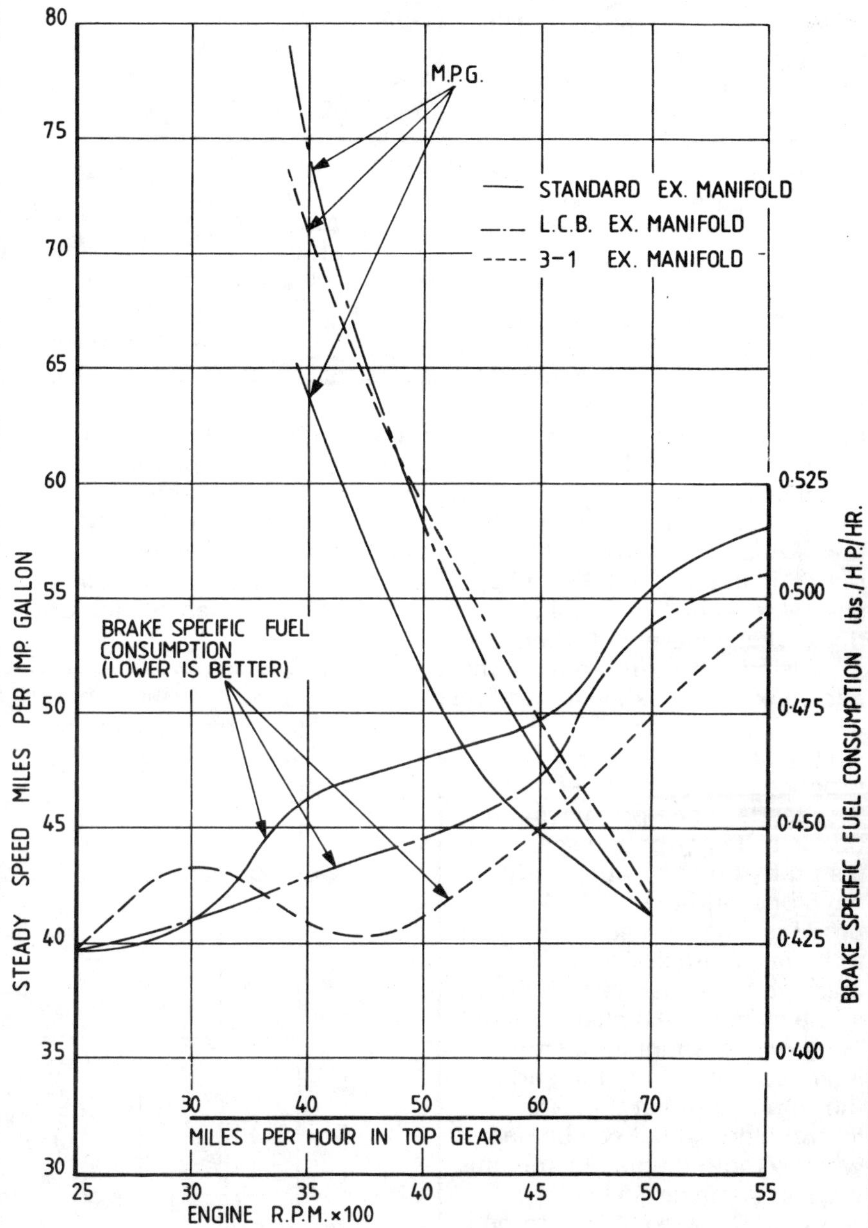

Fig. 10.6 Comparison of fuel consumption using various manifolds
The manifolds used in the test are those enumerated in the previous illustration.

the intake system. On the test 1275cc engine the two most popular alternative types of configuration were tried. **Fig. 10.5** shows the results of tests comparing the standard manifold with an LCB and a 3-into-1 system. The curves in the graph **Fig. 10.5** tells the story, but not one that presents a true picture as far as the advantage of one system over another. **Based on these figures, one would assume that the 3-into-1 manifold would be a last choice, since all it did was produce less torque than the standard manifold.** Before going any further I should point out the need to relate these figures to the state of tune of the engine. Whenever a short period cam is used, the engine is less sensitive to tuned lengths and long tuned lengths at that, than would otherwise be the case.

Obviously the lengths of the 3-into-1 are not paying off with the particular valve timing delivered by the standard cam in the rpm range used. **On the other hand, the LCB system did show a gain in top end power. From 3700 or so rpm upwards it produced better than standard output.** Probably these gains are much less than many might expect or indeed hope for. At present it would seem that a clear-cut victory must go to the LCB over the other types of manifold. However, the graph **Fig. 10.5** only shows power and torque gains or losses as the case may be. If a motor is being built for road use, fuel economy may have a high priority. An investigation of the fuel economy aspect, revealed a significantly different story. The LCB shows better economy than the standard manifold at low speed under part throttle conditions, but by the time the engine reaches the sort of power output required to propel a car along the road at some 70mph, it's a shade worse than that produced with the standard exhaust manifold. On the other hand, under full power conditions it's slightly better over most of the range. The 3-into-1 system doesn't have quite the low-speed economy of the LCB but it maintains an improvement over a much wider operating range than the LCB. Sorting through the numbers in the graph **Fig. 10.6**, should reveal which manifold does what, where and when.

Using all this information the choice of which exhaust manifold to use is somewhat difficult, simply because the LCB produces the best power results, the standard one produces the best results per pound spent, *i.e.* it's pretty good if you already have one, and the 3-into-1 produces the best economy. Also to be fair, there are some improvements that can be made in the standard exhaust manifold. They only account for a small increase in hp at the top end of the rpm range, but they are improvements that can be had with no more than a few minutes' work with the grinder.

First, it's a good idea to open up the ports in the exhaust manifold to the size of the gasket. Secondly, make sure there are no abrupt steps at the outlet of the exhaust manifold. The standard exhaust manifold outlet can also usefully withstand a little enlarging, but it shouldn't be taken too far otherwise it will crack.

With the introduction of the Metro, Leyland themselves brought out what can best be described as a cast iron LCB. On all 1275 Metros an exhaust manifold is used which has the pipe lengths of a typical LCB. If you refer back to the intake manifold chapter, there is a photograph of a cast iron Metro intake manifold attached to its exhaust manifold. From this you will see that the centre pipe goes to one outlet and that the two end cylinders are joined and are then routed to another outlet. These two outlets are connected to two pipes which pass down the back of the engine and join up just under the floor. **Changing**

Mirror images

If you look at the photographs of Mini LCB systems, which are mainly Janspeed or Maniflow systems, you will notice that the pipes from the outer two cylinders are mirror images of each other, and they join at what is virtually the dead centre of the engine. The centre pipe, as viewed from the rear, then moves off to the right. Some exhaust systems are made up by having one continuous pipe from one outer cylinder with the other outer cylinder joining it. Usually these join slightly off-centre, and, in many installations, such exhaust manifolds can foul the mechanical pump. If you want to be sure of avoiding this problem, look for exhaust systems that have two end pipes mirror imaged.

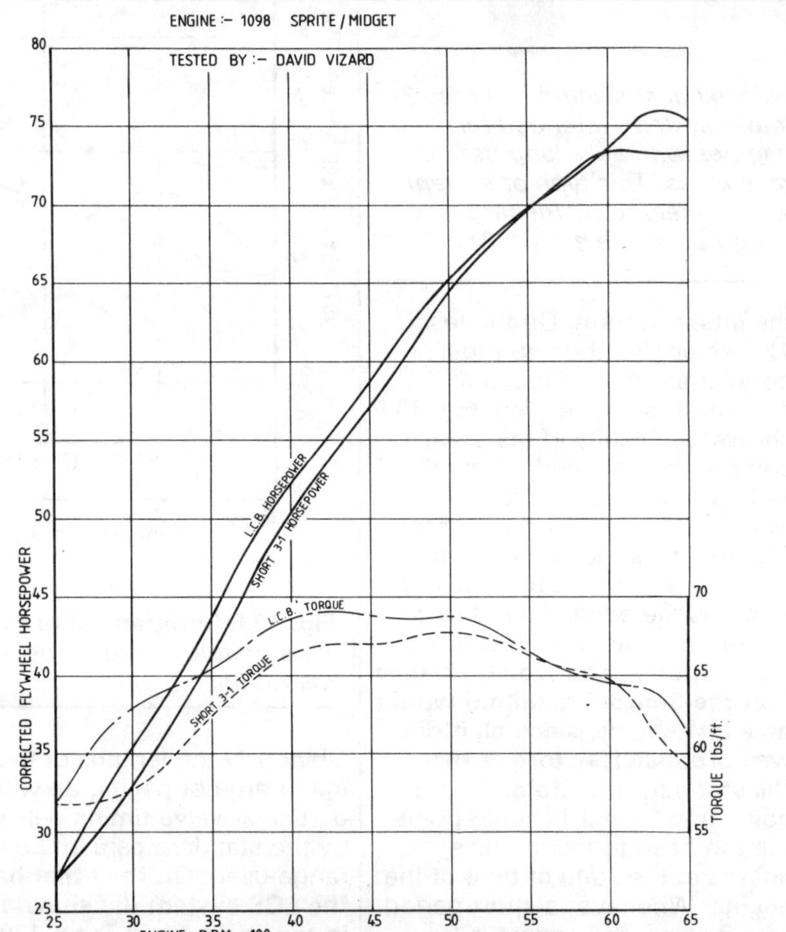

Fig. 10.7 Comparison of short Cooper 'S' 3-1 and LCB exhaust manifolds

this exhaust manifold for an LCB is almost a waste of time. The performance differential achieved by using an LCB of equivalent size is minimal. Porting the cast iron LCB manifold for higher power is practicable though very involved.

Although the cast LCB is a good exhaust manifold, it's not the only effective Leyland exhaust manifold. Cooper 'S' engines used a short, 3-into-1 design which, for engines in a moderate state of tune, works well. Should you be restoring a Cooper 'S' and the engine compartment must look totally authentic, you can rest assured that only a minimal amount of hp is being sacrificed in comparison with, say, an LCB. Certainly, the LCB has an advantage, but it isn't that much until we consider engines over the 80hp mark. **Fig. 10.7** demonstrates just the sort of differences you can expect to see between the Cooper 3-into-1 and an LCB. The particular engine concerned was a 1098cc engine with a single 1½ inch SU, a cleaned up head rather than a fully ported head, and a C-AEG-643 cam. That's Leyland's 388 thousandths lift by 288° period cam. In essence it's a hot, road-going profile. From the curves in **Fig. 10.7**, it can be seen that the power figures with the short, 3-into-1 system were generally within 2 to 2½hp of the LCB system.

Exhaust Systems Part 2

Systems For Hot Motors

One often asked question concerns the selection of the best exhaust manifold for a relatively highly tuned motor. To be honest, it's not possible to say which is best without quoting operational circumstances with some degree of precision. Certain parts combinations will favour one type of manifold whereas other parts combinations will favour another. As a general rule the longer the cam period becomes, the more likely it is that the 3-into-1 system will produce more hp than the LCB. Even so, with cams in the 300° range, there isn't a great deal of difference between them as far as top end hp goes, though the 3-into-1 tends to show better bottom end hp. A back to back test on a 1275 engine utilizing a single HIF SU and a C-AEG-643 cam showed that the two exhaust manifolds could produce positive advantages in different areas. **Fig. 10.8** demonstrates just how much difference there is between

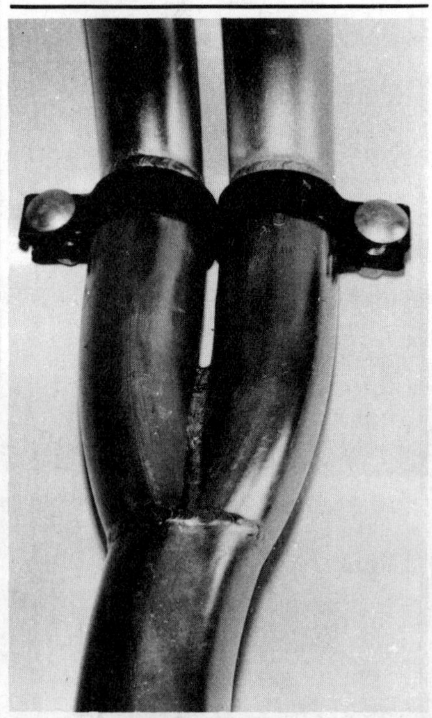

There is hp to be gained or lost at the junction between the down pipes and the tail pipes. It is important to have a smooth transition from one to the other at as narrow an angle as is feasibly possible. This Maniflow manifold shows what the join should look like for an LCB. For a ...

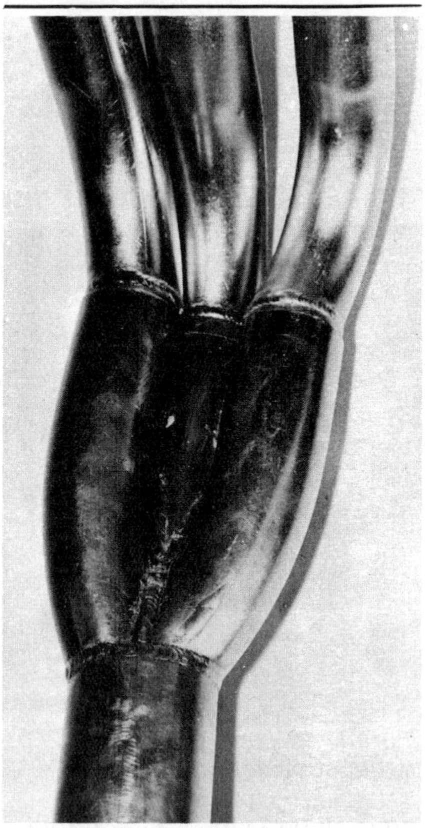

... 3-into-1 system it should look like this. Too abrupt a joint at this point can cost a couple of hp.

bottom end power given by the LCB and the 3-into-1. From 3000 - 6000 the LCB has a slight advantage. From 6000 - 7500 it's pretty even between the two.

Of course a test like the one shown in **Fig. 10.8** really begs for a test to be done between a conventional exhaust system and a Controlled Vortex type system just to see if there is any worthwhile difference between the two. **Fig. 10.9** shows the power potential difference between a conventional LCB and an LCB in CV form. Notice that up to 3000 rpm there is a dramatic increase in torque, thus making this relatively big cam (C-AEG-643) motor pull from low rpm just as if it had an everyday street cam in it. From 3000 up, the advantage changes from one to the other, all the way up the rpm range.

From these figures we can conclude that the CV exhaust system's strong point is probably not seen in a back-to-back test like this. CV exhaust system will, however, allow the engine builder to use a slightly bigger cam before the bottom end torque drops to the point of being insufficient. It can be said for instance, that a certain amount of torque is required at 2000 rpm to make the engine concerned street drivable. With a CV exhaust system, a 300° race cam could have been used whereas with a conventional exhaust system, a 288° cam will be about as big as can be used and still hold the torque at 2000 rpm. A bigger cam would allow the engine to produce more top end hp and, of course, rev higher. It's only when we consider combinations like this that the CV system looks like it has any real advantage over a conventional system. Still, so far we have only dealt with the CV version of the LCB. Let's now look at a CV version of the

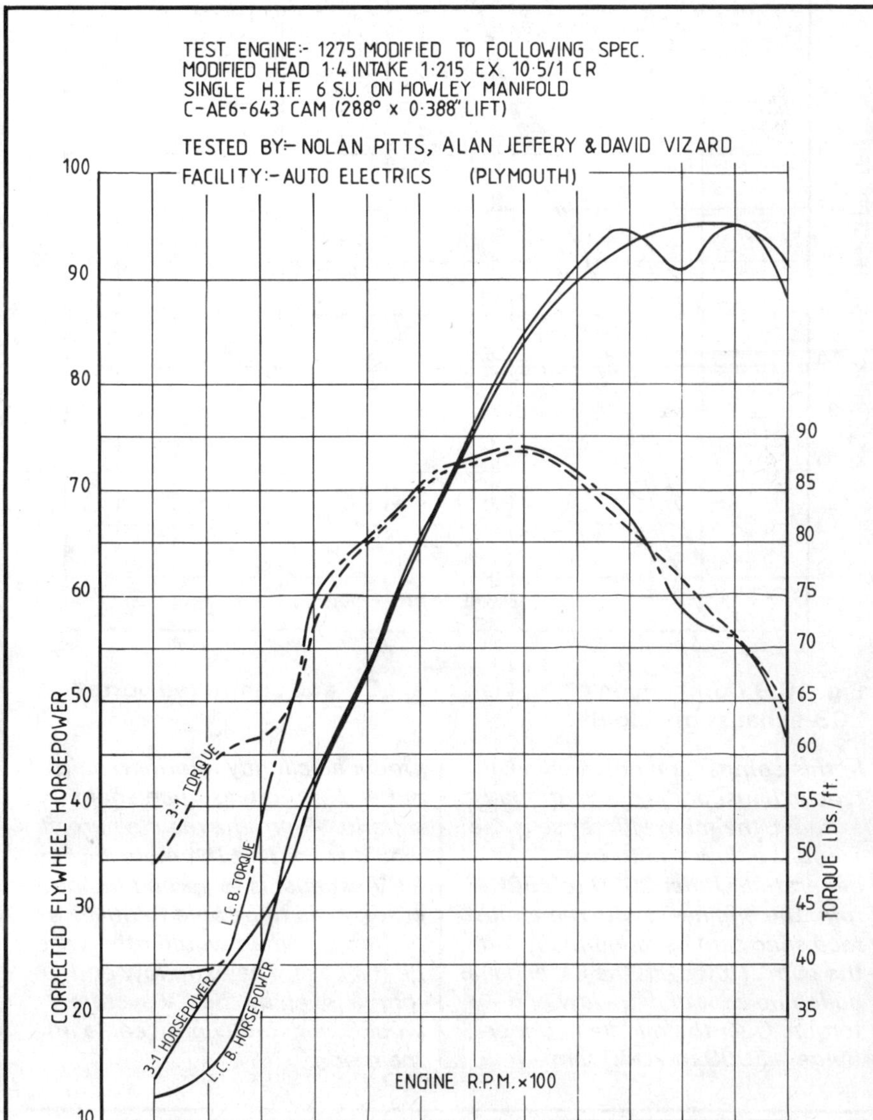

Fig. 10.8 Comparison of LCB and 3-1 exhaust manifolds

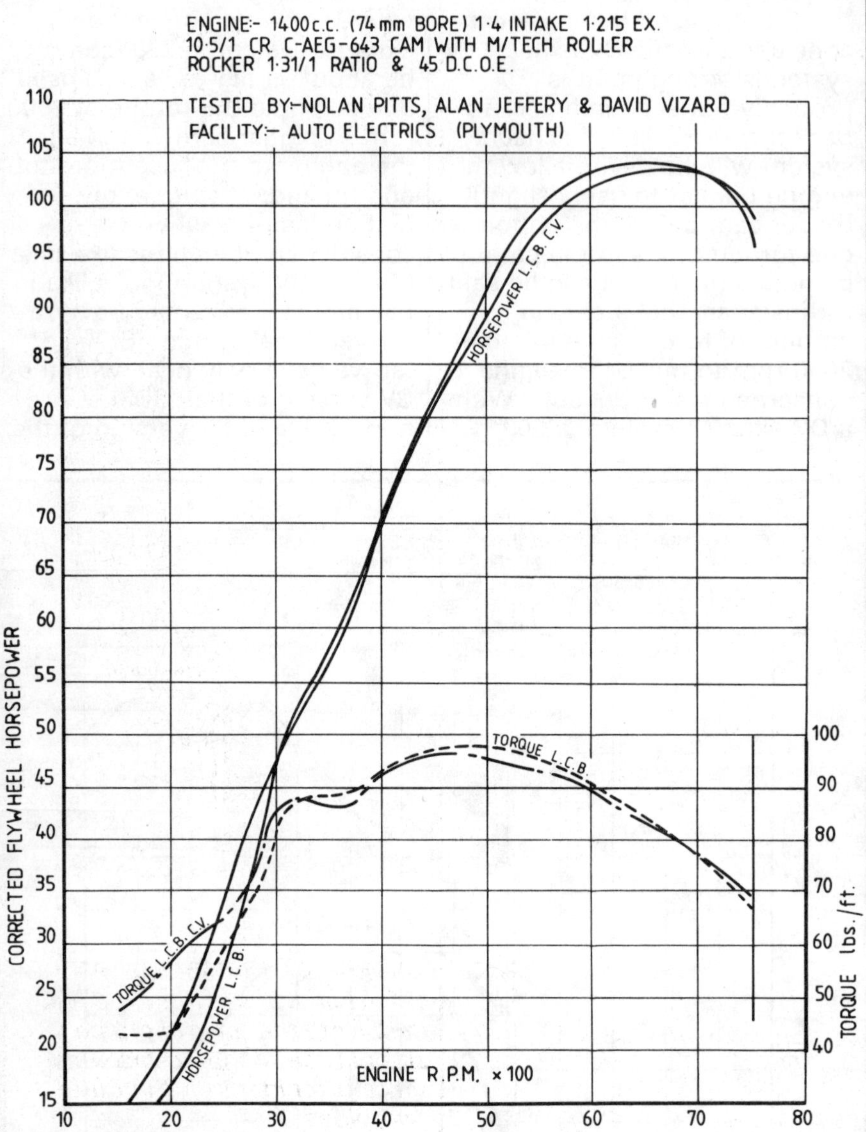

Fig. 10.9 Comparison of conventional LCB and controlled vortex LCB exhaust manifolds

In this comparison between a CV type exhaust and a conventional exhaust, the main difference is the amount of low end torque generated. Under 2000 to 2500 rpm, the engine is, with the almost race spec cam used, normally 'off the cam'. However, the CV exhaust pulled up a lot of extra low end torque. Over the middle to upper range – 3000 to 7000 rpm – it produced slightly inferior results but at 7500 it was once again superior. From the results here, it would seem that the main asset of a CV exhaust is to spread the power curve so as to be able to utilize a bigger cam than would otherwise be the case. The extra low end torque given by the CV system would be instantly noticeable on the road.

Here is an early prototype CV exhaust system made by Janspeed. Note the vortex controlling stubs are mounted on a separate plate. This made it easier to optimize their length and form during the experimental stages.

This was my 1400 dyno donkey engine with the Janspeed CV exhaust system on just before we went to the dyno with it. It looks pretty here, but after 70 hours on the dyno, having various things tested and replaced, it really did look well worn. (Photo Norman Hodgson)

3-into-1 exhaust manifold and see how that fares against its conventional counterpart. **Fig. 10.10** shows the figures.

As you can see, there is not such a clearcut difference here. The 3-into-1 is already better at pulling up low end torque than

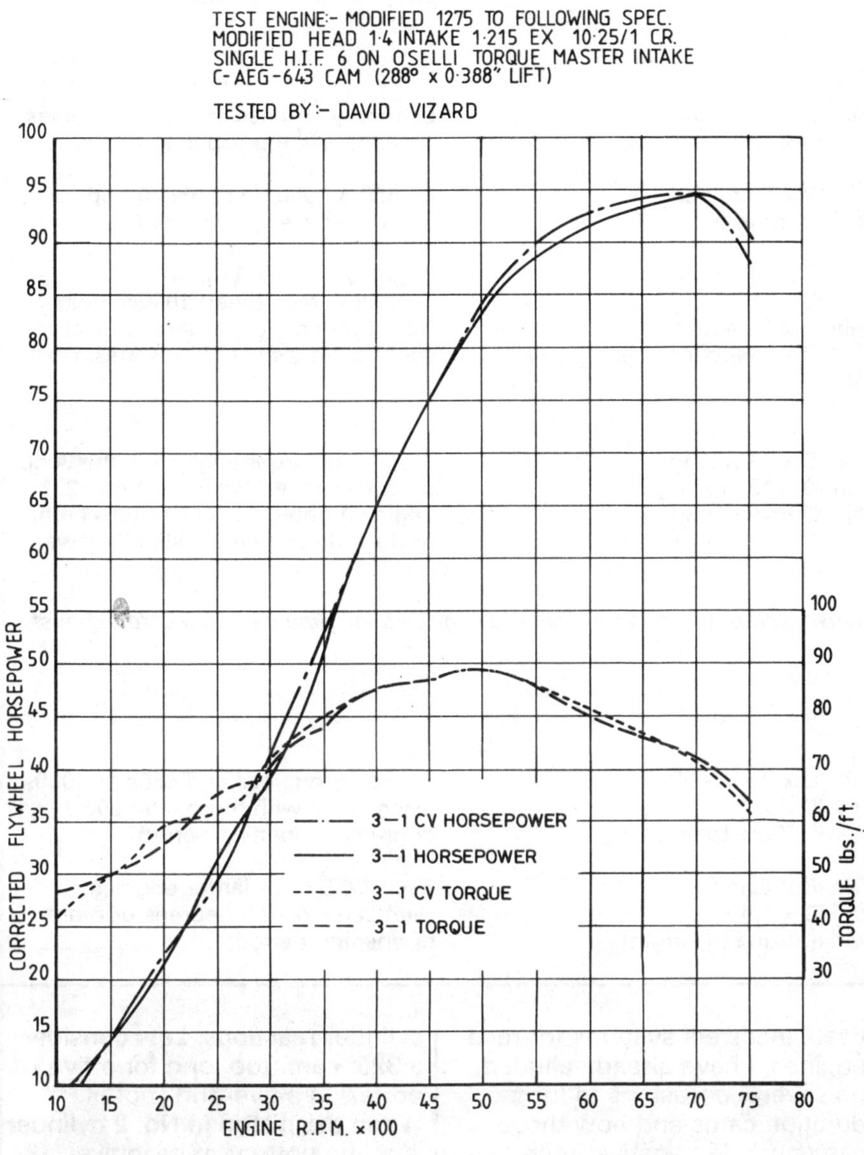

Fig. 10.10 Comparison of conventional 3-1 and CV 3-1 exhaust manifolds

From the graph we can conclude that a 3-into-1 CV system produces little more benefit than a conventional 3-into-1 form. The 3-into-1 manifold in conventional form may well be promoting a more uniform direction of exhaust flow because of the long pipe lengths involved. If this is the case, the CV system has little, if anything, to cure. As a result, it shows little improvement. On a motor with a 300 - 320° cam, the story may be a little different.

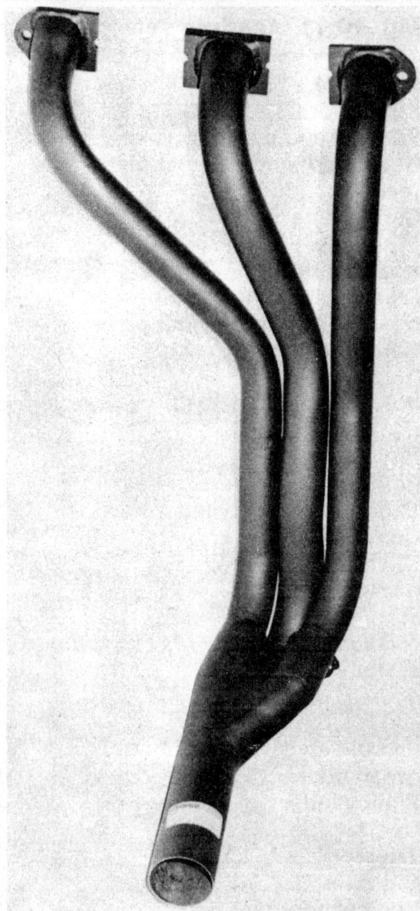

The pre-production Janspeed CV exhaust system looked like this. Note the enlarged pipe sizes at the manifold face. It's more prominent on the outer two pipes.

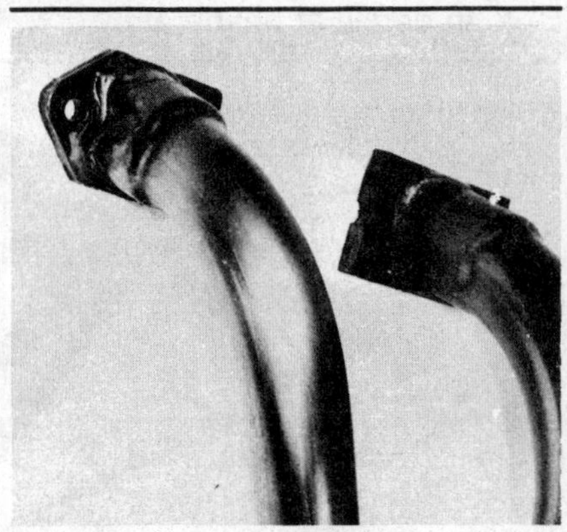

The expansion of the pipe for a CV-type system is clearly seen in this shot of a system for an inline engine.

its LCB counterpart and the CV exhaust system doesn't have so much room to be able to make improvement. Again, it's asset lies more in its ability to be used with a bigger cam.

Earlier, I referred to the subject of exhaust pipe bore sizes for various stages of tune or power level. It's important to remember that big does not necessarily mean best. Fitting an exhaust system with too large a pipe diameter will cut

Fig. 10.11 Manifold selection chart

LCB Style		
Small bore:		
Primary end cylinder pipes	$1\frac{1}{8}$in ID x 11in long	Economy engines and other engines
Secondary end cylinder pipe recommended	$1\frac{1}{4}$in ID x $15\frac{1}{2}$in long	up to 65–70 bhp especially
Centre pipe	$1\frac{1}{4}$in ID x 25in long	on 850 & 1000 up to 55–60 bhp for
*Tail pipe	$1\frac{3}{4}$in ID (tune to length)	whatever use they are put to.
Medium bore:		
Primary end cylinder pipes	$1\frac{1}{4}$in ID x 11in long	Engines with 70–110bhp output
Secondary end cylinder pipes	$1\frac{3}{4}$in ID x $15\frac{1}{2}$in long	including race tuned 1000cc units
Centre pipe	$1\frac{3}{8}$in ID x 25in long	works well on longer stroke units
*Tail pipe	$1\frac{7}{8}$in ID (tune to length)	such as 1100's & 1275 engines. Do
not use on engine with a power expectancy below 70bhp.		
Big bore:		
Primary end cylinder pipes	$1\frac{3}{8}$in ID x 11in long	1275 cc or larger engines in the 90hp
Secondary end cylinder pipes	$1\frac{1}{2}$in ID x $15\frac{1}{2}$in long	or more range. Works best on 1275
Centre pipe	$1\frac{1}{2}$in ID x 25in long	engine if potential is over 105–110hp
*Tail pipe	$1\frac{7}{8}$in ID (tune to length)	or on 1400 engines of slightly lower
power output.		
3 into 1 style		
Small bore:		
End cylinder pipes	$1\frac{1}{4}$in ID x 25in	Economy engines and 850s or 1000s
Centre pipe	$1\frac{3}{8}$in ID x 25in	using cams with more than 305
Tail pipe	$1\frac{3}{4}$in ID (tune to length)	degrees of opening period
Big bore:		
End cylinder pipes	$1\frac{3}{8}$in ID x 25in	Race 1275 cc or larger engines.
Centre cylinder	$1\frac{1}{2}$in ID x 25in	With cams of 305 degrees or more
*Tail pipe	$1\frac{7}{8}$in ID (tune to length)	of opening period.

** Tail pipe normally fits over pipe to exhaust system but on large capacity engines tail pipe can go inside exhaust system pipe.*

the hp and low speed and mid-range torque. Since we have started on the subject of incorrect exhaust systems let's look, in detail, at deliberately used, incorrect systems for race engines. I have already alluded, on several occasions to long duration cams and how these can generate negative inter-cylinder reactions. Lets consider a 320° cam (too long for a five porter) in a five-port motor. When the piston in No. 2 cylinder is at the bottom of its intake stroke the intake valve will still be over 80° away from the closure point. Pressure measurements show that if the cylinder was isolated as per a four intake port system the port and consequently the cylinder

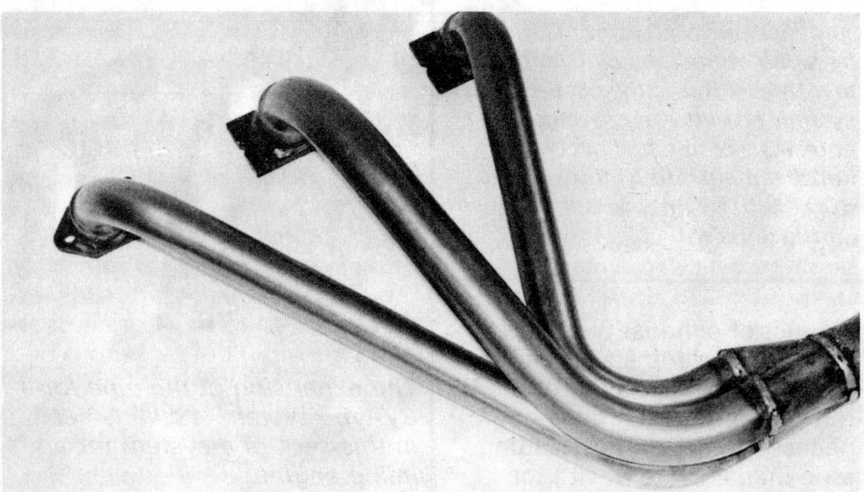

Many inline engines present an easier installation of the exhaust system than do transverse engines, at least until the pipe lengths are such that they have to join up under the floor and then a flat junction rather than the 'one over two junction shown here has to be used.

pressure would be rising due to pressure wave super charging and would be just starting to go above atmospheric pressure. With an isolated cylinder this pressure could, with a well tuned exhaust, be 2–4 psi above atmospheric some 40° before valve closure. Just prior to valve closure it can have risen to as much as 7 psi above atmospheric. The pressure wave mechanics are there to do this with a five-port layout but for the interaction of the centre cylinders with the outer ones. (No. 2 with No. 1 and No. 3 with No. 4).

Here is what happens: during the overlap period of the outer cylinder (in our example No. 1) a good exhaust system will pull the pressure at the exhaust valve down to as low as 6 – 7 psi *below* atmospheric pressure. This super low scavenge pressure (which is in effect 7 psi of suction or vacuum) sucks out the combustion chamber residuals and starts the intake charge into the chamber through the intake valve which is about 125 to 150 thousandths open at TDC. With the valve this far open we have good communication between port and No. 1 cylinder. Right next to this high vacuum source is a port and cylinder (No. 2) which would, left to its own devices, be in the critical *pressure* ramming phase of its intake stroke. The intake valve for this cylinder is about 200 thousandths short of being closed and the piston is coming up the bore. So here we have a port with its pressure rising and it would like to ram fill this cylinder. At the end of this port are not one but two cylinders. One with its pressure starting to rise (No. 2) and one with its pressure well below atmospheric – No. 1. So which cylinder is the flow likely to go to? – right – No. 1 with its low pressure. No. 2 will be robbed of most of its ram fill. This is a very important fraction of the intake charge that could potentially have gone into the No. 2 cylinder. At high rpm, a race cammed motor with a well tuned exhaust system gets about 15% of its total cylinder fill *after BDC from the ram effect that can take place*. With a long cammed five-porter those centre cylinders are deprived of the benefit of the ram fill and unfortunately it is not made up for by better filling of the outer cylinders. Our own tests indicate that the outer cylinders benefit only to the tune of 2–4%.

The absorption and subsequent killing of the ramming pressure wave of the inner cylinders by the outer ones can be compensated for but not cured by taking a number of steps. The first is *avoid long period cams*. Assuming a typical lobe centre line angle cam of around 102–104 degrees we find that cams with more than about 290 degrees start to exhibit measurable inter cylinder robbing. With cams of more than 300 degrees the robbing effect looses as much or more power than the longer duration has the potential to make so total output starts to drop. This effect, has inadvertently been countered by the almost universal use of long 3–1 exhaust manifolds when very long period cams are used. This is fix No. 2. The 3-1 system is essentially an LCB with the outer branches extended to the length of the centre branch. What this does is to give cylinders Nos 1 and 4 a less than optimal length so they do not scavenge as well thus cutting their ability to rob cylinders Nos 2 and 3. So, because of the five-port layout we have had to adjust the pipe length to a dimension that is sufficiently *wrong* to reduce its tuning effect! Putting it another way, we have had to *mis-tune* the outer cylinder exhausts to compensate for the selection of a *less than optimal cam*. This proves to be one of those rare occasions where two wrongs put together actually adds up to a better result than either one singularly!

So the justification of a long race style 3-1 exhaust system is only to compensate for what is essentially an over zealous decision of how much duration the cam should have. A mistake made in one area of the engine spec then has to be compensated for by deliberately using a mismatched exhaust system. Personally I am not enthused with balancing out one mistake with another and this brings me to fix No. 3. I contend that we should try to reduce the robbing problem at its source – that is by deliberately phasing the valve event timing on each pair of cylinders differently so there is less tendency for the robbing to take place. This will in turn allow us to use a more functional LCB style of exhaust manifold. The scattered pattern cam, which you will read about in the cam section, does just that. There is also a trick that can be done with the crank which, since it is relevant to cam timing I also deal with in the cam section.

At this point I may have left you with the impression that the selection of the best exhaust system looks complex. It is at the design level, but hopefully I have done your homework for you. Short of dynoing your own engines to find out exactly what exhaust system works best on them, I would suggest you follow the guidelines in **Fig. 10.11** as to the sizes to be used.

Megaphones

An area that always seems to capture people's imaginations is in tail pipe tuning. Is there a magic shape that will extract those few vital extra hp from the engine? Opinions differ widely but I have not found a great deal of difference in power output by trying various lengths of tail pipe or megaphones or whatever on 5-port engines. At best there seems to be a potential gain of 1 or 2hp by optimizing the exhaust

It's difficult to see in this shot, but here is a Metro cast iron LCB system installed in a Mini. The entire Metro system, right back to the first silencer, will fit in any Mini already fitted with a big bore motor. For small bore engines, the down pipes need to be shortened by 1/4 to 3/8 inch.

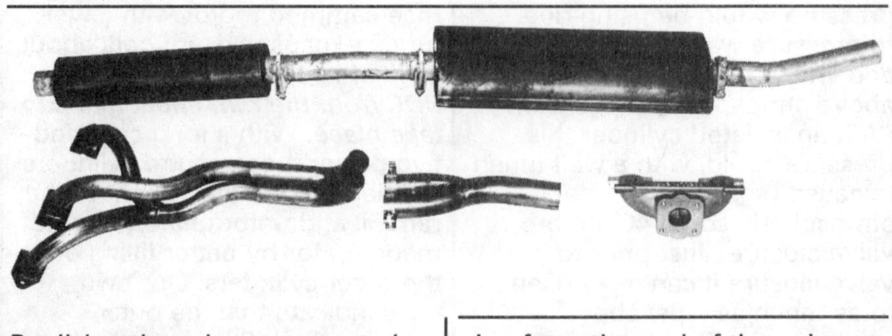

Realizing there is economy to be gained as well as power by improvements in the exhaust system, Oselli marketed this economy kit. It consisted of a medium bore LCB, a high efficiency exhaust system and an intake manifold. On a typical 1275 GT Mini it produced about an 18% improvement in economy along with a 10 to 12hp increase in engine output. That's having your cake and eating it.

pipe from the end of the exhaust manifold. Sometimes a megaphone will show an increase over a straight pipe and other times it won't. In most instances a piece of exhaust pipe that's a slip fit over the exhaust manifold cut to about 25 inch in length seems to be pretty close to the mark. This is, of course, assuming that a fairly racey cam is used. I have to admit I have never tried tuned exhaust pipe lengths and open exhaust on an engine with a standard cam; the need for it has never arisen. If you feel that a plain pipe lacks that certain something, then by all means use a megaphone. Something similar to the dimensions shown in **Fig. 10.12** should be appropriate in most cases.

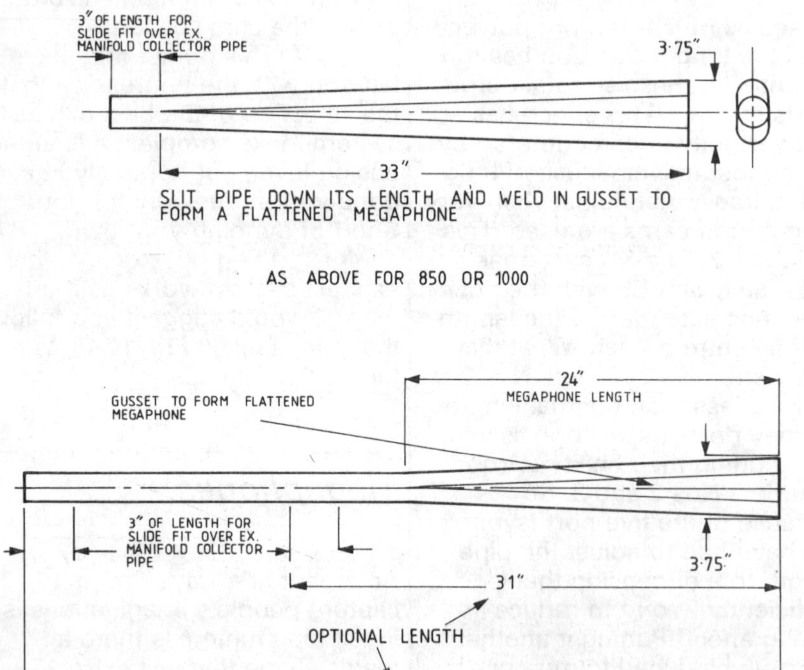

Fig. 10.12 Megaphones and tail pipes
The length given assumes a 300° cam. If a longer period cam is used length may need to be shorter, except on stroked 1275 units where engine speed and breathing potential become a limiting factor.

Eight-Port Exhaust Systems

An engine having individual intake and exhaust ports per cylinder, such as an eight-port or 16-valve head can benefit greatly from exhaust tuning. Actually stating that it can 'benefit greatly' is, if anything, an understatement of some sizable proportions. As previously mentioned the effect of exhaust tuning can, if done in an appropriate manner, boost output over a band as wide as 4,000 rpm and sometimes more. The exhaust, not the intake, in a race tuned system is the primary factor toward achieving volumetric efficiencies as high as 115% or more. Without an appropriately tuned exhaust the full potential of any high performance or race engine will be compromised. Since eight-port eight or 16-valve installations are less than common it is very likely that you will in many cases need to get a special system made up. If KAD or Jack Knight do not have a system made to your requirements then you will need to seek the services of such companies as Manx Racing, Janspeed or Maniflow. These and other specialist manifold manufactures can all produce a system to suit. This now raises the question as to 'what suits?' In the designing of a system it is necessary to not only have the right lengths but also diameters. Also we have to decide whether a 4-2-1 'interference' system or a 4-1 system is most appropriate. Obviously the best way to arrive at a close to optimal system is to start with an intelligent estimate of what is required and then employ the trial and error method of fine tuning on the dyno. The dyno time involved here although it can be costly is not necessarily the major issue. Owning a dyno does not make exhaust system optimization cheap. The real cost lies in the fabrication of as many as half a dozen systems each of which must, in themselves have a degree of adjustability. Unfortunately the testing of a suitably wide range of manifolds is only viable for those more well-funded racers – a group to which I have unfortunately never belonged. As a consequence of this I have had to find ways and means of making more intelligent estimates. The techniques I have employed involved the use of flow bench data and computer simulations to more accurately home in on what is required before any metal is cut. If results are anything to go by the methods used have produced better than average results so I feel justified in passing them on. Let us first deal with exhaust pipe diameters.

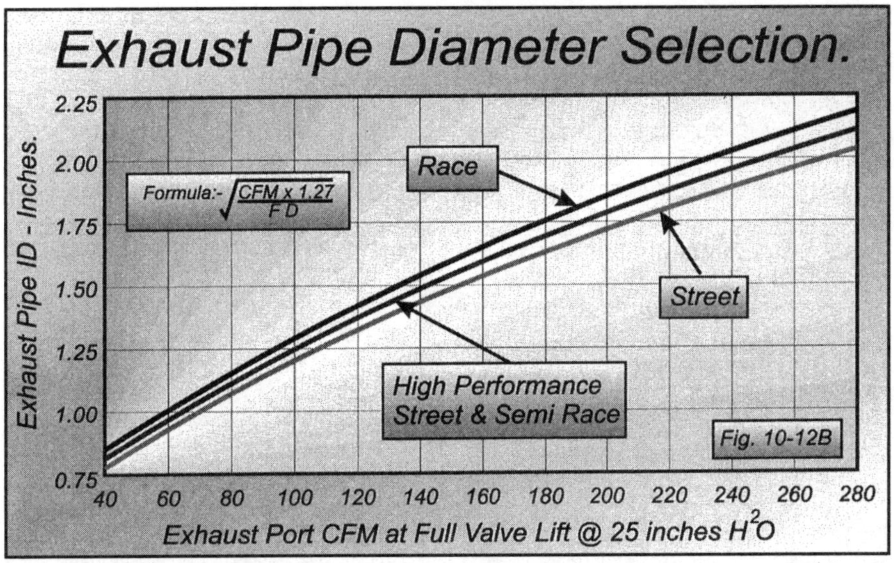

Fig. 10-12B

Pipe Diameters

If the primary exhaust pipe diameter is too large the resistance to the flow will be minimal but the all important extraction effect will be severely compromised. If the pipe diameter is too small the extraction effect may be good in the lower rpm range but top end power will suffer because of too much flow restriction. In my experience determining the best primary pipe diameter is largely a question of factors dictated by application requirements and available exhaust port flow at full valve lift as measured on the flow bench. If given some thought sizing the pipe in relation to the cylinder head's available exhaust port flow makes sense. Having an exhaust pipe with a huge amount of flow by virtue of a large diameter is hardly going to have a positive effect if the exhaust flow is ultimately limited by something as fundamental as a much lower exhaust port flow capacity. To size the exhaust pipe using my method you will need to know the exhaust flow at full valve lift when using 25 inches of test pressure. Given the flow of the cylinder head's exhaust port **Fig. 10-12B** can be used to determine the primary pipe diameter. We will deem that step No. 1. The next move is to decide what style of exhaust system we should use.

System Style

There are two system styles that we could choose for our eight-port exhaust manifold. These are shown in **Fig 10.12C**. The choice between one or the other system is largely dictated by the rpm range the engine is required to run in. When the rest of the engine spec is such that peak power is expected at 8,000 rpm or less the 4-2-1 system is the best choice. It will produce an appreciably higher torque peak and a wider usable torque band than the 4-1 system. However if peak power rpm is expected to

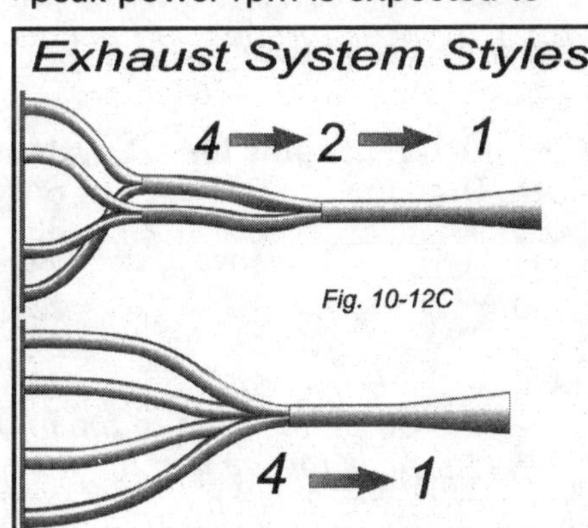

Fig. 10-12C

If you want to go one step better than the charts shown nearby and produce results close to having unlimited dyno testing, then I can recommend the use of a computer program called Dynomation. I have made extensive use of this program and found it to be an extremely powerful tool for exhaust system development by very substantial proportions. An example may prove worthwhile here. After designing a range of cams for a well known UK cam company I was asked to make dimensional exhaust system recommendations for the race cams I had developed. This was funded in part by the cam company involved but at the end of the day this one-man business (me, – I was out of pocket by about 20,000 1985 dollars. I got hold of a copy of Dynomation in 1992 and after using it found it predicted to satisfyingly close limits what had cost so much to achieve on the dyno. Had I had this program in 1985 I estimate I would have saved about $15,000 worth of testing. That was with the '92 version of Dynomation. The 1998 version can accurately be described as enormously more powerful.

From the results personally achieved I can say that even though it sells at some $700 it does, for the serious racer, provide some very cost effective results.

be above 8,000 then the 4-1 system is the way to go.

Pipe Lengths

Pipe lengths are significantly influenced by engine specification. Most pipe length equations attempt to take into account the effect of cam duration. This is a step in the right direction but little more than that. If the situation is analyzed in greater detail it will be seen that a more important factor, and one more complex to deal with, is the effective breathing area given by the combined duration/lift of the valves and their flow capability. While we are dealing with one type of cylinder head be it two or four valves per cylinder, the effect of breathing efficiency is small compared with the effect of cam spec. However when we go from two to four valves per cylinder the low lift flow of the muti valve head is so much greater that it appears, when compared to a two-valve head, to have a cam duration of far greater length than in actuality. This, in effect, leads to a four-valve head having shorter optimal pipe lengths for a given cam duration, than a two-valve head. For example, the pipe lengths that may be optimal for a 320° cammed two-valve head may well be optimal for a 300° four-valve head. The bottom line is that pipe length equations that do not address cylinder head flow capability are not accurate enough to predict what is required in a race world which is demanding greater precision from our calculations if we are to stay out in front. If all this sounds complex then rest assured – you are right in your assumption. However all is not lost, especially if you are building a high output engine for street use. My experience in dyno testing exhaust systems is that if the first part of the manifold is close to what is required then the tail pipe can be used to adjust the overall characteristics very close to what is required. Given that dyno time is a practicality, be it engine or chassis, then

Fig. 10-12D Lengths for 4-2-1 exhaust systems.

Cam Duration	Primary	Secondary	Tail Pipe	Cr	Peak Torque	Peak hp
280	10-12	15-17	26-28	12	4300	6700
290	9-11	14-16	21-24	12	4600	7100
300	8-10	12-14	16-20	12	5000	7300

The cam duration refers to the intake duration. The CR is the ratio used to approximately duplicate the rpms for peak torque and hp. If less CR is used these rpm figures will come down slightly.

Fig. 10-12E Lengths for 4-1 systems

Cam Duration	Primary	Secondary	Cr	Peak Torque	Peak hp
280	23	30	12	4500	6700
290	17	25	12	4900	7100
300	16	23	12	5100	7500
310	15	22	12.5	5300	7600
320	14	20	13	5500	7700

The cam duration refers to the intake duration. The CR is the ratio used to approximately duplicate the rpms for peak torque and hp. If less CR is used these rpm figures will come down slightly. The lengths for cams from 290° on up are for maximum average power over the rpm range used while racing.

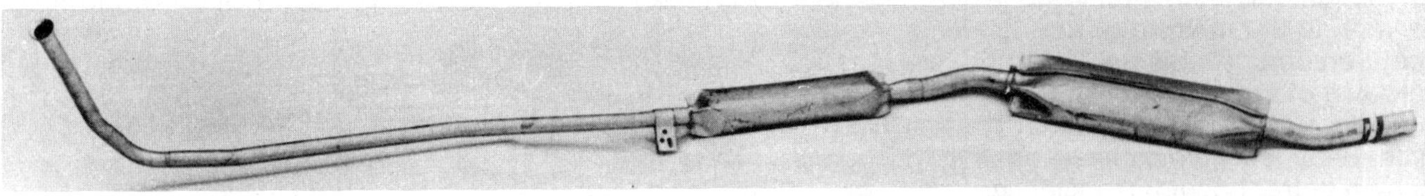

Shown here is the standard Leyland system for Minis employing the standard short, cast iron 3-into-1 exhaust manifold. Although the locations and angles may vary, the same style of silencer boxes are used in many other Series 'A' powered vehicles. The flow figures in Fig. 10.16 show that it's poor in terms of flow efficiency. However, many people . . .

. . . fall into the trap of thinking that the Cooper system, because of its big bore pipe, is much better. Some of the early ones were, but currently available Leyland systems use the same inefficient silencers and the use of a big bore pipe makes only marginal improvements to flow, as the restriction occurs at the silencers, not in the pipe.

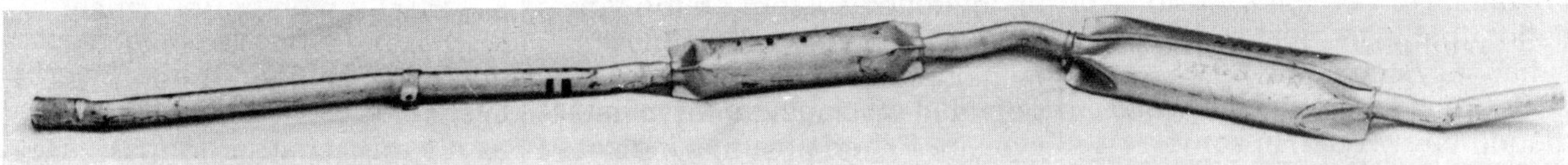

adjusting the tail pipe for best results is about as simple as it comes.

Still on the subject of lengths, if the system of choice is a 4-2-1 type then we find that the selection of primary and secondary pipe lengths is not as critical as a 4-1. This makes the selection of a 4-2-1 system even more desirable for a street driven motor. However the selection of optimal pipe lengths can be severely compromised by the choice of exhaust system used. Although we can deal with this as a separate issue you must be aware of this at the planning stage of the exhaust system so that it can be successfully accommodated on the vehicle. The effective silencing of exhaust noise is mandatory for a road vehicle and, for various reasons, but most usually cost, we have accepted that the penalty paid for low noise is reduced power. For a race engine this is not an acceptable situation unless it applies to every vehicle in the event but yours! Power loss due to a need to suppress noise is not an inevitability. We will deal with this subject a little later, suffice to say at this point that the key is not to destroy the pipe length tuning. As far as lengths are concerned refer to the chart **Fig. 10.12D** for 4-2-1 systems and **Fig. 10.12E** for 4-1 systems.

Eliminating Noise

On road vehicles (and increasingly on competition machinery) the compliance with anti-noise regulations will often mean that the ideal theoretical exhaust system can be compromised. Not only should you consider your own ear drums, but also those of the people in whose area you live or drive through. Even putting this consideration aside, it will be found that a noisy exhaust system attracts unwanted attention from the traffic police. **Fortunately, a quiet exhaust system does not necessarily mean the sacrifice of power, at least if you choose the right exhaust system for the job.**

Earlier in the chapter I raised the question of whether the silencer or manifold should be changed when modifying an exhaust system on a tight budget. Let's pick up that theme again, looking at silencing systems this time rather than exhaust manifolds.

Firstly, it's worth reiterating that the big enemy is back pressure. I have on occasions seen in print that engines need a certain amount of back pressure in order to function properly. Take my word for it, whether for economy or hp the existence of back pressure reduces both. An ideal exhaust system is one which has zero back pressure. Published power curves showing the better hp when silencer-induced back pressure exists in the system are usually the result of a bad test procedure or engine specifications. Changing the silencer often affects the calibration of other components such as the carb and ignition timing. A good friend of mine who designs intake manifolds, once put it very nicely by saying 'muffler back pressure is reflected in carb calibration. Change the back pressure and you change the world the carb sees.' If an engine is set up with ignition and mixture optimized for maximum power with a silencer on, then removing that silencer will generally cause the mixture to go lean. A lean mixture takes more time to burn, so the ignition is now retarded in relation to the mixture's burn requirements. At this point the removal of the silencer can cause a power drop. That's how it is with many tests which have shown better hp with the silencer giving average back pressure as opposed to the almost zero back pressure of an open exhaust. For the tests presented here, a great deal of effort was made to optimize for each circumstance, so the true picture could be seen.

As far as the back pressure is concerned, there is only one exhaust system that was any good that came as standard equipment on a Series 'A' engined vehicle, and that was the *genuine* Leyland Cooper 'S' exhaust system (now generally unavailable). To be accurate this system was only half-way decent but anything else that was or is available as original equipment works about as well as a cork and ends up costing hp. Measuring back pressure presents no real problems for the enthusiast. This means you can check out your own system and often as not give yourself a scare. **Fig. 10.13** shows a typical setup for measuring exhaust back pressure. The technique is to drill a hole in the exhaust pipe, just prior to the first silencer. Braze a length of copper pipe into this hole. 3/16th O.D. pipe will do just fine. The copper pipe needs to be about 18in long so that exhaust heat will be dissipated in the copper pipe prior to reaching the coupling pipe. Because the copper pipe has dissipated the heat, some simple screen washer hose will suffice for the coupling pipe. In fact, a screen washer hose can be used to make a pressure measuring device, namely a manometer. The actual fluid used for indicating the pressure can be either mercury or water. However, it's worth noting that with most standard exhaust systems, the back pressure is so high that a water manometer would, in some cases, have to be as tall as 15 feet. It doesn't take much to realize the impracticality of hauling around a 15 foot manometer in an Austin Healey Sprite. This means that for most applications, mercury will have to be used to test the pressure. Because some systems have high back pressure, it is practical to check out the back pressure by utilizing a standard fuel pressure gauge as a means of measurement.

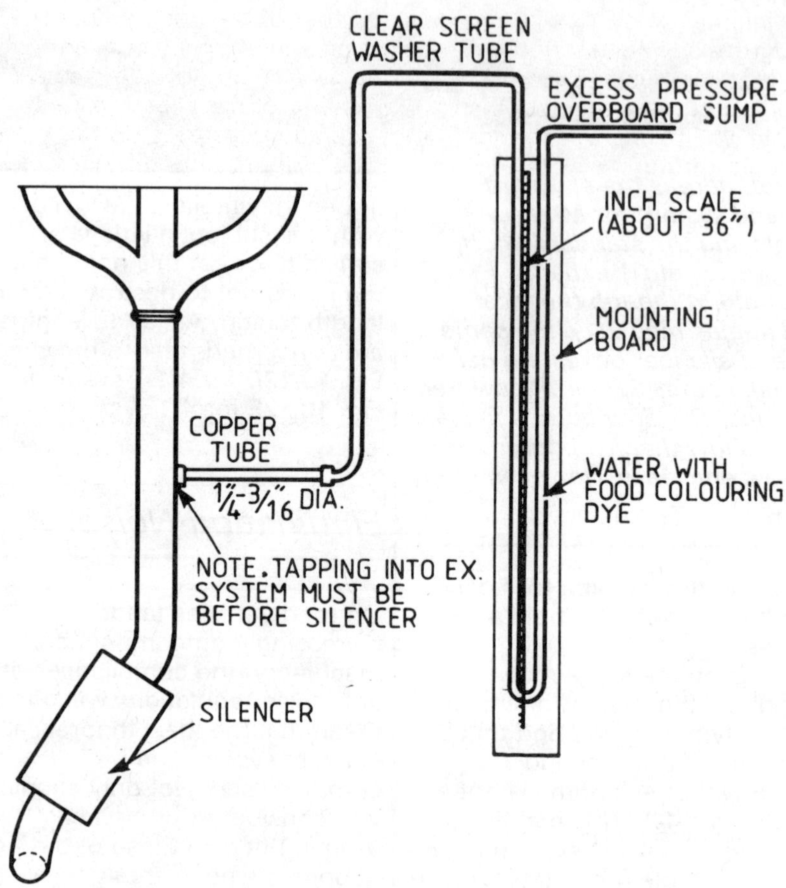

Fig. 10.13 Manometer for back pressure testing

It will be fairly obvious that the cause of back pressure is inadequate flow capability of the silencer. Just as we can check out the performance potential of an induction system on the flow bench, so we can, in a similar fashion, get an idea of the potential efficiency of an exhaust system. **Experience has indicated that a flow rating in the region of 2.2 to 2.5 cubic feet per minute per hp will allow the engine to produce close to or the same power as it does on open exhaust.** A flow rating any higher than this does not necessarily mean more hp or hp any nearer that of a straight, open pipe, simply because flow capability alone is not the sole criterion. As noted before, it's possible to get more than adequate flow in a silencing system by having a big bore. Unfortunately the larger bore a system becomes, the harder it is to eliminate noise, especially if it's a straight through type silencer. In high performance exhaust systems for small engines a straight through silencer is the most satisfactory, as it's much harder to make a baffled system to the same efficiency levels and, anyhow, a straight through silencer is sufficiently quiet. As engine sizes progress above that of the Series 'A', that is engines of 2-litres or more, so the reverse begins to apply, *i.e.* it is harder to make a straight through silencer flow adequately as well as reduce the noise level to an acceptable point by comparison with a baffle or three pass silencing system. This is because a straight through silencer functions on an absorption principle. In other words, a slug of gas passing up the core of the silencer dissipates sound by expansion through the core holes into the pack. Apart from this, a measure of noise reduction is also achieved by virtue of gas friction between the walls of the core and the gas itself. If the core diameter becomes too large, a portion of the gas slug can simply pass up the middle of the core without having any appreciable energy taken from it. The net result is a noisy exhaust system.

The cross-over point between an effective straight-through silencer and an effective three pass silencer is about 2 inch diameter. Fortunately, the biggest exhaust system likely to be used on a Series 'A' engine is around 13/4 inch diameter. Just like the intake, it's necessary to keep the exhaust speed up so that it has some inertial extraction effect back at the cylinders. However, it is the exhaust pipe diameter that is the most important aspect for inertial extraction. It is quite permissible to go up in pipe size at the silencer itself. In other words, it's quite permissible to have, say, a 1½ inch exhaust pipe up to the first silencer box and then a step up in diameter so that it now goes through a 1¾ inch silencing system at the larger size.

The first part of the pipe will keep the exhaust speed up to provide whatever measure of inertial extraction may be had, but at the same time, assuming the pipe diameter isn't ridiculously small for the power output of the engine, that straight pipe will offer minimal amount of resistance to flow. **Unless we are talking of the most efficient silencing systems on the market, the biggest impediment to flow in the exhaust system is the silencer itself.**

Still on the subject of ways of extracting noise from the exhaust gases, I will explain some of the reasons why the standard exhaust system can be a power loser. Original equipment manufacturers put in a great deal of effort toward producing a product which is very much made down to a price. Exhaust systems can be made to silence using three techniques. The first one I have already spoken of and that is absorption. Usually the absorption takes place into a glass-fibre pack, although this is not essential. Air can be used, but it requires a greater volume of air, so, to produce a silencer of practical size, glass is often used. **Apart from absorption, sound can be reduced by reflection as in a baffle-type silencer or by constriction.** Constriction for the high performance enthusiast is an ugly word. Again, baffle-type silencers often use constriction silencing by virtue of small pipes and by setting up tortuous gas routes through and around baffles.

Of the methods discussed, the cheapest silencing method is by constriction. As a result, this is a technique used as much as possible by original equipment manufacturers and manufacturers of systems which would otherwise have little chance of meeting noise regulations. When designing an exhaust system for high performance use which must not only reduce noise but do so without costing a significant amount of power, then advanced techniques have to be employed. **Such systems tend to cost a little more than a regular replacement unit, but in terms of money spent per hp gained, they represent exceedingly good value.** If you then add the improved economy to the equation, it will be evident you will get back much, if not all, of the investment in a quality, hi-tech silencer.

Testing Some Systems

Having discussed silencer design it is now time to look at some performance facts and figures. **Fig. 10.14** shows the hp and torque output as well as the back pressure in the exhaust system of an engine tested with both a standard 1275 GT exhaust system and a **Competition Silencers** R.C. 40 system.

The first point to note is that **the replacement of the standard silencer with the high performance Competition Silencers' unit resulted in a bigger increase in hp and torque than was the case when exhaust manifolds were changed.** When

Undergoing the test here on the flow bench is a TI tailbox unit. To obtain the figures in the accompanying chart, sometimes the silencers were blown through, and . . .

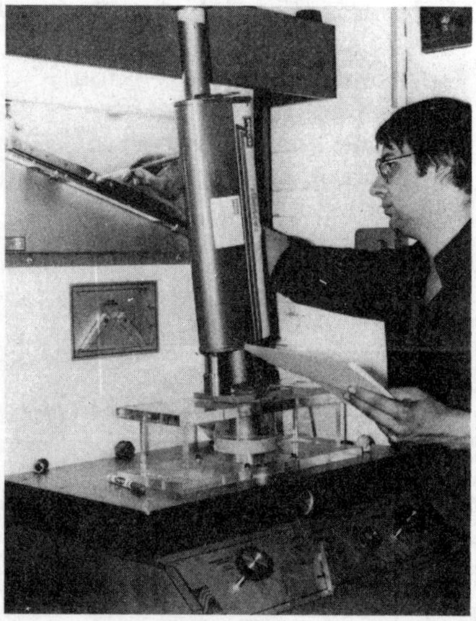

. . . sometimes they were sucked through, but always the flow was in the right direction. In both directions a streamlined entry ring into the silencer was used to eliminate 'edge effects'.

the tests for the exhaust manifolds were conducted, they were done with a Competition Silencers unit in place, so they would have the best possible chance of showing their true worth. When a high performance exhaust manifold is used, together with a restrictive silencer, much of its power potential is lost. So from the figures presented, it's possible to see that a good silencer does give excellent results in terms of power gain. If you take a look at the back pressure on the graph, you will see that it goes up to nearly 7 inch of mercury, that's equal to some 3½ lb psi back pressure. These sort of readings are fairly typical of earlier type exhaust systems. Many of the

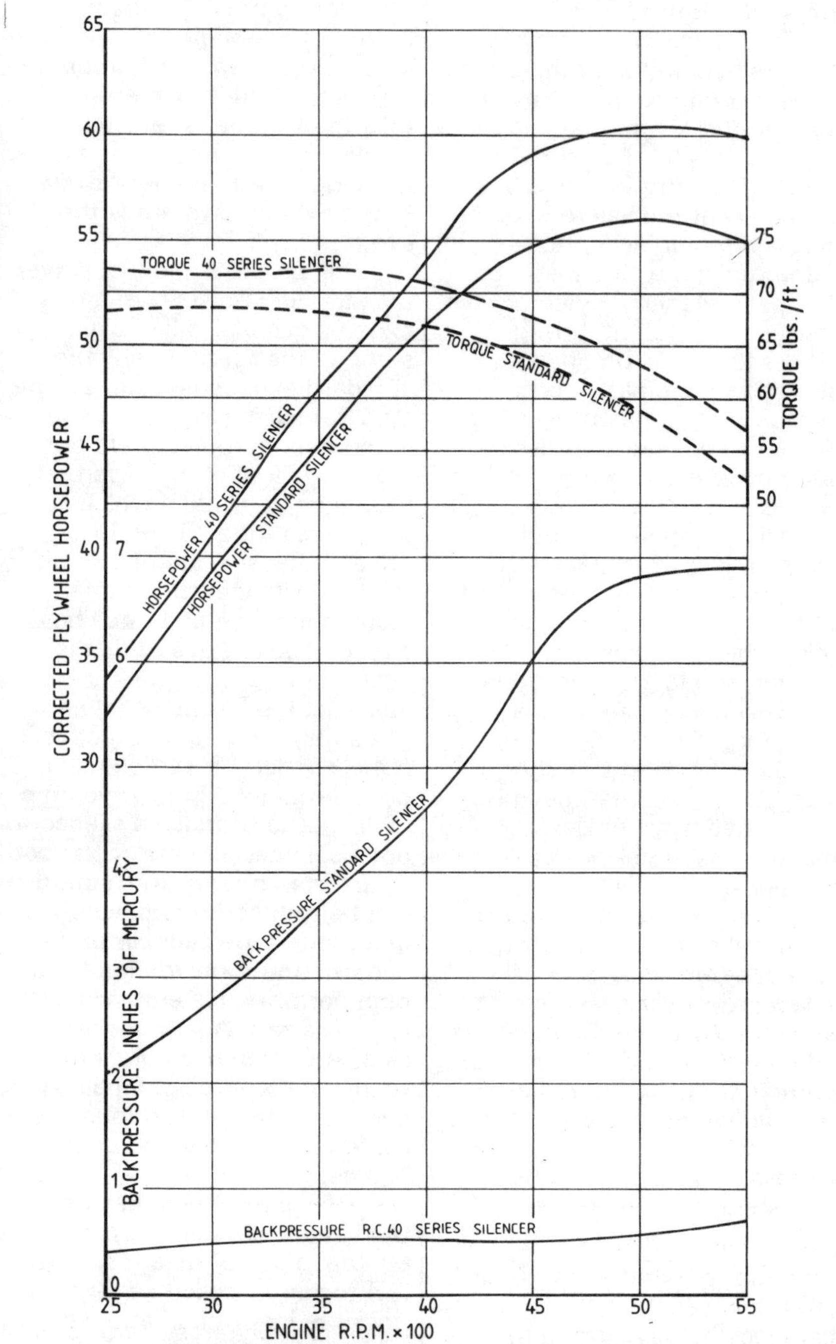

Fig. 10.14 Comparison of Competition Silencers' 40 Series silencer and standard Mini 1275 GT silencer

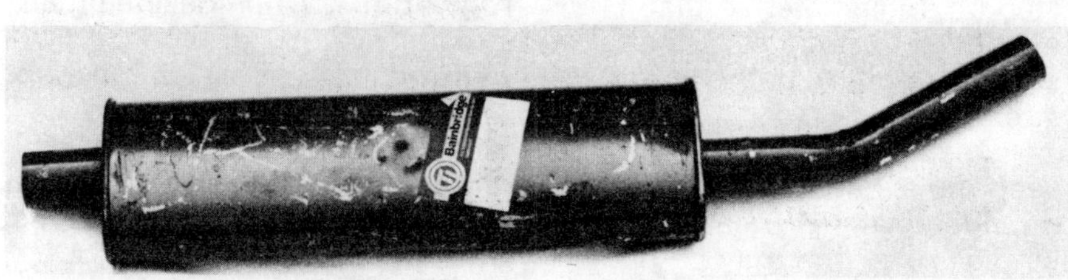

This TI AN-181-T unit has been popular with many enthusiasts for use as a single tailbox silencer. If you don't mind a slightly louder exhaust than standard, then it's entirely practical. However, its flow efficiency is over-rated by many.

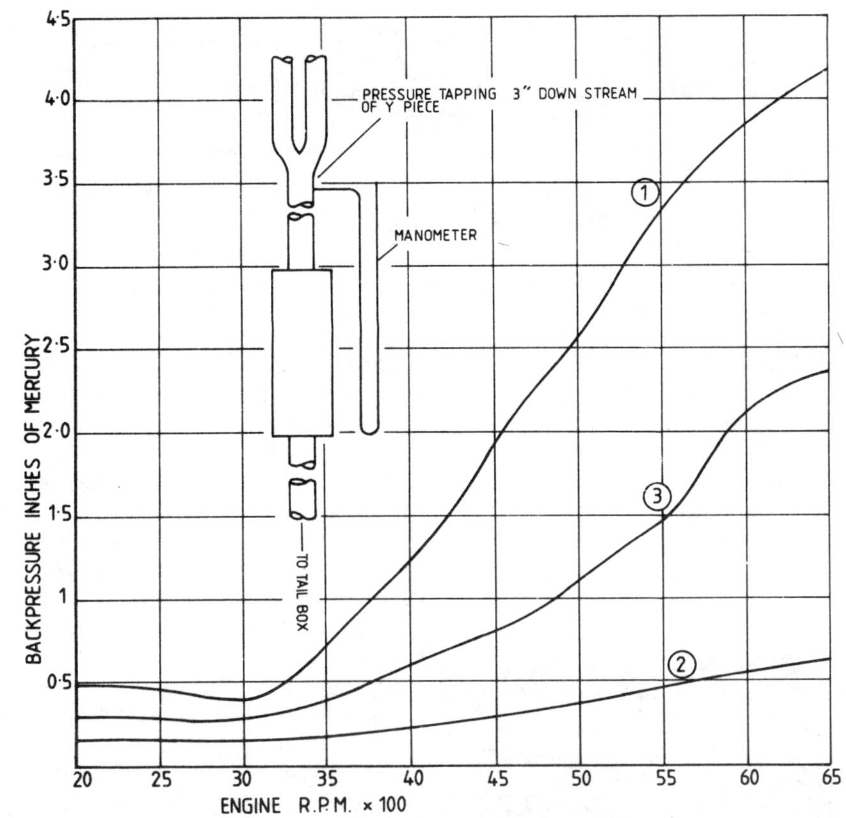

Fig. 10.15 1275 Metro exhaust system back pressure

This graph shows the back pressure generated in a standard Metro exhaust system on a standard engine (curve 1). Compare this with curve (2). Here the tail box of the standard system has been replaced by a Janspeed straight-through unit, but all else remains unchanged. Note how much the back pressure has dropped. For curve (3) the power of the engine has been raised from 60hp to about 70hp due to an intake manifold, cylinder head and exhaust manifold change. The silencing system is as for curve (2) but note how much the back pressure has risen due to the torque and power increase.

later systems, such as that fitted to the Metro, often seem to have better flow capability. Production tolerances however, tend to cloud the issue somewhat and it's possible to pull a unit off the shelf and find that it flows very much worse than the norm. Also, many production systems consist of two silencer boxes. Usually the first box is nothing more than a resonator, sometimes empty, sometimes having a perforated core surrounded by a glass pack. Such resonator boxes or primary silencers tend, due to their straight-through design, to be relatively good in terms of flow. However, they do little other than take the raspy bark out of the exhaust. The bulk of the silencing chore is left to the main box and these are often the real point of flow restriction within the system.

To demonstrate the point, the graph, **Fig. 10.15**, shows what happened to the back pressure in a 1275 Metro exhaust system when the back box only was replaced with a Janspeed straight through main box. The drop in back pressure was significant. In fact it came down to below 1/5th of the original level. As a rough rule of thumb with the Series 'A' engine, for every inch of mercury back pressure that exists, the engine gives away between 1¼ and 1½% of its power. Dropping the back pressure from some little over 4 inch to .7 of an inch gained almost a 5% increase in power. On a 1275 Metro that's approximately 3hp.

Moving to the third curve shown in **Fig. 10.15**. It can be seen what happens when the power output is raised. The original engine as tested put out 60hp. By dropping the back

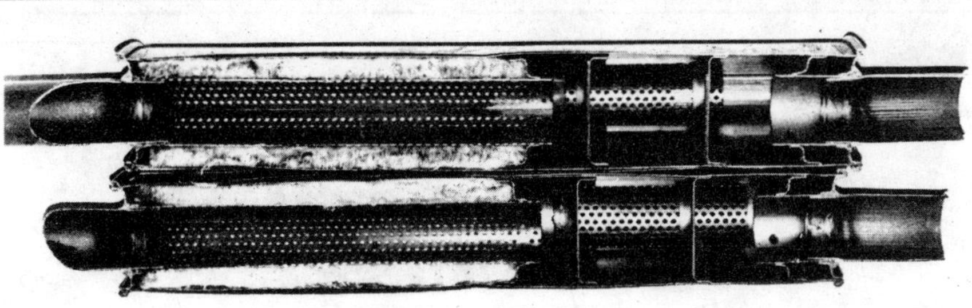

A section through the TI AN-181-T shows that it has a relatively complex internal make up. Part of the fact that the exhaust follows a tortuous route coming in the single pipe on the right, into a chamber, having to split up and go through two pipes, one of them severely restricted, pass through a buffer chamber into a third chamber from where the gas has to make the right-angle turn through a very limited space into the tail pipe, we can see that it's hardly surprising this unit does not flow particularly well.

These two Peco silencers for use on Minis appear well made and the company has a reputation for producing performance silencers. However . . .

pressure in the system with the new tail box, the power went up to 63hp. At this point, some changes were made to raise the power level even further. Along with this, a Howley tunnel port intake manifold, together with a Janspeed LCB exhaust system was installed. Because of the Metro's design of exhaust manifold, it's fairly evident that most of the power gain came not from changing to an LCB so much as the changes made in intake. This, combined with an open intake, resulted in an engine producing 72hp. Dumping this through the same exhaust, caused back pressure curve number 3. **You will note that just a small increase in power has caused the back pressure to go up quite dramatically.** This should make the point that a system which may be adequate on a relatively low hp engine, can very quickly become overloaded on a tuned engine. When we start talking about street motors with a high power potential, such as are rapidly becoming very popular, then it's necessary to ensure that the exhaust system is well designed, otherwise a substantial amount of power is going to be thrown away. The power lost due to the back pressure curve number 3 is around 3% on a 72hp engine. That same exhaust on a 100hp

. . . a look inside this unit, the Peco MO/2 and . . .

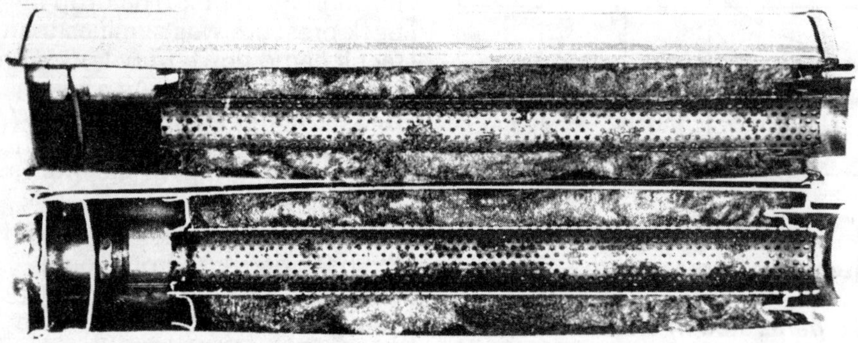

. . . the Peco MO/4 reveals what could hardly be called 'performance innards'. Both these units failed to produce high performance flow figures, as the chart Fig. 10.16 shows.

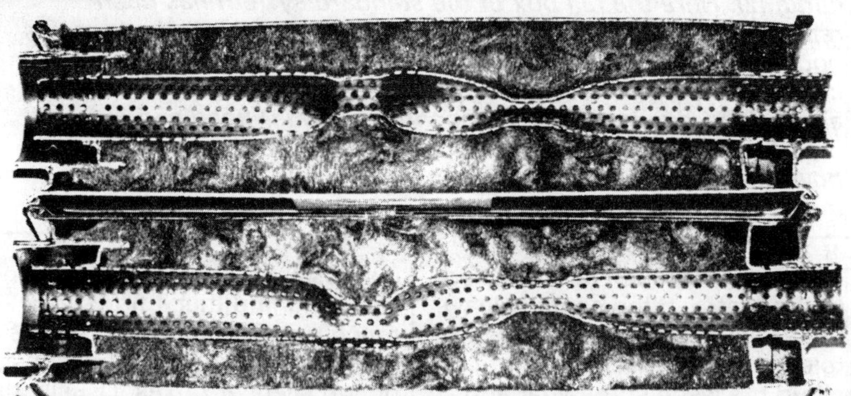

This RC 40 unit from Competition Silencers was built specifically for the Series 'A' engine and was designed to be a high flow unit with excellent noise reduction characteristics. The characteristics of the silencer are built to handle the odd noise signature generated by Series 'A's' three exhaust ports. It produces noise levels equal to or below standard silencers. The flow figures in Fig. 10.16 speak for themselves.

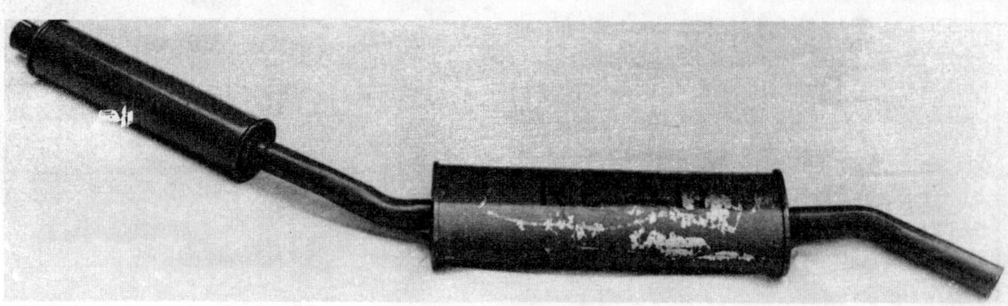

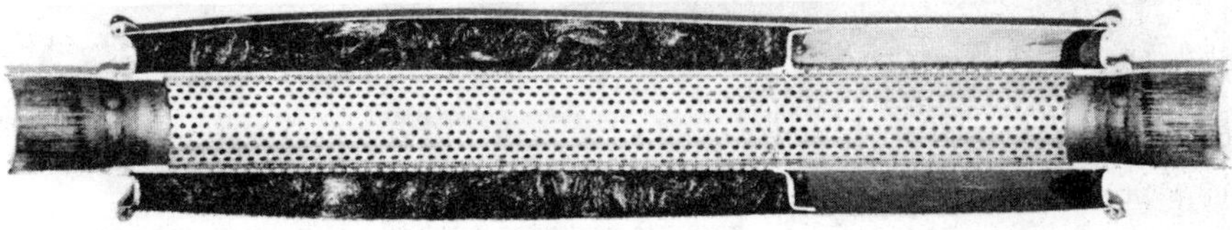

Here's a section through a Competition Silencers' RC 40 second box. A little more flow can be obtained by using only the second box of this system but, since the unit is very efficient to start with, an engine may not produce any more power if the primary box is left out of the system. However, the noise absorption was designed specifically to work with a primary box and the noise levels without the primary box are considerably higher.

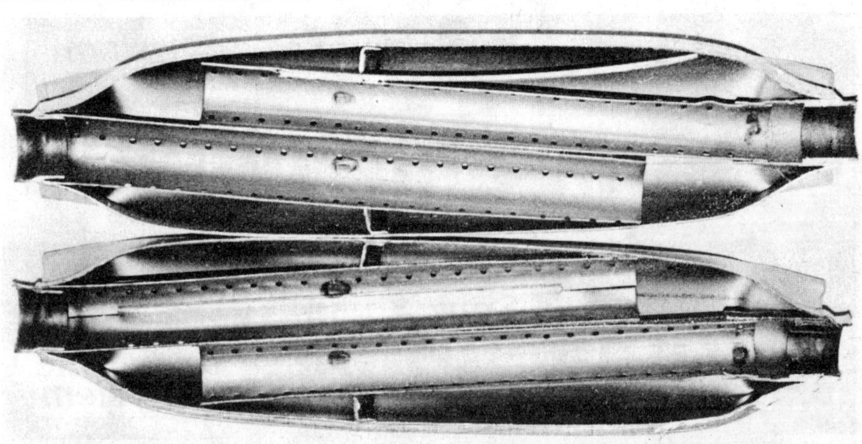

In case you wondered why the standard Leyland system flowed so badly, here is the answer. When I sawed this unit in two, I couldn't believe that the flow figures were as high as they were because apart from those small holes in the sides of the pipe, there is nothing to link one end of the silencer with the other. It seems difficult to conceive of a more inefficient design. However, the tail box unit . . .

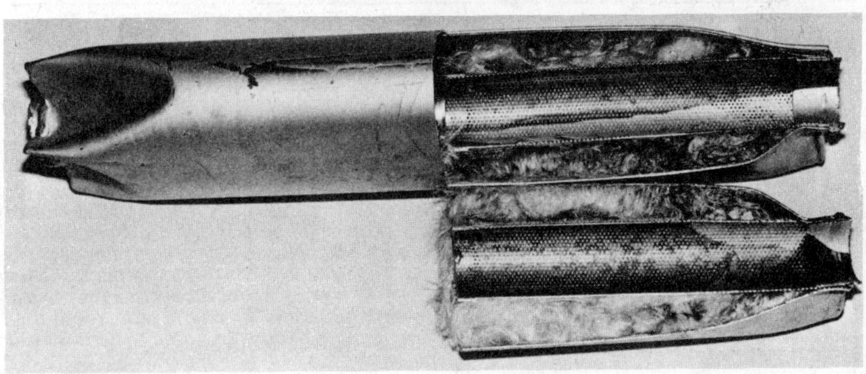

. . . shown here, is much better in terms of flow, but it's not effective enough as a noise reducer to be used on its own, at least not to meet the legal noise limit.

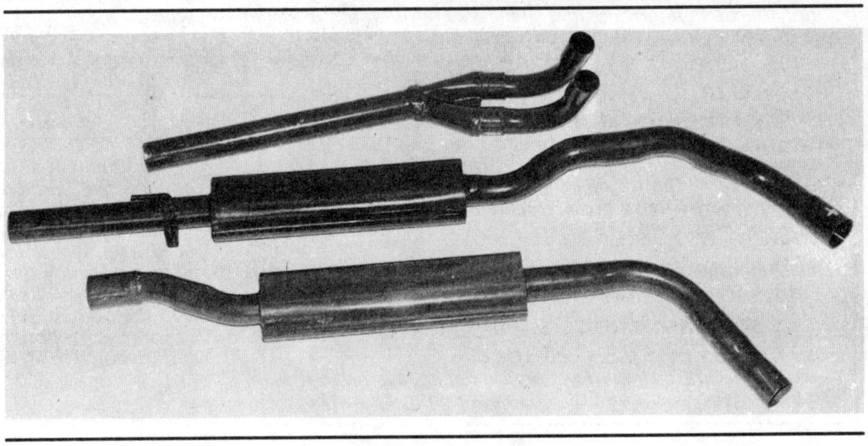

Shown here is a Competition Silencers system for a Metro. This system, as did the Janspeed tail box for the Metro, effectively removed the whooshing exhaust noise given by a standard Metro. Both transformed it into a pleasant but subdued sporty note. Power increases at the wheels are typically around 5hp.

engine could cost as much as 8–10hp. Indeed, in the past I have seen silencing systems of a type commonly used on many rally cars, which effectively chopped off 15hp from the engine. This is a loss that could have been avoided had a well designed system been used in the first place.

Unfortunately, there don't seem to be many systems on the market which meet noise regulations and allow the production of hp. As I stated earlier, the old Cooper 'S' system wasn't bad and 'look alike' but not 'work alike' systems were

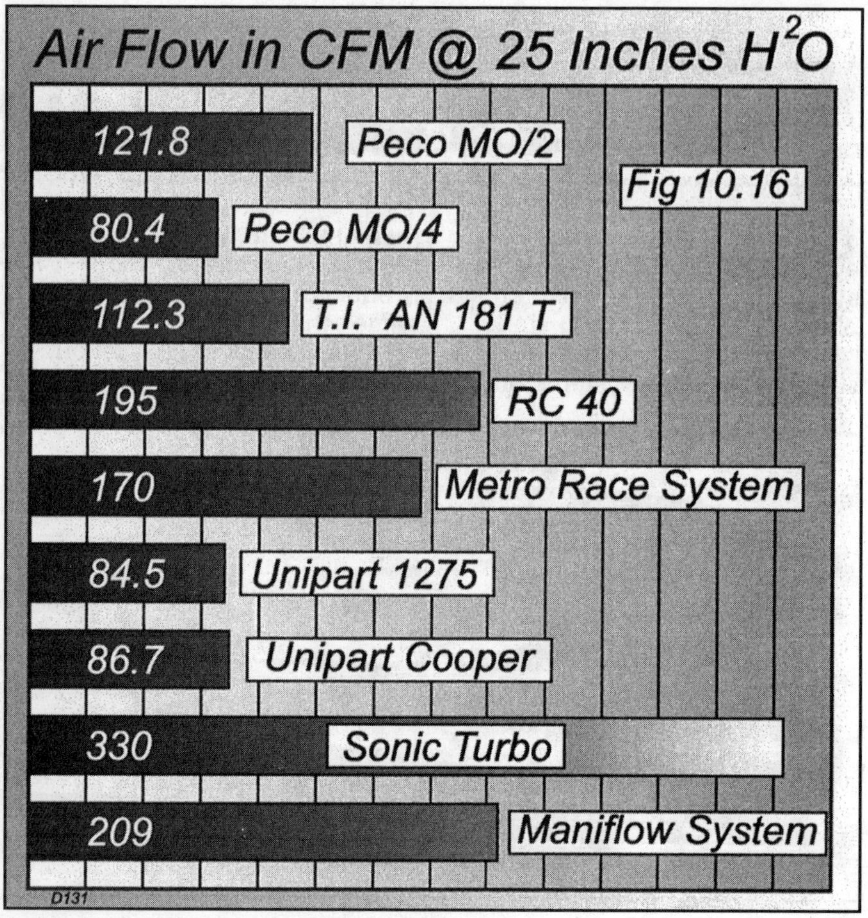

Fig. 10.16 Silencer flow chart

1. Peco part number MO/2 is for the 1275 GT & Mini Cooper and Cooper 'S' range. It is a centre mounted rear box only, having a single big bore outlet pipe. Noise level is noticeably higher than standard.

2. Peco part number MO/4 is for vans, Countryman and estates. It is a centre mounted rear box only. The inlet pipe is small bore (1¼in) to suit size used on 850/1000. Outlet is large bore. Noise level comparable to standard system.

3. The Tube Investments silencer which is also sold under the Bainbridge banner, is a side mounted unit, comprising a tail box only. It is normal to use this unit with a resonator or intermediated silencer. However, it has been popular to use a tail box only because intermediated silencers are often the principal cause of power loss. When used in a tail box only situation, this TL silencer is much louder than the full system, but is quieter than a Cooper 'S' tail box only system.

4. The RC 40 from Competition Silencers is a full system, and in most instances it allows the engine to produce 'open pipe' power with noise levels at, and sometimes below, that of the standard system. The RC 40 is particularly good at taking out mid- and high-frequency noise, thus leaving a deep, mellow sounding exhaust note.

5. This system which exits at the side of a Metro, was designed around some standard Leyland silencers. It is a two box system and is not suitable for road use as it is considerably noisier than a standard system.

6. This is a standard Unipart system which installs on to the short (non-Metro cast iron LCB) 3–1 cast iron system. The particular unit tested was for a 1275 GT Mini, but systems installed on 850 and 1000 Minis, 1100s and 1300s Austin, Americas, Allegros, Marinas etc., are of similar specification.

7. This is a current production Cooper 'S' system. It employs the same silencer boxes as those in 6. The principal difference is that a 1¾in bore exhaust pipe is used. Normally this would allow greater flow but this pipe size is pinched down to 1¼in bore at the silencers. The extra pipe size is in this case largely redundant because it's the silencer boxes that are the prime restriction NOT the pipe. From the figures it can be seen that increasing the pipe area by almost 100% only increased flow by 2½%.

8. Though primarily designed for V-8 use, this silencer will work well on a Series 'A' engine. The 'centre in, side out' Sonic Turbo fits easily into the centre of the rear sub-frame of a Mini for a side exiting tail pipe. It is often the case that the RC 40 is difficult to obtain in the US. The Sonic Turbo, therefore presents a very good alternative. It makes a little more noise than an RC 40 though is still more than acceptably quiet. As far as power is concerned, it shows no real advantage over an RC 40 until about 125–130hp. On a nitrous oxide injection engine of around 180hp, about a five hp advantage over an RC 40 is achieved which at that level is out of its designed power range.

9. Shown here is the flow for the big bore Maniflow mini system. This is a stoutly built system comprising two silencers as per the RC 40 system. It also has skid gussets added between the pipe and the body of the rear silencer to reduce damage from a ground strike. It narrowly beats the RC40 for flow so makes a good choice in all respects.

The art of installing an exhaust system is to hang it, not fit it. This RC 40 system is going onto a Mini with rubber suspension straps at key points. If an exhaust system is fitted solidly, it will transmit engine vibrations and noise into the car and either the bracket or the exhaust system soon break due to the rocking of the engine.

fitted to 1275 GTs from around '77 onwards. Such systems were designed to cope with engines having a power output up to 55hp and not much more. If used on engines capable of outputs above that power level, they start to cost a lot of hp.

A popular move in the past has been to use just the Cooper 'S' tail box. That's the main part of the silencer. This improves the system's efficiency from the gasflow point of view but unfortunately it also increases the noise situation beyond acceptable bounds. I can quite categorically say that your neighbours will not like the sound level put out by your car if you use only a Cooper 'S' tail box. Apart from that, there are genuine Leyland tail boxes and there are others. Sometimes it's not always easy to distinguish between them. Consequently, you may well be using a tail box which is not only loud but does nothing for the power output either. The chart, **Fig. 10.16** will give you a good general idea of what flows and what doesn't flow, but remember, just because they are on the chart, doesn't mean they will meet road legal noise levels. For instance, the Metro competition system is included so that you can see just how

Competition Silencers' R.C. 40 compares. You will notice that the Metro competition exhaust system flowed 170cfm, good for power but there's no way it will pass a noise test. On the other hand, the **Competition Silencers** unit flows 195cfm, some 20% more and it does this while allowing no more noise than the standard exhaust system. At present, the Competition Silencers system is considered by many to be the best available, and **Figs. 10.17** and **18** demonstrate its capability on some engines in full race tune. You will notice in both cases that the power difference between an open pipe and that given by the silenced system is minimal. The question is, just what is the top line capability of the **Competition Silencers** unit? Just how much exhaust can be put through it before it starts to lose hp? Fortunately a friend of mine found that answer for me. He used the Mini system behind an MGB engine and tested with open exhaust and with the R.C. 40. Actual hp went from about 130 at the wheels down to about 125–126 when the unit was fitted, so we can say that in all probability the R.C. 40 type unit (be it one for a Metro, Sprite, Mini or whatever) is good for power levels up to about 125 flywheel hp and at those levels it absorbs between zero and $2^{1/2}$hp.

If you need more flow than the RC 40 gives then read the following section on the Sonic Turbo.

For The Track

A race car silencer/muffler that has been popular for many years, especially in the USA where in certain classes the silenced engine is required due to race regulations, is the Supertrapp muffler. Although available in a variety of configurations, the Supertrapp is basically a megaphone with some stacked conical-shaped diffuser plates in the end of the pipe. The main marketing ploy that Supertrapp have is that it is a tuneable unit

Because it works well, the Cyclone Sonic Turbo has become one of the most popular high performance mufflers in the USA. Units with the appropriate size sleeves to suit the existing pipe are supplied through A.P.T.

Fig. 10.16a Silencer efficiency comparison

Though the Super Trapp race car silencer is popular, its performance leaves much to be desired as these test results show. By the time noise has been reduced to the level produced by an effective Turbo silencer/muffler, the power output of the Super Trapp equipped engine is way down.

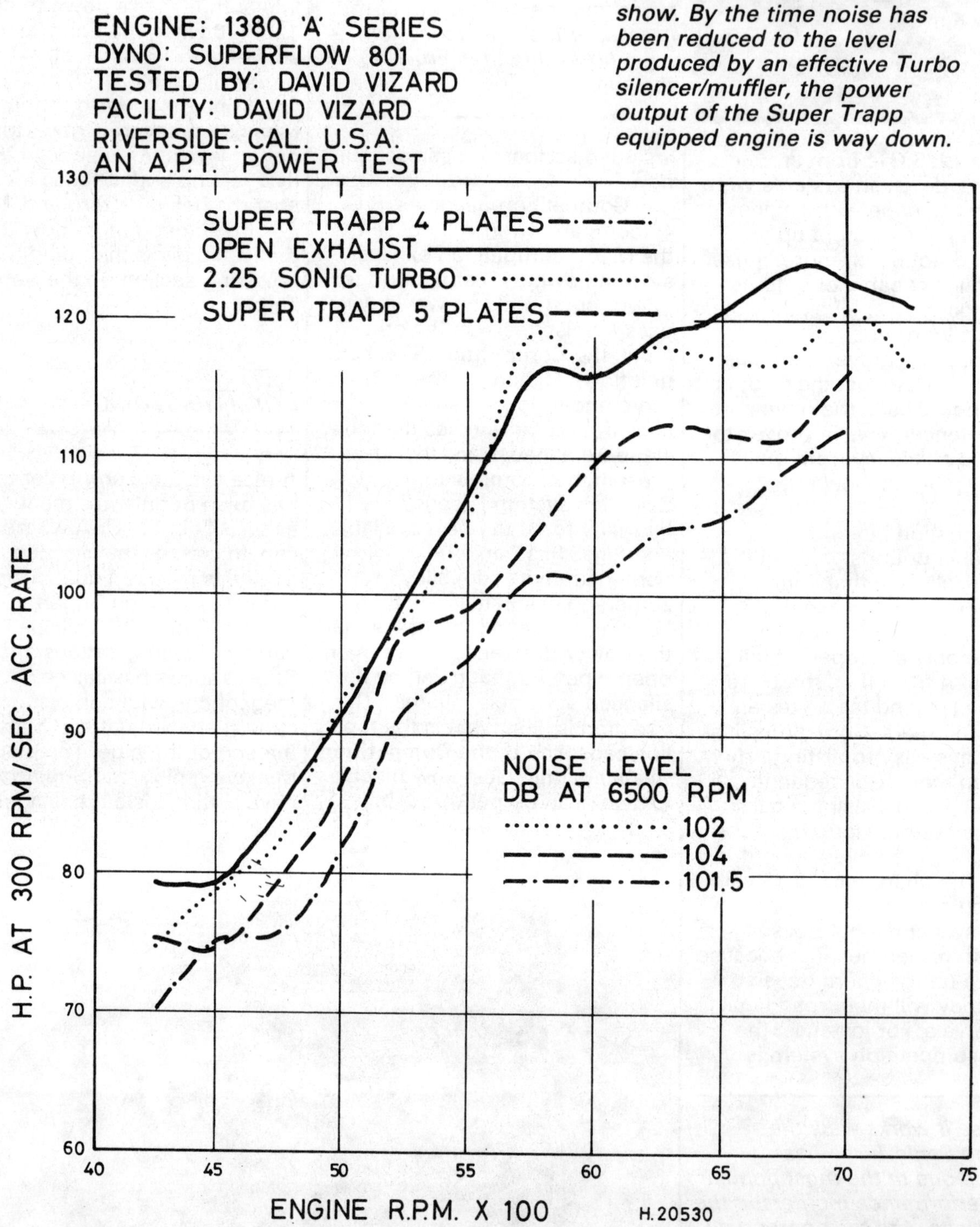

Flow testing a Sonic Turbo always produces a good solid 320cfm-plus. On the other hand . . .

and can be tuned for best results. However, in general they don't qualify that statement by indicating in what area the best results can be obtained. My initial testing of the Supertrapp some years ago was on the flow bench. I found that the unit flowed so badly, unless a large number of plates were put in, that it looked like a waste of time to even try it on the dyno. Eventually, due to pressure from many fellow enthusiasts, I got around to a full blown dyno test. **Fig. 10.16a**, shows how things worked out on the dyno. To make the

. . . the Supertrapp race car muffler cannot boast such flow figures until so many discs are in the system that the noise reduction achieved is very minimal. Unless mandated by regulations, the Supertrapp is not an item I would fit onto any of my vehicles.

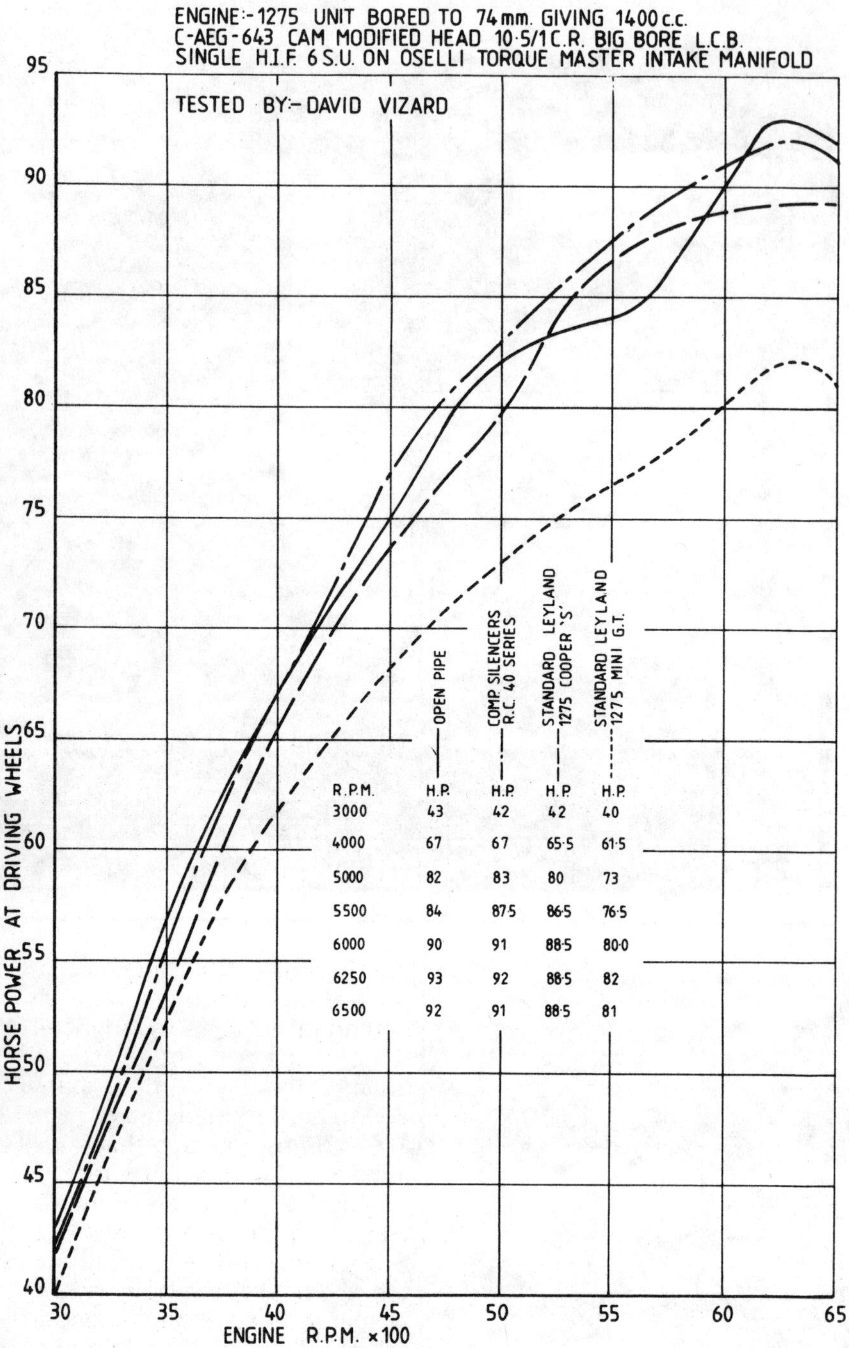

Fig. 10.17 Comparison of silencers on an extensively modified engine

Shown here are some typical figures when different silencers are used on a highly tuned engine. Note the Competition Silencers unit, and to a much lesser extent, the standard Cooper 'S' system, actually gain power over an open exhaust pipe in certain parts of the rpm range. This is probably due to enhanced exhaust gas resonation taking place which is improving the breathing over the rpm range concerned. The presence of the two silencer boxes is probably accountable for this phenomenon. The 'dip' in the power curve between 5500 and 6000rpm is something which is often seen on engines running with open exhausts. It could be that at these rpm a negative pressure wave is actually detracting from the power output and the use of the silencer fixes this problem. Whatever silencer is used, it needs to be an efficient one or the back pressure generated will substantially reduce power output. The 1275 GT unit is a prime example of how to lose power. This type of exhaust system uses a smaller exhaust pipe at 1½in OD as opposed to the 1¾in OD of the other system tested here. An important point to note is that the Cooper 'S' system tested here is a 1975 and earlier type. Current production replacement Cooper 'S' systems use the same silencer units as the 1275 GT unit tested here. The fact that the pipe size on the current 'S' units is larger is of little consequence because it is the silencer units themselves which cause the bulk of the power loss, not the small exhaust pipe. A very important factor to note is that all the silencers tested here cut noise to road legal requirements.

comparison valid, it was necessary for me to silence the Supertrapp-equipped engine down to a level given by a muffler of known performance. The particular unit chosen here was a 2½ inch Cyclone Sonic Turbo. In spite of being a 'three pass' silencer/muffler, this unit will flow in the region of 330cfm. If you inspect the curves on the graph you'll see that the Sonic Turbo-equipped engine produced very close to the same output as an engine equipped with an open exhaust. Only from 6000rpm upward did the power deviate significantly below that of the open exhaust. Of notable interest is the fact that at 5750rpm, the Sonic-equipped engine produced more power than the open exhaust. This suggests that it is not the lack of flow of this muffler

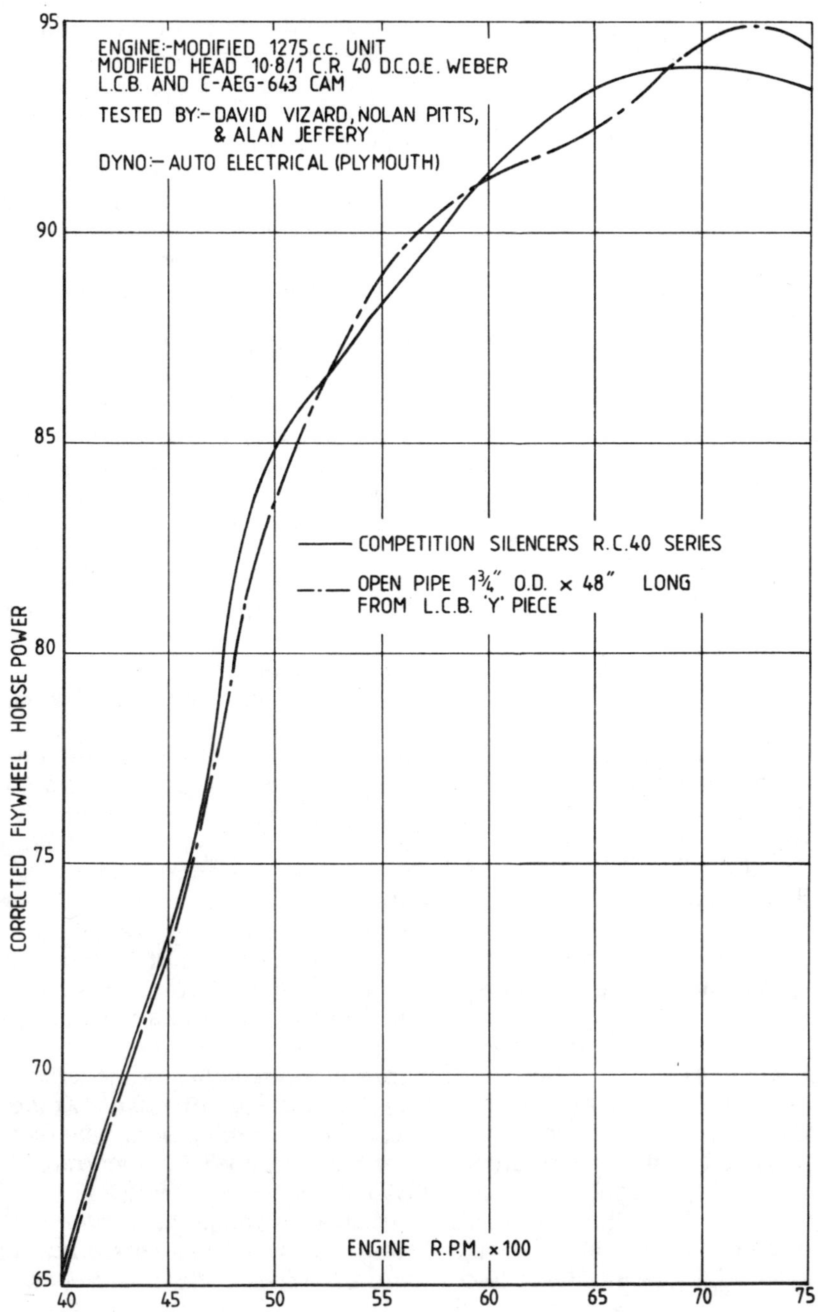

Fig. 10.18 Comparison of tuned-length open exhaust and Competition Silencers' RC 40 series silencer

Over most of the rpm range the RC 40 unit produced as good and sometimes better power than a 48in tuned length exhaust. In this instance, the purpose of the test was to find out exactly what difference the presence of the silencer boxes made. For that reason, the tuned length open exhaust pipe used was $1^{3}/_{4}$in OD instead of the $1^{7}/_{8}$in OD often used. The bigger pipe seems to be worth about 1, at most 2hp in the 6500 to 7500rpm range, but on the debit side it can lose about the same in the 4000 to 5000rpm range.

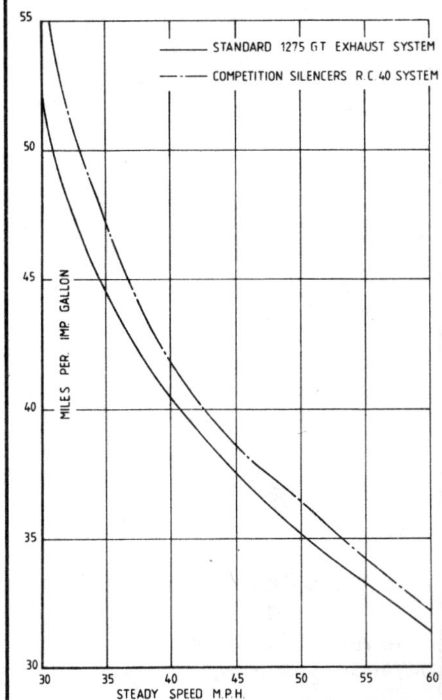

Fig. 10.19 Comparative fuel consumption

The test used a standard BL 1275 GT silencer system and a high efficiency system

that is causing the reduction in horsepower, but that it may be setting up adverse resonances in the exhaust system. As at 330cfm, the flow is more than adequate for an engine of 160–170hp or so and that's well above the output of the test engine. The next part of the test was to equip the engine with a Supertrapp and keep on reducing the plates until the noise at a predetermined engine speed in the middle of the usable power range was the same. As can be seen, with five plates the Supertrapp was just over the noise given by the Sonic Turbo and with four plates it was slightly under. However, it never matched the power output of the Sonic Turbo in either circumstance. Even though it made more noise, the five plate Supertrapp setup produced 8hp less at 5750rpm than the Sonic

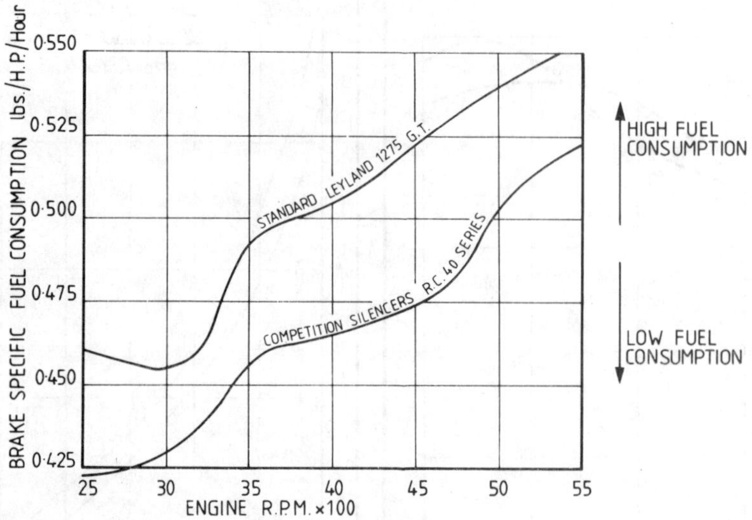

Fig. 10.20 Comparative fuel consumption at full throttle

The test used a standard BL 12 75 GT silencer system and a high efficiency system.

Turbo and close to 19hp less when in the four plate configuration. Tuneability of the Supertrapp is there as the manufacturers claim, but it is hardly the sort of tuneability we want. When you tune in the horsepower, you also tune in the noise. When the noise is tuned out, so is the horsepower. As far as muffling efficiency goes in terms of noise reduction versus power capability, the Supertrapp leaves a lot to be desired as these tests show; it is in fact inferior to a high performance silencer.

Back Pressure & Economy

Right at the beginning of the chapter, I mentioned that high back pressure and low flow capability would cost an engine mileage. If you wondered just how much mileage could be sacrificed, **Fig. 10.19** should provide some answers. This shows some *steady* speed mpg figures for a Mini. These figures were not produced on the road, but by a dynamometer simulation technique. The reason the dyno was used was simply to eradicate the variable effects of wind and rain, road surface gradients, etc. In the dyno cell, all possible variables can be held to very close limits. The simulation technique is quite simple and was done as follows: the first step was to produce a road load power curve for a Mini. In other words, measurements were made to establish the amount of hp required to propel the car at a given speed. Together with this the engine rpm was also established. Knowing these two factors it was then easy to simulate 'on the road' conditions in the dyno cell. For instance, at 30mph the engine rpm of the guinea-pig Mini, was 2000 and the hp required was 5. That's 5 flywheel hp, not driving wheel hp. On the dyno, the engine was set up to turn at 2000rpm and the throttle adjusted to produce 5hp. Under these conditions, very accurate fuel measurements can be made, knowing full well that the car wasn't about to run out of road or suffer a head wind or anything like that. As a result, it was possible to measure fuel consumption figures to a third significant decimal place, a feat very difficult to achieve by any other means. All these figures were then re-converted back to mpg. The graph **Fig. 10.19** is the result of just one series of mpg tests. Installing a high efficiency silencer, again an R.C. 40, the 1275 Mini produced worthwhile gains in economy, especially at the lower speeds. In both cases, the ignition timing and mixture were optimized for each installation.

When it comes to full power conditions, it's easiest to determine the fuel consumption in pounds of fuel used per hp hour. The lower the number is the more economical the engine is or rather, the more fuel efficient it is. Using the same silencer, accurate figures of full power fuel consumption were done. In general, for a drop of 1 inch of mercury back pressure, fuel economy improved approximately 1%. The tests depicted in **Fig. 10.20** show that fuel usage at full power was generally around 6% better. Not bad, considering that the silencer was also producing nearly 10% more hp. That's having your cake and eating it.

11 Camshafts, Cam Timing & Valve Trains Part 1

The camshaft plays an important part in shaping the characteristics of any engine, and the Series 'A' engine is no exception. In fact, **the Series 'A' engine has a few quirks as far as cam specifications are concerned,** that you need to be aware of **if you are to get the most from modifications in this area.** However, first let's look at a few basics concerning cam design.

It's important to understand why the cam is significant. Whether your requirement is for the maximization of hp, torque or economy is important. Let's look at the hp aspect first.

Not surprisingly, the characteristics that the cam designer tries to build into a cam are closely allied to breathing, or to be more precise, the engine's breathing ability and rpm. For a moment consider just the intake valve. The classical situation has the intake valve opening at TDC and closing at BDC. In other words, there is no overlap and no delayed closure of the valve.

In this simple example, starting at the top of the induction stroke, as the piston moves down the bore, so the intake valve opens. When the piston is half-way down the bore, the valve has reached full lift. As the piston approaches the bottom of the bore, so the intake valve closes. This looks simple enough, but as engine revs rise, the engine runs into a breathing problem. Starting to open the valve at TDC means that as revs climb, so the ability of the valve to flow enough air becomes limited, even at the beginning of the induction stroke when piston speed is relatively low. Remember that the airflow speed through the valve is going to be very high because a large piston is drawing on a very small valve area, so even early on in the induction stroke, the higher the valve lift in relation to the piston movement, the better the cylinder filling will be.

A similar situation presents itself at the other end of the induction stroke. Let's assume that rpm has reached such a point the valve cannot supply air fast enough to fill the cylinder by the time the piston reaches BDC. In fact the cylinder may only be about 75% filled. If the valve closes when the piston is at the bottom of the stroke, no more air will be drawn in. A moment spent studying the motion of the

This rollerized rocker assembly may look unusual, but after the cylinder head, the cam and valve train it is the next most important aspect as far as hp and the Series 'A' engine is concerned.

piston in relation to the crankshaft will reveal that around the bottom of the stroke, the crankshaft can turn quite a considerable number of degrees without actually moving the piston any significant amount. Under such a situation, if the valve was held open 20° longer, the piston may have only moved up the bore by 1 or 2% of its stroke. However, the intake valve now has something like 11% more time to fill the cylinder. The net result is that cylinder filling at the end of the induction stroke is likely to be better. This is borne out in practice. Because piston motion is slow at the top and bottom of the stroke, we find that it is possible to open the intake valve a considerable number of degrees before TDC and close it a considerable number of degrees after BDC with no really bad side effects.

The same sort of situation can be applied to the exhaust. If the exhaust is opened before the piston reaches BDC on the power stroke, there will be time for the cylinder to 'blow down', that is to release the residual combustion pressure in the cylinder. There will also be more time for the valve to reach a reasonable lift, so that when the piston has to forcibly expel the remnants of gas in the cylinder on the exhaust stroke, it will have a reasonably free-flowing aperture through which to do it. Remember, **if the exhaust valve flows badly, it takes hp to push the exhaust out, and that means less flywheel hp available to drive the vehicle along.**

Closing the exhaust valve after TDC gives the piston the best chance possible of pushing the largest amount of the exhaust out of the cylinder. In fact a certain scavenging effect can take place here (as discussed in the exhaust manifold chapter) if exhaust velocity through the exhaust port is sufficiently high. If this is so, as the piston slows towards TDC, the velocity of exhaust can still be high enough to draw some of the residual exhaust from the chamber. This reduces pollution of the incoming charge by the exhaust and the result is more hp. If the scavenging effect is high enough, then the intake charge can be started on its way into the cylinder before the piston has actually begun its downward stroke. However, this sort of effect doesn't take place to any marked degree, if at all, with the relatively short period cams normally used for road work. With an engine like the Series 'A' which is not renowned for good breathing capabilities, the scavenging effect is difficult to achieve even on a race engine.

Extended timing over and above the TDC/BDC opening/closure, dictated by the classic Otto four cycle sequence, can help power considerably, but if carried to an extreme, will first cause a loss of low rpm output and then high speed output. Imagine a situation where intake valve closure is delayed to say, 80° after BDC as would be the case for a race cammed engine. The piston, under these circumstances, will have moved a considerable distance up the bore. At low rpm when the gases have ample time to fill the cylinders, we find the new charge being pushed back into the intake port, because the intake valve is still open and the piston is now on its way up the bore. Putting it into technical terms, the low rpm 'trapping efficiency' of a race cam is bad. Also, there's a nasty side effect that is brought about by having a lot of valve overlap. If opening of the intake valve is started very early so that it's well open when the piston starts downward, we find gases from the previous exhaust stroke can now pass into the intake port as well as the exhaust port. Remember, at this point in the cycle the exhaust valve is still open because its closure has also been delayed. In fact, this period around TDC when both valves are open is called 'overlap'. When a lot of overlap exists, we find that the engine suffers a loss of low speed tractability often referred to as race engine 'camminess'. At low rpm such an engine will miss and cough and generally splutter until the rpm reaches such a point when the gases are compelled by momentum to travel in the correct direction. If you refer back to the cylinder head chapter, you will see that I made a special emphasis of having valves which actually tend to bias the flow in the correct direction. The Rimflo valves possess just this tendency. If, when the valve sequence is in the overlap period, the intake valve flows badly in the reverse direction it will tend to suppress the bad effects of a high degree of overlap.

Having shown how extending valve timing effects the power output of the engine, let's have a look at lift. From the flow figures in the cylinder head chapter, you can see that by and large, the airflow capability of the valve/port continues to climb right up to relatively high lifts. In fact, figures of 0.550 – 0.600 of an inch can be achieved before the port's flow capability is exceeded. Valve lift must be considered in relation to what the valve/port combination is capable of. As a guide we can say that the higher the valve lift the better. There is little point, however, in lifting the valve much past the point where the airflow curve levels out. Also, it takes more hp to drive a high lift cam and valve train setup than it does a lower lift one. There comes a time when the hp absorbed to drive the camshaft and valve train, exceeds the power gain achieved by having the higher lift. A further modifying factor, and one linked with the last, is that **the bigger the engine is, the higher the valve lift required for the engine to make its maximum hp.** A good example is related to big capacity engines. Let's assume you have a camshaft which works very well in a 1275cc engine that you intend to stretch to 1400. You will then have a larger engine which has to breathe through the same valves. At any particular point in the rpm range the bigger engine will try to draw in a proportionally larger amount of air. If the cam lift valve lift/ timing characteristics are left unchanged, the valve will represent a greater restriction to the flow of gases. To offset this, ideally the valve lift rate should be increased and together with this, the total lift. **The benefits to be gained by increasing the lift rate and the valve lift are greater on a bigger engine than on a smaller one.**

Looking at the other end of the scale, we find that on a small engine there is a limit to how much valve lift does any good in terms of producing extra power. Unless flow is woefully inadequate to start with, small valves on small motors don't respond to high valve lift by producing much extra hp.

In Particular

All I have said so far applies to engines in general rather than the Series 'A' engine in particular. Each engine has its idiosyncracies so I will now relate cam timing and lift to the Series 'A' engine.

First, a golden rule: the smaller the Series 'A' engine, the more sensitive it is to losing bottom end power when longer cams are used. Here's a good 'for instance': a 290° cam can be made to pull strongly from around 2800 rpm upwards in a 1275 engine. Below 2800 rpm the engine can still function, although its torque output is lower than would be the case with a shorter cam. Trying the same cam in an 850 will produce a very cammy motor. The engine will hardly take full throttle at much below 3500 rpm. This, of course, makes the assumption that the 850, like the 1275, is fully prepared in other respects, i.e. both have big valve heads, Weber carburation etc. This phenomenon, of a big engine pulling on the cam sooner, is quite normal, as the gas speeds are higher in the bigger engine than in the small one. That's basically because each engine is breathing through ports of similar size. The 1275 ports may be bigger, as are the valves, but they are not proportionally bigger in relation to the capacity.

When considering large displacement Series 'A' engines such as 1400s, 1430s, 1480s and on up to 1600, we find that since ports and valves have already been enlarged for maximum flow, there is physically no more room for further development. As a result, the cam characteristics are all tamed down. In other words a 320° cam in a 1600 Series 'A' will still pull reasonably well from rpm as low as 3000. On the other hand, that same cam in an 850 will produce a motor which won't even run much below 4200, let alone pull. A 1600 engine can be given close to full throttle below 3000 and still produce respectable torque. The 850, with that same cam, will simply fall flat on its face and sign off, just as surely as if the ignition were turned off.

The conclusion is that we have to be far more critical in the selection of cams for small engines, than for big engines, especially if that motor is expected to be used for day to day transportation. Another factor which needs to be taken into account if the engine concerned is for use on the road, is fuel economy. The rule here is very simple: the greater the cam timing, the less likely it is to aid economy. Again, this applies more in a small engine than a big engine.

There is also a limit on smaller engines concerning how many degrees of cam timing can usefully be employed and still produce extra power. For instance, a 998 engine equipped with the Leyland 649 race cam which has 300° intake, 300° exhaust timing, will produce a better power curve than that same engine fitted with the Leyland Super Sprint race cam which has 320-320 timing. Maximum hp is slightly better on the Super Sprint cam by about 1½ to 2 hp, but only at the very top end of the rpm range, i.e. around 7500 - 8000 rpm. Below this, the 649 cam generally produces markedly better hp. In other words, the power figures suggest that the Super Sprint cam has just gone over the top for the requirements of a race 998. On the other hand, the Super Sprint Cam used in a 1600 engine, will produce quite a bit more hp than the 300°, 649 cam. Undoubtedly the reason stems from the fact that the cylinder head has totally inadequate breathing in relation to the size of the engine. As a result, hanging the valves open longer to help fill those woefully inadequately supplied cylinders produces more hp.

Choosing The Numbers

If you're contemplating a cam change for your engine, a whole variety of options is available to help in selecting a suitable profile. Just looking through the specifications of other Leyland cam profile within the range may give an idea for a cam which fulfils your requirements. On the other hand you may prefer to select one of the specialist cam grinder's standard profiles that appears to fit your needs. The last choice is to have a cam profile ground to your own specifications. Provided that you are sufficiently knowledgeable about your engine, this will allow you to make small but subtle changes to the cam timing and adjust the cam characteristics to exactly what you want for the specification of motor that you are building. Before having your own cam designs made you must be very confident of your grasp of cam design theory and an understanding of the previously mentioned idiosyncracies of the Series 'A' engine. If not, you will just be wasting money. However, having said that, I will explain the effects of changing various parameters of the cam, so you have at least a usable understanding of what happens.

We have already looked, in simple terms, at what happens when the cam timing is extended. Now let's look at a few of the more subtle details such as the cam's timing angle when it's installed and the intake to exhaust lobe displacement angle. Firstly let's look at cam timing. When the manufacturer, be it Leyland or a performance cam grinding company, designs a cam they design a profile which has a certain amount of duration. This process is done for both the intake and the exhaust and then the cam is ground with a certain relationship between the intake and exhaust. By trial and error, the cam is either advanced or retarded in the engine until it produces the best results for the required application in mind. For what it is worth, a cam set at 2 - 4° of advance is most likely to run best. The next procedure is to grind another cam which has the intake and exhaust profiles in a slightly different relationship and then repeat the procedure to see if this produces any better results. Finally, the cam designer produces a camshaft which gives the best all-round results within the rpm band he designed the cam for. If the original homework was done right, then assuming you are building a typical engine, the camshaft you fit should be installed at the timing that the designer intended. In other words if a camshaft is supposed to be timed in to give an intake valve opening at 50° before TDC and closure at 70 after the exhaust opening 75° before BDC and 45 after TDC, then that's what it should be timed in at. Of course, this makes one assumption, that the power curve produced is what you want. In reality, it may be close to what you want but you may be able to improve performance for your particular needs by doing a little re-timing of the cam. For instance, assuming the original timing for the cam was average, we find that advancing the cam (that is making all the valve events occur earlier) tends to pull up the bottom end power slightly at the expense of the top end. Retarding the cam tends to increase the top end power at the expense of some bottom end power. That may sound like a simple rule and, indeed, the issue can be somewhat clouded. When the cam designer designs the cam, he has to make a guess at the typical engine spec of the engine in which the cam is likely to be installed. Sometimes it's possible to re-time a cam and achieve an improvement in power and torque throughout the rev range. In other words, no bad side effects. This happens because the new cam timing just happens to suit better the particular combination of parts that your engine has. However, we must make the assumption at this point, that in most cases timing figures at or near those quoted for the cam, are most likely to work best.

Let's now consider the lobe displacement angle (in U.S.A. more often known as the lobe centre-line angle), a topic which is not often discussed in Series 'A' engine tuning circles. The lobe displacement angle is the angle between the high lift point of the intake and the high lift point of the exhaust, and it's measured on the cam. By that I mean that it's expressed in cam degrees, not crank degrees and basically it's the only valve event that is normally expressed in terms of cam degrees. **Fig. 11.1** clearly illustrates the angle involved. If you are having cams ground by one of the specialist cam grinders, you have an opportunity to experiment with lobe displacement angles on the type of engines you are building. Reducing the lobe displacement angle or increasing it, may produce a power curve more favourable to your requirements. The effect on power of narrowing, or widening the lobe centres can very soon become complex, as there are numerous factors which tend to influence the situation. Such things as the bore to stroke ratio, valve sizes, the rod length to stroke ratio, port velocity and the flow characteristics of the valves, especially at low lift, all play a part in influencing what lobe displacement angle will work best.

Bearing in mind all these variables. let's lay some basic ground rules. Leyland's favourite lobe displacement angle for production cams is 107½°. This applies right up to and includes the C-AEG-643 cam which is an early Formula Junior profile. In modern terms, this cam can best be described as a very hot road cam, sufficiently hot for an engine to fall off the cam at the lower rpm range. This being the case, it's a good example to demonstrate the effect of changing the lobe centres. If the lobe centres were reduced to, say, 102½°, this would have the effect of increasing the overlap. This means that both intake and exhaust valves are further open when the piston is at TDC and about to start its induction stroke. As a result the engine is now more prone to the passage of exhaust gases out through the intake. This makes it more cammy in the low rpm range, but when the engine rpm reaches the point at which all the gases flow in the right direction the engine will come 'on' the cam much more suddenly. From that point on, if the engine needed more valve lift at TDC to get better volumetric efficiency, the engine will produce more torque and more power. The penalty has been that it is now less flexible. If we take the opposite tack, we widen the lobe centres, we find that the amount of overlap is decreased. As a result, the engine becomes more flexible. However, the effect of scavenging may be reduced if, in fact, any took place in the first instance. This will be so because the valves are much nearer closed when the piston is at TDC and about to start its intake stroke. On the other hand, since the intake valve is now held open longer after BDC, the engine is very likely to be able to rev further up the rev range to achieve its peak power. Up to that point it sounds quite good, but widening the lobe centres can produce less power and less torque throughout the rev range. All it does in fact do is to make the engine much more drivable, and it's an easy way to allow a longer period cam to be used without necessarily throwing away drivability. Because of these effects, there is a tendency to use wider lobe displacement angles on road cams than on race cams, simply because the wider lobe displacement gives better drivability for any given opening period of cam timing.

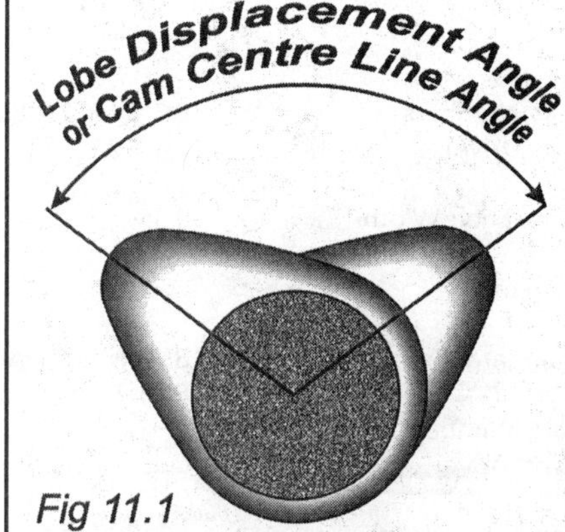

Fig. 11.1 Lobe displacement (cam centre line) angle

Cam Advance/Retard

Throughout this camshaft chapter you will hear references to cam advance and retard. This will be a completely meaningless factor unless you know what the reference points are to determine whether a cam is advanced or retarded. Take a look at **Fig. 11.1** and you'll see the explanation of the lobe centreline angle. Let's take a particular cam and say it has 106° lobe centreline angle. If that cam is then timed in so that the intake valve reaches full lift 106° after top dead centre, ie, the same as the lobe centreline angle number, then it is said to be timed 'straight up'. There's no advance or retard in the cam. If a cam is 4° advanced this means it will reach full lift at 102° after top dead centre on No 1 intake. That's 4° earlier. If it's retarded by 4°, it will reach full lift at 110° after top dead centre on the No 1 intake stroke.

Fig. 11.2 Camshaft chart *All linear measurements in inches*

	1	2	3	4
Part Numbers (Spider Drive)	12A 1065		126 726	AEG 577 AEG 323 AEG 538
Cam Lobe Width	$\frac{3}{8}$ in		$\frac{3}{8}$ in	$\frac{1}{2}$ in
Markings			2 Rings	
Standard Use	Auto Mini, Auto 1100		Auto Mini, Auto 1100	Midget Mk III Sprite Mk IV 1300. 12 H Engines Marina 1.3
Part Numbers (Pin Drive)	8G 712 2A 297 2A 571		12G 165. AEA 630	
Cam Lobe Width	$\frac{3}{8}$ in		$\frac{3}{8}$ in	
Markings			2 Rings	
Standard Use	850 Mini & All 803 & 948 cc Engines		1100 Sprite Mk II Midget Mk I	Cooper S Midget Mk II
Part Numbers (Slot Drive)				
Cam Lobe Width		$\frac{1}{2}$ in		
Markings				
Standard Use		850 & 998 Metro		
In. Opens BTDC	5°	9°	5°	5°
In. Closes ABDC	45°	41°	45°	45°
Ex. Opens BBDC	40°	49°	51°	51°
Ex. Closes ATDC	10°	11°	21°	21°
In. Period	230°	230°	230°	230°
Ex. Period	230°	240°	252°	252°
Cam Lift In/Ex	0.200/0.220	0.250/0.235	0.250/0.250	0.250/0.250
Valve Lift. STD Rockers	0.280	0.318/0.300	0.318/0.318	0.318/0.318
Valve Lift. 1.5 Rockers	0.330	0.375/0.360	0.375/0.375	0.375/0.375
Lobe Displacement Angle	107.5°	107.5°	107.5°	107.5
Tappet Clearance STD 1.5	.010-0.12/ .012-.014	.012-0.12/ .014-.014	.012-.012/ .014-.014	.012-.012/ .014-.014
Checking Clearance STD/1.5	0.019/0.023	0.021/0.025	0.019/0.023	0.016/0.019
Full Inlet Lift At*	110° ATDC	106° ATDC	110° ATDC	110° ATDC

***This is a factory–recommended figure. If more performance is required read the text concerning cam timing.**

	1	2	3	4
Overlap	15°	20°	26°	26°

5	6	7	8
	AFG 542	C- AEG 567	
	$\frac{1}{2}$ in	$\frac{1}{2}$ in	
	1 Wide Ring	AEG 568	
	Cooper 'S'		
	AEG 510	88G 229.2A 948 12A 122	
	$\frac{1}{2}$ in	$\frac{3}{8}$ in	
	1 Ring	1 Ring	
	Cooper 'S'	997 Cooper	
$\frac{1}{2}$ in			$\frac{1}{2}$ in
$\frac{1}{4}$ in Groove After 3rd Lobe			
Metro, Marina, Allegro, MG Turbo Metro			MG Metro, Vanden Plas Metro
9°	10°	16°	16°
41°	50°	56°	56°
55°	51°	51°	59°
17°	21°	21°	29°
230°	240°	252°	252°
252°	252°	252°	268°
0.250/0.250	0.250/0.250	0.250/0.250	0.250/0.250
0.318/0.318	0.318/0.318	0.318/0.318	0.318/0.318
0.375/0.375	0.375/0.375	0.375/0.375	0.375/0.375
107.5	107.5	107.5	107.5
.012/.012-.014/.014	.015/.015-.018/.018	.015/.015-.018/.018	.015/.015-.018/.018
.021/0.025	0.016/0.019	0.016/0.019	0.021/0.025
106° ATDC	110° ATDC	110° ATDC	110° ATDC
26°	31°	37°	45°

Production Road Cams

Ignoring profiles for a moment, there are, amongst factory produced cams, three basic types. The principal difference between these types relates to the drive to the oil pump. The earliest type of cam is the pin drive. In the early '70s the spider drive cam appeared on the scene, and with the introduction of the 'A+' engine, the slot drive cam was introduced. Although it's possible to swop them around, and I'll deal with that in the relevant section, each drive type needs the relevant oil pump and/or oil pump and adaptors, if it is to be fitted into an alternative block. Essentially, any cam can be used in any block if the right modifications or parts are used.

Shown here is an example of every type of camshaft-drive. The top cam is a slot-drive. The one nearest the camera is the original pin-drive and the one farthest away is a spider-drive item.

The cam chart **Fig. 11.2** details the production camshafts available. The cam in column 1 is Leyland's original one-cam-for-every-engine profile. It was used in all 850 Minis, most of the early inline engines, including the 948 Sprite, and, I believe, it was even used in early A30 engines. Merely by looking at the timing figures, one might think it a pretty tame cam which wouldn't

Early cams had ⅜ inch wide lobes whereas the later cams of the higher performance variety have ½ inch wide lobes. When having a cam re-profiled, the wide lobe one is the best type to use.

produce much in the way of results. However, it's a lot better than the specification would suggest especially in smaller engines. In spite of its short timing, if the engine in which it's installed is equipped with other free-breathing components such as carburation, intake manifold, modified head, etc., this cam will allow the use of rpm usefully up to 6000-plus, even though the specification suggests it ought to die by 5000. Its main asset is that it is a cam which produces good flexibility and, another important factor, good economy. Its main drawback is that it has insufficient lift, even for an 850.

Another aspect of this cam is that even though Leyland's specification calls for the cam to be installed so that it reaches full lift on the intake valve 110° after TDC, the cam very often produces better results when it's advanced to give full lift around 105 - 107°. When this is done, an improvement in torque is gained around the 2000 - 3000 rpm range, with at worst only a very minimal loss in power above 5000 rpm, and at best a small gain produces more power output in the most often used range, thus improving top gear overtaking capability. This cam advance applies almost

Virtually all standard profile cams are short on lift. Installing offset bushes in the standard rockers and mounting the rocker and shaft assembly into pillars with a suitable offset in the shaft location bore is an effective way of picking up hp. Depending upon the state of tune of the engine, 2 to 3 hp can be expected from this move without necessarily incurring any loss of low end performance.

Over several years of development period, Dave Mountain has found that most 850 Mini-X engines will produce better mid-range power without losing any top end power, when the cam is timed to 105 to 107° rather than the standard 110°. In other words, the cam is advanced 3 to 5°.

It's not necessary to have an expensive switchable magnetic stand for doing cam timing. This simple magnetic base and dial indicator holder is what I used for years for checking and installing cams on my own engines.

irrespective of the engine it's installed in, although it does prove more beneficial to engines displacing less than 1-litre.

Another characteristic of this cam is that it responds well to being used with high ratio, high lift rockers, especially on bigger engines. This is, of course, due to the insufficiently large 'area' under the lift curve on these bigger engines, because the cam was designed for use in a small engine. If you are installing one of these cams because you have to, i.e. that's all you have available to you or you are building a competition motor for which the rules call for a standard cam, it's well worth noting that these cam profiles can vary quite considerably because production tolerances were wide. It's also worth noting that later cams, that is those ground after about '79 or '80, are reputed to have computer smoothed profiles. In other words the cam profile was optimized on the computer and although the cams have the same timing, the valves early in the lift cycle tend to reach higher lift values more quickly although max lift is often less than is the case for some earlier cams.

The next cam on our list is the Series 'A+' cam for 998 engines, also used, I believe, in export 850 cc Metros. The 998 'A+' cam, commonly found in 1-litre Metros, has a timing of 9-41 - 49-11. It would seem that this cam was developed with a view to producing better all-round power and economy than had been given by any previously used cams of similar duration. It will give more hp than the baseline 850 cam in column 1, and you can expect to get slightly better mpg at typical highway cruising speeds. In any engine this cam is used in, more power will be produced if it's used together with high lift rockers, but watch out on smaller motors for loss in mileage when this is done.

In column 3 we have a cam profile that has been around a long time. I first tried it in the early '60s when it came out in 1100 Midgets and Sprites. This cam still uses the 230° duration of the basic Mini cam but it has more lift at 318 thousandths. The exhaust timing, however, is extended to give a total of 252° duration. Effectively 11° are tacked on to each end of the basic exhaust period of the baseline cam in column 1. The extended period cuts pumping losses, and since the exhaust valve is further open towards the end of the exhaust stroke, more residual gases are cleared from the combustion chamber prior to the start of the intake stroke proper. The extended exhaust period is of particular significance. My experience with camshafts, has indicated that whatever duration of intake timing is used the engine seems to go better if the exhaust opening duration is longer, typically by 10 - 15°, but sometimes by as much as 20. This isn't to say there aren't some cams that produce good results with equal intake and exhaust timing. However, I often ask myself if they might not do better with a shade more exhaust timing.

In column 4, essentially the cam is the same as in column 3, the only difference being that this cam is ground on a ½ inch wide lobe instead of a ⅜ inch wide lobe.

In column 5 is the 'A+' cam as used in the ordinary Metro, i.e. the non-performance derivatives. The timing of this cam may look quite different from anything we have looked at so far. It is, in reality, the same cam as in column 3, though it may have been computer smoothed, but the whole cam has been advanced by 4°. This cam advance has improved the whole power curve on the 'A+' engine. Not only does the engine run well at low rpm, but it runs significantly better at high rpm, so this represents a useful tip. If you have a 5-45 - 51-21 cam in a 1275 motor, the motor, almost certainly, will run better if it's timed with 4° of advance to give the 9-41 - 55-17 timing that the 1.3 Metro has.

Moving on to column 6, here is a cam which first saw use in the Cooper 'S' engine and very rapidly established itself as a cam with excellent characteristics for a road motor. This cam differs from the profiles in columns 3 and 4 by virtue of having 5° of extra duration tacked on to each end of the inlet phase, thus giving a cam of 10-50 - 51-21 when timed to Leyland's specification. Although the cam works well timed at 110° after TDC it can often be made to work better if a free-flowing exhaust (an LCB for example) is used, with a high performance exhaust system, by being timed in 3 or 4° advanced. At 4° advance, this would give a timing of 14-46 - 55-17. This cam, like most of the cams having a standard lift of .318, produces significantly better results when used with a 1.5:1 rocker, and generally speaking, the increase in power is of a greater proportion on 1275 engines than on the smaller engines. While on the subject of lift and engine sizes, it should be pointed out that for cams up to about 270°, the small bore engines, and in particular 850s and 998s, don't seem to respond well in providing extra power when valve lifts go over about .375. In other words, above .375 lift the losses outweigh the gains. 1098 engines can withstand a bit more lift when we consider cams in the 270° range, especially if they are using big valve heads, but as I said, the real benefit of 1.5 rockers and high lifts, comes to light on the bigger motors. Having said this, on small bore engines with 1.5

Identifying Your Cam

Establishing which sort of cam drive is used on any particular cam is quite simple. Merely look at the end of the cam to see whether it's a pin drive, spider drive or slot drive cam.

The identity of this cam can be established using two factors. First, the two rings tell us it is either a 12G 726 spider drive or a 12G 165 or AEA-630 (they are the same) pin drive cam. Look at the drive to ascertain which it is. The nearby chart will tell you the timing figures for this cam.

The cam shown here, distinguishable by the single (arrowed) ring, is of the 2A 948 group. In the early '60s it was one of the most popular performance street cams available.

The three machined rings on this pin drive cam distinguish it as a 731 cam.

The number 649 stamped into the shaft of this camshaft indicates that it is a C-AEA-648 race profile cam.

Identifying the profile of the cam can be done in most instances by measuring the lobe width and noting any identification on the shaft. For instance, an AEG-510 cam can be identified by the fact it has ½ inch wide lobes and has a single ring machined which is left proud of the main part of the shaft. Most camshafts within the range can be identified either by a ring or rings on the shaft in conjunction with lobe widths. The exceptions to this are camshafts which actually have a part number stamped on them. For instance, the C-AEA-648 camshaft has '649' stamped on the shaft. This business of having a number one digit higher on the shaft than the part number of the cam is fairly common within BL part numbers. Other competition cams as you can see from **Fig. 11.3**, also have a number one digit higher than the number stamped on them. All cams with the prefix C are special limited production, competition cams, as opposed to cams for production line vehicles.

rockers and longer period cams, say 300° and over, extra power of about 3 - 4 bhp can be found but only at very high rpm.

The cam in column 7 was produced both as a production cam in pin drive form with a ⅜ inch wide lobe or a competition cam in a ½ inch lobe width and in spider drive. Since the demise of Leyland's Special Tuning Department as such, anything less than a full race cam is now very hard to obtain, but this profile nonetheless has been very popular for the simple reason that it works well. It is especially good in 850s and 1000s as it represents about the hottest cam you can use without sacrificing any noticeable power in the rev range used normally for day-to-day driving. The idle is still smooth, and the engine will respond well to the throttle in high gear at 1000 rpm. It gives a very useful power increase over the standard cam and in both 850s and 1000s works well with offset bushes to increase the valve lift to around 340 - 350 thousandths. On 1100s and 1275 engines, it's especially useful for adding that extra something in the way of top end power, again without noticeably sacrificing power low down. On 1275 engines especially, this cam works well with 1.5 rockers. Again, like many of the cams we have discussed so far, if the engine is being used with a freeflow exhaust system and an LCB, the cam seems to respond to being timed a few degrees advanced. Personally I favour this cam being timed at 105 - 107°.

The introduction of the MG Metro brought with it the hottest production Leyland cam yet. This cam, with timing of 16-56 - 59-29 uses the intake lobe of the cam shown in the previous column of **Fig. 11.2**, together with the exhaust lobe of the 731 or 800 cam depending on whether it has pin drive or spider drive. Even though this is a relatively new cam, many engine builders are finding it excellent in virtually all sizes of Series 'A' engines. It works very well in 1275 engines, especially with high lift rockers. This cam can be used to demonstrate a point which I made earlier on, that the Series 'A' engine likes to have more exhaust timing than intake timing. This rule seems to hold good provided simple symmetrical cams of less than 300–310° period are being used.

Though taking things out of order, I would like to discuss now the use of the 731/800 cam profile. The 731 cam profile has long been considered a half-race profile, but it simply doesn't produce results in a 1275 motor, and it only produces results in 1100 engines when used in conjunction with very particular specifications. I have found that the 731 cam really only comes into its own in 850s and 1000s and even then it's not really a sparkling cam. As one very well-known engine builder said: 'Install a 731 cam, lose all the bottom end power and gain almost nothing at the top end.' That just about hits the nail on the head. My own testing has shown that the 731 produces less power than the MG Metro cam right up to 4000rpm. Above this speed it shows an increase up to a maximum of 3hp. The trade-off of top end gain for bottom end loss is simply not worth it, but we'll talk about the possible reasons why this is so later. The MG Metro cam under discussion, can well be likened to the cam in column 7, but with extended exhaust timing. As such, it works, but the 731 cam which has the same exhaust timing and extended intake timing to match, does not often produce good results. Having discussed briefly the 731 cam I will move on to the Leyland Competition profiles. Most genuine factory cams of these profiles are now very scarce and, therefore, if you intend to use some of these profiles, you may have to use reproduction profiles from speed equipment cam grinders. **Ripspeed** in the UK have, with the help of BL, had the C-AEG-800 cam (731 profile) remanufactured to OE standards on camshaft blanks. Unfortunately, many of these reproduction profile cams are little more than junk, especially (and it irks me to say this) if they have been ground in England. When you buy a reproduction factory profile, or indeed any profile from a speed equipment cam grinder, be absolutely sure that you are buying from a well-known and reputable source, such as Kent Cams as good cam grinders in England are very few and far between. Series 'A' enthusiasts in the States are probably in a far better position to get a good, accurate cam profile than their counterparts in the UK, as the standard of cam grinding in the US performance industry is second to none. In the UK, Kent, and Burton have proved satisfactory. Like wise Crane, Isky, and Comp Cams in the US.

It may seem that, since the quality of after-market copied cams is often poor, the simple answer is to fit a genuine Leyland cam. However, even genuine Leyland competition cams have in the past, left a lot to be desired in quality. The figures quoted for the Leyland performance profiles in the cam chart **Fig. 11.3** are the official factory timing figures. When measured, some of the cams come pretty close to these but some profiles don't achieve the timing figures quoted. For instance the 300/300 C-AEG-648 profile, popularly known as the 649 cam has, in my experience, rarely measured 300/300. Typically when checked they show 4–10° short of this figure.

The next cam on the list, in column 2, is the old Formula Junior profile. It first came out in pin drive form as the 544 cam. Although it was originally designed as a race cam in Formula Junior engines, that is the 1000 and 1100cc units, it is an excellent road cam for bigger Series 'A' engines. It works very well in 1275s and is even better suited to 1400s, 1430s and 1480s. Although bottom end power is down by comparison with, say, a shorter timing cam, the power isn't less than a standard 1275 in these bigger motors, but once the engine gets to 2000, the timing is just right for the larger displacement, and it puts the power right where you need it most, between 3000 and 6000. The only criticism I have of this cam is that ideally it needs a little more lift than the 388 thousandths it has. This is particularly so if you have a well ported cylinder head, that is suited to flow in the higher lift ranges. For this reason I would thoroughly recommend the use of offset bushes for the rockers, but better yet, 1.5 ratio rockers which would typically put the lift figure up to around 440 thousandths.

Although this is a super cam for the big engines, don't write it off for small engines. I have used it very successfully in 1100s, 1000s and 850s and it produced good results in all of them. In the small valve engines, it doesn't respond significantly to having much higher lift than is given by the standard ratio rockers. For instance, using offset bushes on an 1100 engine with $1^1/_{16}$ inch exhaust and $1^1/_4$ inch intake, only results in about $1^1/_2$ hp increase on an engine already giving a little over the 80 mark. As for flexibility, it's still surprisingly drivable for use as

Fig. 11.3 Camshaft chart *All linear measurements in inches*

	1	2	3	4	5
Part Numbers (Spider Drive)	C-AEA-800	C-AEG-643	C-AEG-530	(Sprint Cam) C-AEG-597	(Super Sprint) C-AEA-595
Cam Lobe Width	½ in	½ in	½ in	½ in	½ in
Markings	3 Rings	707/1473	AEG 530	AEG 598	AEG 596
Standard Use					
Part Numbers (Pin Drive)	C-AEA-731	C-AEA-544	C-AEA-648		
Cam Lobe Width	½ in	½ in	½ in		
Markings	3 Rings		AEA-649		
Standard Use					
Part Numbers (slot drive)			STR-930		
Cam Lobe Width			½ in		
Standard Use					
In. Opens BTDC	24°	34°	50°	60°	60°
In. Closes ABDC	64°	74°	70°	80°	80°
Ex. Opens BBDC	59°	69°	75°	75°	85°
Ex. Closes ATDC	29°	39°	45°	45°	55°
In. Period	268°	288°	300°	320°	320°
Ex. Period	268°	288°	300°	300°	320°
Cam Lift In/Ex	0.252/0.252	0.306/0.306	0.315/0.315	0.315/0.315	0.315/0.315
Valve Lift. STD Rockers	0.320/0.320	0.388/0.388	0.394/0.394	0.394/0.394	0.394/0.394
Valve Lift. 15. Rockers	0.378/0.378	0.458/0.458	0.470/0.470	0.470/0.470	0.470/0.470
Lobe Displacement Angle	107.5	107.5	102.5	102.5	102.5
Tappet Clearance STD 1.5	.015/.015- .018/.018	.015/.015- .018/.018	.015/.015- .018/.018	.015/.015- .018/.018	.015/.015- .018/.018
Checking Clearance STD/1.5	0.016/0.012	0.016/0.019	0.016/0.019	0.016/0.019	0.016/0.019
Full Inlet Lift At	110° ATDC	110° ATDC	100° ATDC	100° ATDC	100° ATDC
Overlap	53°	73°	95°	105°	115°

a road cam in 1100 engines, although in 1000s and 850s the bottom end power starts to diminish quickly. In a 1000 'S', which has bigger valves than the equivalent small bore engine, the gas speeds are slower and this means that this particular profile exhibits much more camminess than it does in a 998 engine or indeed even in the 850. With a 1000 'S' engine, the 544 profile causes the engine to come 'up on the cam' distinctly around 3000rpm. Below those revs it runs a little more like a tame race cam than a hot road cam. As a result the engine is a top end screamer.

Moving on to column 3, we have the good old 649 profile. I say 'good old' with reverence rather than humour. This cam has been around a long time and it has proved over the years to be one of the most versatile race cams ever produced for the Series 'A' engine. Earlier I was complaining that the actual timing figures rarely concurred with those specified but this does not detract from the fact that this cam profile does work exceptionally well in the Series 'A' engine. **If you have built an engine which has a combination of components that are half-way right, a 649 profile will perform well, whereas many of the other racier cam profiles are very sensitive in this respect.** The only criticism of the 649 profile is that it's short on lift and this really applies throughout the range of engines. Even 850s, if equipped with state-of-the-art race heads, can use a little more lift than the 394 thousandths this cam is supposed to give. Offset bushes in the rockers will generally add around 30 thousandths.

Fitting 1.5 rockers on the smaller engines doesn't seem to do much, but on bigger engines, especially 1275 and above, the 1.5 rockers with the 649 cam do start to pay off handsomely.

In column 4 is Leyland's sprint profile. **This cam features 320° on the intake and 300 on the exhaust. I have never been able to make this cam work for me,** no matter what the specification of engine. The 649 has always produced more power for me than the sprint cam. Many other engine builders have expressed the same sentiments towards the sprint cam. Kent Cams, for instance, gound some experimental 'reverse' sprint cams (300° intake, 320° exhaust) and found that, in the right combination, these produce better results than the 649, sprint or super sprint cams. It seems to be that the sprint cam has been designed back to front;the 320 intake period is too long for the 300 exhaust period. **I wouldn't recommend the use of this cam in any**

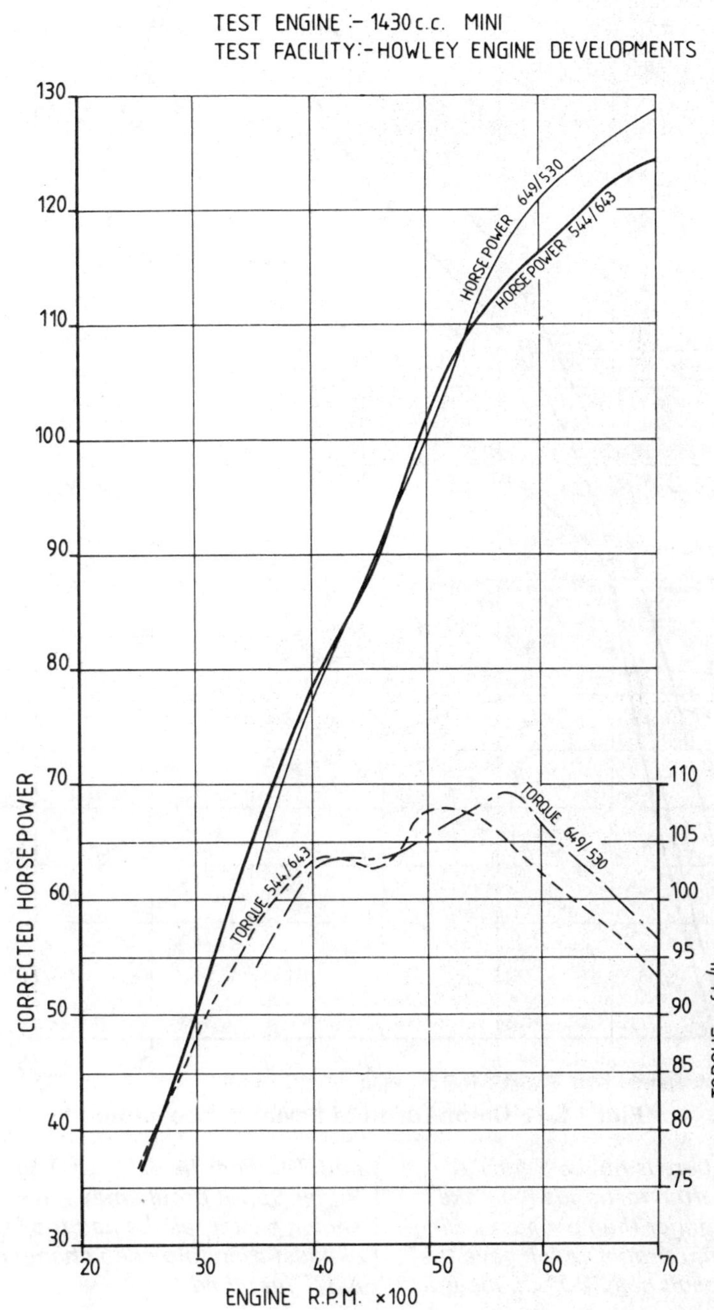

Fig. 11.4a Cam comparison

Cams used were 300°/300° 649/530 Leyland race profile and 288°/288° 544/643 Leyland race profile

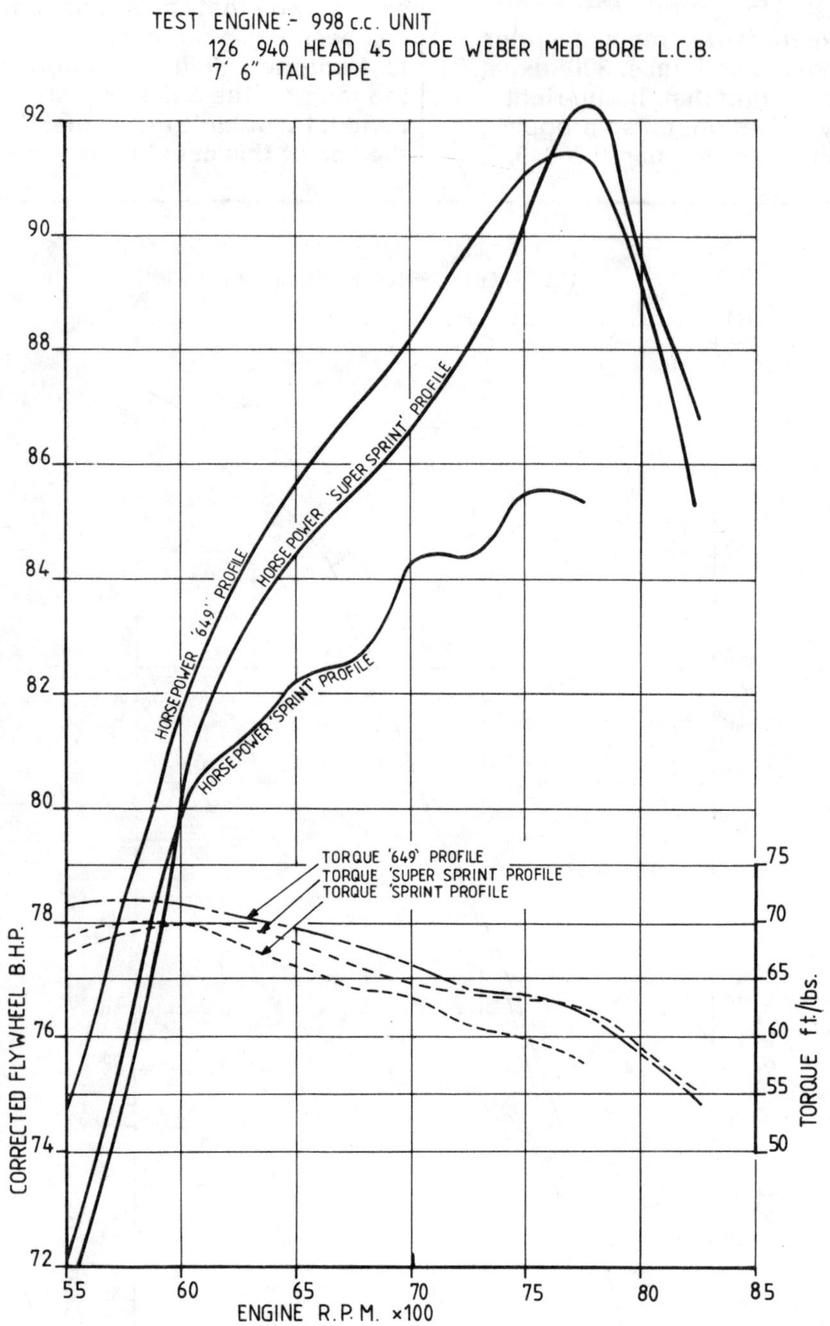

Fig. 11.4b Comparison of Leyland race cams

Shown here is how a Series 'A' engine often responds to intake timing longer than exhaust timing. The Sprint profile which gave the worst result has 320° on the intake and 300° on the exhaust. The Super Sprint profile may have shown better results had the LCB exhaust manifold been changed for a 3-1 manifold.

Series 'A' engine.
In column 5 is the super sprint cam which has 320/320 timing. It is my view that this cam is too radical for the Series 'A' engine, and is very difficult to make work. There are other, better, ways of skinning the cat: more lift and less timing achieves a better result. Provided that careful matching is carried out on such items as the exhaust system, cylinder head and induction lengths etc, some aspects of the super sprint performance can match those of the 649 profile. If we consider only the top end hp, then the 320 cam can produce a bit more than the 649, but at the expense of power throughout most of the rest of the rev range. This is especially so in 850s, 1000s, 1100s, 1150, etc. However, as we get to the bigger engines, so the situation between the 649 cam and the super sprint cam tends to even out. On really big engines, i.e. 1430 and above, the engine is struggling for breathing capability, even if a really good cylinder head is used. When this situation applies the only way to fill big cylinders is to hold the valves open longer, so under such circumstances the super sprint cam can pay off over something like the 649. However, a 649 cam with high lift rockers is more than a match for a super sprint cam with standard lift rockers. How would a big 1600 engine go with a super sprint cam and high lift rockers? Unfortunately that's a combination that I have not tried. The basic reason for this is that I've bypassed the use of such cams with cams of special profile which have more of what I know the engine needs, so I haven't had to make do! I understand, however, that Ripspeed have built big Super Sprint equipped units for rallycross which give good low down torque with good power at the top end. However, I'd be prepared to bet that these engines would have worked better with the 290°× .530 inch lift profiles that I've been using.

Having discussed the factory cams, let's make a few assessments, especially in relation to race cams. Experience has shown that it is easy to make the mistake of putting too much duration into a camshaft for a Leyland race engine. Increasing the duration to that of the 544 profile produces substantial amounts of extra hp in a Series 'A' engine. Going from the 544 profile to the 649 profile does not produce such a big increase. **Fig. 11.4a** demonstrates the sort of increase that occurs on a 1430 Series 'A' engine. To get the best from a 544 cam, it's necessary to pay particular attention to its timing. A mistimed 544 profile will lose substantial hp and torque. Install the cams at 105°–106°, no more, no less, on virtually all sizes of engine, unless you know you have a combination that needs more cam advance. On my own engines, which I know have good heads, manifolds, etc, I will put a standard 544 in at 106°. However, I often have a 544 profile ground on a 102° lobe centreline angle instead of 107.5°. These I put in at 101–102°. On 1275 engines which have had a stroke increase together with an overbore to bring them up to 1430, but more particularly 1400cc engines and upwards, it sometimes helps to advance the cam a couple of degrees, but only if the engine is being used with a relatively long induction system and Weber carburation. If it's only a single carb induction system with a typical high performance intake manifold and an air filter, then stick to the 106 setting.

I think the moral is: if you don't know any different, time any Leyland profile in to the specification quoted. Let us now talk about lobe centres. The 544 profile is the hottest profile that used the production 107.5° lobe centres. All the rest of their competition profiles use 102.5° lobe centres. Whereas 102.5° lobe centres do work well, it is very worthwhile, especially on bigger motors, to experiment with tighter lobe centres. Of course, if you want to experiment in this way, it's going to be necessary to have a Leyland profile ground by one of the performance cam companies. My latest experiences with lobe centres is that the bigger the engine the tighter the lobe centre needs to be. This flies in the face of my previous findings which just goes to show that generalizations supported by insufficient test data are dangerous. My latest experiments, though, are based on literally hundreds of cam tests in 1275–1400cc engines. With this many tests upon which to base conclusions I feel that I've now got a good 'handle' on the situation. Indeed, I've now done so much research work on Series 'A' cams that I can quite categorically say that to get the best output from a given profile or pair of profiles it's necessary to have different lobe centre line angles for each like 'pair' of cylinders. I know at first sight that may sound ridiculous but when I get to the reason later in the cam section all will become clear. While still on the subject of LCAs I have experimented with angles as tight as 98°. Many engines, but in particular those above 1400cc, respond to having LCA on 100°. Apart from experimenting with lobe centres, I feel there is a distinct area for improvement in cam and valve train design for the Series 'A' engines by looking at relatively short period, high lift cams. My own work with Crane and Kent cams has backed up this supposition time and time again. I have already said I feel the sprint cam has gone over the top. What I think would probably work best in, say, a typical 1275 race engine is a cam producing .550–.575 inch lift at the valve with around 290° to maybe 300° intake period and about 305–310° of exhaust period. These timing figures, along with LCAs to suit each individual cylinder requirement, looks like the route to go. Producing such cams can present a few problems which we'll look at in a moment, but for now let's look at some of the wide range of the 'off-the-shelf' alternative cam profiles available from specialist cam grinders.

Camshafts, Cam Timing & Valve Trains Part 2

Choosing the right cam specification for the application is the entire key to success when it comes to choosing a performance cam. Opening duration, which is so often the only factor considered, is not the sole influence on the cams delivered performance. Valve event timing is, and that means overlap and lobe centre line angles are more influential than duration. If the lobe centre line angle is inappropriate for the motor it will never make good torque so the quest for maximum performance is lost right there. However before we delve into this let's take a look at off-the-shelf cams and what they can deliver.

Formed in the early '70s by Bob Fry and the late Bob Munt, Kent Cams is a British company that has grown in prominence relatively quickly. This is due mainly, it would seem, to their producing some of the highest quality cams available in the UK. The Berko cam grinding machines that all of their cams are ground on, are based on design specs laid out by Harvey Crane of Crane CamDesign in Daytona Beach, Florida. Since leaving Crane Cams in 1990 Harvey has started a cam design and consultancy business. He is recognized world wide as a leading authority on cams and their manufacture. In addition to consultancy work Harvey also runs a highly acclaimed 'cam school' for professionals.

Kent Cams started out with good quality equipment, thus avoiding the pitfall that many UK cam grinders fall into, *i.e.* inadequate equipment for the job. During late 1984 Bob Munt

To avoid knocking the cam bearings as you install the cam, turn the block on end and lower the cam in as seen here.

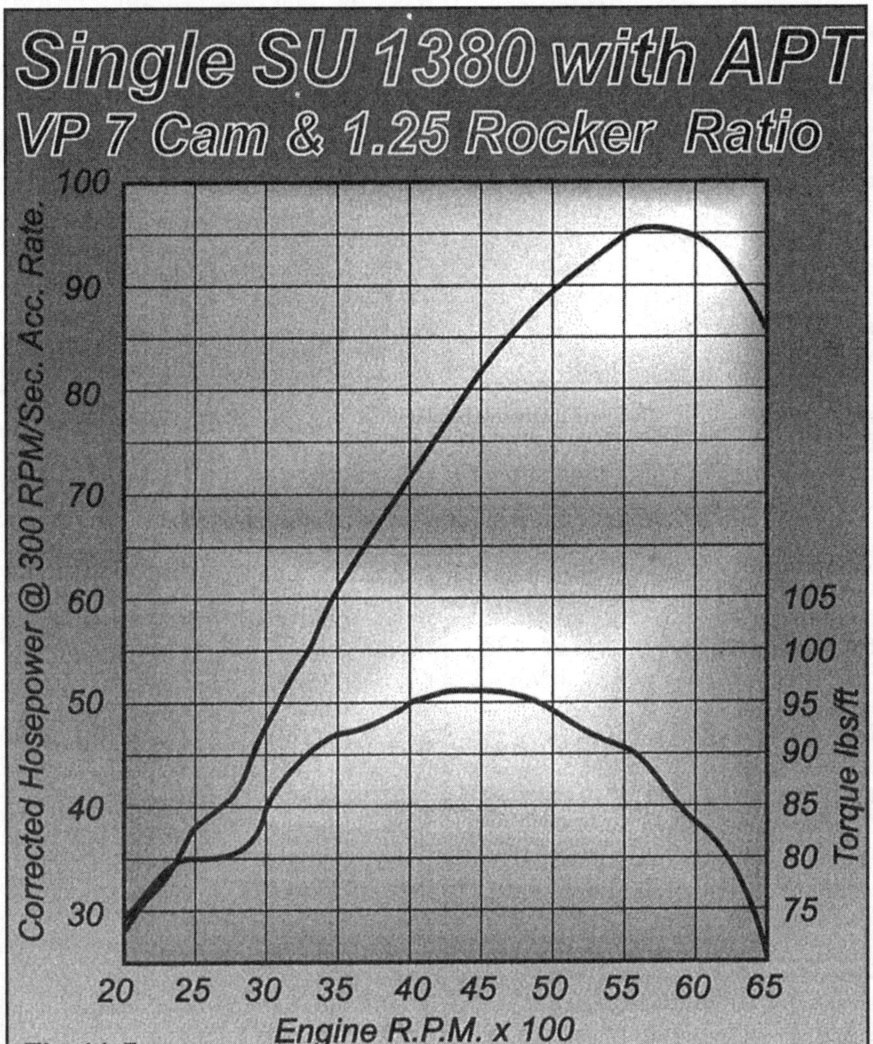

Fig 11.5

Here is a prime example of what can be done even on restricted funds. This motor has the most basic of valve trains in as much as it still retains the standard 1.25:1 ratio rockers. The point here is that if the cam has the appropriate event timing then it will allow a good combination of parts elsewhere to produce a strong motor. This particular motor, with its 73.5 mm Omega pistons and fully modified head sported a 10.3:1 CR. Induction was via an air flowed 1/3/4 SU on a Howley manifold and the exhaust dumped through an LCB and an RC 40 exhaust system. The result was a motor that would pull smoothly from 1,250 rpm in 4th and produce 96 ft/lbs of torque and 96 horsepower. If the 1.25 rockers were replaced with 1.5s at a later date this big bore motor would produce about 6hp more. To put this in perspective this motor makes more hp and torque than a Turbo MG Metro and does so with much less complication and cost.

of Kent Cams made me an offer that was difficult to refuse. In a nutshell, the deal was that Kent Cams would supply engines and grind on cam blanks any combination of profiles that was within their profile library. They would pick up the tab for testing the cams until I was satisfied that, within reasonable engineering limitations, the range of profiles was as near optimal as possible. From my point of view this was almost a licence to kill, but Kent Cams were as good as their word and from January of 1985 to June of 1985 an intense cam development campaign was embarked upon. If my memory serves me correctly, during that time several thousand – yes thousand – combinations of cam and valve train setups were tried and analysed in detail. Because of the sophisticated equipment to hand I was able to 'home in' very rapidly on near optimal cam timing figures for a given power band. The normal way of checking cams is to run one against the other to see which is better. The problem with this is that if one cam is worse, there's not a lot that can be learned about why unless full data of the engine's many other characteristics are taken. Most dynomometers are simply not equipped to do this. With my facility I was able to find, within reason, why a particular cam did not work, thus allowing intelligent changes to be made to intake timing, exhaust timing, lobe centreline angle, lift characteristics, etc. So the second time an attempt to improve on a particular cam was made, it was invariably much nearer what was needed. Usually by the third or fourth attempt the sought for power increases were being realized. Again, like the cylinder head situation, filling you in on all the details on the cam project would take up too much space. The printed test results from that one session alone (and there have been others since) amounted to over 400 typed pages of numbers. The explanation as to why various things happened, or didn't happen as the case may be, would simply have taken another 400–500 pages. In essence, what the end user is more interested in is what the positive results were and this is what I intend to outline here.

Firstly, the project was to **deal with the valve train as a whole.** After all, the function of the cam is to open and close the valves in the most favourable way possible. Yet it must function in conjunction with a suitable valve train. Although we're dealing with cams in this section, it is not possible to separate the cam from the valve train and talk about a specific item when we're trying to look at the power results obtained. So here we're going to look at the effects of altering certain components in the valve train namely the rockers. Although I deal with rockers in particular later on, here I'm going to deal with rocker ratios in general and hopefully prove beyond a shadow of a doubt that the 1.5, or yet higher, ratio is what the Series 'A' engine needs. If you're serious about getting horsepower from your engine, don't even contemplate running with the standard 1.25 ratio rocker. To put you in the picture it is going to be necessary to cover all the whys and wherefores of the valve train. It may look a bit late in the cam chapter to do this, but it is an appropriate time because up till now I've only spoken of what is available off-the-shelf in general terms. Now we'll look at the Series 'A' valve train in terms of its various design limitations.

Although it may not at first seem obvious, one of the principal limitations on any pushrod valve train is the cam follower. The diameter of the cam follower dictates just how much velocity can be put into the follower by the cam profile. The highest velocity occurs when the cam contacts the follower as near the edge as possible without actually edge riding. The simple formula that closely approximates (assumes flat follower face instead of 50 inch radius) the follower's maximum velocity in terms of thousandths of an inch per degree is given by:

$$\frac{D''-25}{114.6} \text{ thousandths/degree}$$

Where D = tappet diameter in thousandths of an inch. As can be seen with the 0.811 inch diameter follower, the Series 'A' engine is limited to a lift rate expressed in terms of degrees of cam rotation per increase in lift of approximately 6.85 thousandths per degree. The easiest way to envisage how the velocity is governed by the size of the follower is to imagine that the cam is actually an arm, the length of which is determined by the distance from the centre of the rotation to the contact point. Obviously, the longer that arm, the faster the peripheral movement. When the cam is on the base circle, the contact patch is in the middle of the follower, excluding any offset in the follower for the moment. The easiest way to envisage how the acceleration of a lifter takes place is to imagine how fast the contact patch moves from the centre of the follower to the outer edge in relation to the rotation of the cam. If it takes only a few degrees for the contact patch to move from the centre to the outer edge, then the acceleration on the follower is high as it takes only a few degrees to go from zero thousandths per degree lift to maximum. On the other hand, if the contact patch moves out slowly, this is because the acceleration of the lifter is slow. To make the Series 'A' engine perform, both follower acceleration and velocity are important. To get horsepower from the Series 'A' engine it is necessary to get that valve moving quickly to as fast a speed as possible. Since there are clearly defined limits on what can be done at the lifter/follower, this leaves us with a situation whereby an increase in velocity and acceleration at the valve must be catered for by **increasing the rocker ratio**. Here we find what is probably the saving grace of the Series 'A' valve train. It was, from the outset, designed with too low a rocker ratio, as explained in a later section. This means that using a high lift rocker is an **essential ingredient toward building horsepower.** By using a high lift rocker, many assets can be accrued. For instance, say you have a certain horsepower target. That target can be achieved easier with a high lift rocker setup utilizing a shorter cam than would be the case with the 1.25 rocker and a longer cam. The shorter cam will give more torque, plus a wider power band than a longer cam with low lift rockers. Essentially, the testing I did showed that even moderate timing cams with the lift, velocity and acceleration maximized, make for perfectly driveable characteristics yet produced close to the horsepower normally associated with long timing race or near race cams.

So much for theory, now let's look at some of the basics of what was achieved in this test series. Firstly there were some constraints applied to the tests. The cams developed had to be either single or dual pattern cams. Those of you familiar with cams may think that covers everything. Normally yes, but for where the DV/A.P.T. shop was with cam technology for the Series 'A' engines, it didn't cover everything. Since virtually every cam that's been available on the market is either a single or dual pattern cam, this constraint did not hold matters back anymore than anyone else with endeavours in this field. For those not familiar with the terms, let me explain. A single pattern cam is one which uses the same lobe on both the intake and exhaust. A dual pattern cam has one lobe for the intake and one for the exhaust. If you think about it, for any given engine there is no guarantee that both lobes should be the same. The Series 'A' engine certainly presents a situation where exhaust duration needs to be different from intake

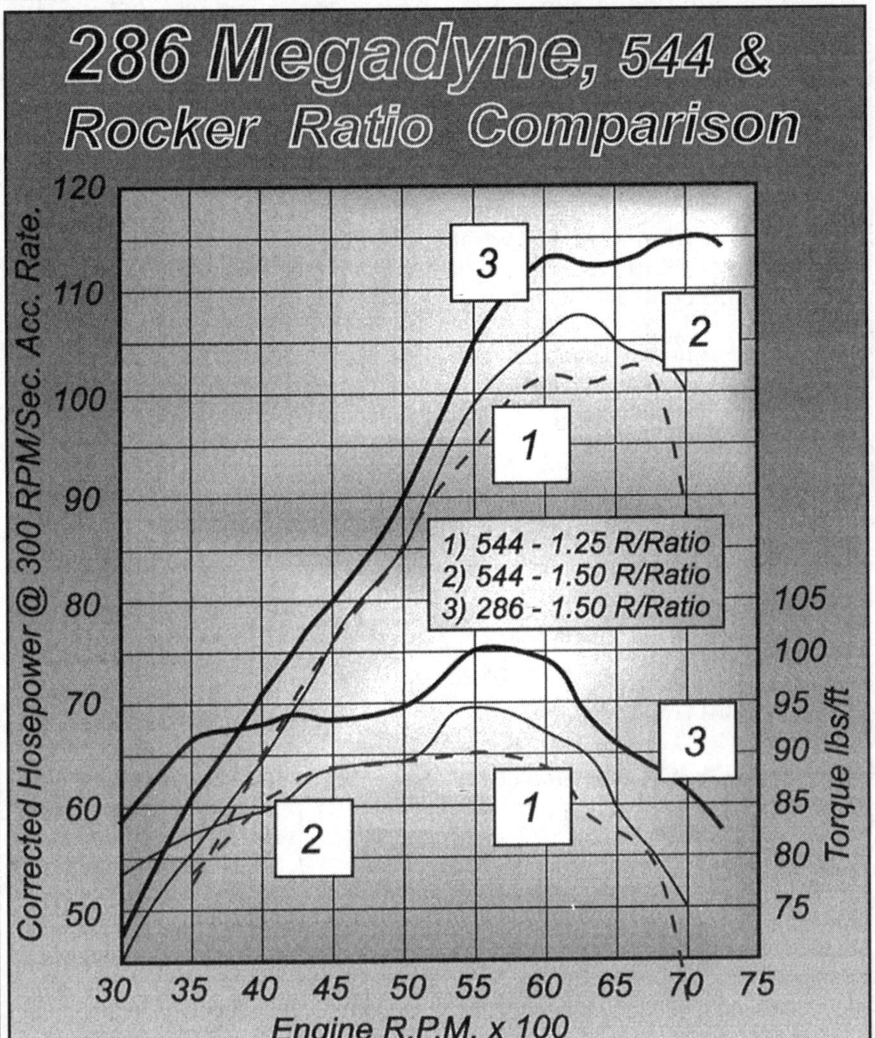

Fig 11.5a

Fig. 11.5a Comparison of rocker ratio influence on power
This test, performed at A.P.T., is based on what can best be described as a typical 1380cc enthusiast motor where everything is a little more on the conservative side than I would necessarily build. For instance the CR is 10.2:1 whereas I would have used 10.5:1. Water temperature is 185–190°F (85–87.8°C) whereas I strive to keep this down to 170°F (76.7°C). The cylinder head is a good one but not exceptional and so on. All this makes it very representative of the motors most commonly used on the street. A number of factors are evident from these results. Some are more subtle than others but the effect of adding lift in a bigger motor is blatantly obvious. As can be seen from the curves labelled 1, the standard rockers with a 544 cam simple cannot supply the air demand of a 1380cc motor. The result is that the top end turns over with the torque falling right off the graph. Adding lift by way of 1.5 rockers (curves 2) makes a considerable difference to the top end output. However the valve timing is still not right for the motor's requirements. All we have done here is give the valves more lift. If this is now combined with more appropriate valve event timing brought about by tightening the lobe centre line angle and adding some extra exhaust duration we make further and very significant gains as shown by the curves labelled 3. If the technology behind the development of this cam appeals to you then it will be worthwhile checking out the results shown in **Figs 11.5e** *and* **11.5f**.

duration. As I've stated previously, it tends to prefer longer exhaust duration than intake. It was with this in mind that most of the cam designs worked on came into being.

The 286 Megadyne

The first cam I'm going to deal with in the Megadyne range (as they became known) of cams from Kent Cams is the 286. I consider this cam to be about the hottest practical cam for most road use. Some people have been installing 286 cams into 1275s and then expect to drive satisfactorily in London traffic jams. Unless they are prepared to be patient, this is not the cam for those sort of situations. Firstly, a 286 cam should be used in no less than a 1275 engine and, strictly speaking, it was designed to go into 1380 or bigger engines as 1380–1400s are becoming extremely popular. Secondly, it is not meant to deal with day-to-day heavy traffic especially if the compression ratio has not been raised sufficiently to restore most of the lost low end output. It is meant to be the sort of cam where most of the driving is done on relatively empty roads at speeds over 40–50 mph. It will idle and drive in traffic when the engine is set up correctly, but is not the most docile cam for these conditions. As a baseline from which to compare what was to be the eventual design of the 286 Megadyne, I started off with the 544 profile (288°x288°x107.5° LCA, 0.388 in lift). The first thing was to establish, as suspected, that this cam, though it's an excellent cam, has insufficient lift for the job at hand on a large displacement Series 'A' engine. The first test was to run the engine with standard rockers and then with 1.5 rockers. In each case, the timing of the cam was adjusted to give the best possible power curve. On this, as in all the tests done, an adjustable belt drive was used on the engine so that the cam timing could be very quickly optimized. It literally took a minute to advance or retard the cam as necessary. The first cam put into the motor was the standard 544-643 cam ground on 107.5° lobe centreline angles. The graph, **Fig. 11.5a,** shows the effect on the power curve of simply optimizing this cam on both 1.25 rockers and 1.55 rockers. On our particular test engine, in both instances best power was produced on 106° valve setting. That is, 106° after top dead centre to full lift on No 1 intake. On the face of it the 1.5 rockers appear only to help the horsepower from 5000rpm upwards where they then show a considerable advantage. However, this graph can be very misleading because it does not give a complete picture of what's happening. The 1.5 rockers actually produced a considerable increase in volumetric efficiency throughout the rpm range, but in so doing, **made the exhaust timing a mismatch to the intake.** Much of the potential to produce extra power was lost because now the exhaust valve timing did not match the intake timing when the 1.5 rockers are used. More air is going into the engine, but the volume of exhaust has increased by a greater proportion, because of the temperature, than the intake. Now the engine is suffering from undue pumping losses. The next step was to grind a cam which utilized the same sort of intake parameters as the original baselined cam, but with a longer exhaust. This is exactly what the 286 Megadyne is. The uppermost curve on **Fig. 11.5a** shows what happens when the exhaust timing was increased and the lobe centrelines of the cam set to a figure that the engine more nearly wanted. As can be seen, power increased throughout the rpm range, although the volumetric efficiency actually dropped slightly up to about 6000rpm over the standard 544 profile with 1.5 rockers. The fact that the VE dropped slightly is not a point in question here. What's happening is the engine is utilizing the air that it is drawing in much more efficiently. Its brake specific air consumption has improved to a far greater extent than the drop in VE. The reason the power has come up is because it's able to expel the exhaust easier and therefore cut pumping loss. Above 6000rpm the longer exhaust timing plus the more favourable lobe centreline angles improved the VE slightly. You'll notice that the cam and valve train combination went on to produce significantly more power

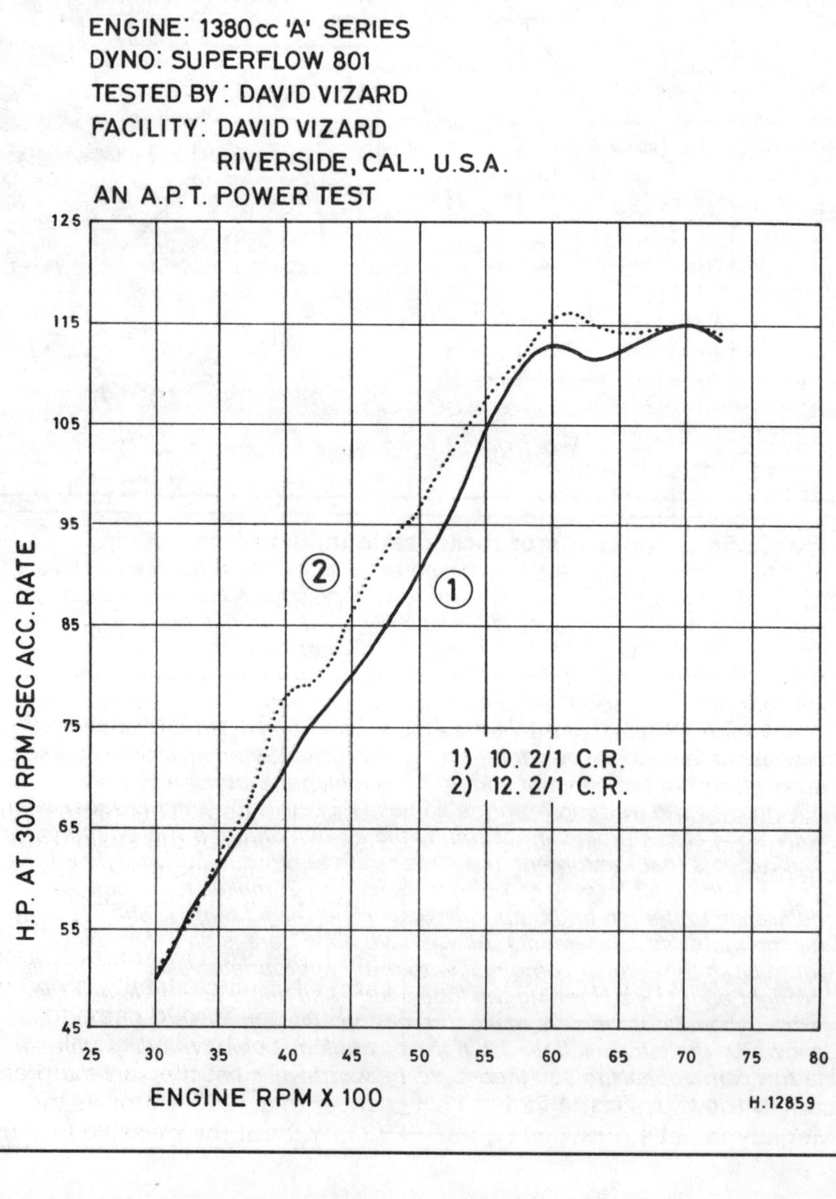

Fig. 11.5b The effect of increasing the CR

The differences between power curve 1 and 2 are typical of the effect the compression ratio has on engine output. The extra power is developed by virtue of a substantial torque increase throughout the rpm range. If fuel octane values are sufficient to allow a suitably high compression ratio, then it's definitely the route to take.

than the regular 544 profile. In spite of producing figures as good as a 649 with standard rockers, the cam still pulled strongly down to 3000rpm, and in a well set up motor (emphasising well set up) will go on to pull *down* to as low as 1800rpm. The particular engine used to test in **Fig. 11.5a** is consistent with about the hottest spec you would expect to use on the road. Since these tests were run there has been a trend of improved motor octane number in the fuels available at the service station pump. This has been more so in the USA than in the UK but it does mean that a little more CR can be used than the 10.2:1 used for this test. With modern fuels and attention to running temperatures 10.5 is probably on and more if you are prepared to use a good octane booster. (Aldon Automotive's will pump up the octane by about 2.5–3 numbers per recommended dose – while most other brands are hard pushed to make 1 number increase).

If you're prepared to use an octane booster, then there is much more potential in the engine with this cam. It is of a long enough period to warrant the use of a high compression to get the best from it. Shorter cams will function with a lower compression ratio without such great loss in horsepower. The graph, **Fig. 11.5b,** shows how much extra power was made when the compression ratio was raised from 10.2:1 to 12.2:1. Notice how the mid-range torque went up considerably due to more effective scavenging and less dilution of the incoming charge because of the smaller chamber.

Fig. 11.5c Kent Cams Application Chart

	Type	Timing*	Duration	LCA+	Lift 1.25 R/R	Lift 1.5 R/R
1	256 Megadyne	21-53-53-21	254/254	106	.320/.320	.380/.380
2	266 Megadyne	24-56-61-29	260-268	106	.320/.330	.380/.390
3	276 Megadyne	29-61-66-34	270/280	106	.360/.380	.430/.455
4	286 Megadyne	34-66-71-39	280/290	106	.388/.400	.460/.468
5	296 Megadyne	39-71-76-44	290/300	106	.400/.420	.470/.495
6	310 Megadyne	49-81-81-49	310/310	106	.420/.420	.495/.495

**Kent quote zero cam advance timing, but up to 3° advance can be beneficial in most instances.*
+These lobe centreline angles are known to work in typical displacement engines, i.e., 1275 to 1400. For larger or smaller engines Kent Cams can grind those cams on tighter or wider LCAs to suit.

Type	Application
256 Megadyne	Smooth idle, good mileage. Pulls from very low rpm. Excellent traffic cam, excellent towing, increase in low, mid and top end power.
266 Megadyne	Smooth idle, pulls from 1200rpm, towing okay. Good traffic cam. Small increase in low end, big increase in mid and top end.
276 Megadyne	Slightly lopey idle, okay in traffic. Strong mid and top end increase.
286 Megadyne	Lopey idle. About maximum for most street usage. Okay in light traffic. Very strong mid and top end power.
296 Megadyne	Lopey idle. Race cam needs free flow exhaust, CR, etc. to achieve best results. Mid and top end power only.
310 Megadyne	Drag race/sprint cam only. Use in fully prepped motor. Strong top end power.

The 266 Megadyne

Having looked at a killer street cam in the form of the 286 Megadyne, let's now have a look at something which retains all of the drivability and idle qualities you would expect of a production cam. This represents the other end of the scale from the 286 in as much as civility is a prime requirement for this profile. As a starting point for this cam I took the Leyland MG Metro cam, which incidentally is one of the best street/road cams to come out of Leyland. It will provide almost the same top end power in a 1275 motor as the 731, without the massive loss in low end power normally associated with the 731. The reason it works well is that it has essentially the right ingredients. However, the concept of the cam can, as my own testing was to prove, be fine honed to produce better results yet. Again, using a 1380cc unit as a mule motor, a standard Metro cam was installed and optimized. The valve train, with 1.25 rockers, was standard. The normal setting for the MG Metro cam is 110° after top dead centre to full lift on No 1 intake.

This is the timing used for most factory cams. However, this setting does not give optimal horsepower, and generally speaking, for most applications if the cam is advanced 2 or 3° it will produce a superior power curve. The test results in **Fig. 11.5e** we see the power curve with the MG Metro cam optimized, and under these conditions it produced 2–3ft lbs extra throughout the rpm range. At the top end this amounted to about a 3½hp difference. By changing the cam lobe centreline angle to 106° timing it in with 3° of advance and using it with 1.5 rockers to get the lift required, the power curve took a substantial upward leap with virtually no penalty down the low end. In fact, it's worth comparing with the power curve of a standard 544 cam used together with 1.25 rockers. The 544 cam was considered, at one time, to be the most potent road cam available and as such, sacrificed a degree of road manners down the bottom end. Here we have a cam and valve train in the form of the 266 Megadyne, which is not giving away anything in terms of drivability, yet is matching the output of what was once considered a race cam. A.P.T.'s own VP2 cam is very close kin to the Kent Cam's Megadyne 266 and the only changes we contemplated were softer closing ramps on the exhaust to minimize valve seat recession from extensive use of lead-free fuel in the USA. All the cams in the Kent Megadyne range had their design features settled upon as a result of the dyno testing in the DV/A.P.T. workshop. Within the Megadyne range I also developed some cams for smaller engines, *i.e.*, up to 1100, and for people who felt they could not afford the cost of high lift rockers. As far as possible, these all embodied the same design parameters as the cams intended for 1.5 rockers. Incidentally, though most of the cams I've worked on were developed to make full use of a 1.5 rocker, they all produce positive results when used with standard 1.25 rockers. This gives the end user, who has limited finances, the chance to install such a cam at the time of an engine build and realize a performance gain which can be increased to its full extent when finances allow the installation of 1.5 rockers.

In fact, the same religious development procedure was applied to all cams right through to the 310 Megadyne. The 310 Megadyne is right on the limit as to how much timing a Series 'A' can tolerate. This cam loses a lot of torque down the low and mid-range and it really is a cam to be used in a very high rpm engine. It must be used with compatible components elsewhere in the engine otherwise success is anything but assured. For instance, I would not recommend anything less than a 7-inch intake manifold. A compression ratio of 13.5:1 minimum, maybe as much as 14.5:1 and some experimentation will be needed because of fuels. It's also worth trying both an LCB and a three-into-one exhaust manifold to see which works best. There's no hard and fast rules as to what will apply here because factors dictated by the grinding of the cylinder head can have as much influence as anything on what happens at the end of the day. If you are going to cam an engine with the 310 be prepared to put the engine on a dyno and spend time optimizing the cam timing, exhaust pipe length and exhaust system. Pick the right parts and you can expect 140 plus hp from a 1400.

A.P.T./Crane Cams

About the time I was halfway through testing cams for Kent Cams, I was approached by Harvey Crane. I found out from Harvey that he had gone into production with billet cam blanks for the Series 'A' engine. By starting from scratch on the cam blanks, he was able to use the widest and largest lobe possible consistent with going through the cam bearings. This, along with superior material would allow a camshaft to be produced with faster opening rates and more valve lift in relation to the amount of cam timing involved. This presented an exciting prospect, so I spent some time discussing with Harvey the timing requirements I thought best suited the Series 'A' engines. I confirmed to him that valve acceleration was as he had anticipated, extremely important and that too much timing was detrimental to the power curve of this type of engine. Using a relatively new programme, Harvey produced some profiles which he referred to as 'tight lash' profiles. In other words, they used very little tappet clearance and produced very rapid acceleration off the seat. To give you an idea of how much more quickly these profiles opened the valves than conventional cams, let me make a comparison with a Leyland 649. A properly ground 649 profile has 300° of seat timing with 244° timing at 50 thousandths tappet lift. The Crane 'tight lash' profile has 286° of seat timing with 252° of timing at 50 thousandths tappet lift. From this, we can see that the profile is 14° shorter on the seat, but 8° longer at 50 thousandths lift which means it's opening that much quicker. To find out part of the secret of how come this was possible, I put one of the Crane profiles on my cam checking machine and found that the basic ingredients for the short timing were virtually no tappet ramp on the start of the profile and extremely rapid acceleration rates off the seat. The check of the cam profile was made after I had used these cams for some time, so the reliability of the cams was not in question.

However, there were certain precautions that had to be taken. Firstly, they cannot be used with standard Leyland lifters; only one type of cam follower is compatible with this type of cam. It must be a chilled iron follower and to make sure you're getting the right one, I would advise getting them from A.P.T. or a source that gets them from A.P.T. Unfortunately these followers are becoming scarce. In anticipation of their eventual demise, David Anton and I have been working on some alternatives. At the time of writing we have what appears to be a follower with even better wear properties than the chilled iron items we've found so effective to date. If you attempt to use anything but a chilled iron follower (or our new one assuming it delivers its currently indicated promise) one of two things can happen. If the cam is sprung correctly, as it approaches its designed rpm it is likely to break the follower through frictionally induced side loading. If the cam is sprung lightly, turning too many rpm will cause valve float and the cam will wear out or break the followers from impact relatively quickly. Since the quality followers are little different in price than any other high performance follower, it makes sense to take the time and trouble to get them.

As you would expect, A.P.T. has a range of cams for the Series 'A' engine ground on the standard camshaft. Most require a selected cam to start with as they need as much material as possible to take the new profile.

Many of these cams were developed during our tests for Kent Cams and are ground to our specs either by Kent or Iskenderian. The A.P.T. cams cost a little more, but that's easy to understand when you consider that our race cams have to be ground on 'as new' Leyland race cams. Many of these cams are now offered by A.P.T. in their Crane equivalent. For instance, one of the most popular A.P.T. cams is the SPVP3; this is what I consider our top-of-the-line street burner. The VP3C is the Crane equivalent. Timing is 276° intake, 286° exhaust and it sports almost 500 thousandths lift at the valves on 1.5 rockers. In terms of performance, the A.P.T./Crane cams are for normal single and dual pattern cams state-of-the art. They will produce more horsepower, more torque and a

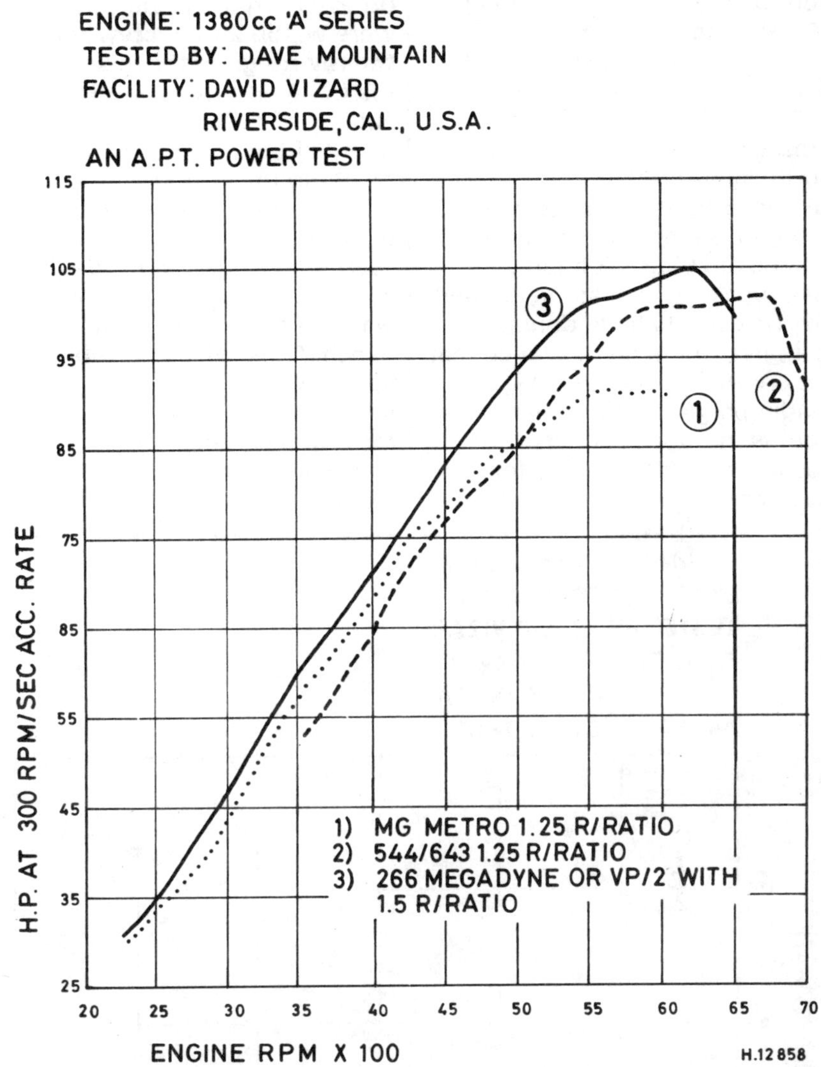

Fig. 11.5e Cam comparison
This graph demonstrates the basic effect of timing and lift on the Series 'A' engine. Curve 1 is the basic 252/268 MG Metro cam with nominally 0.320 inch valve lift Curve 2 is the result of changing to a cam having 288°/288° x 0.380 inch lift. The result is that power below 5000rpm dropped, but above 5000 it increased. Curve 3 utilizes a shorter timing cam with 1.5 rockers to give a valve lift of 0.375 inch in conjunction with a more favourable set of timing figures. Result, more horsepower throughout the rpm range than either the standard cam or the pseudo race profile of the 544/643 cam. As a matter of reference, A.P.T.'s VP2 cam produces a virtually identical power curve to the Kent Cam's 266 Megadyne cam.

wider power band than any cam currently available ground to normal specifications **(See Fig. 11.5e)**. In fact, the Crane profiles enabled, as far as I know for the first time, the passing of the 200psi brake mean effective pressure (bmep) barrier in a normally aspirated Series 'A' engine. Engines built to my specs have, on several occasions seen bmeps of 203psi. A figure, incidentally, which some learned engine experts have doubted, claiming, in the first instance that the DV/APT dyno must be reading high. Just to quell those doubts, not too long after one such test I ran an engine for Cosworth which had already been on one of their many accurate dynos. The figures I obtained on my Superflow dyno were within 2hp **everywhere in the rev range** of those seen on the Cosworth dyno. On the subject of calibration, since so many of the figures I use from the dyno become public domain by virtue of all the articles and books I do, it is imperative that I keep the calibration of the dyno as near spot on as possible, as with the calibration of the flow bench. Otherwise the published figures will be difficult for anyone to relate to. In other words, are the modifications I'm doing worthwhile or is my dyno reading high? Generally speaking, my dyno reads lower because essentially I don't so much as sell engines as hard core data. My clients, which are mostly big companies, want hard, raw facts, not the inflated numbers that are often fed to racers so they can trot away happily thinking they have some wonder engine. To give you an idea of how much numbers can be inflated, a client who moved his engine across town to run it on my dyno experienced a 50 horsepower loss over what he was seeing the previous day on a well-known and internationally respected engine builder's dyno.

The strong point then of the A.P.T./Crane cams is they bump the torque considerably. Over the years, this has been the hardest aspect to improve on the Series 'A' engine due to its limited breathing capabilities. Right now our dyno test engines, which are not always the best to take as a criterion since they receive so

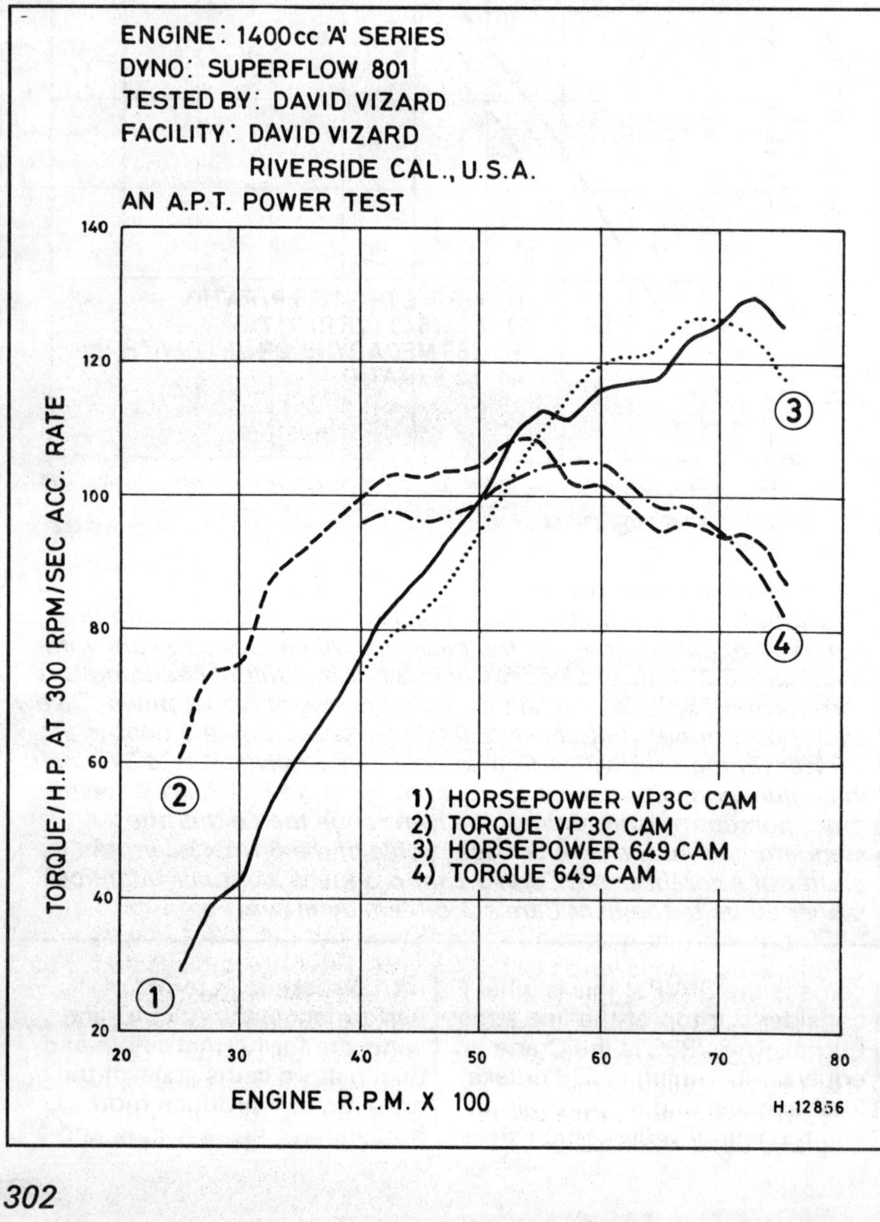

Fig. 11.5f Cam comparison

Here's what giving the engine what it wants does for output. The VP3C cam is 276°x286°x0.510 inch lift. The Leyland 649 with the same 1.5 rockers is 300°x300°x0.480 inch lift. The VP3C cam will take full throttle all the way down to less than 2000rpm whereas the 649 'falls off' the cam at about 3000rpm. Except for the small range between 6000 and 7000rpm, the VP3C beats the 649 in every other respect. It produces more peak horsepower, 130 as opposed to 126 for the 649 and more peak torque with 109.2lb ft versus 105.4. A few years ago a street cam versus a race cam situation would have produced expected results, with the race cam winning every time. However, buying the latest technology now means the performance of yesterday's race cams for the street today.

Fig. 11.5g Cam Comparison

What you see here is about as positive proof as you can get that the Series 'A' engine prefers high lift, short duration cams to long period cams with a lesser lift. In most of their years of race cam production, Leyland's now defunct competitions department failed to appreciate this important, but subtle fact. The results shown here are for two cams, both on 102° LCAs; the 649 is 300°x300°x0.480 inch lift and the VP8C is 286°x286°x0.520 inch lift. The 'area under the lift curve' for the two cams is similar, but the power, torque and power band width are all superior with the VP8C. Why? The higher lift and faster opening rates produce higher VE figures. The shorter seat timing cut the shared intake port time, so reduces robbing between the siamesed ports and the earlier closing improves trapping efficiency. At the time of writing the VP8C is probably the **best conventional** *race/rally cam available. With the right combination of parts, over 140 steady state horsepower is possible from a 1400. The 1380 shown here would pull 137 at a steady 7250rpm and this engine was in need of new rings!*

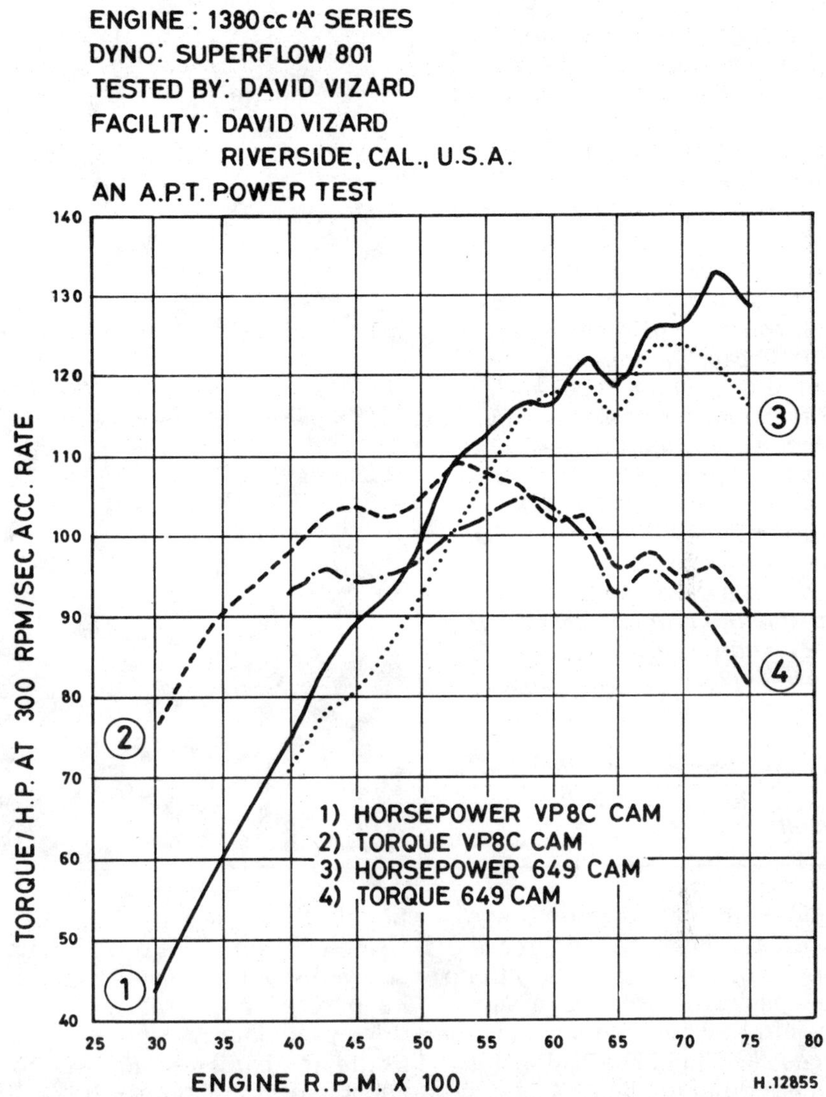

much abuse on the dyno, have consistently seen 1.39ft lbs per cubic inch. That's 85ft lbs per litre. When you consider there are many successful four valve per cylinder engines that only make 87 – 90ft lbs per litre, you'll see that's a pretty creditable torque output. In terms of horsepower output, the VP8, which is a 286°x286°x520° lift cam, in a 1400 with a high flow offset exhaust valve head has beaten anything I have tested in the way of a conventional single or dual pattern cam to the tune of 6hp and 5ft lbs. Most of the cams that have run close to this are our own cams. Yet the big thing about these short period cams is they allow the production of a lot of horsepower without the associated rpm. For instance, we've run some cams which have around 310° period, which have had to turn as much as 1500rpm more to make an almost always smaller peak.

While on the subject of power, it may do well to relate some successes used on engines that I've *not* built myself. During 1985 my daughter's race car in England used a 286°x286° VP8 cam in her Mini. Mike Parry did all the engine building and Dave Mountain did all the dyno work. Even though the car still sported a helical gear box, on Dave Mountain's rollers it put out 126hp at the wheels. To give you a comparison, some of the best 1700 crossflow Fords for the street put out 120–124hp. At the time, Dave Mountain said it was not only the highest power output by a Mini he had seen by a substantial margin (12hp), but also was the fastest Mini he had ever driven, race weight or not. The car was standard street

Seen here are our experimental set of 1.62 ratio Titan roller rockers. The workmanship in these rockers is extremely good, as is their attention to design detail. If you look carefully at the tips of the rockers you will see that each roller is offset to take care of the small amount of offset that exists on a 1275 between the pushrod and valve stem centres. Most rocker manufacturers ignore this requirement and satisfactory alignment is only achieved by juggling the position of the rocker on the shaft spacers and having a wide enough pad at the end of the rocker so that it spans the end of the valve in a satisfactory manner. Tests in the A.P.T. shop have shown that these 1.62 rockers have the potential to produce more horsepower, but they are likely to require a cam specification change to produce optimum results.

As many cam tests are performed at A.P.T., it was necessary to perfect a time-saving technique. After removing the rocker gear, the cam followers were 'lifted' from the cam profile by magnets affixed to the end of aluminium tubes. The followers were then held in the raised position using clothes pegs. Pulling the cam . . .

weight and my daughter drag raced it to finish second in a championship and that's with no concessions for the fact a Mini uses front wheel drive. In half the races she ran in she had to square off in the finals against V-8 powered cars. On slicks the car was capable of 0–100mph in under 15 seconds, 0–60 in 5.6. Using our computer programme showed that had this engine been installed in a rear wheel drive such as an Austin Healy Sprite, the car would have been capable of running low 13s, maybe even high 12s, under ideal conditions with trap speeds

. . . was simply a matter of removing the belt from the adjustable belt drive, pulling off the cam sprocket and seal, then sliding the cam out. Incidentally, the spider drive for the oil pump was brazed into the oil pump so that it didn't drop down into the sump of the engine.

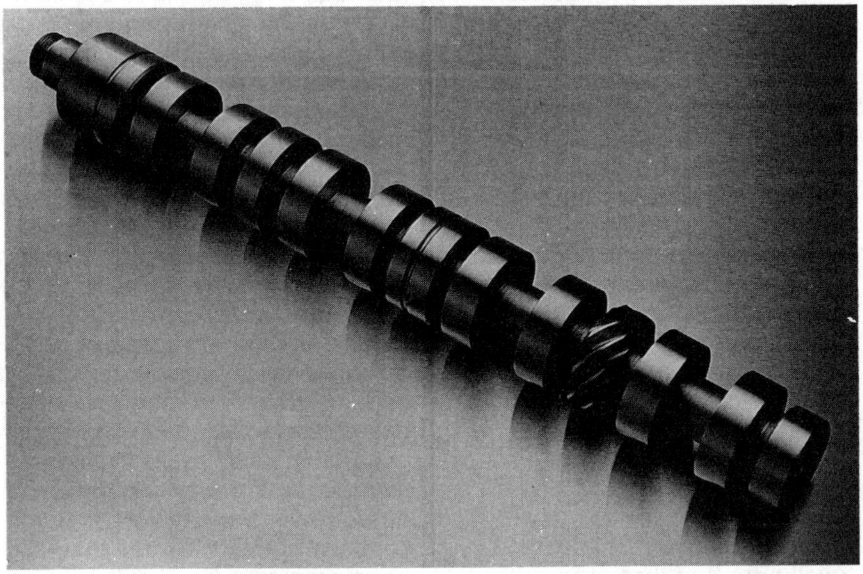

After seeing how effective Crane's billet cams have proven to be, Kent Cams followed suit by tooling to produce their own Series 'A' cam blanks. Seen here is a Kent Cams blank ready for any profile you care to have, including a roller follower type profile.

in the region of 104mph. Consistent performances by Samantha Vizard on the drag strip with trap speeds in the 96–97mph range, on a track that was less than ideal, indicate that the engine was making 144hp at the flywheel.

Fig. 11.5h A.P.T. Cam application chart

Type	Timing	Duration	LCA	Lift 1.25/1	Lift 1.5/1	Application
VP3C	34-62-69-41	276/288	106	394/414	465/489	1
SPVP3	See	Chart	Fig 11.5k			2
SPVP4	See	Chart	Fig 11.5k			3
SPVP5	See	Chart	Fig 11.5k			4
VP6	44-64-68-40	288/288	102	386/386	451/451	5
VP7	32-56-64-32	268/276	106	375/375	440/440	6
VP8C	44-64-68-40	286/286	102	432/432	510/510	7
VP9C	42-66-74-42	286/296	104	432/439	510/523	8

1) Based on a steel billet this cam is an extremely effective street performer. Its short duration high lift format produces a high output at the top end with little compromise to low end output. Idle and drivability are what you would expect of a street cam but power is not. In a 1275 or larger motor this cam will produce a greater output than Leyland/Austin Rover's 649 profile (see Fig. 11.5f). In smaller motors (850–1000) 1.25 rockers are recommended although in 1100s some benefit is seen with the use of offset rocker bushes.
2) This cam is an excellent street cam with all the qualities expected of a cam required to idle smoothly and make good low end. It is designed primarily for 1275 motors on up and though it still works well with 1.25 rockers it is at its best with 1.5s.
3) A.P.T.'s basic vintage race cam. Based on the short timing, high lift concept, this cam is easy to get good results from. Power band in a 1000cc motor is generally from 3,750–7,750rpm and in large motors 3,000–7,500rpm. Needs to be used with a fully prepped motor spec and the CR must be over 11/1 especially in smaller motors. If used the silencer/muffler must be high flow low restriction type.
4) This particular cam is for an all out drag/sprint type unit of 1400cc or more. It needs 13.5:1 compression ratio minimum and delivers a power band 4,000 to 8,000rpm. It produces less power to 6,500rpm than a SPVP4 but more from there on up. Motor spec must be well sorted to make this cam work so it is best recommended for professional builders.
5) A high performance street cam intended specifically for smaller motors and larger ones where the 1.25 rockers are to be retained. Works well in 850, 948 and 1000cc units. It delivers a slightly lopey idle with a wide power band in all motors – good for solo type events.
6) A high performance street cam intended for use primarily with 1.25 rockers. It has a slightly lopey idle in smaller units but pulls strong from 2,500–7,000rpm. It has a smoother idle in 1100s and 1275s and produces excellent mid and top end compared to factory cams including the 1275 Metro cam.
7) This is the short timing high lift concept maximized and, due to its high lift valve spring set up is critical. This rally/race cam is, unlike Leyland/Austin – Rover's Sprint/Super Sprint cams, very easy to use as it is very insensitive to motor spec. It produces outstanding results even in totally conventional race units. It is well suited to 1430cc units on up and functions best with high flow exhaust ports as it is a single pattern cam.
8) Extended exhaust timing on this cam allows larger displacement, ie 1430cc on up, motors to make the best of their capacity. Short intake timing with high lift maximizes breathing into a large cylinder while extended exhaust timing clears the cylinder ready for the next charge. This is an excellent rallycross/rally/race cam and will produce a strong power band from 3,500 to 7,500rpm.

Currently, it looks like the VP8 could be the cam that everybody is using, but there is a drawback . . . cost. Cutting a cam from a solid billet is expensive and the price of one of these billet cams is about 250% more than its Leyland cam equivalent.

Now we turn to the home of the hot rod cam industry, the U.S.A. **Iskenderian Cams** have a catalogue profile which is really their version of the 649 cam and which produces very similar results, though generally the cam is more accurately ground than the original Leyland one. Apart from 3 or 4 catalogue profiles, Iskenderian have about half-a-dozen special profiles that are worth investigating if you know what your particular application needs.

Good cams can also be had from **Crane**, and from **Huffaker** who produced the original 649 and can also produce a cam from a brand new billet, though it's rather expensive, and **Elgin Cams** have some good working profiles for the Series 'A'. Of special interest to those of you running big engines, is their super high lift profiles which give some 460° lift with a standard rocker ratio and a whopping great 550 or so with a high lift rocker. Since Dimitri Elgin, the boss at Elgin Cams, has had a lot of experience racing Austin Healey Sprites, it's not surprising that he worked overtime on the development of cam profiles that work in the Series 'A' engine, so anyone buying a cam from Dimitri at Elgin Cams, does get the benefit of the boss's personal experience.

Scatter Pattern Cams

You'll note that throughout the previous text I've alluded to cam types by the terms single pattern and dual pattern. The reason I've done this is there's another type of cam that David Anton and I have worked upon extensively at A.P.T. This is known as the scatter pattern cam. This type of cam is anything but simple in its nature. Basically it addresses the problem that the Series 'A' engine is not one four-cylinder engine but rather – for the purposes of cam design – two, two-cylinder engines. The scatter pattern cam has different timings for each pair of cylinders and seeks to take advantage of producing the most horsepower per cylinder. A point I'll make right up front is that grinding a cam like this is it's

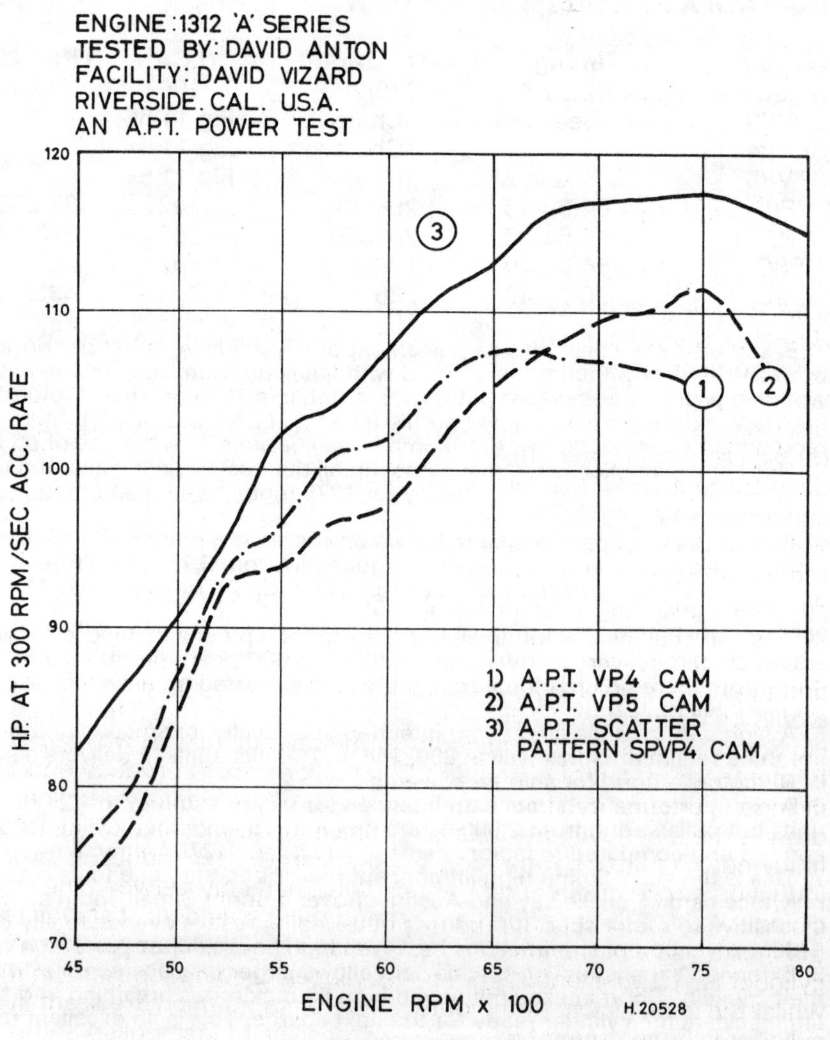

Fig. 11.5i Cam comparison
The superiority of the scatter pattern cam in a race engine can be clearly seen here. The scatter pattern produced more of everything that was useful. The engine had a wider power band, coming on the cam sooner, going off later. It also produced more of our No 1 requirement, namely, torque and horsepower. One of the most important effects of using a scatter pattern cam is the more equal charge distribution between inner and outer cylinders resulting in reduced jetting compromise as compared with a long duration, non scattered, cam. This tends to indicate that mixture variations between inner and outer cylinders is not a function of manifold design but cam timing.

Fig. 11.5j Cam comparison
This graph shows how the scatter pattern cam fared in a highly tuned road going engine. With 290° period, the A.P.T. SPVP4 is really a race cam, but in a well set up engine where other factors are tailored for the road, it can be used successfully without too much aggravation so long as traffic conditions are relatively light. As can be seen, by scattering the pattern of the VP4 in the appropriate manner, both power and torque improved and, important for the road, the scatter pattern cam improved drivability. As an aside, don't expect a high lift race style cam to last 100,000 miles in a street motor. Too much time is spent idling which is below its mechanical stress design parameters. At low speed the nose of the cam is highly loaded and wears faster.

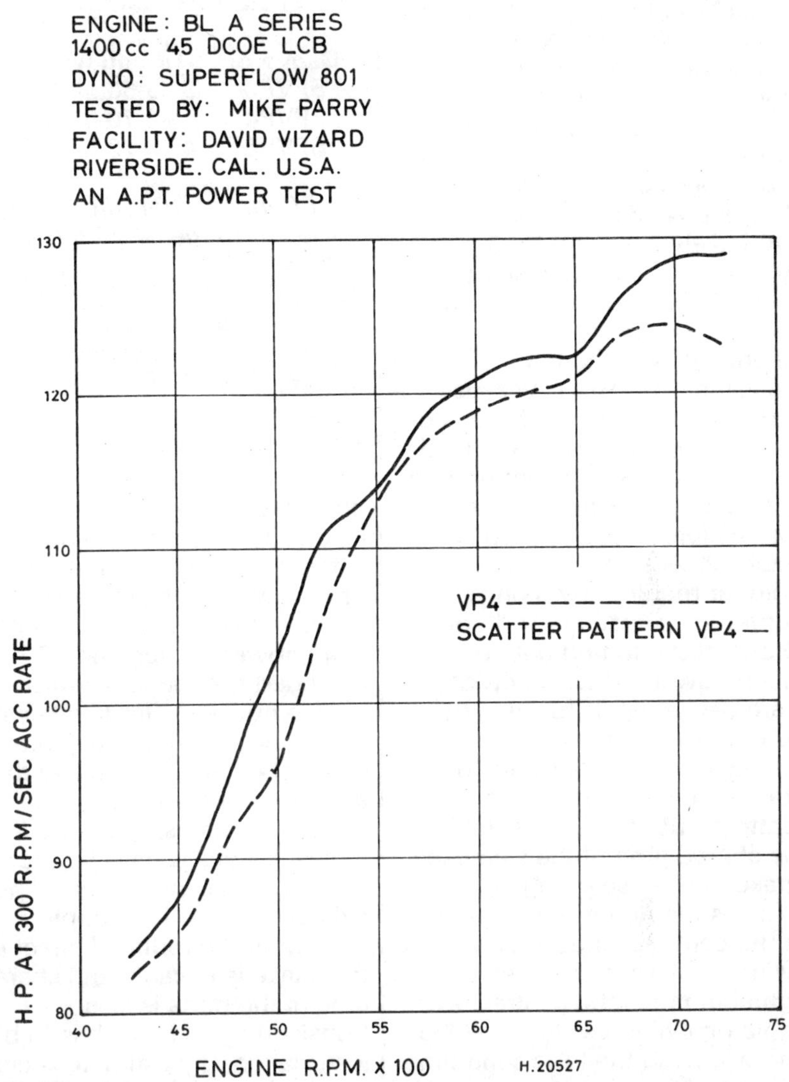

expensive. The scatter pattern cam takes longer to grind than a conventional cam because the machine needs to be reset to grind each set of lobes as if it were a separate camshaft. However, during our cam development campaign with the late Bob Munt of Kent Cams, we discussed scatter pattern cams. He was so pleased with the development we made with conventional cams that he was willing to delve extensively into scatter pattern cams and grind whatever was needed in terms of camshaft profiles. The concept of a scatter pattern cam is that timing on No 1 and No 2 cylinders would be different with cylinders No 3 and No 4 as mirror images. Remember that out of each pair of cylinders, the first cylinder to draw is the inner cylinder. This is followed, 180° later, by the induction stroke of the outer cylinder. With a long period cam, the intake valve of the inner cylinder is still substantially open whilst the intake valve of the outer cylinder is open during the overlap period. This causes robbing of the inner cylinder by the outer. It appears that this effect is exaggerated by the scavenging of the last part of the intake charge of the inner cylinder by the scavenging action of the outer cylinder. This, I feel, is the fundamental reason why super long cams don't produce good results in the Series 'A' engine. Since the induction conditions at the start of each induction stroke differ markedly between the inner and outer cylinders, it stands to reason that the chances of each cylinder needing the same in the way of cam timing are scant at best. When I first looked at this problem it was not clear whether No 1 cylinder would rob No 2 or that the momentum built up by No 2 cylinder would help the filling of No 1 cylinder. Optimizing two different sets of cam timing in one engine is complex enough, but if we were going to get everything as near optimum as possible, two sets of exhaust timing figures also had to be considered. The exhaust timing that is required for the centre cylinders will be different from that required by the outer cylinders because the siamesed port on the centre cylinder has such different characteristics to the outer ones. This meant that any cam that sought to optimize cam timing for all the cylinders would have to have an intake lobe timing for No 1 and No 4 cylinders, another intake lobe timing for No 2 and No 3 cylinders, a specific exhaust timing for No 1 and No 4 cylinders and a specific exhaust timing for No 2 and No 3 cylinders. Fortunately No 2 and No 3 cylinders exhaust 360° apart so there appears to be little need to phase these two cylinders differently even though they share the same port. This is because the exhaust pulses are evenly timed.

Fig. 11.5k A.P.T. Cams Scatter pattern series cams

Cam No.	Timing Nos 1 & 4	Timing Nos 2 & 3	Lobe Lift
SPVP3	31.5-56.5-58.5-33.5	28.5-59.5-61.5-30.5	312/319
SPVP4	45-70-67.5-42.5	42-73-70.5-39.5	324/321
SPVP5	47.5-72.5-72.5-42.5	44.5-75.5-75.5-44.5	339/339

Kent Cams scattered cams are very similar. Part No. 286 SP is eqivalent to the SPVP3, the 296SP the SPVP4 and the 310SP the SPVP5.
The SPVP3 and 4 are intended for road/rally and rally/race use where a reasonably wide power band is required. If used for rally applications the silencer/muffler must be a low back pressure design. The SPVP5 is an all out race design and should only be used as such.

Finding out what was needed for a scatter pattern cam proved to be a complex process of elimination. It was necessary to establish what timing each cylinder required by trying numerous camshafts and doing 'Morse tests' to find out how much power individual cylinders were delivering. After running hundreds of tests and studying the results it was possible to home-in on a trend. Essentially, a scatter pattern cam cuts down the shared time of the siamesed intake port. In so doing, it reduces the negative interaction of the centre cylinders by the end cylinders. This is achieved by tightening the lobe centreline angle on cylinders No 2 and No 3 and widening the lobe centreline angle on No 1 and No 4 over a comparable single and dual pattern cam. This is the basics of a **simple** scatter pattern cam. If a multi-scatter pattern cam is used, things get complex in a hurry. In its ultimate form, the scatter pattern cam utilizes an exhaust timing similar, and sometimes shorter, on No 1 and No 4 exhaust than its equivalent intake. On No 2 and No 3 it utilizes exhaust timing longer than the intake timing of the other cylinders. All this sounds very complex and the question is, how effective are these cams.

To date we have only used cams profiled from original Leyland camshafts which has precluded the use of the more radical and effective Crane profiles because of the expense involved in prototype work on the latter. Yet, the scatter pattern cams have proven incredibly effective in terms of producing extra power, torque and an extended rpm band in which the engine operates. To date tests have indicated that the shorter the cam gets, the less effective the scattering of the cam's intake pattern becomes, but there is almost always justification for variations in centre to outer port exhaust timing. The simple reason for the reduced effect on the intake is because the shared time of the ports is less. Considering that cost has to be a justification, my estimate is cams much under 260° intake don't really justify having their patterns scattered. To date, our experience has shown that scatter pattern cams pay off in all sizes of Series 'A' engines. About the worst situation for a scatter pattern cam is a small valved engine. In a SCCA Sprite engine (where valve size must remain standard), in which we had a particularly heavy hand in determining the specification, the power output went from 72hp at the wheels to 75hp. Torque increased throughout the rev range and the power band was widened. On bigger valved engines the increase can be greater as the graph **Fig. 11.5i**, for a 1312cc engine and **Fig. 11.5j** for a silenced (Sonic Turbo) road going motor show. One aspect of scatter patterns noticed to date is that there is no clear cut figure to time the longer race cams into. Generally speaking, they seem to like running with No 1 intake timed between 2 and 4° advance, but this is not a hard and fast rule. A full 30% of the cams tried have fallen outside this band width. For that reason, **I recommend the use of an adjustable cam timing device** so that optimal cam timing can be achieved. It would appear the specification of the engine is the prime factor affecting where a scatter pattern cam runs best. As can be seen from the graphs mentioned above, the use of a scatter pattern cam is well worth the effort.

Scatter pattern looks like it's an exciting concept for the future. I ask myself where do we go from here. The obvious answer would be to use Crane profiles to develop even more exotic scatter pattern cams. However, this would be costly, not only because Crane blanks are expensive, but also the fact that it's more labour intensive to grind a scatter pattern cam than a single pattern cam. A good guess at this stage would be a scatter pattern cam based on Crane's profiles and blanks could net 1.5ft lbs per cubic inch and a comparable increase in horsepower over the more conventional profiles ground on original Leyland camshafts.

10 Steps to Successfully Spec a Cam for Maximum Engine Output

Step 1 Decide on the fuel octane that the engine will use.
Step 2 Determine the required dynamic compression ratio as measured using the trapped volume above the piston at valve closure for cam duration to CR compatibility.
Step 3 Determine the overlap required of the application.
Step 4 Determine the Lobe Centre Line Angle (LCA).
Step 5 Estimate the static CR the engine is likely to need and provisionally use this number to correct the LCA.
Step 6 Determine the duration that satisfies Steps 4 and 5.
Step 7 Determine the static CR that will result from the chamber volume to give the desired CR at the valve closure point. Go back to Step 5 and use this figure instead of an estimated static CR.
Step 8 If your original estimate of the CR required was way off, re-do Steps 5 and 6 using the last static CR figure developed.
Step 9 Determine the highest lobe lift possible for your engine at the duration established by Step 6.
Step 10 Install at the appropriate timing.

Now you know it's that simple, read how it's done.

Valve Trains for Power and Torque

Luck – a Selection Element?
A problem confronting many engine builders is making the best choice of cam and valve train components for their engine. Picking up a cam catalogue and browsing through doesn't necessarily help because of the overwhelming number of choices possible. On a single page you may see several cam specs given and many may seem likely candidates. Maybe one has a little more lift than another but a little less duration. Another may have slightly less duration but be on a tighter lobe Centre Line angle than yet another possibility. Sure you can ask someone what has worked in their motor. Although that may seem a simple, sure-fire solution let me assure you it can fail all too often to produce the best results and if you are racing, nothing but the best will do. Why is this so? First ask yourself if the person you are seeking advice from has successfully built an 'A' Series engine like you intend building and secondly, do you intend to closely follow the specification that was used? If there is any doubt on either count then *you need to be armed with the relevant information to make your own decision.*

Basic Five
There are five basic parameters that need to be addressed when selecting a cam and valve train for best performance. These are *lift, duration, the Lobe Centre Line Angle (LCA), overlap* and the cam's *advance/retard* setting in the engine. Just how each of these affects the output of an engine depends on the design of the engine. Some engine configurations respond better to added lift than increased duration and some the reverse. Some engines need a tighter LCA than others and so on. When we have even a basic understanding of the reasons for these differences much of the mystery of making a good cam selection is removed. Investigating in depth reveals that required cam characteristics are *intimately tied up with the cylinder head's air flow capacity and the size of cylinder it has to supply*. This implies that a *flow bench* is not only a tool for *developing heads* but also *for aiding the selection of a compatible cam* design. To see how the primary factors of flow and displacement affects the cam spec required let's start with a seemingly disconnected issue of fuel octane.

Step 1 Choose Your Fuel Octane
The fuel octane influences the CR that can be used and an engine's CR must, for best results be tied to the cam duration. The key factor here is the *dynamic CR based on the intake valve closing point.* Obviously if the valve closes very late, as per a long duration cam, the piston will be well up the bore and – at the lower rpm at least, will trap less air, therefore producing a lower dynamic CR and consequently less torque. The drawing **SB11-1** illustrates this point.

Essentially we need the CR to be as high as possible without detonation occurring. Detonation is brought about by a combination of time, heat and pressure. For most engines detonation occurs at low rpm because of the greater length of time the end gases are exposed to the advancing flame front. When a long cam is used the compression pressures at low rpm are reduced thus not only allowing but requiring a higher CR to be used.

Step 2 Dynamic CR
The *static* CR must be based on fuel octane and intake valve closure point. This means establishing the *dynamic* CR required. The following chart, **SB11-2**, shows the *CR at valve closure* needed for three different intake temperatures. Choose appropriately and then, using the research octane, which is typically four numbers higher than the R+M/2 value, determine the dynamic CR that is needed for the application.

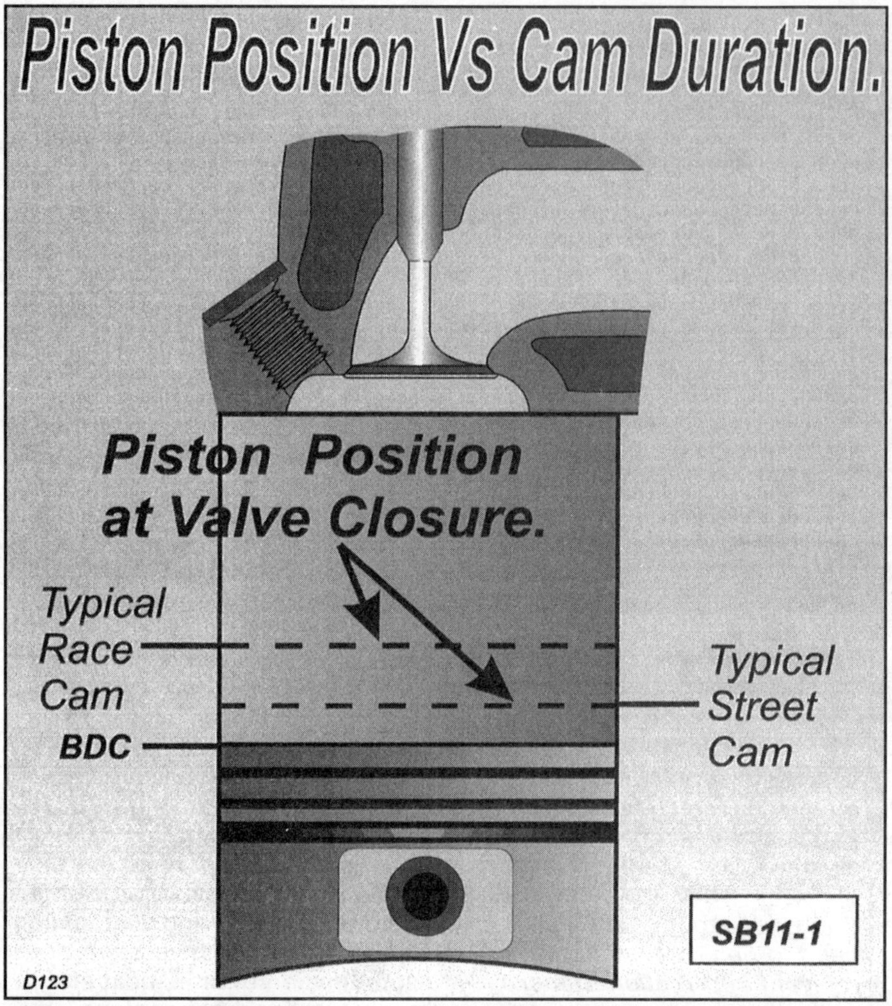

practice what appears to be only an extra *25%* of overlap actually ends up being some *80%* more. The bottom line here is don't be tempted to use too much overlap. The effect *overlap through flow area* and the cylinder displacement has on the power curve are closely linked. The greater the through flow area is the higher the rpm the engine needs to turn before it starts to perform. The bigger the engine is for a given cylinder head or valve size the lower the rpm at which it starts to function. This means the selection of the overlap must be based largely on the lowest rpm that the engine is required to run from. The following chart, **SB11-3**, gives a good indication of what is required for a two-valves per cylinder engine equipped with an exhaust system with near zero back pressure and an unrestricted intake. To put numbers to these we are talking about an exhaust with less than ½ inch of mercury (0.25 psi) and less than 1° inch of intake manifold vacuum at peak power. If exhaust back pressure increases the amount of overlap must be decreased. If the intake

Step 3 Required Overlap
Overlap is the time around TDC that both the intake and exhaust valves are open. This cam characteristic is crucial towards obtaining high output from a race engine. Though good for power it's very destructive of low end output, drivability and mileage. This makes the decision as to how much should be used for a given application critical. It is tempting to add a few more degrees in the hope that it won't cost much at the bottom end but in reality it is like playing Russian Roulette with numbers. Increasing overlap from say 30° to 40° may only appear to be an extra 25% but this is not really the case. The critical factor is the *overlap area* brought about by the combined degrees and lift. Adding more overlap also adds lift during the overlap period. In

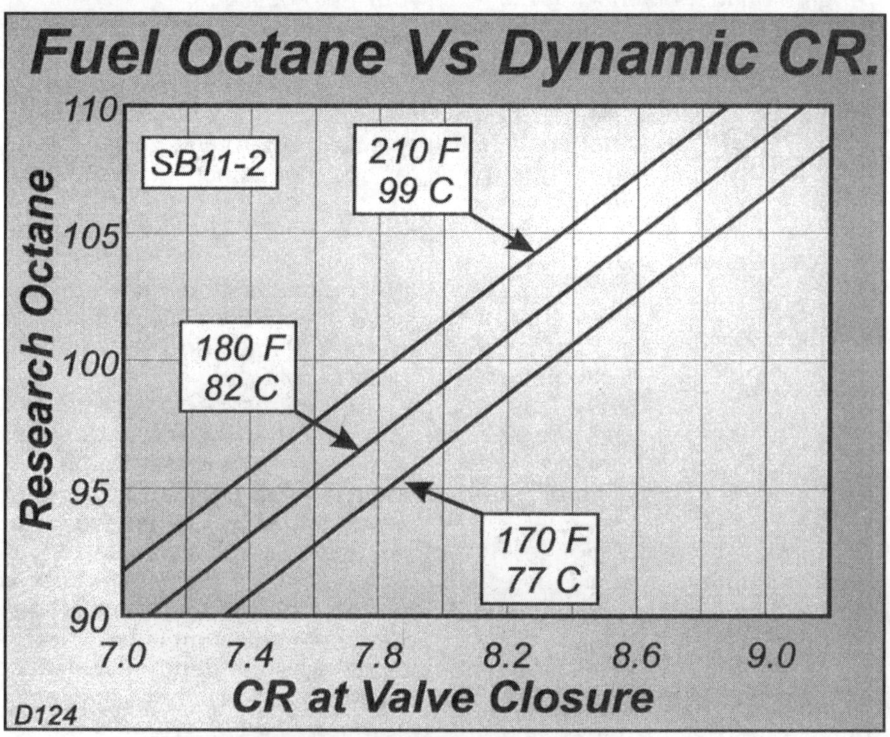

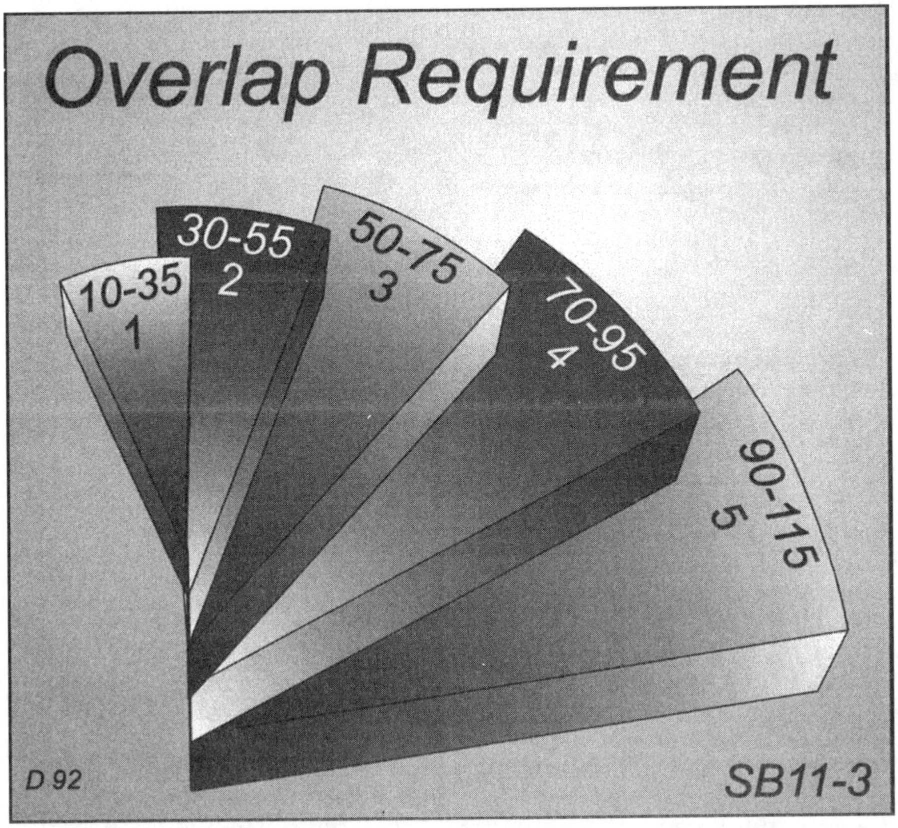

is restricted the situation becomes complex and is influenced to an even greater extent by the exhaust used. In the overlap chart (**SB 11-3**) section 1 is for working trucks and vans where low end torque and mileage is an important requirement. Section 2 is for daily driven cars where a strong emphasis is still on idle and low speed performance. Section 3 is for hot street machines where performance is now a primary requirement. Section 4 is for oval track, all out road race and heavier sprint/drag racing cars. Section 5 is for 'no compromise' sprint/drag race engines with seven or eight port heads.

Step 4 LCA Selection

The camshaft Lobe Centre Line, the only cam spec quoted in cam degrees rather than crank degrees, is seemingly the most difficult aspect of cam spec for an engine builder, or cam grinder for that matter, to optimize. The subject is complex enough to warrant a complete paper in its own right but we must forego that here. The intention is to present a basic understanding of what's required and a simple method to determine it so you can make the best cam choice for your engine.

One basic rule that can be applied across the board is that the more an engine becomes *under valved* the *tighter* the Lobe Centre Line Angle needs to be. Normally, the LCA required is based on the cylinder head's flow capability and the size of cylinder to be serviced. Unfortunately only a minority group are in a position to get cylinder head flow figures. This means adopting the next best method that still produces creditable results. This can be done using valve size as a criterion instead of flow capability. Going this route makes life considerably easier because the principle element involved can be reduced to the simple curve shown in the following graph, **SB11-4,** which is then corrected by the CR to be used. To use the graph it is necessary to establish the *capacity to valve area ratio* in terms of ccs per square cm of valve area. The simple formula for valve area is *diameter*^{2} *x 0.785*. Using a 1400cc 'A' Series engine as a working example will illustrate what we are attempting to do. Big valve heads for this engine will typically have 36.6 mm (1.44 inch) so for this engine the area will be 36.62 x 0.785 = 1,050 sq mm or 10.5 sq cm intakes. The ccs per sq cm of intake valve equals cylinder volume divided by valve area. That is 350/10.5 which equals 33.3ccs per sq cm.

Now we have this figure, go to the graph and locate the 33.3 point on the lower scale. Next move up vertically to intersect the graph line, then from here move across horizontally. The result, indicated on the left-hand scale, is 108° for the Lobe Centre Line Angle and applies to a well-ported head with one intake port per cylinder. If you are building

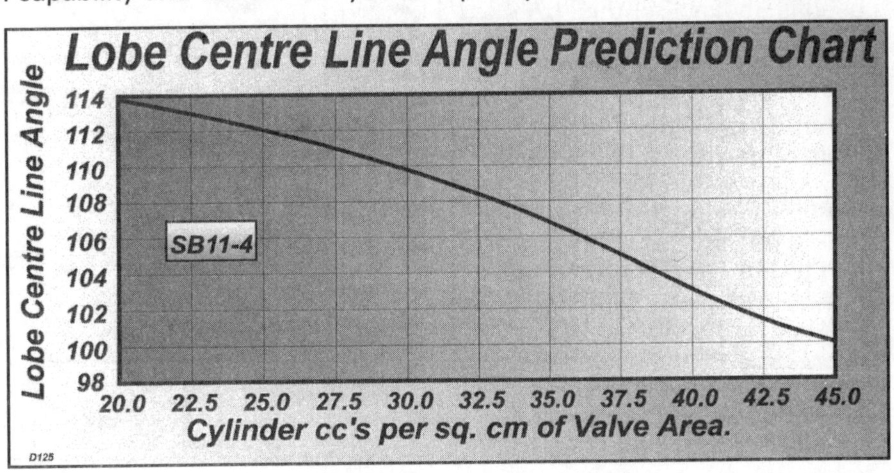

an engine with a five-port head then it is necessary to compensate for the five-ports configuration's reduced ram effect. This is done by tightening up the LCA by three to four degrees over that required for a seven or eight-port motor. Also, if you think your cylinder head might not be up to par (often the case if not done by a reputable, flow bench equipped professional) then you should consider tightening the LCA by an additional 1 maybe 2° over that predicted by the chart. Just for the record power is lost far faster if the LCA is too wide rather than too tight. Three degrees too wide can cost 6-10hp where as 3° too tight costs little or nothing but the motor comes on the cam more abruptly. For our example 1,400 five-porter, an LCA of 104-106 will get the job done.

If you are building a street motor and idle quality is important then you should consider widening the predicted angle by 1 or 2°. However you will only need to go this route for a quality idle if you have been over enthusiastic in selecting the overlap to be used.

Step 5 Correcting for the CR
Using the graph above you can calculate the LCA for maximum power for virtually any combinations but with a few provisos and here they are. First these LCAs apply to an engine of 11 to 13:1 CR. If the CR being used is above 13:1 the LCA needs to spread by about 1° per ratio. If the CR is below 11:1 then the LCA needs to be tightened by about ½° per ratio. If you are building a street motor then be warned, the figure predicted by the graph makes no concession to idle quality. It optimizes power and torque only. However *idle quality and slow speed drivability will not be a problem if the overlap* does not *exceed* the *limits given by the earlier overlap graph*. If idle and vacuum are important factors add between 2 to maybe as much as 4° to the predicted LCA. This will improve the vacuum at idle and cruise and can sometimes add a little to the output right at the top end. But the price paid is reduced torque throughout the low and mid range rpm. Obviously a judgement call is needed here depending on how important the idle quality is to you.

Step 6 Duration
Adding valve opening duration has always been an accepted technique for making more power upstairs but it has its pros and cons. Adding duration is easy on the valve train and makes for reliability in that department. But the price of added top end is almost a direct trade off as it reduces low end output. Just how much it reduces low end output depends on the ratio of valve area to cylinder volume. Small displacement big valve engines rapidly loose low end output compared with under-valved large units. The bottom line here is that if you are using a smaller version of an engine you need to be more conservative with duration. If the engine has been stretched by boring, stroking or both, then a little extra duration proves an asset. By extending the cam timing we cause the torque curve to move up the rpm range. More cam timing *does not mean more torque* just more rpm at which it occurs at and this translates into more HP. The following graph, **SB11-5,** shows the effect of more duration (in a non-siamesed intake port motor) while the CR *remained the same* throughout the test. Cams Nos 1 to 3 are nominally 270, 285, and 300° of seat duration. It does not take much studying of these curves to see how much of direct trade off added duration can be. However because the CR was not increased for the longer cams these tests show a 'worst case' situation. Adding the appropriate CR would redeem about half of what was lost at the low end

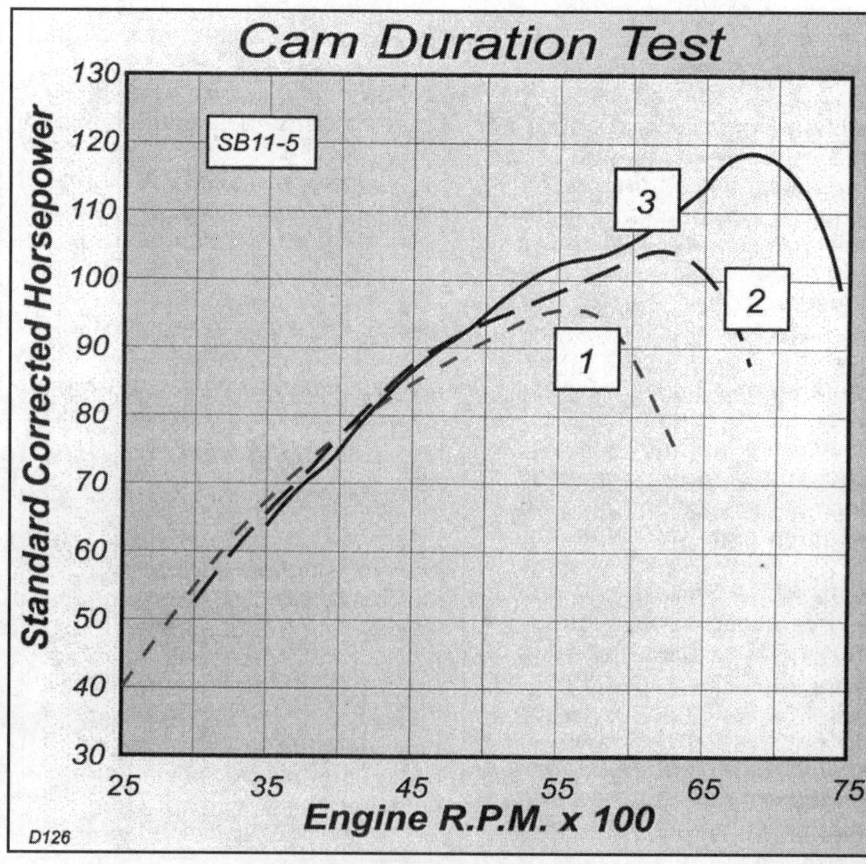

while adding to the top end output.

Before deciding to use a long duration carefully consider the rpm range you want the engine to have. Also be aware that longer duration cams become more fussy as far as the rest of the engine spec is concerned. This means it is far better to err on the short side, especially on a street motor, as you will like the results far better.

As far as the actual intake duration is concerned we have in reality already set its value because we have determined the maximum overlap that will be used and the lobe Centre Line angle. From these two numbers *we get the duration by dividing the overlap by two then adding this to the LCA and then multiplying the result by two.* As an example let us assume we have determined an overlap of 50° suits the application and the engine spec calls for a 106 LCA then the duration will be (50/2) + 106 = 25 +106 –131. Multiply this by two and we get 262. You will now choose a cam lobe profile that will a) have the highest lift possible while giving an opening duration as near the calculated 262° as possible. (This is opening duration at the valve clearance point) and b) a profile that will not run off the edge of the tappet diameter that is being used. Normally an 'A' Series is 0.812 inches but a useful trick to allow more tappet velocity to be used on the profile is to increase the tappet diameter to 0.875 inches.

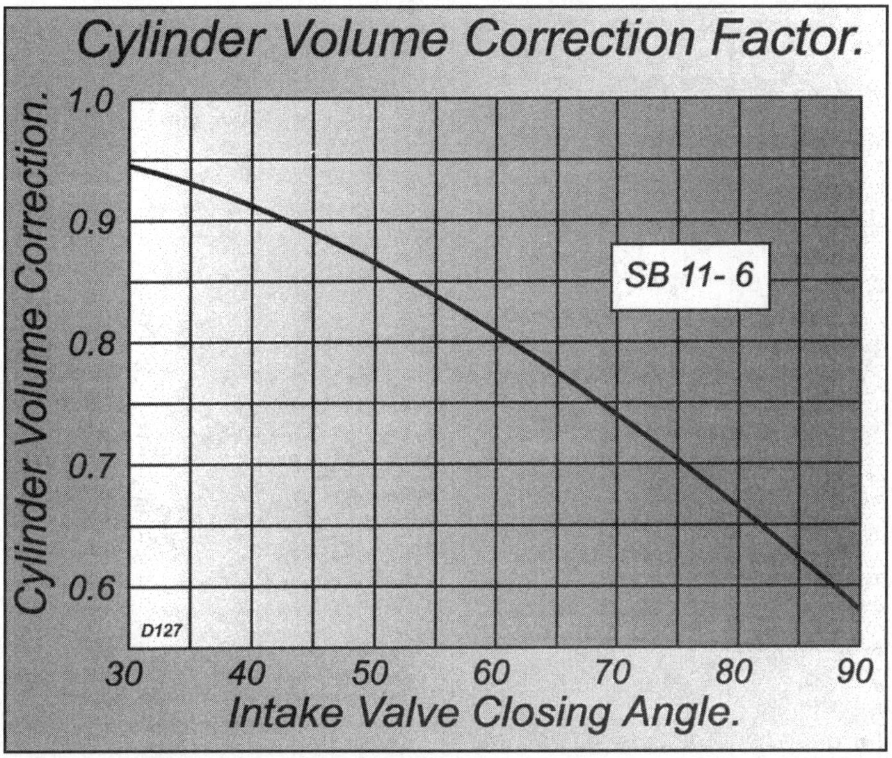

Step 7 Determining Static CR
Once the LCA and the duration have been determined we can now easily calculate the intake valve closure point. This will be the LCA + half the duration – 180 if the cam is timed 'straight up', *ie* neither advanced or retard. In our example this works out to 106 + 131 –180 = 57, that is the intake closes 57° after BDC. In a five or eight-port 'A' Series engine best power is almost always made at 4° of advance so, in our example the 57° will become 53° when 4° of advance is introduced. Once the intake closure is fixed we must determine what the combustion chamber volume must be to give the desired CR at the valve closure point. First determine the volume there is above the piston at valve closure. Use the following chart, **SB11-6**, for this. Find the valve closure point on the bottom scale then go up to intersect the graph line then across to the vertical axis. Multiply the cylinder volume by this number. Example – a 70° intake closure on a 350cc cylinder reduces its effective capacity to 0.766 x 350 = 271cc. This is the *effective cylinder volume* and is the number used to compute the *combustion chamber volume*. To compute combustion chamber volume required subtract one from the *dynamic CR* determined from the first chart (**SB11-1**). Next, divide this into the *effective cylinder volume* we have just obtained. Let us use the 350cc cylinder as an example once again. The corrected, or effective cylinder volume was 271cc. If we had determined from the octane of fuel being used that a 7.8:1 ratio could be used then we will subtract 1 from this number leaving 6.8 and then divide this into 271 which equals *39.8cc for the chamber volume*. Calculating the CR by the conventional method gives a ratio of (350 + 39.8) divided by 39.8 = 9.8:1. In reality the numbers predicted are on the conservative side to make allowances for those people who feel bound to run their engines too hot and with too much ignition advance. *A competent engine builder can add a half to three quarters of a ratio to the calculated numbers and still be out of the detonation area*. Use this static CR and return to Step 5 to verify the LCA to be used.

Step 8 Fine Tuning the Static CR
If your original estimate of the CR required was *way off* re-do Steps 5 and 6 using the last static CR figure developed.

Step 9 Lift Vs Output
The number one question concerning lift is how much is

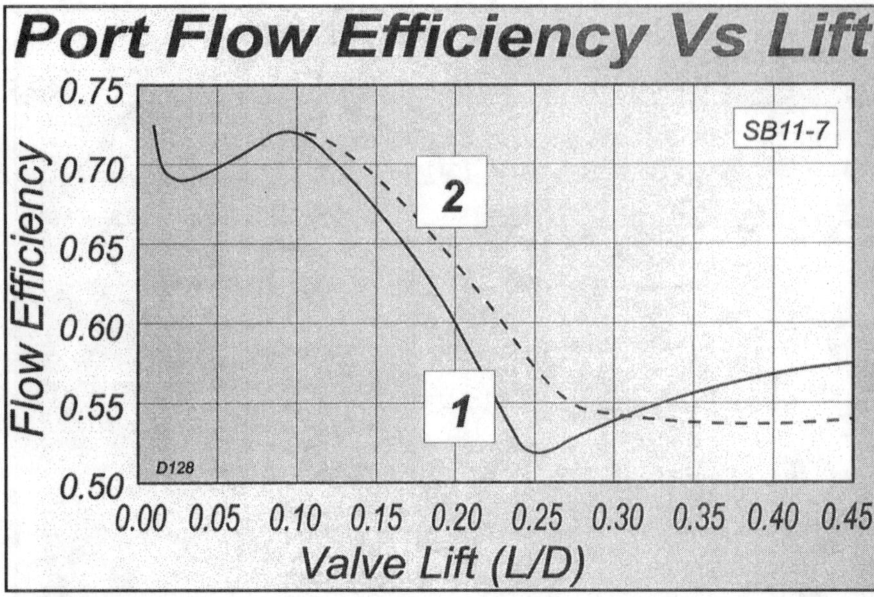

needed before any further increase becomes pointless. This might look tricky to answer but fortunately there are some simple calculations that help simplify things. Let us first consider a two valves per cylinder layout with parallel valves. The flow bench shows us that the flow efficiency of a well developed head follows the form shown in **SB11-7.** Curve 1 is for a typical parallel valve head and curve 2 for an inclined valve hemispherical or pent roof chamber style head with the ports steeply down drafted. The flow efficiency in either case is high as the valve comes off the seat. But for the parallel valve head efficiency drops more quickly reaching its lowest value at a lift approximately equal to a **quarter** of the valves **diameter** (0.25 D). It is at this point that shrouding caused by the cylinder wall or combustion chamber is greatest. If lift continues beyond this we find, if the port itself can support sufficient airflow, that the efficiency starts to rise. This is due to the valve moving out of the sphere of influence of the valve seat in the head. This brings about a change in flow pattern where the flow is predominantly across the back of the valve rather than around it as shown in **SB11-8.** For a typical *functionally modified* port we find that flow will peak out at 35 to 45% of the valve diameter. This means that high valve lift is of significant importance to airflow into an engine equipped with a two valve head. If we add to this the fact that a two valve 'A' Series engine is almost always under valve *ie* has more than 25ccs per sq cm of intake valve, then it follows that lifting the valves fast and high should pay off. Also the more capacity the engine has under a given cylinder head the more important valve lift becomes specially if there is no more room for bigger valves. It is revealing to compare the output of two cams having the same lift area under the curve. Although each produces about the same hp the short, high lift, design tends to produce its power with less rpm. This means it makes more torque. The difference

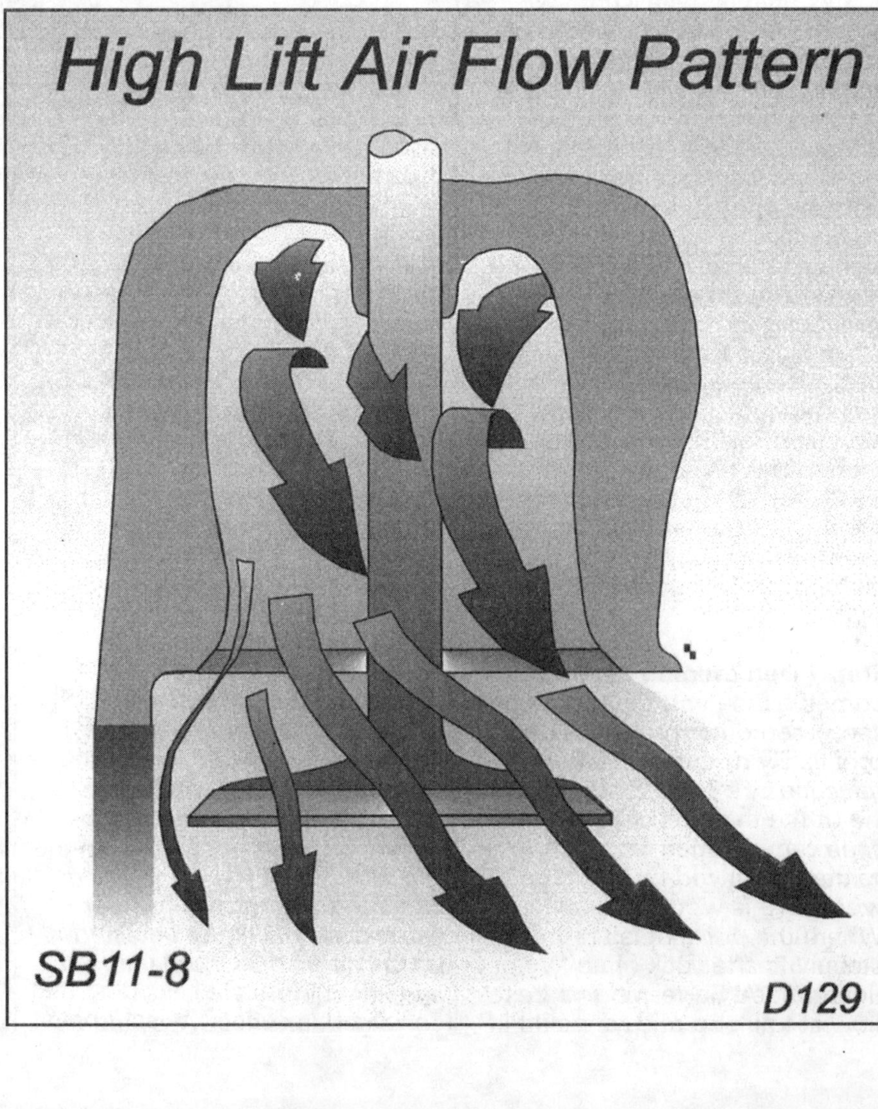

between two such cams, as tested in a non-siamese port motor, can be seen in the graph **SB11-9**. Curve 1 is for a short high lift cam whereas curve 2 is for a longer cam with less lift. From these curves it can clearly be seen that cam No. 1 produces the most usable output especially for street use.

To establish the maximum useful lift required for a two valves per cylinder engine is simple. We just apply whichever of the following formulas predicts the *lowest* lift.

$$\text{Valve lift} = 0.40 \times D$$

or

$$\text{Valve lift} = \frac{\text{CC of one cylinder}}{D \times 0.65}$$

Where D is the valve diameter in mm.

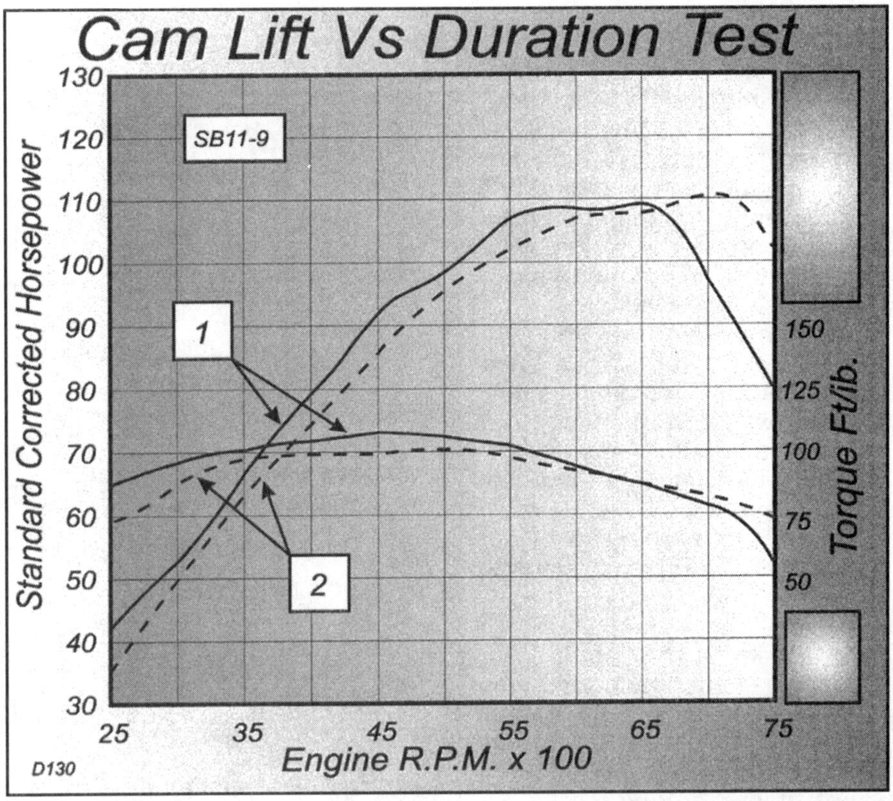

For a street or even a semi-race application this produces numbers far higher than can be achieved with available hardware. Only for opening durations in excess of about 290° does it become possible. This alone should indicate that *maximizing lift, especially with a street cam, should be considered a high ranking priority*.

Assuming you have or are going to have a decent head on your engine then going for a little more duration and as much extra lift as possible is the way to go. For pushrod engines this means a serious look not only at the cam profile but also *the rocker ratio which on a standard 'A' Series is woefully low at 1.25:1*. For a given cam duration there comes a time when no more lift can be accommodated on the profile. From there on it's *a question of higher ratio rockers*.

If the engine has *a hemi chamber configuration or four valves per cylinder the situation with valve lift is a little different*. Where the port angles more steeply to the back of an unshrouded valve we find that flow at low and mid range lift is stronger than with a parallel valve head but it tapers off more quickly at the higher lifts. Unless it is for dynamic reasons associated with rapid opening and closing, lifting much above 30% the valve diameter becomes a futile exercise in such heads. The name of the game here is to make the most of rapid valve opening especially if the engine concerned is over head cam.

Attempting to establish which of two similar duration profiles has the fastest opening is difficult but in nine out of ten cases it will be the cam with the most lift so, once again, lift becomes a predominant factor.

Putting together a valve train that generates high lift is not without its problems. Avoiding spring coil binding must rank a number 1 priority. *To make a system reliable it also helps to cut valve train weight wherever it proves practical*. It is also important to *use the best spring available* as described under the springs and retainer heading that follows this section. Also it pays to optimize the valve train geometry. This usually means getting the pushrod length right for maximum valve lift. None of these problems is insurmountable and when properly taken care of can produce an engine with a very satisfying power curve.

Dual Pattern

Many cams have different intake to exhaust duration. For *high compression, high rpm applications* this is desirable because of the reduced time period within which the exhaust must be expelled. Usually about 2/3 of the extra opening period, usually from 4–10°, is on the opening side of the exhaust thus giving the cylinder more time to 'blow down'. On a purpose-built performance street motor *with one port per valve*, this extra blow down can cut mileage and low end torque by as much as 5%. From mid range rpm on up it starts to help power. For a five-port motor the extra timing, which is needed mostly by the centre siamesed exhaust cylinders pays off from about 2,000rpm up. In addition to this many 'off-the-shelf' street cams are speced this way because the cam designer anticipated their use in engines with standard and consequently less than optimal exhaust ports and systems. For a true street engine of moderate compression but with an effective modified head and a free flow exhaust system a single pattern cam will, as often as not, *be the best choice if mileage is a factor*. It is not until we start dealing with engines having compression ratios of 9.5:1 or more and higher rpm capabilities that a dual pattern really starts to pay off.

Step 10 Timing the Cam In
The valve event timing produced by the cam is as you would expect, an important player in the game. Surprisingly enough there is not, at least *among most conventional, normally aspirated engines*, a great deal of difference in what is required to work. For most two valves per cylinder engines 2–4° of cam advanced works well. For four-valve or hemi style engines zero to 2° usually gets the job done. Having said what is needed let's define exactly what is meant by cam advance or retard as the case may be. To work out what cam setting will produce a given advance you will need to know the cam's LCA. If the cam is set in the engine to produce full lift at the same number of degrees after TDC on No. 1 inlet as the cam's LCA, it is said to be timed 'straight up'. This means it is neither advanced nor retarded. For example if a cam on 106° LCA was timed in to give full lift on No. 1 intake at 106° after TDC it is in 'straight up'. This can be easily appreciated when you consider that at this setting the exhaust valve reaches full lift 106° before TDC.

If the the cam is *advanced* the number of degrees after TDC for the intake to reach 'full open' gets smaller and when *retarded* gets *bigger*. For instance if we were to time our 106 LCA cam in 4° advanced then it would deliver

Calculating the Lobe Centre Line Angle from cam catalogue figures

Not all cam catalogues specify the LCA and buying a cam without that number means you are really putting your trust in the cam company's ability to know what your motor needs. If the subject does not crop up they will sell you what they have and this may not be what your motor needs. For those occasions it is omitted here how to calculate the LCA.

The first point to appreciate is that cam duration/timing figures are usually quoted at two distinct tappet lift points. The first is at a low tappet rise and is often, but not always, the point at which the tappet has taken up the designed tappet clearance. In case you find the need to check your cam spec at the valve remember that this lift figure is at the tappet and any figure measured at the valve must take into account the rocker ratio used. For instance the seat to seat opening duration of a 649 profile must be checked with a 15 thousandths tappet rise which equates to almost 19 thousandths with a standard rocker or 22.5 with a 1.5:1 rocker.

The second duration figure is the 50 thousandths tappet rise number. It is given because the rate of tappet movement in relation to the crank degrees turned is sufficiently high to allow much more precise measurement of the cam's position in the motor. Either of these sets of figures can be used to determine the LCA but if you are having to measure the cam's spec from the motor because its origin is unknown, then the 50 thousandths figures will produce more accurate results.

A typical set of catalogue cam timing figures could read 34-66-76-28. In layman's terms this means the intake opens at 34° before TDC and closes 66° after BDC. The exhaust opens 76° before BDC and closes 28 after TDC. The first move towards establishing the LCA a catalogue cam has is to determine the duration of both the intake and exhaust. To do this add each pair of intake or exhaust numbers together plus 180. The intake in this case will be 34 + 66 + 180 = 280 and the exhaust 76 + 28 + 180 = 284.

The next step is to determine the full open position of both lobes. This is often called the intake or exhaust centre line. The centre line for the intake will be half the duration minus the amount of opening before TDC which in our case is 140 – 34 = 106. The procedure for the exhaust is similar except the degrees to closure after TDC is used. For our example cam half the exhaust duration is 142 and subtracting 28 from that gives us 114. Just to make it clear these figures are the position that the intake valve is fully open after TDC and the position the exhaust valve is fully open before TDC. The final step to get the LCA from these two figures is to add them together and divide by two. For our example this is 106 + 114 = 220 divided by two = 110. This is the LCA. Let's take things one step further. Because this cam called for a 106° intake centre line that means the cam manufacturer is calling for it to be set into the motor with 4° of advance ie 110 – 106 = + 4.

full lift on No. 1 intake at 102° after TDC.

How important is it to get the cam timed in right? It is a job you should do when installing a new cam but there is a certain amount of leeway within which you can work without losing any significant amount of output. In practice if you hit the mark within about 2° you are unlikely to feel the difference although a stop watch might. If the cam is retarded there tends to be a loss of low end output but possibly, but not always, a small gain in top end results and visa versa. In most cases *performance is lost faster as the cam becomes too retarded* than it does *too advanced*. For a high performance street machine it *is definitely better* to have the cam slightly too far advanced than too far retarded. A benefit of having the cam slightly more advanced is that as the timing chain wears so it retards the cam. Another factor that is worth noting concerns cam timing and the LCA used. *If the cam's LCA has been selected to optimize power then be aware that timing it into the engine optimally becomes more critical*. If we had a cam of say 116° LCA retarded from optimum by say 6° it may lose 5ft/lb at 4,000 rpm. The same profile cam on an optimal 108 LCA could easily loose double that amount. *The more optimally specced the cam the more critical timing it in becomes*.

Camshafts, Cam Timing & Valve Trains Part 3

Springs & Retainers

Having chosen the cam, the next most important thing to consider is the valve spring type to be used. **The golden rule with any Series 'A' engine is not to 'overspring' the valve.** All overspringing generates increased frictional losses and as a result more hp is used up just turning over the engine. Except for the very early engines which used a single coil spring which allowed valve bounce at about 5200rpm, almost all engines utilizing a standard cam are adequately sprung. If you are not sure about the spring type, and the engine is high mileage it's worthwhile changing springs. The single coil springs available through a Leyland/Austin Rover distributor are of stiffer spring pressure than in the past. For instance, the ordinary single coil spring as used in the early '60s was only good for about 5200, though a really good set of springs may have run to 5500. The outer Cooper spring, which was a design which became standardized for single spring applications, would allow an 850 Mini to run to 6500 before valve

Though a bit more expensive than most springs, these Iskenderian valve springs are well worth the extra money. For race purposes they hold their delivered force indefinitely and, because they occupy less room, they will go to a higher lift before valve spring coil binding occurs than their British counterpart. Kent Cams have these springs. The standard form retainer is not compatible so use Kent S33 retainers. A.P.T. lists the same springs as part No. VPS–01 to be used with retainer 903-0000. Isky also has a street variant of this spring which has a different inner.(A.P.T. No. VPS-02) Kent has an 80 thousandths spring shim available but if you want finer setting of the seat force then A.P.T.'s VSS03 at 0.030in thick will be what you need.

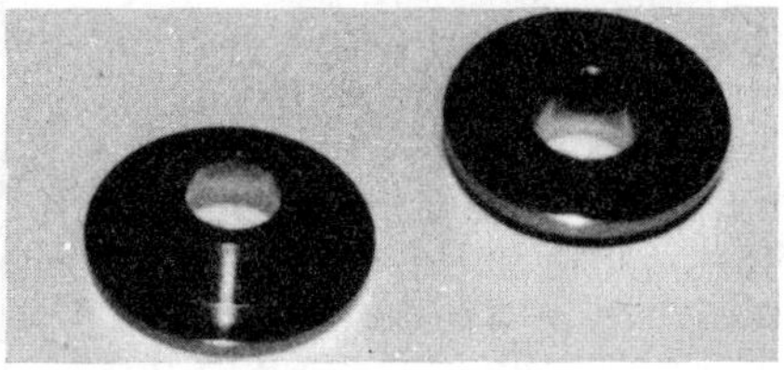

Although effective in terms of rpm the aluminium Isky retainers tended to be a little unreliable. One trip to valve float and they needed to be replaced if a breakage was to be avoided. The best way to go for light weight retainers is with Kent or A.P.T. titanium items.

crash speed was reached. **If your engine is equipped with a standard Leyland profile, the simple rule here is to use the springs intended for use with that profile or, at most, the next heaviest set of springs in the range.** The only real justification for using slightly heavier springs is if the engine has been modified in other areas, such as the top end breathing, and consequently hp at high rpm has been improved. Under such circumstances an engine which revs, say, to 6500 may prove more useful if it were to rev, say, to 7000. This would apply in a vehicle employing wide ratio gears as is typical of standard cars, where it needs to be able to rev 700 to 1000rpm past peak power so that when the gear change is made, the engine drops back to a more favourable point in the rpm range. The spring chart **Fig. 11.6** lists the production springs available, together with stronger Terrys springs, normally used with Leyland's competition profiles. These springs, however, are not the only ones available, nor are they an efficient spring package. Although acceptable with a standard valve train, at lifts exceeding .450 inch coil binding is likely. A high lift valve train must have a better spring package that will achieve the correct open/closed loads without binding. **If you are installing a cam produced by a cam specialist, you should seek their advice as to exactly what springs should be used with**

Fig. 11.6 Valve spring chart

	P/N BLMC	P/N Terry's	Application	Load at check length	Free length	Rate	Fitted load Closed	Open
1		451–452	Standard on early 850	37.5 lb at 1.297"	1.625"	113 lb in	37¼ lb	78 lb
2	AEA 311	451–458	Standard on 997 Coopers and 850 Minis. Standard on single carb 1100's.	88 lb at 0.973"	1.75"	113 lb in	51 lb	92 lb
3		451–557	Heavy poundage for 997 Cooper and 1098 single carb.	100 lb at 1.011"	1.687"	148 lb in	57½ lb	111 lb
4	AEA 311 AEA 401	451–580 Pair	998 Cooper and twin carb 1100's	88 lb at 0.973" 30 lb at 0.847"	1.75" 1.687"	113 lb in 35 lb in	51 lb 12½ lb	92 lb 25 lb
5	12G 1015	451–662	Standard 1300 & 1300 GT from 1969 on	124 lb at 1 065"	1.953"	140 lb in	91 lb	142 lb
6	AEA 767 AEA 768	Pair	Standard on Cooper S	47 lb at 1.297" 17½ lb at 1.165"	1.75" 1.672"	105 lb in 37 lb in	47 lb 17½ lb	85 lb 31 lb
7		451–595	Heavy poundage for 997 and 1098	128 lb at 1.011"	1.937"	138 lb in	88 lb	138 lb
8	AEA 311 Non std inner	451–578 Pair	Suitable for all units	88 lb at 0.973" 20 lb at 1.011"	1.75" 1.75"	113 lb in 31 lb in	51 lb 18 lb	92 lb 29 lb
9		451–579 Pair	Suitable for all units	100 lb at 1.011" 62 lb at 0.887"	1.687" 1.53"	148 lb in 96 lb in	46 lb 35½ lb	99 lb 71 lb
10		451–613 Pair	Suitable for competition units	115 lb at 1.00" 63 lb at 0.926"	1.83" 1.76"	139 lb in 75 lb in	74 lb 37 lb	124 lb 64 lb
11		451–614 Pair	Suitable for competition units	134 lb at 0.989" 71 lb at 0.926"	1.954" 1.866"	139 lb in 75 lb in	91 lb 52½ lb	141 lb 79 lb

Where there are two springs, the heavier outer spring is on the first line. The second line is the lighter inner spring.
**Assuming fitted length of outer to be 1.297 and inner 1.165 inch. Open load assumes typical hi-perf street cam i.e. 0.360in valve lift.*

that profile. A good cam manufacturer will supply the correct springs to go with their cam. A good source of high quality springs in the States is **A.P.T. Iskenderian,** and **Crane.**

The A.P.T. and Isky springs are made by the same manufacturer and are the most popular with racers in the UK. These springs (and a less expensive road version) are available through Kent Cams. They hold their delivered force for longer than most U.K. produced springs and allow about 0.09 inch more lift before coil binding occurs.

When increasing spring force or going from a single to a dual valve spring setup, it is important to make sure that a spring retainer of sufficient strength is used. Taking things one step at a time, let's assume first that you are going from a single to a dual spring installation, in which case it's important to install the inner spring locater. This ring sits around the valve guide and prevents the bottom of the inner spring from wandering. Many of the spring retainers fitted to run-of-the-mill production vehicles are made of ordinary mild steel and, as such, are not up to taking spring poundages much over 150lb. When using a typical race cam, Cooper 'S' retainers or equivalent must be used. For lightweight application, Kent cams and Isky produce a hard anodized aluminium retainer which stands the rigours of race poundage springs. These super lightweight aluminium spring retainers are about one-third the weight of steel retainers. Experience shows them to be reliable with a typical 220lb valve spring, **as long as a gear is never missed which promotes valve float.** If this happens, the retainers start to pull through and need replacing. On road motors having spring forces less than those typically used on race engines, the retainers seem to have an almost indefinite life. However, a little spring preparation is necessary to ensure this state of affairs. Because aluminium is soft, any sharp edges on the spring can easily dig into the retainer. To avoid this, it is a good idea to radius any sharp edges from the inside diameter of the spring. An alternative to these sometimes unreliable aluminium retainers are the titanium items sold by Kent and A.P.T. These are virtually 100% reliable and increase the rpm to loss of valve train control by about 150 rpm.

Now to the selection of collets, or keepers as they are known in the States, of which two types have been produced for the Series 'A' engine. There are collets that fit in the square groove valves and those that fit in the round groove valves. In about 1972, the square groove design was discontinued in all Series 'A' vehicles. For any high performance use, the half round groove design should be considered THE one to use.

There is a difference in the type of materials used for the collets/keepers. Cooper 'S' items are machined from solid steel, whereas many of the later types are of sintered steel design. Although I have not experienced any problems with the sintered steel design, the Cooper 'S' type are to be preferred.

Rockers

The golden rule for rockers, and the rocker shaft if it comes to that, is never use worn items. These contribute more to valve train noise than any other component, and also, rockers which have a worn pad cause the valve guides to wear quicker. This is especially so if the rockers were used with a standard camshaft, but are now being used with a high lift cam. The reason for this is that the lip formed at the edge that corresponds to the highest lift position tends to dig in to the valve and put a lot of side load on it, when it goes to the higher lift generated by the new high lift cam. The result is rapid guide wear and, depending on how extreme the situation is, the guides can be worn out in anything from a few hundred to a few thousand miles.

Also if rocker shaft bearings are worn, more oil finds its way up into the top of the engine. Which means more oil for the oil seals to control. The greater the amount of oil that goes down the guides, the greater the octane reduction of the fuel will be, and in turn, this will increase the motor's tendency to detonate as oil reduces the fuel octane.

Basically, there are three distinct types of rocker. The original pressed steel type is made up of two thicknesses of stamped out material which is spot-welded together. Over the years this type of rocker has been much maligned. When the forged Cooper 'S' rocker came out, this was hailed as THE rocker to have. During that era I could not afford THE rocker so I had to make do with the pressed steel rockers on my race engines. Over a period of some 5 years I used nothing else, and never broke one. This doesn't mean they are unbreakable. I have seen one or two break, but breakages were rare, and usually rather than fail completely, they bent first as an early warning, and as a result, no untoward damage was ever done. In all instances failures occurred because the rockers were used as standard rather than being strengthened. The easiest way to beef up pressed steel rockers is to put a bead of

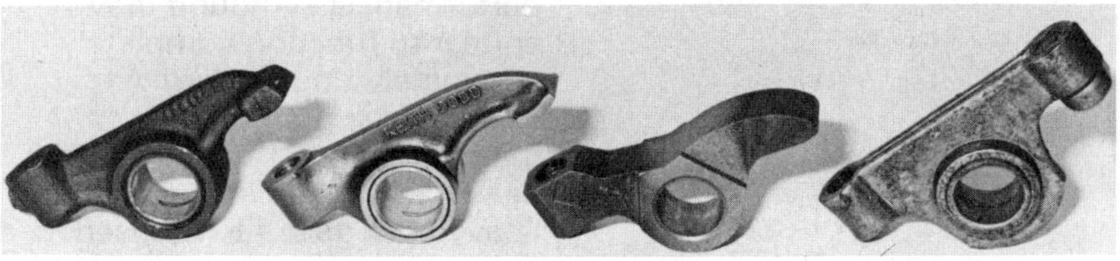

Here's an example of each of the non-rollerized rockers you are likely to come across. On the left is the original pattern Cooper 'S' forged rocker. Tough but notoriously inconsistent as far as ratio is concerned. Next is the beautifully made Mini Spares rocker which has a ratio of 1.5:1. Next is the current sintered 'A+' rocker. It's relatively heavy, most certainly ungainly and is something of an unknown factor when it comes to heavier-than-standard valve springs; sometimes they survive and sometimes they don't. On the far right is the original pattern pressed steel rocker. It's a lot more functional than many people have given it credit for.

arc weld along the top edge. This stops the edges spreading and allows the rocker to preserve its strength as a beam. Under these circumstances, the rockers have shown satisfactory reliability when used with typical race cam profiles such as the 649, and they are quite adequate for engine speeds up to 8000rpm. Subtle changes have occurred to the pressed steel rocker during its life. 1300 rockers give slightly more lift than do the ones used originally on the small bore motors such as 850s, 1000s, etc. However, on 'A+' engines, whether it be a small or big bore unit, the factory quoted rocker ratio is the same, namely 1.245:1. However, production tolerances can cause that figure to vary and it would seem possible to get rocker ratios as high as 1.28:1 fairly often, and some engine builders claim that with standard parts it is possible to get up to 1.32 rocker ratio with everything still within production tolerances. The essential point is that rocker ratios do vary, and **if valve lift is important because every last bit of hp is being sought, then it's necessary to check the valve lift to see if the rockers are giving something reasonably close to maximum possible lift.**

With the introduction of the Cooper 'S' came the forged steel rocker. This rocker, as far as competition was concerned, appears to be able to take any punishment that can be meted out to it. They do break on rare occasions. However, breakages appear to be due more to material or manufacturing inconsistencies rather than inadequate design. It's often found that by comparison with the pressed steel rocker, the Cooper 'S' rocker gives slightly less lift. To some extent this situation can be rectified by replacing the standard bearing bush in the rocker with an offset bush. Generally speaking the offset rocker bush will produce 25–30 thousandths extra lift at the valve. Such a bush can also be installed in the pressed steel rocker and will increase lift to a similar degree.

When installing offset rocker bushes, it will be necessary to assemble the valve train on the engine and check the position of the end of the rocker pad in relation to the end of the valve stem, as their use will cause the pad to move nearer the valve. This can lead to the rocker pad being very close to the far edge of the valve as it opens the valve. The easiest way around this problem is to run a $^{11}/_{32}$ inch drill through the stud holes in the rocker pillars and position the pillars back against the studs to bring the rocker pad back to its original position relative to the valves.

The question arises of what is the extra lift worth in terms of power increase? This depends on the engine and the components fitted. When discussing cams, I made particular reference to which cams were likely to work well with higher lifts and which wouldn't, but basically we can assume that if you have a standard or near-standard cam in any engine, higher lift will produce higher power, especially in the bigger engines. In such engines, even when a conventional high lift cam is used, **more power is achieved from more lift due to the use of high lift rockers, but be sure to check that spring coil binding doesn't occur.** To give you an idea of just what it's worth on a relatively standard motor, **Fig. 11.7** shows the power increase achieved by simply installing offset bushes into a set of standard pressed steel rockers. This mod is well worth doing in terms of the power increase it gives for such an apparently simple modification. However, installing the bushes can be a little more time consuming than at first appears to be the case. The procedure involves pressing out the old bushes, installing the new ones and drilling oilways as necessary and then reaming the bushes to fit. The bush material is very tough and reaming them to size is very time consuming. However, they do last a long time and you can be sure that instead of the bushes wearing out, the rocker shafts wear out.

The third type of rocker is that fitted to 'A+' motors. These rockers are definitely made to a price, rather than to a

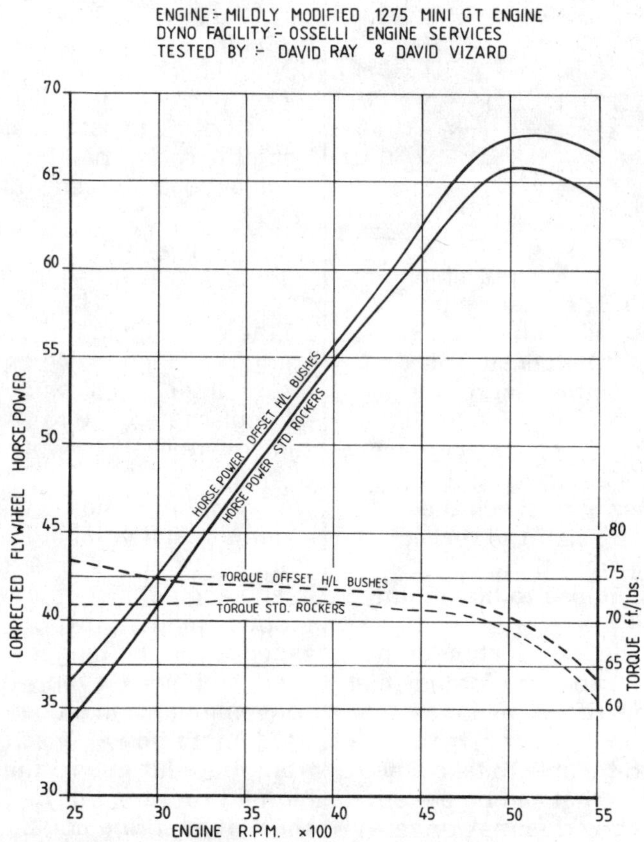

Fig. 11.7 Comparison of power output between standard rockers and standard rockers with offset bushes

Here we see the effect that offset bushes, to increase rocker ratio, can have on power. Note that even as low as 2500rpm the power and torque are substantially better, thus indicating that even on a standard motor the lift is inadequate.

By installing offset bushes valve lift has been increased from 0.325in to 0.350in.

performance. Although they appear to function completely adequately on standard 'A+' engines, the method of manufacture makes them rather heavy and a little suspect if heavy springs are to be used. The 'A+' rockers are made by a powdered, sintered metallurgy process and as such, their integrity under heavy load, is a little less than one would desire for high performance use. On valve springs which don't have excessive load because they are not being used with race cams, these rockers appear to be adequate to about 7500rpm, but past that I'm not sure I would want to use them. They may be able to take much more, but the disadvantage with this type of material is when it breaks, unlike the pressed steel rocker, it gives no warning. Apart from that, unless the heat treatment has been done absolutely to specification, the pad areas can be prone to wear. This wear can be exaggerated on those engines using heavy valve springs.

High Lift Rockers

If we bear in mind that the Series 'A' engine was designed many years ago it will be apparent why some of its design characteristics are a little less than perfect. For instance,

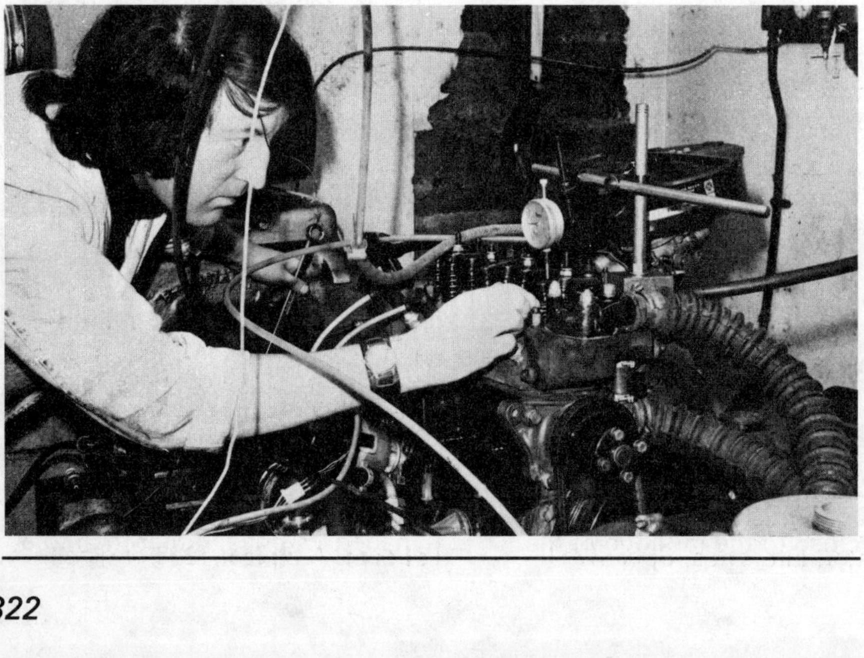

Here I am checking the effect on valve lift of installing a set of higher ratio rockers on a 1275 engine. The brickwork showing in the background behind the air filter was due to a small fire I started in this dyno cell of Oselli's. Hopefully when words of this minor indiscretion get around, it won't deter any other company bosses out there from letting me use their dyno cells!

through many years of experience with pushrod engines it has been established that rocker ratios between about 1.5 and 1.7 are generally optimum for an effective valve train. By having a high ratio rocker, the cam follower and pushrod do not have to be moved as far or as fast for a given amount of valve lift as would be the case with a lower ratio rocker. Another advantage is that the spring has a greater mechanical advantage over the pushrod and cam follower. Let's look at an example to show how this can affect the spring force required to control the valve train. Let us say we have 1.25:1 ratio rocker. If this is operated with a spring giving 200lb open load at full lift, the load to return the pushrod and cam follower down the closing side of the cam will be 200 x 1.25 *i.e.* 250lb. With a 1.5 rocker ratio, the load at full lift available to push the cam follower and pushrod down the closing side of the cam will be 200 x 1.5, *i.e.* 300lb. Since the main mass of the valve train is in the cam follower and pushrod, we find that **the higher ratio rocker can control the valve train with a given poundage of valve spring better than the lower ratio one.** This is a strong argument in favour of obtaining high valve lift with a high ratio rocker. If we add to that the fact that, generally speaking, very few of the high performance cams lift the valves high enough for maximum hp then we can see there are two good reasons for going to high lift, high ratio rockers. Firstly they have the potential to increase the power output and secondly they will also allow the engine to rev to higher rpm on a given poundage of spring before valve crash speed is reached.

Strangely enough, when first introduced by **Speedwell** in the U.K. 1.5 ratio rockers, or high lift rockers as they were simply known, met substantial sales resistance, and as such, were not a great success. The irony is that about the time they became generally accepted as providing better results than the standard ones, Speedwell stopped selling them. For a period of a few years, a set of Speedwell high lift rockers commanded a considerably inflated price as no substitute was available. Now four companies in the UK produce high lift rockers. **Avonbar Racing** sell what could best be described as a replica of the Speedwell rockers. The standard of manufacture is more than adequate and they are sold at a very acceptable price which is similar to the price of a set of standard rockers, pillars and shafts. Note that when 1.5 rockers are used, the pivot point is moved so much further away from the valve that it is necessary to use special pillars, supplied with the kit which locate the rocker shaft in the appropriate position.

High lift rockers don't work every time. Tests I did with this 1000cc engine equipped with a Piper Magnum 270 showed a massive drop in hp. As much as 7hp in the lower rpm range were lost when 1.5 ratio rockers were used. The engine simply had too much valve opening during the overlap period. To make high lift rockers work with this cam, I suspect the lobe displacement angle needed to be widened by as much as 5°.

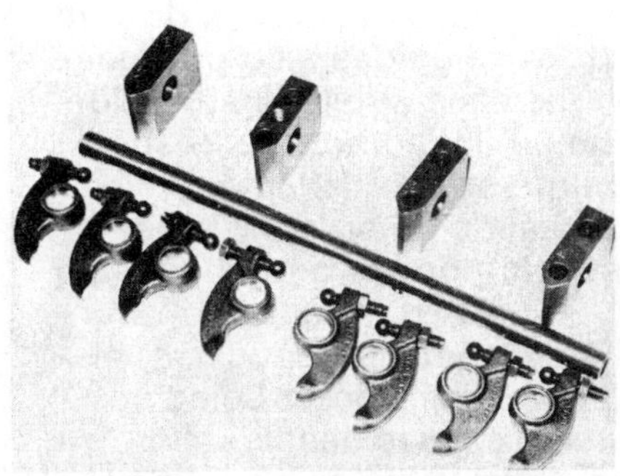

The Avonbar high lift rocker kit provides a 1.5 ratio rocker which bears a distinct resemblance to the early and now defunct, Speedwell rocker. The price is virtually the same as the Mini Spares item.

Another source of high lift rockers is **Mini Spares Centre.** The quality of the Mini Spares rocker is one of the highest I have seen for a conventional rocker. Also, in its standard form it is about 15% lighter than a standard 'S' rocker, and about the same weight as a lightened 'S' rocker. It has a wide pad on the valve end so the business of having to line up the rockers spot-on over the centreline of each valve, becomes a redundant operation. With the 'Speedwell' type rocker from Avonbar, it is best to line up the rockers over the centre of the valve. Usually this is only a ten minute job and simply a matter of juggling shim thickness between pillars and rockers. The job is not complex, but it is one which, nonetheless, needs to be done.

Although their use as a race car item is becoming more and more prevalent, the primary purpose of these high lift rockers is for use on engines where no cam change has been made, the extra lift giving approximately the same effect as moving up one specification of cam. A typical 1275 engine

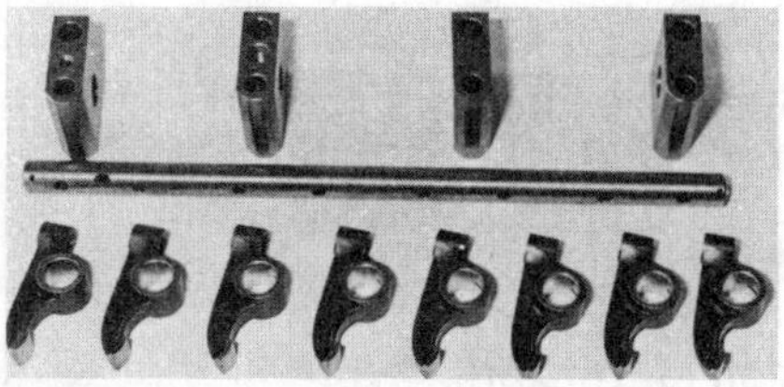

Here's the parts you get from Mini Spares if you buy their high lift rocker kit.

will pick up in the region of 4 or so hp. In fact a power gain will occur throughout most of the rev range with any standard cammed vehicle with the exception of those using the hotter MG Metro profile. With these cams, little bottom end power is gained. In fact sometimes the engine will actually lose the odd ft/lb or so of torque down very low. However, it's virtually unnoticeable, but the extra power at the mid and top end is noticeable.

Roller Rockers

Although the roller-style rocker has been around for many years as a high performance item for pushrod engines, its use has been confined mostly to American V8 engines. However, in the early '80s roller rockers began to find their way onto the Series 'A' engine with, might I add, considerable justification.

The first and most obvious point of recognition for a roller rocker is the roller on the tip of the rocker itself. By having a rolling action across the end of the valve, any side loads produced during the valve opening and closing phase are much reduced. Although side loads are not very great when valve lifts of 300, 350 even 400 thousandths are used, they do start to attain sizeable proportions when valve lifts of 450, 475, even 500 thousandths are used. These side loads can increase very quickly in relation to lift, because high lifts require heavier poundage springs, and this, in turn, increases the frictional side loads on the guide. The roller rocker is a relatively effective way of reducing side loads on guides and thus contributing toward better guide wear rates over a longer period. This in turn, means the engine will maintain its hp output because the valve sealing due to tipping valves, is not lost. It pays to remember worn guides mean worn seats, and worn seats mean a drop in hp. Worn guides also produce an increase in oil consumption. The oil dilution of the fresh charge drops fuel octane which can cause detonation, thus reducing power and engine life.

Apart from the roller on the tip, some roller rockers, the **M-Tech** roller rocker being an example, have needle roller bearings. Whether or not having a needle roller bearing is justified depends upon the application. Since the rest of the valve train need only have a fraction of the amount of oil that's normally supplied, the needle roller bearing gives the engine builder the option of cutting the oil supply so there is less oil to control in the top end of the engine. Ultimately this means less work for oil seals to do, but that particular aspect is really only of minor consequence. Essentially, needle roller bearings, though effective in this instance, are not really designed for reciprocating motion which is what they are subjected to when used in rockers. To be quite honest, a plain bearing is within a hair's breadth of being as good as a needle roller bearing and it's nothing like as fussy to use. For instance, a plain bearing, when it wears, is not likely to cause any reliability problems on the engine. If the cage of a needle roller bearing should break needles will be shed into the engine. Another point to consider is that needle roller bearings cannot tolerate any dirt or grit going through them, whereas plain bearings are not nearly as fussy in this respect. To sum up, needle rollers are nice if you can afford the extra cost, but a plain bearing will get you 99% of the way there if you cannot afford a needle roller.

As far as roller rockers availability is concerned, three come to mind, the **M-Tech** rocker and, from the UK, the Windsor and Titan rockers. The **M-Tech** rocker can best be described as the blueprinted ratio roller rocker. In other words, it was made for SCCA racing when high lift rockers were expressedly banned by the regulations. The rules have changed to allow the use of any

If you really want to move into the big league in the rocker department, then these SCCA-legal roller rockers from M-Tech, provide the highest legal ratio for SCCA racing.

This close-up of an M-Tech rocker shows the sturdy pin on which the tip roller is mounted. Also can be seen the needle roller bearings on which the rocker runs.

ratio. In producing a rocker for this type of racing, M-Tech made allowances for all the tolerances working in favour of generating the highest lift possible and yet still stay SCCA legal. Generally this produces a rocker with around 1.3:1 ratio, as opposed to the factory-quoted standards ratio of 1.245:1. On a typical race cam, the **M-Tech** rocker gives about 25 thousandths extra lift. I have used these **M-Tech** roller rockers on my own engines for thousands of miles and hundreds of hours on the dyno, and they have produced totally trouble-free operation. As bought from **M-Tech**, the roller rocker kit consists of a hardened steel shaft and a set of roller rockers. You have to install your own adjuster screws and make provision to supply your own rocker pillars. In spite of the fact the rockers are blueprinted to the standard ratio, the standard rockers pillars will place the rocker shaft too close to the valves, thus putting the roller over the edge of the valve. As a result, it is necessary to reposition the shaft in relation to the valve, using pillars with the shaft suitably relocated. Some companies make up steel pillars to reposition the shaft when offset rocker bushes are used in standard rockers and these pillars also are suitable for use with the M-Tech rollers. Example sources of pillars for rockers having a ratio of around 1.3:1 (*i.e.* offset bush standard or **M-Tech** roller) is **Richard Longman & Co** and **Avonbar Racing** also **Mini Spares Centre** can provide a bush and pillar set.

Because of the increased ratio over standard, the M-Tech rockers will inevitably need the rocker shaft centreline moving away from the valve stem centreline, so that the roller tip ends up centrally over the end of the rocker. If this is not done, it is possible for the rocker to cause extremely rapid guide wear, especially if the roller just runs off the tip of the valve onto the valve tip chamfer.

The **Windsor** roller rockers differ from the **M-Tech** ones in that they are 1.5 ratio items and the rockers come as part of a kit which includes a special hardened steel shaft and special three-post pillars. These pillars support the rockers from each side, so that the stiffness of the valve train as a whole is substantially increased. As you can see from the **accompanying photo,** the whole setup makes a very sturdy assembly.

The Windsor roller rockers are made from a one-piece forging and can be had either with a plain bearing bush or with a roller bearing. The Windsor rockers are made to run with the special Windsor pillars which support the rockers each side for a super rigid setup.

The high lift rocker sets from **Titan** look a little more conventional than the Windsor rockers, but are manufactured to very high standards, as are all the Titan components I've dealt with. Because of the more compact nature of the Titan setup and its overall dimensions, it fits, usually without interference, under valve covers. Functionally, the 1.5 Titans can be expected to produce the same results as the 1.5 Windsors. However, A.P.T. have been doing some experimentation with rockers ranging from 1.6 to 1.7 from Titan. At the time these were specials but are now available over the counter. The indications to date are that once cam timing/profiles are developed for the higher lift and greater accelerations produced at the valve, then these higher ratio rockers will show a marked increase in output. Based on experience to date it would seem the use of these high ratio rockers almost precludes the use of a single pattern cam. With the additional ratio available, a valve which lifted 0.520 inch on a 1.5 rocker will be lifting over 0.560 inch on a 1.62. This has the potential of improving the volumetric efficiency to the point where more time to expel the exhaust is needed. Hence, the requirement for longer timing on the exhaust. It's too early to tell yet just what 1.62 rockers will do, but the potential is certainly there.

I have already discussed in general the benefits of using high lift rockers with standard or near-standard cams, but nothing equals some actual test figures to demonstrate the effectiveness of 1.5 rockers on bigger engines. For the test in **Fig. 11.8**, I used for a baseline the M-Tech blueprinted ratio roller rockers versus the Windsor 1.5 roller rockers. In other words, it's a roller rocker versus a roller rocker and the highest ratio standard rocker you are likely to get versus a straight 1.5 rocker. The particular engine concerned was equipped with a '649' cam, and by using 1.5 ratio rockers, some 453 thousandths valve lift was achieved. As can be seen from the figures in **Fig. 11.8** hp was noticeably increased and this was on an engine that sported 'S' size valves only. Had

PIllars for the Windsor rockers are made as beefy as possible to give the maximum rigidity to the valve train. However, you may find that the valve cover is a tight fit over this valve train, due mostly to the end pillars intruding into the area the valve cover takes up. Normally it takes only a few minutes with a file to relieve the end pillars to give them clearance. With some valve covers there is no problem.

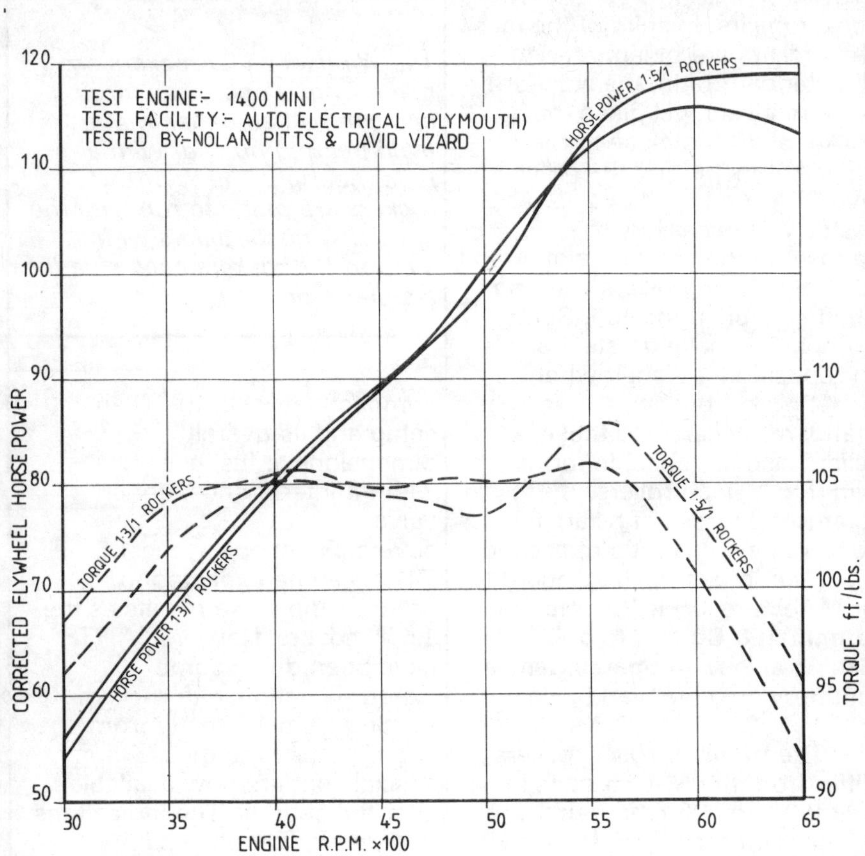

Fig. 11.8 Comparison of power output between 1.3:1 roller rockers and 1.5:1 roller rockers

Even though high lift, high ratio rockers don't increase the seat-to-seat timing (i.e. 300° cam stays 300°), they do increase effective breathing, sometimes to the detriment of low speed output. This is usually due to insufficient exhaust timing which prevents the expelling of the increased volume of exhaust gases generated. Usually the best results are obtained by increasing the exhaust timing, but other avenues for improvement are open. It can sometimes pay off to use a slightly shorter cam or spread the lobe centres of the existing profile by 2°–4°. Once rpm has climbed to the 5500 mark, the effect of the increased breathing of the 1.5 rockers starts to pay off. The further up the rev range the power figures are taken, the greater the advantage of the 1.5 rocker.

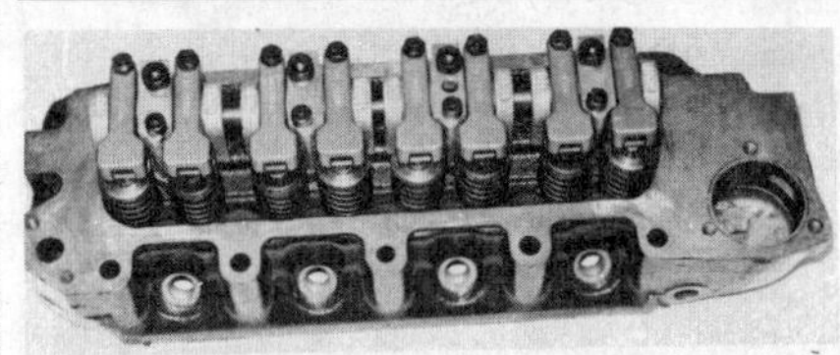

This is what the entire Windsor roller setup looks like when it's assembled. In this particular shot, the Allen bolts have been used simply to hold the system down on the cylinder head for the purposes of this photograph. On the engine, of course, the setup would be held in place by the studs as for the standard rocker shaft setup.

it been a super big valve head, *i.e.* 1.48 intake and 1.25 exhaust, then I would have expected an increase even greater than this.

Before we branch off onto other subjects, there is one point I would like to make that I feel is of real importance to both the cam grinders and to the high performance enthusiast. **The days of the low lift cam and valve train assembly for the performance Series 'A' engine are numbered.** Since cam blanks are very few and far between, the cam grinder must grind a high lift cam out of a standard cam which inevitably is of a low lift variety to start with. The only way to get a high lift profile from a low lift standard profile is to reduce quite dramatically overall profile size. This means that not only does a lot of material have to come off the base circle but also the smaller a cam gets, the sharper the profile is over the nose of the cam. Whenever sharp curves exist on a cam the surface stresses increase and, as a result, the life of the cam lobe is reduced. **Because the standard cam was designed to operate with a 1.245 or so ratio, if a switch to 1.5 is made then, for any given amount of lift required, less material has to come off the standard cam blank to produce a hot profile.** By making the change to a 1.5 rocker, the cam grinder has greater scope for producing a more radical race profile from a standard cam. In many ways it is far cheaper to provide a given valve opening motion by using 1.5 rockers and grinding a new profile from a standard cam profile than trying to grind the profile from a brand new unground billet. If added to this is the fact that the 1.5 rocker ratio also requires less spring pressure, then it can be seen that the all-round advantage lies with the 1.5 rocker rather than the 1.245 rocker.

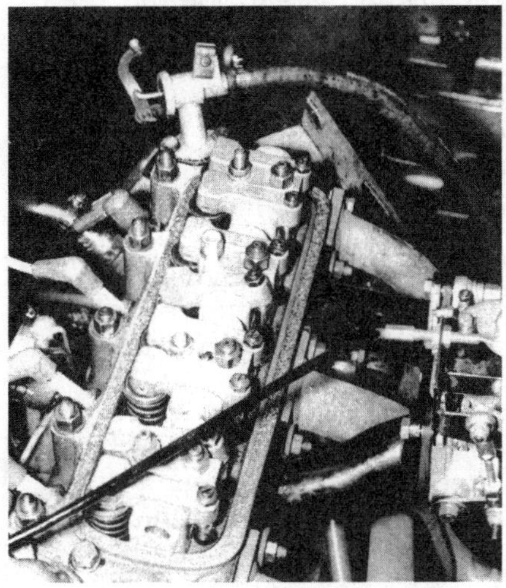

Time spent dyno testing the Windsor roller rockers and carbon fibre pushrod was time well spent. Both the rpm potential and the top end hp of the engine was increased. However, some mid-range performance was lost, indicating that these high lift rockers may fare better with cams designed to match.

Rocker Shaft Spacers

It has been conventional practice over the years to space rockers with tubular spacers. Various reasons have been given for this, the usual one being that it reduces friction. The difference in friction, however, between tubular and spring spacers is so small as to be immeasurable for all practical purposes. On the other hand, springs do perform a useful function that spacers don't. First they take up all the slack in the system, and thus reduce noise, and secondly they tend to damp out valve train vibration due to the pressure of the rockers against the pillars. This in itself is a good argument against using spacers. To conclude, don't bother with spacers, two springs will work better.

Rocker Shafts & Pillars

There's not a great deal to be said on this subject, but what you do need to know could well prevent a minor disaster. If you have an early type engine (those up to 1966–67) which has never been torn down and completely rebuilt, it may still be equipped with the early type of rocker shaft and pillar assembly. In these early engines, the tapered grub screw which locates the rocker shaft was in the rocker pillar nearest the thermostat housing. This rocker shaft was replaced by what at one time was known as the competition shaft. The competition rocker shaft was productionized because of its supposedly superior design and strength. On the 'competition' item, the location of the shaft in relation to the pillars was achieved by having the tapered grub screw in the number 2 pillar. Thus, if you swop from the early shaft to the late one (which will be the case if a new shaft is installed as the early types are no longer available) you will have to move number 1 rocker pillar to the number 2 position. However, the number 1 rocker pillar is the one which has the oil hole in it to feed the rocker shaft assembly. If you are to use early rocker pillars because only the number 1 pillar had an oil hole it will be necessary to drill an oil hole in the number 2 pillar which is now in the number 1 position. In current production rocker pillars all pillars have an oil hole. The easiest way to drill a pillar is to align the number 2 pillar with the number 1 pillar, using a bolt through one or other of the stud holes. The next stage is to drill through from one pillar to another using the existing hole as a guide. Having spotted the position of the hole complete the drilling operation and chamfer the orifice to ensure alignment with the hole in the cylinder head.

Three types of pillar have been fitted, the particular type depending on the year of the engine. Early Series 'A' engines were equipped with aluminium pillars, and this was standard practice on most non-high performance cars up until 1967–1968 or so. Indeed, some of the twin carb engines such as MG 1100s, Sprites, etc., were also equipped with this cast aluminium pillar. The problem with this type of pillar is that if the cylinder head is torqued down just a few lb ft over the limit, and there is justification for this in some instances, then the pillar starts to collapse. The result is that clamping pressure is lost on the cylinder head. This can lead to a head gasket failure as well as a reduction in valve train mechanical integrity. Early aluminium pillars are best scrapped.

When the Cooper 'S' was introduced, it came with cast steel rocker pillars which were just fine for any application. However when it became apparent that cast aluminium pillars were a liability, they were replaced by a sintered iron pillar. The sintered iron pillar was a big enough improvement to the point where it was practical to use this pillar in all applications, be it for road or competition. The cast 'S' pillars may be stronger, but their strength is superfluous to requirements in almost all instances.

Having decided on the pillars to be fitted, it is necessary to give the rocker shaft some thought. Although the shaft is strong enough for all applications, it does suffer from the fact that being soft, it wears very rapidly, thus causing a noisy valve train. This can be overcome by using the Windsor hardened steel rocker shaft or tuftriding the standard rocker shaft. If the standard rocker shaft is to be tuftrided then it's necessary to gut it first, as the tuftriding salts can accumulate inside the shaft. After tuftriding, scrub the bore clean, then replug the ends of the shaft. **Don't omit the post-tuftriding cleanout. Flushing with paraffin/kerosene or any other cleaner is insufficient to get the job done.** If any salts are left in the shaft, they will find their way around the rest of the engine; they are very abrasive. If you don't want to go to this trouble many Series 'A' specialists such as **Richard Longman & Co.** supply a strengthened shaft of thicker material.

Pushrods

It's well-known that valve train rigidity can contribute to an engine's power output. One of the advantages of an overhead cam engine is that because the valve train is so rigid the valves can be opened faster and further in a given time. This means that the cam need not have as great an opening period to produce the same breathing capability. The shorter timing also helps the bottom end power. This is one of the reasons why an overhead cam engine can produce a wider power band than a pushrod engine. Of course other factors such as ports, combustion, exhaust, etc., are involved. However, the stiffer the valve train, the more likely is the possibility of getting power at both ends of the rev range. Unfortunately, the real weak link in achieving rigidity in a pushrod engine is, in fact, the pushrod itself. Pushrod flexibility often means the cam designer has to moderate the valve train accelerations imposed by the camshaft simply because too rapid an acceleration rate at the cam will cause the pushrod to bend, rather than cause the valve to follow the opening profile of the cam. To be fair to standard pushrod design, we are now looking at refinements rather than essentials. In most instances, the standard pushrod copes adequately. Only when maximum power is required is it necessary to consider improvement of the standard pushrod. It is possible to fabricate and install tabular pushrods. It's necessary to trim these to length and put the inserts in to suit the ball at one end and the tappet or cam follower at the other. This type of pushrod, however, still represents a relatively heavy item with not necessarily a great increase in stiffness. However stiffness as measured statically is not a true indication of the pushrod ability to transmit motion from the cam to the valve. A factor that I'm beginning to appreciate the significance of, since the last update of this book in 1987 is the lateral vibration frequency of the pushrod. The higher this is the less the pushrod is likely to resonate with the rest of the system. Result – more rpm before loss of valve control be it either over the nose, follower toss, or at the seat in the form of seat or valve bounce. Follower toss at peak lift can help output but can be destructive to the valve train. Seat bounce is not only destructive to the valve train but also power output. The bottom line here is that although the tubular pushrod may be of

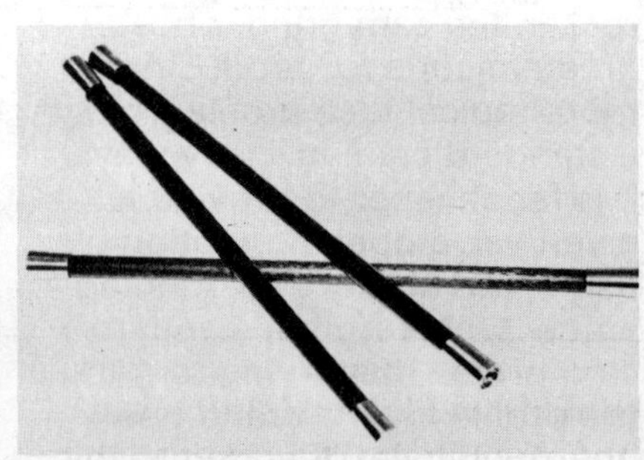

These Windsor carbon fibre pushrods represent a massive reduction in valve train weight over their equivalent steel counterparts.

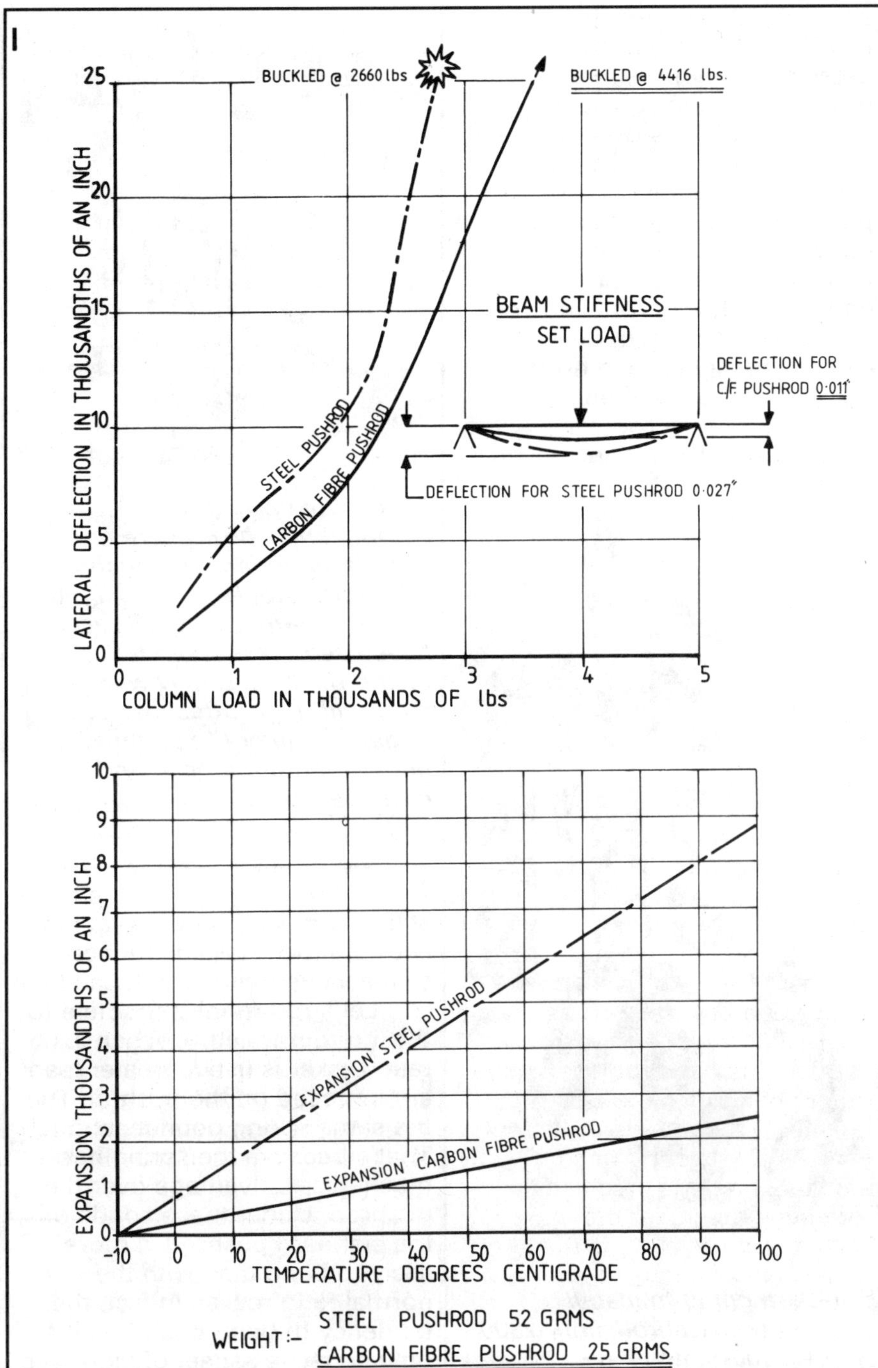

Fig. 11.9 Comparison of stiffness of steel and carbon fibre pushrods

Although this is a crude way to test a pushrod, it did allow me to get some comparative figures of the stiffness of a carbon fibre pushrod versus a steel one. The load on the pushrod was calculated from the pressure read out on the pressure gauge. The dial indicators measured the deflection of the pushrod at the centre of its length.

When the standard steel pushrod buckled like this, it stayed that way and it had no ability to hold the load whatsoever after buckling had taken place. This shows quite conclusively that even a slightly bent pushrod can get into trouble in a high rpm engine.

similar weight and stiffness as the standard solid pushrod it has a much higher lateral frequency resulting in better performance. For this reason I strongly recommend the use of either the APT 1275PR-BB or 905-0003 depending on the length required.

The need for stiffness and low mass are the primary reasons for considering carbon fibre pushrods. Along with this must be added the fact that they have both a high lateral resonant frequency and good vibration absorption qualities. However they are not without their problems. But before we go into that let us look at carbon fibre pushrods in terms of stiffness and mass. This type of pushrod is less than half the weight of the steel pushrod, yet exhibits something like 2½ times the stiffness. **Fig. 11.9** shows just how much better it is throughout the normal working range than the standard steel 1275 pushrod. Although the specification of the carbon fibre pushrod may look impressive on paper, in reality on a typical 1275 motor running up to 7000rpm, there is no detectable power difference, even when a typical race cam such as the 649 is used. It is not until very high rpm that the carbon fibre pushrod starts to pay off with an otherwise conventional valve train. However, if the valve train is employing 1.5 rockers and you intend to use custom designed cams with higher than normal acceleration rates, then the carbon fibre pushrod will pay off. Reduced weight and increased stiffness means faster acceleration rates can be applied to the valve train.

Let's return for a moment to the 1.5 rocker setup. When a 1.5 ratio rocker is used, greater loads are imposed on the pushrod for the same spring poundage. That's because the spring has a mechanical advantage over the pushrod. Under these conditions, the standard pushrod is more likely to bend than with the normal ratio rocker. In fact, the tendency to bend pushrods is the only negative aspect of high ratio rockers. Fortunately for us, it is a very small disadvantage in relation to the number of advantages that high ratio rockers offer. **When 1.5 rockers are used, the requirement for a stiffer pushrod is increased, and**

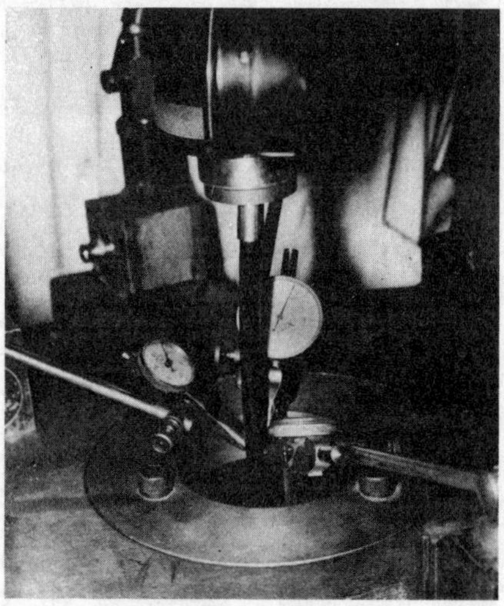

It took more than double the load to bend this carbon fibre pushrod to this extent when compared with a steel one. Relieving the load from the pushrod caused it to return more or less to its original form. Even though it had suffered structural damage because it was tested to the limit, it would still hold a load comparable to a standard pushrod.

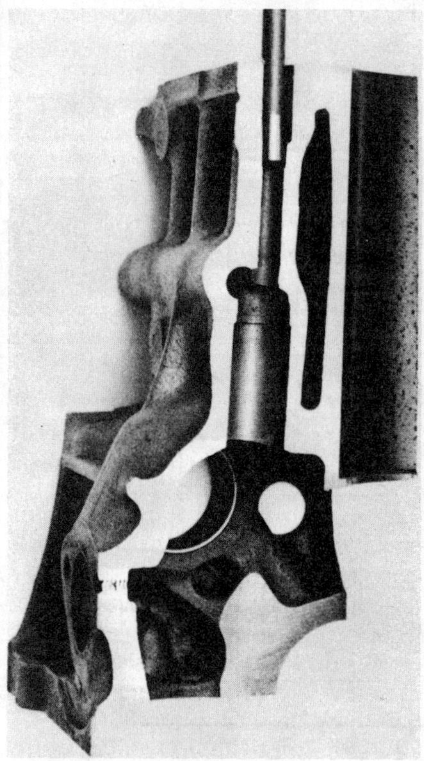

If you are going to install carbon fibre pushrods into 1300 or 'A+' blocks, it will be necessary to drill out the pushrod holes to obtain clearance. As can be seen the greater bulk of a carbon fibre pushrod will not pass through the pushrod hole in the block.

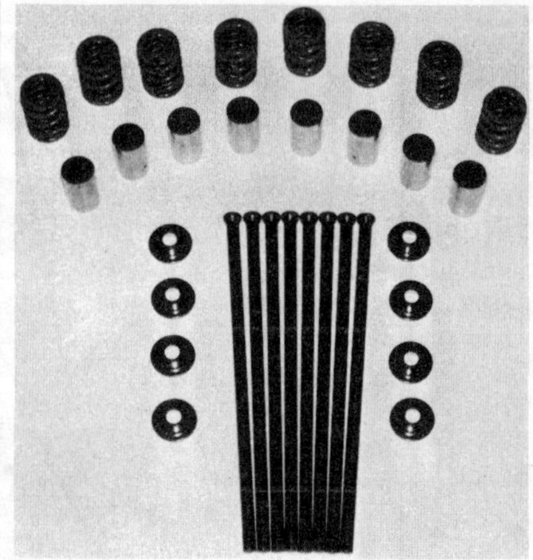

This set of valve train components shown here are all from Iskenderian. The pushrods, which are tubular, are no lighter than standard. Many US racers have used these Iskenderian pushrods successfully with higher lift cams. Because of their larger diameter the pushrod holes in both the head and block must be elongated (not drilled) to suit.

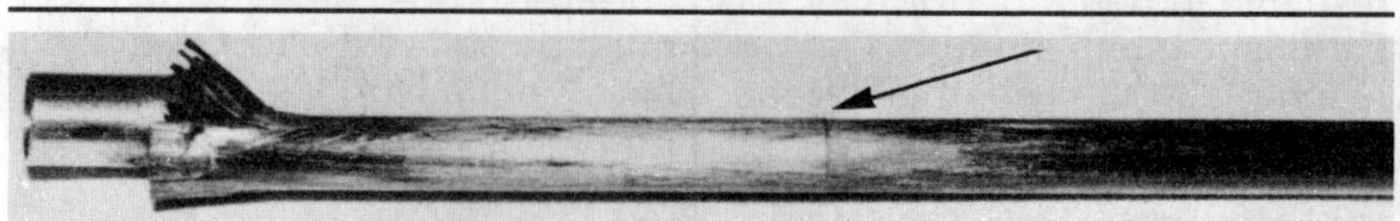

Here's something that Nolan Pitts and I found out the hard way about carbon fibre pushrods. If the carbon fibre pushrod touches the cylinder head or block at any point in its travel, as it did in this example here (point of contact indicated by arrow), the pushrod will, in all probability, fail at the end as shown here, not, as one would expect, at the point of contact.

carbon fibre pushrods have the potential to fulfil this requirement.

Apart from their performance potential, carbon fibre pushrods do have another asset, in that, in comparison with steel, carbon fibre does not support vibrations very well but acts as a noise damper. Steel has a great tendency to transmit noise throughout the engine, the valve train being one of the prime sources of noise generation. Even when very wide tappet clearances are used, the carbon fibre pushrod will damp out most of the excess noise. In fact, from the noise point of view, carbon fibre pushrods are the next best thing to having hydraulic tappets.

Now for the disadvantages of carbon fibre pushrods. Such disadvantages as there are basically fall under one heading – design refinement. Because carbon fibre itself does not make a very good bearing material, it's necessary to cap the carbon fibre pushrod with a steel sleeve to bear on the cam follower at one end and the rocker adjusting screw the other. So far, so good. The problem arises in as much as carbon fibre is not a solid lump of material: it is in fact lots of very, very fine strands of material bonded together. Unless the loads borne by the carbon fibre part of the pushrod itself can be transmitted evenly by both end caps, the carbon fibre material will break down. Imagine the end of the carbon fibre cut off at an angle, yet the cap which sits over the end of it is square inside. The result is that instead of the load being distributed throughout the carbon fibre, it's concentrated just on one side. This causes local overloading of the carbon fibre which then breaks down. My experience with carbon fibre pushrods is that they are not entirely reliable without further design refinement the only way to get a working set of pushrods is to start off with, maybe, ten and expect one or two to break, then replace them. If a carbon fibre pushrod is going to fail, it does so relatively early and usually without damage to anything else. Often it's just a question of running the engine for a half an hour at reasonable high speed and checking the pushrods. If all is well, it will probably remain that way. If not, it's simply a matter of replacing those that appear distressed. It seems that about one pushrod in sixteen fails. As I had indicated earlier, some design refinement is needed if the carbon fibre push rod is to succeed. The principle problems stem from the difficulty in transmitting the load from the tip to the fibres themselves. Also, the bursting stress of a carbon fibre column is low and the loads produced in a valve train area close to or in excess of that which it will tolerate. This means there are two inter linked problems to solve. A technique which I have had success with is to encase the carbon fibre in a thin wall titanium sleeve. The carbon fibre is ground to produce about a 2 thousandths clearance in a 15 thousandths wall tube and installed with the highest locking grade of Loctite. After this has set, the ends are precision ground to form as flat a face as possible. These are then installed with a high compressive load 3M epoxy adhesive into the ends. The pushrods are kept under a load of 200 lbs while the adhesive sets. When spin tested such a pushrod was able to run reliably to 9,200rpm with a spring of 500 lbs. at a full valve lift of 675 thousandths. When will they be available? They won't be – there isn't enough demand but if I ever decide to go back mini racing that is how I will make up my pushrods. If you want some you know how to do it!

Using Short Pushrods

There is a practice amongst some engine builders of using the shorter pushrods from the small bore engine in the bog bore units which require shorter than standard pushrods due to block and head machining. The use of these shorter, small bore pushrods often necessitates the use of longer tappet adjusting screws. Such a practice results in reduced valve lift because the further out from the rocker the tappet screw is adjusted, the greater the lever length. Unfortunately the increased lever length is on the pushrod side and as a result, it reduces the effective rocker ratio. Also, tappet adjusting screws which are adjusted a long way out of the rocker, experience a bending load and can break.

Tappet Clearances

The tappet clearances, valve lash, or whatever you choose to call it are usually quoted for use with rockers having a standard ratio and with stock, steel pushrods. When either of these two factors change, the tappet clearances need to be changed accordingly. Let's look first at the effect that high ratio rockers have.

When the tappet clearance is set, it is measured at the valve end of the valve train, but in fact what we are doing is setting the clearance required between the camshaft base circle and the cam follower (tappet). Most Leyland performance cams are designed to have a working clearance of between 10 and 12 thousandths of an inch between the cam profile base circle and the cam follower. Because measurements are made at the valve, they are multiplied by the rocker ratio, this produces 12–15 thousandths clearance at the valve. If a 1.5 rocker as opposed to 1.25 is used, the clearance set at the valve needs to be increased. For instance, if the valve clearance required was 12 thousandths with a standard rocker ratio, then it will need to go, in round figures, to 15 thousandths, *i.e.* 9.6 thousandths clearance at the tappet times 1.5. If it's a race cam, most of which employ 15 thousandths clearance with a 1.25 rocker, we find that that's 12 thousandths at the cam follower. Translating this back to valve clearance means 18 thousandths with a 1.5 rocker.

The use of carbon fibre pushrods modifies the clearance further. Carbon fibre pushrods expand at a far lower rate (as can be seen from **Fig. 11.9**) than do steel pushrods. As a result, if the valve clearances are set cold, then the expansion of the carbon fibre pushrods is insufficient to bring about the correct clearances when the engine has reached operating temperature. Generally speaking, the carbon fibre pushrods expand 2 to 3 thousandths less than the steel ones. This lesser degree of expansion is, of course, on the cam follower side of the rocker, so it's necessary to close the gap on the valve side by a factor eoual to 2 to 3 thousandths times the rocker ratio. An example using 3 thousandths less expansion as a working figure should clarify the picture. For standard rockers the gap would need to be closed by 3 thousandths times 1.25. For 1.5 rockers it would need to close by 3 thousandths times 1.5. This works out to 3.75 and 4.5 thousandths respectively.

Adjusting the valve train to tenths of a thou isn't practical, so we can say that all that's necessary with carbon fibre pushrods is to close up the valve clearance by about 4 thousandths in comparison with steel ones. The foregoing assumes that the valve clearance quoted is the optimum. It's worthwhile noting that a little experimentation with valve clearance can be beneficial. Firstly, do not fall into the trap of thinking that more valve lift and longer duration achieved by tightening the valve clearance will pay off in terms of hp because in 99 cases out of 100 it won't. Basically, the Series 'A' engine likes rapid valve accelerations from the seat. Clearance ramps on most cam profiles have notoriously slow acceleration rates to keep valve train noise and stresses down. The longer the follower is actually running on the clearance ramp to produce valve opening the less the hp output is likely to be.

If you can spend some time on a dyno, try tightening up those tappet clearances by 3 or 4 thou. Usually you can watch the odd hp or two disappear. On the other hand, opening up the valve clearances can often pay off in terms of a little extra power, but again it pays not to go too wild on this because wide clearances cause violent accelerations from the seat, leading to excessive stresses on the cam followers, pushrods and rockers. The exception to this is, of course, the A.P.T./ Crane 'tight lash' profiles. Since they have no opening tappet ramp to speak of they can often benefit from tight clearances. We have used as little as 0.004 inch. If you have dyno time experiment with clearances up to about 25 thousandths on a race engine using a conventional profile, but don't go any further. Also, if you are doing a bit of experimenting, especially if you are running a radical cam, do periodic checks to see how the valve train is coping. It could be that with the particular profile you are using excessive clearance is causing rapid wear.

Valve Train Geometry

During the building of a high performance engine, there are many variables to contend with in the valve train and there are numerous modifications which affect the valve train geometry. For instance, skimming the block to make the pistons flush, skimming the head to raise the CR, reprofiling the cam. These are just three changes which upset the original valve train geometry. If we allow for the fact that the manufacturer's criteria in designing the valve train is probably different to our own, then we can see that there is more than just the odd variable or two to contend with.

Let's assume first that you are putting together an engine substantially of standard parts, rockers being included in that category. If the head and block have been skimmed substantially, the pressed steel rockers can foul the springs or retainers just on the curved section nearest the spring. The cure for this is to put packing strips underneath the rocker pillars to raise the pillars. The most commonly available form of packing is the lock tab strip that locks the grub screw into the shaft on number 2 pillar. Of course you need 4 of these, 1

Whenever the cylinder head and block assembly is extensively skimmed attention must be given to getting a working valve train geometry. In this particular instance, so much metal has been removed from the block and the head that the rocker angle has altered sufficiently for the rocker to catch the valve spring at the point indicated by the arrow. The simple solution to this problem was to put some packing under the rocker pillars.

under each pillar, and be sure that the plate you are fitting doesn't block off the oil hole which transfers oil from the cylinder head to number 1 pillar.

There is another, much better, solution which involves shortening the pushrods to suit. If the head and block have had a combined 150 thousandths skimmed off them, then the pushrods need to be shortened by a like amount. However, this assumes that the original geometry was optimum, which in most instances it isn't. If you are using a reprofiled cam, it is likely that the base circle is smaller than standard. If nothing else changes, this means that ideally a longer pushrod is required, so cams with smaller base circles begin to offset any skimming that has taken place. However, the reduction in base circle diameter is usually relatively small in comparison with the amount that is machined from cylinder heads or blocks. Apart from that, reducing the cam base circle diameter is undesirable from the point of view of the cam dynamics and the wear properties of the cam, so it shouldn't be looked upon as a means to cure a problem in the valve train geometry.

Given the variables involved there is no simple solution to optimizing the valve train; it's necessary to adopt a straightforward, if time consuming procedure.

First, it will be necessary to make up an adjustable length dummy pushrod. The easiest way to do this is to take a standard pushrod and cut it off about 1½ inch from the end that goes into the cam follower. Make up a sleeve which has a male threaded portion on the end, about ¾ inch long. The sleeve needs to fit over the end of the pushrod and can be 'Loctited' in place. Using some easy to machine material such as mild steel or aluminium, make up an internally threaded end-piece that simulates the part that sits in the cam follower, and which can be threaded onto the end of the pushrod. Make it of such a length that the pushrod can be adjusted about ½ inch in length. Also leave room for a slim lock nut. What you now have is a pushrod that can be adjusted in length and used as a setting pushrod.

Acquire some old rocker pillars (they can be of any type, the cheaper the better) of which you will need, say, 6. Measure the height of a standard rocker pillar and then machine three pairs, each pair progressively 20 thousandths shorter than the first. To use these bits and pieces, it's necessary to substitute them for the real parts during the dry build. Instead of using the valve springs that are required with the particular cam involved, load up number 1 intake and exhaust with light springs. You can use the same sort of spring as is used between the rockers. Two of these on each valve will be more than adequate for the job.

The procedure at this point is to start off with the standard rocker height on number 1 and 2

Using a special form tool and by running the lathe at very slow speed, it's possible to shorten and remachine the end of the pushrod to . . .

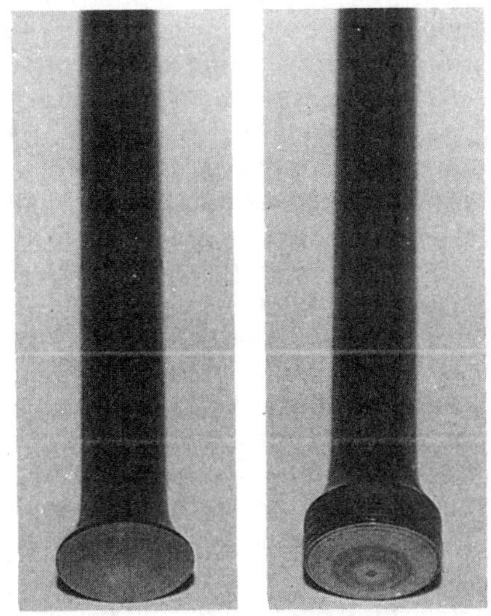

. . . restore it to its original form. Seen here is the difference between the end of a shortened pushrod and a standard pushrod.

pillars only. Install the pushrod, adjust it provisionally to a length equal to the standard pushrod, minus the approximate amount that has been skimmed from the head and the block. Install the pushrod, rocker and pillars in just one end of the cylinder head, then turn the engine over and check the valve lift given when the tappet has been adjusted to the correct clearance. Now remove the rocker (that's easiest on either of the outer ends because it's only a question of sliding it off the shaft after the tappet adjusting screw has been backed off). Adjust the pushrod shorter, re-assemble the rocker and check the lift. If you are looking for maximum lift, continue this operation until you find where the lift is at its greatest. At the same time keep an eye on the rocker pad. Check that the contact area stays as near as possible to the centre of the valve. The least wiping action possible across the top of the valve means the least side loads on the valve. Try this with several heights of pillar to see which gives optimum results.

On the face of it, that sounds a pretty tedious job, but it isn't really. Once you get the hang of it, in less than an hour's work you can determine the length of pushrod that's required to optimize the valve lift. By the way, when you come to measure the pushrod, don't measure the overall length from the top of the pushrod to the ball but take the measurement from the pushrod's ball end to the base of the cup. Use a dial gauge with a ball-ended probe. Stand the master pushrod up under the dial indicator and record the reading or set the gauge to zero. The working pushrods can then be adjusted to exactly the same length one at a time.

Going to the trouble of adjusting the valve train to very precise limits can mean the addition of the odd 1 or 2hp as well as extended reliability but, for engines not intended for racing, the value of the effort involved is questionable.

Here are four Leyland followers. The one on the left is the standard follower without the oil hole. The one next to it is the standard follower with an oil hole. The cam follower second from the right has had a clean-up cut taken down the bore of the follower, to remove any sand inclusions or stress risers from the rough cast finish. This lightens the follower slightly. The follower on the right has been machined so that the wall thickness is about 30 thousandths. Followers lightened to this extent can break, and super-lightweight followers do not contribute much extra rpm potential.

Cam Followers

The standard cam followers' greatest drawback is that it's a bit untidy inside. **Lightening the cam follower to achieve extra rpm is a task which achieves little.** A fully lightened cam follower may add, at a guess, about 50rpm to the engine's rpm capability but, on the debit side, is also prone to breakage due to the side thrust on its walls. **When a breakage occurs, the fracture start point is often at the oil hole.** My advice is simply to clean the excess material and the rough cast finish out of the bore, leaving a 50 thousandths wall thickness minimum to retain the followers' reliability without it carrying undue weight. If you can use the type of cam follower without the hole in it then so much the better, as the oil hole doesn't really appear to be conducive to extra life. Plenty of oil finds its way around the cam follower without that hole.

It's sometimes possible to get cam followers which are sold through engine reconditioning companies and which appear not to be made at Leyland. I found Mini Spares were selling the type of follower shown here, and if appearances were anything to go by, they look to be of higher quality than the original Leyland followers.

ALWAYS fit new cam followers when you change cams. Even 50 mile old ones are too old if you make a cam change.

Having spent a lot of time developing cam profiles and camshafts for the Series 'A' engine, I can say quite honestly that I've just about run the gambit on cam followers. During the testing of camshafts one thing became very evident, there are plenty of cam followers on the market which will not stand heavy spring poundages and the rigours of rapid acceleration race profiles. Added to this, 1.5 rockers will aggravate a cam/follower problem if the followers are not up to par. All this led to a hard look at cam followers, especially when the A.P.T./Crane cam profiles came along. As a result, the only way to minimize cam follower problems was to use the A.P.T. chilled iron lifters/followers.

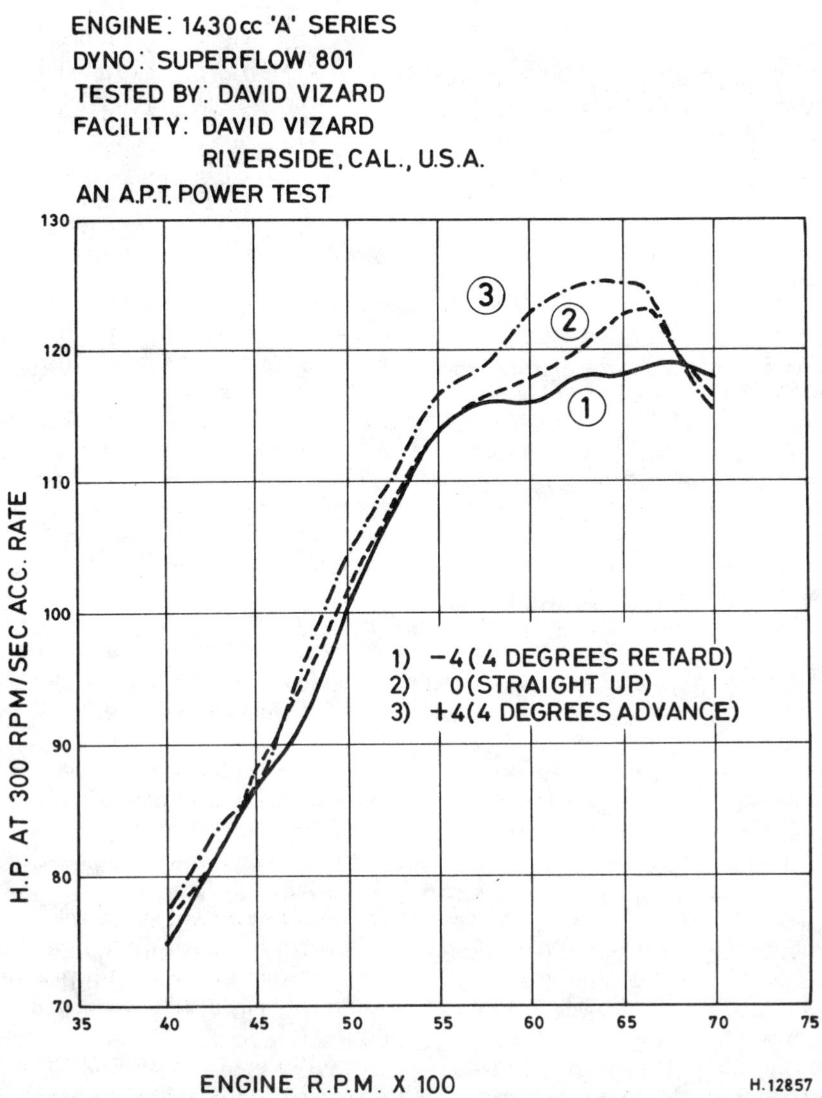

Fig. 11.10 Cam timing comparison

The particular curves shown here are not ones selected to show the best effect of timing the cam correctly, but are, in general, typical of the changes in power that will be seen on a 1400 Series 'A' engine. The curve, No 3, represents the optimum timing for the cam. Any more advance than that given showed a reduction in horsepower.

These are no longer available so we have had to take steps to remedy the situation. As was mentioned previously tests are being done virtually as I pen this manuscript. A follower with a hardening, low friction heat treatment so far appears to be even better than the chilled iron item that we were forced to find a replacement for.

Camshaft Drives

The camshaft is driven from the crank by a variety of methods depending on engine year. Early Series 'A' engines had a simplex (single row) chain the tension of which was maintained by two neoprene rings inset into the camshaft sprocket. The neoprene rings stood proud of the base of the gear teeth and as such, should have taken the slack out of the chain. **Unfortunately, the neoprene chain tensioners work efficiently only for 5000-10,000 miles.**

With the introduction of the Cooper 'S' came the duplex timing chain, which had no tensioner. It relied on the fact that the centre-to-centre distance of the camshaft and crank was held to close enough tolerances to allow the chain to be installed tight. The two-row chain meant the wear on the links were less and, as a result, the chain stretched less. With a Cooper 'S' engine, the duplex gear set was made of steel and it is still accepted as THE item to use for racing, at least when using factory parts.

As supplied off-the-shelf, the cam sprocket is a rather heavy item. To make things accelerate more easily when the throttle is applied, it's worthwhile lightening the cam sprocket by facing down the area between the centre hub and the outer section the gears are cut into. Also, lightening holes can be drilled in addition to the two holes that are already in the face of the gear.

Though very good, the Cooper 'S' gear set is expensive. When the non-'S' 1275 engine was introduced it came equipped with a cheaper type of duplex timing gear set made using a sintering process. Although not quite as good as the Cooper 'S' gear set, it is, nonetheless highly functional and generally reliable. It can be used as a race item, although the gear teeth do wear faster than the ones on the Cooper 'S' steel gears. I don't recommend lightening of the sprocket used in the sintered gear set.

Although the duplex setup has a much better life than the original simplex setup, eventually it does wear and timing chain rattle starts,

Setting The Cam Timing

If you are new to engine building, you have probably heard the expression 'timing in the cam' but did not quite know exactly what it meant. Well, here for the benefit of those who are not quite sure, is all you would like to know about cam timing, but were afraid to ask.

By merely installing the cam, and lining up the timing dots, you are not likely to realize the maximum benefit from the cam change. If you are going to do the job properly it is necessary to 'time the cam in'. This becomes even more important with some of the speed equipment cam grinds because, for one reason or another, during the process of grinding the cam profile, the profile itself may be moved around one way or another in relation to the keyway. Because of this there could be considerable error in cam timing by just lining up the dots. For instance, **I have used a particularly effective profile from one of the speed equipment performance cam grinders that on one occasion was no less than 16° retarded when installed with a standard keyway.** With this much retard, the cam is unlikely to produce as much power as a standard cam. If this factor hadn't been taken into account, then I would have been very disappointed with the performance of the engine. However, I make it a practice to time in the cam on every engine I build. This is the only way to ensure that the true potential of the cam is achieved.

If you think, to get the best from a high performance cam you can get away with just lining up dots as seen here, then you are in for a rude awakening.

To do a proper timing check, you will need a dial gauge indicator and a magnetic stand, plus a 360° protractor. The technique is as follows. Set up a pointer on the block, then using the dial indicator on the centre of the piston, establish exactly where TDC occurs, install the protractor on the crank nose, then set the pointer to read zero on the protractor. You now have your true TDC setting.

One of the most important aspects of timing the cam is first to establish accurately the position of TDC with a dial gauge positioned along the centre line of the piston.

Install a pushrod into number 1 intake follower and turn the engine over until full lift is achieved. Set the dial indicator to zero. Now turn the engine backwards so that the dial indicator drops something in the order of 100 thousandths below

Having established the TDC position of the piston and crank mount a pointer in the fashion seen here to indicate zero on a protractor. By the way, the Castrol engine timing discs are available from most performance orientated accessory shops. Castrol find it a cheap way to advertise their oil because people like me use the disc for photographs in my books. I wonder how many they had to print to make sure I picked one up?

Once the dial gauge has been used to set the TDC position, it must be mounted like this so that the position can be determined where full lift on number 1 intake lobe takes place.

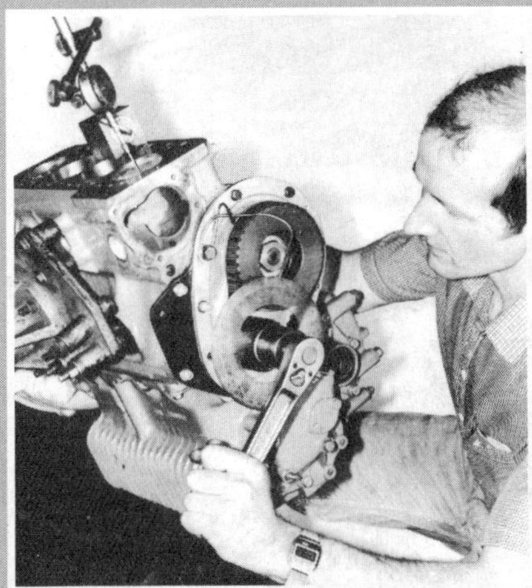

If you are building a motor which is using a belt drive, the technique for timing the cam remains unchanged.

full lift on the cam lobe. Stop and turn the engine in the correct direction of rotation toward full lift. When it is, say, 3 thousandths short of full lift, stop and note the angle on the protractor. Having done that, continue to turn the engine over in the correct direction of rotation until the dial indicator indicates full lift and then drops back down to 3 thousandths below full lift. Again note the angle on the protractor. Add the first angle to the second angle and divide by 2. The figure you are left with is the number of degrees that the valve reaches full lift after TDC. The reason two readings were used to establish where full lift is reached, is that performance cam profiles have a slow valve motion over the nose, and indeed some profiles actually have dwell. Because of this, it is difficult to tell where the exact centre of full lift occurs. By taking a measurement short of full lift and a measurement after full lift has occurred, it's possible to establish to a relatively accurate degree, where full lift or at least, the centre of full lift, occurs. For checking purposes, I chose 3 thousandths as a nominal figure, but there is nothing stopping you using 5, 10 or even 20 thousandths before TDC. The only exception to this is in the rare instance when an asymmetric cam profile is used, in which case the nearer the figure is to TDC, the more accurate will be the determination of the full lift point.

No doubt, **having established where full lift takes place, it will almost inevitably not be where you require it to be.** How much error in cam timing is acceptable? Generally speaking, up to 4° with most engines makes no real difference. However, the Series 'A' engine is less tolerant than many other types of engine, and sometimes as little as 2° off makes for a measurable change in the hp. It's for that reason that I would recommend that you set the cam timing as per the cam manufacturer's figures (or as indicated elsewhere in this book) to within 3° at most and preferably only 2°. If you are building a race engine, there is no excuse for not getting it within 1°.

To time the cam accurately, steps must be made to allow alteration of the cam sprocket in relation to the crank. Keys with 2, 4, 6 and 8° are available and represent the simplest solution to correcting errors. A simple but more expensive approach is to fit a duplex timing chain setup using the adjustable duplex cam gear produced by Kent Cams. When using a chain drive to the cam timing the cam in 1° advanced of the nominally ideal setting is a good idea, because it won't take long for the chain to wear sufficiently to retard the cam timing by 1°. This doesn't apply when using a belt-drive; aim for the cam to be timed precisely.

Precise timing of the belt drive system can be greatly simplified by using the **Steve Harris** modified belt-drive set up. This incorporates a Ford type offset dowel pin to allow quick changes in cam timing to be made. If adjustments need to be made simply and quickly either on the dyno or at the track, the vernier belt-drive arrangement offered by **APT** or **Titan** will be of great value.

Before installing the timing chain, check that the cam and crank gears are aligned. If they are not aligned, shim the crank gear as necessary to get alignment. If the gears are out of alignment, power will be lost to friction and the timing gears and chain will wear far more quickly than necessary.

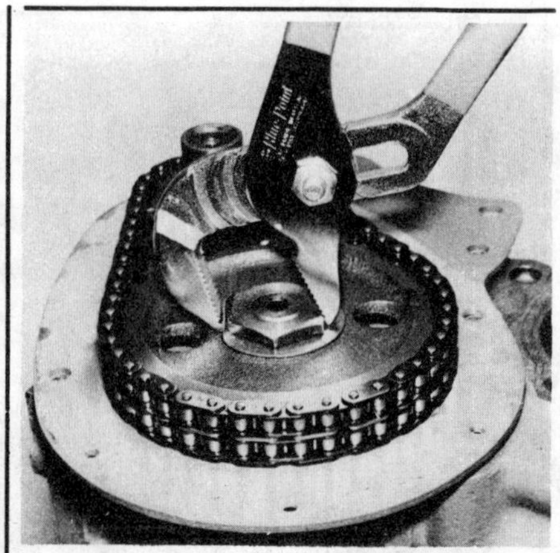

Here I am just pulling up the cam sprocket tab washer with a set of Snap-On pliers made for the job. The particular gear set being used here is not Cooper 'S' type, but the much cheaper 1300 setup.

though it's always far less severe than that found with the simplex setup. To overcome this problem, Leyland introduced a simplex gear with a 'strip' timing chain tensioner of the type seen in the **accompanying photo**. This setup is generally the quietest one available and a single row chain is more than adequate for any street-type cam. However, if you want to have the greatest reliability and also reasonable quietness, it's possible to utilize a duplex timing gear set together with a chain tensioner. To swop from a non-tension chain to a tension chain, you will need the backplate, strip tensioner and the timing chain cover from an 'A+' engine. Although this setup results in a reasonably quiet cam drive system, it does produce a further problem. the timing chain links wear. As they wear, so the chain becomes slack causing a variation in cam timing. As the chain wears, so the cam timing retards. The complete answer is to utilize a belt-drive system, marketed by Keith Carver, late of Mini Spares Centre, which involves replacing the backplate with a cast backplate having an oil seal in it for the crank and cam. A special toothed belt sprocket is installed on the cam and the crank, and a wide toothed belt is used to transmit the drive from crank to camshaft. These toothed belts are the non-stretch variety. they produce little in the way of noise during operation, (less than a brand new timing chain)

If you are changing from a simplex timing chain to a duplex timing chain, it will be necessary to countersink the two holes arrowed and install countersunk-headed socket screws.

If you want to swop over to the later and quieter running simplex setup with the timing chain tensioner, these are the parts you will need. This setup, with the exception of the gears, can also be used with a duplex timing chain installation.

cam timing more accurate than chain drive, and the complete assembly is lighter than the chain setup.

Having used a belt drive on many engines, I am very reluctant to go back to a chain drive, simply because the engine runs so much more quietly and as a bonus, feels more responsive. Also, a competition engine will hold its power longer because it doesn't suffer from cam timing retardation which is caused by chain wear. Finally, belt drives tend to damp vibrations from the crank to the cam and vice versa, so in fact the belt drive also acts, to a limited degree, as a vibration damper.

For the ultimate in accurate cam timing, low power loss and quiet running, this cogged belt-drive installation, though cost is high. This setup is available from Caltrac Automotive Design in the UK.

After installation of the working parts, the tooth belt cam drive is shielded by the arrowed cover. On this particular engine destined to go in a drag racer, sprinter and hill climber, the ancillaries, such as the water pump, are also driven by a cogged belt.

Variable Timing Cam Drives

A variable timing cam drive is a device which allows the adjustment of the cam timing without having to resort to changing of dowels, keyways, or anything else that makes retiming a long winded operation. Obviously a variable timing cam drive will be more expensive than a simple offset key, dowel or whatever, so how can they justify their existence? In such a context as we're considering here, variable drives are justifiable on two specific counts. The first is timing errors, so let's go into detail as to why an error in cam timing should exist in the first place. The simple answer is a stack up of tolerances. The timing of the cam is not only dependent on how the lobes are ground on the camshaft, but also the position of the keyway in relation to those lobes, the keyway position in the gear in relation to the teeth, the relationship of the camshaft in the block relative to the crankshaft centreline, the position of the keyway in the crank and the slot-to-gear-teeth relationship in the crank gear. Whether tolerances are additive or not dictates whether they cancel out. If they don't (usually the case), they can produce a substantial error in the timing of the camshaft. It is not uncommon to find a standard camshaft timing in with as much as 4° of error. However, if certain other factors which have already been alluded to are added to the equation, attempting simply to install a cam by lining up the dots would be ludicrous if maximum performance is being sought. Which brings us to the second justification for a variable timing cam drive.

Today's cam profiles are, from necessity, being ground from shafts that were not necessarily intended to carry high performance profiles. To put these profiles on shafts whose existing profiles are milder means some lobe juggling has to be undertaken to find the position where all the lobes will clean up at the new profile. Often the new lobes do not fall in the correct position in relation to the existing keyway. This means even the most conscientious cam manufacturer may be producing cams which do not reflect the original lobe- to-keyway relationship. Because of this, some substantial errors in cam timing can result. This alone is justification for checking and correcting as necessary the timing of any high performance cam; a job which is greatly simplified by using an adjustable cam spocket. If adjustment alone were the only criterion, an offset keyway would still suffice although it would take longer to do the job. Where the real justification for an adjustable cam timing device lies is when it comes to setting up the engine for maximum horsepower. Sure, we can all time the cam to the manufacturer's recommended spec, but that specification assumes a typical engine. The manufacturer of the cam cannot know precisely what specification your engine is going to be and the figures set down are those which, on average, will work best. Very often an engine will produce better results by power-timing the cam. That means either getting on a dyno or trying out various cam settings on the race track or drag strip. The dyno, of course is the best choice and fine tuning a cam means nothing but extra horsepower. I've personally used cams that have been required to be set 4 and 5° different to the manufacturer's recommended settings. This is not because the cam manufacturer didn't know what

Here's the variable timing belt drive set-up used on the A.P.T. dyno engines. Basically, it's a conversion performed on the normal belt drive kit. Changing the cam timing takes literally less than a minute.

he was doing, it was because he didn't know my engine. It was sufficiently different from the norm to warrant alternative valve events and power-timing proved to be worth a useful amount of extra horsepower as the graph a few pages back in the section 'Setting the Cam Timing' demonstrates.

The Vernier Gear

Probably the hottest selling adjustable cam timing device is Kent Cams' vernier gear. It doesn't cost significantly more than a proper set of steel duplex timing gears, yet it grossly simplifies the setting of the cam in the first instance and speeds up the resetting of it if some fine tuning is needed on the dyno in the second. The only disadvantage of the vernier gear drive over the belt drive is that it does require the timing cover to be taken off to make the adjustments. Since the belt drives have now become so expensive, I personally only use a belt drive on a 'cost-no-object' engine and this is invariably a belt drive that we've modified at A.P.T. so as to be fully adjustable by virtue of slots and clamping screws. On anything less than such engines, it's now become standard practice to use the Kent Cams' vernier spocket set.

The duplex vernier cam drive from Kent Cams which uses slots and Allen screws to provide infinitely variable ajdustment.

12 Forced Induction

The techniques for producing more power from the Series 'A' engine have so far been directed at ways of increasing its air consumption so it can burn more fuel. The techniques discussed concern the production of extra power while the engine is in a normally aspirated form. A normally aspirated engine, in case you are not familiar with the term, is an engine which fills its cylinders with air which is, at best, at normal air pressure. If by some means, more air is crammed into the engine, then we have introduced a situation known as 'forced induction'. To force the air into an engine, a supercharger of some sort is needed. This can take a number of forms such as a mechanically-driven supercharger or an exhaust-driven supercharger, commonly known as a turbocharger. Forcing air into the engine causes it to consume a greater quantity of air. This, along with the appropriate amount of additional fuel, allows it to produce more power. **With a supercharged engine, the extra power shows up as an increase in torque.** The reason for this is not difficult to understand. If we consider for a moment a supercharger giving a boost in air pressure of 7.5psi (that's approximately half an atmosphere above normal air pressure) it follows the engine will draw in a weight of charge 50% denser than would otherwise be the case. This means that the pressure generated in the cylinder upon combustion will be 50% higher, thus the torque can be expected to be 50% higher, as a result, the power at any given rpm will be 50% higher. At least this would be the case if things followed simple theory.

Of course in the real world, all sorts of interactions take place which complicate the outcome.

Turbocharging & Supercharging: The Differences

The type of superchargers likely to be seen on Series 'A' engines or at least those that are practical from the everyday standpoint, are those that are belt driven. Most fall into the positive displacement category, that is, if the supercharger turns over once, it displaces a certain amount of air from one side of the blower to the other and into the engine, ignoring for the moment leakage.

If we had a supercharger which, when rotated once, displaced 1000cc of air and it forced the output of that air into a 500cc cylinder, then that cylinder would be supercharged at 2 atmospheres. Generally speaking, the amount of boost given by a mechanically driven, positive displacement supercharger is governed by the ratio of the drive from the crankshaft to the supercharger drive pulley. The faster the supercharger is turned in relation to the engine, the higher the boost.

Although it is a supercharger, a turbocharger is not a positive displacement supercharger. It is, in fact, a centrifugal supercharger. It is the energy that the rotating vanes put into the air, which is then centrifuged to the outside of a case, that causes the supercharge pressure. At low rpm the intake turbine (compressor) of a turbocharger gives little or no supercharging effect. However, as the speed of the turbine increases the boost increases roughly with the square of the speed. If we were to gear drive a centrifugal supercharger from the engine, that is, have it turn at a speed directly related to the engine speed, at low rpm the engine would have no boost and as the speed went up, so the boost would very quickly climb. This gives an engine which produces an extremely spikey power curve. It was this type of supercharger that was used for the V16 BRM back in the '50s. Certainly, it made an enormous amount of hp from the 1½ litre engine, but the power band was very narrow. The only easy way around this problem is to drive the supercharger at such high speed that it delivers sufficient boost at, say, some mid-point in the rpm range, and then dump excess boost. This has the disadvantage that power from the engine had to be used to compress that air in the first instance. Throwing it away, by dumping it back to atmosphere without deriving any benefit from it, means that cylinder pressures have to be higher than would otherwise be the case, to generate the same power output. The other aspect is that the speeds at which the turbine impellor has to turn can be so high that considerable stresses are induced in the turbine when the engine is revved up, simply because it has to be geared to give at least 3 times (often 8-10 times) engine rpm.

The solution to this problem is to drive the compressor turbine in such a way that it is not rigidly coupled to the engine. An obvious method, originally thought of in 1903, is to use exhaust gases to drive a turbine which is coupled to the same shaft as the intake turbine. This is the essence of a turbocharger. Exhaust energy is used to power the supercharger. The best way to envisage the working of a turbocharger is as follows: the exhaust gases emanating from the exhaust ports are collected in the exhaust manifold and directed towards the exhaust turbine of the turbocharger. Attached to the self-same shaft but in a separate housing is the intake turbine or, to use the correct turbo terminology, the compressor. At idle, the exhaust gases impinging on the exhaust turbine will have very little energy, probably not enough even to cause the turbine to windmill. There are plenty of routes through the blades of the turbine for the exhaust to escape. Within the exhaust manifold there is no real back pressure at this moment. Imagine the car is now being driven down the road, and there is a greater amount of throttle opening and more air is being drawn into the engine, not forced in, but drawn in, just as occurs in a normally aspirated engine. This extra volume of air produces an extra volume of exhaust and we will assume at this point that there is enough exhaust impinging on the exhaust turbine to cause it to windmill over at an appreciable speed, but still nothing like enough to give any supercharged pressure.

At this point the throttle is depressed fully, which in turn opens fully the butterflies in the induction system. The engine is now able to draw in a relatively large charge of air, approximately the same, in fact, as a normally aspirated engine would at full throttle at the same rpm. This causes the amount of exhaust to increase quite substantially. The exhaust gas impinges on the exhaust turbine and causes it to speed up. In so doing, the intake turbine also speeds up and starts to draw in just a little more air than the engine would normally draw in of its own accord. In other words the induction pressure has just started to rise. The cylinder is now filled with more air on the induction stroke, and when this is burnt together with its fuel, more exhaust is created. This in turn will cause the exhaust turbine to spin faster, the compressor turbine will spin faster; as a result the cycle will continue leading to even more boost. Eventually the inefficiencies of the various ports and valves, etc. will finally defeat the turbo and the boost pressure will climb no higher. As long as that boost pressure isn't above what is required, all is well. However, the higher up the rpm range the engine goes, the more exhaust it creates and the greater is the possibility that the boost pressure will go too high. In fact, above a certain point in the rpm range, dependant on the size of components chosen and the engine characteristics, the boost pressure will run away and blow the engine to bits. As a result, it is necessary to control the boost level, and this can be done in a number of ways.

Probably the easiest method is to use a blow off valve to dump excess boost pressure. Though simple, this is not necessarily the best answer as, once again, power has been expended to compress the air and it's wasteful to simply dump it. The second way is to induce deliberately a restriction to the gas flow, within intake and/or exhaust, and thereby limit the boost. A third way is to have an exhaust waste gate. In this system the exhaust waste gate senses the intake supercharged pressure and when it reaches a predetermined level, the waste gate opens and diverts some of the exhaust down a separate exhaust pipe to by-pass the exhaust turbine. If the boost pressure drops below the predetermined level, the waste gate closes and directs all the exhaust through the turbine.

On the face of it, driving a centrifugal supercharger by means of an exhaust-driven turbine gets around the problem of supplying boost throughout the rev range. That, however, is not quite the case. At low rpm there simply isn't enough exhaust to drive a turbine to enough speed to give boost pressure. As a result, turbo motors are generally poor in respect of low end power. Generally speaking turbochargers don't really produce much boost until about 2500rpm and maybe sometimes as high as 3000 though this depends to some extent on the size chosen for the exhaust housing. On the other hand, belt-driven superchargers can produce a measurable amount of boost, even at very low engine rpm, e.g. 1500. This would make a positive displacement supercharger seem far better than a turbo. A positive displacement supercharger, however, tends not to be as efficient at compressing air as a turbine supercharger, and because of this tends to heat up the air more than a turbine supercharger. In other words, the pumping efficiency of a turbine supercharger can be appreciably higher than a positive displacement supercharger. Unfortunately the engine compartment of a turboed engine can get very hot, simply because of the vast amount of heat that is dumped into the exhaust turbine housing. So **even though a turbo supercharger may be more efficient, the net result is the air fed into the engine can be at substantially higher a temperature than would otherwise be expected**.

Another snag with a turbocharger is that it suffers from what is commonly called turbo lag. A more precise definition would, of course, be boost lag. Flooring the throttle does not immediately give boost to the engine. It takes a little time for the boost to build up, due to the cycle of events described previously. On the other hand, flooring the throttle on a positive displacement supercharger produces instant boost, thus giving instant throttle response.

As far as power output is concerned, the turbo seems to have the edge over a positive displacement type supercharger. On the other hand, the throttle response of a positive displacement blower is better, and as a result, the car can respond quicker out of tight turns and this may offset the slight power advantage the turbo may have. Excellent results have been achieved with both types of forced induction, it would need a very comprehensive engineering programme to decide which was the best approach for your particular requirements.

Suck-Through Versus Blow-Through

Assuming the engine is to be fitted with a carburettor (rather than fuel injection) there is an important question concerning whether the supercharged air should be drawn through the carb first (suck-through system) or blown through the carb under pressure (blow-through system). The argument seems to be more important, or at least more hotly argued, when turbocharging is considered. Those who favour the blow-through system argue that the throttle response is better and, of course, finding every way possible to reduce throttle lag to a minimum is important on a turbocharged engine. However, since boost lag or throttle lag does not occur on a positive displacement type supercharger, the question of whether it should be suck-through or blow-through is less important. Almost all installations using a positive

This Janspeed turbo installation, designed for a Metro, utilizes a blow through system. The turbo supplying the air ...

... is situated down by the clutch cover. Power output is about 110 hp+

displacement supercharger suck through the carb.

A 'suck through' system is entirely practical with a turbocharged setup. Those who favour a suck-through system point out that the carb is operating in an entirely normal mode when air is drawn through it. This differs from a blow-through mode because here the air density is increased upstream of the carburettor. Carbs do not recognize changes in air density, metering fuel flow as a result of the velocity of the air through the venturi not the density. If the air is squeezed down to half its volume, then it need only travel relatively slowly through the venturis. In doing this it will not draw a sufficiently large amount of fuel to match the increased oxygen content in the air unless a major carb design change is made specifically to cater for such an event. If this is not done the resulting carburettor miscalibration would mean that the benefit of turbo-charging would be largely lost. There are, however, ways of overcoming this by providing the carburettor with density information using mechanical components.

To continue the suck/blow discussion in relation to turbochargers let's look at the problem of throttle response. If a suck-through system is used, then the carb, as stated before, functions as normal. It's only a matter of putting in the right needles/jets, etc. to provide the correct mixture. However, when the mixture leaves the carb, it goes into a centrifugal compressor. This compressor tends to centrifuge the fuel out to the walls of the turbo housing. As a result there can be a considerable amount of fuel flow along the walls of the compressor and on into the intake manifold. In fact, the amount of wall flow of fuel into the engine can be substantially higher than would otherwise be the case, and it's reasonable to suppose that some of that fuel may never be re-emulsified in the air sufficiently effectively to be ignitable. A way around this is to design intake manifolds which have sharp corners in them to induce turbulence and give every possible chance of re-introducing the fuel into the air.

With the blow-through system, the situation concerning wall fuel flow is reduced to that of a normally aspirated engine simply because the fuel/air mixture does not get centrifuged by the intake turbine. However, to make this system work, it's necessary for the carb to know by how much it's being boosted so that the carb can deliver extra fuel in proportion to the increased air density. The easiest way to do this is to restrict very slightly the flow of air into the carb so that the pressure at the mouth of the carb is higher than just inside the carb. If the difference in air pressure is communicated to the float chamber it will cause the pressure in the float chamber to rise slightly above that in the venturi. Thus, as the boost pressure rises, so the pressure in the float chamber rises faster than the pressure in the intake manifold downstream of the carb. The result is that increase in fuel delivered by the carb as the boost goes up. To clarify this, **Fig. 12.1** shows a schematic illustration of how it works. Leyland's own turbocharged engine, the turbo Metro, uses a blow-through system and works on this principle.

If the carb is used in a blow-through mode, other factors apart from mixture calibration have to be taken into account. Firstly it's necessary to seal the throttle spindles, otherwise a small amount of fuel/air mixture will be blown out through the throttle spindles. In a normally aspirated engine, of course, air would be sucked in at this point.

A further consideration is the supply of fuel to the float chamber. Fuel must be delivered at a pressure which is constantly above the supercharger pressure. It's no good supplying 3psi of fuel to a float chamber that has 7psi of boost applied to it because there will be 4psi pushing the fuel back whence it came, i.e. the fuel tank. To compensate for this it's necessary to have a pressure regulating valve which is sensitive to boost pressure. The pressure regulating valve has to regulate the fuel pressure from a high pressure pump so that the fuel delivered to the float bowl is always a pre-set level above the boost, typically 2½ to 3psi above boost.

At this point the blow-through system appears to be a little more fussy in terms of installation than the suck-through system. Not so, because we haven't looked at the whole picture yet. With a suck-through system, although there are no real carb calibration problems with regard to catering for boost pressure, a situation occurs during part throttle or idle use where a vacuum exists

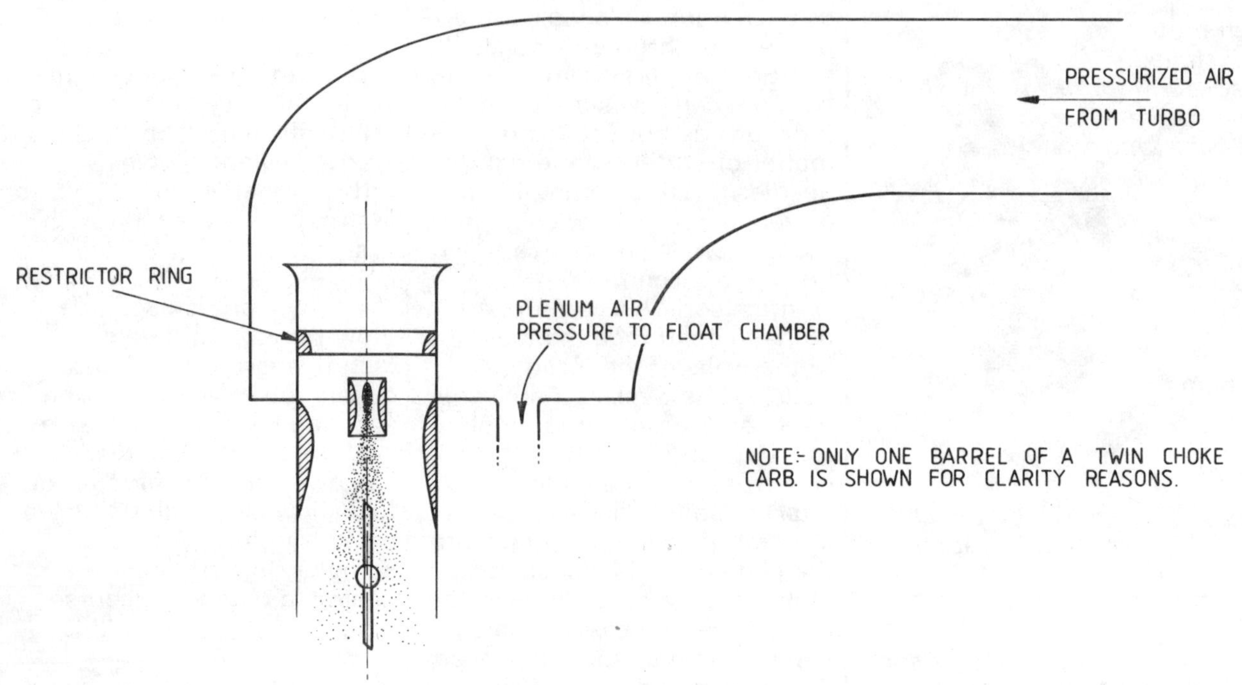

Fig. 12.1 Method of making the carburettor boost sensitive

To compensate for the increase in boost density on a blow-through system, it is necessary to make the carburation 'boost' sensitive as well as velocity sensitive. Carbs for normally aspirated engines are largely velocity sensitive. In other words, they meter fuel more in proportion to the velocity through the venturies rather than any other factor. By putting a restriction of some sort at the carb mouth, (the restrictor ring shown here being just an example) the carb becomes sensitive to boost pressure. By having some form of restriction at the carb mouth, the air pressure in the float bowl will increase faster than the air pressure in the carb barrel. This will cause a greater pressure differential across the main jet and as a result, more fuel will flow. Because of the limited restriction to flow offered by the restrictor ring (or whatever else is used) at low rpm little fuel enrichment will occur, even if boost pressures are relatively high. This restriction technique for boost-orientated enrichment works best in the higher rpm ranges. The greater the restriction in the carb mouth, the greater the density orientated enrichment achieved.

downstream of the carb. When this happens the compressor is running in a partial vacuum. Under these conditions, oil can be drawn out of the turbo bearings into the induction system. Because turbos were originally made to supercharge diesels, no provision was made

In this shot, the exhaust pipe has been removed from the turbo exhaust housing so that a clear view can be had of the wastegate and turbo. The particular items used in this instance are from the California-based Rotomaster company.

Much thought has been put into the installation features on the Janspeed Metro turbo kit. The waste gate and turbo are easily accessible and to avoid the problems of cracking exhausts due to temperature differences, a stainless steel bellows has been added between the exhaust manifold and the turbo.

for them to run in a situation where vacuum existed. Any turbo for a suck-through system needs to have a special seal fitted in the intake turbo housing to prevent oil being drawn out of the bearing into the intake system. Without such a seal, this vehicle will consume vast quantities of oil and emit immense clouds of smoke from the tail-pipe.

These days there are so many turbo applications, both blow-through and suck-through, that it's possible to get most turbos with or without a carbon seal on the intake turbine housing.

In conclusion, the suck-through and blow-through systems each have advantages and disadvantages and are fairly equally matched both in the technical difficulties of installation and the performance that can be expected to be produced. The choice is yours.

Turbo Sizing & Matching

If you intend fitting a supercharger of any type, one of the first decisions concerns the amount of boost required. In a roundabout way, it's really a question of how much hp you would like to see from the engine and then relating this back to the amount of boost required to give such a power level. With a belt-driven supercharger, setting the boost is easy enough. It's a matter of picking the right size of supercharger for the engine and then arranging that it be driven at the right speed. A turbocharger installation is, in principle, no different.

The first thing we have to do is size the turbocharger to the engine. In other words, we have to pick the turbo, at least as far as the intake side is concerned, such that it can supply the amount of air the engine needs to make the sort of hp we want. Sizing a turbo undersize will, of course, limit the hp. Sizing the turbo oversize does not have such dire consequences though there are disadvantages. The turbo is capable of producing a great deal more airflow than the engine requires and, of course, a bigger impellor has a greater mass which does not assist quick throttle response. In practice, however, a turbo as much as 100% oversize is not of any real consequence. For instance, if we want to see 110hp from the engine, it's not that much of a disadvantage to have a turbo which can deliver enough air for 220hp. Obviously it would be better if it was sized a little nearer, but sometimes it is not always possible to come up with a turbo exactly the size we anticipated.

Just for reference, the turbo used on the Metro is essentially the same turbo as used on the Lotus Esprit, and *that* turbo is capable of supplying enough air for well over 230hp. The turbo Metro puts out 90!

Once a turbo has been selected the 'drive' characteristics of that turbo have to be matched to the engine. Referring back to a mechanical analogy, we have to 'gear' the turbo correctly to the engine. Of course, as such there is no mechanical connection; the turbo is driven by the exhaust. However, by selecting the correct area/radius ratio (A/R ratio) on the exhaust housing, we can, using our mechanical analogy, 'gear' the turbo to the engine, so that it's running at the sort of speeds required to give boost in the desired rpm range. The A/R ratio of a housing is the area of orifice leading into the exhaust housing to the radius that orifice is from the centre of the exhaust housing. The smaller the A/R ratio is, the higher 'geared' the housing is. In other words, if we have a, say, 0.39 A/R ratio, this will cause the engine to come on boost sooner than a 0.5 A/R ratio, and that will come on boost sooner than, say, a 0.68. If the A/R ratio selected is too small, the engine will come on boost low down and boost will become excessive by the time the engine is only about three quarters of the way up the rpm range. On the other hand, if a large A/R ratio housing is used, the engine may have to turn 5000 before effective boost is provided.

In practice it is a case of selecting the housing A/R ratio in an attempt to put the boost in the best rpm range available to us, because it is not always possible to develop boost at lower levels in the rev range. At low rpm there is insufficient exhaust going through the motor, so it is impractical to get boost.

If we select a small A/R housing, the boost will come on early and, without any method of control, it will reach excessive levels before maximum engine rpm is reached. However, if a waste gate is used it is entirely practical to select a smaller A/R housing than would otherwise be the case, to bring on the boost sooner, then deliberately bleed off excess exhaust through the waste gate, once the desired boost level has been achieved. By doing this a much flatter boost curve is achieved. As noted earlier, there are other techniques that can be used to limit the boost. The simplest method of control is to use a restrictive exhaust system downstream of the turbo. This reduces the amount of pressure drop across the turbo and prevents the boost reaching excessive levels. In a similar fashion, it is also possible to restrict the intake so that after a certain point the turbo is incapable of drawing any more air through the system.

Restrictive controls have advantages and disadvantages. The prime advantage is that they are cheap. The disadvantage is that they are inherently less efficient than a waste gate. Restricting the exhaust can cause an undesirable temperature build-up on the piston crown and exhaust valves. Also, too much exhaust restriction can produce an engine which has a slower throttle response than would otherwise be the case. Too much restriction on the intake can drive a turbo into 'resonance'. This is a situation where the turbine takes a gulp of air, which then leaves too much of a vacuum upstream of it, and the air it has just taken in tends to bleed back from where it came, simply because the source can't feed quickly enough the vacuum in front of the turbo. When this happens, it can be destructive to the turbine, so it is a situation which needs to be avoided.

From a practical standpoint the enthusiast no longer need worry about sizing and matching turbos; there are enough turbo kit manufacturers on the market with working kits to have alleviated this problem. Of course, if you have modified the engine, you may have to make some small adjustment. At most a change might be required of exhaust housing size to the next larger A/R ratio. Installation of turbochargers is a complete subject in itself, and one that cannot possibly be covered here. There are, however, a number of books on the subject, amongst which can

Here's what the Janspeed installation looks like with the exhaust pipe mounted on the turbo. Note that it fits snugly to the block. By the time it gets down to the area of the final drive, it's in pretty much the same position as a standard exhaust pipe. Also seen in this shot is the pipe coming from the waste gate, joining into the downpipe of the exhaust system.

be recommended the following. 'Turbochargers' by Hugh McInnes (published by **H.P. Books**), 'Turbocharging & Supercharging' by Alan Allard (published by **Patrick Stephens Ltd.**), 'Super Power' by Ganahl, Atherton & McFarland (published by **S-A Design Publishing Co.**).

Leyland's Turbo Metro

The Metro hadn't been out for long before it became apparent that the car's handling could cope with more power. The provision of power for additional performance might have been found by following the route previously taken for Minis, but the Metro was a much more refined car, and having a raucous engine in a Metro would mean losing much of this refinement. The obvious answer was to turbocharge for quiet power. Literally within weeks of the Metro being introduced, companies were developing turbocharged versions. However, it seems Leyland already had this in mind. Peformance derivatives of the standard Metro were already being developed. It was intended that the MG Metro should have a normally aspirated engine, producing around 10hp more than the ordinary 1300 Metro, and on top of that, a 90hp turbo Metro was to be developed. The requirements of a production line turbocharged engine are different from those of a speed equipment company whose product is essentially an add-on. A production line turbocharged engine must have 'body drop-on capability' and therefore the installation has to be very compact. Leyland's engineers looked at numerous ways of turbocharging their engine, and the method they finally settled on was a system blowing through an SU HIF6 carb. The carb compensated for density changes by having a slight restriction at the carb mouth. Unlike most turbocharged engines, this particular engine utilizes quite a high CR at 9.4:1. Most turbocharged engines have CRs between 7 and 8:1. A problem can easily arise with a CR as high as that of the turbo Metro engine if detonation is to be avoided, which begs the question of why so high a CR was decided upon. **Whenever the CR is lowered, the part-throttle economy of an engine is reduced.** The high CR is maintained with the turbo Metro engine to improve part-throttle economy and going from 8:1 to 9.4:1 was worth 6 to 8% in mileage on the fuel economy cycle. The other aspect concerns low speed torque. The high CR pulls the low speed torque up, thus reducing the 'turbo feel' to the low end of the rpm range. Of course, having a high CR like this works fine until the boost comes on, and then the effective compression, that is the geometric CR, plus the compression the turbo puts into the engine by virtue of the boosted charge, can cause the onset of detonation very easily. The factory engine avoids detonation by having an advance curve which retards the ignition timing from optimum at operating conditions where detonation may be a problem. Apart from this, detonation is also suppressed in certain critical areas in the speed range by mixture enrichment. This mixture enrichment is over and above what is normally needed for the engine in all other respects. In other words, a little surplus fuel is put through the engine, just to cool it. The extra fuel is introduced into the system by means of the shape of the needle in the carb. Since it performs this extra 'detonation avoidance' function, the needle profile in the SU of a turbo Metro is critical. Should you make any alterations which raise the boost level on a turbo Metro engine, be sure the mixture is well and truly rich and start calibrating it from there back down to leaner levels. **Don't ever let a turbo Metro engine run lean at or near full throttle otherwise detonation will rapidly melt the pistons.**

Another aspect of the turbo Metro engine is that it has a modulated boost system in which the waste gate is set at around 4 to 4½psi. What the factory engineers actually wanted was a boost of some 7 to 7½psi but they found with a boost level that high, the transmission reliability dropped below what was considered an absolute minimum for the vehicle. Here you must bear in mind that the factory tests the car undergoes are far more brutal than the average driver will subject the car to out on the road.

It was determined that for long-term reliability the gearbox can cope with no more than 95ft lb of engine torque. So there would be no problem with excess warranty claims, it was decided that the maximum torque of a turbo engine would be limited to 85ft lb. One of the aspects of a turbocharged engine is that it has a tendency to produce massive mid-range torque figures. So it was with the turbo Metro. With 7½lb of boost the engine produced well over 95ft lb of torque between about 4000 and 4500rpm. The idea of the boost modulator was to chop the mid-range torque and here's how it works:

The waste gate is set at some 4½psi. This means if the modulator does not come into action, the boost figure goes up to 4½psi and stays there. However, at this boost level the power produced will be below that required. The waste gate senses when to open by virtue of a pipe routed from the intake manifold to the waste gate diaphragm. It is the intake boost pressure which opens the diaphragm. In the line between the manifold and the diaphragm, the boost modulator is installed. This boost modulator is an electronic device which is coupled up to the ignition system of the engine so it senses rpm. At a certain rpm, about 3300, the boost has reached 4½psi so the waste gate starts to open. However, at about 4500 rpm the engine has passed the point at which it will produce excessive torque if the boost were 7½psi, so the modulator starts to fool the waste gate into thinking that the boost in the intake manifold is less than it really is. It achieves this by rapidly opening and closing so as to bleed off some of the 'signal' from the intake manifold. This fools the waste gate into thinking the boost in the intake manifold is dropping, so it starts to close, thus causing the boost to go higher. This sequence of events continues right up until the modulator is dumping about 3lb of signal pressure from the manifold. In other words the waste gate thinks there is only 4½psi in the manifold, whereas in fact there is 7½ and 3lb of that signal is dumped back into the intake system. The graph **Fig. 12.2** shows schematically what the torque and boost curves would be like with and without the modulator.

The Leyland engineers were worried about failure of the drop gears and also they were concerned with the cracking of the centre web in the gearbox. The centre web cracking was due to an extra tapped hole in Metro gearboxes that had not been used a great deal previously. If any gearbox prior to about 1979 or after 1981 is used, the gearbox centre web problem is minimal. Gearboxes produced in the period 1979 to about the end of 1981, are the ones to avoid. What the performance enthusiast regards as reliable and what the average motorist regards as reliable are quite different things. Running the standard gearbox near the 95ft lb limit is not out of the question provided it is not of the suspect type just mentioned. In fact, if straight-cut drop gears are used in the transverse Leyland transmission, along with the standard helical gears, it is not out of the question to put 110 to 115ft lb of torque through the gearbox relatively reliably. If an entire straight-cut transmission is used, and again not in the case with the extra drilled hole in the centre web, then outputs of 125ft lb can be accommodated.

There are many ways of improving the power output of a turbo Metro engine. The obvious way is to increase the boost. However, since the engine is very close to detonation, this cannot be done without taking some precautionary measures to prevent damage. Fuel enrichment and ignition retard have been used to guard against detonation virtually to the maximum possible extent. Increasing the boost and then retarding the ignition further to stay out of detonation will probably not yield anything like the hp increase that could be extracted from the engine if the boost was increased and detonation suppressed by water injection. The easiest way to make up a water injection system is to have an oil pressure switch sensing manifold pressure and use it to trigger a windscreen washer pump to inject water into the intake manifold. Most windscreen washer pumps will produce about 45psi if small jets are used. In this particular instance, small jets are all that's needed and it can well be that it's impractical to drill a hole small enough to give a water

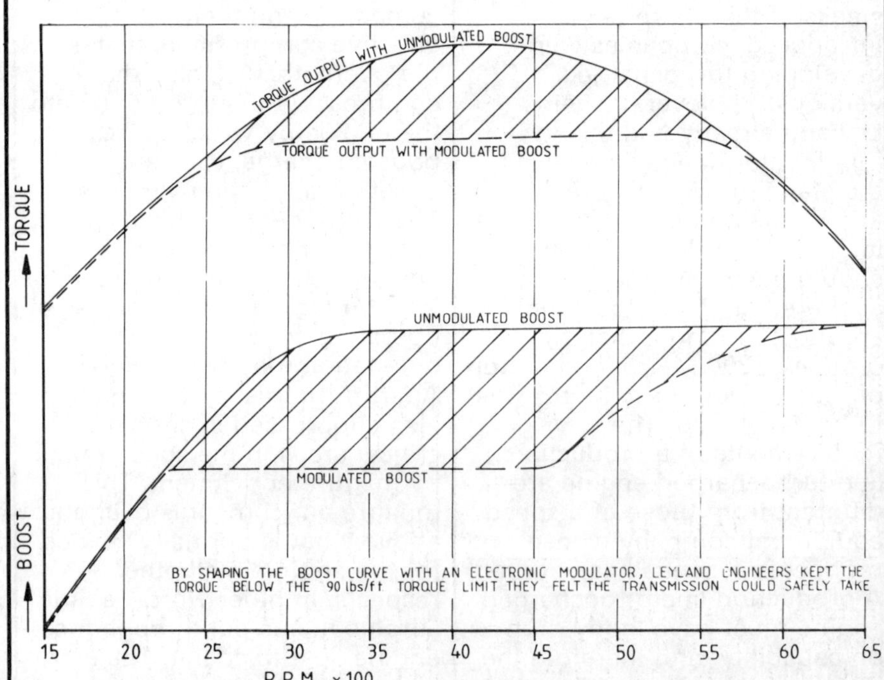

Fig. 12.2 Comparison of boost and torque curves of modulated and unmodulated boost in Turbo Metro installation

flow that only just suppresses detonation. It's difficult for most people using home tools to drill a hole much smaller than about 15 thousandths. Through an orifice this size a pump giving 45psi will give enough water flow to suppress detonation up to about 10psi of boost. Although one water injection nozzle suitably positioned will supply enough water to quench detonation, there may be a problem of mal-distribution of the water amongst the cylinders and, for this reason, using an injection nozzle in each intake port, is probably a better arrangement.

Obviously a system such as just described is not going to give proportional water flow into the engine, but nevertheless it will suppress detonation satisfactorily. Naturally, there is a small cost in hp as a result of water injecting the engine, but detonation costs far more hp and usually costs engines into the bargain. If you would like to use a more sophisticated water injection system, then good systems are made by **Spearco, Holley** and **Edelbrock.** All three of these systems are calibratible with electronic control on the supply of water. All the systems sense manifold pressure and engine rpm, and meter water into the engine proportional to these two functions. This means they can be set to give just the right amount of water to suppress detonation and no more. That is the optimum situation in terms of power output from the engine.

Freer Breathing

Increasing the boost to make more hp, as discussed for the turbo Metro engine, is only one of the numerous power increasing options available. Turbo engines, as with normally aspirated engines, respond well when breathing improvements are made. More hp can be obtained from the turbo Metro engine by gas flowing the cylinder head. Intake ports and combustion chamber shapes need to be modified along the same lines as for the normally aspirated engines shown in the cylinder head chapter.

The exhaust port can also stand improving, but note that it is not a good idea to cut too much metal away from the guide boss because the guide boss is needed to conduct heat away from the valve guide, which in turn is needed to conduct heat from the valve stem of the sodium-filled exhaust valve that's used in these engines. If the exhaust valve is allowed to run any hotter than it already does, the problem of detonation will occur sooner than would otherwise be the case. At the present, no big sodium-filled exhaust valves are available for the Series 'A' head, so it's necessary to revert to a conventional valve. However in the interests of keeping the valve head cool and having it able to conduct heat away to the stem, it is best to continue with a 5⁄16 inch stem and with a valve which has a relatively bulky underhead shape. Use this in conjunction with a bronze valve guide.

As far as the intake valve is concerned, the turbo Metro uses a 1.3 inch diameter valve and it can benefit from being increased to 1.4. Also, this engine will respond well to the fitting of 1.5 ratio rockers. The increased valve opening rates allow the engine to breathe better and the higher lift given puts the valves much nearer the high flow regime that is achieved at lifts over 400 thousandths. Incidentally, the cam in the turbo Metro is the standard Metro cam with 320 thousandths lift; changing to 1.5:1 rockers typically puts the lift figure up to around 380. Depending on how effectively the head work is done, adding 1.5 rockers and a modified head to an otherwise standard turbo Metro typically gives a power increase of around 10 to 12hp. This, together with an additional 2½psi boost, plus water injection to suppress detonation, is good for power levels around the 110 to 115hp mark.

The simplest method of increasing the boost pressure is to build another device into the line which senses the manifold boost pressure and signals the waste gate diaphragm. As I said earlier, the modulator fools the waste gate into thinking there is less manifold pressure than is actually the case. The easiest way to fool the waste gate further is by using an adjustable valve to dump some of the signal pressure from the modulator back to the air filter case. The simplest device for this is an adjustable valve of the type used for balancing the flow of the air to aerator stones in aquariums. These little plastic valves are cheap and functional. If you cut into the line downstream of the modulator valve, put in a T-piece and re-couple the modulator line to the diaphragm of the waste gate, then it's a simple matter to restrict the leak from the leg of the T by putting in the fish tank valve in a line from there back to the air filter case. With the valve shut off completely, the boost curve will be unaffected. If the valve is opened slightly, and I mean slightly, the effect it will have is to raise the entire waste gate controlled part of the boost curve. This means instead of the modulator limiting the boost to 4½psi it will limit it to some higher level and since we are dealing with a fixed orifice rather than a modulated orifice, it will have the effect of increasing the low end boost from the point where the waste gate shuts fully a little more than it will the top end.

If reliability is not your first concern and you are not worried about the transmission breaking, it's a simple matter to bypass the modulator altogether and install a stronger spring in the diaphragm of the waste gate so it gives a higher boost figure before it opens. Fine control of this boost can then be achieved with the fish tank bleed valve as previously described. Getting rid of the modulator valve will, of course, give a huge increase in mid-range torque. In fact with 10lb of boost, the engine should make around 110ft lb of torque at about 4000rpm. To put things into perspective, that is the sort of torque output one would expect from a good normally aspirated production 2 litre type engine. In other words, a turboed 'A' Series vehicle will feel as if it has a 2 litre engine.

Blow-Through Side Draft

The Janspeed turbo Metro was the first machine fitted with a turbo 'A' series engine with a blow-through setup I drove that had REAL power. In fact I think this was the turning point that convinced me it was actually possible, with careful engineering, to produce a blow-through setup that really worked well. I am referring to the sort of power output which is not possible with an engine in normally aspirated form. I would estimate the power output of the particular vehicle I drove to be somewhere in the region of 115 to 120hp. This power output was achieved from 1275cc and it was totally road drivable. 1000rpm in top gear could be used just as effectively in this car as it could in a standard Metro. There was no boost gauge on the car, but Jan O'dor informed me that the boost had been set to around 12 to 13psi.

The particular carb used in this installation was a Dellorto carb which had been specially developed by the factory in Italy for use on the turbo Esprit which, incidentally, is also a blow-through system. On the first examples of the Janspeed blow-through system, the typical blow-through, high-speed lean out was experienced, but this was overcome by juggling around with carb calibrations and making sure the boost pressure over the fuel in the float chamber was just a tad higher than that within the carb barrels.

As I have touched on the subject of float bowls there, it might be wise to point out at this stage that if high boost pressures are used in a blow-through situation a conventional float can collapse. A carb for blow-through installation must have a float of the foam filled type so that the structural rigidity of the float remains unaffected in the boosted environment.

As far as performance potential goes, it's difficult to say whether the Janspeed turbo kit has an edge over the Leyland blow-through SU set-up. Either way, it should be possible to buy all the parts to make up your own turbo motor, but the Janspeed might well be the setup to choose. First, the airflow capacity of the twin choke Dellorto is much higher than the SU and so there is less loss of boost pressure across the carb. Secondly it's much easier to analyse what's going on and change a few jets, air correctors and emulsion tubes on the Dellorto than it is to spend hours trying different needles in an SU. The third factor concerns the exhaust plumbing. The exhaust manifold on the Janspeed turbo kit is much more streamlined and the waste gate is separate from the turbo itself. This means there is much less heat in the area of the exhaust turbo housing, and thus less heat is passed through to the intake turbo. Consequently, at any given boost pressure a Janspeed

Here's the Janspeed Metro turbo conversion installed in a car. As is usual with most Janspeed products, the installation is very neat and workmanlike. In the top LH corner of the under-bonnet area, the Microdynamics turbocharger ignition control module can be seen.

Here are the parts in the Mini turbo kit from Seven Enterprises. I have driven a car equipped with this kit, hp was one thing it had in abundance. Much more, in fact, than could successfully be put on the ground, even with a car equipped with slicks.

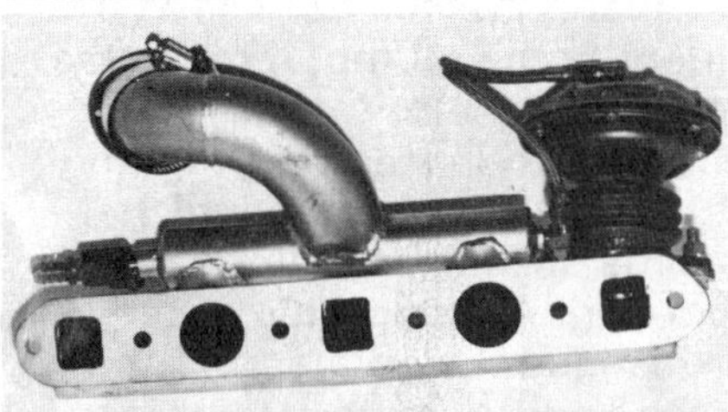

This is the Seven Enterprises turbo kit intake and exhaust manifold. Note that the intake manifold makes no pretence at being streamlined. Its log-type manifold is intended to keep the fuel and air as turbulent as possible to prevent separation.

setup should run slightly cooler. Also, divorced waste gates tend to be more free-flowing and therefore more efficient than waste gates that are integral with turbo housings.

Against the Janspeed setup is that it is bulkier. However, for all its bulk, as can be seen **from the photos,** it is a very neat workmanlike installation.

The most impressive turboed Series 'A' I have ever driven, from the sheer power point of view, was the suck-through system developed by aircraft engineer John Kendrick of Florida. This kit, primarily intended for left-hand drive Minis, is now sold by **Seven Enterprises Ltd.** in the U.S.A. A conservative estimate of the hp would be about 180. However, I think 200 would be nearer the mark. The engine was equipped with many parts found in normally aspirated engines. For instance, the cam used was a 649 profile; the cylinder head was modified in a conventional manner and sported 'S' size valves. The block used was a Cooper 'S' item, and the crank and rods were also 'S' items. Although officially red-lined at 7000, this engine was regularly taken to 7500rpm and with the 18 to 20lb boost used, it had, not surprisingly, a limited life. It would appear that the unreliability was essentially in the main bearings which had a life of only a few hundred miles, but the bearings used were the grooved type as opposed to the sort developed for the turbo Metro engine. It would seem probable that an engine turboed to this power level would have a far greater main bearing life when built with a turbo Metro block, which has superior main bearings with a greater load carrying capability.

With an engine built to the Kendrick specification, it was difficult to tell where turbo lag started or finished and where camminess came into the picture. As I said before, the engine utilized a 649 cam, as a result anything approaching full throttle below 3000rpm caused the rough running one would expect from a full race cam. From 4000rpm upwards, the power and torque came on exceptionally strongly. In practical terms, this engine with its super high boost and race cam was a little extreme for a road-going vehicle.

A close-up of a turbo. This particular example used on the Seven Enterprises turbo kit, is a Rayjay unit.

Reliability

The first problem area is that of head gaskets. With small bore engines it's vital to use the competition 998 head gasket.

All the studs should be changed to the latest Y-studs and the nuts used to hold the cylinder head down should be the flanged variety. If more than about 10lb boost is used, the cylinder head securing nuts should be torqued down to 60ft lb.

If a big bore engine is being built the same rule applies as far as head studs and nuts go. From 1983, the best head gasket currently available for a big bore Series 'A' engine is the copper-asbestos competition head gasket. However, development on the turbo head gasket may change this picture. If you are turboing or supercharging an engine, don't indulge in big overbores on the block. With a 73½mm bore, bore flexure is already a problem in a normally aspirated engine. In a highly boosted engine it can be a greater problem still which is just not worth the hassle for those extra few cc, especially as the engine could drop more power by bad ring seal than it gains from the extra capacity. Also, big overbores cut the reliability of the head gasket. If supercharging is your intention and the engine is to do any amount of mileage, then a 60 overbore should be considered the maximum permissible. If high boost pressures are used along with high rpm, pay particular attention to the main bearings. With big bore engines it's a good idea to convert to the ungrooved, turbo Metro main bearing shells.

To allow the later bearing shells to be fitted a tang groove must be filed into the main bearing caps. **For high boost applications, because of the fact it has less rigidity, avoid using the 1275 block which has a hole over the centre main bearing. As soon as the power approaches a respectable level, the flange is likely to break.** If starting from scratch, use a turbo Metro block. On highly boosted engines or engines which are expected to run at full throttle for an extended period of time, the pistons can get very hot and it has become the common practice to oil-cool the underside of the pistons. This is achieved on Leyland's racing engines by the installation of jets fed with oil from the main gallery which direct the oil towards the underside of the piston. An easy way for the amateur engine builder to achieve much the same effect is to drill at an angle through the main bearing webs from the oil groove, aiming the drill at a point which corresponds to the centre of the piston when it's at TDC. The hole needs to be drilled clear through from the main bearing housing out through the sides of the block above the main bearing housing. Next, using a clearance drill for a 3/16 inch thread, drill down to within about 3/8 inch of where the drill breaks through into the crankcase. Using a suitably long tap, thread the remaining part of the hole, but don't take the full thread all the way through, so that when a grub screw is fitted, it locks up against the part of the thread that is not fully formed. Now drill a 1mm hole through the grub screw. Screw the grub screw into the hole that you've drilled and lock it against the end of the thread. This technique can be done as described for the front three cylinders. The flywheel end has to be attended to in a slightly different way. Because there is no oil groove, it will be necessary to drill a hole from scratch, through the housing, aimed toward the centre of the piston. The start of this hole should be adjacent to the oil hole feeding the main bearing. After the hole is drilled, grind a slot between the main bearing oil feed hole and the hole you have just drilled. Then drill and tap the new hole as per the other three.

Because it now has to feed these four cooling spray jets, the oil pump must be up to scratch. In fact, if the engine is not to have poor oil pressure at idle, it will be necessary to use a particular type of oil pump. The pump is of the type which has a steel plate top cover on it, and has about a 20% higher capacity than other pumps available for the Series 'A'. To identify this pump, refer to the chapter on lubrication.

Another area on a turbocharged engine which needs special attention is cooling. Although it's of no consequence on a sprint or drag race engine, if the power is to be kept on for any length of time, as in a circuit race or rally car, the cooling capacity of the radiator system must be adequate. The extent to which a cooling problem could occur depends upon installation of the engine and the power level it's being boosted to. Some vehicles have an inadequate radiator in the first place, Minis falling into this category. Other vehicles have plenty of radiator but the water is not being circulated in the most effective manner to cool the engine, the Metro being an example here. A turbo Metro has a radiator at least 50% bigger than it need be for the turbocharged engine it carries. If the vehicle has insufficient radiator in the first place, then the problem of cooling the engine is obviously more severe. A solution will require the addition of an extra radiator or the production of an entirely new oversize radiator. Even if a radiator of adequate size is available, as it is in the Metro, highly boosted engines can suffer internal block cooling problems. The cylinders at the opposite end to the thermostat tend to overheat. One solution to the problem is to take water out of the cylinder head from the heater take-off point, pass it through an additional radiator, then dump it back into the original cooling system at the thermostat housing or the top hose. If the engine is used with an installation that has ample radiator capacity, then all that may be necessary is to take water from the heater hose end of the cylinder head and pass it back into the thermostat housing or top hose. Doing this improves the volume of water flowing through the engine so that hot water stays less time in the block and passes more times through the radiator.

Another move which helps on 1275 blocks is to cut away the water jacket from the water pump mounting face around the back of the cylinders. A photo in the cooling chapter shows how this modification is done. As far as fans are concerned, **an electric fan will not cure an overheating problem. It just doesn't have enough power to move enough air. However, if you have added an auxiliary radiator, adding an auxiliary electric fan in addition to the standard fan, will help matters.**

Power Potential

Although it's pleasant talking about, and even working on, very high hp turbo motors (such as that shown in **Fig. 12.3a**), the majority of turbos will be fitted to engines which will produce a lot less than the 160 to 200hp discussed in this chapter. Most users will be more than satisfied with a power level equal to a normally aspirated race engine but without the temperament, noise and inflexibility that such an engine displays. A 1275 race engine will typically develop around 115 to 120hp. Such power outputs are entirely practical for road driven engines if a turbo is used as part of the tuning aid.

At this point, let's have a look at the effect a turbocharger

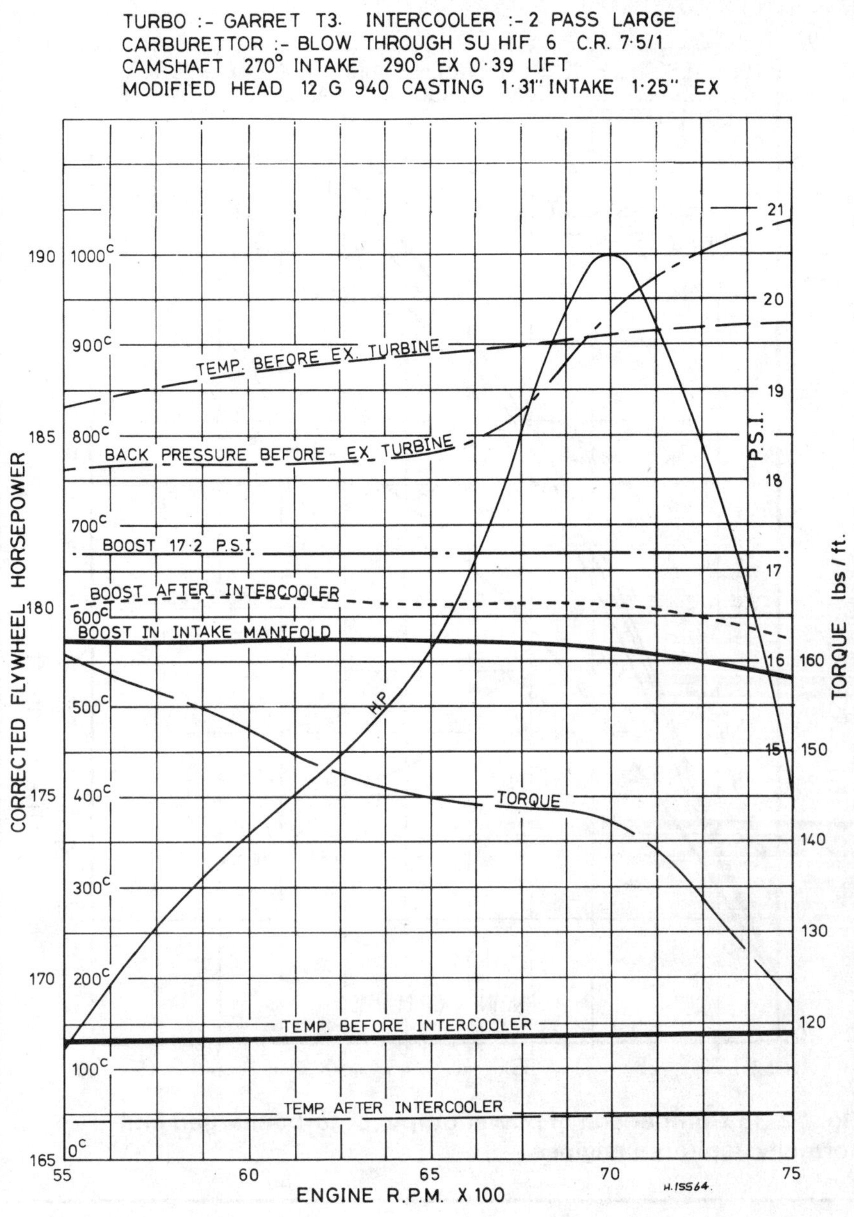

Fig. 12.3a Turbocharged 1293cc Series 'A' engine

Here's what you can do with a Series 'A' engine when it's developed to make the most of turbocharging. 190hp at 7000rpm is the sort of power expected of a 2000cc engine. The graph, although somewhat 'busy', shows many important points concerning turbo applications, relevant not only for the Series 'A' engine, but almost any engine. Important points to note are that the mid range torque is high and that the power starts to drop quite steeply when the back pressure in the exhaust manifold starts to climb. The effect of the intercooler is also shown here to good advantage. For about 0.7psi drop in boost the intercooler drops the intake air temperature from an average of 133°C down to 42°C. This is worth far more increase in power than the possible loss due to the 0.7psi reduction in boost pressure.

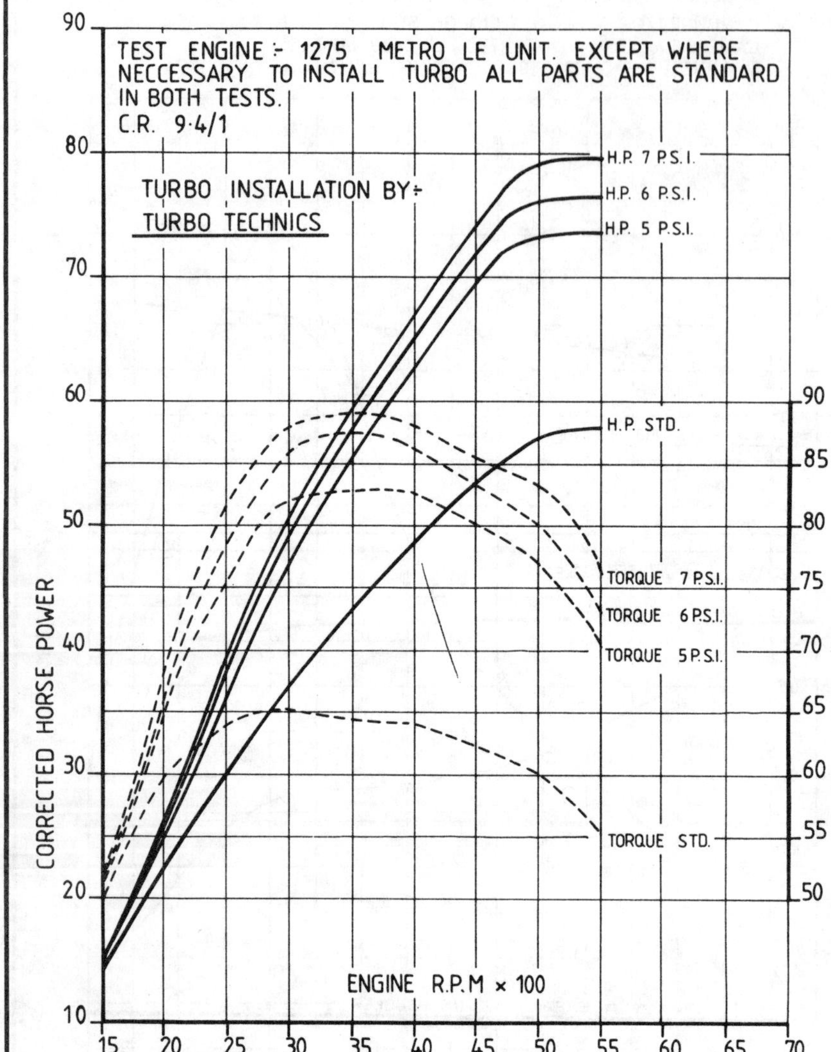

Fig. 12.3b Comparison of power output of turbocharged and normally aspirated engines

Here's the Seven Enterprises kit which, incidentally, was designed by John Kendrick of Florida, installed in a Mini. My only criticism is that it doesn't have an air filter. If any foreign bodies get ingested, it's going to be pretty hard on the intake turbine, let alone the rest of the engine.

kit has on the power output of an otherwise standard 1275 engine. The particular kit used in this instance is the **Turbotechnics** system. This is a draw-through design utilizing an HIF6 SU. As is most often the case these days, this kit incorporates a waste gate. Both turbo and waste gate are AiResearch items and the waste gate can be adjusted over quite a wide range to give varying amounts of boost. For these tests, the baseline engine was an ordinary 1982 Metro unit which is of 'A+' specification.

Fig. 12.3b gives the comparative outputs of the Turbotechnics kit on an otherwise standard 1275 Metro engine. The main point to note on this is that a substantial increase in torque is achieved by the turbo engine. At 7psi boost, the torque figure is raised from some 65lb ft to around 90lb ft. That is the sort of torque increase that might be expected from an engine if stretched in capacity to around 1500. Getting such a torque increase by means of a turbocharger is, of course, a much cheaper means to the same end.

Looking at these curves raises the interesting question of whether the power increase gained is proportional to the boost pressure. Not quite; there are always bound to be losses in the system. If the engine's output had been increased in direct proportion to the boost pressure, 96ft lb of torque might have been expected, instead of the 90ft lb measured. From the curves it can be seen that, generally speaking, each extra pound of boost does not produce the same amount of power as the previous one. In other words, a law of diminishing returns is in force. Two factors have a hand in this. The first is that there is a considerable increase in back pressure when a turbocharger is installed on the engine. The increase in back pressure means that the pumping losses are higher. However, they are not quite as bad as they seem at first, because some of the pumping loss is recovered by the fact that pressure on the piston during the intake stroke means that the intake stroke is also a partial power stroke.

The other aspect is that any time air is compressed, it heats up, and obviously hot air takes up more room than cold air. This means that although a supercharger may have increased the air pressure supplied to the engine the air density has not increased proportionally. Assume that the engine is boosted to half an atmosphere. Though the air pressure is now 50% higher than it would normally be, the air temperature increase will prevent the full 50% increase in density being realized. Depending upon the temperatures involved, the engine may only see a 40% increase in density. Each time the boost pressure increases, the hotter the air gets. Finally, although the boost pressure increases, the air density changes little or not at all. Long before this point is reached, there will be an obvious need to rectify the situation, and it's at this stage of the proceedings that intercoolers need to be considered.

Although intercoolers will increase efficiency at even moderate boost pressures, they are bulky. If the boost pressure is not above 7psi the return measured in terms of the small power increase is not really worth the installation problems that are likely to occur. The higher the CR of the standard engine, the sooner an intercooler will be needed because high inlet temperatures also cause the onset of detonation earlier than would otherwise be the case. If we assume for the moment, we are supercharging or turboing an engine which has a relatively low compression, i.e. 8 to 8.5:1, then it's fairly safe to assume that an intercooler won't be needed until supercharged pressure reaches 12 or 13 pounds. If high octane fuel or water injection is available, then it's quite practical to go to 18lb of boost without an intercooler, although it should be stated that an intercooler at this stage, if there's room for it, can provide a very useful power bonus or the same power at lesser boost.

Returning to the 1275 turbo engine, you can see from the power curves that the extra power is achieved without any extra increase in rpm due solely to the big torque increase. This is the beauty of supercharging over tuning an engine in normally aspirated form. Basically the engine reacts just as if its capacity had been increased. However, just because the engine now has forced induction, do not run away with the idea that the air is going to be pushed into the cylinder, come what may. Turbocharged engines still respond to air flowing techniques just as do normally aspirated engines. If the intake valve is very restrictive, the boost pressure in the manifold will be reached sooner in the rpm range and much more of the boost energy will be lost in overcoming the flow restriction at the valve. Remember, **the object of a supercharging exercise is not to generate boost, but hp.**

Exhaust flow on a turbo engine also can be important. In practice, if the exhaust valve and port are made to flow better, more of the energy that is normally lost due to restrictions in these areas will be available to drive the turbo. The result is the ability to use a larger A/R exhaust housing or the same housing and boost earlier in the rev range.

Another aspect worth considering is that turbos respond very well to heavily pulsed flow at low rpm. Let us assume a basic quantity of exhaust flowing through a turbo exhaust housing, say, 100cfm. If that 100cfm flows through at a steady rate, the turbo would produce less boost than if it went through in, say, 5 shots of, say, 200cfm with equal periods between those shots of no flow at all. In other words, after 1 minute, the flow through the turbo had been 100 cubic feet but it was in 5 pulses of 200cfm, each of a period of 6 seconds, and then 6 seconds of no flow. In each case, the same TOTAL quantity of air passed through the turbo (i.e. 5 shots × 6 secs = 30 secs. ½ min. × 200cfm = 100cfm). At low rpm, pulsing in the exhaust can be increased by making sure the exhaust valve is as efficient as possible, especially at low lifts, so that when the exhaust valve opens, a more sudden pulse of exhaust goes into the manifold. Likewise a cam which opens the exhaust valve a little faster and sooner, can also enhance this effect. It may only be a small advantage, but it's one worth seeking.

Turbos & Silencers

Continuing the subject of exhaust for a moment, **an aspect that is important to consider on a waste gated engine is the efficiency of the silencing system as this can have a distinct effect on the boost obtainable.** If the exhaust system presents any measurable back pressure, it will reduce the amount of boost achieved. This is because excess back pressure in the silencer will reduce the pressure drop across the exhaust turbine. This means there is less pressure differential available from the exhaust to drive the turbine, and hence drive the compressor into boost. On any turboed engine, where a waste gate is used to control boost, an efficient exhaust should be used.

From the noise point of view, a turbocharged engine has an advantage. The fact it has a turbine in the exhaust system means the amount of noise emitted at the exit of the turbine is reduced quite substantially in comparison with that produced at the unsilenced exhaust pipes of a normally aspirated engine. An American friend of mine describes a turbo as a 'rotary muffler' and as such, it's the only muffler available which really adds power to an engine. The noise reduction brought about by the use of a turbo is not enough to make an engine road-legal, but enough to make the silencing of the engine far easier.

The high performance, dual box exhaust system will produce very low noise levels because of the presence of the turbo. If you are only seeking to silence the car down to the level normal in, say, a standard 1275 'S', then a high performance single box system, such as typically produced by Competition Silencers or Janspeed will be adequate for that purpose.

More Power - Same Boost

Using the same engine as was used to develop the power curves in **Fig. 12.3b**, tests were done to see what the effect of improving the breathing efficiency of the engine would be in terms of power output. The particular head used was a **Longman** big valve head, having a chamber volume of 29cc as opposed to the standard head which has around 21½ to 22½cc. This cylinder head was used along with Powermax pistons in which the dish had been machined to 9.3cc. The combination of parts used resulted in a CR of 8.1:1 as opposed to the standard Metro CR of 9.4:1. The 9.4:1 compression, when used with 7lb boost will, if the ignition timing is optimized for power, be very close to or at detonation level, hence the lower CR on the modified engine. To improve the breathing further, a cam having the 510 Cooper 'S' profile on the lobes, was installed in the engine. This operated the valves through rockers equipped with offset bushes, thus giving a valve lift of around 345 to 350 thousandths. To see what these modifications did in terms of power increase, look at **Fig. 12.4**.

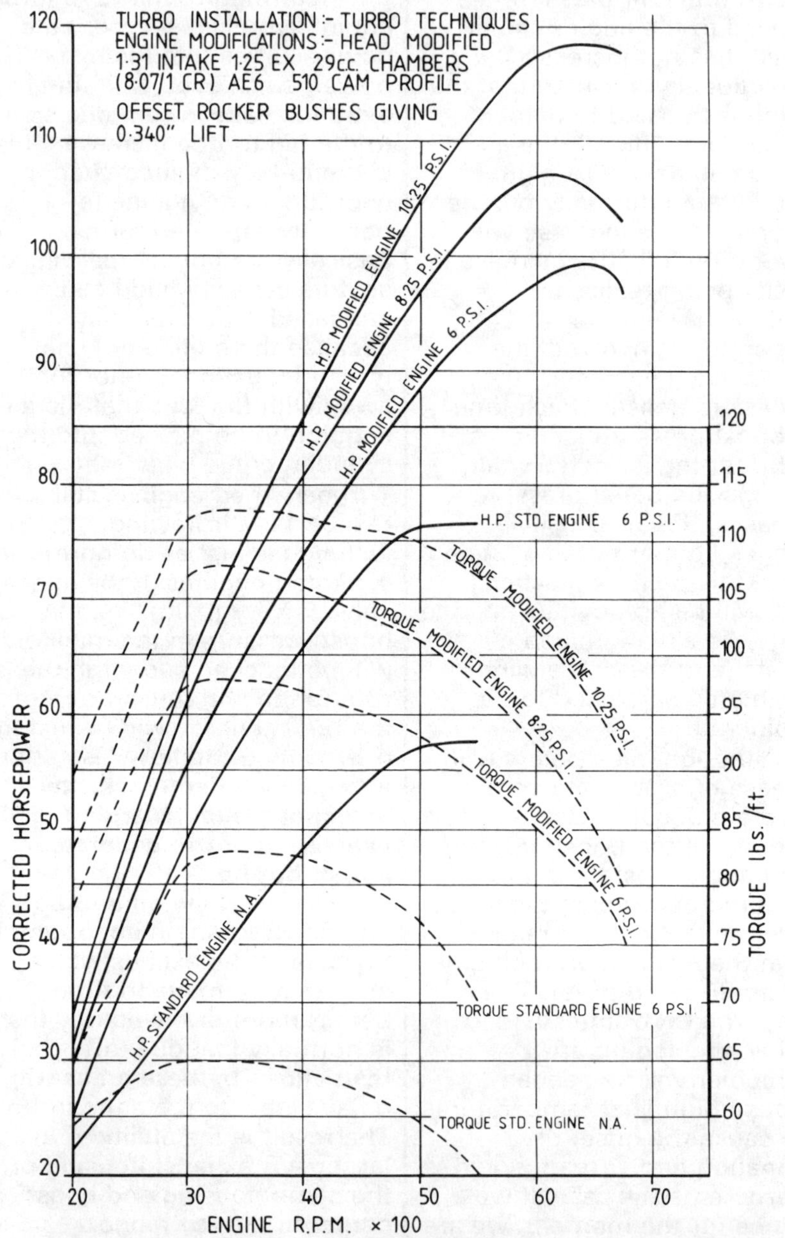

Fig. 12.4 Effect of improved breathing on power output of 1275 turbocharged engine

On this graph is the output achieved with 6lb of boost on a standard engine. Note that with 6lb of boost on the modified engine, both torque and power have increased throughout the measured rev range. In other words the same amount of boost has produced more power. When the boost figures are increased, we find more power increase per lb of extra boost pressure on the modified engine than on the standard engine. The engine responds better to the turbo, principally because the flow of gases through the engine is more easily accomplished. With 10¼lb of boost pressure the test engine produced 118hp at 6500rpm. It takes a very good 1480cc normally aspirated engine to produce 118hp at 6500. As far as torque is concerned, this turbo engine managed 115ft lb. Such a torque figure is difficult to achieve from a Series 'A' engine other than a well developed large displacement unit.

Positive Displacement Supercharger

Whenever superchargers of any sort are considered the question arises of whether a conventional supercharger is better than a turbocharger. If you read to the end of this book, you'll find that, though it need not necessarily be the case, for a conventional turbo setup turbo lag does exist and this is the prime argument for looking at mechanically driven positive displacement superchargers. The question is, can such a supercharger produce 'turbo results?' Those who favour turbos often put forward the point that since the turbo is a more efficient compressor, it should produce more hp than a positive displacement type supercharger. On the other hand, turbos do generate a lot of heat in the engine compartment and they may be self-defeating from that point of view simply because they draw pre-heated air into the engine. So to represent the other camp it's worthwhile to look at some supercharging exercises which were done utilizing belt driven positive displacement superchargers. The particular supercharger setup I have in mind was built by Mike Broad of **Autopower Services** in Daventry. It utilized a **Magnuson** blower and was installed as a suck-through system. As I said previously, this is the conventional way to configure a carburated, supercharged engine when a belt-driven blower is used.

Although there are installations which blow-through, they are very much in the minority. Of course it's possible to fuel inject. If this is the method used, then it's much simpler to port inject, but when injection is used, the question of blow-through or suck-through is removed because it is irrelevant to the fuel injection metering system where it's injected as long as calibration is maintained. From the point of view of power with the fuel injection system, it seems best to inject as close to the intake valves as possible, hence port injection is to be favoured.

Boost on the Magnuson blown engine is adjusted by changing the crankshaft sprocket to blower drive sprocket ratio. The blower unit, which displaces approximately 1000cc per revolution, was tested with zero overdrive (that's a 1:1 ratio), 22% overdrive and 37½% overdrive. These ratios produced the boost curves shown in **Fig. 12.5.** A point worth noting here is that the Magnuson blower is capable of producing substantial amounts of boost as low down as 2000rpm. Even when geared at 1:1 it produced 2lb of boost at 2000 whereas the turbocharged engine was only just beginning to register boost at 2000rpm. When driven at 37½% overdrive, just over 5lb of boost was possible at 2000rpm. Because a positive displacement blower is much better at producing supercharged pressure at low rpm, one can expect the torque output to be better than that of a turboed engine. Couple this to the fact that a positive displacement type supercharger has no throttle lag, it's easy to see that it will appear to be a more 'lively' engine with superior lugging power at low rpm.

Seen here is Gerry Cookson's blown 998 Berkley. This car is used for drag racing and runs in the 13s.

As far as total torque and hp is concerned, making a comparison between the positive displacement supercharger and the turbocharger is not as easy as it would at first appear. Numerous factors are brought into play, not the least of which is the original spec of the engine. Because of this outputs in **Fig. 12.6** cannot be directly compared to those in **Fig. 12.4** for the turbo engine. From

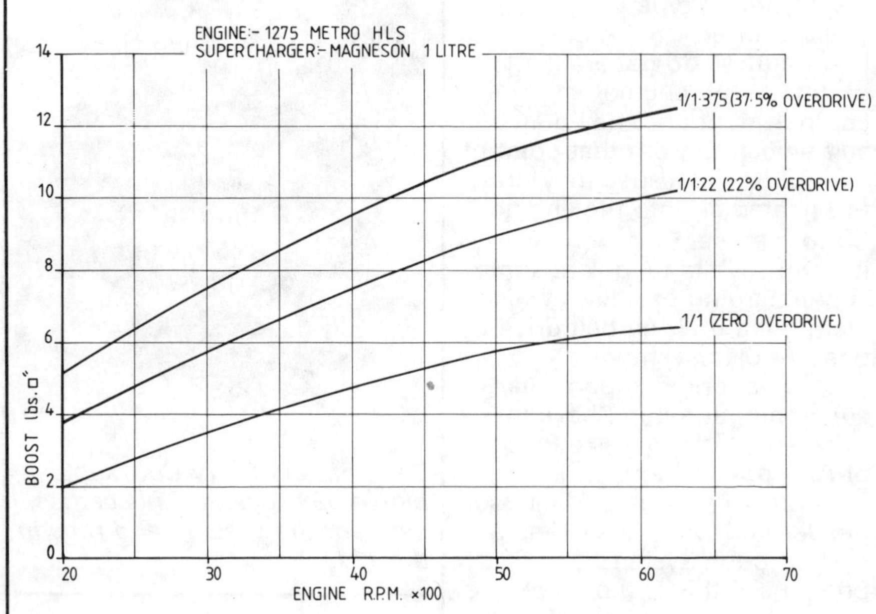

Fig. 12.5 Boost curves of supercharger using various drive ratios

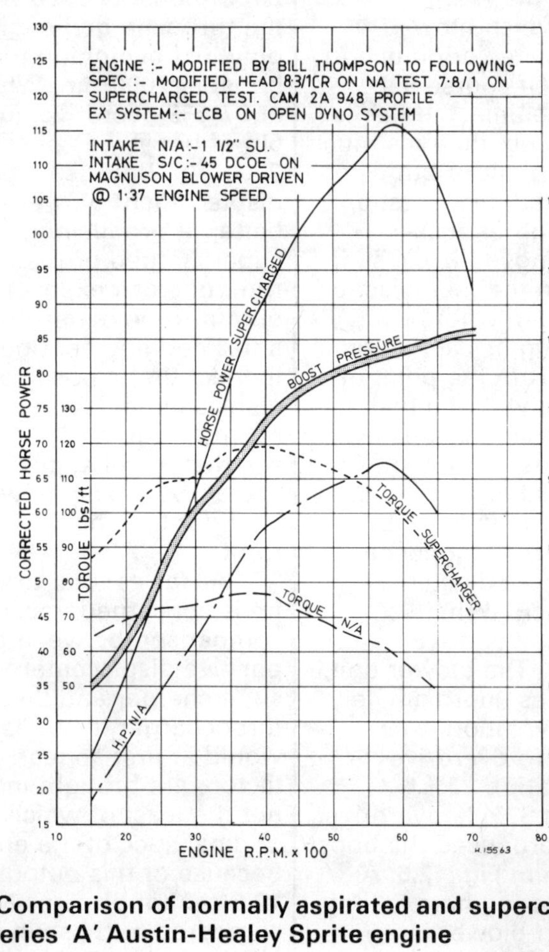

Fig. 12.6 Comparison of normally aspirated and supercharged 1330 cc Series 'A' Austin-Healey Sprite engine

these graphs you can only view trends. Which method of supercharging is chosen depends on the type of response you want from the engine. In competition, both positive displacement and turbochargers have been used effectively.

To aid selection it might help to give preferences of people who have tried both routes in different situations. Generally speaking, for use on loose, off-road situations such as autocross, grass track, rallycross and rally-style events, the positive displacement type supercharger is more popular. This also applies to tight, twisty courses such as often seen on hill climbs and slalom events. The turbo seems to be favoured most for road race courses. I don't want you to run away with the thought that it's all cut and dried. This is not so. The Magnuson supercharger in particular seems to be breaking new ground as far as positive displacement type superchargers are concerned and it does seem to be making in-roads into situations which were previously considered to be dominated by turbocharged engines. Certainly in the field of forced induction on the Series 'A' engine, a lot still has to be learned with both types of supercharging. Developments in both areas keep on making the job of deciding which way to go as difficult for the expert as it is for the novice.

Matching Components

To the inexperienced at building supercharged engines there are likely to be two areas which may cause problems relating to the selection of compatible parts. These concern camshaft selection and ignition timing. As far as ignition timing goes, it should be remembered that the more boost pressure increases, the more it's necessary to retard the ignition. Highly boosted engines do not need nearly as much ignition advance as normally aspirated engines. The more boost that is used, the less advance is needed. However, as cam specifications entailing more opening period are used, so the ignition timing will need to be advanced. Also, the advance will need to come in faster so the distributor will have to be 'curved' to suit. Usually engines with race cams are so equipped in an attempt to make the most hp from the boost available, consequently race cams tend to be used together with high boost pressures. As a result, ignition timing figures on high octane race fuel tend to end up around 24 to 28°.

I only mention ignition timing here as a passing reference; a more comprehensive look at what's required for ignition systems on supercharged and turbocharged engines is covered in the chapter on ignition systems.

As far as camshafts are concerned, the rule of thumb to follow here is relatively simple. If the engine is intended for road use and low speed drivability is of importance, it's worthwhile remembering that the engine can probably make the hp being sought without the use of a particularly long period cam. With a well setup supercharged engine, a 260° cam will produce useful power up to about 7000rpm. A 270° cam will extend that to around 7300 or 7400. 300° cams generally can make 8000rpm usefully available, as long as the rest of the engine is capable of coping with it.

As far as lobe centreline angles for a turbocharged engine are concerned, better performance is usual with lobe centreline angles about 105 to 108° for a race engine. In some cases, mostly for road use, 108 to 112 may be beneficial. These larger lobe displacement angles are better when the exhaust back pressure is quite a bit above the intake boost pressure because smaller A/R housings are being used. The nearer these numbers get to parity, the better the smaller lobe displacement angles seem to work. With a mechanically driven supercharger which has no self-induced exhaust back pressure, lobe displacement angles around the 100 to 104 mark are beneficial. This is because the increased overlap period brought about by the tight lobe centreline angles improves intake charge scavenging of the combustion chamber. Let's say we have 4psi in the intake manifold and only 0.1psi back pressure in the exhaust. By opening the intake valve sooner and having a greater amount of valve opening at TDC, the intake pressure can purge the residual exhaust from the combustion chamber, thus providing better filling of the cylinder. If the cylinder over-scavenges, ie, blows fresh charge out of the exhaust port, the cam lobe centreline angle needs to be increased and/or the compression ratio lowered. Remember at this point the piston has not started to travel down the bore to draw in a fresh charge. When it does so, it will draw in a fresh charge in addition to what's in the combustion chamber. With a turbocharged engine this cylinder scavenging does not take place because in most instances the exhaust back pressure is higher than the intake pressure, so if any gases are going to flow, it's going to be exhaust into the intake and that's not good for hp.

Another aspect that's well worth considering on a cam profile for a turbocharged engine is the slightly earlier opening of the exhaust valve than would be the case for a normally aspirated engine or a positive displacement supercharged engine.

13 Blocks, Cranks, Rods & Pistons Part 1

All the power the engine will ever produce is made above the piston and it is the function of the components discussed here to transmit that power to the flywheel as efficiently as possible. None of the components we are looking at here actually produces additional hp. Efforts to perfect these components are really efforts in improving their efficiency so that they lose as little hp as possible. A situation which must, of course, go hand in hand with reliability.

From the performance aspect, **a factor which is becoming increasingly important in today's highly competitive motor sporting scene is the weight of the rotating and reciprocating parts within the engine.** The weight of reciprocating parts, i.e. pistons and connecting rods, needs to be kept to a minimum, not only for performance, but because cutting the weight of these components relieves the crankshaft of additional stresses. Likewise, lighter pistons reduce the stresses that are likely to break connecting rods. However if lightening is taken too far, these components may break purely because of inadequate strength. There is a need to tread a fine dividing line. All the material carried within the components concerned needs to be there to do a job. Nothing should be along for a free ride, and with those words of caution in mind, let's start by looking at pistons.

Pistons

I will describe what to look for in piston design so you can ascertain for yourself what type of piston should be used. Firstly, split skirt pistons are out. Fortunately, this type of piston is now a rarity although it was the original design of piston used in engines such as the 948, 803 and 850. These pistons were so designed to be quiet in operation. At high rpm the skirts flex and eventually fatigue breaks them. Also, the fact that the skirt flexes causes the piston to rock in the bore and compression and oil rings tend to be less effective at their respective functions.

Many Series 'A' engines had pistons that had an oil control ring below the gudgeon pin as well as above. A good, modern, two rail and expander type oil control ring set is perfectly adequate for oil control. The extra ring of early pistons contributes measurably to increased internal engine friction.

Avoid also the W type piston. This piston, pioneered by **Hepworth & Grandage**, is designed to run quietly but be less susceptible to fatigue breakage than the split skirt variety. As such, it's a reasonably strong piston and it does function well within the realms of its design parameters, but it is not a high performance or race piston. Inspection of this piston will quickly show that the major part of the skirt is not actually connected to the crown of the piston. **At high rpm the crown of this type of piston can separate from the piston pin bosses, and after such an event, performance will be found to be relatively poor.**

The type of piston you should look for is a solid skirt one with the entire circumference of the skirt connected without a break, to the ring belt and thence to the crown of the piston. These types of piston are inevitably the strongest. On many pistons, a slot is machined into the oil control ring groove so the oil scrapped from the bore is passed into the groove and fed back to the inside of the piston.

Here's one each of just about every popular piston for the Series 'A' engine. No marks for identifying each one by part number.

Many early Series 'A' pistons were of the split skirt variety as shown here. These pistons ran very quietly but they had a tendency to break up, if used at too high an rpm and too heavy a power loading.

The slot method works well (and it's cheap to machine a piston this way) but it does weaken the piston simply because the crown of the piston is largely separated from the skirt. **On a true high performance or race piston, oil is returned to the crank case from the oil control ring by drilled holes,** and this type is to be preferred over the slotted ones. Having said that, there are, nonetheless, some slotted pistons which have proved to be effective in high performance use. Even so, if you have the choice, drilled oil return holes are much to be preferred.

Now let's discuss cast pistons versus forged pistons. Cast pistons are used for many production engines simply because they can be made more cheaply, but that in itself need not rule out these pistons for use in Series 'A' engines. **A good cast piston can be most satisfactory for high performance use if expressedly designed for that purpose.** The cast piston has improved immeasurably when compared with similar items made 15 or 20 years ago. One of the advantages of a cast piston is that by the very nature of the manufacturing technique, metal can be placed exactly where it's required, whereas with a forged piston it is necessary to consider how the forging die is to be removed from the completed piston. For instance, there can be no complex pin boss undercuts in a forged piston, but in a cast piston the problem does not arise.

Although ideally forged pistons are to be preferred, a little bit of over-engineering in a cast piston can produce a component virtually equal to a forged piston in every respect but cost. A cast piston always scores over a forged one in that respect.

For competition purposes, some of the Hepolite Powermax pistons are cast. However, this company has, for some time, been using a process known as 'squeeze casting' which allows to be combined many of the advantages of forging and casting. Just before the material in the mould solidifies, the casting is impacted, thus imparting some of the properties of a forged piston to the finished product. I have used Powermax pistons manufactured by this method in high performance and race

Series 'A' engines and have found them to be trouble-free.

Another company that manufactures an excellent cast piston is **Omega Pistons Ltd.** This company produces cast pistons from a relatively complex piston alloy which, can best be described as a LO-EX alloy, developed by modifying the fairly common LM13 piston alloy. By putting their special alloy cast pistons through a series of heat treatment processes during manufacture, Omega manage to bring the hardness and strength of their cast pistons to much above what is usual. Some of the cast pistons that Omega produce are the type used for making big bore Minis, i.e. 73.5 or 74mm. I have used numerous sets of these pistons in high performance and race tuned engines. Over the last few years I have built many dyno test engines with their pistons and, during experimentation on the dyno, these pistons have been subjected to the most arduous treatment. **Even with the injection of very high loads of nitrous oxide not only did the Omega cast pistons survive for many hours of testing, but they showed no signs of distress until subjected to power levels in excess of 170 bhp.** When you are choosing a high performance piston for the road, remember cast piston are generally harder than forged ones and this leads to longer ring groove and skirt life.

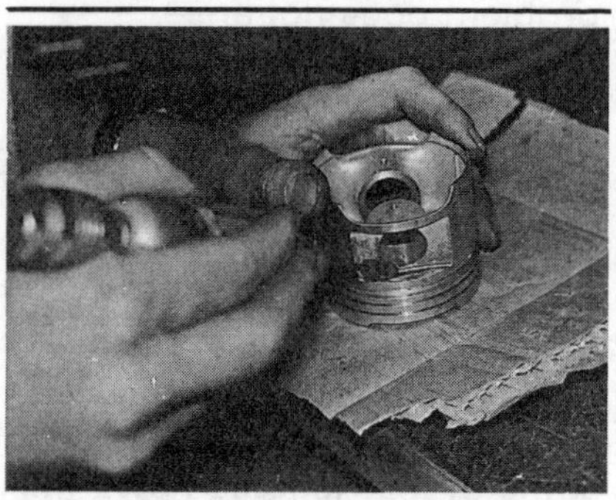

The Omega cast pistons have proved an incredibly reliable piston and have only showed signs of failure at extremely high power levels, usually in excess of 170 hp. The usual point of failure are cracks appearing in corners of the flats on the gudgeon pin boss. The chances of this occurring can be reduced by polishing out the radius as seen here and then masking the piston and having the radius lightly shot peened.

Naturally, the rings as well as the piston are also important. On engines which are not necessarily subjected to high rpm usage, the rings can be made relatively wide and of a material that's not necessarily very sophisticated and ring life can be expected to be more than adequate. However, **as soon as high rpm considerations are taken into account, the rings become more critical in both material content and design. Wide rings are not suited to high rpm applications.** For very high rpm usage, ring widths need to be reduced to about 40 thousandths (1mm) and stainless or chrome plated or moly filled top rings are a definite asset. Both stainless and chrome rings have a disadvantage in that they take a long time to bed in. This, however, can be overcome largely by careful selection of the bore finish.

If you are considering an engine that is overbored to any extent, i.e. boring out to 73.5 or 74 mm, then a new problem begins to appear on the horizon, namely that of oil control. On engines which are overbored substantially, bore rigidity is reduced. As a result, the oil control ring has a harder time coping and sometimes it's necessary to pass on some of the responsibility of controlling the oil to the second compression ring. Here, a napier scraper ring is often the answer. Again, in overbored engines, plain ductile iron rings seem to have a limited life, if the engine is used continuously at wide open throttle at relatively high rpm.

Where bore flexibility is evident, as soon as the ring loses some of its tension, power-robbing blowby can occur. What appears to happen on most test engines is that the ring holds out until a certain point in the rpm range, and then suddenly the blowby increases dramatically. At this point, power increase with increasing rpm does not follow the usual trend. Although I have never tried it on a big piston, a 'Dykes' ring would seem like a good idea. The Dykes ring is an L-shaped ring which has a wider face pressing against the bore than a conventional plain ring, whilst the narrow section of the ring sits in a narrow groove. Gas pressure reaches behind the foot of the L during the combustion stroke and forces the ring out onto the bore to produce a better seal.

When choosing a piston, refer to the chart **Fig. 13.1**. This details most of the popular pistons of the Series 'A' engine and gives a good idea as to their specification and suitability as far as competition or high performance use is concerned. However, there is additional important information not on

From left to right we have the four pistons as in Group 1 of the chart Fig. 13.1. The piston to note here is the Carl Schmitt piston, second from left. For a 'standard' type piston, it is excellent for high performance use. So is the piston on the far right. This is a Powermax piston with a flat topped crown, solid skirt and little weight.

Fig. 13.1 Piston selection chart

	Brand	Type	Bore size	Engine	Number of rings & thickness	Compression height	Crown style	Total weight	Pin weight	Std CR	Comments
	Hepolite (15037/K)	Cast 'W' type	2.477	848	5 (3 compression 2 oil control) .07/.07/.07/.125/.125	1.498"	Shallow dish 1.5 cc	304 grm	53 grm	8.3	Not suitable for high perf. Use for std rebuild only
1	Karl Schmidt	Cast Steel ex. insert	2.477	848	3 (2 compression 1 oil control) .07/.07/.125	1.498	Shallow dish 1.5 cc	302 grm	52 grm	8.3	This piston seems most successful in 850 mini-cross racers
	Hepolite (14396/K)	Cast 'W' type	2.477	948	5 (3 compression 2 oil control) .07/.07/.07/.125/.125	1.342	4.6 cc dish	286 grm	53 grm	8.3	Not suitable for high perf. Use for std rebuild only
	Hepolite (18519/KR)	Cast solid skirt	2.477	948	3 (2 compression 1 oil control) .062/.062/.125	1.342	Flat top	267 grm	54 grm	9.35 or 9.0	**Powermax design.** Suitable for road & race
	Hepolite (20754)	Cast solid skirt	2.542	998	4 (3 compression 1 oil control) .062/.062/.062/.157	1.333	Flat top	305 grm	60 grm	9.6 in Metro	Piston is longer in skirt than power max piston below. Very robust for std piston rings too wide for very high rpm
2	Hepolite (20950 KR)	Cast solid skirt	2.542	998	3 (2 compression 1 oil control) .062/.062/.157	1.333	Flat top	275 grm	60 grm	9.6 in Metro	**Powermax design** Excellent high performance street usage but rings are a little wide for 8000-9000 rpm race engines
	Hepolite (20773)	Cast solid skirt	2.542	998	4 (3 compression 1 oil control) .062/.062/.062/.157	1.333	6 cc dish	279 grm	60 grm	8.3 in Metro	Has deep oil return slot in oil ring groove O.K. for moderate perf. use up to 6000 rpm
	Hepolite (16179)	Cast 'W' type	2.542	1098	5 (3 compression 2 oil control) .062/.062/.062/.125/.125	1.196	3 cc dish	282 grm	60 grm	8.9	Has too many rings not suited to more than 6000 rpm
3	Hepolite (20398)	Cast solid skirt	2.542	1098	4 (3 compression 1 oil control) .062/.062/.062/.125	1.194	13 cc dish	270 grm	60 grm	7.5	Has horizontal slot in skirt, ok for moderate performance
	Hepolite (20397)	Cast solid skirt	2.542	1098	4 (3 compression 1 oil control) .062/.062/.062/.125	1.194	7 cc dish	268 grm	60 grm	8.5	Has horizontal slot in skirt. OK for moderate performance
	Hepolite (18520 KR)	Cast solid skirt	2.542	1098	3 (2 compression 1 oil control) .062/.062/.125	1.194	Flat top	262 grm	62 grm	9.5 or 10.0	**Powermax design.** Suitable for use to 8000 rpm
	Hepolite (20395/V1)	Cast solid skirt	2.780	1275	4 (3 compression 1 oil control) .062/.062/.062/.156	1.497	16 cc dish	380 grm	93 grm	8.0	Robust for cast piston. OK for Turbo to 10 lbs boost
	Hepolite (20755)	Cast solid skirt	2.780	1275	3 (2 compression 1 oil control) .062/.062/.157	1.527	12 cc dish	393 grm	93 grm	9.4	Robust for cast piston. OK for hi perf use to 7000 rpm
	Leyland	Cast solid skirt	2.780	1275	3 (2 compression 1 oil control) .062/.062./156	1.527	8 cc dish	389 grm	101 grm	10.0	Leyland Metro piston. Has proved very satisfactory in high performance street 1275 engines
4	Hepolite (20394/V1)	Cast solid skirt	2.780	1275	4 (3 compression 1 oil control) .062/.062/.062/.156	1.495	10 cc dish	389 grm	94 grm	8.8	A robust cast piston OK for street perf up to 7000 rpm
	Hepolite (20747/KR)	Forged solid skirt	2.780	1275	3 (2 compression 1 oil control) .046/.046/.159	1.497	6 cc dish	363 grm	88 grm	9.7	**Powermax design** excellent high perf/race piston
	Leyland Cooper 'S'	Cast solid skirt	2.780	970-1275	3 (2 compression 1 oil control) .062/.062/.157	1.497	6 cc dish	357 grm	93 grm	9.7	Excellent high performance street piston
	Howley Omega	Forged solid skirt	2.800	1293	3 (2 compression 1 oil control) .046/.046/.157	1.510	Flat top or dished to suit	356 grm	90 grm	11.0*	Excellent race piston approx 345 grm when dished
	Longman	Forged solid skirt	2.800	1293-1310	3 (1 steel compression 1 SG iron, 1 low drag oil control)	1.497	6.5 cc offset bowl	350 grm	85 grm	9.6	Forged in RR58 increased squish, high-tech rings
	Cosworth	Forged solid skirt	73.5 mm (2.894)	1380	3 (2 compression 1 oil control) .040/.040/.156	1.500	Flat top	377 grm	87 grm	11.0*	Excellent race piston
	Hepolite (20659/KR)	Cast solid skirt	73.5 mm (2.894)	1380	3 (2 compression 1 oil control) .060/.060/.156	1.481	10 cc dish	403 grm	94 grm	9.25	Excellent big bore high perf street piston
	Omega	Cast solid skirt	73.5 mm (2.894)	1380	3 (2 compression 1 oil control) .046/.046/.156	1.497	7 cc dish	373 grm	91 grm	10.1	Big bore street/race with std stroke crank
	Omega	Cast solid skirt	73.5 mm (2.894)	1380	As above	1.497	11 cc dish	360 grm	91 grm	9.2	Big bore street/race with std stroke crank
5	Omega	Cast solid skirt	73.5 mm (2.894)	1380	As above	1.426	7 cc dish	366 grm	91 grm	†	Big bore street/race with long stroke crank
	Omega	Cast solid skirt	73.5 mm (2.894)	1380	As above	1.426	11 cc dish	354 grm	91 grm	†	Big bore street/race with long stroke crank
	Omega	Cast solid skirt	74 mm (2.913)	1399	As above	1.497	7 cc dish	377 grm	91 grm	†	Big bore street/race with std stroke crank
	Omega	Cast solid skirt	74 mm (2.913)	1399	As above	1.497	11 cc dish	369 grm	91 grm	†	Big bore street/race with std stroke crank
	Omega	Cast solid skirt	74 mm (2.913)	1399	As above	1.426	7 cc dish	373 grm	91 grm	†	Big bore street/race with long stroke crank
	Omega	Cast solid skirt	74 mm (2.913)	1399	As above	1.426	11 cc dish	361 grm	91 grm	†	Big bore street/race with long stroke crank

* In flat top configuration

† Depends on crank stroke

the chart which may be of use. Firstly, if you are looking for a high performance piston for either an 850 or a 948 engine, the Sprite flat topped, Powermax piston from **Hepolite**, part number 18519/KR is a pretty good bet. The piston is of a somewhat dated design in that the rings are a little wider than is ideal and a double rail, cast iron ring is used rather than a 3-piece stainless ring for oil control. However, in spite of this it's still a good working piston in this size of engine. Of course if you use it in an 850 engine, it will be necessary to machine about 150 thousandths off the top of the block to get the piston to come flush.

When building an 850 Minicross engine where the regulations require that either a standard or standard replacement piston is used, then the **Carl Schmitt** piston for 850s is a good bet. It is slightly heavier than many pistons, but it is very robust and skirt flexure is minimal. This piston is unusual amongst most of the pistons listed in the chart **Fig. 13.1** in as much as it's the only steel strut piston listed. This company, however, does produce pistons other than those for the 850 and can supply pistons for the bigger Series 'A' engines.

The pistons shown here, left to right, are in the same order as those in Group 2, top to bottom. The piston on the left is worthy of some note since it is standard in all high compression 998 Metro engines. The piston is tough and good for high performance use. Its only drawback (and this is relatively minor in nature) is that it's 20 to 30 grams heavier than a more competition orientated piston.

Moving on to the bigger small bore engines, the original 998 Cooper piston had a raised D-section on the crown. If this type of piston is used in an engine which has a non-open chamber type cylinder head (as per single carb 998) that raised D-section will hit the beak of the combustion chamber. An obvious solution is to either machine a slice off the piston to make the D-section smaller, or to relieve the beak in the combustion chamber. Production of this type of piston ceased in 1982 so examples are going to become increasingly hard to find. Current 998 Cooper pistons are flat topped. This means replacing raised crown pistons with the current flat topped items will reduce the CR.

When building a 998 racer, two choices of piston are available. Either the Powermax piston 20950/KR or the Leyland competition slipper piston intended for use with 998s equipped with Cooper 'S' rods. Broadly, the slipper piston seems to have the edge on the 998 piston as far as power is concerned. However, this piston/rod combination is heavy and its use normally entails approximately 200 to 230 grams extra reciprocating mass. On short track events, the extra power produced by the slipper piston/'S' rod combination may not offset the weight penalty because in the lower gears that extra power is largely absorbed accelerating the extra weight. It's not until the speed builds up that any real difference is likely to be noted.

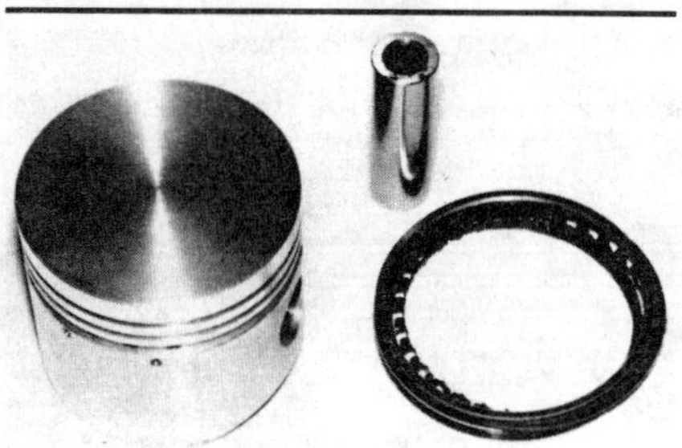

Here's the Hepolite 20950 KR Powermax piston for a 998 engine with which I've had a great deal of success. I have used them in 998 Miglia type engines with power levels up to 95 hp without any sign of unreliability.

Just in case you never get to see any in the flesh, here are the, now rare, 998 Mini Miglia 'slipper' pistons.

Regarding power output variations. According to Steve Harris of **Steve Harris Race Engines** the slipper pistons are worth 1 - 1½ hp over the Hepolite piston. Also, the slipper piston/'S' rod combination is considerably more expensive than using 998 rods with Powermax pistons.

For the 998 engines, the Hepolite Powermax piston 20950/KR is a good working piston. However, in comparison with some of the high performance pistons in the Series 'A', especially the big bore engines, this piston has its

This time, from right to left, the order of these pistons corresponds to the top to bottom of Group 3 in the chart Fig. 13.1. Intended primarily for use in 998 and 1098 engines, the only thing I have to say here is that the order of preference for use in performance engines starts on the left with the Powermax item which is the most preferable, and works across to the W piston with five rings on it, on the right, which is the least preferable.

rings fairly close to the piston crown; the top face of the top ring is only about ¼ inch down from the crown of the piston. In normal circumstances this is fine but if this piston is used in an engine running close to or at detonation level for any length of time the lands tend to become hammered down. This pinches the top ring and sometimes the second ring, so destroying ring seal. Of course this isn't so much the fault of the piston as it is the engine builder for being a little over-ambitious with the CR. It pays to note the fact that octane levels are not totally consistent. This in itself is justification of being a tad conservative with the CR when this piston is used.

The foregoing also applies to the Hepolite Powermax piston 18520/KR. This piston is intended for use in 1098 engines and the top ring, which is of the Dykes type, is very close to the top of the piston; the main part of the body of the ring is ¼ inch from the crown, but the top edge of the L-section of the Dykes ring is some 50 thousandths closer. This piston functions well, too, as long as an excessive compression ratio isn't used. Another bonus is that this particular item is very light which is probably fortunate since it has to be pushed and pulled through a 3.3 inch stroke.

While we are still on the subject of small bore engines, it might be worth mentioning the fact that the latest 1000 Metro piston (1980 on) is a fairly stout item. If you are modifying one of these engines, you need have no fears concerning piston strength.

Big Bore Pistons

Virtually all the Leyland cast pistons for high performance motors such as those of the 'S' are a good basis upon which to work as are the latest 1275 Metro pistons which are proving very robust. Building a race engine with either type is quite feasible. I have used standard cast 'S' pistons on a race engine which was turned regularly to 8000+ rpm and they survived for more than 2 years. However, these standard pistons do not have a great deal of leeway as far as running them close or in detonation or running them with lean carb settings. In other words, they are reliable when used correctly, but are less than satisfactory when used incorrectly, but then, the responsibility for incorrect usage lies solely with the engine builder.

If you are building a high performance road engine using standard pistons, I would recommend an extra ½ thou of piston clearance. If the pistons are to be used for competition purposes, then go for an extra thousandths on piston skirt clearance.

On the left is a 1300 piston as used commonly in pre-Series 'A+' engines. On the right is its modern counterpart, a high compression piston from a 1275 Metro which is one of the best production pistons used in Leyland engines.

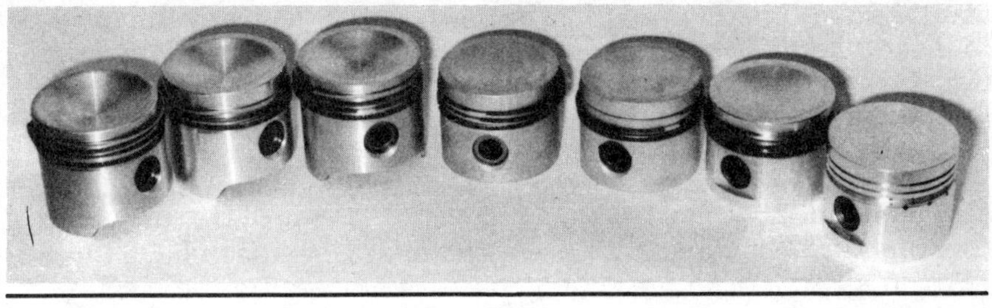

Here are most of the pistons for 1275 engines. Even the worst of the standard pistons is a pretty tough item, so in this line-up there are pistons which range from good to downright excellent. From left to right they are listed in the chart from top to bottom in Group 4. Not shown is the Longman item.

This line-up represents the sum total of pistons available for making a big bore engine out of a 1275. Most of this line-up is from Omega, but Cosworth and Hepolite are represented. The order from left to right corresponds with the chart section Group 5, top to bottom.

If for any reason a change is made between 1275 Metro pistons and 1275 'S' pistons, it's worthwhile noting that there is a difference in compression height. The Metro piston has nominally 30 thousandths more compression height and comes closer to the top of the block than the 'S' piston. This is how the Metro's extra CR is achieved.

If a high performance piston is needed for a big bore motor, for bore sizes no more than normal rebore sizes, many options are open. An obvious choice is the Hepolite Powermax piston, part number 20747/KR. This piston has narrow rings suitable for high rpm application and, considering it's a forged piston, it's pretty light at 363 grams.

For many years a minor controversy has existed regarding dish versus flat topped pistons in big bore motors. Firstly, a dished piston consistently shows a power advantage over a flat topped piston. This is probably due to one of two reasons or a combination of both. It would be difficult to differentiate between the two, but basically the more a Series 'A' cylinder head is skimmed to reduce the depth of the chamber, the better it breathes. I detailed this in the cylinder head chapter. Why this is so, is not exactly clear, but nevertheless it's a measurable phenomenon. Testing has shown that if the combustion chamber is concentrated more on the spark plug side, better combustion results. As a half-way house, a dish piston probably produces more effective combustion than a flat topped piston. As a result, a dished piston with a shallow chamber produces better hp than a flat topped piston with a deep chamber when each have the same compression ratio.

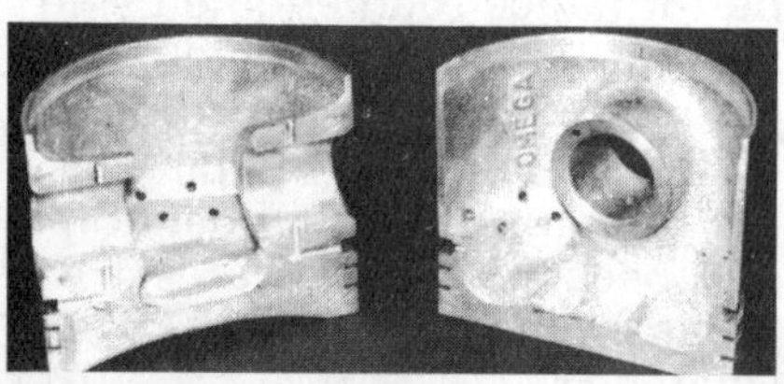

Here are two cutaway sections of a big bore Omega piston. One at 90° to the pin, the other through the pin. The sectioned half on the right is from a flat topped piston. On the left is a dish piston. Note how sturdy the piston is in the region of the crown and pins.

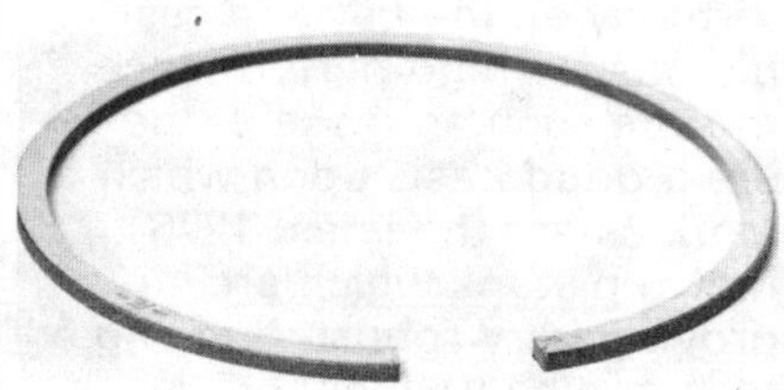

The rings used in the Series 'A' engine are very important. Make sure you choose top quality rings and that they are correctly gapped for installation. Finally, ensure the rings are put in the correct grooves and around the right way. To stop dry scuff, always lubricate the rings in their grooves adequately before assembling them.

Piston Crowns – Critical Configurations

I spent a lot of time in the camshaft section dealing with high lift cams and rapid valve opening rates. These two factors bring along some problems associated with piston crowns. Firstly, on all but the bigger engines, we find **high lift** cams can require valve cut-outs to be made in the pistons. In a 1275 engine this is a sure fire requirement for a typical race engine configuration. Obviously if compression ratios are kept low, the requirement for cut-outs in the pistons to clear the valves becomes less. For 1380s and 1400s valve cut-outs are really dependent on how much compression the engine's going to run and just how radical the cam is. Certainly, if valve lifts in the region of 0.480 inch-plus are used and compression ratios are around 12.5, some sort of valve cut-out in the piston is almost always going to be required. The point is, if valve cut-outs are not dealt with correctly, power will be lost. The most obvious way to do a valve cut-out is to use a circular cutter and just put a pocket in the top of the piston a little larger than each valve. Unfortunately, dyno testing at A.P.T. has shown that this type of pocket in the piston crown causes an adverse effect on the flame travel and reduces horsepower quite substantially. The problem appears to be the thin edge left at the outer rim of the cut-out towards the edge of the piston. During our own

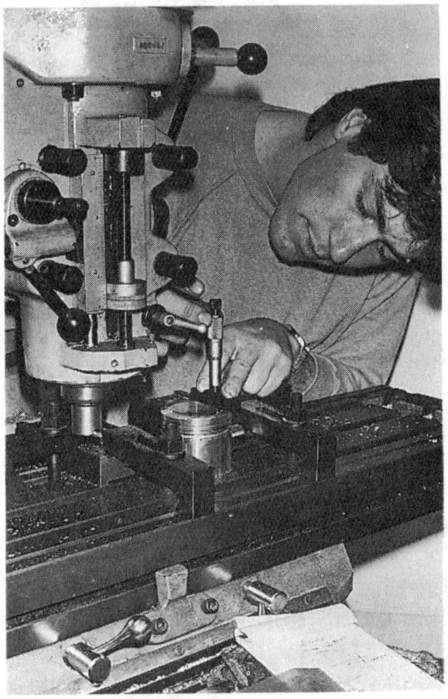

In an effort to find out what worked in the way of piston crown shapes for high lift cams, many configurations were machined up. The semi-circular cut-outs so often used . . .

. . . proved measurably inferior to a simple slot across the piston as shown in these two examples.

After much experimentation in this area, the best dish form I've found to date is the wedge dish shown here. Because it's being used in conjunction with a high lift cam, ie 0.500 inch lift, the piston also has a valve clearance slot machined across it.

To improve piston strength and heat retention in the combustion chamber, we have been having our pistons coated on the crown with a ceramic coating. This has great insulating properties and effectively cuts piston temperature, thus allowing the piston to operate in a lower temperature/higher strength region. Power increase in the region of 2% have been experienced due to lower heat rejection.

testing, this edge was progressively cut further and further away because it was suspected of being the problem source. Eventually it was found that when cut away to the point of being a parallel slot across the top of the piston, best results were achieved. So this is the way I recommend doing valve cut-outs for the Series 'A' engine. Just check the nearby photos and see why putting a parallel slot across the piston is the easiest way to produce a cut-out. Not only does it produce best results on the dyno, but also it is far easier than putting two circular-type pockets in the top of the piston.

The next aspect I want to deal with is the dish in the top of the piston. For large displacement engines with compression ratios suitable for the street, the options concerning the dish are relatively limited. Under such circumstances, a dish as much as 11cc is required in the top of the piston. A dish this size is almost certainly going to be a centrally located deep dish. This is because it is simply too big to put anywhere else but in the middle of the piston. Such a dish extends right out to as close to the edge as possible, bearing in mind the thermo-mechanical limitations imposed by thin sections at the edge of a piston. If dishes any smaller than about 7cc are contemplated, then there are avenues for increased horsepower by having a dish of a rather unconventional form. My own work in this area has indicated that a superior type of dish for the Series 'A' piston is of an offset wedge form. This design, now known as the 'V' wedge, proves to produce more horsepower than either the simple offset dish or the centrally located dish. It proved especially successful when used with the piston slots for valve cut-outs. After an extensive number of tests with different piston configurations, it was found that pistons with an offset dish and slots for valve clearance produced a minimum improvement of 2% in virtually every aspect. Two per cent more horsepower, torque 2% better, brake, specific fuel and air consumption 2% better, etc.

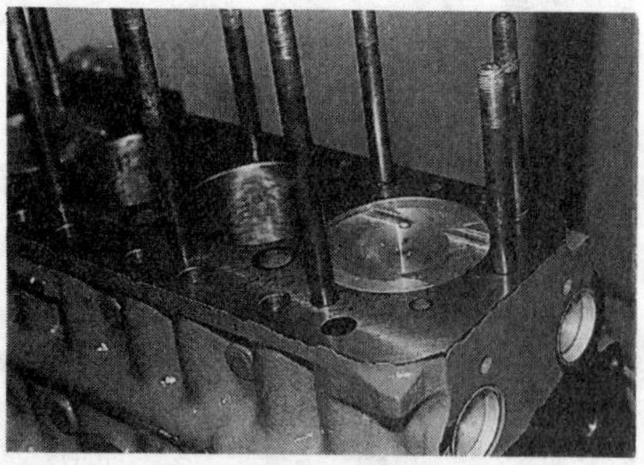

Whatever piston configuration you use, it's important to remember that the position of the piston at T.D.C. has an influence on horsepower. Pistons in standard engines sometimes fall short of the top of the block by as much as 0.060 inch. This can cost the engine 2 or 3 hp. Whatever piston you use, machine the block so that the piston comes flush to the block deck. So long as clearances throughout the engine are held to tight tolerances, it is possible, once you know the combination of the engine you're working with, to build as much as 0.015 inch 'pop-up' out of the block before the pistons hit the cylinder head. I would only recommend going to a positive deck height for those engine builders who work consistently with a known combination.

In fact, a minimum of 2% right across the board. Now 2% may not sound much, but it is 2% better than anything else that was available. However, the difference between having valve cut-outs wrong and a dish that is not conductive to power compared with one that was right, can prove substantial. Getting the dish and slots right with the wedged dish and parallel slots allowed us to realise 128 hp from the test engine. A conventional dish piston with semi-circular valve cut-outs (not the worst by any means) only made 122 hp, the difference being the circular valve cut-outs cost horsepower, whereas the parallel slots across the piston did not.

Blocks

Regarding the 803cc block, unless original restoration is the goal, don't use one. This being the case, let's move onto the next size block up, which is the 850cc Mini block.

If you intend building an 850cc Mini engine and it is to utilize the standard stroke, any block will serve your purpose. If you intend putting in a high performance cam install cam bearings in the block, as only the front cam bearing has a proper tri-metal bearing. At the other locations cam journals are run directly in the cast iron. Normally this is no great problem, as long as high revs or excessive spring pressures are not used, but once you start building a high performance engine problems do arise.

If the intention is to fit a 998 or 1100 crank into an 850 block, remember the main bearing width on 850s is wider than on 998 or 1100 blocks, so the centre main bearing housing must be machined narrower. As far as 948 inline engines go, these are much of a muchness. During the early days of Formula Junior racing there were some special strengthened blocks produced, some of these having special fitting and drive take-offs so they could be dry sumped, but these blocks are extremely scarce. Of course, the outgrowth of these blocks was the big bore 'S' block, but it really is too much of an outgrowth to say that it's interchangeable with a 948 block.

The transverse 998 and 1100 blocks are essentially the same, at least if we consider the early ones and just look at the machining. Any part that fits into a 998 block will also fit into an 1100 and vice versa. However, it's worth noting that early 1100s are (as relatively reliable rumour has it) nothing more than bored out 850 blocks. This may not be important if you are simply rebuilding the engine, but it can be if you intend to overbore the block. Also there are thick flange and thin flange blocks, and this really applies across the board on transverse engines. The flange referred to is the surface that mates with the gearbox at the base of the block. On long stroke, high revving engines, the bottom of the block tends to flex or conversely the clamping loads between the block and the gearbox are insufficient. Whatever the case may be, the thin flange blocks tend to leak at this point more easily than the thick flange blocks. If you are building a high rpm engine, it's best to base it on a thick flange block to cut down the likelihood of oil leaks. Even with thick flange blocks, drill out all the 1/4 UNF screw thread holes in the block that are used to bolt the block to the gearbox casing, and tap them to 5/16 inch UNF. You should then substitute standard bolts for 5/16 inch socket-headed bolts.

Of the small bore, Series 'A' range, probably the optimum in blocks is the 'A+' in small bore configuration as used in 998 Metros, etc. It can easily be recognised by the stiffening ribs on the outside. These follow the same pattern as the stiffening ribs in the big bore, 1275 blocks.

Now to Cooper 'S' blocks. Most people are aware that these are the first generation of big bore blocks. They are readily identifiable by having the pressed steel tappet chest covers on the back (exhaust) side. This is in contrast to the later 1300 blocks which had the solid tappet chest. The point made about thick and thin flange blocks applies equally to the 'S' block. In fact, it's more important on a 1275 'S' if it's for performance use, to use a thick flange block, as the thin flange ones are far more prone to distortion.

If you intend you engine to produce high power outputs and it is to be used in a transverse installation, it's a good idea to drill out the 1/4 inch fine thread holes that secure the block to the transmission and tap them 5/16 inch. Be sure to tap each hole square.

Another point worth noting here is that 970 and 1071 blocks are shorter than 1275 ones. However, all blocks were machined from the same type of casting. The basic deck casting thickness on a 1275 block is nominally 5/8 inch. This substantial deck thickness was deliberately produced so that all 'S' blocks could be machined from the same casting. If a 970 block is needed, and they are getting scarce, it's a simple matter to machine down a 1275 'S' block to produce the identical end product the factory would have produced.

All 'S' blocks have good bottom end reliability. The blocks themselves are stout, and the main bearings are held in position by high tensile studs, rather than the bolts used on more mundane 1275 blocks. As for using an 'S' block as a basis for a big bore engine, beware. Failure to bore successfully over 2.9 inch is common and failures do occur at 73.5 mm bore (that's 2.894 inch). Also it seems that some of the early 1275 Sprite blocks don't take kindly to boring to too large a size.

Having got onto the Sprite engine, it's worth pointing out that again you need to think in terms of a thick flange block as being better for high performance work. In fact, the best 1275 block to use is the 'A+' block as used in the Ital (last of the Marinas).

Returning to transverse engines, in the late '60s the non-'S' 1275 engine was introduced which looked initially to be a good substitute for an 'S' engine. From the point of view of the block the main advantage was that it had a solid tappet chest, i.e no removable tappet chest cover; this was important for block stiffness. The bad point was that there was a large casting hole above the centre main bearing and on numerous occasions this hole has proved to be the starting point of fatigue cracks. I had experience with three of these blocks breaking the centre main housing from the rest of the block, taking with it the crank and most of the rest of the engine and transmission. Strangely, I have known race engines built on these blocks which have been entirely reliable, but to be absolutely safe give this block a wide berth. The tendency to crack can be significantly reduced by grinding the area around the centre main bearing hole and polishing as best as you can. This will help prevent a fatigue crack.

Around 1976-1977 the casting hole was deleted from 1300 blocks. These later blocks had more rigidity than the 'S' blocks because of the solid tappet chest and the increased strength in the main bearing area. They can be used as race blocks and the bolt retention of the main caps rather than the studs of the 'S', causes no noticeable reliability problem. The degree to which these blocks can safely be bored out is difficult to quantify as success rates differ between engine builders. The variation in success ranges widely from 50-90%, the mean falling somewhere between the two. Obviously boring to 74mm is likely to have a greater chance of success than going to 74.7mm. As much as anything this is a subject you should discuss first with your machinist, after all he's the one who is going to be doing the boring.

1300 blocks are quite common, whereas 1275 'S' blocks aren't. This being the case, if you own a 970 'S' and you haven't been able to find a 1275 block, it can be very tempting to simply skim the top of a 1300 block. Be warned, the deck on a 1300 block is only about 450 thousandths thick, and so after 1/4 inch has been skimmed off the deck is only about 200 thousandths thick, and that's not really enough to prevent head gasket problems. In other words, using a 1300 block to put together a 970 is not impossible but it is risky.

The latest blocks are, of course, the Series 'A+' blocks and these, as already mentioned, are available as inline blocks and transverse blocks. These are the strongest Series 'A' blocks made to date, and they will stand overboring to 74.7mm at least as successfully as the previous 1300 blocks. However, if you contemplate putting stroked cranks into these blocks, beware; the longer strokes, i.e. 86mm and up, cause clearance problems and when grinding to find clearance one often finds fresh air. When building a 91mm stroke engine, use the block type immediately prior to the Series 'A+' block.

Another point to consider if swopping from a pre-'A+' block to an 'A' block is that the distributor mounting and drive are different. A change of block will require a change of distributor assembly. A point worth noting here is that it's

The ribs arrowed in this shot distinguish the 'A+' block from the earlier blocks. Note also the thick flange which is seen here at the bottom of the picture.

virtually impossible to drop the distributor driveshaft into the engine while installing it on an 'A+' block, whereas it is always a hazard on blocks prior to the 'A+'. Of course if you drop the distributor driveshaft into the sump of an inline engine, it's no big deal, but it's very annoying if you drop it into the sump of a transverse engine and the engine is in a fully assembled state.

Probably the best high performance block, at least if stroke lengths are kept within the confines of what the block will take, is the latest turbo Metro block. This has a different main bearing cap which uses a non-grooved bottom half to the main bearing. The load capability of this bearing is said, by Leyland engineers, to be about 80% higher than the grooved main bearing.

If you find yourself with a 1982 transverse Series 'A+' block, the chances are it's linered because of boring problems at the factory. These blocks are okay at standard and ordinary +20 or +40 overbores, and they are okay for moderate performance applications, but I don't advise their use in all-out race engines.

Another point to check: on many Metro blocks, a mis-alignment exists of the block main bearing oil hole to the main bearing shell. On many blocks these holes were drilled outside positional tolerances at the factory. This is not a problem with a standard engine, but it could easily cause catastrophic failure in a high performance engine. A simple 'fix' is to chamfer the hole in the block so it aligns with the hole in the bearing shell.

Shown here is the main bearing cap of a turbo Metro block. As you can see, the bearing retention tang is in the centre of the cap rather than to the side. This is because . . .

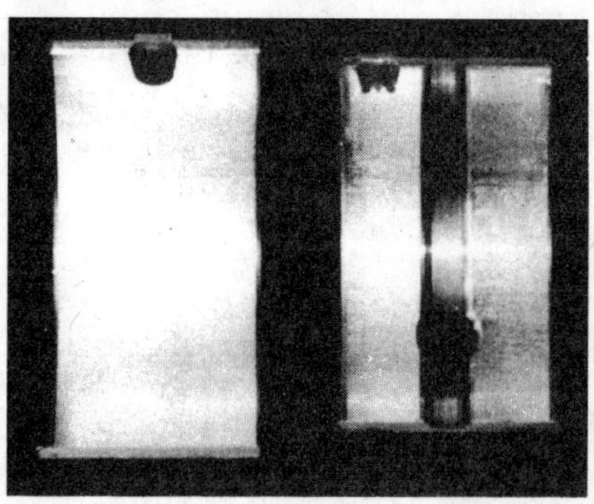

. . . it takes a bearing combination having a slotted shell in the top of the block and a plain one in the bottom. If these shells were reversed, no oil would be fed to the bearing, hence the 'one way only' assembly design.

Block Preparation

Just how the block is prepared is important. It should go without saying that cleanliness ranks high on the priority list. If you are like me, and can rarely afford to have a new block upon which to build your high performance or race engine. it is doubly important to be meticulous about how clean the block is. Ideally the block should be put into a boil-out tank, but if you have to resort to more normal cleaning techniques, then the best plan is to remove all the oil gallery plugs (and I mean all) and then drill and tap as necessary the end of each gallery to accept a socket screw plug. However, **leave all the plugs out until all the machining on the block is done. If the block has to be decked, bored and honed, there will be a considerable amount of grit accumulated.** Also you may be working on an 850, and new cam bearings may need to be put in. There are 1001 machining operations, all of which produce swarf. Because of this the gallery cleaning and the installation of oil plugs should be one of the last jobs done on the block.

When using an old block run a tap down each thread, especially the threads in the deck of the block, and make sure that all the threads are clear. Also, it is important to chamfer the block threads. The chamfer should be about 1/16 inch larger in diameter than the diameter of the thread. If this isn't done, then the top threads can pull up. Normally an operation like this is done on the block at the factory, but if the block is decked it doesn't take much decking to remove that chamfer.

Depending upon the type of block you are using, and the combination of the various available oil pump/cam drives, it may be necessary to drill and tap extra holes to secure the pump properly.

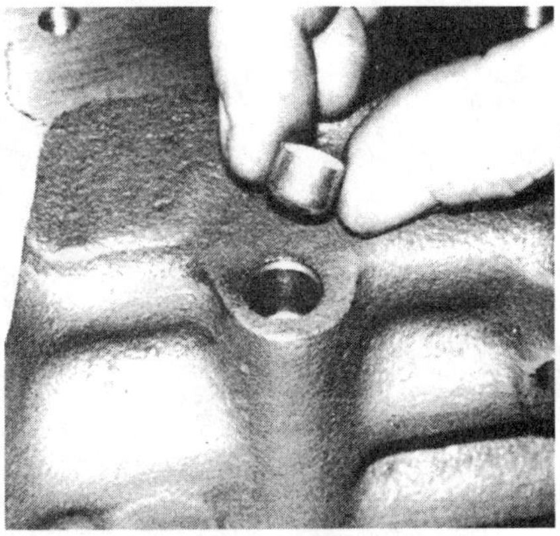

Never assume all the oilways in your block are clean, even if it's a new block. Plugs to replug the galleries, like the ones shown here, are available from your dealer. If you want to pull them from the block, the easiest way . . .

. . . is to tap the plug with a self-tapper or course threaded bolt and pull it out, by placing the sleeve or some large nuts over the hole and tightening the bolt down on the nuts.

Of course the other job I mentioned that's well worthwhile doing on transverse engines, is increasing the size of the flange bolt holes from 1/4 inch UNF to 5/16 inch UNF.

Centre Main Bearings

On all blocks with the possible exception of 'S' blocks, it's a good idea to consider reinforcing the centre main bearing. It used to be normal practice on 850 blocks, that is 850 blocks that were to be used for racing, to replace the standard centre main with a steel centre main bearing cap. Of course this would necessitate having the cap line bored. This job needs skill on the part of the machinist because there is no spare metal in the block half of the housing. As a result, it is necessary for the machinist to align the boring tool so it follows exactly the position of the block half of the original centre main bearing bore so the surface of the new cap is in an

If you want to be able to clean out the oil galleries easily each time the engine is rebuilt, then it's a good idea to tap them and fit taper Allen plugs. The same basic procedure can be used . . .

. . . here, first drill, then tap, then . . .

. . . install a taper Allen plug preferably with a little sealer on it. The emphasis being on 'little'.

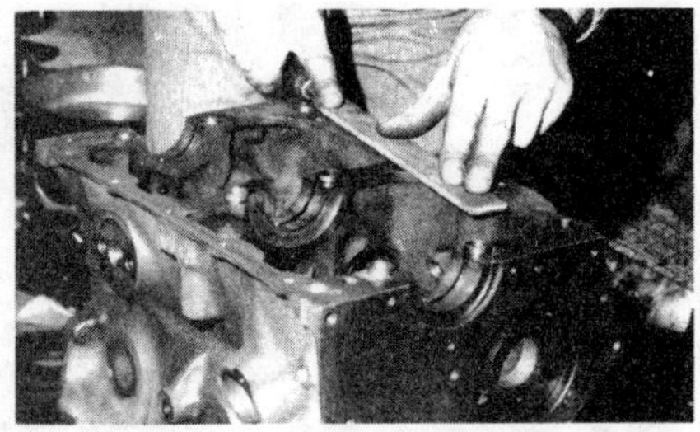

If there are any dings, dents or nicks in the bottom face of the block, take a flat file and remove them. This is specially important in the area upon which the main bearing caps seat.

If a high CR is not a limiting factor, then flushing the pistons with the deck of the block aids power by helping suppress detonation. Note on this block that all the holes in the block face are neatly chamfered off. This is especially important on the stud holes as it prevents the top thread pulling and causing a gasket problem.

There is no such thing as a block which is too clean. After all the machining is done, go through the entire engine with a stiff brush and paraffin (kerosene) to make sure all foreign bodies have been removed.

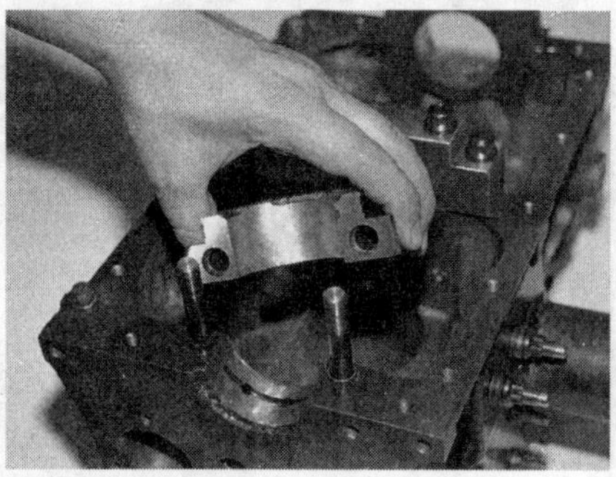

If you are building an all out motor it is worth homing the mains to assure their accuracy. To do this it is necessary to grind the caps a couple of thousandths, not that this block is equipped with ARP studs.

identical position to that of the cap it replaced. A good half-way house on high rpm 850s is to use a centre main bearing strap. To do this, the cap is machined flat so that a suitably drilled steel bar can, with the aid of longer bolts, be clamped onto the centre main bearing cap. This technique is relatively standard practice on all highly tuned engines within the group with the exception of the 'S' range.

If you are considering building a highly tuned 948 engine, its reliability can also benefit from the use of a steel

Whenever really high power outputs are envisaged, a centre main bearing strap should be considered. This is especially relevant on almost all blocks except perhaps 'S' blocks and the latest 1275 'A+' Series blocks. For all practical purposes, these may not need additional support in the centre main bearing cap until power outputs over 120hp are achieved.

For high output or endurance race motors a 4-bolt centre main is worth considering. This ties in the outer walls of the block to the centre thus allowing more load sharing. To clear the longer studs used on all the mains cap the . . .

centre main bearing cap. Retention of the bottom end integrity can be measurably influenced by the quality and strength of the fasteners used. Since the late '80s David Anton and I have been working with ARP in the States to develop a line of super high tensile fasteners with maximized fatigue life specifically for the 'A' Series motors. All of these fasteners are made from aircraft quality

. . . gearbox on Minis will need some clearancing as seen here.

These $^{5}/_{16}$ths ARP 12-point high tensile bolts should be used to secure the gearbox to the block in a superior manner. Tightening 12-point bolts is far easier than tightening the often used hex socket Allen bolts.

certified material, the same as used in this company's F1 fasteners. The sidebar **Fig. 13.1a** deals not only with main bearing cap fasteners but also with the critical problem of stock rod bolt failures so be sure to read this section. As far as securing the main bearing caps are concerned studs are to be preferred over bolts. However it is important to install the studs correctly and the first step toward that is to understand they should under no circumstances be torqued into the block. Nor, for that matter should they be Loctited in unless you install a cap and torque them up while the Loctite sets up. If this procedure is not done then the studs will loose some of the pre-load as the Loctite settles. For most practical purposes the simple installation of studs is more than adequate and this involves no more than installing them finger tight in the block.

Block Machining Accuracy

When the time comes to assemble your engine, do not assume that all the machining, whether done by the factory or your local machine shop, has been done accurately. Always check. Most of the checks can be done by simply installing parts to see that they do not bind. For instance, it's a simple enough job to install the crank, torque up all the bearings successively and check that the crank spins freely between the tightening of each bearing cap. With typical 30 weight engine oil on the bearings, you should be able to take hold of the crankshaft between two fingers and spin it freely. If the crank doesn't turn freely then you have one of two problems: either the crank is bent or the block main bearings are not aligned. Neither problem is insurmountable, in fact the factory will let cranks go out from inspection with a bend as much as 3 thousandths. It's claimed by the factory that a 3 thousandths bend in the crank does not make any noticeable difference to power output, so that's why this limit is set. However, for a high performance engine we are not looking for noticeable power; we are trying to find any measurable power, and if a 3 thou bend causes the loss of ½ a hp, then if you are really determined to produce a good motor, you should worry about any bend or inaccuracy.

Another aspect to consider is bore sizing, but I'll deal with that when I talk about piston clearances. To ease assembly the tops of the bores should be chamfered. If they are not, it can be difficult to get the rings to go into the bores, and it's possible to damage them during assembly if a small chamfer is not present. On the other hand, it's important not to have too large a chamfer, especially on big bore engines, as this can cause a head gasket problem. A 60 or 75° chamfer on the top of the bore about 20 thousandths larger than the bore diameter, will considerably ease assembly. Finally, take care with the bore finish. A 20 to 30 micro-inch finish seems to work well with either ductile cast iron rings or stainless or chrome plated rings. As a warning, don't try for too fine a finish for use with the stainless or chrome plated rings as the rings will take a long time to bed in. In fact they may never bed in.

Although a block must be honed after it's bored, the more accurately it is bored in the first place, the greater the likelihood of achieving an accurate end product. Boring too close to the finished size does not give the hone time enough to correct any errors. At least 2 thousandths of material should be left in after boring.

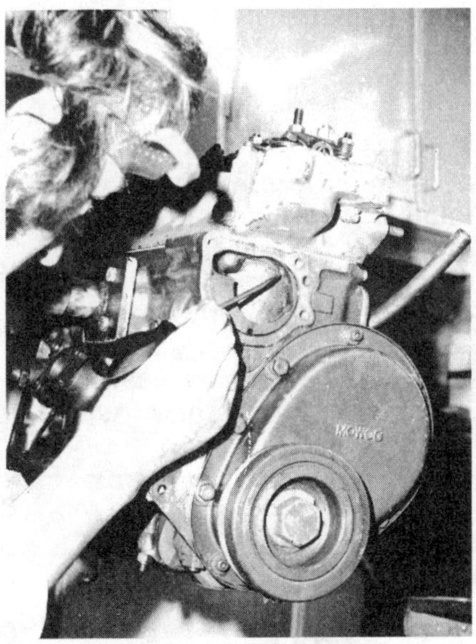

Though a rare occurrence in a block and head which have been skimmed a lot, the bolt may project through the hole indicated here by a screwdriver. If it does, it can foul the water pump impellor blades. In fact it will stop them turning, so check this part before installing the water pump and shorten the bolt as appropriate.

Blocks, Cranks, Rods & Pistons Part 2

Connecting Rods

All Series 'A' connecting rods, with the exception of the 970 'S' rod, are 5.75 inch between centres. Because of this, in any given type of block, i.e. small bore or big bore, an increase in stroke length is made up for by a reduction in piston compression height. If it weren't, of course, the piston would jut out of the top of the block.

Between small and big bore engines, major design differences exist between rods. The small bore big ends have a diagonal split-line; arranged this way so that the rod will pass down the bore to enable the rod to be assembled onto the crank journal. The little end and beam of a small bore rod is of relatively light construction, but the small bore rods have excessive material around the big ends, making them, for what they are, a very heavy rod. Fortunately, they are easily and substantially lightened. As a general rule all Series 'A' connecting rods can be considered grossly overweight in standard form.

In the foreground is an example of an 850 Mini rod. For what it accomplishes, it weighs a ton. The one in the background has been lightened by about 25%. A similar technique can be applied to the fully floating rods found in the 1000 and 1098 engines.

The original 803, 948, 850 and 997 Cooper engines employ a clamped little end type of rod. The piston pin has a central notch in its length in which the shank of the little end bolt locates, thus positioning the pin and rod within the piston. In the early days of racing and high performance use of the Series 'A' engine, this rod developed a reputation for having little end problems. Whether this was due to bad assembly practices, or the fact that the design is just not compatible with 8000 rpm, is not certain. Personally, I have had no problems with this type of little end but nonetheless, I would advise changing to the later, fully floating, rod as it is a better design. This rod has less weight in the little end and the pin has no securing bolt groove in it so this stress raising point is removed.

To use the later type of rod in an early engine, it is necessary to either use a piston with circlip type pin retention or modify an existing piston to take circlips. Another alternative is to use Teflon buttons to stop the pin ends running against the bore and gouging grooves in it.

Since its introduction in the 998/1100 engine, the fully floating pin type rod for small bore motors has remained relatively unchanged throughout production. Unchanged, that is, except for a 1982 factory version of this rod. In an effort to cut weight, the big end was trepanned from each side to form a rod that looked like it had had a mild competition lightening job. Unfortunately, the trepanning operation left some sharp edges on the cap. Bending loads in the cap causes these edges to stretch. The very thin edge proves to be prone to fracturing, and once a crack starts at the edge, it very quickly propogates through to the rest of the rod. The result is a major rod failure. Unfortunately most of the production of 1982 was this type of rod, so **if you tear an engine down and find the trepanned type of rods in it, remove them from the engine and throw them away, as you can never be sure how much of the fatigue life is left in them.**

From about the end of 1982 or the beginning of 1983, the rod design reverted to the original style, and if anything, these later rods have shown an improvement in their capability because manufacturing quality has been increased.

I have not personally weighed one of the trepanned rods so I can't say how much weight was saved, but just for the record, a typical 998/1100 rod weighs around 645 grams and the 850/948 etc. rod with the clamped little end weighs 15 to 20 grams heavier. As you will see when we look at other rods, in spite of the lightweight construction of the beam and little end itself, overall this rod, in standard form, is no featherweight.

Big Bore Rods

Basically there are five types of connecting rods for big bore motors. Firstly, there is the Cooper 'S' rod (AEG-521). The rod is made of EN24V and it's very tough and equally heavy, though not the heaviest in the range. As far as strength is concerned, the Cooper 'S' rod is the best Series 'A' rod available, but running a close second is the 1275 Midget/Sprite rod (AEG-625). This rod is made of slightly less complex material, EN21, and can be used in every situation in which a Cooper 'S' rod can be used. This is because the Midget rod as well as the Cooper 'S' is over engineered. So in spite of the fact the Midget rod is made of a lesser grade material, it's still capable of taking everything that can be meted out to it in the way of rpm and hp.

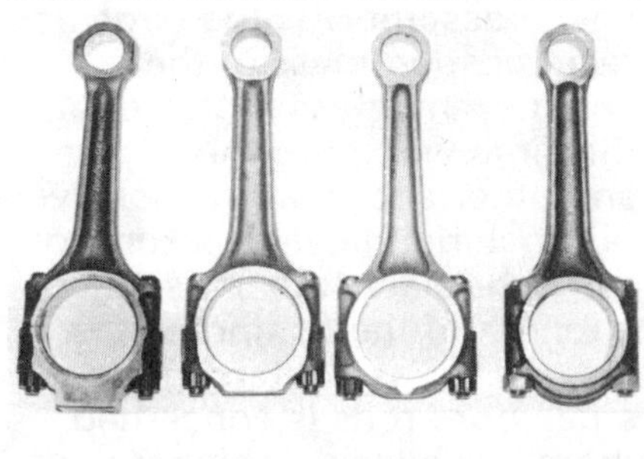

Big bore rods, left to right. First is the 1275/1300 rod as used in 1275 GTs and all '1300' saloons including the Marina. Second from left is definitely the 'good-guy' rod. This type of rod was used in the Innocenti Mini. It's one of the lighter standard rods and it's tough. Second from right is the latest Series 'A+' rod. It's the lightest of all but the material spec is down. In the far right is a Cooper 'S' rod. This is made of a high grade material and is almost bullet-proof. The 1275 rod is made in virtually the same forging but to a slightly lesser material spec. The nuts holding the cap onto the rod in this photo, are only shipping nuts. The nuts used at the time of engine build are, of course, high tensile type nuts.

Both Cooper 'S' and Sprite/Midget rods with the 1⅝ diameter big end, weigh about 690 grams. Having dealt with the Sprite/Midget/Cooper 'S' type rod, I have now covered all the small journal rods. All other rods in the Series 'A' range are the big journal type rods, that is, rods which utilize a 1¾ inch big end. It's worthwhile pointing out here that the big end size was increased on cranks in 1275 engines because it was found to be cheaper to make a bigger journal crank of a lesser grade material than to continue with the very expensive EN40B cranks that were used in 'S' type motors. Along with big journal cranks came the obvious rod change. These bigger journal rods were also made of a lesser grade material. **The best of the big journal rods are the 1300 GT/Innocenti rods produced from EN19. They are easily recognizable by the fact they have a small strengthening web between the two beams,** as shown in the **accompanying photo**. This is a good rod in spite of material downgrading. The balance pads are not of excessive size and weight is typically around 675 to 680 grams. These rods are relatively common and exceptionally good for a high performance motor.

Next on the scale of rod desirability is the standard 1300 rod, that is the rod used prior to

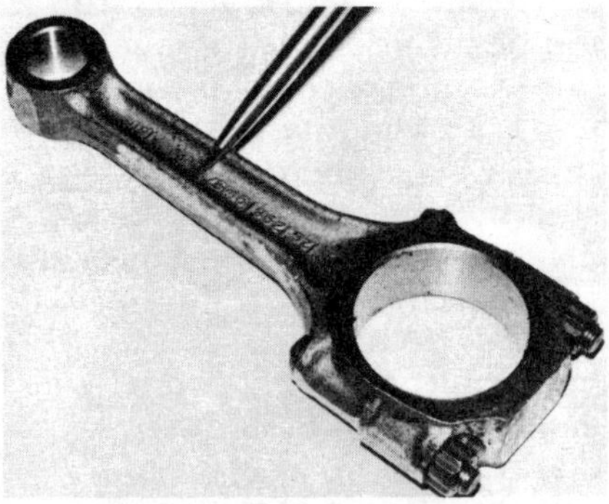

The Innocenti rod shown here, can be recognized by this miniature beam half-way down the shank, tying in the two flank beams. It doesn't actually need to be there, but it is a sure way to identify this rod.

the 'A+' engine. This holds the record for being the heaviest Series 'A' rod ever made, typically being 740 to 750 grams. They are recognizable by the massive balance weights they have on the big end and the little end. Originally these rods were also made of EN19 material, but as time went by (I don't know exactly when) this material was changed to EN16. I would guess that rods prior to about 1973 were made of EN19 and most of those after were EN16. This type of rod generally went out of production in 1982. The next rod to be produced was the 'A+' rod, and here **a genuine effort was made at the factory to design a rod which didn't carry excessive weight as it had been realized that the big bore Series 'A' rod was over-designed.** The result of this redesigning was a rod which weighed 650 grams. Along with this redesigning came a further reduction in material quality, and EN15 was used. However, though the material quality of the rod was reduced, these rods are still, for most practical purposes, over-engineered. They are capable of taking greater loads and stresses than the standard engine puts out, and there is still no reason why even the worst of the big bore rods won't live perfectly adequately in engines of 100 bhp with a 7500 rpm potential. The reason I am stressing the point of weights and materials is that if you are building a super exotic engine, it would pay to select the best of the type of rods for the particular application you have in mind.

Connecting Rod Lightening

Increased acceleration is the result of reducing by as much as possible the weight of the reciprocating components in the engine. The connecting rod and piston assembly is the largest reciprocating mass in the engine. Any mass reduction in this area will make the engine smoother and more responsive but material must be taken from the right place, otherwise the strength of the component is lost. As far as lightening the small bore rods is concerned, there is a considerable amount of scope and the weight reduction possible is prodigious. In fact, these rods can be lightened by 25%. The techniques for achieving this can be understood by studying the **accompanying photos**. The example shown in the photos weighs around 480 to 490 grams. Note that these rods use a cap-headed bolt rather than the conventional big end bolt. Using this type of high tensile big end bolt means tightening to 40 ft lb instead of the normal 35 to 38. No lock washer is needed with this system.

Although not obvious from the photos, it is very important when lightening rods to make sure there are no sharp edges anywhere and that when the beams are ground, they are ground up and down the length of the beam, not across the width. Lateral grinding marks would invite dangerous stress points. The key to successfully lightening rods and retaining adequate fatigue life is to make sure there are no sharp edges from which fatigue fractures can start. After a rod has been lightened it should be checked for straightness, because very often removing the skin of the rod will cause the rod to move slightly. If you intend to have the rods shot peened or tuftrided, then the straightness checking should be the last operation. While on the subject of shot peening or tuftriding, either of these processes are to be recommended if the rods are to be used in a race engine because they increase the fatigue life of the rod substantially. Rods prepared in the manner shown in the **accompanying photos** are good for up to about 8500 rpm and at that rev limit you can expect them to last for a typical season's circuit racing, i.e. 350

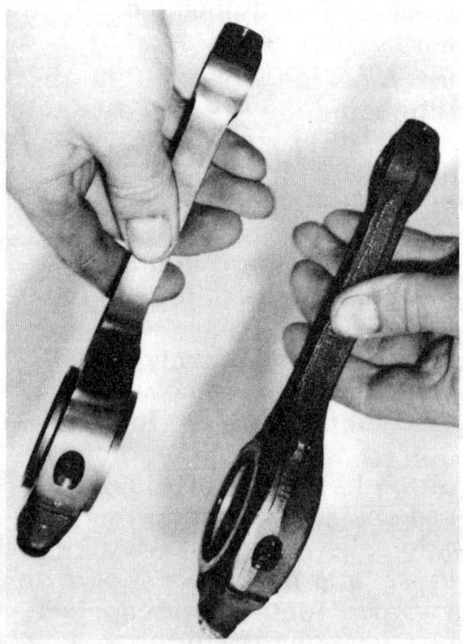

The big end area of a small bore rod carries a lot of surplus metal. Most of the weight reduction of this rod takes place in this area.

When lightening the rods, it is important that the edge indicated here with a pen has a reasonably generous radius on it, otherwise a fatigue crack will start in this area. The radius needs to be at least 30 thousandths and polished with the polishing marks along the length of the rib, not across it.

to 500 miles. If you are prepared to change the rods on a more frequent basis, then the rods are safe to about 9000 rpm.

The floating pin rod for small bore engines, has only one real fault, that being the piston pin is too small. Although I have not tried it myself, it would seem a good idea to remove the bush in the little end of the piston, hone the bore out so it has 2 or 3 tenths clearance on a ¾ inch diameter pin, and then machine out a piston with suitably strong pin bosses to take the ¾ inch diameter pin. Increasing the pin size like this would give a more rigid pin and a reduction in weight because reduced wall thickness can more than compensate for increased diameter. For a given pin stiffness, the larger the pin diameter the lighter it can be made. Incidentally, using modern oils, a steel-on-steel bearing as would be the case for the 'big pin' modification is entirely satisfactory for as much as 500 race miles. It is the pin rather than the rod that wears, so replacing the pin restores the clearance. I mention connecting rod lightness and pin rigidity mainly in relation to 998 Mini Miglia engines. Because racing in this category is so close, the engines are approaching the point of being overstressed. On high rpm 998 units, the standard con rod, piston and pin assembly can be a problem. If the circlips come out of this type of piston, and they sometimes do, the pin will very quickly gouge the bore. This can be prevented either by resorting to a press-fit pin or putting Teflon buttons in the end of the pin to prevent bore contact. For many years, the press-fit pin was usual. This technique normally meant the utilization of a Cooper 'S' rod.

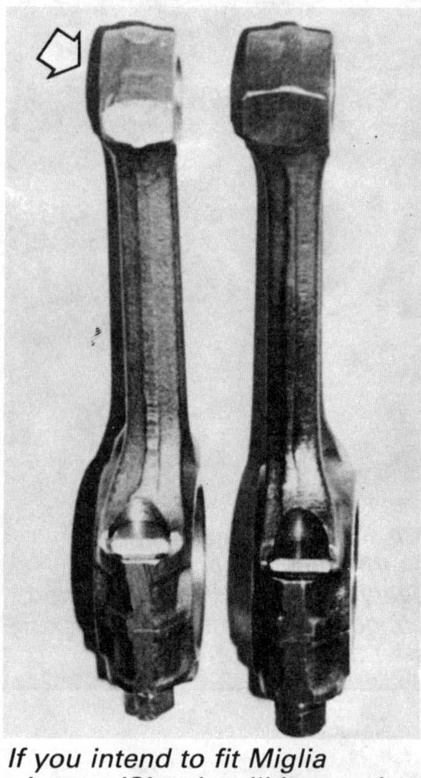

If you intend to fit Miglia pistons, 'S' rods will be required in your 998 motor. They need to be modified by cutting down the offset. This involves machining about 90 thousandths off the face arrowed on the rod on the left which has had this operation done.

So that an 'S' rod can be installed into a 998 engine, it is necessary to machine 90 thousandths off the side of the little end that is opposite the side of the largest offset of the big end. When this type of connecting rod is used, it's necessary to use the special forged Miglia pistons. These pistons were sold by the now-defunct Special Tuning department. This type of piston is still around, but they are rare and getting rarer. Although a Miglia piston shows more hp on the dyno, typically about 1 to 1½ more, the penalty paid is a substantial increase in reciprocating weight as it is difficult to reduce the weight of the 'S' rod much below about 605 grams. This, plus the heavier Miglia piston, increases the reciprocating mass to about 150 grams more than if a modified 998 setup is used. Personally, I prefer to use a lightweight 998 setup because, in addition to the weight advantage, it's much cheaper.

As far as lightening big bore rods is concerned, again the photos should reveal what is necessary. The particular example shown in the photos is a 1300 rod. Of course, this has the greatest scope for lightening, simply because it's the heaviest to start with.

One of the first lightening operations on the heavyweight 1275/1300 rod is to machine out the centre of the balance pad on the big end cap.

Quite a bit of lightening on the big end can be performed simply on a 1275 rod by facing off material in the lathe. However, this represents the bulk of the machining and . . .

Nevertheless the same techniques can be applied more or less across the board, to the other rods. Because of its initial lightness care should be taken not to over-lighten the 1275 'A+' rod. If you do intend to

. . . the rest of the work has to be done by hand, as seen here.

Here's a 1300 rod which has had all the simple lightening procedures done to it. Its weight has dropped from 750 grams to 660. The amount of time to do this lightening is relatively small. However, to go from this down to about 600 grams, which is about as light as you can safely make these rods, requires hours and hours of work on each rod. Still, if you have the time . . .

lighten this rod be sure that shot peening and/or tuftriding is carried out.

When lightening big bore rods, you will probably notice a mis-match at the split line between the cap and the main part of the rod. When looking at the rod, it appears that the machining of the side outboard of the bolts has gone slightly askew. The radius in this area is not symmetrical with the rest of the rod, but is so machined that the rod can pass through the bore. It's recommended, to remove the cap overhang, as it contributes nothing towards rod strength, and its removal will save about 5 grams.

It is important not to leave a sharp edge at this point. Everything needs to be rounded off, otherwise a fatigue crack can start at this point.

Rod Bolts

When originally introduced, the 1275 Midget and Cooper 'S' small journal rods were equipped with $^3/_8$th rod bolts that, when correctly torqued, gave close to zero problems at the rpm and power levels of the day. That was over 30 years ago. Since then many substandard bolts have been produced and sold as replacements of equal quality. The bottom line is that some are but many are not. With that in mind let us go through rod bolt basics so you can avoid what will always be a catastrophic failure in this area.

First if you are performing

'A+' rods employ a higher specification rod bolt and nut than earlier rods. With the improved quality control on the spot facing area of the rods, these flanged nuts with their respective bolts are showing excellent fatigue life.

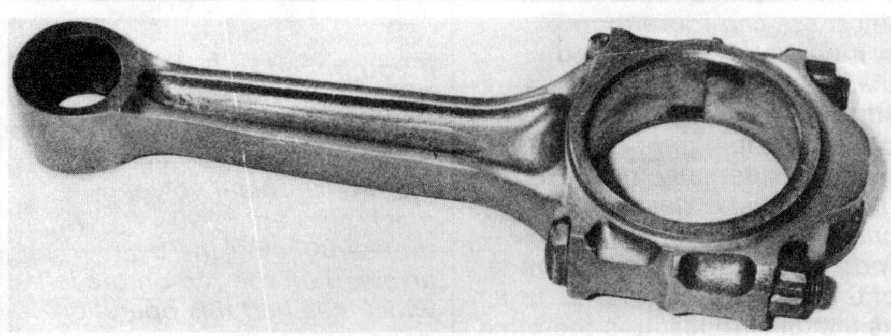

Here's an 850 Mini crank, prepared for Mini-X racing. The key word here is 'lightness' to give the best possible throttle response out of the turns.

only a mild performance build the stock bolts may be OK. To check these out use a five times power or more magnifying glass to check out the threads. If there is any sign of bolt distress, thread distortion or the rods have been previously used for racing, replace the bolts. They are relatively cheap compared to a new engine and transmission. If the bolt appears to be OK it can be re-used but always use new nuts. The ARP 12-point nuts should be your number one choice.

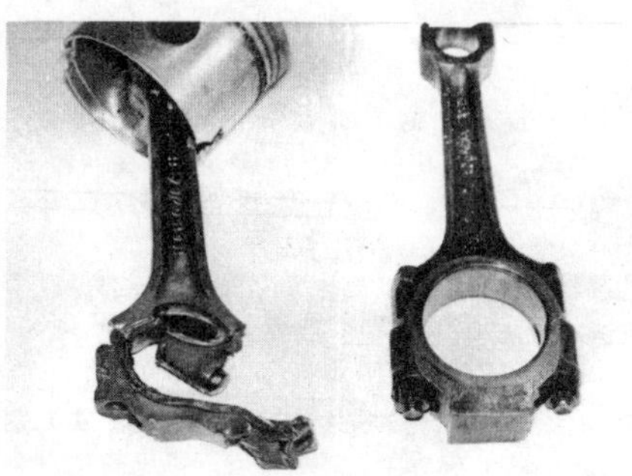

A photo to bring to your attention the importance of good rod bolts. Not only did the engine concerned need a new rod, but also a new block, crank, cam and gearbox, so for the price of checking out and replacing the rod bolts as necessary, the owner of this engine had to suffer a very expensive rebuild.

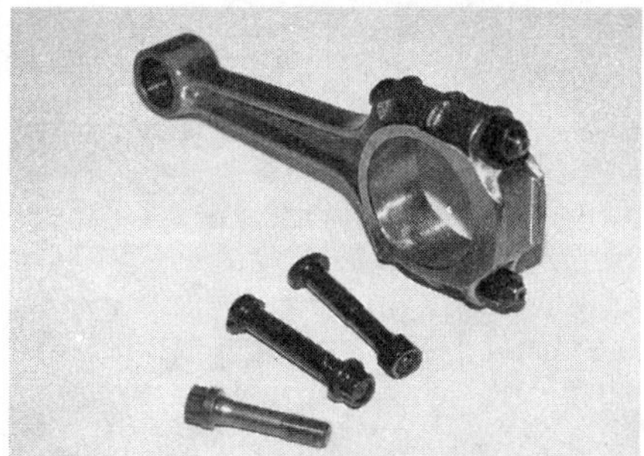

If rod bolts are not up to the job the rods will not survive. Seen here is an example of each of the ARP rod bolts for various 'A' Series rods. In the foreground is the 12 point bolt for the small bore rods as used in 850s, 998s and 1098s. The other two bolts are for the 1275 and 'S' rod. Note the 12 point nut on the middle bolt. This is an option you should consider because this nut is about 40% lighter than the regular hex nut.

If your motor is, or will be capable of rpm significantly above that of a standard unit or you are going to go racing you should install the best rod bolt available – this means ARP items.

The bolts used on the bigger journal rods are smaller at 11/32 but generally don't give any trouble under normal circumstances if correctly installed. If the rpms turned are to be higher than standard the least you should do here is to install the A+ rod bolts and the flanged 12-point nuts as they are significantly better than the earlier type. Because the bolt is smaller than the 'S' bolts racing requires that you have the best available which once again means ARP items.

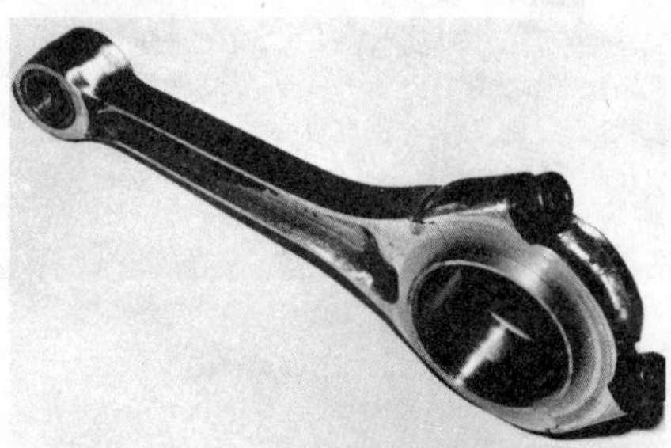

Instead of using the standard bolts, high tensile Allen bolts can be used for big end bolts. They should be torqued up to 5ft lb over the standard setting and as you can see, are used without lock tabs.

On the Series 'A+' rods, a great deal of attention was paid to improving the bolt/nut spot-facing on the rods. This, in turn, improved the clamping efficiency of the nut/bolt combination by virtue of reduced friction just as the nut seats onto the rod itself. If you are using this type of nut and bolt, be sure to use a good, high pressure lubricant on the nut face and threads before assembly, especially if you are working with earlier rods. If you intend to fully prepare a set of earlier rods, *i.e.* pre-'A+', it is a good idea to check the spot facing. If the rod/cap shoulders are not flat or square or have sharp edges in the corners, have the spot facing re-done at a reputable machine shop.

Alternative Rods

Until now the only rods I've talked about are those available based on factory forgings. But for a price, there are alternatives, though you may ask why, since the factory rods are reliable. There are sound reasons for considering alternatives. For instance, you may want to go to a specially made rod either to save weight or to change the rod/stroke ratio to something you feel may be more favourable for the combination you're putting together. A rod centre to centre length versus stroke length ratio of around 1.8 to 1.9 was deemed to be about optimal for a two-valve motor. However in light of tests done in the last 10 years this now seems not to be the case. The less breathing capability the

Lightweight Engine Internals

Amongst the uninitiated, there are two schools of thought concerning lightweight engine parts. The first school, and one that prevails mostly amongst the more expert engine builders, is that lightweight engine parts DON'T contribute to power output. The second school, and one which prevails mostly amongst amateur engine builders, is that lightweight parts DO help power output. The bad news is that both are wrong, but the second group are less wrong than the first. If this is the case then, you may well ask why we bother to lighten engine components. The fact of the matter is, if we measure the output of an engine first with lightweight components and then with the standard weight components fitted on a typical engine dyno no change in hp will be seen. At this stage it looks like the professionals are right: no increase in power due to lightweight parts. However, let's stop and consider what we are doing when we measure the power output on a typical dyno. The engine is held at a steady speed by driving against a steady load, and the power measured is a result of how much load it can drive against. When used in a vehicle, the only time the engine sees a steady load is at the maximum chosen speed. Under all other conditions, it is accelerating, so the conditions on a typical dyno don't actually simulate what happens to the engine when in use. My own Superflow dyno is able to measure power output while the engine is accelerating as it would when installed in the car. It is a dynamometer of this type that reveals exactly what happens when lightweight parts are used in the engine. Under normal conditions, the combustion pressures within the cylinder have to accelerate not only the mass of the car but also the mass of the engine. Since the engine is geared in relation to the road wheels, the rate of acceleration of the parts within the engine is much higher than the car. Although lightening internal components may only shed a matter of a few pounds, the effect of doing so is out of all proportion to the small weight reduction. This is the more so, the lower the gear ratio employed by the vehicle.

If a comparison is made of the power output of an engine taken under steady state conditions, and accelerating conditions, we find that the engine power output whilst accelerating is considerably less than it is during steady state conditions. My own dyno can measure the power output of an engine while it is accelerating at rates as high as 2000 rpm per second. The power output of an engine measured thus, is typically 25% - 30% less than the power output measured at constant rpm, so where did that missing power go? Quite simply, the power is absorbed by the accelerating internal mass of the engine, and only the 'surplus' power is left to accelerate the vehicle. The lighter those internals are, the less power they absorb during acceleration. By using lightweight parts, a greater amount of surplus power is left to accelerate the vehicle. The longer the cam period used, the more the engine becomes a 'high rpm' engine. Such engines usually require lower final drives for maximum performance. In such situations, lightweight engine components become even more desirable because of their effect on improving acceleration. Although lightweight parts are important at any stage of the game, super high revving 850s and 1000s really need them, especially in the flywheel area. On the big motors which have a very long stroke, it's necessary to make particular efforts to pare the weight of connecting rods and pistons, simply to relieve the crank of as much inertial load as possible.

It is possible to calculate the effect of lightening components directly attached to the engine. Calculating the effect of lightweight rods and pistons can be very difficult, simply because of the numerous factors involved, but calculating the effect of the lightweight flywheel is much easier. I will give you an example so you can calculate for yourself just how much difference a lightweight flywheel can make to your vehicle.

At this point let me introduce an engineering term; 'radius of gyration'. It sounds very grand, but what it means is the radius which we could assume all the mass of a rotating component could be compacted into to give the same effect as it has in its more spread-out state. In other words, for a flywheel, we could say that the radius of gyration is the radius around which all the mass of the flywheel acts. The bigger the radius of gyration of a flywheel, the harder it is for the engine to spin it to speed. This is due to the mass being concentrated further from the centre and therefore absorbing more torque to accelerate to a given rpm. Conversely, the smaller the radius of gyration of the flywheel, the quicker it will accelerate. For reasons of simplicity, the formula I will show will make the assumption the radius of gyration of the flywheel and clutch assembly doesn't alter during any lightening operations. By making this assumption, it's possible simply to calculate the effect of a lightened flywheel, with reasonable accuracy. The formula is:

$$\frac{0.5 \times n^2 \times r^2 + R^2}{R^2}$$

In this formula, n = total gear ratio (the total gear ratio is the gearbox ratio × final drive ratio). r = radius of gyration of flywheel. In transverse engines, this is about 3.75 inches. For inline engines, it's about 4 inches. R = Radius of wheel (this is half O.D. of tyre in inches).

For anyone with a typical calculator, that formula is not too daunting to slot numbers into and come up with an answer. The question is, what does the answer give? Basically it gives the weight equivalent that would have to be removed from the car to give the same effect as the weight removed from the flywheel. Let me explain in a simple, though not totally technically correct, manner. A gearbox is really nothing more than a complex mechanical lever. When the engine accelerates the vehicle in first gear, it doesn't really see the weight of the vehicle as the amount of mass it's accelerating so much as the weight of the vehicle divided by the total gear ratio. Let's say for the sake of argument, that the vehicle in question is an Austin Healey Sprite which, with driver, weighs 1500 lb. If first gear is 3.5:1 and the final drive ratio is 4.1:1, then we have an overall gear ratio of 14.35. What the engine actually sees is the weight of the car divided by this overall ratio of 14.35. In other words, the engine only 'sees' a little over 100 lb, but it is also looking at the weight of the flywheel plus a few other drive train parts which will typically add up to around 35 - 40 lb. So in total, the engine sees something in the region of 140 - 145 lb. If you manage to lighten the flywheel by 14 lb, then as far as the engine is concerned it has been relieved of 10% of the mass it has to accelerate. Of course when you make a gear change into second, the effect of 14 lb off the flywheel won't be so great. It will be even less in third and still less in top. Lightening a flywheel won't make one bit of difference to the top speed of the vehicle, it won't go even ½inch an hour faster. What it does do is allow the car to accelerate faster, and let's face it, it is the acceleration mode which is most important in a vehicle because virtually all performance driving except Bonneville and motorways are in accelerating or decelerating modes.

Having looked at the basic principal, let's go through an example to see what the formula produces. Let us assume the guinea-pig vehicle is a hill-climb car, powered by a 1000 cc engine which runs to 9,500 - 10,000 rpm (it's a very highly developed engine). Let's assume the car is a space-frame machine with a glass fibre body and it weighs in at 900 lb. With a close ratio gearbox, the first gear will typically be 2.6:1. Because it's a very high rpm unit, it will more than likely have a final drive ratio of around 6:1. Let's assume also that the flywheel weight dropped from 20 lb to 10 lb. Now let's slot all that into the equation. First the total gear ratio will be 2.6 × 6 = 15.6, therefore n = 15.6. r, we'll say, is 4 (because I have assumed an inline engine) and R we'll say is 11 inches. That would be about the typical tyre size on a modern slick for, say, a Sprite. Inserting all that into the equation, we have:

$$\frac{0.5 \times 15.6^2 \times 4^2 + 11^2}{11^2}$$

The answer we get is 17.09 lb. What our formula has told us here is that every pound off the flywheel is equivalent to 17.09 lb off the weight of the car in total. The actual drop in flywheel weight was from 20 lb to 10 lb. Though only 10 lb has come off the weight of the car in first gear, it looks like 171 lb, i.e. 10 × 17.09. In first gear, then, the engine 'sees' the car as being 19% lighter than in the previous situation. Assuming that with the gear ratios given, the tyre and suspension work well enough to hook the car up, you can see that that lightweight flywheel can mean considerably more acceleration from a standing start, and in any short course form of motor racing, the rate of acceleration right off the line or out of a tight turn is much more important than top speed over the finish line.

If having lightweight components works out so well, why do original equipment flywheels from the factory end up being so heavy? The answer to that is quite simple; when a lightweight flywheel is installed, smooth idling almost always suffers severely. That is particularly the case if a near-street timing cam is fitted. If a longer period cam is used, the situation may not be quite as bad. With such a cam, the valves close much later in the cycle, so it doesn't take as much flywheel energy to run the piston up on the compression stroke and the idle is by nature smoother.

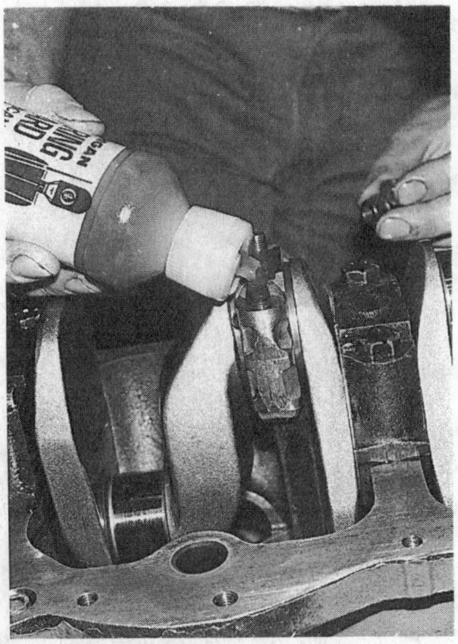

Maximum rod/clamping loads can only be achieved by careful attention to reducing friction both on the thread and clamping face of the rod/nut/bolt assembly. To achieve optimum results a good high pressure lubricant should be used on the thread and nut face.

head has in relation to the displacement of the cylinder it has to feed the longer the rod needs to be. There are two reasons for this. The first is that a longer rod has less side thrust and so the friction is lower. Secondly a longer rod reaches a lower peak velocity which also occurs later in the induction stroke. This means peak air demand happens later when the intake valve is further open. If a 6 inch rod is used with a 3.2 inch stroke the peak piston speed occurs about $1\frac{1}{2}°$ later so the intake valve on a typical race cam is about 8-10 thousandths further open. To get a rod around 6 inches centre to centre into a standard stroke motor will require some juggling of the pistons ring pack but it is do-able. If you are stretching the motor by virtue of a longer stroke you should, if specialty rods and pistons are on the agenda, seriously consider maximizing the rod length as this will give better results.

As the displacement of the motor becomes less under a given cylinder head the need for larger rod stroke ratios decreases. A standard 5.75 long rod in a 1000cc 'S' produces a rod stroke ratio of 2.39 which, if anything is a little too much. If I was building a cost-no-object 1000 then I would attempt to bring the rod stroke ratio down to 1.9:1. This would require a piston with a large compression height even after the block had been decked as much as possible. In reality it is practical to bring this figure down only to about 2.1 or so. By shortening the rod the amount of charge trapped at valve closure is increased slightly so the torque output is increased.

If you go for a specially built rod, all of these options are opened to you. Carrillo in California can produce a connecting rod for the Series 'A' engine to almost any length that can be feasibly accommodated in the engine. With a typical rod length, the connecting rod is somewhat lighter than the Leyland equivalent. A typical $5\frac{3}{4}$ inch Carrillo connecting rod weighs in at about 570 grams.

Crankshafts

Taking things in approximate chronological order and starting way back with the A30, we have, as I've said before, a crank which resembles a bent piece of wire. This you can see from the top photo opposite. Once again, I am not going to say much about the A30 bottom end, but move straight on to the 948 engine.

Fundamentally, the crank of the 948 engine is a fairly robust item. Under most conditions it's safe to about 7000rpm though that does rather depend upon the state of tune of the rest of the engine. Obviously the more power the crank has to transmit, the less life it can be expected to have. However, the standard 948 crank responds to all the same modifications as other cranks within the range, so if you follow the general rules I have outlined later in the chapter, you will find that this crank is totally acceptable in most applications.

When the 948 engine was in its racing heyday, special cranks were made at the factory. There was a red colour-code crank (of EN40B) and a black colour-code crank. These cranks were not made in any great numbers, and finding one these days would be very difficult. If you wanted a high rpm crank specifically for racing, which would be as near 100% reliable as a race crank can be expected to be, then a billet crank should be made from EN40B. Failing that, run modified standard cranks and change them at regular intervals to ensure engine reliability.

Early Mini engines did develop a few crank problems, a position not really unexpected on a setup which was as

An A30 crank, not the stoutest Series 'A' crank ever made.

A standard 850 Mini crank. Points to look for here are that there is a degree of overlap between the big end and the main bearing journals, thus making the crank stiffer than its long stroke counterparts, and the fact that there is also some deliberate counterweighting opposite the big ends.

revolutionary as the Mini was in 1959. The original Mini cranks had an oil-fed primary gear. While the seal was working, everything was fine; no oil got onto the clutch. Unfortunately the reliability of the seal was far poorer than anticipated and, as a result, most of the cars equipped with these early cranks ended up with slipping clutches. This oil fed primary gear crank was used on all 850s and 997 Coopers until January 1963. The shortcomings of this type of crank/primary gear combination were very quickly realized and the 'deva' metal primary gear was introduced. The deva metal bearing in the primary gear needed no special lubrication so the oil-feed on the original pressure fed primary gear type crank could be blocked off. This eliminated the source of clutch slip. It is unlikely that many of these cranks still exist, but if you should find yourself in possession of one, I would suggest you simply don't use it. A far better choice would be to go to a breaker's yard and obtain a crank from a newer engine.

While still on the subject of early Mini cranks, it's worth noting that there are 'thin tail' cranks and 'thick tail' cranks. The tail on the 'thin' cranks is $1\frac{3}{8}$ inch diameter; the thick ones were $1\frac{1}{2}$. Under certain conditions, usually when a heavy flywheel and no crank damper is used, these thin tail cranks will break. If a crank damper is used with a thin tail crank, along with a lightened flywheel, reliability is considerably enhanced, however, since 850 cranks are so cheap in the breaker's yard, it's hardly worthwhile using a thin tail item. The thick tail cranks were made from September 1964 and for any serious competition work this is the crank to use. Once again, I should add that special competition cranks were produced by Special Tuning for racing. However these are no longer available and it's unlikely you will be able to acquire one. However, the standard crank, correctly prepared, is totally functional for race use.

Although they are made from a different forging, cranks for 998s and 1100 engines follow the same sort of pattern as 850 Mini cranks. In essence they are long stroke 850 Mini cranks. In fact the cranks are interchangeable between all three engines as long as due allowance is made for the fact 850 Minis have a wider centre main bearing and bearing saddles than the two bigger engines.

Early 1275 Sprites and Midgets had nitrided cranks, but these were quickly discontinued. Later cranks were made of EN16T and tuftrided, and the last of the Sprite/Midget cranks for the 1275 engine, made in 1974, were EN16T and non-heat treated. Of course the Sprite finished production in 1971, leaving the Midget the only sports car in Leyland's range using the Series 'A' engine, but even that came to an end in 1974 when the 1500cc derivative of the Triumph engine was adopted.

Vehicles other than the MG Midget/Sprite use the inline engine. A large number of these engines went into the Marina range of cars. The inline engine, along with the transverse engine did, in the '80s, develop into the 'A+' engine, but I'm getting ahead of myself here; let's go back to 1275 transverse engines. The original 1275 engine was the 1275 'S' engine which, as mentioned, employed a nitrided EN40B crankshaft. This crank was difficult to mass produce. It supposedly had a few minor problems with 'growing' during the heat treatment process, and production was stopped because of the production expense entailed. Production of the Cooper 'S' ceased in 1969 and during the last year to 18 months of production the cranks used were tuftrided EN16T. When the nitrided crank was dropped from production, many engine builders (myself included) were concerned that the EN16T replacement crank would simply not be strong enough for

Don't assume the stroke length is anything like correct on your crank. A 10 thousandth variation in stroke from one big end to another is common, but I have seen ones which have varied as much as 30 thousandths on stroke. Here, a crank is being stroke checked and corrected as necessary to bring all the throws to the same dimension. By the way, if the stroke varies on each cylinder, a balancing operation will not truly balance the differing inertial effects of one cylinder to another.

Fig 13.1a
This section on fasteners has been put here because rod bolts, closely followed by head studs are the most important in the motor. But they are far from the only ones and you will make life easier for yourself by reading what is said here about ARP fasteners and the 'A' Series.

ARP Bolts and Studs

Part No. AMS-1 This is a dome end stainless steel in/ex. Manifold stud kit. It comes complete with stainless washers and 12-point nuts. ARP stainless is rated at 170,000psi. These studs and nuts do not, unlike the standard items corrode. This means regardless of what the vehicle has been through or how long they have been fitted they can be removed as easily as the day they were installed. Also it takes a smaller 3/8in socket for assembly which makes getting in to tighten everything much easier. Torque: With oil – 30ft/lbs or with ARP lube, 20ft/lbs.

Part No. BEB-08 This 12-point rod bolt suits 850, 948, 997, 998 and 1098 and is immensely stronger than the standard items and is lighter. Torque: With oil – 65ft/lbs or with ARP lube, 45ft/lbs. Lock tabs/washers are not recommended. See Note No. 1.

Part No. BEB-1 This 11/32in rod bolt/nut suits all 1300 engines. In an 'A' Series this 190,000psi material bolt has an almost indefinite fatigue life and is virtually indestructible. Torque: Measure stretch if possible which should increase 0.0067–0.007 inches or torque: With oil – 55ft/lbs or with ARP lube, 38ft/lbs. See Note No. 1.

Part No. BEB-6001 This 190,000psi material 3/8in rod bolt/nut is for all Cooper S or 1275 Midget rods. Torque: Measure stretch if possible which should increase 0.0065–0.007 inches or torque: With oil – 60ft/lbs or with ARP lube, 40ft/lbs. See Note No. 1

Part No. FB900 These 12-point hi-tensile flywheel bolts for in-line engines virtually eliminate the occasional shearing seen with the standard flywheel bolts. For 1275s use six and for 948s and 1098s use four. Torque: With oil – 60ft/lbs or with ARP lube, 40ft/lbs. (**Note:** *Flywheels should be doweled with preferably two dowel pins*) Lock tabs not required.

Part No. HSA9 Basic 190,000psi head and rocker stud kit for all 'A' Series engines. Includes rocker studs and all hardened washers and 12-point nuts. These head studs give significantly greater clamping forces than the standard stud and should be considered a must if head gasket failures are to be avoided. The large head studs should have the nuts torqued to 60ft/lbs with oil or 40ft/lbs with ARP lube. The small rocker shaft stud torque requirement is 30ft/lbs for oil and 20ft/lbs for ARP lube.

Part No. HSA11 This kit is as per HSA9 except it has the two extra studs to suit engines where the head and block have been drilled and tapped to accommodate them. The front stud should only be torqued to 40ft/lbs with oil or 30ft/lbs with ARP lube because the block is open around the water pump. All other torque settings are as per HSA9.

Part No. HSA11-B This kit is as per HSA11 except has a 5/16in bolt instead of a stud at the extra fastener location by the thermostat. The torque settings for this bolt are 30ft/lbs for oil and 20ft/lbs for ARP lube. All other torque settings are as for HSA11 or HSA9.

Part No. MSA33 Main stud kit for Cooper S or 1098 2in main engines. Studs are 190,000psi. These studs are shouldered for proper location of the caps. Torque: With oil – 95ft/lbs or with ARP lube, 75ft/lbs.

Part No. MSA44 Main stud kit for all 'A' Series engines with 1.75in dia. mains which includes 848, 948, 998 and early 1098. This kit contains 12-point nuts and hardened washers along with the 190,000psi studs. Torque: With

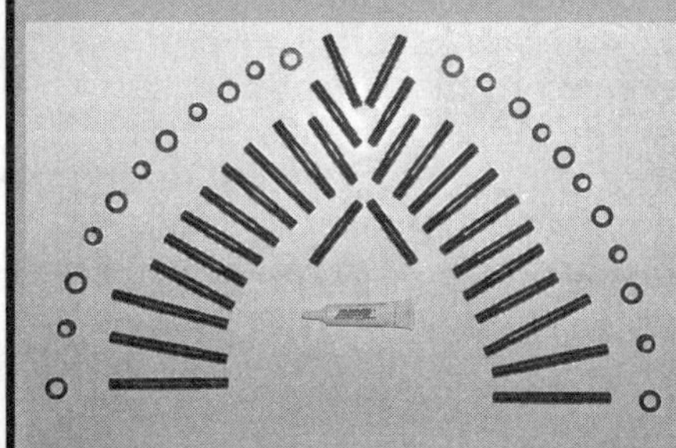

Who says there is no art in a tech book. Here are the ARP studs for the mains and head of a 1275 motor. The quality of these parts is such that each is a work of art in its own right.

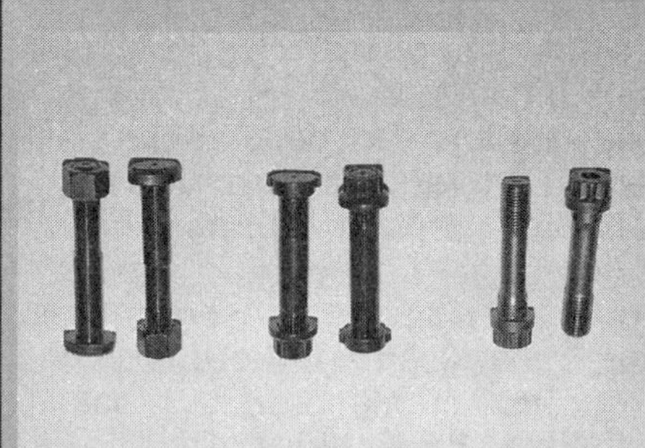

If you have limited finances the number one requirement is to buy a set of ARP rod bolts. The failure of a standard part in this area will clean out what ever may remain of your bank balance from building the motor in the first place.

These ARP stainless studs for the thermostat housing eliminate the effects of fastener seizure due to electrolysis corrosion at this point.

oil – 95ft/lbs or with ARP lube, 75ft/lbs.

Part No. MSA54 Main stud kit for all 1275 engines except Cooper S. When using these in a Mini it is recommended that the washers be left off the studs on each end. In order to clear the gearbox webs (which should always be checked) it may be necessary to either shorten the studs, or remove the top two or three threads from the block so the top of the stud is flush with the top of the nut. Always check for clearance during trial assembly.

Stainless ARP studs to secure the intake and exhaust manifold are not a luxury but a high convenience item. Not only does the 12-point nut allow easy access with the spanner or socket but also the fact it is stainless means it does not seize due to repeated heating and cooling

Part No. MCS-1 Extra two 7/16in dia. studs for use when using a four-bolt type centre main cap. Use MSA 54 for the other six studs. Torque: With oil – 85ft/lbs or with ARP lube, 70ft/lbs.

Part No. THS-1 This is a stainless steel thermostat housing stud kit which for ever ends stud corrosion. The kit is for use on housing up to 1.00 inch in thickness which means it suits most 'A' Series engines except A+ units. Torque: With oil – 25ft/lbs or with ARP lube, 20 ft/lbs.

Part No. L99 ARP lubricant. This is a special high pressure lubricant design to maximize a threaded fastener's clamping force. Its use is strongly recommended.

Part No. THS-2 As for THS-1 but for A+ thermostat housing – 2 inches total height. Torque: With oil – 25ft/lbs or with ARP lube, 20ft/lbs.

Important Fastener Tech – do not omit reading.

Always wash the fasteners before use. The grease they are covered with is a protective coating *not* a lubricant. Always lubricate the threads and mating surfaces of the washers/nut flanges – preferably with the special ARP lubricant made specifically for the job and torque to the recommended settings. If torqued to the recommended settings the bolts/studs are preloaded to 75% of their yield strength. Be sure to use a very generous chamfer on all stud holes. When screwing studs into the block be sure to put them in hand tight which is about 4ft/lbs max. If a thread locking compound is used on the stud threads be sure to perform the final torque well before the compound cures. Bear in mind that for probably 99% of applications thread locking compound on the studs is unnecessary. Do not use lock tabs. They are usually made of material which is too soft for the loads generated by higher tensile bolts. This eventually starts to 'squeeze' out and pre-load is lost.

Higher clamping loads are generated by finer pitch threads and smoother thread finishes. Adequate lubrication on fine threads is more critical toward achieving the maximum clamping loads. Be sure to tap all female threads so they are clean and the thread surface is smooth so as to minimize friction.

ARP lubricant produced more clamping load from less tightening torque.

Note 1
For all connecting rods be sure to check the bolt hole chamfer is sufficiently large enough to clear the bolt under head radius and correct if required. This is *very important*.

All ARP fasteners enquiries for Europe should be directed to:
Kent Cams Ltd, Units 1-4 Military Road, Shorncliffe Industrial Estate, Folkestone, Kent CT20 3SP
Phone: 01303-248666
Fax: 01303-252508
E-mail: kentcams@BTinternet.com
For all other enquiries contact Advanced Performance Technology (APT), 595 Iowa Ave- Suit 'C', Riverside, CA 92507 USA. (Tel 909-686-0260 or 800-278-3278 (800-APT-Fast) Fax: 909-686-2831
E-Mail: techinfo@aptfast.com
Web: www.aptfast.com

competition use, but this proved not to be the case. Although the EN16T crank was not as good as the EN40B one, it proved entirely adequate, and undue crank problems were not experienced. The then newly introduced 1300 GT also shared the last type of 'S' bottom end in as much as it had a tuftrided crank together with 'S' or Midget rods. When the 1275 GT was introduced in 1969, this as such, did not have an 'S' engine. However, early ones were often built using existing factory stock and a number of early 1275 GTs were equipped with either a tuftrided 'S'-type crank (some of which had 1300 centre main bearing widths) and 'S' rods, an 'S' crank and Midget rods, or the then newly introduced big journal cranks and EN19 rods. The big journal cranks and EN19 rods are the most common combination, and generally referred to as 1300 cranks and rods. The rods I have already described. The 1300 rods are the ones with the big balance pads on, and the 1300 cranks were made from EN16T and non-heat treated, the extra strength being made up by having a 1¾ inch big end instead of a 1⅝ inch big end.

When the Metro was originally envisaged, it was deemed necessary for two reasons that the power level of the Series 'A' engine would have to be raised to keep the performance of the new car in line with then current standards and to compensate for the fact that the Metro would, in all probability, be a heavier car than the Mini. This requirement led to much development work being undertaken in terms of improving long term reliability of the engine at these new, higher power levels. Much of this work was done on the crankshaft. With the start of the Series 'A+' engine development, **cranks were initially taken straight from the 1275 GT line, but under endurance tests these cranks developed fatigue problems in the fillet radius,** that's the radius between the big end or main bearing journal and the cheek of the crank web. Testing revealed that if the cranks were not made exactly to drawing spec, they would fail in the fillet radius. The most common failure was between number 4 big end and the main bearing nearest the tail of the crank. The second most common failure was between number 1 big end and number 1 main bearing. The third most common failure was from number 1 main bearing through the web that joins number 1 to number 2 big end. Tests showed that almost all these failures were the result of too small a fillet radius. Usually it was too small because it was outside production tolerances. Unfortunately a lot of cranks left the factory like this. However, I think I should make the point here that the tests Leyland conduct on their cranks to see if they have adequate fatigue life are very arduous. It amounts to driving the engine at maximum rpm under full throttle for 100 hours or more. This is roughly equivalent to 10,000 miles of flat-out motoring at maximum rpm on the motorway. Cranks that did break tended to do so near the end of the test. On the other hand, cranks that were made to drawings would complete the test without failure. These would often run as far as 50,000 miles before they finally broke. The obvious solution to such crank breakage (as the factory engineers well knew) was tuftriding, however, though a known cure to the problem, the tuftriding process, at the time, was causing controversy because of the atmospheric pollution it causes. Very stringent regulations were being laid down relating to when and where tuftriding plants could be built and operated. As a result of the environmental problems Leyland decided to investigate the potential of rolled fillet radius cranks. This was a process which had been used successfully by Ford, BMW and many other big manufacturers to increase crankshaft fatigue life. For instance, on a 2000 overhead cam Ford engine, rolled fillet radii proved to extend the fatigue life by a factor of at least 100%. So a pilot plant was put into operation and rolled fillet radii cranks were tried. In the experimental stages, they appeared to be totally successful. Unfortunately, when the scheme was put into production, all did not function smoothly, and many cranks did not have the rolled fillet radii correctly applied. The result was, a giant step backwards as far as reliability was concerned.

This meant that until rolled fillet radii production techniques had been corrected, cranks would have to be tuftrided. Fillet rolling techniques were eventually sorted out satisfactorily and cranks produced from mid-'83 on, are generally pretty tough pieces.

When the turbo Metro was introduced, there was much ado made about the fact it had a nitrided crank. Neither I nor anyone who has torn down any of these engines has managed to locate a turbo Metro with a nitrided crank. Who knows, maybe Leyland intends to produce them, but at the time of going to press, none exist.

Blocks, Cranks, Rods & Pistons Part 3

Improving The Crankshaft

I'll start the ball rolling in the crankshaft 'tweaking' department with the inline engines because it makes life easier. Basically, the inline engines can be split into two groups, small bore and big bore, and the only small bore cranks worthwhile dealing with, are those used in the 948 and 1098 engines. Big bore inline cranks should be dealt with in much the same way as the transverse cranks, which I'll get to later.

On 948 engines, if you want a crank for high performance use, you have only two options, either prepare the standard crank to the best of your ability to cope with the hoped for extra power or get a billet crank made from EN40B. The last mentioned is a little expensive, but the end product is near bullet-proof. Billet cranks such as this can be obtained from **Race Engine Components**, but you can expect a long wait during manufacturing time, as virtually all are done to special/order.

As far as preparing the standard crank is concerned, all the basic rules detailed later will apply. If you have a 1098 engine, then at least you have some sort of choice as to what you should use, because there are two 1098 cranks, a small main bearing one and a big main bearing one. The small main bearing crank has 1¾ inch main bearings whereas the big one has 2 inch. The 2 inch crank should be selected if high outputs are envisaged. The small main bearing crank, however, properly prepared, is quite adequate for an 85 hp or so road-going engine.

Whichever type of crank used, and this applies throughout the range, you must remember it is not an instantaneous overload due, say, to a missed gear, that breaks the crank, but fatigue over a long period at moderately high revs. Fatigue stresses are caused not only by reciprocating loads, but also torsional vibrations, a situation aggravated by heavy cranks and flywheels and the lack of an adequate crankshaft damper. Dampers, especially on the bigger engines, are very important. If the engine is to spend any length of time at high rpm, the crankshaft damper becomes A1 important. If the flywheel is lightened, the amplitude is decreased of the vibrations involved and the rpm is increased at which the principle unwanted vibrations occur. By using a lightweight flywheel and clutch assembly the torsional vibration frequency can be moved up the rpm range high enough to reduce its severity as a problem.

Moving onto the big bore inline cranks, we find that a very favourable situation exists in terms of fatigue life. Firstly, if you have an EN40B crank from the early Sprites, then you can expect to have a very long life crank, but if you don't have one of these cranks then the life tends to be better on an inline crank than a transverse engine crank simply because the combined flywheel weight and crank length is less. This leads to greater torsional stiffness. As a result, torsional vibrations occur further up the rpm range, so fatigue life is enhanced because the engine does not operate in an undesirable rev range quite as much. As the inline engine went through the phase of becoming the 'A+' engine, so we find it developed into a large journal crank, i.e. 1¾ inch as opposed to 1⅝ inch. These cranks, fully prepared, are intrinsically very reliable units.

As far as transverse cranks are concerned, the pros and cons on 850 cranks have already been touched upon. If you have a thick tail crank for an 850, then in it's fully prepared condition, it's a pretty reliable unit and can be used up to as high as 9000 rpm. 998 cranks by and large, also seem rarely to be a problem. By the way, if you are putting together a 998 engine from bits, it could be that you may confuse a 998 crank with a 1098 crank. Well if the paint is still on, there's a pink blob for 998 and a yellow one for 1098, and if you remember that the oil discolours these, you may have to make allowances for a change in colour. You should be able to identify it within a fair degree of certainty. An alternative way of identifying cranks is as follows. Lay the crank down so that you are looking straight down at the journals. On 998 cranks the big-end journal slightly overlaps the main bearing journal. Even when the journals are ground undersize the overlap is still discernible. On the 1100 crank, however, the journals do not overlap. Subsequent regrinding makes this more apparent. Anyway, back to 998 cranks. These rarely seem to be a problem as far as torsional vibration is concerned, as long as a crank damper is fitted. 1098 cranks with their longer stroke do have more of a problem and although it's possible to rev engines with this crank to 8000 rpm, they will break after a while simply because they fatigue. My advice is, if you want it to live for any length of time for competition purposes, redline an 1100 crank at around 7000. If it's for the road, make that 6500.

With 1275 cranks it has been found that in conjunction with a typical weight of clutch and flywheel assembly they go through two critical vibration periods which will be in the rpm range used for high performance work. These vibration periods coincide with 5750 rpm and 6100 rpm. With lightweight flywheels, the rpm at which the vibration occurs is moved up, so from that aspect at least, the lighter the flywheel used, the better. It's difficult to lighten the flywheel/clutch assembly enough to move the vibration frequency completely out of the usable rpm range, so this means living with torsional vibration. It also means the crankshaft damper is important. Whilst on the subject of crankshaft dampers, it's worth mentioning that two crankshaft dampers are used: a Holset and a Freudenburg. At regular intervals you should inspect the crank damper as they are critical to the fatigue life of the crank. **If the damper shows any sign of distress, namely the outer ring moving on the rubber, either don't rev the engine or replace the crank damper.** Having said that the crank damper is critical, let me make one provision: if you are putting together a really high rpm sprinter or a drag race engine, then the chances are the fatigue life of the crank is longer than a lot of other components. In other words, the probability of wear or breakage in other areas is greater than a fatigue breakage of the crank. Under these conditions, it's okay to run without a damper, but remember, at 8500 to 9000 rpm the crank may only have a life expectancy of 250 miles.

Having said that for a Sprint or drag race engine you can run without a damper, we need to look at the other side of the coin. Namely, the effect a damper has on power output. At first sight it would seem that a damper would have no effect on power output because it is simply a mass of material being hurled around the end of the crankshaft. This proves not to be the case. Some extensive testing I did on Chevrolet engines showed the crank damper to have a significant effect on engine power because vibrations in the crankshaft can be transmitted directly to the camshaft. Since the camshaft has the important function of operating the valves at very precise moments, any disturbing element is unlikely to do anything other than reduce horsepower. My tests on a Chevrolet engine with no damper, several different types of rubber dampers and a viscous damper showed that the damping effect on the crankshaft was important to power output. The mass of the damper had less effect on reducing the accelerated power output than the effect the damper had on improving power output due to damping crank vibrations. A conventional rubber damper only damps vibrations at its tuned frequency. Unfortunately, with all the modifications we do to the Series 'A' engines, we're not in a position to determine where crankshaft vibrations peak. This means we have to accept the operational frequency of some standard damper and in effect, it may not be operating very well at the point the crankshaft vibrations are maximum. The tests with the Chevrolet engine showed that a viscous-type damper, which is not frequency sensitive, not only has the capability of damping vibrations as much as 400% better, but also will damp the vibrations **wherever** they occur in the rpm range, so it has no need to be tuned to the vibrational frequency of the assembly. At present, nobody makes an off-the-shelf viscous damper for the Series 'A' engine, but Mike Parry at Racing Engine Technique can produce a 5-inch viscous damper to special order. It's expensive, but it's the next best thing to insurance as far as crankshaft life is concerned. Not specifically relating to Series 'A' engines, my own experience has been that using a viscous damper compared with an effective rubber damper has the same effect on crankshaft life as moving up one grade of steel. Though it's too early to say for sure, it would appear a viscous damper has the capability to more than double the fatigue life of the crank. As far as power output is concerned on V-8 engines, I've typically seen 11-14 hp more than that given by the best metalastic-type damper. On a Series 'A' engine I would estimate that 2-3 hp improvement is on the cards. Finding a horsepower improvement along with an improvement in reliability is definitely an asset in any high performance engine.

Torsional vibrations obviously cause cranks to break and it is surprising the number of places they can fail. I have already discussed failures through fillet radii earlier on, but that's not the only place breakages occur. During manufacture, two cut-outs are put into the crank webs to locate the cranks for machining of the journals. Even though these cut-outs appear to be a long way away from any highly stressed area, cracks can occur from the corner and spread into the web. The likelihood of a crack forming here is not as remote as might be thought. However, there is a simple solution to it which, even if you are just putting together a very basic motor, is worthwhile doing. Using a round needle file, file the corner of these cutouts (**Fig. 13.2**). If the crank is tuftrided or shot peened, the likelihood of a crack starting here is far more remote, but there again, shot peening and tuftriding cost money, filing a radius doesn't.

Another point where a fatigue crack can start is the C-clip groove in the tail of a transverse crank. There should be a small radius between the walls of the groove, and the bottom floor of the groove. If there is a radius generally there won't be a problem, except possibly on cranks which have been stroked. Again we can take action to reduce the likelihood of a crack emanating from here by getting into the slot, polishing the corner with very fine emery, i.e. 600 grit, and/or tuftriding the crank. In this particular instance, shot peening is not really satisfactory as it's difficult to get into the groove to do an effective job. If you are in a position to select cranks, look through a batch of cranks to see which has the largest radius in the bottom of the groove. A radius of 0.005 inch or more is ideal.

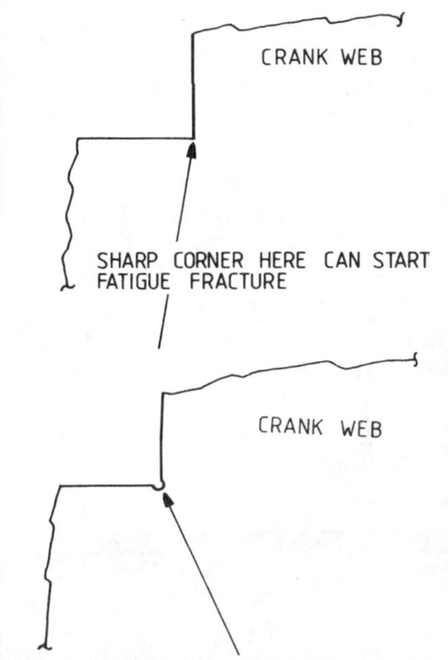

Fig. 13.2 Radiusing crank web cut-outs

'So You Want A Billet Crankshaft'.

If you buy a crank from Paul Ivey's Race Engine Components, you will find that it's considerably more expensive than an off-the-shelf Leyland one. If you see how it's made, you will understand why.

A special crank starts off as a parallel billet of steel, usually expensive EN 40B, recognized as ***the*** *steel for crankshafts. The first operation is to machine this tough material so that the main bearing diameters are in position. 13-51*

The next operation, using a special offsetting fixture, is to machine the big end throws so they are in the correct position in relation to the main bearings.

After all the lathe operations have been done, a great number of milling machine operations are performed so that the crank now actually looks like a crank. Here's one of the last operations to be done, drilling the oilways. After this operation, the crank goes through a complex series of heat treatment processes which, from start to finish, takes the crank off the machine shop floor for about ten days. When it comes back ... 13-53

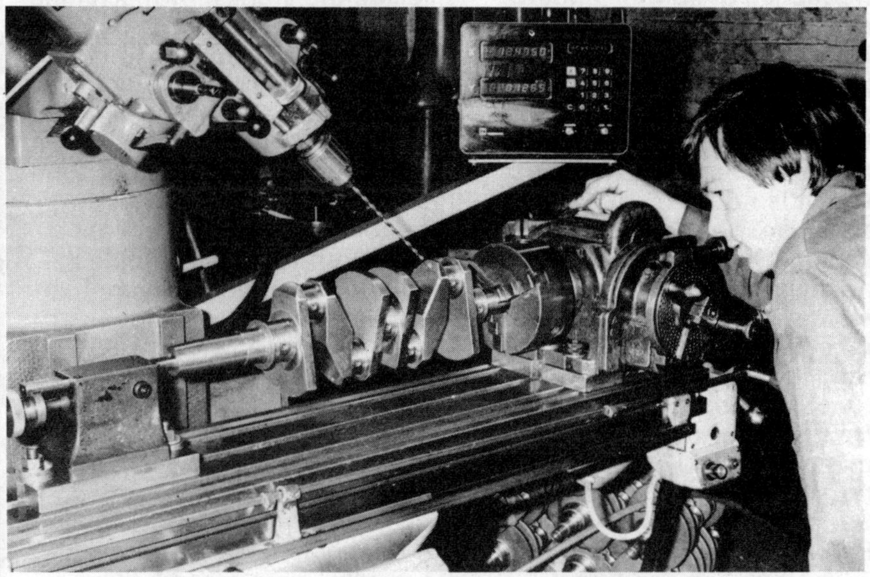

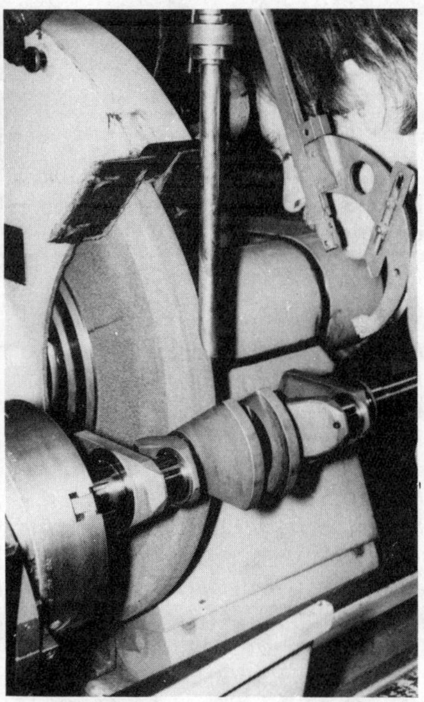

... the crank goes through a series of grinding operations. Remember, it is now too hard to do any machining other than grinding on it. Grinding operations involve grinding all surfaces down to their finished size. This includes the tail of the crank, big ends and mains and all thrust faces. 13-54

Apart from oiling and packing, balancing is the final operation before the crank leaves the shop. One of these billet cranks is balanced to fine enough tolerances such that only the parts that are added on to it need be balanced. 13-55

A crank forging starts off life looking like the item at the top of the photo. From this virgin forging, an 86mm stroke crank like the one seen in the lower part of the photo, can be made. The reason the journals aren't shiny on this crank is that I didn't wipe off the protective grease from all the surfaces, as I was about to pack the crank back in the box.

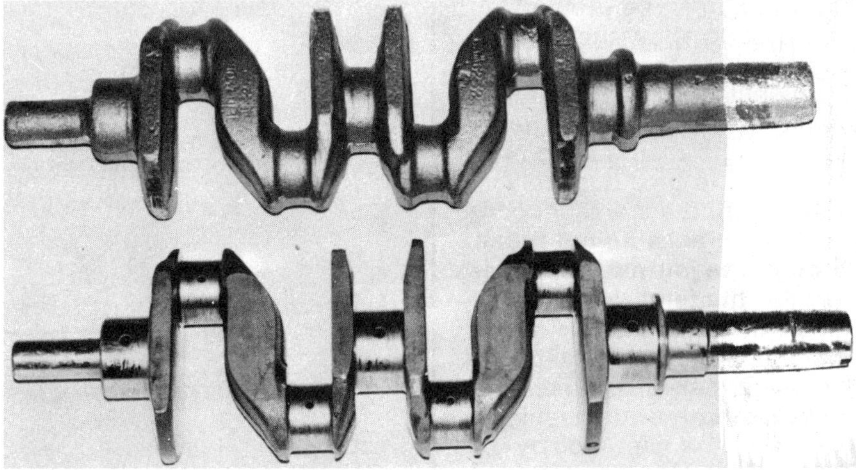

The function of the crankshaft damper is very important. Without one, crankshaft life, due to fatigue, is considerably shortened. For 850s and 1000s utilizing lightweight engine components and clutch and flywheel assembly, the damper can be machined as shown here. This moves the optimum damping frequency up the rpm range, putting it in an area more closely resembling the rpm range the engine will run in.

Flywheels with a lot of mass toward their outer edge, like this 1300 saloon flywheel, can at certain rpm, cause considerable crankshaft flexure. The rpm range this occurs at can often be right in the middle of what is normally used for high performance application. Lightening the rotational parts ...

... hung on the crankshaft can move the critical vibration frequency of the crankshaft assembly much further up the rpm range. The parts shown here are typical of a lightweight clutch and flywheel assembly for a transverse Series 'A' engine. Lightening these parts can reduce fatigue induced crankshaft breakages.

Resisting Fatigue

It's time now to move on to the journals and fillet radii. This is where a crank is most likely to suffer a failure. Although the factory calls for a radius of some 30 thou in the corner of all the journals, many cranks have been produced with what amounts to a sharp edge in these corners. Such cranks are prime candidates for breakage if they are used on a high performance engine. Note that a fillet radius is important on all cranks, not just 1275 transverse cranks. If a crank is to be prepared for high performance use, it must have a generous radius in the corner between all journals and the adjoining flank. 30 thousandths is about as big as it need be for good results. 20 thousandths is just about acceptable and 10 thousandths is too small. Whatever crank you use, take a look at the fillet radius with a magnifying glass. **If the radius appears small or non-existent, ditch the crank. If it's about 10 thousandths or more, and the crank is still at standard or near-standard size, then the crank may be redeemable by grinding.** At this point it will be necessary to seek the services of a top quality motor machine shop.

The crank must be ground undersize to allow the forming of the radii in the corners. Because it would appear that a smaller journal diameter has less strength than a large one it may seem to be a dastardly thing to grind a crank undersize. However, **cranks do not break through the journal. They break through the fillet radii.** Reducing the size of the crank journal by a few thousandths at this stage makes no difference to the strength of the crank, but increasing the fillet radii by a few thousandths can improve dramatically the fatigue life of the crank. The technique to 'fix' cranks with too small a fillet radius is to have them reground to produce as large a fillet radius as possible. Just how big a radius can be made (and remember 30 thousandths is the maximum required) will depend on the size of radius that is there at present and how much is taken off the journal. You will probably have to talk to the man who is operating the crank grinder and get his opinion. It is important to remember that when producing a larger fillet radius from an existing small or sharp radius, all the original radii must be removed. It's no good having a little 'nick' in the new radius that is part of the old one because a crack will simply start from that nick. So, if the crank is reground, it's necessary to exercise a little judgement as to just how big a radius can be squeezed in and **to clean up totally** the old, small radius. If a crank is being 'stroked', as is often the case with 1300 cranks, the radius is VERY important if the crank is to have any real life. In fact if a 1300 crank is stroked, it's a good idea to adopt a belt and braces technique, by making sure the radius is as large as possible in the first instance, and then having the crank shot peened or better yet, tuftrided.

Rolled fillet radius cranks present a few problems, not that the idea is bad, but there were several thousand cranks produced which were done wrongly. If you have an 'A+' crank with a rolled fillet radius, inspect it very carefully, especially if it's '81 to '82 vintage. The first thing to appreciate is that the rolled fillet should undercut the journal. In other words, the bearing area of the crank should be proud of the rolled fillet radius. The next thing to do is to see if the radius will pass the fingernail

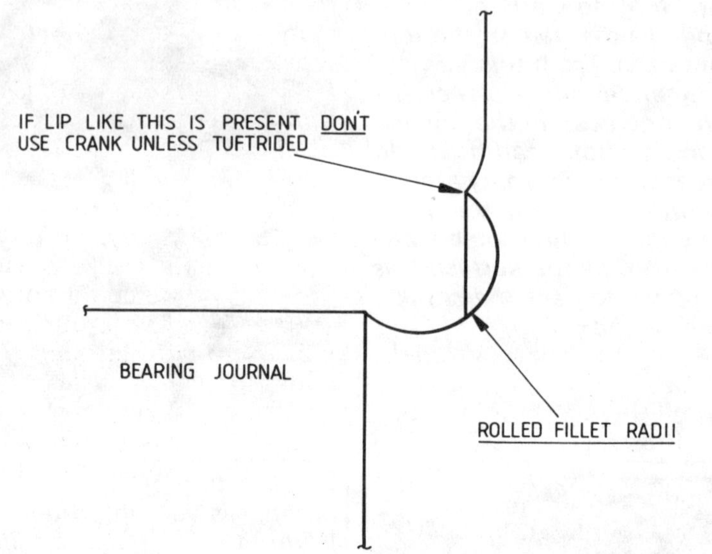

Fig. 13.3 Bearing journal rolled fillet radii

One of the most important aspects of improving a crankshaft's fatigue life is to minimize the stresses in the corners of the journal as arrowed. Here's a crank prepared in my shop. After ensuring that the fillet radii were as large as possible, all the bearing surfaces of the crank were masked up leaving the radii exposed. The crank was then sent off to be shot peened. Properly shot peening the fillet radii is the primary reason fatigue life of the crank is improved.

test. Many cranks were produced wrongly, and the rolling left a lip and this is evidence that the crank has been done wrongly. These cranks will break if used to any reasonably high rpm for any length of time. The easy way to see is to stick your fingernail into the radius and pull it up the web of the crank. If you feel a small lip or ridge as shown in **Fig. 13.3** DON'T use the crankshaft unless it's tuftrided.

Your problems, however, don't end there. The rolling of the fillet radius causes a considerable amount of stress to be induced into the crank, and tends to distort them, so the cranks have to go through a straightening procedure after the fillet radii have been rolled. When a crank is tuftrided the process also stress relieves the crank. The stresses induced during the rolling and straightening operations will settle out and the crank will revert to roughly the shape it was before it was straightened. If this is the case, the crank journals will have to be reground after tuftriding. In certain cases the crank may bend so much that it is not worthwhile using it. Again, your local machine shop will be able to advise you whether or not they can save the crank. Fortunately most can be salvaged as the distortion doesn't usually exceed 6 to 8 thousandths over the length of the crank, and note that in this case the grinding has to be done after tuftriding. If the crank does not have a rolled fillet radius, then it can be tuftrided before grinding and the bloom on the journals merely polished off afterwards.

If the crank has rolled fillet radii that are in good condition, then it's not advisable to tuftride as the rolled fillet radii will achieve the same end as tuftriding. If you tuftride such a crank and it was acceptable before, beware, you may involve yourself in a grinding operation just because the crank, being a rolled fillet radius type, has distorted.

Just because you have a Series 'A+' engine, do not assume it has a rolled fillet radius crank in it. Because of production problems with this type of crank, Leyland reverted to earlier spec cranks several times, and you could find that your engine has a tuftrided crank or a non-tuftrided crank with big fillet radii simply ground on. If you have a large fillet radius crank, this is probably the best crank to work with as it gives you scope to have the crank tuftrided and the tuftrided crank with the large fillet radius seems to be the best possible combination to have. A crank with a good radius in the corner and tuftrided, will survive 120 hours at a continuous 6200 rpm on a 90 hp engine and tests indicate that such cranks can be run on a more or less continuous basis at 7000 rpm. However, a crank prepared to that spec certainly isn't the ultimate. There is still more we can do to them to make them more suitable for high performance use.

Crank Modifications

Looking at cranks across the whole range of Leyland Series 'A' engines, it can be seen that some of the cranks, especially those in the smaller capacity range, have larger counter weights in proportion to their stroke than the 1275 cranks. By contrast, 1275 cranks seem to be under counter-weighted. In other words, they are too heavy on the big end journal side of the crank or conversely, not heavy enough on the counter weight side opposite the big end. If an inspection is made of a 1275 crank that has been in a Series 'A' engine for a high

Here's a crank that has been set up in the mill for a wedging operation on the crank webs. I have put a black line on the web so you can see how much material will be taken off by the time it's down to the size required. 13-47

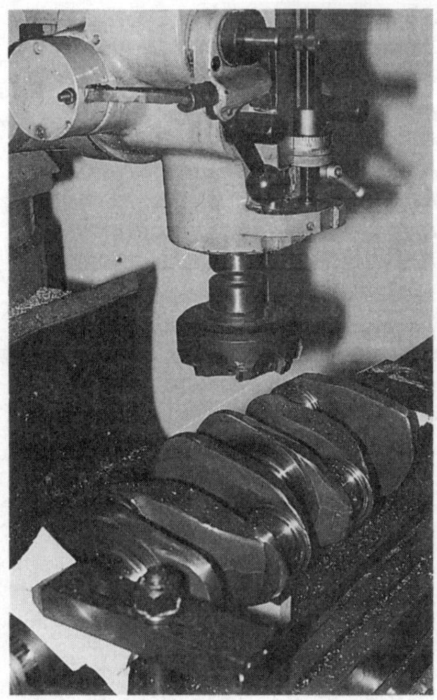

Here's the set-up I use for wedging cranks in my shop. After doing a number of cranks in this fashion, one important aspect became evident. The toughness of the material various cranks are made of differs quite markedly to the point whereby the toughest required the tool bits replacing at much more frequent intervals. 13/49a

Mini cranks aren't the only ones to come in for the wedging treatment. The crank that David Anton is working on here is destined for use in an SCCA Sprite. Although a lot of work is involved, it is worth extra high rpm horsepower plus improved reliability and throttle response.

This is what an APT prepped crank looks like after being reworked. First it is wedged then ground. During the grinding process attention is given to the fillet radius to maximise it. Also the crank is indexed and stroke corrected with each throw delivering within one thousandths. After grinding, the crank is balanced then put through a proprietary heat treatment and surface hardening process. The final operation is to micro polish the bearing journals.

Here's the finished crank out of the DV shop after many hours have gone into it. If time is what you've got to spare, you can do a lot of this sort of preparation work yourself. If you must pay someone to do it, it could be cheaper to get a billet crank made. A compromise would be to have someone do all the heavy machining and then do all the hand finishing yourself.

mileage, it will be found that the main bearing wear is concentrated more on the crank pin side of the adjacent main bearing journal than the side opposite. In other words, taking the big ends nearest the centre main, you will find the centre main bearing has worn more on the big end journal side. The two main bearings at each end of the crank tend to wear on the adjacent crank pin side, this being the side opposite the centre main bearing. This wear pattern suggests that for a standard rod at least, more weight is needed opposite each big end. Well, adding weight is difficult, and not necessarily a good thing for a high performance engine. Taking it off

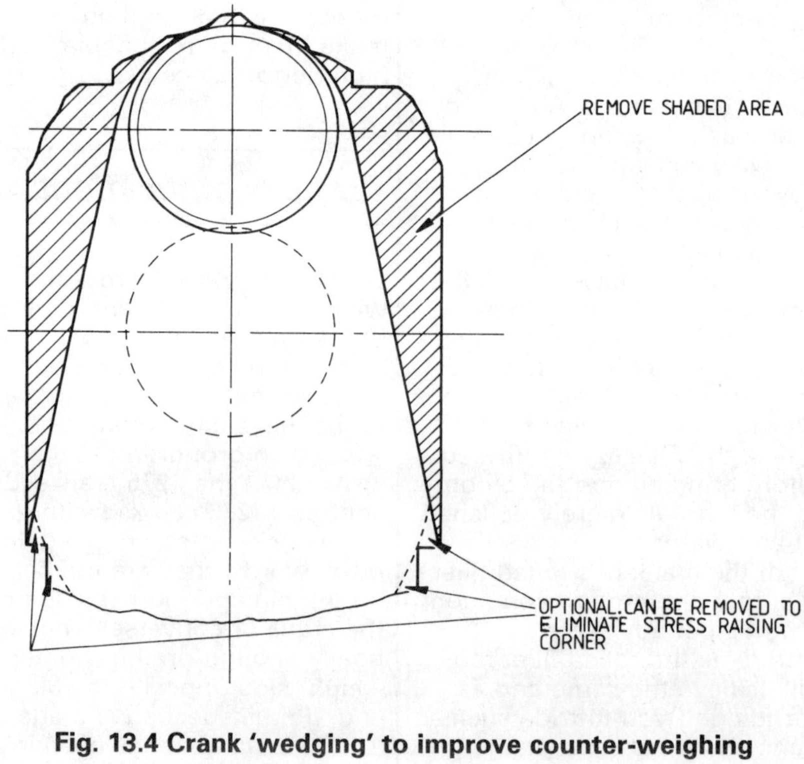

Fig. 13.4 Crank 'wedging' to improve counter-weighing

After the wedging operation has been done on the mill, the crank comes in for a good clean-up with an emery drum. The idea is to put a generous chamfer or radius on all sharp corners, to cut down oil drag, and then generally clean up the rest of the crank to make it look tidy.

If this crank looks strange it is hardly surprising. To eliminate rubbing between cylinders one of my Australian friends, Simon Gardner, built himself this crankshaft for his Sprite. With the pistons in each half of the motor going up and down . . .

the opposite side is easier and also lightens the crank. For want of a better word, I'll call this operation 'wedging'. Basically 'wedging' involves machining the crank as shown in the accompanying photos and drawing (**Fig. 13.4**). Whilst all this machining is going on, it is a good idea to round off all the sharp edges on the corners of counter weights and webs so as to cut down the oil and air drag of the crank as it rotates. It is not really necessary to knife-edge the crank. Certainly this is the best procedure, but a generous radius on each sharp corner will produce 90% of the effect of a knife-edge and it's much quicker to do. Apart from that, a knife-edge or anything approaching it can produce a stress point, whereas a nice radius does not.

A 1275 crank modified to this sort of spec, will drop about 1 lb in weight. Not a great deal on an item normally weighing around 26 lb, but it's better than adding 1 lb in weight.

Although I have talked about wedging cranks on a 1275

. . . at the same time the motor balances like two 360° firing parallel twins joined at the centre journal. Primary balance for this motor is as per normal but there is a larger secondary out of balance and a large end out of balance couple. In terms of induction pulses this crank produces an induction pulse every 360° at each intake port so there is no inter cylinder robbing. The centre exhaust however now has a long dwell period and an overlapping exhaust phase from cylinders nos 2 and 3.

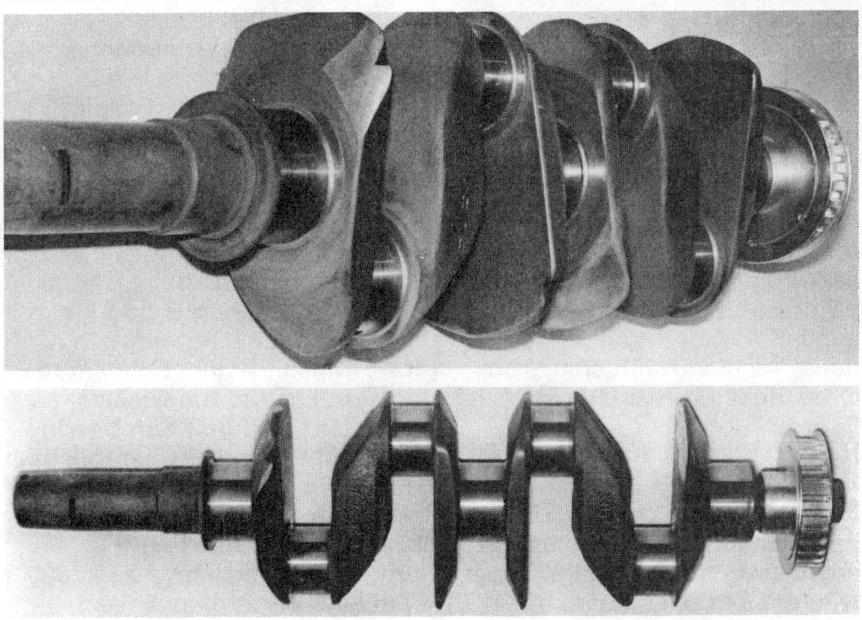

Here are two shots of the crank that you saw being machined earlier in its finished state. The dull finish on some of the surfaces is due to the fact that after machining, it went through the tuftriding process. Tuftriding doesn't necessarily make cranks look pretty but it does make a dramatic improvement in the fatigue strength of the item.

engines, this technique can be applied to other engines and it's especially useful on 850s, not because it increases the power to any marked degree, but because the weight reduction is a great help in assisting more rapidly. Indeed, on 850s, anything which reduces the rotating mass and, helps acceleration is worth doing.

Crank Lubrication

As is standard practice with virtually all engines, oil to the big ends on Series 'A' cranks is supplied from the main bearing journal. The faster the engine revs, the more centrifuging action there is forcing oil to the big ends. Under extreme conditions, the rpm can be high enough to cause the oil flow to the big ends to equal or even try to outstrip the oil flow to the main bearings. This tends to starve the main bearings of oil. Result: premature main bearing failure. In an effort to cut down this centrifuging action, competition cranks are cross-drilled. This operation involves plugging the normal exit in the big end journal and drilling a hole a right angles to it so that instead of the oil exiting at the maximum radius possible, it exits at a smaller radius, thus reducing the centrifuging action. This process, called cross-drilling, can be applied to those cranks not so done. All the EN40B cranks are cross-drilled as were most of the Special Tuning competition cranks.

Although it's better to have a cross-drilled crank, I personally have not found it a problem on engines running up to 8000 rpm with the 81mm stroke of the 1275 engine, when the cranks used are non cross-drilled type cranks. Nevertheless, cross-drilling gives a bigger margin for safety as far as lubrication on the main bearings goes and it is worthwhile. Many high performance engine specialists can modify the non cross-drilled cranks to the cross-drilled spec of the Cooper 'S'.

Bearings

In U.K. at least, the two principal types of bearing used in the Series 'A' are Vandervell and Glacier. Of the two, the Vandervell bearing tends to be thought of as THE high performance bearing particularly by professional race engine builders. Certainly their 'lead indium' bearing proves very successful in race engines. However, let us not run away with the concept that because a bearing is good in a race engine. It's necessarily the best in a road engine. A different set of circumstances prevail. A race engine is not expected to run for the period of time a road engine is, and just because a bearing lasts well in a race engine, which may only do 1500 miles before tear-down to replace all wearables, such as bearings, rings, etc., it may not live for 50,000 miles in a road engine, whereas a bearing that is totally unsuitable for a race engine, may live in a high performance road engine for 100,000 miles. However, do not draw the inference that just because the Vandervell bearing is good in a race engine, it's not good in a road engine. What I do want to point out is that **the Glacier reticular tin/aluminium bearing, which has not really established itself as a race bearing, has proved to be an excellent bearing when used with a tuftrided crankshaft.** For a 7000 rpm high performance road engine, the previously mentioned combination has proved to be very successful for long life. For race engines, I have used both Vandervell and Glacier and with the Series 'A' engines I have built I can only recall four catastrophic bearing failures. Two of these used Vandervell bearings and two used Glacier bearings.

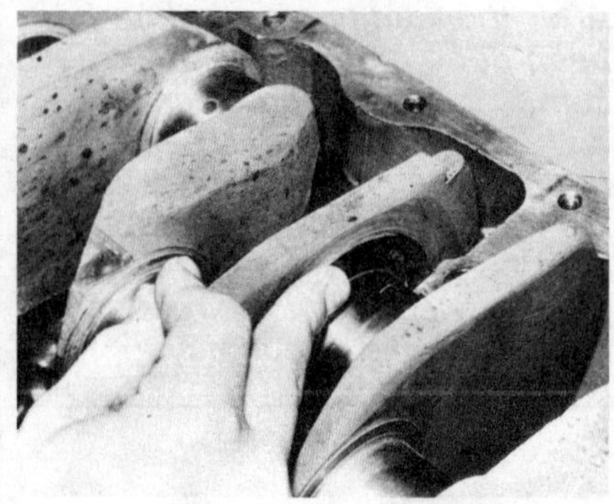

If you expect to run the engine consistently over 7000 rpm, then the thrust bearing clearances need to be increased. On small bore engines, go 2 to 3 thou over top limit. On big bore engines, especially those that have been stroked, go 4 to 5 over top limit.

In the main I have used Glacier bearings in my road engines, simply because there is a main supplier just down the road, and over the years I have had no worries about their ability to do the job.

Generally, big ends are not a problem and, for most practical purposes, nor are main bearings. However, if high engine rpm is envisaged with any of the Series 'A' engines, the first bearing to fail due to overloading is the centre main bearing. This problem can be due to two factors. First, inadequate strength in the main bearing cap, and second the bearing is simply overloaded. In the section on blocks I talked about adding a strap to the centre main bearing or a machine cap, whichever the case, and this is something you should seriously consider on all blocks except, perhaps, the latest 'A+' blocks of the turbo Metro type. These blocks use a different main bearing shell in the bottom half of the bearing. Instead of having a groove in the

Crank journal sizing is important. Different sizes are needed for different applications. Read the text before having your crank ground.

shell, they are plain and this increases the load capacity by about 80%.

Not only is it important to consider the type and design of bearing used in the Series 'A', but also the bearing clearances. **Often, a bearing can fail because the local temperature of the oil gets too high. This is not the temperature that is read off a gauge measuring the temperature in the sump, but the temperature between the journal and the bearing itself.** If bearing clearances are very small, even though metal-to-metal contact may not occur, during high rpm operation, the oil may break down because it's local temperatures are simply too high. To counteract this, it is necessary to increase bearing clearances. However, **when bearing clearances go up, the load supporting capability of the bearing, not surprisingly, comes down. As a result a compromise is necessary.** A basic rule of thumb I use when building my own engines is as follows. If the engine is to be a quiet running, smooth road engine, where performance is of secondary importance to long life and quietness, the bearings fitted are as near to the middle limit as possible on the specified factory sizes. If the engine is a moderate road performance engine, then I have a machine shop size all the journals on bottom limit. If it's to be a higher performance road engine, where performance is more important than other aspects, then the crank is sized 3 to 5 tenths of a thou below bottom limit. If it's a circuit race or rally engine, 7 tenths to 1 thousandth below bottom limit, and if it's a drag race, sprint or hill climb engine, then they are sized 1.5 to 2 thousandths below bottom limit.

By sizing the crank journals undersize, it's possible to extract a little more hp from the engine due to reduced bearing friction. It also makes more of an allowance for mis-alignment of bearings and distortion that can take place in both the crank and the block at high rpm. In other words we are giving the crank a chance of not binding up. These bearing clearances are not necessarily optimum, but because I have to operate on a limited budget and cannot afford to blow up engines very often I tend to err on the conservative side.

On the subject of binding up, it might be as well to talk about connecting rod side clearances. Consider the fact that at high rpm, the crankshaft flexes to a surprisingly large degree. The fact that at 9000 rpm a piston which has 15 thousandths clearance when a dry build is done, can touch the cylinder head, shows how much flexing is taking place. A little of this is rod stretch, but the larger amount is caused by crankshaft flexure. With a three bearing crank, crankshaft flexure can cause pinching of the connecting rods on the journals. Typically, side clearance recommendations for a standard engine are 6 to 10 thousandths of an inch between a rod face and the cheek of the crank. To make sure pinching doesn't take place, it's a good idea to set the side clearance to

Wherever a crankshaft is reworked in an effort to lighten it or to correct any original machining errors in the stroke it is going to be essential that the crank be rebalanced.

10 thousandths on a high performance road engine and 12 to 14 thousandths on a high rpm race engine.

Balancing

All sorts of wonderful things are claimed of balancing an engine, mostly by people who have balancing machines and can sell you the service. **One of the most popular claims of balancing is that it will increase the rpm limit of your engine by about 500 rpm. This is gross exaggeration.** In all probability it will increase the failure rpm of your engine by about 5 to 10 rpm and possibly at most, 50 if it happens to cure some particularly bothersome vibration frequency. Certainly if you are looking to balancing to improve the rpm capability of your engine before destruction occurs, then you are going to be disappointed. Really, you should think of balancing as producing a smoother running engine and little else. It will be fairly obvious that this will be so if you consider what balancing entails.

Firstly, it's a question of getting all the reciprocating parts to weigh the same, so that the two pistons going up the bore as near as possible counteract the inertial loads of those coming down. As it happens, due to geometric reasons in a four cylinder engine, this can never be so, but at least if all the parts weigh the same, it will be as nearly so as possible.

Balancing the pistons is no problem. All that's necessary is to see that each piston weighs exactly the same as the others. Connecting rods are a little more of a problem because the little end of the connecting rod is totally reciprocating. The big end is totally rotating so it's necessary to balance the rods not only for overall weight but also end for end. In other words all the little ends of the rod must weigh the same and all the big ends must weigh the same and if this occurs then it follows also that the total weights will also be the same.

To balance the crank, a dynamic balancer has to be used, as static balancing achieves nothing. This means a special machine must be used. Balancing is done by spinning the crank up

Balancing has its place in a scheme of events toward building high performance engines. All of the crankshafts for the engines that I build are balanced because I am trying to achieve perfection. Here, my good friend Jerry Goodale of SMS in Riverside, balances one of my cranks until a zero error is shown.

on the machine and having the machine detect where the out-of-balance is occurring and hence where the heavy spot is on the crank. By drilling and grinding away at the heavy spots, the crank is brought into balance. After the crank has been balanced, the crankshaft damper is added and its effect is taken into account.

Any out-of-balance forces now seen by the balancer are due to the crank damper being added so it is drilled as necessary to bring it into balance. The flywheel is then added and so on until all the engine's crankshaft assembly is balanced.

If you are building an engine on a budget, it's sometimes difficult to decide where to spend your money and except for achieving added smoothness, money spent on balancing doesn't actually improve performance. So when should you balance and when should you not? A simple rule is that if you are on a very tight budget, don't disturb the factory balance by lightening any of the parts unless you can balance the pistons and rods yourself by borrowing the necessary equipment, *i.e.* a set of scales or a very accurate set of spring balances. If you are in a position to balance your own pistons and rods, then it's not totally necessary to balance the crankshaft if it has not been touched or indeed any of the parts that rotate with it have not been touched, i.e. flywheel, crank damper, clutch, etc. If all these parts retain their original balance, then they will be no worse than they were when they left the factory. On the other hand, if the budget is up to spending the necessary cash to have an engine balanced, then it does make it that much nicer to run to high rpm when it does so smoothly.

14 Big Motors Part 1

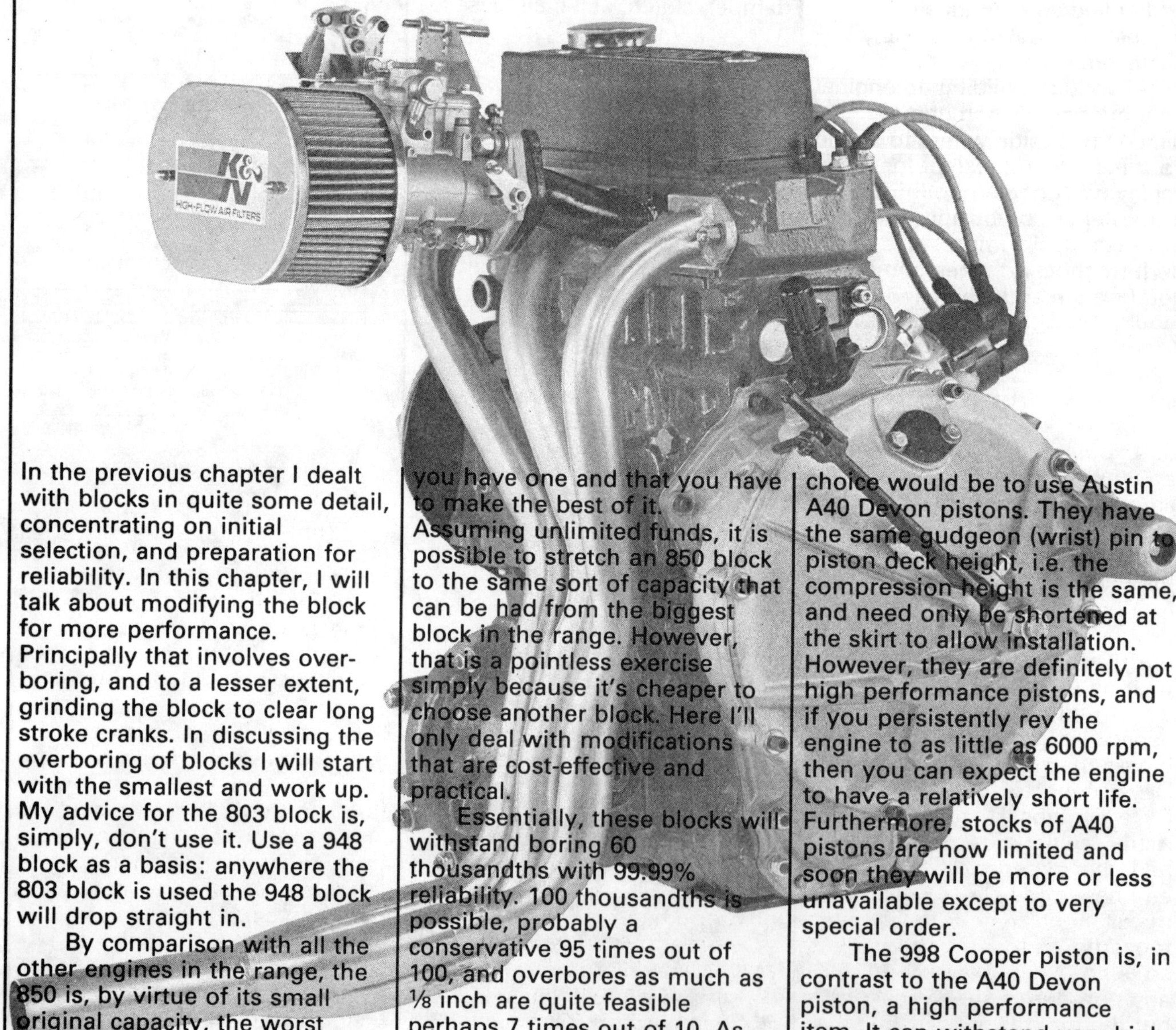

In the previous chapter I dealt with blocks in quite some detail, concentrating on initial selection, and preparation for reliability. In this chapter, I will talk about modifying the block for more performance. Principally that involves overboring, and to a lesser extent, grinding the block to clear long stroke cranks. In discussing the overboring of blocks I will start with the smallest and work up. My advice for the 803 block is, simply, don't use it. Use a 948 block as a basis: anywhere the 803 block is used the 948 block will drop straight in.

By comparison with all the other engines in the range, the 850 is, by virtue of its small original capacity, the worst (practical) block to choose as a basis upon which to build an engine. However, let's assume you have one and that you have to make the best of it. Assuming unlimited funds, it is possible to stretch an 850 block to the same sort of capacity that can be had from the biggest block in the range. However, that is a pointless exercise simply because it's cheaper to choose another block. Here I'll only deal with modifications that are cost-effective and practical.

Essentially, these blocks will withstand boring 60 thousandths with 99.99% reliability. 100 thousandths is possible, probably a conservative 95 times out of 100, and overbores as much as ⅛ inch are quite feasible perhaps 7 times out of 10. As for pistons for such overbores, there are a number of choices. At first sight, the simplest choice would be to use Austin A40 Devon pistons. They have the same gudgeon (wrist) pin to piston deck height, i.e. the compression height is the same, and need only be shortened at the skirt to allow installation. However, they are definitely not high performance pistons, and if you persistently rev the engine to as little as 6000 rpm, then you can expect the engine to have a relatively short life. Furthermore, stocks of A40 pistons are now limited and soon they will be more or less unavailable except to very special order.

The 998 Cooper piston is, in contrast to the A40 Devon piston, a high performance item. It can withstand very high rpm, far higher than is likely for all normal purposes, even racing. The only snag in using

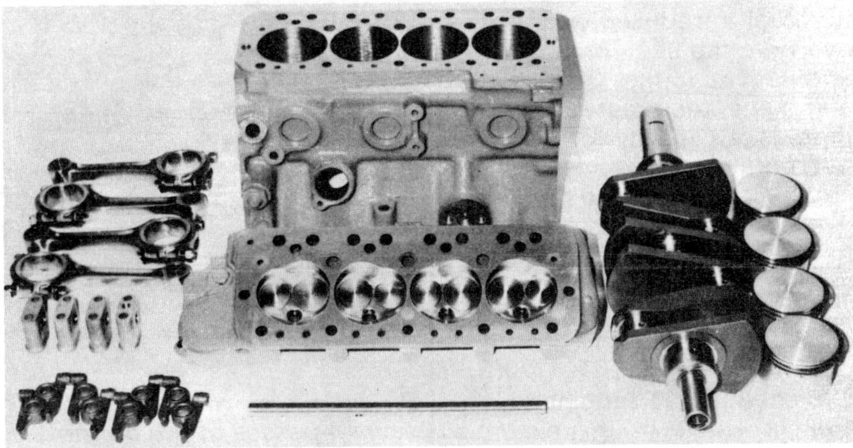

Just how big is big for a Series 'A' engine? The parts shown constitute what's necessary to build a 1600cc motor; that's about the limit.

the 998 Cooper piston is that the piston compression height is some 131 thousandths (nominally) less than the 850 Mini piston. This amount will come off the top of the block with no untoward effects whatsoever. Using this piston gives the option of oversizing the 850 Mini bore by 66 thousandths with a standard Cooper piston, or up to 60 thousandths on top of this by using one of the rebore size Cooper pistons. Using the biggest Cooper pistons will stretch the engine to 936cc and, if the block is good, will produce a very nice short stroke Mini engine. On the other hand if it is a short stroke Mini engine you are trying to build, then it would be better to start with a 998/1100 block and fit the Mini crank to it and machine the top of the block as necessary. After all the 998 block is really only an 850 block with the bores stepped up in size, so this means that with their correspondingly thicker coring, the 998 block can be taken out to a bigger size without any problems. In other words, the bore can be taken to 2.6 inch with no untoward side effects.

An interesting point especially in relation to 850 Mini blocks, has been made to me by Dave Mountain of **Mountune Race Engines.** Dave builds many racing 850s and it has been his experience that engines bored to +40 seem to make slightly better power than those bored to +60. Why this is so is not clear. There could be any number of subtle reasons, but the most likely cause is that the bore walls are flexing too much with a +60 bore. It's for this reason that if maximum hp is being sought, rather than the slightly better torque a bigger bore will give mid-range, that I recommend using the 998 block. Although the safe speed capability of Cooper 998 pistons is, for all practical purposes unlimited, if you happen to obtain a set of these pistons with the raised D-crown, when the block is decked the 'D' will hit the 'beak' on closed chamber cylinder heads. The D-section raised crown pistons, however, are no longer in general production so can only be obtained from old stock. From 1983, Cooper pistons, including the Powermax ones, are flat topped.

The inline 948 cc engine shares the same bore size that the 850 Mini does. Its extra capacity resulting from a longer stroke, i.e. 3 inch as opposed to 2.687 inch. This means that since the rod length is the same (5.75 inch) the piston compression height has to be less because the crank/rod/piston assembly fits into the same block height. Like the 850 Mini, this block can be bored up to ⅛ inch. The most obvious piston to use for an overbored version of this engine is, once again, the 998 Cooper piston.

For use in 948 engines, the Cooper piston is 25 thousandths too tall, but will withstand this amount being removed. However, it's not necessary to remove a whole 25 thousandths, simply because in standard form, the pistons rarely come to the top of the block. If the engine is to be built so the pistons are set flush to the block, then you may find that the block needs a few thousandths off it, or the pistons need a few thousandths off them. It just depends on the way the tolerances fall, so the 998 piston makes an excellent piston to use in this particular case.

A topic closely related to piston selection concerns connecting rods. The 998 Cooper engine uses a fully floating pin. 850 and 948 engines and, in fact, all early engines, used a split little end with a clamping bolt which locates in a groove in the pin. Obviously if you are going to retain that type of rod, you will need to change the pins in the 998 Cooper pistons to suit. However, a better bet is to swop the rods for the later, fully floating items. These rods are pretty much interchangeable from one engine to another. Any small bore, fully floating rod can be used for this application. As far as size is concerned, overboring the 948 motor to 2.6 inch (that's +60 on a standard Cooper piston) will increase the capacity to 1050 cc. In other words, a nominal 10% increase in capacity.

A popular move when it was first introduced was to overbore the 997 Cooper engine. The longer the stroke of an engine, the greater increase there is in capacity for a given overbore size. With it's long stroke, it doesn't take much overboring on a 997 to stretch it to 1100 cc. Hepolite used to manufacture a piston specifically for this job. Utilizing a 2.579 inch bore, the Hepolite piston (part number 18512/KR) stretched the 997 engine to 1095 cc. However, this piston is no longer manufactured, so once stocks are used up it will be necessary to seek alternative ways of stretching this engine. Fortunately there are plenty.

The difference in compression height between the 997 piston and an 1100 piston is only 48 thousandths of an inch, the 1100 piston being the shorter of the two and, this being the case, it's a simple job to bore a 997 engine out to accept 1100 pistons. Like overbores on 948s and 850s, doing this conversion with 1100 pistons, you will need to accommodate the pinch bolt little end of the 997. You either scrap the 997 rods and use a fully floating rod or you use gudgeon (wrist) pins consistent with the type of connecting rod you used as standard.

When using 1100 pistons in the 997 Cooper, a small interference problem will occur. If you retain the 997 rods, which are the pinch bolt type, you will find that the corner of the boss into which the pinch bolt screws just fouls the piston under-head area. To alleviate this, it's necessary to grind a little material off the rod to produce some clearance. Of course, a better bet is to swop to the fully floating pin rods.

The 1100 engine is the biggest of the small bore engines. It also has the longest stroke of any production Series 'A' engine, and as such, overbores on this engine

Big overbores may necessitate several passes of the boring machine, so the cost of the work may be a little more expensive but it's still a cheap way of getting power. 14-2

produce the biggest increase in capacity. Unfortunately the range of pistons available to produce extra capacity is limited, the obvious overbore size being a simple +60 which stretches the engine to 1150 cc. If you want to go larger than that, then things get a little complicated. Before 1275 engines became common, a popular conversion with the 1100 engine was to bore the block to take Hillman Imp pistons. Standard size Imp pistons at 2.677 inch diameter would stretch an 1100 out to 1215 cc and if the block could withstand a 40 thousandths overbore, this would bring it to 1251 cc. Basically the use of this piston for overbores as large as this does present a few problems, though none are insurmountable, but the success of boring to this size is not guaranteed.

Firstly, the compression height on the Imp piston is nominally some 135 thousandths less than the 1100 piston. This means that approximately 135 thousandths has to be machined from the top of the block to get the block deck height acceptable. That's no real problem. However, what is a problem is that squeezing in bores of this size necessitates a little juggling on bore centres, so that everything has the best chance of working out right. **Fig. 14.1** shows how the bores must be re-centred. The end bore (number 4) you will note is re-positioned only 15 thousandths, the reason for this being that it is possible to break into the diagonal oilway which passes to one side of the cylinder. By offsetting only 15 thousandths, the likelihood of this happening is reduced, though by no means eliminated. If this oilway is broken into because it 'wandered' during the drilling at the time of manufacture, then it is

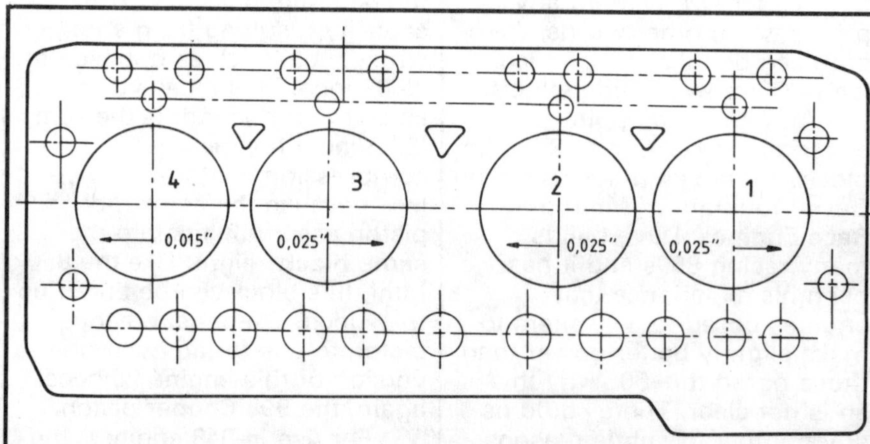

Fig. 14.1 Bore recentring for overbored 1098 engines using Imp pistons for 1215 cc plus

necessary to sleeve the hole. The sleeve, of course, needs to be a good press-fit, though not so tight that it causes the bore to crack inwards when the cylinder block is rebored to machine the side of the sleeve to match the bore itself. To make an overbore like this work, it's necessary to use a cylinder head gasket that doesn't overhang the bores. This means the standard 998/1100 head gasket is not usable. The situation requires the use of one of two types of head gasket. My first choice is the competition head gasket for 998s; part number C-AEA-647. Alternatively you can use a gasket from a big bore engine and the best gasket to use here is the copper asbestos competition head gasket for the 1275 engine. This is part number C-AHT188.

Having told you how to do this 1215 cc conversion, let me throw in a few words of caution. Although it's by no means established, it would seem that 1100 blocks prior to '64 and maybe early '65, were in fact nothing more than bored out 850 blocks. This being the case, they will not take kindly to being bored to the Imp bore size, simply because the coring around the cylinders is inadequate. If you intend to overbore an 1100 to this size, be sure the block you use is no older than mid-'65.

If it's an inline engine you are dealing with, i.e. that fitted to Sprites, Midgets and Marinas, then the situation is a little different. Generally it would seem that these blocks can be overbored with a fair degree of success.

Stroking Small Bore Motors

Overboring a block is one way to gain capacity. Increasing the length of the stroke or stroking, as it's commonly known, is another. However, within the small bore Series 'A' group, there is little point in going to all the time, trouble, effort and expense of a stroke increase on a crankshaft, or indeed having a special crank made, for one simple reason. If you have one of the small engines, it's cheaper to get the block and crank assembly from one of the larger engines and use that. The 1100 engine with its 3.3 inch stroke is as long a stroke as you should consider using, bearing in mind the bore size the engine has. If you need more capacity, then one of the big bore blocks must be used.

Generally, cranks are interchangeable from one block to another amongst small block engines. However, there is one exception, this being amongst 1100 Sprites and Midgets. The Marks 111 and 1V from 1964 on used a 2 inch main bearing crank which, of course, has to be fitted in the relevant block. With this exception, all the other cranks used the more common 1¾ inch main bearing.

Results

Increasing the capacity of any of the small bore engines produces very positive results, especially if this extra capacity is achieved by virtue of an overbore. Remember, all the engines we are dealing with are essentially long stroke, small bore engines. Anything which makes them nearer 'square' helps in extracting more power. The extra bore size not only gives extra engine capacity but also it allows the exhaust or intake valve to be unshrouded on the side adjacent to the bore, and thus makes a small improvement in breathing. Whether or not the benefits of this improved breathing are realized depends on how well the combustion chamber is matched up with the bore along the areas where their edges should be co-incidental or nearly so.

Generally speaking, any capacity increase as a result of overboring on a small bore engine will give a power and torque increase on a par with the percentage increase in capacity, and on some occasions it will give a little more, **but it's important to remember that whenever the engine capacity is increased, flow restrictions due to the valve lift, cam timing, etc., will have greater effect. Usually this results in a power increase less than on a pro rata basis with the capacity increase.** Torque increase, however, is not generally affected so much. If the engine has 10% increase in capacity, torque usually goes up by 10% because the breathing ability of the cylinder head is not generally a limiting factor in the mid rpm ranges. Whenever a stroke increase is made, it should be remembered that piston inertia will put more loads on the crank. Long stroke engines cannot be revved as high as short stroke engines. However, having said that, even 1100 engines using a small main bearing crank, can be made to rev to as high as 8000 intermittently and still retain a practical reliability level.

Bigger Big Bore Blocks

Essentially there are two reasons for overboring a big bore block. Either you are intent on producing a big bore, short stroke engine that falls into a given capacity class, or you are intent on producing, within reasonable limits, as large an engine as possible. Let's look at the first case. Except for the 970 and 1071 'S' engines, all the big bore engines, in spite of their bigger bore, are still under-square engines. A long stroke means increased crankshaft loadings and these ultimately limit the rpm potential of the engine. A short stroke produces an engine which has more rpm potential and a bigger bore into which bigger valves can fit. In other words, the bigger the bore is, the more room there is to put in large diameter valves for improved breathing. A popular move for racing, when the rules allow changes in bore and stroke, is to have special short stroke cranks made up and overbore the block 73.5 or even 74mm. 73.5mm is usually favoured simply because it leaves the block just that much stiffer and the other ½mm doesn't really seem to produce anything except possibly bore flexure, to the point where a little power can be lost. To make a short stroke, big bore engine work, it's necessary to become ultra-critical on cylinder head design, otherwise the engine rpm capability will not be utilized.

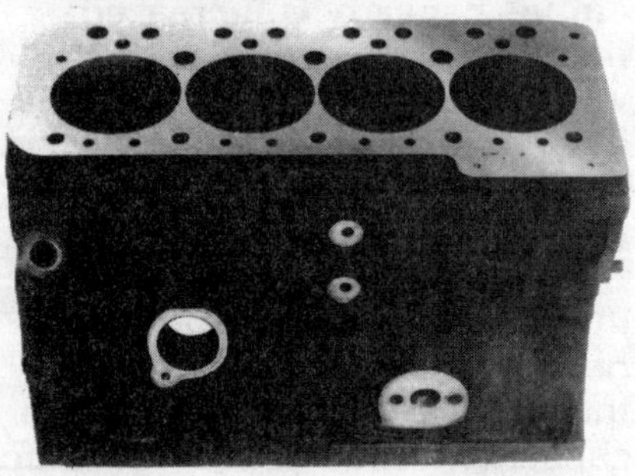

Here's a 1300 block that has been bored to 74mm. You can clearly see that the wall thickness between the centre two cylinders is down to a minimum.

To give you an idea of just what sort of stroke reduction is typical, let's take a Series 'A' engine for a 1300 cc class. The standard stroke for a 1275 'S' engine is 81 mm. If we build a 1299 short stroke engine with a 73½mm bore, then the stroke comes down to 76 mm, a very worthwhile reduction if the advantage is fully utilized by compatible breathing efficiency from the cylinder head's induction and exhaust systems. For most people, however, overboring represents a way to extra capacity, rather than a change in engine configuration to meet a capacity limit. At one time it was necessary to hunt through catalogues and find piston replacements from other engines that would do the job of overboring for extra capacity. These days, however, there is less need to do that because there are very good pistons available. Much of which was discussed in a previous chapter. What I intend to deal with now is the idiosyncracies of each type of block you are likely to encounter, and how well suited it may be to boring.

I have had reasonable success with over-boring 'S' blocks to 73.5mm, but the experience of others is less happy. This is a topic worth discussing with your pet machinist before going ahead. In view of this, if you are starting from scratch and choosing a block to bore oversize, 1300 blocks and Series 'A+' blocks are a better bet as they tend to have thicker cylinder walls. Often when boring 1275 blocks it is found that the boring operation has uncovered a section of metal which is a different colour half-way down the bore at the siamese section. Don't panic. The metal is a piece of mild steel which is used to tie in the cores from one side of the block to the other and it's simply that the big over-bore has machined into this piece of mild steel. You can safely ignore its presence as in virtually 99% of cases it causes no problems.

When engine displacement goes up due to overboring and stroking, the problem arises of getting the CR down. To prevent CRs becoming unacceptable , pistons that are heavily dished are needed, such as this Omega item.

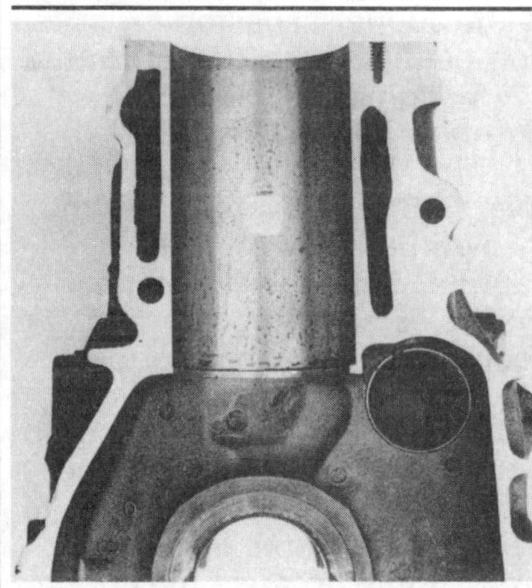

Big overbores on some blocks will cause a light patch to show half-way down the bore. This is no cause for alarm. The block is not about to fail. The light patch is caused by the boring uncovering ...

Boring out 1275 blocks, or at least those of the non-'S' type, to 74mm is usually no problem provided off-set boring is carried out. This is done to keep the wall between the centre pair of cylinders roughly the same as those between the outer pairs of cylinders. The off-setting is achieved by spacing each pair of cylinders

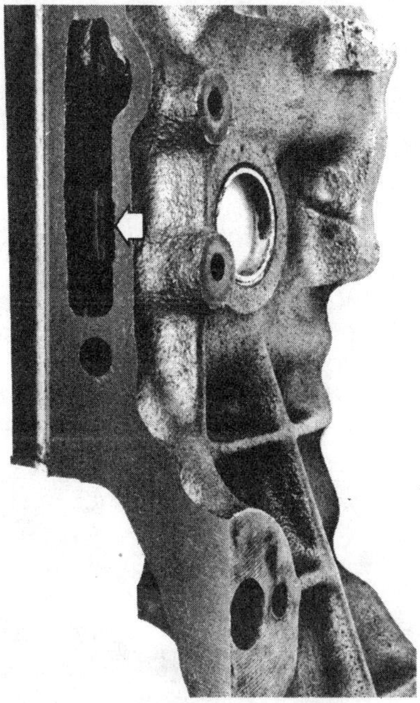

... a steel insert (arrowed) which ties in the sand cores from one side of the block to the other. When the sand is removed from the casting, the steel insert remains.

Sectioned here is a 1275 Metro block bored to 74mm. As you can see there is still adequate cylinder wall thickness, and at these overbore sizes, there is little problem with porosity, but ...

... that doesn't mean that problems can't occur. It is possible to break into one or both of two oilways close to the bottom of the block. Here the bore has just broken into the diagonally orientated drilling that is the continuation of the oil pipe fitting union hole on the distributor side of the block. This is the pipe that takes oil from the number 4 cylinder down to the oil filter. About this same point ...

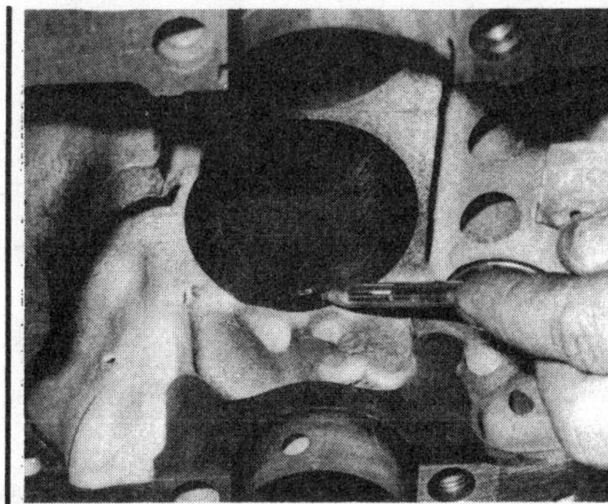

... it can also break into the horizontal drilling from the pressure release valve. Either hole can be sleeved to cure the problem.

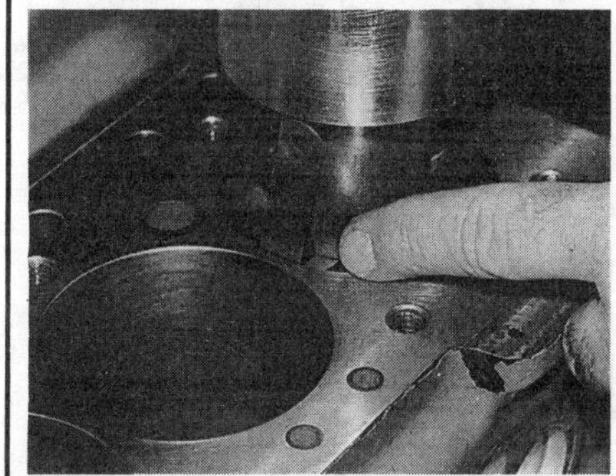

If you're contemplating boring a block to 73.5 mm leaving the 74 mm piston size for a 'last re-bore' before the block's scrap, then it is advisable to offset bore the block as if it were going directly to 74 mm. If you don't the new offset needed when finally boring to 74mm will not clean up the old 73.5mm bore.

out from the centre by 10 thousandths. This leaves a situation where the centre-to-centre distance of the centre pair of cylinders is increased by 20 thou, whereas the centre-to-centre distance of each outer pair of cylinders remains the same. If a block is bored to 73.5mm there is no need to offset-bore unless a further rebore to 74mm is anticipated after the bore has become worn. It is difficult, sometimes impossible, to offset-bore a block already at 73.5mm; so make your mind up before it's too late.

When blocks are being bored to this size, it is possible to break into oil galleries. The most obvious one is the oil gallery that continues down from the oil pressure release valve. There is another gallery that goes across the bottom of the bore on number 1, and this also can be broken into if the drilling has wandered. Apart from the fact that it may leak oil, it's of no great consequence because this part of the bore never gets subjected to gas pressures, so simply sleeving the block there and re-boring it will cure the problem.

If the block is to be bored more than 74mm, then a twilight zone is entered where success is by no means guaranteed. The normal overize used when going to bores bigger than 74mm, is 74.7. This is the diameter of a Triumph

Precision machining of the block is important. Not only must the bores be accurate prior to honing, but also they must be square to the crankshaft centreline. If the bottom of the block is truly parallel to the main bearing bores, then the block can be bored using the bottom face as a reference. Otherwise the block should be bored with reference from the main bearings. 14/54

Although a block looks a substantial piece of cast iron, it's not as stiff as you think. Bolting the cylinder head on distorts the top part of the block. To simulate cylinder head and securing bolt torque loads, I prefer to use a 'torque plate' on my blocks while they are honed. This ensures that the cylinder bore is round when the cylinder head is on. Depending upon the over-bores involved, bores can distort by as much as half a thousandth for the first one inch down from the block face.

piston normally used in the 2.5 litre 6 cylinder engine. These pistons, because they have a very low compression height, can only be used with cranks of 81mm stroke length or more. With a standard 81mm stroke, it's necessary to mill a large chunk from the top of the block to get the correct deck height. However, the longer the stroke used, the less the amount that will be required to come off the top of the block.

Personally I don't like to use the 74.7mm bore. There are

Here's a close-up of the 'torque plate'. The smaller numbers give the tightening sequence and the larger ones are the Allan bolt tightening torque to simulate the torque loads produced by the cylinder head nuts. Remember, the threads in the block are coarse whereas the threads on the holding down studs are fine, hence, the difference in the torque value used.

several reasons for this, the principal one being that the cylinder walls are too thin. On a high powered engine, there is always a lot of blow-by, and controlling oil consumption at high rpm and heavy load conditions is difficult. In fact, at 74mm bore, this problem is already evident and is only reduced to tolerable proportions by using a good piston, such as the Omega, for which the ring pack, normally supplied with such a piston, is designed as far as possible to cope with flexing bores. Boring out another

Bore finish is important, not too coarse or too fine. Using the torque plate during the honing operation and with the ring combinations normally used, a 20 micro-inch finish seems to give excellent overall results. 14/57

0.7mm merely exaggerates the problem. This doesn't mean to say it can't be worthwhile in terms of power, but it's definitely getting to the point of not being a nice engine.

Stroking 1275s

The standard stroke length on a 1275 engine is 81mm. Whether this can be changed depends, simply, on which 1275 engine you have. Amongst 1275 engines there are basically two configurations of crank other than the inline or transverse crank types: the difference being in the big end bearing. Cooper 'S' and 1275 Sprite and Midget engines share a common 1.625 inch big end bearing. 1300 and 1275 GT cranks, and in fact all non-'S' transverse engine cranks and Marina cranks, have a 1¾ inch big end. The larger diameter cranks are made of lesser material; BL's engineers concluded that a little more section thickness would offset the fact they are not made of such strong material. As it happens, these big bearing cranks still have ample strength, even when their bearing journals are ground to the 1.625 size. Because of this **by offset grinding a 1300 crank to 1275 big end sizes, it is entirely practical to produce a stroke approximately 3mm longer. This, together with a 74mm piston produces an engine of 1445 cc.** Obviously the use of such a crank will necessitate the use of 'S' or MG Midget connecting rods. These can be quite expensive items, so actually stroking the crank is going to mean the price of the work on the crank, plus a set of rods. Having acquired these rods, a little modification will have to be done. 1300 rods are narrower in the big end area than 'S' or Midget rods. Since their place is being taken by a wide rod, it will be necessary to narrow the 'S'/Midget rods to the width of a standard 1300 rod so it will fit between the cheeks of the thrust face on the big ends.

On a high performance engine, aim for a rod side clearance of around 12 thousandths, that's generally more than the factory call for but these cranks flex more at longer strokes and high rpm.

Having seen how easy it is to acquire an 84mm stroke crank, you may be anxious to know if a still longer stroke can be easily and simply attained without an expensive billet crank. It is possible to get an 86mm stroke out of a standard, unmachined virgin forging. **Cranks machined on a virgin forging are cheaper than billet cranks and an 86mm stroke crank with 74mm bore will produce 1480 cc,** but if you are prepared to pay out for a billet crank, then it's possible to squeeze in stroke lengths as long as 91mm. This will give 1556cc with a 74mm bore or 1596cc with a 74.7mm bore. For stroke lengths of this magnitude, an 'A+' block is not the one to use. In strengthening the Series 'A+' block, BL engineers put metal in high stress places and took it away from others where it is, in fact, needed if these very long strokes are used. If you are going to build yourself a real monster Series 'A' engine, go for the pre-Series 'A+' blocks. This means a 1275/1300 block or a Marina or 1275 Sprite or Midget block. Avoid the early Sprite and Midget blocks; try to use a casting that is post-'68 as these have more metal around the cylinder bores. Long strokes will go into 'S' blocks also but don't try using them with bore sizes much over 73½mm.

When long stroke cranks up to 91mm are used, it's necessary to grind a little clearance into the block. This is what it takes for rod clearance with a 91mm stroke. A similar amount of clearance has to be ground into the gearbox if the engine is for a transverse installation.

Keeping It All Together

At these big displacement sizes, the Series 'A' engine is obviously being stressed to a far higher level than was ever intended. The longer the stroke of the crank the more problems there are associated with its use. This is especially so if the engine is tuned to the limit in other areas. In other words, building a full race engine based on a big bore and long stroke, can present problems. I haven't built many engines of the super big displacement category and when I have, I have been careful not to run them into an rpm regime where they are likely to break, simply because I cannot afford the breakage. However, Steve Roberts, a racer of modified sports cars of some repute, has considerable experience of long stroke cranks in big displacement Series 'A' engines. After finding the breakage of 'conventional' cranks seemed to be the rule rather than the exception, Steve looked at the possibility of increasing crank strength. A successful formula proved to be a billet crank with 2⅛ inch mains rather than the 2 inch. The block, main bearings, saddles and caps were inline bored to take Ford 1600 crossflow bearings, the bearings being drilled to get the oilways aligned. This configuration of crank was tried with big end bearing sizes at standard 'S' size and an extra oversize which utilized Vauxhall Viva bearings in a specially made connecting rod. To give you an idea of how successful this type of crank was, when standard 2 inch 'S' sized mains and standard big ends were used, the crank life before it broke, was about 1500 miles and that's using a 7500 rpm limit on an out and out race engine. With the big main bearing and a standard 'S' sized big end, or the extra oversized big end, crank life is in excess of 4000 miles and at the time of writing neither of these sizes of crank had broken.

Another move to improve crank strength concerns the increase in the size of the crankshaft tail when used in a transverse configuration. These big tail cranks have no C-clip in them. To use a big tail crank requires a special **Jack Knight** straight cut primary gear. To go with this, a special splined hub, also produced by Jack Knight, has to be used for the clutch plate, so that the clutch plate can be modified to suit the primary gear. The last and most obvious modification, that the flywheel centre has to be machined out so that the taper suits the taper on the crank.

Of course, a special technique is necessary to install such a crank having no C-clip, so this is how it's done. Firstly, install the flywheel on the crank without the primary gear and measure the distance from the end of the crank to the end face of the flywheel, that's the face that the flywheel securing bolt clamps down on. It's not necessary to do the bolt up super-tight. Just sufficiently to make sure the flywheel is seated on the taper. Having done this, strip and install the primary gear with the thrust washer on the block end of the tail only. Re-measure the distance between the end of the flywheel and the end of the crank, and then machine the back side of the flywheel to give 50 thousandths clearance. When the flywheel is bolted up hard, the clearance will come down to 25 thousandths. Again, the correct technique is needed to put the flywheel on so that it stays on. Firstly lap the flywheel to the crank to ensure they are a perfect fit. When installing the flywheel, heat it for one hour in

Here's a long stroke crank with standard 'S' size bearings. It's readily apparent from this view that there is very little overlap of the journals. As a result, the crank is fairly flexible. Such flexibility may (though it is by no means certain) lead to an early fatigue failure if high revs are used for any length of time. By comparison ...

... this big journal crank with its 2⅛ inch main bearings and 1300 size big ends, has noticeably more overlap, thus making the crank more rigid and more able to cope with the long stroke it employs. This particular crank used in Steve Roberts' Mod Sports racer has done two years of racing with revs up to around 8000 rpm.

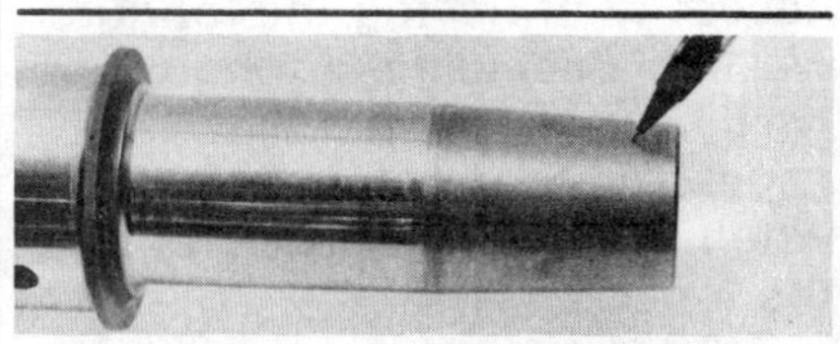

Here is the tail of a special, big journal crank. Note the lack of a key slot. Also note the high quality finish on the taper of the crank, although the flywheel has, in fact, been on and off on numerous occasions. Although having to transmit the power of a very high powered Series 'A', there is no fretting on the crank, a condition which would indicate the flywheel was working loose. The text tells how the flywheel was installed to achieve this state of affairs.

The secret of making a long stroke crank survive at high rpm for a respectable mileage is to increase the size of the bearings. The main bearing indicated here is 2⅛ inch diameter.

an oven set at 425° F. After heating, fit the flywheel in place. Using a new securing bolt each time, torque it up just as tight as it will go. Tight enough, in fact, so you feel that the nut is just on the point of letting go. At a guess, this is around 250 ft lb. A point I should make here is that with such a combination, it is not necessary to slot the end of the crank. The grip of the flywheel on the taper is more than sufficient to transmit the power from the crankshaft to the flywheel and out through the primary gear without the necessity of having the tongued washer. Because the tongued washer is not required, it's not necessary to do any alignment when fitting the flywheel onto the taper.

After everything is bolted down hard, check that the primary gear is not being clamped. To do this, relieve the pressure on the pressure plate by backing off the clutch bolts and check to see that the transmission can be spun over when it's in gear. This will prove that the primary gear is free to rotate.

Flywheel Removal

The foregoing procedure works very well indeed from the point of view of reliability, but to strip the assembly down requires something of a technique as well. Firstly, the ⅜ inch flywheel bolt holes used to attach the puller, are inadequate. These must be drilled out to at least ½ inch and tapped accordingly. It will also be necessary to make up a frame so that a hydraulic ram can be placed between the end of the crankshaft and this frame. A 30-ton hydraulic ram, at least, will be required to remove the flywheel from the taper; even then the application of heat to the flywheel may be needed. **When the flywheel does come off, be sure you are not standing in the flight path.** Though this all sounds pretty brutal, Steve Roberts has found that the combination of big crank/flywheel etc., has been very reliable.

Dampers For Strokers

Whilst on the subject of reliability, crankshaft vibrations in long stroke cranks must be considered. The longer the crank stroke, the greater the weight of crankshaft and the more critical the use of a crank damper becomes. If lightweight flywheels are used, there is less of a problem with crankshaft vibrations as the vibration frequency goes up to a higher level. However, a longer stroke crank itself brings down critical vibration frequencies to a lower level in the rpm range. Big bearing cranks are inherently stiffer than small bearing cranks, and as a result their vibrational frequency is generally moved up the rpm range. It is probably for this reason that the big bearing cranks survive, whereas the small bearing ones don't. It is fatigue which finally kills a crank. However, **to get the best reliability from whichever crank you have, it's necessary to pay particular attention to the crank damper.** Generally speaking, big bearing cranks are relatively trouble-free when a standard 1275 Mini GT crankshaft damper is used. However, it's a good idea to Loctite the crank damper onto the crankshaft with bearing-fit Loctite.

If a stroked crank is used with small bearings, the crankshaft damper has a lot more work to do, and often the amount of energy it has to absorb, causes damper distress. You can usually spot this by virtue of the fact that the outer ring of the crank damper is slipping on the rubber core. When you see this, scrap the damper. If the damper isn't functioning properly, your crank may not be long for this world. Richard Longman & Co can supply a large diameter Freudenberg damper specially developed for long stroke engines. Alternatively, if you wish to fit a viscous damper contact Mike Parry at Race Techniques.

Big Motors Part 2

Power Potential

We have all heard the expression 'there's no substitute for cubic inches', and generally it is true, but it is possible to over-step the mark by making the modifications in one area too extreme. Usually this comes about by machining standard components to such a point that they are too weak or incur some other disadvantage which ultimately detracts from the power output. **Before investing a lot of money in a monster Series 'A' engine, stop and consider just what advantages are achieved at various points along the road to increasing capacity. The biggest is not necessarily the best.**

No doubt many Series 'A' enthusiasts have toyed with the idea of building a monster motor, the sort, say, in which is used a crank with a 91mm stroke and bore of 74mm plus in diameter. This combination, as you can see from **Fig. 14.2**, stretches a 1275 motor all the way out to around 1600cc, and at first sight, it seems like an exciting proposition. The question is, are 1600 engines worth the effort or not? Certainly, many of us, in Europe at least, have seen them in competition, especially on television in rallycross, and they seem to pay off in terms of performance, but **do not forget that nearly all the 1600 motors seen in top events are also driven by top drivers, many of which have demonstrated their winning capabilities even with engines down on power.**

To try to establish the real worth of a 1600 unit in terms of power potential, I, along with Nolan Pitts of Auto Electrical (Plymouth) and Keith Dodd of Mini Spares Centre, set about building two engines of as near identical configuration as possible with the exception that one employed an 81mm stroke (standard) and the other a 91mm stroke. In other words, it was going to be a shoot out between a 1400cc and 1600cc engine.

In building the engines reliability was a fundamental requirement, for if an engine were to break half-way through a test, it would be reliability or lack of it that was being tested, rather than the power difference. To minimize the breakage factor, it was decided not to build a pair of out and

Fig. 14.2 Engine capacities at overbore sizes and using various pistons

Standard Bore size	Standard engine capacity	Capacity at standard oversize			Alternative oversize pistons			
		+0.020"	+ 0.040"	+ 0.060"				
2.477	848 (68.26 mm Stroke)	862	876	890	A40 Devon pistons			
					Std	+0.020"	+0.040"	
					919	933	947	–
					998 Cooper pistons			
					Std	+0.020"	+0.040"	+0.060"
					894	908	923	937
2.477	948 (76.2 mm Stroke)	962	978	993	998 Cooper pistons			
					Std	+0.020"	+0.040"	+0.060"
					998	1014	1030	1046
2.457	997 (81.3 mm Stroke)	1010	1027	1043	1098 Sprite/Midget pistons			
					Std	+0.020"	+0.040"	+0.060"
					1065	1082	1098	1116
2.543	998 (76.2 mm Stroke)	1014	1030	1046	–	–	–	–
2.543	1098 (83.72 mm Stroke)	1115	1132	1150	Chrysler/Talbot/Hillman Imp pistons			
					Std	+0.020	+0/040	
					1215	1233	1251	–
					Oversize bore diameter			
					73.5 mm	74 mm	74.7 mm (Triumph 2.5 pistons)	
2.780	970 (61.01 mm Stroke)	984	998	1012	1050	1065	–	
2.780	1071 (68.26 mm Stroke)	1086	1102	1117	1160	1176	–	
2.780	1.275 (81.3 mm Stroke)	1293	1312	1330	1380	1399	1426	
Bored & Restroked Big bore Engines		Stroke Length in mm						
		58.9 (For short stroke 1000)			999			
		76.5 (For short stroke 1300)			1299			
		75.5 (For short stroke 1300)				1299		
		84			1425	1445	1473	
		85			1442	1463	1491	
		86			1459	1480	1508	
		87			1476	1497	1526	
		88			1493	1514	1543	
		89			1510	1531	1561	
		90			1527	1549	1578	
		91			1544	1556	1596	

out race engines, but to build two relatively high powered road engines.

To say that the engines could be built identical other than for stroke length is, in fact, easier said than done. It seems simple enough, but certain complications arise which may or may not affect the validity of the tests. Personally I say they do not, simply because where those differences lie, they are unavoidable and a natural inheritance of building a 1600cc rather than a 1400cc engine. I'm sure that point is going to need some explanation, so let's look at it in more detail.

Components which are normally considered to have the greatest effect on an engine's power curve were, in fact, identical. For instance the cam

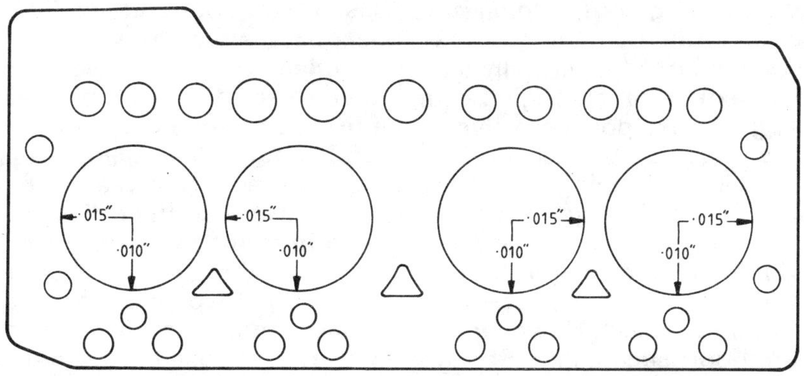

Fig. 14.3 Bore recentring for overbored 1275 engine using 74.7 mm pistons

and valve train in both engines were identical. In this instance, parts were swopped from one engine to the other so they did, in fact, constitute an 'identical' valve train. This also applied to carburation, intake and exhaust manifolds. Ignition systems were also identical, although ignition advance curves were optimized for each size of engine. Ancilliary components fell into the identical category also.

Two components which did prove difficult to make identical in type were the pistons and the cylinder head. Both engines used Omega pistons. The 1400 unit had dished pistons. On the 1400 fitting these pistons, together with the cylinder head having minimal metal out of the combustion chambers, was all that was required to achieve the desired CR. On the other hand, achieving the same compression for the a 1600cc engine presented more of a problem, and to do this it was necessary to utilize a combustion chamber which had the floor deepened, as well as machining the piston crowns to remove the dish and leave the piston short of the top of the block at TDC. This meant that the combustion chamber configuration of the two engines did differ. The valves, too, were different, those used on the 1600 being bigger. This, however, was a necessary move. When the combustion chamber floor is moved upwards, it causes more of an abrupt turn in the port. This, in turn, necessitates some sharper turns for the incoming air to negotiate, and results in less efficient airflow. Although the cylinder head on the 1600 head sported bigger valves, the flow figures between this head and the one on the 1400 were, in fact, comparable. The only difference was that the gas speed through the valves in the big valve head would, for the same cfm, be slightly lower. As to how these differences affect the power, I can only guess. From previous experience, I can say that although these differencies would affect the power, they would be minimal by comparison with the 200cc difference between the two engines. However, **it's important here not to lose sight of the fact that the differences in combustion chamber and piston shape are modifications which will almost inevitably have to be done to build a 1600. This means they legitimately represent an accountable factor in a comparison between a 1400cc and a 1600cc unit.**

The crank used in the 1600 engine was supplied by Paul Ivey's Race Engine Components company. As was to be expected, the crank was very nicely produced, but it didn't employ any of the special trick features common to Steve Roberts' cranks. This crank was simply a straightforward 91mm crank, having 2 inch mains and 1¾ inch big ends. Remember, this particular engine isn't an all-out race engine and it was being dynoed at relatively low rpm for the sake of reliability.

Again on the subject of reliability, machining the top of the Omega pistons to get the 10:1 CR on the 1600 engine, left the top ring very close to the piston crown. Initially there was some cause for concern here, and I did wonder just how long the piston would last. Normally if the top ring of a piston is too close to the piston crown, the land above the piston overheats, collapses and traps the top ring. At this point the engine loses compression and consequently loses a lot of hp and suffers extreme blowby. Fortunately, on our test engine over the entire time of dyno testing, which represented a considerable amount of full throttle work, the close ring to piston crown proximity caused no apparent problems. As discussed in the piston chapter, rings which are as close to the top of the piston as possible, are, in fact, better for hp so long as no reliability problem is incurred, so at least this situation is on the positive side as far as the 1600cc engine goes.

Pros & Cons On the Numbers Game

As I said earlier, it's easy to assume that the 1600 engine must make more hp than the

1400 simply because it has an extra 200cc representing a 14% increase in capacity. But let's consider what's on the debit side that may make that extra capacity less than proportionally effective. To determine whether or not that is going to be the case, it will be necessary to consider some theory, starting with the cylinder head.

Virtually all big bore, Series 'A' engines, that's everything from 970 'S' through to 1275 and more if it has been stretched, utilized the same cylinder head. This means that all the air the engine consumes has to pass through the same ports and the same valves. This is irrespective of the size of the engine the cylinder head may be used on.

Let's now consider the fact that the power developed in the cylinder of the engine is proportional to the amount of air consumed. Very basically, the amount of air consumed depends on (a) the flow capability of the cylinder head, and (b) the engine rpm. Since all our big bore Series 'A' engines can utilize the same cylinder head, we should find that it is the cylinder head which is the limiting factor to power output, not the capacity of the engine. For instance, in theory a 1000cc engine should develop the same power as a 1300, but at proportionally higher rpm, i.e. 30% higher rpm. Most of us are aware, however, that theory in this instance does not follow what happens in practice. 1300 engines DO make more hp than 1000s, although it should be pointed out here that the 1000cc engine tends to make more hp per litre than the 1300, so where does the discrepancy lie between our theory and practice?

The discrepancy lies in the inertia of the air that the engine is drawing on. After every intake stroke, the air in the port adjacent to the valve has to come to a complete standstill, and the energy put in to start its motion at the beginning of the induction stroke is virtually lost. To an extent, on an engine with a single intake port per valve this energy can be recovered. The sudden stopping of the air causes a shock wave to run back up the induction system and if it's ultimate return is timed right, it can produce a 'shock ram' effect. However, a shared port such as the Series 'A' has, produces such a jumble of shock waves, that trying to utilize useful ones and reject ones that are not, is virtually impossible. On the Series 'A' engine, shock wave tuning, as I detailed in an earier chapter, is like chasing the pot of gold at the end of the rainbow.

Because of air inertia the small engine never gets close to the power of the big engine simply because it runs out of breath due to the more rapid stop-start motion of the air at the higher rpm involved. Since the continual stopping and starting of the air is not conducive to power, we find that turning a big engine at low rpm is likely to give us more total air consumption and hence more total power than turning a small engine at proportionally higher rpm, so from the total air consumption point of view (as opposed to volumetric efficiency) a bigger engine scores positively. However, that isn't the end of the story. The next point to consider in the chain of breathing events is whether the Leyland 5-port head, even when it's really well modified, has enough airflow potential for a big engine. Looking at the question another way, if the engine is too small, then the need for very high rpm defeats its purpose as a power producer. If the engine is too big then the requirement for high flow capability from the cylinder head becomes the stumbling block. This now poses the question, what is about the optimum size block assembly to put under a 5-port head? **Over the years, experience has indicated, and I have to add this is by no means a firm assessment, a short stroke, big bore engine around 1200cc seems to make the best power usage of the cylinder head in terms of hp per litre.**

From the point of view of high rpm, it would appear that a 1600cc engine simply will not perform because it can't breathe at high rpm, but that is not the only reason why it will never be a high rpm engine. A 91mm stroke and a bore that's just a tad over 74mm comes nowhere near being a square configuration. In fact it's about as long a stroke engine as is practical, and there's probably no other engine currently in production in the world, powering a car which has a bore to stroke ratio of this magnitude. Not only does the long stroke impose considerably higher inertial loadings on the crank, but also much of the extra power produced in the cylinders is used up by internal friction in the engine. This is because in a 1600cc engine, the piston has to travel up and down a much greater length of bore, and as a result, more power is lost to friction.

With the aid of some experimental data and a computer, I calculated the frictional hp lost in both 1400cc and 1600cc engines. The results of these calculations are shown in **Fig. 14.4**. You will note that at 3000 rpm the big engine absorbs 3 hp more from the power developed in the cylinder than the small one. At 6000 rpm it absorbs 6.3 hp more of the power developed in the cylinders than the smaller engine. This means that whatever power may have been gained by stepping up the capacity, will be reduced by the difference between the two figures.

From these figures it can be

Fig. 14.4 Comparison of bhp absorbed by internal engine friction of 1400 and 1600 Mini engines

rpm	Calculated absorbed bhp		Difference
	1400	*1600*	
3000	13.6	16.6	3.0
3500	14.9	18.3	3.4
4000	16.3	20.2	3.9
4500	17.9	22.2	4.3
5000	19.5	24.4	4.9
5500	21.3	26.9	5.6
6000	23.3	29.6	6.3

seen that the overall losses are much in line with those deduced theoretically. For figures of what can be achieved in reality refer to **Fig. 14.5** which are the results of dyno testing the two engines discussed earlier. From these figures, it's apparent that if you were looking for big top end numbers, you would be disappointed. A 1600 engine just does not produce significantly different top end power outputs to a 1400. However, looking down the rev range and inspecting the torque figures shows that the 1600 engine does crank out a lot more low end hp and torque. It's obvious that the only justification for a 1600 motor is one which will produce a lot of hp without necessarily turning any high rpm. In certain respects this is fortunate because a crankshaft of this stroke length is not capable of turning very high rpm and staying in one piece for very long. It appears fairly certain that with cylinder heads as used on the 1600 engine, a realistic breathing limit was being approached, and at about the same time as reaching a practical rpm limit on the crankshaft.

Was the 1600 engine dying off in terms of top end hp because of the extra friction involved, or because it was running out of breath? Fortunately, with the numbers produced in **Fig. 14.5**, some comparisons can be made from which some reasonably valid conclusions can be drawn. Taking the power developed in the cylinders of the 1400 engine and comparing it with that developed in the cylinders of the 1600 engine, it can be seen whether or not the 1600 engine is reaching the limit of breathing capability of the cylinder head.

The power developed in the cylinders is, of course, the flywheel hp added to the frictional hp, and if you look at the chart **Fig. 14.6** you will see in column 1, the power developed in the cylinders of the 1400 engine. In column 2 we have the power the 1600 engine should develop in the cylinders if the breathing ability of the cylinder head is just as capable of supplying the 1600 as it was the 1400. In column 3 we have the power that was actually developed in the cylinders of the test 1600. You will note that up to 4000 rpm the figures correlate generally within about 2 hp, which suggests the cylinder head is perfectly capable of supplying air to the 1600 engine in this rpm range. At 5000 rpm the power figures begin to diverge by about 4 hp. At 6000 rpm this trend has progressed and there is about 8 hp difference between what the engine should have developed and what it did develop in the cylinders. This suggests that the 1600 engine is now beginning to gasp because of lack of breathing ability in the cylinder head.

Conclusions

So what has been learned here? The most obvious conclusion is that a 1600 engine for out and out hp production will not be successful. Here, it is worth noting some comments made by Paul Ivey of Race Engine Components. Paul's company is prepared to sell you whatever stroke length crank you want for your Series 'A' engine, but his advice is that cranks between 84 and 86, maybe 86½mm, are about as far as it appears to be worth going. These stroke lengths will produce engines, when a big bore is used, between 1430 and 1485cc. Going much bigger in size than this helps torque but seems to gain very little in the way of extra power. Also the engine loses a lot in terms of rpm potential and ultimately in reliability. In other words, **unless rpm are kept down, a 1600 unit might only be one step from a 4-cylinder, cast iron, Series 'A' grenade.**

Another important point to note here is that if big motors are to produce anywhere near the extra power their added displacement warrants, then the

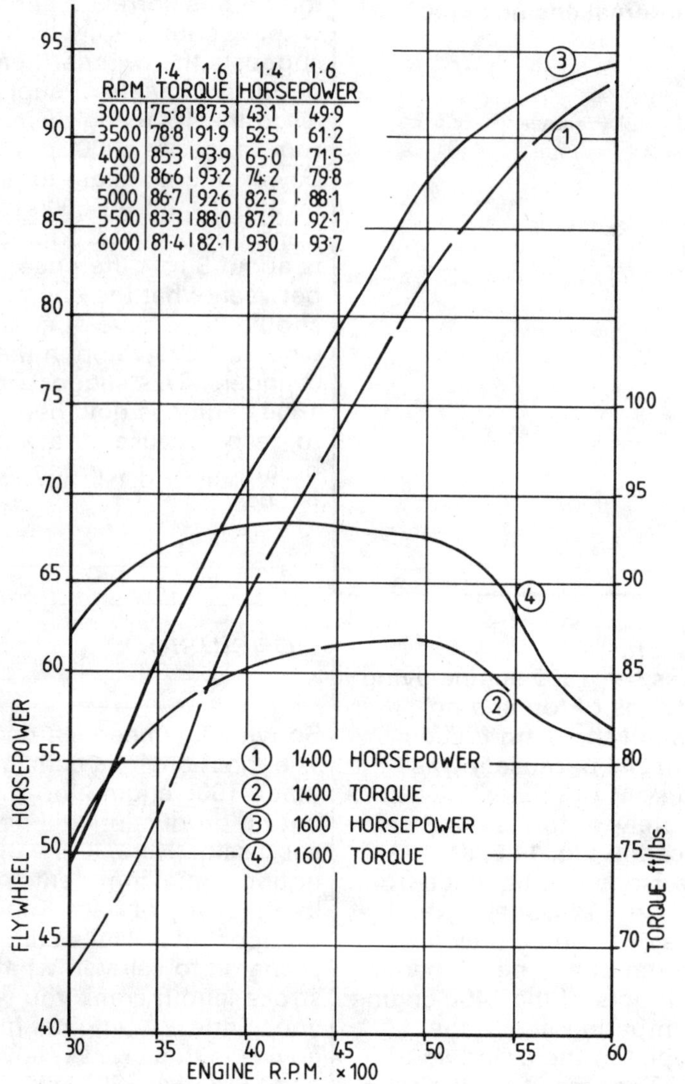

Fig. 14.5 Comparison between 1400 cc and 1600 cc Series 'A' engines

cylinder for a given valve opening period. We can, however, accomplish the same goal by using a cam with more timing, so the valves are held open longer. If the cylinder head is a liability due to the fact that it has (a) small valves and (b) inadequate flow capabilities for the engine's size, then installing a long period cam produces some positive results without necessarily incurring the negative aspects normally experienced if that cam were installed in a smaller motor. By virtue of the small valves and the larger capacity of the cylinder, the gas speed through the valves and ports is higher. As a result, **a big displacement Series 'A' engine will come 'on the cam' much sooner than a small one.** For instance, if you use a 649 cam in a 1000cc 'S' engine, the engine is almost undrivable at full throttle under about 3800 to 4000 rpm. When it hits 4000 rpm there is enough gas speed in the induction system and ports to maintain gas in the correct direction through the engine. That is, in through the intake and out through the exhaust. The effect of back pumping is now overcome and the engine gets up on the cam and really starts to produce power. With a big cylinder head used becomes more critical. **The bigger the engine is the more important it is that the cylinder head has good flow capabilities. This applies to ALL Series 'A' engines from 1275 up.** Whenever the capacity is increased, it's necessary to increase the breathing ability.

It's important not to lose sight of the fact that there is more than one way of getting air into an engine. A high efficiency cylinder head allows more air to be packed into the

Fig. 14.6 Practical and theoretical power comparison

rpm	1 bhp developed in cylinders of 1400	2 Column 1 multiplied by 1.142	3 bhp developed in cylinders of 1600
3000	56.7	64.7	66.5
3500	67.4	77.0	79.5
4000	81.3	92.8	91.7
5000	102	116.0	112.5
6000	116.3	132.0	123.3

The figures in col 2 represent the power a 1600 would give by comparison with the 1400 in col 1 if nothing other than the capacity changed. In other words col 2 is a direct scaleup in proportion from 1400 to 1600.

displacement engine, this effect happens much sooner. A 1600 engine can utilize a 649 cam and yet be tamed at the bottom end to be perfectly drivable for normal road use. In a 1600 engine such a camshaft can be made to pull from as low as 1500 rpm with no real difficulty. Torque output at such low rpm, that is 2000 and under, is usually less than a good 1400 would produce with a shorter, more streetable cam. However, by the time 3500 rpm is reached, the use of a 649 cam over a more normal road cam will show torque increases of around 7 or 8, and sometimes as much as 10 ft lb extra. Peak power with a 649 cam as opposed to a hot street cam, say a 544 profile, is generally up by 4 or 5 hp, but even so, with a 649 cam, a 1600 engine is still no super fire breather. Using a Super Sprint cam adds further to the top end power of a 1600, but begins to seriously detract from the bottom end hp.

Having stretched the engine to 1600cc, it isn't really producing any real extra hp that a 1400 can't produce. A 1400 engine, by comparison, is a highly revvable engine. Of course you may think a 1600 is justifiable if engine rpm are to be kept down, as would be the case for a road engine, but another aspect of a road engine is that it should also run smoothly. Well, here's some bad news: Series 'A' engines with stroke lengths of 91mm or close to that figure, are anything but smooth running. When doing these back-to-back tests, I had originally intended to do some vibrational checks but the 1600 engine ran so roughly compared with the 1400 engine, that there was no need to instrument either engine to find out the difference. I would find the vibration level of a 1600 engine to be unacceptable in a road-going car. A 1600 is a 'shaker' and it would require very sophisticated engine mount technology to be able to isolate all the additional vibrations from the car.

Based on my experiences with big engines, I have concluded that building a 1600 engine for anything other than maximum torque delivery such as might be required in rallycross, is really a waste of time. Even then it seems probable that the torque output of a rallycross engine could be equalled by development in other areas of the engine. The 1400cc engine represents the best value for money in terms of big displacement Series 'A' engines. It is very good on rpm potential, excellent on torque, cheap to produce and reliable. If I were building an engine where cost was not a factor, then I would use the 74mm bore size and increase the stroke to around 86mm to produce an engine of around 1480. This, I feel, represents about the best all-round compromise in terms of reliability, power and torque. These conclusions tend to be borne out by performances on the race track as opposed to a rallycross circuit, as 1480cc engines seem to be more than a match for 1600s and a well-built 1400 is only marginally slower and considerably cheaper.

15 Lubrication Systems & Oils

An oil pump can be likened to the heart of an engine: without an effective oil pump to send the life-sustaining flow of oil to the vital parts of the engine, a seizure would occur in a very short time. Like a living counterpart, the more powerful the engine is (equating to greater physical strength) the more critical the oil pump and lubrication system can become.

On early inline Series 'A' engines the oil pump was marginal for the job. When fitted to transverse engines the pump was found to be inadequate as it often suffered from debris collected from the gearbox because the engine shares the same oil as the transmission. Fortunately for us, the days of the inadequate oil pump have gone. It was probably the Cooper 'S' engine that started the trend towards the adequate to more-than-adequate pump design. Prior to the introduction of the 'S', most Series 'A' engines, and especially the 850 Mini, suffered from oil pump maladies. Whilst the pump was adequate when in perfect condition, the slightest bit of wear inevitably meant that there wasn't enough pump capacity present and many 850 Minis simply went out of operation because of low oil pressure due to a worn pump. With the introduction of the 'S' engine, most engine tuners of that era opted to fit the 'S' pump to most of the other engines in the range. Because of the difference in block length between a small bore engine and a big bore engine, it was necessary to take the oil pump drive shaft out of the 'S' pump and replace it with a driveshaft from the small bore engine pump. These days, such operations are not necessary, since any pump you buy will have a capacity at least equal to that of the original Cooper 'S'. This is because BL figured life would be easier if pumps were standardized across the board. The change to large capacity oil pumps took place in about 1969. Since then there have been numerous changes in oil pumps, some of which have been brought about by changes in camshaft design whilst others were related to the capacity of the pump.

The first significant change in pump design, at least in the area that concerns us, was the change from the original pin drive which had served for many years. On the introduction

Shown here is one each of every type of drive used on the Series 'A' engine's oil pumps. Left to right we have spider drive, pin drive and slot drive. If that sounds like it's round the wrong way, bear in mind that the terminology refers to the part that's driving the pump, that's why the centre pin drive has a slot in the end of the pump because the slot is driven by the pin in the cam.

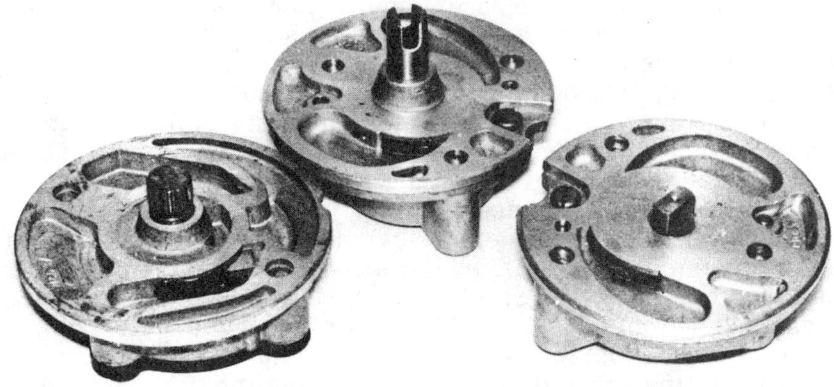

Look at the shaft of the oil pump on the left. Note that it is longer than the one on the right. Left is a pump for a small bore block, right is a pump for a large bore block, such as 1275 'S'.

of the 1300 engine, a new type of drive from cam to pump, popularly known as the 'spider' drive, was fitted. On the introduction of the 'A+' engine, yet another type of drive was introduced, namely the 'slot' drive. All these drives must be used with the appropriate cam.

As far as interchangeability goes, there seems to be endless possibilities; some are useful, others are pointless. However, here are a few tips on the interchanges you might want to do.

The first substitution is to adapt a spider drive pump to a small bore engine. Remember that the spider drive pump was developed for the 1300 engine; that's a big bore engine. You may have a cam profile that is fairly common in spider drive on a spider drive cam, but not readily available in pin drive (here I am talking about original Leyland cams). To be able to put such a cam into a small bore engine it will be necessary to use the spider drive pump along with some adaptation. First an oil pump spacer will be required to pack out the oil pump to the relevant position in relation to the cam. Also it will be necessary to use the spider to actually couple the oil pump to the cam, but of course that will be obvious when you come to assemble it.

Another factor to be considered is that although current pumps (as of 1984) have a three-bolt fixing, early spider drive pumps could have 2 or 4-bolt fixing. You may have such a pump. If your oil pump was in good condition, then

If for any reason a spider drive pump is to be fitted to a small bore block, a spacer will be needed as shown here to the right of the pump. In some instances ...

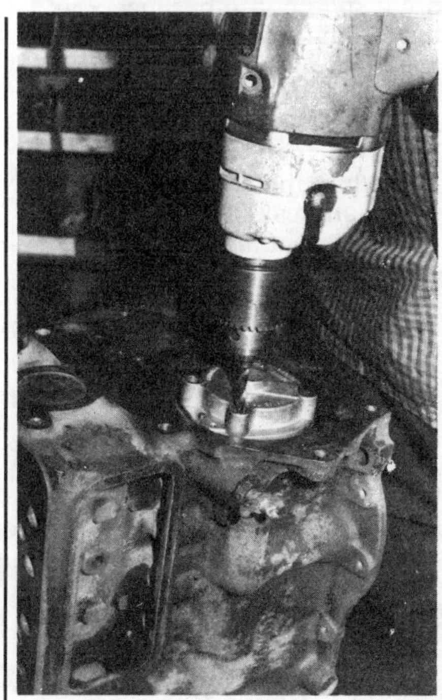

... it will also be necessary to drill extra securing holes for the pump in the block and tap them out to the relevant size for the fixing bolts.

there will be no need to replace it just because it had the wrong number of fixing holes. It's quite a simple job to bolt the pump in place, using one of the two holes that do line up and then drill and tap through on the other two. There's no need to go overboard on either depth or the amount of thread in the other two holes. 5⁄16 inch of full thread will be more than adequate for the job.

Many of the spider drive pumps have three instead of four fixing holes, only one of which will line up, in which case use the same procedure and drill in the extra holes to secure the pump. If you have a Cooper 'S' and wish to install a spider drive pump because you have a spider drive cam, then the operation is simple. It is just a question of drilling any extra holes necessary and bolting it up. No spacer plate will be required. If you have a 1300 block and you wish to use a pin drive style pump because you have a pin drive cam with a profile you want on it, then you simply use a Cooper 'S' pump and again drill any relevant fixing holes as may be needed.

When the slot drive style of pump and cam enters the picture the number of combinations available goes up dramatically. Basically, if you wish to use a slot drive pump with a pin drive cam, you will need to make up an adaptor along the lines of that shown in **Fig. 15.1**. If you are going to use a slot drive cam and pump in a small bore engine, you may or may not need a spacer between the pump and the block, as per the spider drive situation. Conversely, if you were using an 'A+' small bore block, i.e. the 998 Metro block, then the slot drive pump and cam combination will fit.

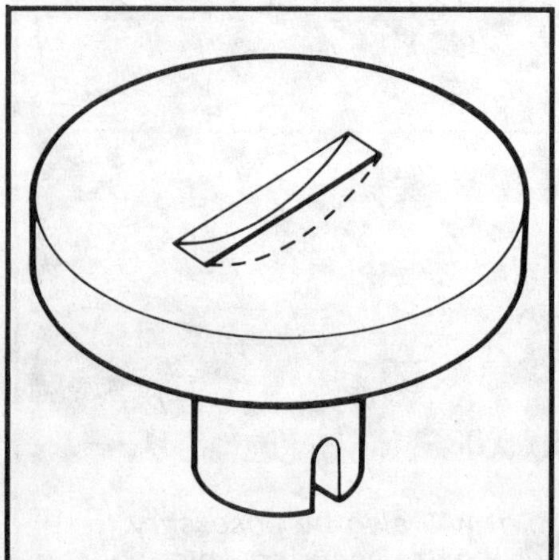

Fig. 15.1 Adaptor for converting pin drive cam to fit slot drive pump

Pump Capacity

It has been Leyland's policy for some years now to source products made outside the factory from more than one supplier. Oil pumps are an item not actually made within the Leyland group. Two different manufacturers supply pumps to Leyland and they are notably different in design. For all normal purposes, either pump is adequate for the job in terms of being able to supply the engine with enough oil, but one pump has about 20% more capacity than the other. Under normal conditions, it is not a good idea to put on an oil pump of more capacity than is really needed because it takes hp to drive it. On the other hand, if you have a turbocharged engine and you have modified the block as detailed in the relevant chapter so that oil jets cool the underside of the pistons, you will need to select a pump with the maximum capacity possible to cope with the extra oil flow requirements. Identifying the pumps is straightforward. The high capacity pump has a black steel plate on the outer face. This steel plate is secured by two cruciform dome-headed screws. The other pump has an all-aluminium case and is the type that should be used for all other applications other than in a turbocharged engine.

Oil Pressure

The oil pressure on most engines as they come from the factory is 40 to 50 psi. Increasing the oil pressure for high performance applications is easy enough. It's just a question of using a stronger spring or some packing under the spring that is currently installed in the pressure release valve. An excessively high oil pressure is not needed, nor even desirable. I have seen many engines which have had the oil pressure release valve set to give 80, 90 even 100 psi. Oil pressure this high does not protect the bearings any more effectively than does oil pressure around 60 to 65 psi. What it can do though is cause more hp to be absorbed in driving the oil pump in the first instance, and secondly it can unnecessarily raise the oil temperature. With some engine installations, controlling oil temperature at reasonable levels can be a problem; high oil pressure may do nothing more than aggravate that problem.

On the SCCA Mini that A.P.T. sponsors, the oil pressure is set by this adjustable pressure release valve. Oil pressure is set to give 60 psi hot - no more, no less.

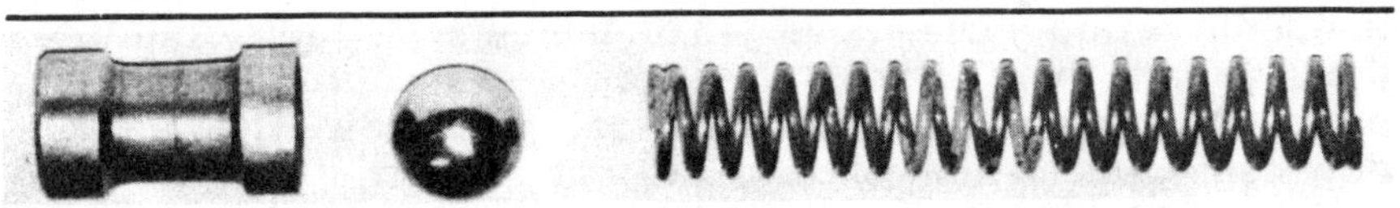

If you want to prevent sticking of the standard oil pressure release plunger, should any dirt or debris from the gearbox get trapped around it, then swop to a ball bearing as shown here. To the left of the ball bearing is the seat that is pressed into the block. If you have just built a new engine and you are not getting any oil pressure, check that the seat is, in fact, installed.

If you find after you have built your engine that you are having the reverse problem, i.e. no oil pressure, and you have used a brand new block, then check that the oil pressure release valve seat is, in fact, installed. Contrary to what is often supposed, the oil pressure release seat is not machined in the block but is pressed in after the block has been machined to accept it.

Still on the subject of low oil pressure, sometimes the oil pressure release valve can become jammed. This happens most often in transverse engine installations, primarily because debris produced by the gearbox is subsequently picked up in the pump. From here debris is passed on to the pressure release valve, and because the standard pressure release valve piston is not a good fit in the bore, it can sometimes jam in the open position. Here's what happens: when the engine is started up from cold, the oil, being thick, does not flow easily through the galleries. The oil pressure release valve, in controlling the oil pressure, lifts a long way off its seat and allows excess oil to go back to the sump. Any grit or particles or debris present could cause the plunger to stick at this very wide open position. As the engine warms up, so the oil becomes more fluid. Under normal circumstances the oil pressure release valve will shut right down to a small opening. However, it is stuck at the large opening developed during the cold start phase. As a result, the oil pump no longer has enough capacity to feed the engine and the wide open oil pressure release valve, and the net result is low oil pressure.

There are two simple solutions to this problem. The most obvious is to make the oil pressure release valve plunger a loose fit in its bore. Reducing its diameter by about 5 thousandths usually cures the problem, Alternatively, you can use a 9/16 inch ball bearing in place of the piston. (The balls in the inner driveshaft joint are 9/16 inch.) This achieves two things at once; it increases the oil pressure and it largely obviates the problem of sticking. You may find that using a ball instead of a plunger, causes the oil pressure to go too high, in which case progressively shorten the spring until the desired oil pressure is achieved under hot running conditions.

Oil Coolers

In the early days, many performance Series 'A' engine builders were almost paranoid about oil temperatures. Based on the sort of oils produced at the time, this was, perhaps, justified. If oil temperatures went just a shade too high, the engine's bearings would disintegrate. It was in these sort of circumstances that oil coolers became standard wear on any race Series 'A' engine. However, modern, high quality oils have a higher temperature tolerance than was the case 10 or 15 years ago. Synthetic oils have, by comparison, an extremely high temperature tolerance, and it is these oil improvements that have lead to the oil cooler being less of a necessity. It doesn't necessarily mean you can do without one, but it does mean that the typical Series 'A' engine up to about 90 hp or so, whether raced or not, doesn't really need an oil cooler, so long as overheating of the coolant system is nowhere in sight.

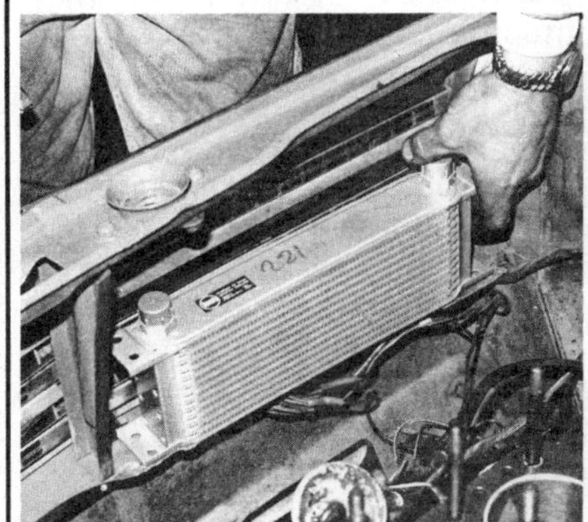

In warm weather, oil temperatures in highly tuned engines can soar. It can pay to check your oil temperature to see if you need a cooler. Oil temperature of 250° are on the high side. 220 is much better in terms of engine reliability and power.

Assuming you are starting with a typical, good quality 10-40 grade oil, the best power output in most instances occurs when oil temperatures reach 210 to 220°F. In reality it isn't so much the oil temperature that allows the production of power, but the fact that high temperatures cause the viscosity of the oil to drop. Most good quality conventional oils will function at this temperature as long as mileage with the oil is limited. Of course if temperatures go much above the 220 mark, there is the possibility that the oil will break down and cause a lubrication problem. Many engine builders like to run engine oil at temperatures around 175°F (65°C). Whilst this provides safe conditions for bearings it is not hot enough for maximum power or economy to be achieved. Also many blow-up products are not evaporated off, making necessary more frequent oil changes. Synthetic oils can have the edge for two reasons. They have a lower viscosity to start with so less power is lost pumping or churning the oil and they have much greater tolerance to overheating. Many top engine builders use only synthetic oils, Mobil 1 being, it seems, the favourite.

Whether or not a conventional oil cooler is used on a competition car, plumb in an oil temperature gauge to measure the sump temperature. Depending on the conditions prevailing on race day, the oil cooler should be blanked off so the oil temperature runs somewhere near the 200° mark.

Actually plumbing in an oil cooler is a simple enough task. If an oil to air cooler is used, then it is plumbed in between the block and the filter and takes the place of the pipe which runs across the block in the upper side near number 4 cylinder, down to the oil filter housing situated between number 1 and number 2 cylinders.

The introduction of the turbo Metro, saw the use of an oil to water cooler that is installed between the filter and filter housing. The use of such an oil cooler is fine, provided the water temperature required is, in fact, cooler than the oil temperature. However, there is one factor that needs to be taken into account. By transferring heat from the oil to the water, a greater load is put on the engine's radiator. If the radiator was of marginal capability prior to using an oil to water cooler, then the engine may run significantly warmer when oil heat is dumped into the water as well. On many vehicles, but principally the Metro, the radiator has far more cooling capacity than the engine really requires, but on vehicles such as the Cooper 'S', the standard radiator cannot cope with even hot summer conditions in England, let alone the temperatures found in summer in the U.S.A. If a condition such as this is likely to be encountered fit an oil to air cooler.

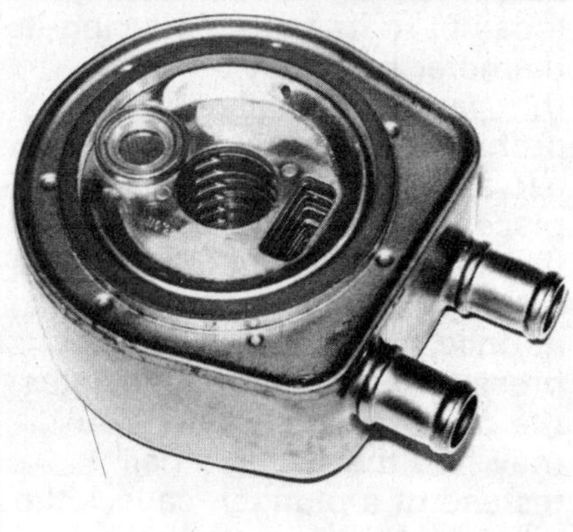

Shown here is the oil to water heat exchanger used on the MG Metro. This unit is positioned between ...

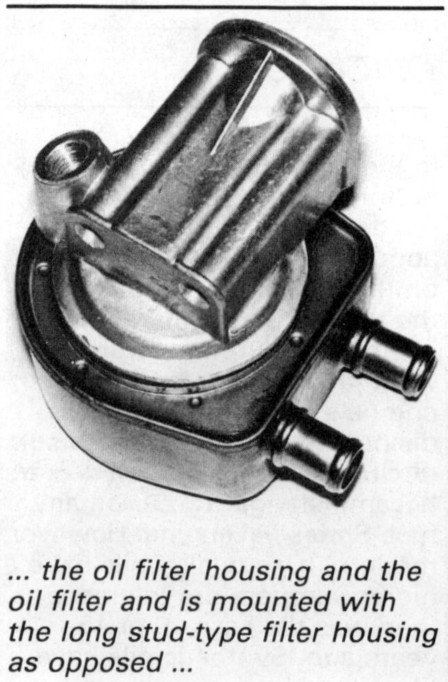

... the oil filter housing and the oil filter and is mounted with the long stud-type filter housing as opposed ...

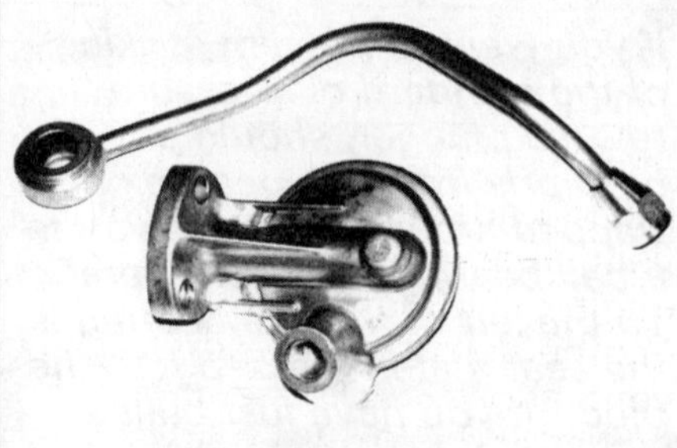

... to the type shown here, which can either be mounted on short studs or bolts. Also worth noting is the fact that with a change of filter housing, a small change in form of the oil feed pipe was also made, although they will interchange, it takes a little bit of bending to make an oil pipe from one type of filter housing, fit another.

Over the years a number of filter housings have been used. Originally filters were secured to the housing by a long bolt, passing through the filter can and a replaceable element was installed in this can. Later filter housings employed a spin-on filter and two types of filter housings were available for use with the spin-on filter. One housing utilized two long studs to mount it to the block and the other utilized two short bolts to fix it. By changing from studs to bolts or vice versa, you can use either filter housing on virtually any block. However, if you are to use an oil to water oil cooler, then it is necessary to use the housing that mounts on the long studs. If you swopped from the short bolt fixing filter housing to a long stud one, so the oil/water cooler could be used, then it will be necessary to either slightly rebend the oil pipe that goes from the top corner of the block down to the filter housing, or buy the relevant oil pipe, as they are marginally different.

Crank Case Ventilation

Having adequate crank case ventilation on a Series 'A' engine is important. If the engine doesn't have good ventilation, it will lose some power and quickly establish every possible point for developing an oil leak. At least two breathers in the region of ½ inch to ⅝ inch bore must be fitted. For a competition engine or a really high performance road engine, three or even four breathers will be needed to keep crank case pressures from going significantly above atmospheric. On road going engines, do not just vent the crank case to open air, as the engine will breathe through this pipe. Not only can pollutants from piston blowby go out of this pipe, but also grit in the air can be drawn in as the engine cools off. To stop this happening and to prevent engine oil from being splashed around the engine compartment, use one of the small crank case ventilation filters. This allows fumes and gases out but won't allow dirt in. Filters are available from **K & N** and also BL.

Series 'A' engines like to be well vented. These special K & N vent filters are ideal for the job. They allow gases to go out but stop all dirt going in.

It is not commonly realised that most Series 'A' oil filters are of the bypass type. If the oil is cold and thick or the filter is starting to clog the least bit, the bypass valve opens and some of the oil bypasses the element. Installing a large remote filter of the full-flow type, as shown here, ensures clean oil under all conditions.

The oil separator shown here in the transmission case of this Mini engine proves to be a handy little device. It will plug straight in to the Leyland plastic breather that bolts on to the case and allows. . .

. . . the connection of a rubber hose from the transmission case to. . .

If you are building a competition engine, another avenue to deal with crank case pressures is open to you. This technique, which has become popular in U.S.A. is known as an 'evacupan' system. **It has been found that if the crank case pressure can be reduced below atmospheric pressure, the amount of oil that can go up past the rings is greatly reduced.** This brings about a number of advantages. First, reducing the amount of oil which goes past the oil rings means a higher CR can be used, simply because the engine does not now suffer octane reduction due to combustion chamber oil pollution. Secondly, as a result of the first factor, tension on the oil rings can be reduced, thus reducing bore friction. Thirdly, reducing pressure in the crank case seems to encourage the oil to de-aerate faster and spend less time being whirled around by the rapidly rotating internals of the engine. Again, this leads to a small increase in power. My own test on big bore engines have indicated that about 2hp extra can be expected, mostly at the top of the rpm range. Another aspect of the Evacupan system is that it is a good engine condition indicator. My current (1987) evacupan systems will create a depression of 25-30 inch of water in the crankcase. This figure starts to drop, say, 15-30 minutes running time before the engine starts to go off tune due to seal loss at the rings or valves. The indications have proved so consistent in this respect that I now regularly use a vacuum gauge connected to the sump t0 check that an engine is still up to scratch for producing accurate test data. The drawing **Fig. 15.2** covers the layout of such a system.

. . . the exhaust extractor and anti-backfire valve assembly situated in the exhaust system. The Evacupan unit shown here was capable of pulling as much as 36 inch of water vacuum in the crankcase of a Mini engine. 15/61

Basically it works like this: at idle the vacuum in the intake manifold draws air out of the crank case, through a PVC valve. When the engine is under load, there will, of course, be no vacuum in the intake manifold, because the butterflies will be wide open. Under such conditions, the high speed passage of the exhaust in the collector part of the exhaust manifold will cause suction on the pipe connected to the anti-backfire valve. This will draw air out of the sump. If a backfire should occur and the pressure in the exhaust line goes up, the anti-backfire valve will, of course, close, preventing the backfire going all the way to the sump and possibly producing a sump explosion. The anti-backfire valve also seals the system at idle so that exhaust isn't drawn into the intake manifold via the crank case and the PVC valve. For this system to work, it is essential that back pressure downstream of the collector is minimal. This means the system must be employed either with an open exhaust or an extremely unrestrictive silencer. If the silencer presents any back pressure, it will close off the anti-backfire valve and no evacuation will occur.

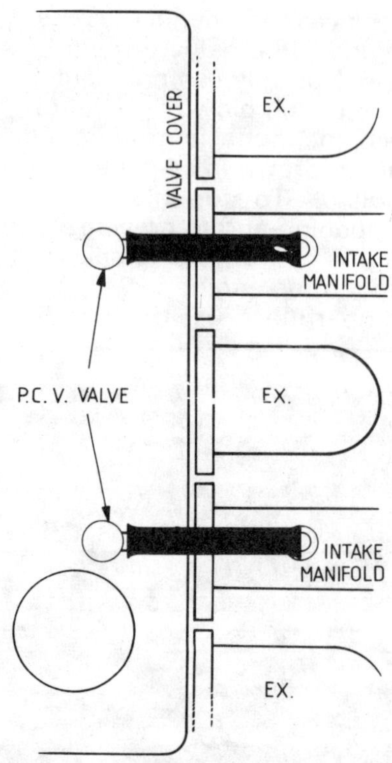

Fig. 15.2a Installation of PCV valve in evacusump system

Turbo Motors

Earlier, the significance of oil temperatures and hp was discussed. If your engine is turbocharged anything beyond relatively moderate boost levels then it's a good idea to use an oil cooler that holds the oil temperature below the figures previously quoted as optimum for normally aspirated engines. The reason for so doing is that you will in all probability be using the oil to cool the pistons. It will do this job better if it's kept at a lower temperature in the sump. As far as a turboed engine is concerned, extracting the last hp is not the name of the game. Once a turbo is on the engine, enough hp can be developed to grenade the engine. The real trick is to keep the engine in one piece, so oil temperature needs to be biased toward generating reliability rather than hp.

Another factor you ought to take into account is that turbocharger centre bearings can be easily wiped out by exposing them to a hot soak. If your turbocharged engine has been run at high speed for any length of time and then rapidly brought to a standstill and the engine turned off, a substantial amount of heat from the turbo housing, which was probably running at or close to red heat, will be dumped into the centre section. The temperature of the oil in the centre section can exceed 600° F. At these temperatures it is literally burned to a powder which is non-lubricating. I shouldn't need to say that turbo centre bearings are precision items, and for them to run dry is disastrous. If, when the next time the car is started, the engine is immediately brought up to relatively wide throttles under load, the turbo will spin up to the typical ultra-high rpm it normally runs at before a fresh supply of oil reaches the bearings, and the result will be wiped out bearings.

The way to avoid this situation is to allow the engine to idle for a few minutes after making any hard runs prior to switching it off, and when you start up your turboed engine, let it idle for 30 seconds or so, so that a fresh supply of oil is delivered to the turbo centre bearing.

Of interest is the fact that some turbochargers are now employing water cooled centre bearings. These water cooled centre bearings keep the temperature well below the point at which the oil is fried to death, so the problem of insufficient lubrication is virtually eliminated. Even so, it's still a good idea to let the engine idle for probably 10 or 15 seconds before pulling away.

Fig. 15.2 'Evacusump' crankcase pressure reducing system

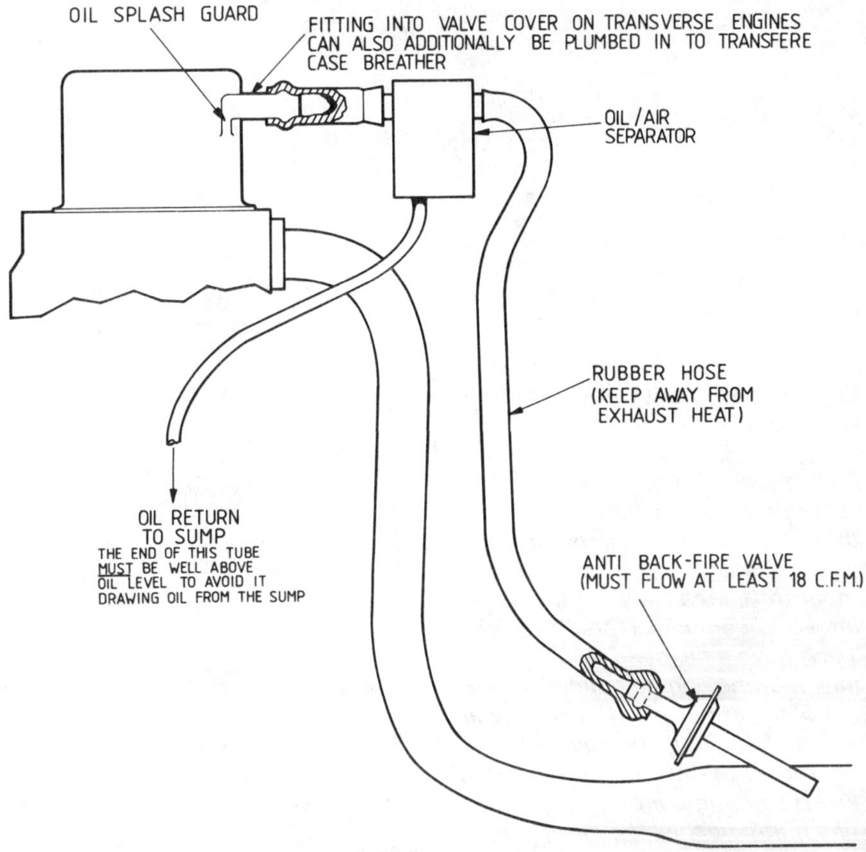

Shown here is an 'evacusump' system. This system draws air out of the crankcase on the engine so as to reduce the pressure in the crankcase. This has the effect of considerably easing the amount of work the oil control rings have to do. With this system, even worn oil control rings can still prove capable of good oil control. A few years ago, when high octane fuel was available, such niceties as this were not really needed. If fuel lost the odd octane due to oil dilution, it was of no great consequence. These days, however, there is no octane margin if service station pump fuel has to be used in high performance and race engines. (This is the case in UK at least.) This evacusump system is effective enough to make a real smoker into a clean burner. Because it's so effective at cutting down oil consumption, tension on the oil control rings can be reduced, thus cutting bore friction and allowing the engine to produce more power. Another side effect is that it appears to cut down windage losses. This means less power is lost churning the air and oil around in the crankcase.

Here's how the system works: at idle, intake manifold vacuum is used to draw a vacuum on the crankcase. Depending upon the type of manifold used, either 1 or 2 PCV valves can be plumbed into the crankcase at the rocker cover, as shown in the sketch here. When an intake manifold vacuum exists, air will be drawn from the crankcase via the valve cover and the fumes will simply be burnt in the combustion chamber. When the throttle is opened, of course, the intake manifold vacuum will go and this is where the second phase of the system takes over. As exhaust speed increases in the exhaust collector, air will be drawn out of the pipe that is plumbed into the collector. The anti-backfire valve located in this pipe is positioned such that air can be drawn out of the system but if a backfire should occur, pressure cannot travel back up the pipe. If this item is not put into the system, the engine can easily experience a sump explosion if a backfire takes place. The pipe from the exhaust evacuation system can be plumbed into the valve cover or the drop gear or transfer gear cover on a transverse engine. If it goes into the transfer gear cover, then it's necessary to make sure there is a good oil separator in the system and one that will drain oil back into the sump, otherwise the evacusump system will draw oil out of the engine. For this reason, I prefer to plumb into the valve cover, and as a precaution against any oil being drawn out, install an oil/air separator with a drip line going back to the sump. This, of course, must be plumbed in well above the oil level. For the system to work effectively, the engine has to be virtually air tight in every respect apart from the air that can be drawn out through the various vents that are associated with the evacusump system. The easiest way to check that the engine is reasonably air tight is to plug all the holes that are supposed to be in it, and then use a bicycle pump with the end of the plunger reversed so it can be used as a vacuum pump. You then connect this up to a water manometer. With this simple setup, it's possible to check that the engine will hold some sort of vacuum. Most engines will hold 15 to 20-inches of water vacuum before air starts to leak in past the oil seals. Another point of leakage will, of course, be the distributor and dipstick hole, and it will be necessary to grease distributor bearings, otherwise oil in the bearings will be drawn out. If you really want to get fussy about this, then of course some sort of seal in the distributor can be used, though I have never tried it this way.

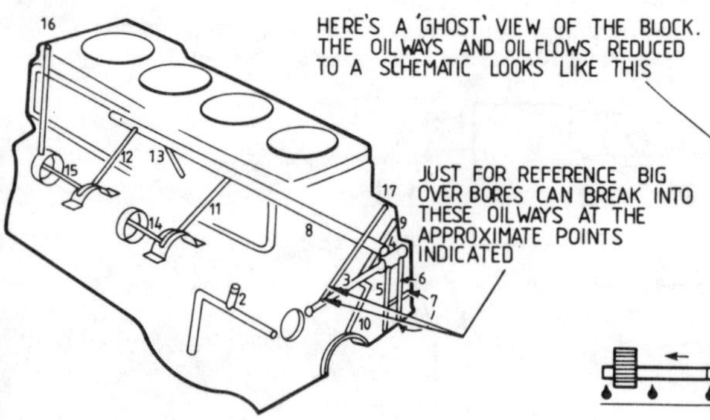

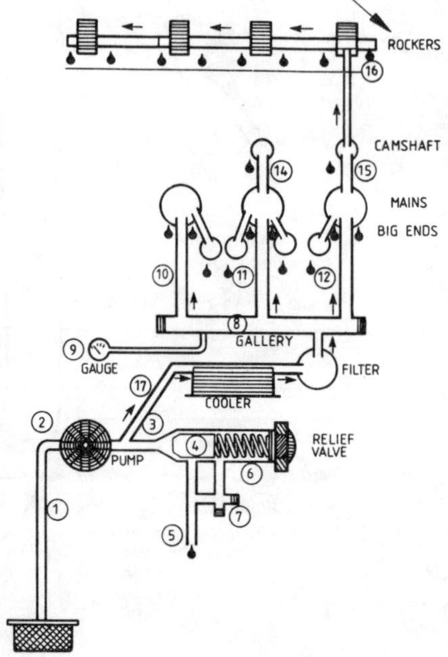

Fig. 15.3 How oil travels through the block

Oil is sucked from the sump through the gauze 'filter'. It transfers via the transmission case and the rear corner of the block (near the clutch slave) to the pump intake. Older blocks (pre 85H, 99H, 10H) have a hexhead plug enroute at (2). Some engine builders modify the block at this point to include an external oil filter, so supplying filtered oil to the pump thus greatly extending pump life.

The pump pushes the oil via intersection (3) through the cooler (if fitted) to the filter. It is worth noting that all the oil passes via the filter before getting to the main gallery (8) which runs the length of the engine on the distributor side, though it is not all filtered due to the 'by-pass' nature of the filter itself. This gallery is a great parking space for bits of metal and gravel and other abrasives which gather there until the block is used for competition. This is good reason to draw out the plugs at each end and clean out the gallery very carefully.

All pre-rod linkage units have filters with replaceable cartridges which, if they get sufficiently clogged, will lift off their seats and allow unfiltered oil to pass into the gallery (8) and on to the bearings. Some engines, of about 1966 to 1969 vintage, had an electric sensor built into the top of the filter housing which switched on a yellow light in the dash. The modern spin-off filters do not have provision for this facility.

Subject to the foregoing, clean oil passes from the filter to the main gallery (8). The oil pressure in the filter and gallery is determined by the relief valve (4) and its associated spring (6). Oil pressure in the gallery (8) is registered on the oil pressure gauge (or sensed by a pressure switch and associated yellow bulb) at (9). Excess oil released by the relief valve falls via drilling (5) to the sump. The drillings at (7) dispose of any oil which passes behind the valve and might otherwise cause a hydraulic lock.

The schematic diagram shows clearly the points at which oil escapes from the pressurized channels. The relief valve, because it normally receives unfiltered oil, can be a source of pressure loss. Worn crankshaft shells are usually an effect rather than a cause of low oil pressure, except when they are destroyed by dislodged debris that was not removed from the gallery during a rebuild.

16 Ignition Systems Part 1

So as to put everything into perspective, let's start off with a simple basic ground rule as far as ignition systems and the Series 'A' engine are concerned.

The Series 'A' engine, what ever type it may be, is not overly sensitive in the ignition department. As long as the spark is adequate to fire up the mixture effectively no further improvement in this area will be necessary. A super spark which falls into the artificial lightening category, does very little to actually increase the power output. This isn't to say it won't increase the power output, but **the gains achieved by a super-sophisticated ignition are generally small when compared with that achieved with a totally functional, standard, contact breaker-type ignition system.**

The foregoing is a basic view looking up the scale from a standard spark to a super spark. Going the other way and regressing from an adequate spark, the Series 'A' engine doesn't have to suffer much spark degredation before both power and economy very rapidly fall off.

As an overview of the benefits of electronic ignition, it would be fair to say, assuming a reliable ignition system is chosen in the first instance, the main advantages will be that you should have a service-free ignition system. The elimination of contact breaker errosion or the elimination of the entire contact breaker assembly (depending on the system used) will mean performance should not drop off with increasing mileage. In addition to peak ignition capability maintenance, there is likely to be a small increase in power and economy, but if you are looking at the economics of the situation, you may find it will take a long time to get your money back in fuel saving.

Do not assume that spark intensity is all that matters when super tuning an ignition system. Because of the trend during the late '70s and early '80s of reduced fuel octanes, the ignition timing and advance curve have become super-critical if the best power curve shape and maximum economy are to be achieved. At this point it is worth expanding on the term 'the best power curve shape'. **Any distributor with any advance curve can be set to produce the engine's true maximum power.** For instance, if an engine takes 32° of full advance (crankshaft) to produce maximum power at, say, 7000 rpm, then any distributor, regardless of its advance curve, can be set to give this 32° advance. However, if a distributor does not have the right advance curve, the engine could, at, say, 3000 rpm, be delivering 5 hp less than it's capable of, or worse yet, it could be over-advanced to the point where it detonates the engine into uselessness. These days the most important single factor associated with distributors and ignition is tailoring the advance curve to be compatible with the current low octane fuels. Because manufacturers are striving to achieve good fuel economy, they are producing engines with as high a CR as possible. In the lower rpm ranges, the octane required to run ideal ignition timing is higher than is available. This means if the ignition timing was set to optimum, the engine would detonate. To get over this factor, the ignition advance curves are developed to keep the engine below the detonation limit, rather than to give maximum output. This poses the question, why such high CRs? The simple answer is fuel economy. High compressions are used because they increase part throttle fuel economy quite dramatically. Cast your mind back to what was said in the cylinder head chapter, namely: high CRs equal high economy. The factory's strategy has been to build an engine which will run at nearer optimum during part throttle use for good fuel economy. The penalty has been a CR that is a little over the top under Wide Open Throttle (WOT) conditions if optimum WOT ignition timing is used.

Once over 3000 rpm, the octane requirement of a Series 'A' engine is not quite as high. As a result, ignition timing can be advanced to a level that will produce maximum power with the high compression ratios used in engines from about 1980 on. It is hoped that from the discussion so far, I have convinced you that there is more to ignition systems than meets the eye. If you hope to get the best overall performance from your Series 'A' engine you will need to pay close attention to the entire ignition system.

Spark Plugs

A good place to start in our analysis of the ignition system, is the spark plugs. There are three factors which come out on top as principal considerations when choosing a plug. These considerations are thermal, electrical and mechanical. If any one or any combination of these three are significantly awry, then at best the engine fails to reach its true power potential and at worst the engine is wrecked. To quote an example: a few years ago I was testing a supercharged Mini engine on a chassis dyno. The engine had been run in (broken in) on Champion N6Ys so it was given a power run on these to see what they looked like after a hard blast. The motor gave bang-on 90 hp at 6000 rpm and the plugs were pulled out and inspected. There appeared to be no sign of over-heating. Nonetheless I thought some retracted nose racing plugs were worth a try, and a set of Champion N62Rs were installed. The result was the engine would make no more than 81 hp at 6000 rpm. Were these plugs running too cold? Probably not, but the next move was done more out of curiosity than anything. A set of N4 plugs were put in just to see what difference they made to the hp. The engine was quickly run up to six grand, and would you believe, it wouldn't pull anything more than 60 hp.

Admittedly, this is an extreme case, but that's how much difference the wrong plugs for an engine can make, and don't think it's always easy to tell by the sound of the engine. This particular engine sounds exactly the same with any one of the three types of plugs in.

This example, establishes fairly well that plug selection is important. In selecting a plug let us consider first the thermal rating. This is, in fact, a most important criterion if we are to avoid destructive engine detonation or misfires due to plug fouling. Looking at the conditions under which a plug operates we find the nose projecting into a combustion chamber, the environment of which continually buffets the plug with flame temperatures which instantaneously reach over 1000° C. That's enough to melt most materials. To prevent its electrodes melting or glowing red hot, thus causing pre-ignition, the plug must conduct excess heat away. Also, the plug is continually being coated by products of combustion which can slowly build up and eventually short the plug out so it no longer sparks. This is exactly what happens when heat is conducted away from the plug tip too quickly and the plug runs too cool.

The result of operating at the other extreme, namely with the plugs running too hot, have far more dire consequences; the electrodes begin to glow and become the source of ignition even before the spark occurs. In other words, pre-ignition is almost certain and detonation can quickly follow, due to the overheating caused by the pre-ignition. It doesn't take long, running under these conditions to melt down or blow holes in the piston crowns. The happy

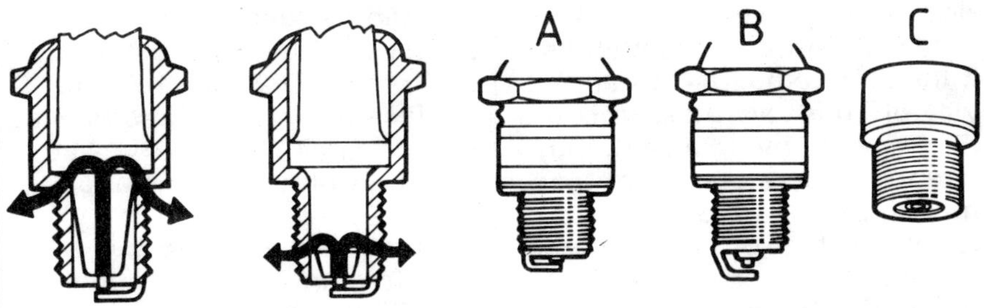

Fig. 16.1 Heat rating of plugs

The heat rating of a plug is determined by the electrode material used and the length of the heat path from the electrode tip to the main body of the plug. The hot plug on the left has a long path and so remains hotter longer than the cold plug on the right.
Plug A is a normal nosed plug such as a Champion N5 but the plug B has as an extended nose which protrudes into the combustion chamber for better cooling and more power. C is a surface discharge plug which is very cold running and needs transistor ignition to fire it.

compromise then, is to select a plug which conducts heat away just fast enough to stay below the upper pre-ignition and detonation inducing temperature limit and yet retain enough heat to burn off the deposits and so avoid fouling.

Choosing a plug which falls between these two limits is mainly a question of selecting a plug with the right heat path length. In the drawing, **Fig. 16.1** you will see two plugs, one with a short heat path and one with a long heat path. The one with the short heat path can conduct heat away faster than the one with the long heat path. Whatever engine is concerned, both plugs should operate at the same temperature. In practice, the difference lies in the source of heat feeding these plugs. Because of this, they require a different heat dissipation capability to retain a similar operating temperature between the two. Generally speaking, you can tell how hot or cold a plug will run compared with another, by looking at the length of the insulator core on the central electrode. The shorter this is, the less distance the heat has to travel to reach the outer shell of the plug. This isn't a precise guide, however, because the materials the plugs are made of can also affect the plug's heat conductivity. Most plug manufacturers produce an equivalence chart, showing how their plugs compare in heat range to other manufacturers. Though not always totally accurate, it is a good idea to consult such a chart if you are swopping from one plug brand to another. As far as different brands are concerned, I have had very good results with NGK plugs in the Series 'A' engine. These results, incidentally, are after running the gamut of plugs. Once in a while, I'll and buy a wide heat range and style of every brand of plug available for a particular engine and run a plug 'scan' on the engine. The result of recent plug testing sessions has shown the NGKs to be above average from the point of view of power output, life and resistance to fouling. To compare a typical range of plugs for the Series 'A' engine, refer to **Fig. 16.2**. NGK plugs have copper core which, because of its high heat conductance, pulls the heat away from the electrode nose faster than a steel core plug. As a result making a heat range comparison between an NGK plug and, say, a steel core Champion, by looking at centre electrode insulator lengths will be very misleading. You will find a copper core plug has a much longer centre than its equivalent steel insulator core equivalent. Generally speaking, it would seem that plugs with a copper core have a wider heat range capability in the Series 'A' engine. As a result, they seem to be less critical as far as making an exact choice of heat

	NGK	Champion	Bosch	AC	Motocraft
14 mm 3/4" reach	BP5ES	N-12Y, UN-12Y N-11Y	W160T30	45XLS 44XLS	AG52 AG42
	BP6ES	N-10Y	W175T30 W200T30	43XLS	AG32
	BP7ES	N-9Y N-8Y	W230T30	42XLS, 42-8XLS 41-2XLS	AG22
	BP8ES	N-7Y, N-6Y	W260T30	41XLS	AG12

Fig. 16.2 Spark plug equivalents

range is concerned.

For most practical purposes, I find that for standard engines throughout the Series 'A' range, right up to units with about a 50% increase in power, the BP6ES plug functions well. For the higher hp engines and road use, a BP7ES works out well. For many competition engines, especially the big displacement ones, a BP8ES seems to do the job. However, when you are putting together a race engine, it is always worthwhile trying out retracted nose plugs. For no obvious reason some engines seem to respond better to the retracted nose plugs. Some informed sources, however, believe that squish (quench), or lack of it, has some bearing on the phenomenon. Champion make a good range of this type of plug, and I would suggest that an N57R, and N60R or an N64R plug should be tried, the N57R being the coldest running and the N64R being the hottest. Sometimes this type of plug causes difficulty in cold-starting so it may be necessary to warm the engine up on standard plugs first. On the other hand, if a high output electronic ignition is used, starting from cold on retracted nose plugs should not be a problem.

Visual Heat Checks

On completion of a newly-built engine it is likely that some testing will be required prior to a rolling road session, simply because the engine needs some bedding-in mileage and some indication will be required that the basic settings are reasonably close to the, as yet unkown, final requirements. One of the first checks is to see if the mixture is reasonably satisfactory, usually done by removing each spark plug and looking at the electrodes. However, a mixture check is only valid within certain limits. If the plug temperatures are way off, this can 'mask' mixture readings by giving a colouring due to temperature rather than mixture. So, how is it possible to tell whether or not the plugs in your engine are over-heating? If the plug isn't reaching high enough temperatures, it will foul and this can be seen by virtue of the combustion deposits collecting on it. However, you must not confuse this state of affairs with that of a rich mixture. We had better assume, at this stage, that your mixture is approximately right, and therefore any adverse plug condition seen is due to temperature.

White deposits and damaged porcelain insulation indicating overheating

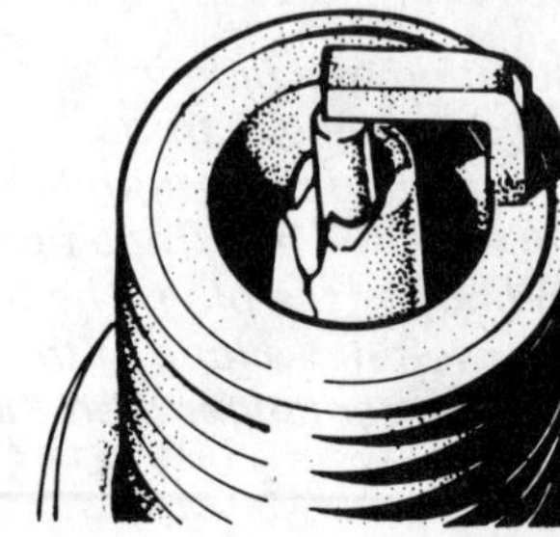

Broken porcelain insulation due to bent central electrode Check piston clearance

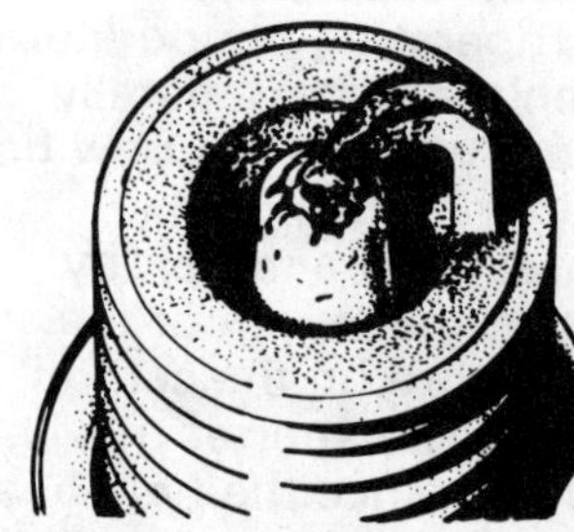

Electrodes burnt away due to wrong heat value or chronic pre-ignition (pinking)

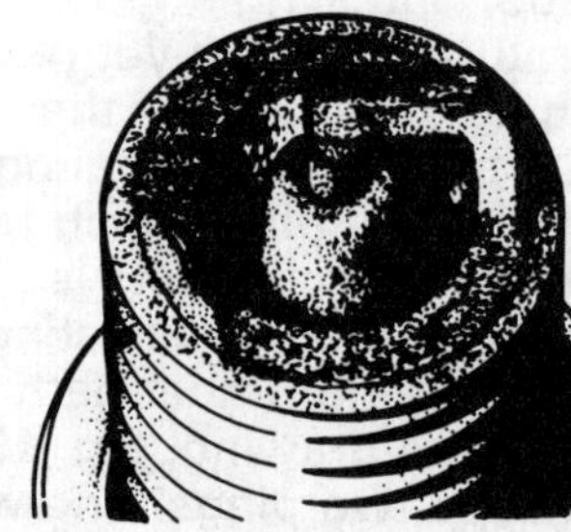

Excessive black deposits caused by over-rich mixture or wrong heat value

Mild white deposits and electrode burnt indicating too weak a fuel mixture

Plug in sound condition with light greyish brown deposits indicating correct running conditions

Fig. 16.3 Spark plug electrode conditions

Now I want to make a point clear here: a lot of people try to give the impression that plug reading is easy. Unless you have a lot of experience, it most certainly isn't. A cold fouling plug is probably the easiest to spot, but a plug which is over-heating needs a little more of a trained eye. For plug overheating, what you need to look for are light colourations of the plug. These will tend to be towards grey or white and if overheating is present, are often accompanied by blistering of the centre electrode. Sometimes you may find tiny particles of metal on and around the centre electrode. This is a fair indication that the engine has been detonating and those are particles of aluminium from the pistons. The easiest thing to confuse plug over-heating with is a weak mixture, as this can often show similar signs to over-heating, except the centre electrode won't be blistered. **A point worth noting is that slight plug over-heating can be caused by a weak mixture, because it burns slower and therefore confronts the plug with radiation from the flame for a longer period.** Also, a weak mixture has no excess fuel to cool the combustion chamber or plug. This is part of the reason why an engine which runs weak at full throttle, runs hotter than one with the correct full power rich mixture. If the plug is of the right heat range, you will find the centre electrode tends to look a mid- to dark brown in colour with the shell tinged with black around the edges, but not a sooty black.

Since mixture conditions can affect the temperature of a spark plug, it's worthwhile practising the subtle art of plug reading so as to be able to assess with reasonable accuracy the mixture state. Quite frankly though this method is nothing more than a guide. Anyone who believes they can set up a mixture spot-on throughout the entire operating range by looking at plug colouring is mistaken. Assuming you want to use plug reading only as a rough guide, a black sooty plug indicates that the mixture is too rich. One which is generally brown, tinged with black around the edge of the shell is pretty near, or one which is light brown and running to grey or even white, is lean, whitest, of course, being the leanest. When it's that lean it's foolhardy running the motor at WOT because it could melt a piston. These days, with fuel economy being of paramount importance, many engines run very lean mixtures at cruise. If you look at the plugs after a long cruise, then for an economy mixture the plug colouration should tend towards the white end of the colour scale.

Electrical Considerations

The most important electrical aspect of a plug is, of course, the electrodes and the conditions under which they operate. First of all let's deal with the spark plug gap. Here there are a few conflicting requirements; the wider the gap the higher the voltage required to cause a spark to jump across the gap. On the other hand, if the ignition system does have sufficient energy, the wider the gap, within reason, the better the spark's ability to fire the mixture. The upper limit of a useful gap is around 50 thousandths. Ordinary coil ignitions generally won't fire a gap as big as this. When the engine's CR is put up, it is harder for the plug to fire and, therefore, the maximum gap the ignition system can cope with at the plug has to come down. Generally one needs to think in terms of a couple of thousandths smaller gap for each whole ratio the compression is raised over 10:1.

Another aspect that should be considered if you're after maximum spark capability for your engine, is the actual

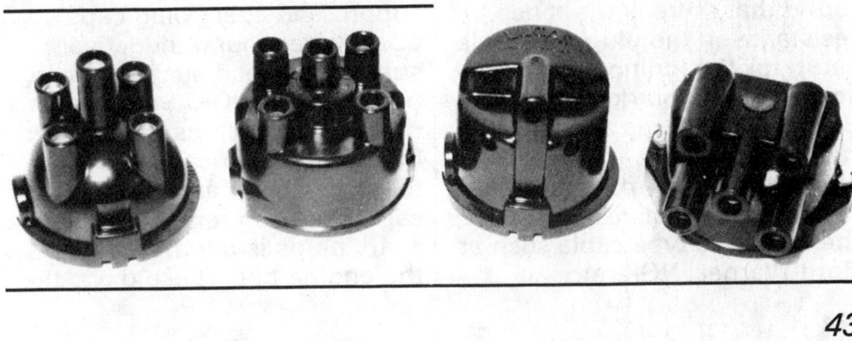

There are a variety of distributor caps that can be used on Lucas distributors. The pair on the left are the straight type, the one on the far left being for use on a 23D or 25D type distributor and the one next to it, on a 43D or 45D type distributor. The pair of caps on the right are 90° items. The one on the far right is intended to be secured to the distributor body by two screws and it requires modifying to be used with a spring clip fit distributor, normally used on Series 'A' engines. This is simply done by removing the screws and filing a semi-circular groove in the two ears on the distributor. This then allows the clips to hold the cap in place, as normal. This type of cap is used for the 43D and 45D type distributors. The cap next to it, most commonly seen on transverse engines, is a 90° cap for the 23D and 25D type distributor. 16-4

condition of the plug electrodes. When sparks jump the gap, they take material from one electrode to the other and this tends to round everything off, but the fact of the matter is, sparks will jump from a sharp edge easier than they will from a round edge. If your plugs have done a few thousand miles and are rounding on the side electrode edges then, using a needle file, dress the face of the side electrode to produce a sharp 90° edge. If you want to be really fussy, file the edge so it is right over the middle of the centre electrode. Of course, preparing a plug like this is only really important if, for some reason, you are forced to use a totally standard ignition. High powered ignitions are not so fussy about the condition of the plug. If the ignition system is a high powered electronic setup with bags of energy potential on hand, then plug condition (within limits) becomes almost irrelevant.

Resistor Plugs

With cars becoming quieter and performance of radio and stereo units becoming better, manufacturers in the ignition business need to pay increasingly close attention to the suppression of radio interference. The resistor plug is a step along that road. Although these plugs can function perfectly well, if they are used with suppressed plug leads of the resistive type, it may be found that there is too much resistance in the plug and cable setup for the ignition system to cope with. If you do decide to use resistive plugs to keep the radio interference to an absolute minimum, then rather than fitting resistive leads, use the inductive type cable such as Borg-Warner, NGK, etc.

Ignition Cables

To get a good spark, especially from a standard coil ignition system, the amount of resistance that exists on the high tension (HT) side of the ignition system is important. Too much resistance and you make life harder for the coil. A coil can deliver a certain maximum voltage but the plug takes only the voltage it needs to fire it. For instance, if a plug requires 10,000 volts to fire it under a particular circumstance, the coil will build up to 10,000 volts, fire the plug and there will be no further rise in voltage, and no more to fire the plug, even though the coil may be capable of delivering 15,000 volts. Problems start when a coil can only supply 15,000 and the plug, together with losses due to resistance between coil and plug, needs 16,000 volts to generate a spark. The result: no spark. **To avoid misfires, we obviously need to make sure that the voltage required to fire the plug is less than the voltage available.**

Aged suppressed leads are one of the most common spark killers. Many suppressed leads usually take the form of carbon impregnated string. Unfortunately this type of lead often breaks down electrically. When new they may be fine, but after a few years they are definitely a liability if used with a standard ignition system.

To avoid the problems brought about by old leads, you have two choices. Either use copper core type leads with suppressed spark plug caps or use a wire wound inductively suppressed lead such as those produced by NGK. Some leads of this type such as those by Sparkrite and Lucas have a tendency to separate from the cap if lead/cap removal from the spark plugs is attempted with the engine hot. These days, the copper core lead and suppressed plug cap are not often used. However, if you elect to use copper core plug leads, don't be tempted to omit all in-line radio suppression. Suppressed plug caps usually have a built-in resistor in them, and if they are omitted and if the leads have no resistance themselves, other than the very low resistance of a copper lead, the spark intensity can actually be reduced. If you are using a standard ignition system or for that matter, many of the transistorized ignition systems, the best spark intensity seems to be produced when there are 5000 to 6000 Ohms of resistance in the system. On the face of it this may seem to contradict what's been said earlier, but it will become clear if you understand what happens at the plug. When the spark jumps the gap at the plug, the resistance of the plug, due to the arc formed at the gap, diminishes to nothing. In fact, it looks like negative resistance. Under these conditions, the current goes up, the voltage drops and the spark becomes self-extinguishing.

With a little resistance in the system, the spark can be increased in both duration and temperature.

To attain the correct resistance most easily use a suppressor of the specified resistance or a suppressed lead between the distributor and coil. If you use a suppressed lead, it usually requires a length of about 6 inch of lead to get the desired resistance. It will usually be necessary to have a joint between this lead and a piece of copper core lead to reach the distributor. You will usually find that resistive lead is around 25 thousand Ohms (i.e. 25K) Ohms per metre when new, and an awful lot more when it's old.

One last point on plug leads: many plug cable manufacturers are making leads with a silicone rubber casing so

Here's the inside view of the two types of distributor cap. On the left is a 23D and 25D type distributor cap. That's the older type of distributor. On the right is a 43D and 45D type distributor. Notice the much deeper trench around the centre rotor arm contact on the cap on the right. Also, the distribution points are at a larger radius. Both of these features reduce the likelihood of tracking and cross-firing with today's higher output ignition systems.

that leads in close proximity to the exhaust system will survive higher temperatures without melting. On a Series 'A' engine silicone plug leads are non-essential since the exhaust is around the other side of the engine.

Contact Breakers

Over the years, distributors used in Series 'A' engines have been progressively refined. The original distributors used way back in the early days (pre-'60) were functional but had little margin left in them for high performance applications. The cam design of the contact breaker was such that it could easily induce points bounce. When this happens, the spark output drops dramatically. **That severe spark degradation occurs is simple enough to understand when you consider that the coil is being charged ready for the next spark only while the points are closed.** The coil needs time to build up the necessary energy to discharge into the plug. Points bounce causes a substantial reduction in the charging time available and this occurs just when every bit of spark energy is needed, namely at high rpm. Over the years, distributors have become better in respect of points bounce. Up until the early '70s most Lucas distributors had an essentially symmetrical contact breaker cam lobe shape. In the early days of Formula 3, the Competition Department at Lucas developed an assymetric cam, one which opened the contact breakers sharply (that's a prime requirement for a good spark) but closed them more gently. This gave a much more controlled action and staved off points bounce to a couple of thousand rpm higher without any increase in contact breaker spring tension. This type of cam lobe has now found its way into production distributors and virtually any modern Lucas distributor will have an assymetric cam lobe.

Depending upon the type of Lucas distributor your engine is equipped with, one of two rotor arms will be needed. The basic difference between them is that one rotor arm accommodates the larger radius of the 43D and 45D type distributor caps. That's the uppermost one of the pair shown here.

Another attempt at producing high rpm ignition capability without fitting sophisticated electronic ignition, which at the time was in its infancy, involved the use of the heavy duty contact breaker set. This type of contact breaker set became standard equipment in the Cooper 'S'. The spring gave 32 ounces of contact pressure and was capable of running up to well over 8000 rpm when used with the assymetric cam. By contrast, most of the contact breaker sets used these days are of the 'quik-fit' type. This type of contact breaker has only 22 ounces of spring pressure but by and large, with the now common assymetric cam, are perfectly adequate to about 7500 rpm. However, if used with a symmetrical type cam, points bounce is likely to occur around 5500 to 6000 rpm; well below the rpm capability of most Series 'A' engines.

To build a distributor with the heavy duty 32 ounce points, it's necessary to have a long pin backplate assembly. The **accompanying Photo** shows the type required. A long pin backplate will accept either the quick fit type contact breaker set or the heavy duty set. However, distributors made for the quick

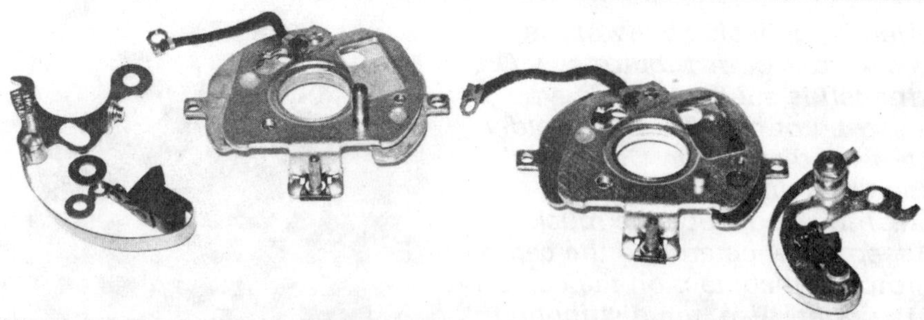

On the left is a long pin vacuum advance backplate. This is the type needed to fit the heavier tension springs as per Cooper 'S'. On the right is a short pin backplate which utilizes a vacuum backplate which uses the 'Quik Fit' points.

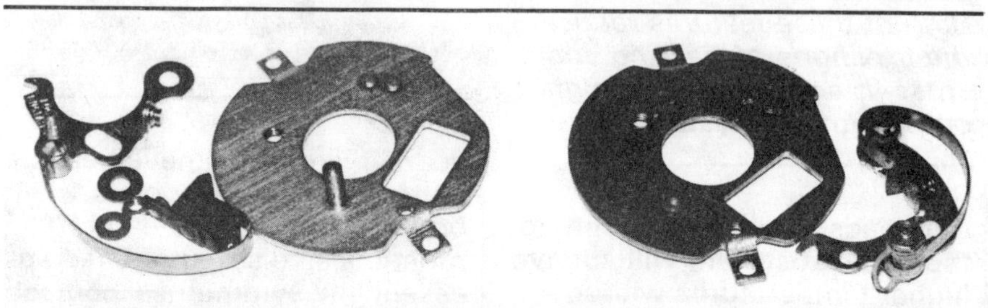

Where non-vacuum type distributors are used, a different backplate is required. For the Cooper 'S' distributor which are still equipped with the original heavy duty Cooper 'S' points, the long pin backplate on the left is needed. The assembly on the right is the short pin variety for Quik Fit points.

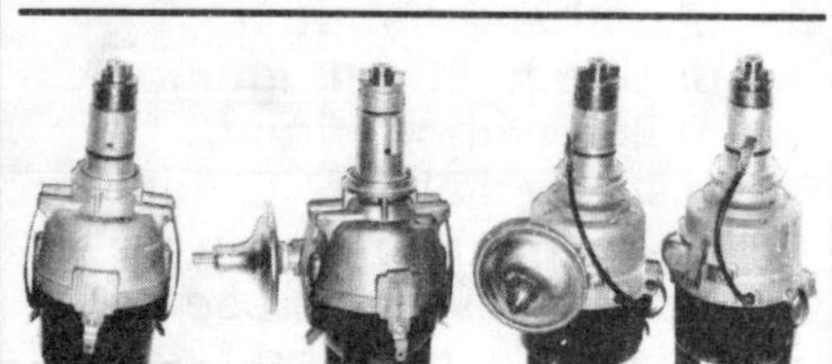

If, as is likely, your Series 'A' engine is equipped with a Lucas distributor it will more than likely be one of the four distributor types shown here. Left is a 43D. Although this particular unit looks like it might be intended for high performance use because it has no vacuum advance, it wasn't. The vacuum advance was omitted on this distributor to cure crank rumble on the 1.3 Marina it was installed in. Next in line is the 45D distributor with the vacuum advance. Second from right is a 23D distributor which was a type commonly used on 'S' engines and on the far right is a 25D distributor. The pair of distributors on the right have different distributor caps, but they are interchangeable, depending on whether you want a straight out distributor or a 90° distributor cap.

fit set have a short pin, and they will not accept heavy duty type points.

The very latest distributors from Lucas, and those from the French company Ducellier used as alternative supply in the latest 'A+' Metro engines, have

32-Ounce Points & Long Pin Back Plates

No doubt having read the main text you would like to know whether you can convert your distributor to the 32-ounce point type setup and the long pin backplate. If you have a Cooper 'S' distributor (i.e. type 23 D4 distributor), there is no problem. This type of distributor, however, is no longer made and buying a new distributor is unlikely to be possible. If you have a vacuum distributor of the same general type it will come under the type number of 25 D4. These distributors are still available, but not that common, and they can be had with a long pin backplate. Both those types of distributors were made in quick fit backplate style assemblies. To convert to a long pin backplate, Lucas part number 54412436 will be required for a type 23 D4 unit, and Lucas part number 422318 for a 25 D4 unit. Since the mid-'70s, most of the distributors were the more modern 43 or 45 D4 units, neither of which can be had with long pin backplates. These, however, are high rpm units anyway. They all had the assymetric type cam and will turn to about 8000 with the assymetric type cam and the latest, Quik Fit type points. All these distributors apply to engines prior to the 'A+'.

Another distributor which is worth mentioning is the 48 D4 sliding points distributor. A few 1.3 Marina engines which were pre-'A+' engines were fitted with this type of distributor. For use up to 8000 rpm it's probably the best distributor of the bunch because it offers a longer time between service.

As far as contact breakers go, the 32-ounce points are available under Lucas part number 5441356 or CS9 or DSB 109.

a setup commonly known as sliding points. Here, one-half of the contact breaker assembly is induced to move up and down so that instead of the contacts touching at the same spot all the time, the wear is distributed over the face of each contact. This greatly extends the life and slows the rate of degradation due to electrical contact erosion. This type of distributor is probably the best type of contact breaker distributor with regard to holding its tune. It produces results over a 12,000 mile period close to a contactless ignition system.

Some non-genuine replacement points can suffer from 'soft' plastic on the actuating arm. The softening occurs sometimes after a period of high rpm operation when 'drag' will occur on the advance mechanism causing artificial retardation and, of course, loss of power. This malady is very difficult to detect because, after cooling down, the plastic returns to normal, and so does the timing. The only way of being sure of avoiding this problem is by using genuine Lucas replacement points. The foregoing only applies to non-'A+' type distributors.

Spark Scatter

One of the problems with distributors, especially those in the pre-'A+' engines, is that the bearings in the distributor can wear quite rapidly, especially if there is any mis-alignment between the distributor shaft mounting bore and the driveshaft and the angle it makes to the end of the distributor. Usually there is enough slack to take this up, but sometimes a little binding can occur at this point. It's a good idea if you are building an engine from scratch to install the distributor and driveshaft assembly prior to installing anything else in the block and just check that it all rotates freely. If it doesn't, use another driveshaft or ascertain where any tightness occurs and try taking a little metal from the offending part. **If distributor bearings are worn and loose, the spark scatter that occurs can cause substantial power loss. Fig. 16.4** shows some power figures I took many years ago on an 850 engine. As you can see, a little thing like distributor driveshaft wear can cost a high percentage of power loss. There are two reasons why this happens. The first is that spark scatter causes an erratic change in ignition timing. For instance, if the optimum ignition timing is supposedly 30°, then with a 4° spark scatter, the spark could occur at random anywhere between 28 and 32°.

The other aspect is that the spark scatter, since it is caused by a sloppy distributor shaft, also causes the dwell angle to change, again at random. At high rpm, the dwell angle can be reduced quite dramatically. Also, points bounce can be induced much sooner than would otherwise be the case. None of the foregoing helps power output, so anything that can be done to the distributor to

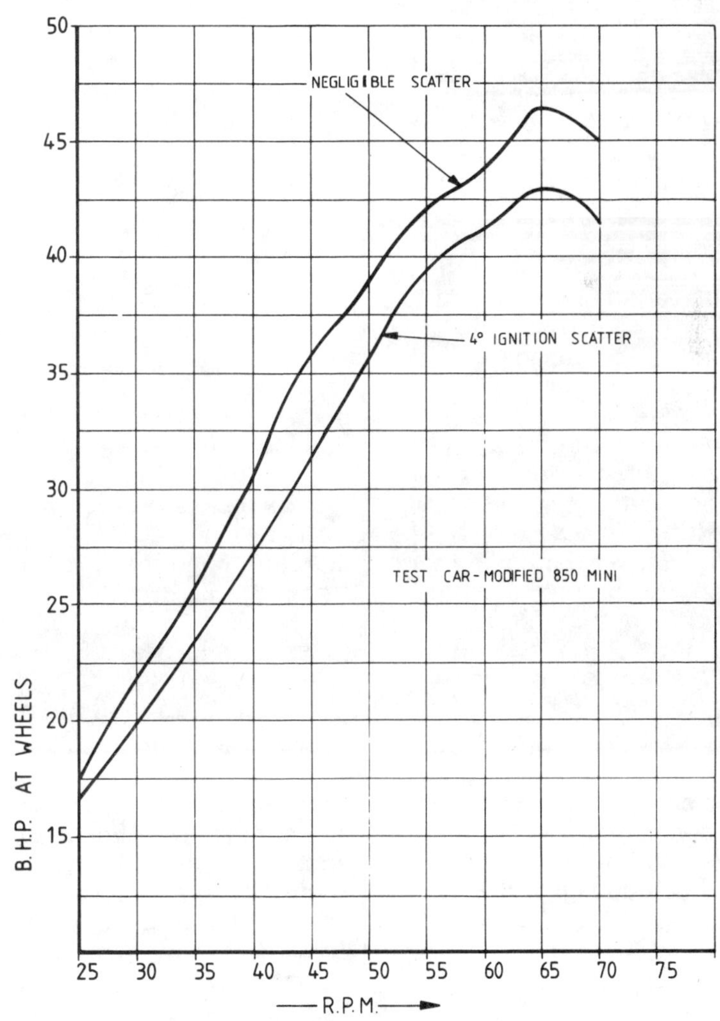

Fig. 16.4 Typical power loss due to spark scatter

Distributor Bearing Modification

Being fed up with replacing distributors, I mentioned to a retired engineer friend, John Blackwell, the possibility of fitting roller bearings into a distributor to cure the problem. John looked into a number of ways of modifying these distributors and decided upon the following as being most feasible.

First of all, it was agreed that the roller bearing would be required in the top bearing only, the one directly under the advance/retard system, as this was the one that took most of the load resulting from the vibrations imposed by opening and closing the points. The bearing used was an Ena or Torrington bearing, type number NA 4901 2RS. This is a roller race with a 24mm outer diameter and a 12mm bore. The bore is just too small to fit over the shaft of the Lucas distributor, this being 0.04798 inch diameter, so it is necessary to grind the bore of the inner race to fit the distributor shaft. To get everything in the relevant position on the shaft, a spacing washer has to be made up ⅝ inch O.D., 0.480 inch I.D. by 0.070 inch thick. This is fitted prior to the installation of the inner race onto the distributor shaft. The inner race is 'Loctited' in place on the distributor shaft, then the nylon thrust bearing and shim are bored to fit over the inner race. Next the stripped down distributor housing is mounted in a lathe and trued up on the original bore of the bearings. From here on the distributor case is machined as in the diagram. Note that the bore is stepped; this is so that the inner race, which is longer than the outer race, is clear of the distributor housing.

The last operation before assembly is to drill a hole in the boss and tap it quarter Whit (or equivalent U.S. course thread) 0.300 inch deep and 0.25 inch down from the thrust face. Using an ⅛ inch bit, drill a hole from the bottom of the tapped hole into the bearing housing. This drilling allows the fitting of an oil nipple to supply lubricant to the bearing. Assuming that care has been taken in the modification and final assembly, distributor bearing problems should be a thing of the past.

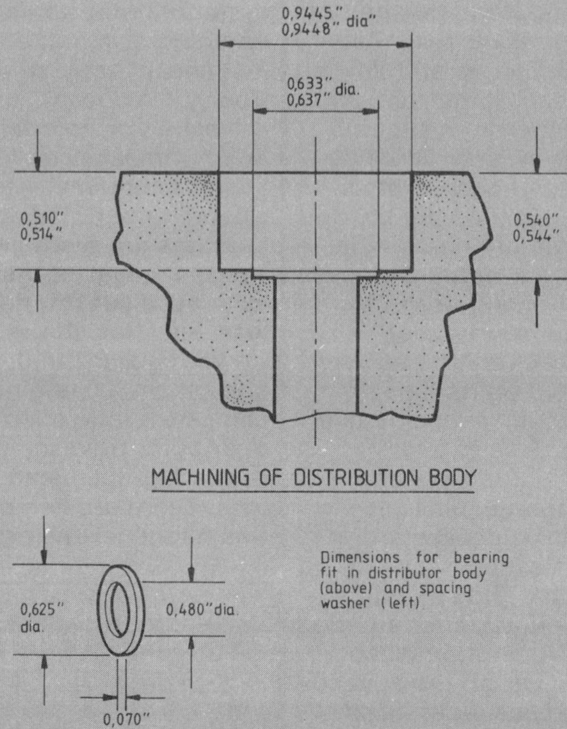

Modifying a distributor to roller bearings

reduce spark scatter and hold more consistent dwell angle figures, the better. A simple mod which does just that is machining of the distributor shaft down to 5⁄16 inch diameter along its shank. This, it seems, adds enough flexibility to the shaft that it reduces the stresses on the distributor driveshaft and in turn, allows the bearings live to longer. As a result the dwell angle stays more constant.

Ignition Systems Part 2

Electronic Ignition

You can see so far that I have dwelt extensively on contact breakers, and not without good cause. The contact breaker itself, though it can function perfectly well when new, does deteriorate rapidly with age. As the dwell angle changes due to the numerous factors discussed, it can alter the ignition timing. To recap, much of this alteration occurs not only due to wear and tear caused by the points action battering the bearings in the distributor, but also by electrical erosion of the points and cam follower heel wear altering the dwell angle, and hence the spark effectiveness. If the contact breaker erosion can be eliminated, the points would last almost indefinitely, though we are still stuck with wear and tear on the distributor bearings due to the contact breaker action. If a change is made to an electronic ignition system, two choices are available. We can fit a Transistor Assisted Contact (TAC) electronic ignition system which employs the standard contact breaker assembly as a low voltage switching system to trigger the higher powered switching operations handled by transistors. Alternatively, the points can be replaced by a magnetic, inductive, or light triggered system. We are only ever going to fit an electronic system for three reasons: increased time between servicing; the elimination of servicing altogether; or improved power and/or economy. If an electronic system gives none of these advantages it's not worth considering. Provided it doesn't self-destruct, an electronic ignition system will be virtually maintenance free, being only necessary to replace spark plugs every 12,000 to 15,000 miles and sometimes at even greater intervals. Just how much extra power and economy is possible by using electronic ignition? This varies from engine to engine, but **one thing you can be sure of is that electronic ignition system manufacturers' claims are rarely, if ever, achieved in practice.** In reality, power gains usually amount to only 2 or 3%. However, to show a balanced picture of the situation, I will start by dealing with the power

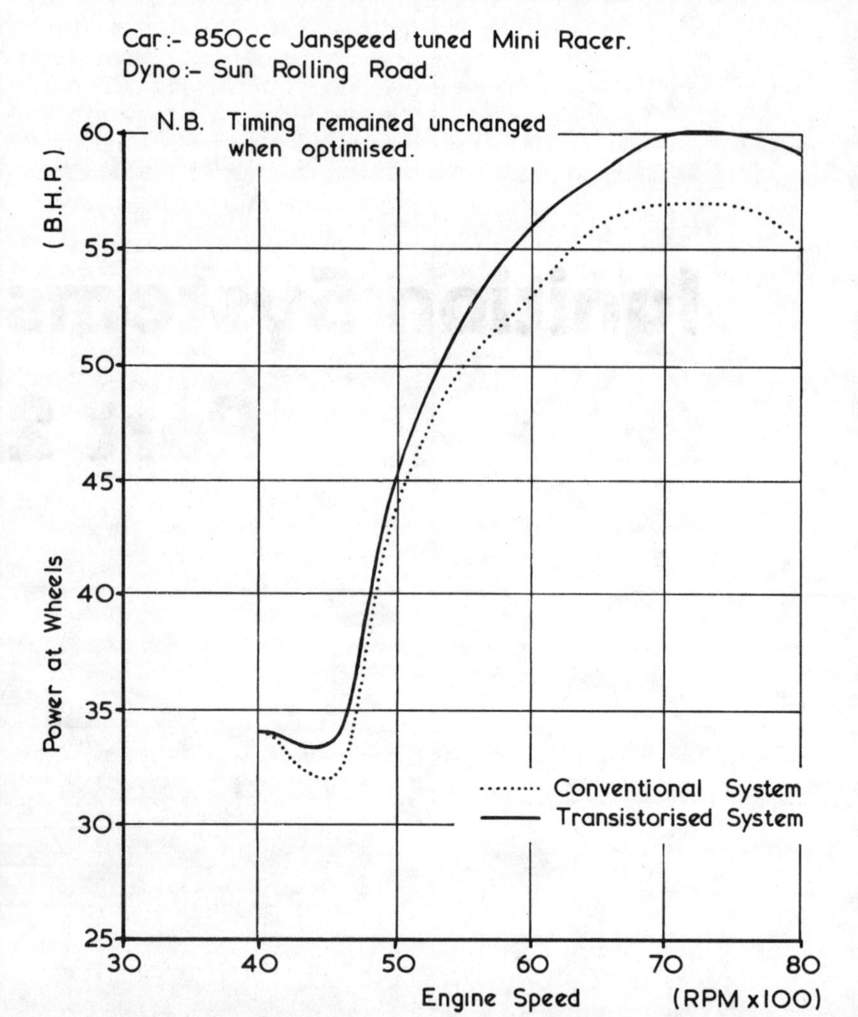

Fig. 16.5 Comparison of conventional and transistor assisted contact breaker ignition system

curve in **Fig. 16.5**. This engine was for a racing 850 Mini and was fully prepared by **Janspeed Engineering**. The ignition system concerned was an early **Sparkrite** system, and it produced very creditable gains in terms of extra power. In fact this engine and ignition system showed the biggest percentage gain of any ignition system I have tried on a Series 'A'. This is not to say that the particular ignition system was any better than any others. It's just a race engine built of this particular combination of parts needed an electronic ignition system more than most. The net gain with the electronic ignition was about 2½ hp and this represented an increase of some 4%. **So if you assume that a 4% increase in power is about the maximum your engine is likely to gain by using an electronic ignition system, then you will be in the realms of reality.** However, an often more realistic gain achieved with electronic ignition is shown in **Fig. 16.6**. An important point to note here is that neither of these power curves were obtained as a result of a 'straight swop-over'. In each case the standard ignition system was optimized and when the change from the standard ignition system to the electronic ignition system was made, it was necessary to re-optimize. **When simply switching from the standard ignition to the Lumenition system as per tests in Fig. 16.6, no measurable difference in power output was seen.** It was only by adjusting various aspects of the engine to suit the higher output characteristics of the Lumenition system that any extra power or economy was produced.

The adjustment procedure required that the plugs be opened up to a 40 thousandths gap. It was also found that slightly less ignition advance was required with electronic ignition than with standard. In addition, to obtain the extra possible economy, the mixture has to be much leaner. This brought about the need for some recalibration of the carburettor under part-throttle conditions. After being optimized approximately 5% gain in steady speed fuel economy was achieved. With the Lumenition system, or indeed any contactless ignition system, part of the gain can be attributed to the fact that a higher energy spark is delivered, and part (a much smaller part) due to the fact that ignition timing accuracy was improved. However, increased accuracy is not achieved without some snag, and the snag is that most electronic ignitions change the shape of the advance curve. However, this aspect of distribution will be discussed later. Now, lets turn to the subject of more sophisticated ignition systems.

Two particular systems I have tested fall into this category. The first system is **Autotronics** MSD system. This is a multi-firing system and instead of one spark, it produces a multiplicity of sparks, one after another, spaced at about 2° intervals. Depending upon the unit, either road or race, the number of

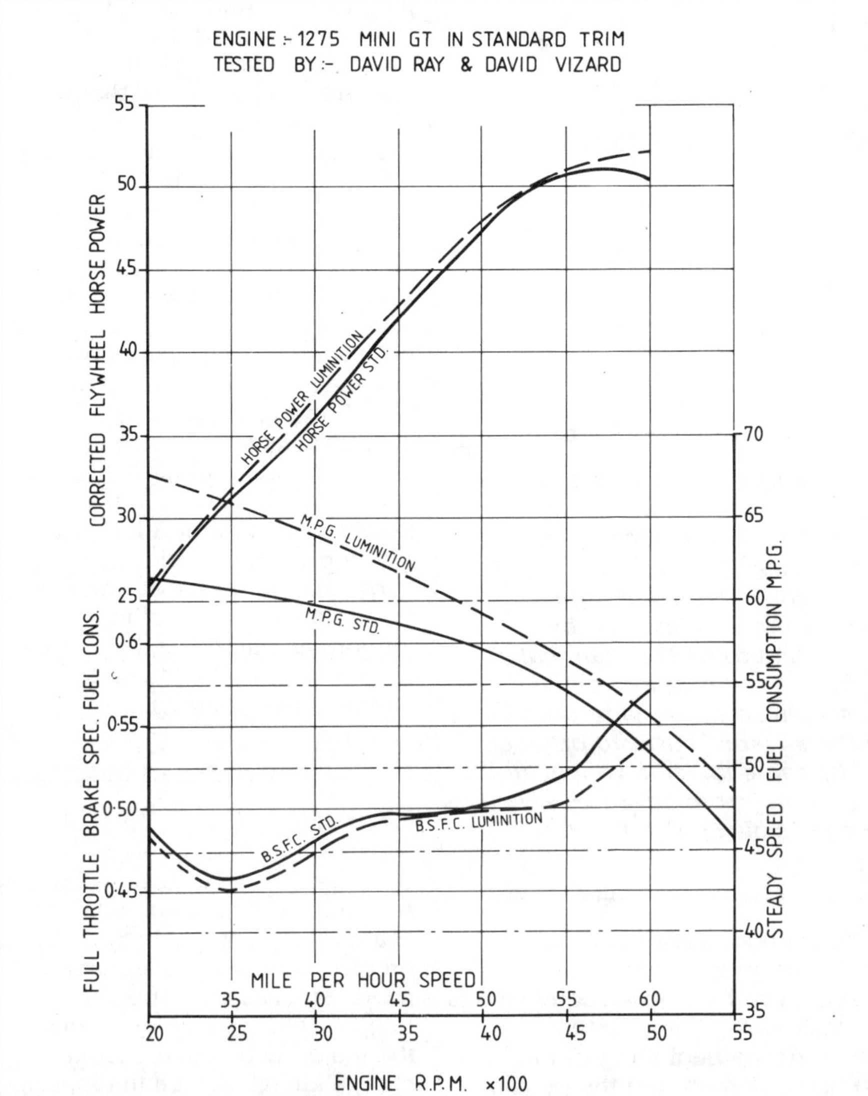

Fig. 16.6 Comparative performance of standard ignition system and Luminition contactless electronic ignition system

sparks produced ranges between about 2 to 6 for the MSD6, and up to 10 from the MSD7. The MSD system is very popular in the U.S.A. for use on V8 engines where flame path lengths across large combustion chambers are long. Anything that speeds up ignition is generally a help.

I have found, as a power producer the MSD7 offers no measurable advantage over a single spark delivered by a high powered electronic ignition. However, all my Series 'A' testing has been done on race engines but I've tested this system on road motors of other types and have generally found that multiple sparks allow much leaner, and hence more fuel efficient, part throttle mixtures to be ignited successfully. In some instances, except in very cold weather, the choke action on the carb has not been required. It follows that further experimenting for economy using multiple spark ignition may very well prove to be worthwhile.

Another system I have tested which gives an extended ignition time is a plasma ignition. I am mentioning it here because future ignition systems may be along these lines, but the particular system I tried, though functional, had several drawbacks from the electronics point of view and for various reasons it's no longer commercially available. However, I did do some quick, if somewhat incomplete tests, with a plasma-type ignition on **Aldon Automotive**'s dyno which produced the results **seen in Fig. 16.7**. If you study the figures, you will understand why the tests could be considered incomplete. Basically, the standard ignition was optimized to produce the figures in column 1. This power curve was achieved at 34° of total advance. When the plasma ignition was switched in, it was found that maximum power

This is an Aldon race distributor. It is based on many of the parts of a Lucas 43D and 45D type distributor, but it has been re-engineered for the sole purpose of supplying the needs of a race engine.

Fig. 16.7 Dyno test of plasma ignition

Engine: 1275cc Metro HLS unit. Dyno facility: Aldon Automotive.

rpm	hp Std. ignition	hp Plasma
2000	20.0	18.0
3000	35.8	35.0
4000	49.5	52.3
5000	61.0	61.2
5500	66.1	68.1
5750	–	68.1
6000	66.1	66.1
6500	61.5	61.7

Ignition timing: Std. ignition 34° total. Plasma ignition 30° total.

In case some of you experienced dyno operators are wondering how we read Aldon's dyno to the first decimal place, let's put you straight. Since we plan to do tests for a good many months, all the figures have been corrected to standard conditions. After multiplying the observed horsepower, the answer is rounded off to the first decimal place.
During our dyno testing, it was found, at least with the relatively weak mixtures our test engine was using, that the plasma ignited charge needed less total advance (30° instead of 34°) than the conventional charge. Retarding the ignition by 4° appears to be the main culprit for the reduced low end horsepower. Some re-tailoring of the advance curve to produce the same initial advance, and less high rpm advance, would be a possible solution.

was delivered with 30° of advance. However, the altered combustion characteristics produced by the plasma ignition, required a different advance curve in the distributor, and time was against doing an ignition plot and removing the distributor and re-curving it. My guess is that the reduced power at the low rpm level is due to the fact that the distributor really needed an advance curve to suit the plasma ignition.

As far as the lean burn characteristics are concerned, some tests done on this same unit by one of the big petroleum manufacturing research centres showed that the ignition system was easily capable of firing mixtures as lean as 28 to 30:1. With a conventional single spark ignition system, it is difficult to ignite mixtures much leaner than 22:1, so a plasma type ignition does promise improved economy. As far as additional power is concerned, there appears to be little or no gain over that produced by a good, conventional electronic ignition system.

At the beginning of this chapter I went into the possible benefits of super sparks at some length. The emphasis was that you keep a realistic perspective on exactly what a super spark will do for your 'A' Series motor. Apart from spark intensity spark duration has also been a consideration. Longer looks like it should always be better but things are rarely all they seem to be. For most of us nuts and bolts mechanical types the technicalities of ignition design and function can leave us wondering exactly where on the intelligence quotient scale we fall. Discuss the subject with half a dozen so called ignition experts and you will only learn one thing for sure. Namely the experts can't even agree among themselves as to exactly what is required at the plug for optimum ignition. If you talk to enough professional engine builders or dyno operators you will find that they almost universally adopt the 'positive result' principle. What this means is that, regardless of how amazingly advanced the theory behind the systems design may be if it doesn't result in more power it's thrown out. By adopting this philosophy it is not necessary to understand how any new fangled, brain taxing, whiz bang ignition system produces its spark. All that is necessary is to relate the type of spark characteristics to the dyno results and look for a trend for future predictions. Not surprisingly an almost universal trend has been that bigger, fatter, longer sparks add to the engine's output and smaller, thinner, shorter ones subtract.

Quantifying

Assuming they are known there seems to be some difficulty predicting to what extent spark characteristics affect power output. Of the variables involved the major factors influencing charge ignitability are the fuel/air ratio and how well the fuel is atomized and mixed prior to delivery into the combustion chamber. Next on the list is combustion chamber burn capability. Although the 'A' Series rarely falls into this category some chamber designs are difficult to light off or are bad burners or worse yet both. If, for whatever reason, a fire lighting problem exists then the motor responds to a super spark by delivering a lot more power. If not then gains are small to intermediate. All the foregoing is no more than to be expected and from what has been said previously you can see where, in this picture, the 'A' Series falls. However when maximum output for the purposes of winning races is required there is no excuse for anything less than the ultimate spark. The question is – what is the ultimate spark? This we will look into now.

Mags and Coils

Up until the early 1960s an ignition system meant either a magneto or the Kettering system comprising a contact breaker and coil. Both designs date back far more years than any of us have been around and what they produced at the plug was just a good ol' fashioned spark. If the motor needed more then options were limited to efforts toward hopping up the basic design. In the early '60s transistors made their presence known and the potential to improve the conventional breaker switched coil designs came about. To say that the semi-conductor revolutionized ignition design would be an understatement. In a relatively short time a number of techniques for performing spark enhancements were developed. Extensive experience testing newly introduced electronics systems quickly established one important fact. That being the power increases their designers claimed would be seen almost always outweighed in reality by several orders of magnitude. But some systems did produce positive results. First came the systems that delivered bigger more intense sparks – they worked. Next came big multiple sparks and/or big long duration (compared with normal duration) sparks. They worked as well. All, to a greater or lesser degree, added to power output and all improved drivability and throttle response.

Arc Sparks

In the early 1980s the design of an apparently novel ignition system was brought to my attention by my friend David Ray who, at that time, worked at Esso (Exxon) research. The simplest way to describe this systems *modus operandi* is that it rode piggy back on a conventional system. To detail how it worked in greater depth it is necessary to appreciate a few physical effects that occur when a spark is generated. For the electrical energy to jump the plug gap and create a spark a certain minimum voltage, known as the ionization voltage, is required. The greater the number of molecules of insulating material between the plug electrodes the greater the voltage required. Air is a good insulator so the higher the cylinder pressure the greater the voltage needed to fire the plug. The resistance just prior to the spark occurring is very high, in the order of millions of ohms. To jump this gap typically needs 20,000–30,000 volts but once the spark makes it across the gap the air between the electrodes becomes ionized. When this happens the electrical resistance plummets and the gap actually becomes a moderately good conductor. The piggy back system actually made use of this phenomenon. After the normal spark ionized the gap the piggy back system applied a DC voltage across the plug creating an output similar to a small arc welder. Depending on the rpm involved the spark could be sustained for as much as 90°. At this point that may sound like a long spark duration but before assuming too much this will need to be put into perspective – that's something we will do in a moment or two but for now let's get back to the plot. Being performance fiends and racers at heart this apparent spark overkill had a certain undeniable appeal. As a result not only did I conduct my own tests on a race spec motor but also convinced David Ray at Esso Research to run some comprehensive tests on several types of motor. After completing the tests the Esso engineers involved opinioned that this suedo arc welder was the most significant step forward in ignition system design they had tested to date. This was a statement of some heavy weight significance considering the number of dynos and the amount of testing done at this establishment. The long burn arc improved every single parameter associated with performance, economy and emissions. The system tested though was not without its shortcomings. Reliability, high current consumption and a few other factors prevented it going into full scale production and so it died a death.

About 12 years later I was approached by an electronics design engineer who wanted to test an ignition concept for race motors. An outline of the concept revealed that the 'arc welder' principle had resurfaced but in an apparently more refined and potent form. Because of the technique's past success I was more than ready to test the concept against the industry's then most powerful ignition systems. In addition I would be testing on a small block Chevrolet V8, a motor that like the 'A' Series has an extremely effective combustion chamber. The bottom line here is that if the test motor showed any real increase in output the long burn arc would have proved its worth.

Test Time

The long burn arc could have been given an easy time to show superiority by testing against an ignition system of average capability but that would fail to prove its worth in the field of competition. Rather than this it

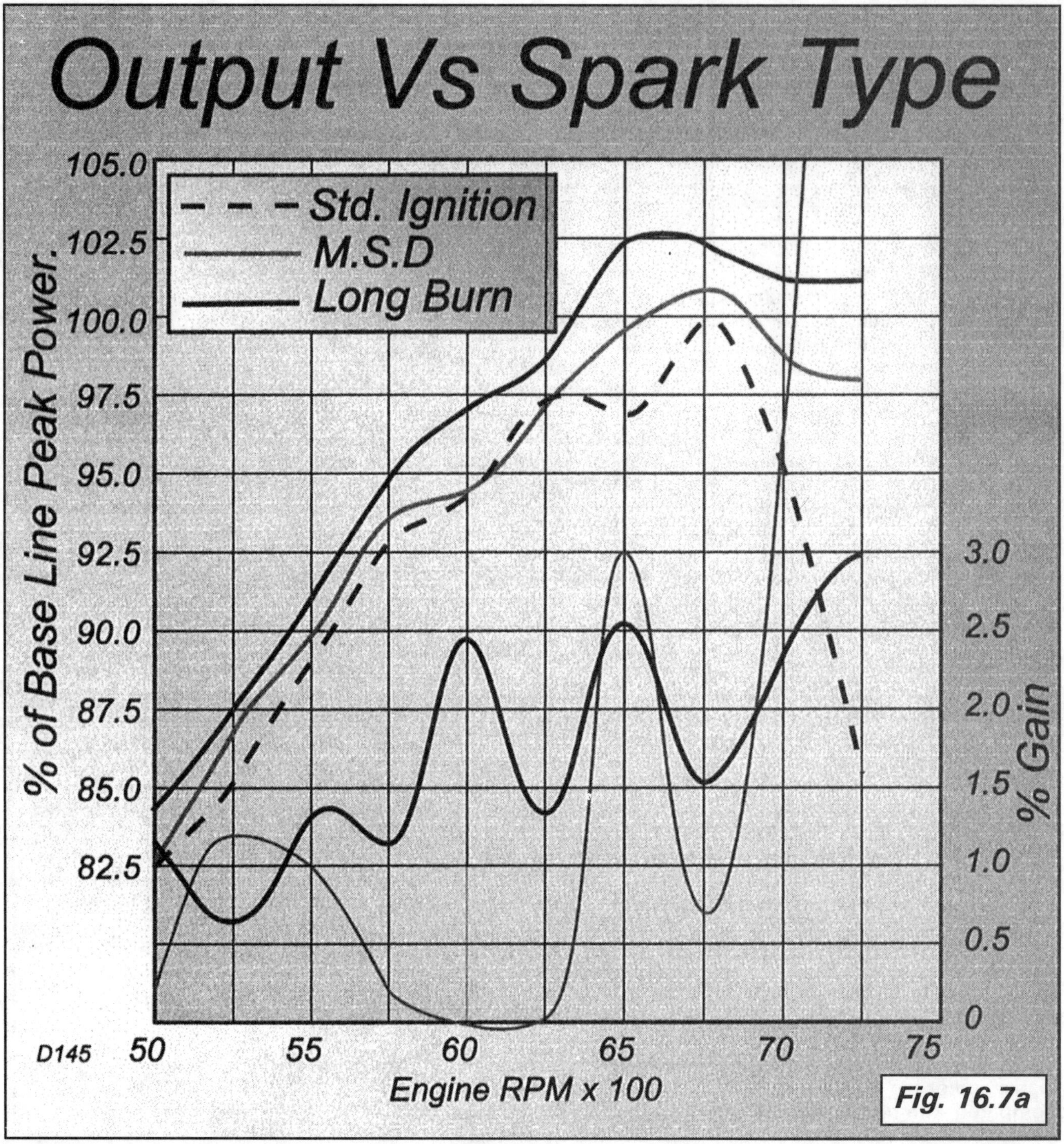

Fig. 16.7a

was tested against a specially built dyno dedicated system based on the then most effective race systems used. The dyno system was capable of firing bolts of lightning across a 6 inch gap in open air – with ease! If you think that sounds like an impressive display of electrical pyrotechnics the long burn arc system will impress you even more. This system could put so much energy into the plug gap that it would literally consume the electrodes. The question is did it work? It did and in no mean way at that as the results in **Fig. 16.7a** show. So where is this system now? Well, for various reasons, safety included, as a shock from it would really shake your neurons, the unit never made it into production. The tests however did establish the usefulness of an intense long burn discharge at the plug. The functioning principle seems simple enough, if the first moment of the ignition discharge doesn't get it the next will. The concept guaranteed to reduce the possibility of a poorly ignited cylinder to an absolute minimum.

Now Short

Bearing in mind the collective experience to date you can understand the scepticism when I was confronted by an engineer who claimed going short on spark duration was also a valid way to increase power output. The gentleman concerned, German Ulf Arens of Nology Engineering in San Diego California, had not so much designed a new ignition system but developed a less than conventional method of delivering the existing system's energy to the plug. The typical spark duration delivered by a modern inductive discharge electronic ignition system is between 1 and 2 milli seconds.(0.001–0.002 sec.) Now let's consider this spark duration in relation to crank rotation and the apparent advantage of a long burn spark. If a conventional spark lasts just one and a half milli seconds then, if the motor is turning 7500 rpm, the spark will last for 67.5°. Isn't this the sort of spark duration, in terms of crank degrees, that a long burn spark was supposed to deliver? Under normal circumstances most of the charge is burned about evenly either side of TDC. This means that if the ignition fired at 32° before TDC the burn should be over with by the time the crank has turned to 32 after. That is about the duration of a normal spark at the rpm we are considering here. These figures prompt the question that if a spark continues even longer, then just how much of a benefit can it be. Also, if the charge was not burned in the expected time does it call for more duration or more spark intensity. To answer that we have to consider that only heat produced at TDC benefits from the full expansion ratio available to the now heated charge. Any combustion which takes place later operates at a lower expansion ratio. In other words a slow and consequently later burning charge operates as if the CR was lower. Clearly this is not what we want. But the long burn ignition showed positive results on the dyno. This provokes the question as to whether it was the duration of the spark or its intensity that got the job done. All the long burn ignitions tested also had very intense sparks so were gains seen by virtue of duration or intensity? What if the existing coils output is dumped into a spark of shorter duration? That's a question that is at least easy to answer. If a given amount of electrical energy is dumped into the plug gap faster the power (wattage) the temperature of the spark goes up. This means that the consequence of maintaining the energy available for the spark but shortening up its duration is a fatter hotter spark – and this at least sounds like what is needed.

Using the alternative Nology Engineering technique for speeding up the process of dumping the spark energy into the plug results in a spark duration of as little as 4 nano seconds. Putting it in more easily assimilated terms this spark occurs quarter to a half a million times faster than a normal inductive discharge spark. If the trade off of time for spark power and temperature is not appreciated going to the short duration spark appears to contradict all experience to date. This was my first impression. The race and high performance business seems to harbour more than the odd crackpot or

The Nology HotWires provide an immense amount of spark power to the point where there can never be a doubt that the spark is more than up to the job at hand. Also seen in this shot is the crank trigger which was used in conjunction with a Motech distributorless programmable ignition system. Although more expensive than a distributor the advance curve could be tailored to the motor's exact requirements.

two. They seem to surface about as frequently as people who believe the 100-mile per gallon carb exists and has been suppressed by the oil companies. (It doesn't and it wasn't!) If it was not for the fact that a handful of racers were successfully using the Nology system I may not have paid any further attention to it or Ulf Arens. However the fact that a Nology assisted magneto was used to make what was at the time the World's fastest 1/4 mile pass prompted a second and longer look at what Mr Arens was proposing. As is so often the case there proved to be more to the short spark concept than at first seemed. Ulf did not reject the validity of the long burn tests done to date. What he was in fact doing was advocating a simpler and potentially better alternative. He reasoned that if the cylinder was ignited aggressively enough in the first place the long burn arc with its associated high energy consumption becomes redundant. Also, if the combustion takes place faster then more of the heat is used at or near the full expansion ratio and more power will be delivered regardless of anything else.

The foregoing seemed reasonable enough but with all the problems of inadequately mixed fuel and air how do we guarantee a good start to the combustion cycle? The long burn cycle keeps the plug alight with a high temperature (about 2,750°C) arc until near complete combustion is assured. However such a technique still does not necessarily mean maximum cylinder pressures are developed. Cyclic pressure variations below the maximum indicate some part of the charge was slower than normal to burn. This is usually caused by air/fuel mixing anomalies which occur on virtually every cycle. Both a richer and leaner mixture burn slower than a maximum power mixture. As previously mentioned when this happens it is as if the slower burning part of the charge is operating at a lower compression and consequently expansion ratio, than it would have been if it had burned faster. The long burn arc may be a good bet for as near complete combustion as is possible but the flame front progression on the way to that goal may still be less than optimum. The long burn arc may move your motor one step nearer optimal combustion but lighting up an already burning charge with a long persistence arc may not be the only way to get the job done. If the initial flame kernel can be started aggressively enough this could prove to be as good if not better source of ignition for the rest of the charge as a long burn arc. From a logical engineering standpoint such a concept seems very reasonable. If enough power is dumped into the plug gap more horse power via a short spark starts to look like a practicality.

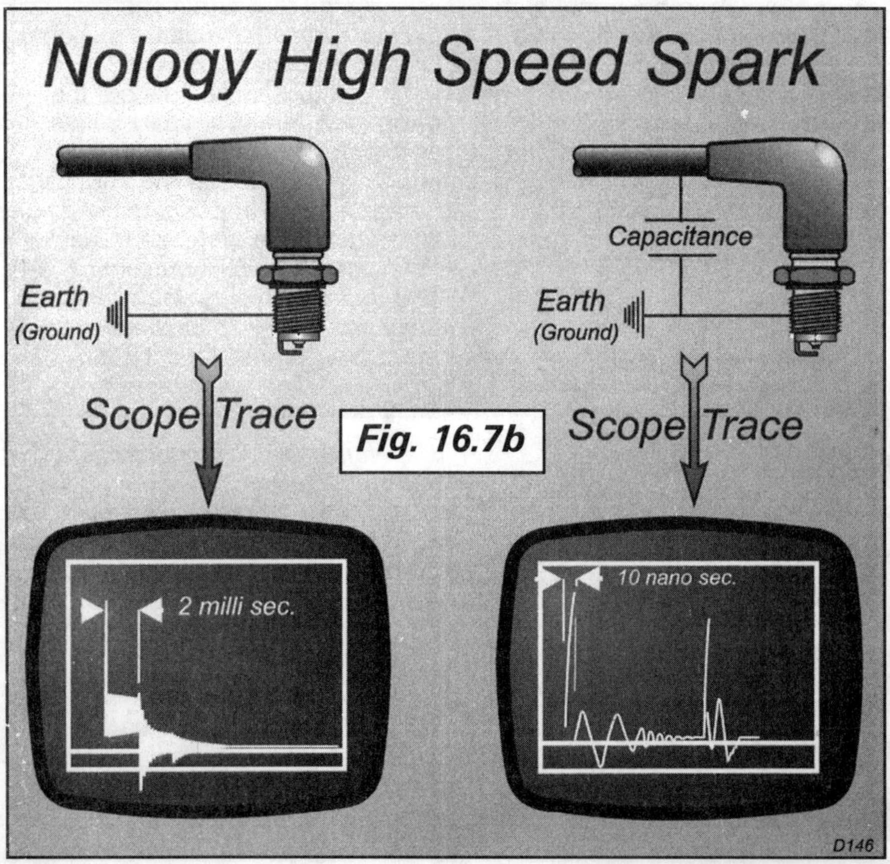

Fig. 16.7b

At this stage we are now confronted with the question of how the spark power can be dramatically increased without incurring the already high current demand seen on today's high output ignition systems. The theoretical solution is simple. Power is the rate at which energy is delivered. The coil has a certain amount of energy available to produce a spark at the time it is required. The faster the coil dumps its stored energy the higher the power involved. By going this route we are not contravening the bigger, fatter, hotter, longer spark concept completely. The only aspect we are in opposition with is the 'longer' part but we are trading heat, power and intensity for duration. By releasing coil energy in as little as 4 nano seconds the Nology system can theoretically create plug gap power figures up

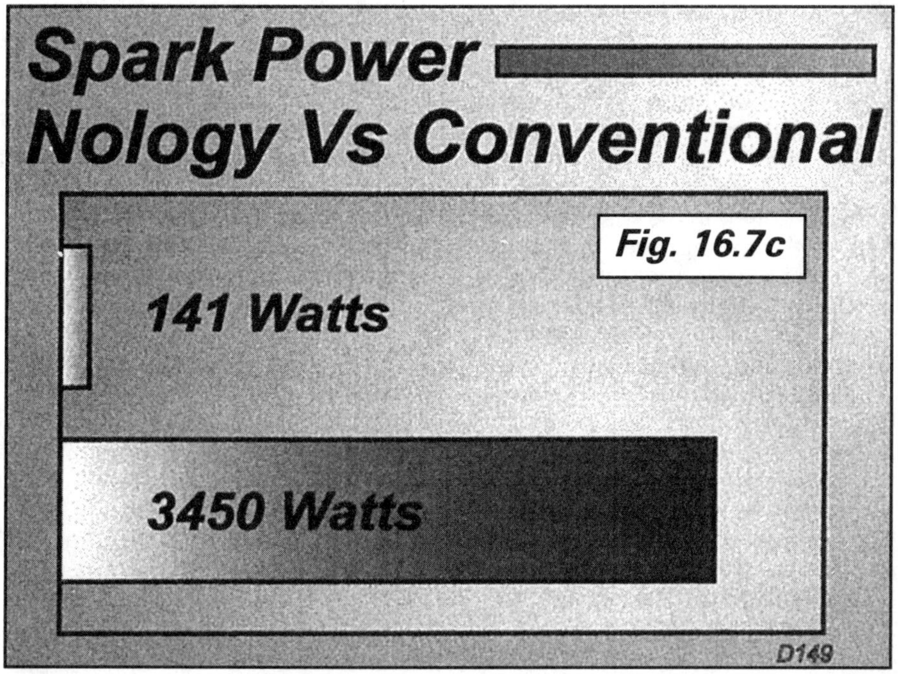

to 100,000 watts rather than the more normal 125 to as much as 1,000 for a super high output race system. Added to this the temperature will soar. It will rise from a typical 1,650°C to as much as 30,000°C.

Easy Installation

The design of the Nology system seems simple enough. Firstly it works with whatever system you have unless you are already using a long burn system (unlikely). Essentially the use of the Nology system involves replacing the conventional plug leads with their special HotWire leads. The fitment of these leads on the motor depicted on the front cover may have caught your attention at the time you purchased this book. The distinguishing feature of the Nology leads is the additional connection other than just to the spark plug. This additional cable is grounded and a capacitance between the inner solid copper core wire and the outer braided copper sheath is formed by the special cable insulation. As far as the function goes the drawing **Fig. 16.7b** should show precisely what is going on. For a conventional system (A) the coil voltage builds to the point of firing the plug and producing the trace shown. The Nology method introduces a capacitance into the system as shown in (B). Here the coil energy goes toward charging the capacitance until the voltage has reached the point it can fire the plug. When that time comes the energy in the capacitor gets dumped into the plug at a far faster rate producing the scope trace shown. If this all appears too simple to be of any consequence let me tell you that dyno numbers and race track wins are proving to the contrary.

Although the Nology method of making a spark was first used on nitro burning dragsters this has hardly any relevance to the 'A' Series motor. However some very comprehensive tests that were done on a small block Chevrolet, which has very similar combustion qualities to an 'A' Series motor are. These tests were done at great length and produced far more in the way of test results than we would care to go through here, so a summary is all that will be given. First the power figures. These are shown in the form of the percentage gain over race grade regular plug leads being fired by a high performance multi fire ignition system. As can be seen the Nology HotWires increased output by up to 2.3%. This compares very favourably with the 3% or so increase given by costly, plug consuming long burn arc ignition. Producing such an increase with no down side is also very much a credit to the Nology HotWires.

Power Source

An electrical test of the Nology leads was interesting. Using a sophisticated (and expensive) oscilloscope I measured the wattage dumped into the plug gap under working conditions with a regular electronic ignition and conventional leads, then repeated the tests with the same ignition and the Nology leads. The results are as shown in **Fig. 16.7c.** as can be seen the power in the plug went up from 141 watts to 3,450 watts. The plug temperature also climbed a similar amount from 1,650°C to somewhere above 20,000°C. The spark happens in such a short time that plug erosion is no different to normal.

At the end of the day power is all that really matters for a race or high performance car but analyzing how it came about is always worthwhile. When the power tests were done a Superflow cylinder pressure cycle analyzer was used to collect combustion pressure data. When the Nology leads were used the peak pressure variation from one stroke to the next was cut by almost 30%. This was the case at part throttle as well as full throttle so the inference here is that the motor ran smoother. This would be as you would expect if the cylinder was lit up more aggressively. Anytime the motor gets on with the business of burning the fuel more effectively we usually find that it burns it faster and this means less advance is needed. When changing to the Nology HotWires a motor may need retarding by as much as 3° because of the faster burn time produced. Another side benefit is that the amount of choke enrichment needed during cold start is considerably reduced. For racers this means less bore wearing fuel slopping around while warming the motor. For a street vehicle it spells easier starting, more mpg and less pollution. For those of you who also own and run two-stroke powered vehicles it means the news is just as good. Along with added output potential and significantly better throttle response the short high powered spark means a virtual end to plug fouling.

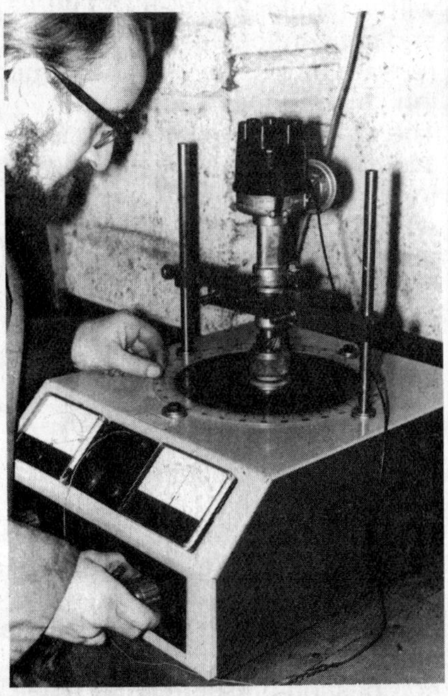

Here Don Loughlin checks out the final advance curve produced by an Aldon distributor on a distributor test machine. You will notice that this distributor is for a 6-cylinder engine, and yes, you would be right if you thought that 6-cylinder Series 'A' engines are very rare

If there is a down side to these leads it is the fact they are far from cheap. Countering this is the fact they open up an avenue of power production that is not available to us by any other means so whatever you have now the Nology leads have the potential to improve on it.

I can only guarantee that you will have the most effective plug lead known to mankind. What I cannot do is guarantee a power increase. Each motor is different. If there is a combustion problem the Nology lead will reduce its negative effect on output and more power will result. If no measurable combustion problem exists no power increase will be seen and if you remember, we discussed all that at the beginning of this chapter. If you always bear in mind that it is not possible to cure a problem that does not exist the reason no measurable power increase will be seen is obvious.

Even if you dyno test and see no measurable power increase do not abandon the use of the leads. After having dyno tested the Nology leads a number of times and recommended them to racers and teams I was working with another aspect that came to light. Several of the teams reported only a small change in output as seen on the dyno but when on the track the differences were far more apparent. How did this discrepancy come about? The reason is quite simple and is more apparent on drag race cars than circuit machines. The easiest mixture to light off is a maximum power mixture of 12.8 to 13 to 1. When a motor is on the dyno there are no G forces to upset the accurate calibration of the carburettor. The result of this is that a near optimum mixture is delivered while dyno testing. On the track things are somewhat different with a carburated motor. The G forces throw the fuel around in the float bowl to the extent that sudden and violent changes in direction cause the fuel level and the action of the needle valve to be far removed from ideal. As a result the mixture can be quite a way off ideal. When this happens the intense spark of the Nology lead does a far better job of lighting off the charge. Result – faster times at the track or on the road and remember we race cars not dynos.

Although produced in San Diego California Nology leads are available in the UK from Nimbus whose telephone number is 0137 721 7888.

Advance Curves

One of the most common mistakes made by the novice engine tuner is to assume, after modifying the engine, that the distributor will be needed from a higher performance version of that engine. Unfortunately, this can be anything but the case. The chances of a distributor from a more highly tuned engine in the same range having the right advance curve for your particular engine is about as likely as the advance curve it presently has being correct. To illustrate why this is so I will run through the various aspects which affect just when the distributor should initiate the spark.

The most obvious point is that a distributor has to deliver the spark to the cylinder at the right time. The timing figure most generally referred to is the initial or static timing. Basically this defines the ignition timing that should exist when it is correctly set at idle with a stroboscopic timing light. If the ignition is set statically, then it is the ignition timing which exists as the engine is very slowly turned over, usually by hand, to the position where the points just open. Since distributors don't generally start to advance the timing until about 800 to 1000 engine rpm, usually the stroboscopically set ignition

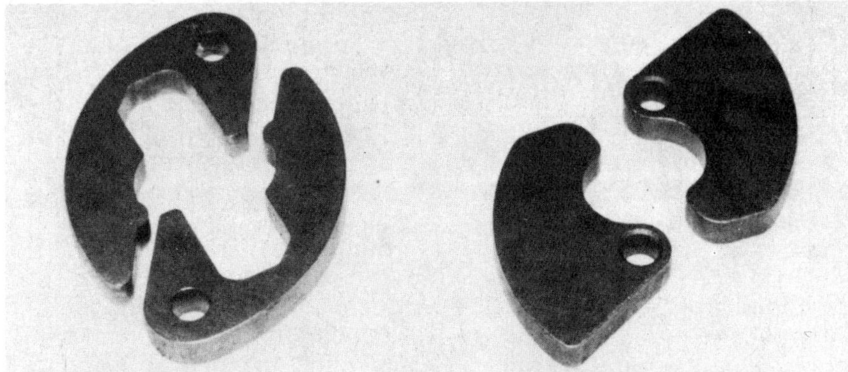

There is a difference in the advance weights of the 23D and 25D type distributors and the newer 43D, 45D and 49D distributors. Seen on the left are the weights for the newer type of distributor. On the right are the short nosed weights of the older distributor. By short nose, I mean . . .

. . . the type on the left as opposed to the 23D 25D type on the right. Both these types of weight are interchangeable and the long nosed ones, being heavier, cause the advance to come on slightly sooner in the rpm range.

timing at idle and the static timing are pretty much one and the same thing. If either of these means of setting the timing is recommended, it is assumed that the advance characteristics of the distributor, both vacuum and centrifugal, are correct for the engine's requirements. If they are, the ignition timing will, within the realms of what can be achieved with a mechanical distributor, be right under all circumstances. **If the advance characteristics of the distributor are not right, then there will be an error between the timing the distributor gives and the ignition timing the engine actually requires.** When an engine is modified it alters the combustion characteristics. Some modifications cause only a small deviation from those of the engine standard and therefore the timing error is proportionally small. Other **modifications can cause a dramatic change in ignition requirements thereby rendering the standard distributor totally inadequate for the job.** Let's look at two engines of similar output as an example and consider whether or not they should require the same distributor. In producing a (supposedly) 78hp Series 'A' engine, the factory decided, as in the case of the Cooper 'S', to increase capacity to 1275cc in an engine with relatively big bores, large valves, twin carbs and tubular exhaust manifold. The cam used was relatively short on timing. Let's say you are going to build an 80hp engine and you decide that as a basis, you are going to use an 1100 engine. Certainly 80hp is possible, but to get that 80hp a hotter cam will be required with much more lift than that of the Cooper 'S'. It's unlikely the 1100 engine will be using valves as big as those of a Cooper 'S' unless extensive head and block modification are involved. To provide effective carburation you may elect to use a side draft Weber. Even at this stage the engine is radically different and we haven't

The heavy springs on the left are the secondary centrifugal advance controlling springs. Those on the right are the primary. The lighter the spring is, the greater the amount of advance it will allow at any particular rpm range. That is until . . .

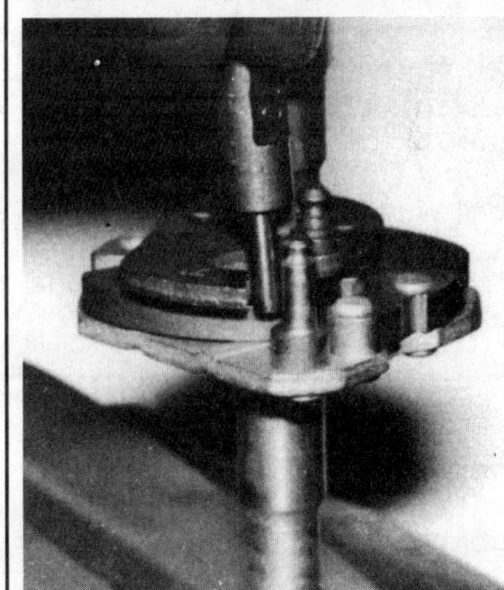

. . . the advance mechanism comes up against the advance stop in the baseplate. The gauge being used here between the tang of the advance weight and the stop, determines the amount of advance that will occur. By adding material or removing it as necessary, the total advance can be increased or decreased.

'S' Distributors – What Are They Good For?

As a distributor for a race engine/the basic 'S' distributor is a good choice. Most race engines require little or no advance from about 4000rpm up. Indeed, on 300° cams, the difference required in timing from 5000 to about 7500 or 8000 in the case of smaller engines, is minimal. Longer cams such as the Super Sprint type profile, may require the advance to continue right up to 8000, but generally the difference in timing between, say, 5000 and 8000rpm, may only be 3 or 4°. On some engines there may be a hole in the ignition requirement curve. In other words, the engine may require less ignition advance at, say, 5500rpm than it requires at 4000, but by the time the engine gets to, say, 6500, it may require more advance.

If there's a dip in the advance curve requirement a mechanical distributor will not cope being unable to retard half-way up the rpm range. This sort of advance curve required must be met using an electronic distributor with a programmable advance curve.

For most race engine applications, where a 300° cam is used, the advance curve below about 3500rpm is not really important other than the fact it makes starting easier. Most Series 'A' engines need around 36° of advance with 100+ octane fuel, and 30 to 32° advance on 96/97 octane fuel. However, they cannot start with this much advance, so it's necessary to have a distributor with 10, maybe 15° of centrifugal advance built in. That's distributor degrees, not crank degrees. 10° of distributor advance produces 20° of advance when measured at the crank. If we have a 10° distributor and require say, 30° of total advance, then the ignition timing can be set to 10° before TDC to give suitable starting characteristics. Once the engine gets to 3500 to 4000rpm, the distributor will be on virtually fixed timing.

If you require a distributor to advance quickly up to, say, 4000rpm and then advance slowly from there on, a unit will be required with a relatively soft primary spring and a very heavy secondary spring, so that full advance is not reached until about 8000rpm.

even though they may be as powerful as the old 'S'. As a result, modern distributors have ignition advance characteristics which take into account the vacuum advance requirements and therefore produce much better economy. An example of this is the MG Metro which uses manifold vacuum picked up at the carb. Also there have been occasions on some Series 'A' engines such as some Marinas, where the vacuum advance has been omitted, not for reasons of economy, but as a 'Bandaid' cure for crank rumble.

When the need for economy is considered in relation to ever decreasing octane levels, we find a great deal hinges on having an advance curve in the distributor which can cope with conflicting requirements. The modern Series 'A' engine in the main uses a substantial CR, 9.4:1 being common and some of them are as high as 10.2:1. All this seems to contradict what is actually needed because of octane levels which have dropped to 96 to 97 (R.O.N.) Trying to power optimize the ignition timing at these CRs on fuel octane levels that are currently available will result in detonation under full throttle conditions in the lower rpm ranges especially if a heated intake manifold is used.

Basically the factory engineers have developed

considered so far that one is a small bore engine as opposed to a big bore engine. Indeed, for all practical purposes they are two totally different engines, so why should a Cooper 'S' distributor curve suit this 'other' engine, even though they are producing comparable power? Not only are the characteristics of that distributor likely to be completely wrong from the centrifugal advance point of view, but also, because design philosophies have changed over the years, the Cooper 'S' distributor would probably be built today with a vacuum advance; something it did not originally have. In the days when the 'S' was designed, fuel was much cheaper and the manufacturer assumed that the car would be driven fast most of the time, by virtue of the market it was going to. It was also not recognized that the car would be the best seller that it was, so again it was expected only to fall into the hands of enthusiasts who would make use of its performance. These days engines in the Series 'A' range of a similar type, have far more attention paid to fuel economy,

Series 'A+' Distributors

If you are working on the Series 'A+' engine, remember it does use a physically different drive shaft and distributor and a forked clamp to secure it to the block. The type of 'dizzy' used in a 49 D4.

Apart from the sliding points, the rest of the distributor is pretty much the same as the 43/45 D4 range.

advance curves which, up to about 3000rpm and full throttle are retarded from what would be regarded as optimum if the octane value the engine needed was provided by the fuel. From 3000rpm upwards, the Series 'A' engine does not require as high an octane level to utilize a given CR, so from this point upwards, the engine can have optimum ignition timing characteristics built into the distributor. **With current octane levels, the modern high performance distributor has to have an advance curve which keeps the engine out of low speed detonation. This inevitably means less initial advance or an advance curve which comes on slower or a combination of both.** The penalty for the octane enforced ignition retard is reduced low end torque. When developing an advance curve the octane limitation factor has to be superimposed on any basic advance curve that may be used in the Series 'A' engine. As to how the advance curve may vary depends on engine specification and can best be determined from the chart **Fig. 16.8.** Understandably there are many variables to take into account. The question is, how do we get a distributor with the required advance curve? Any good engine tuneup shop should, within reason, be able to re-tailor your existing distributor, assuming its in good condition. However, if you have changed the engine's specification substantially, something significantly different may be needed in the way of an advance curve. In the early days, fitting a competition distributor from Lucas was the only method of providing the correct timing. The company, however, was not easily accessible except to commercial engine tuners of

Fig. 16.8 Ignition requirements chart

Item	Effect on total advance	Effect on centrifugal advance	Effect on vacuum advance
Low octane fuel	Retard from ideal because detonation limits advance.	Rate of advance slows down and total advance should be less.	At high vacuum fuel octane makes little difference. As vacuum drops, less advance can be used due to detonation limits.
Raise CR	Needs less TOTAL timing.	Advance rate stays about the same or slightly slower.	Vacuum advance stays about the same.
High energy ignition	Needs less TOTAL timing.	Advance rate stays about the same or slightly slower.	Vacuum advance stays about the same.
High lift, standard duration cams	Needs slightly less total timing.	Advance curve needs slowing down from 2500.	Stays about the same.
Longer period than standard cams	Needs slightly more total timing.	Needs quicker advance up to approx. 2500-3000 rpm.	May need about the same total advance but may need to come in quicker.
Modified heads	Needs less total.	Stays about the same.	Stays about the same.
Manifolds and high flow exhaust system	Usually needs less total but if initial charge temp. drops, due to good scavenging, may need more total advance.	Advance curve may need to go faster or slower depending upon scavenging characteristics.	Usually needs a little more vacuum advance.
Turbochargers	Needs less total.	Usually needs quicker centrifugal curve.	Approx. the same total vacuum but should have pressure retard.
High flow intake manifold and carbs	Needs less total.	Slightly slower from 2500 on up.	Could require more or less depending on manifold characteristics at part throttle.
Water injection	Needs more total.	Remains unchanged.	Remains unchanged.

Care must be taken when assembling a high performance distributor not to over-tension the springs as this can affect the precision of the advance curve. Advance curve accuracy may not need to be particularly good on a standard distributor, but when you are running an engine that is close to detonation, accuracy of advance becomes very important.

competition cars. As a result, high performance road engines were not catered for to any great extent. These days the situation is largely alleviated because **Aldon Automotive** produce tailor-made distributors, based on experience gained from two 'in house' dynamometers which have been run the best part of five days a week for the last fifteen years. During that time, they have dyno-tested engines of the Series 'A' type to almost every conceivable specification. Because of the experience gained, they can build a custom distributor based on the spec of your engine. If your engine is a

Seen here is a Pirahna contactless ignition system installed in an Aldon race distributor. Such conversions should always be done on Aldon distributors at their factory, as installing a contactless system alters the advance curve. They therefore need to build the advance curve to suit the type of ignition system used.

little out of the ordinary, then they are geared up to put your car on a chassis dyno and plot out the advance curve required. From this they'll build a distributor to suit. Such distributors can be built for contact breaker type ignition systems or electronic contactless ignition systems to suit varying budgets and requirements.

This brings me to a subject I alluded to earlier, that is that the distributor advance curve may be changed when converting from a contact breaker system to a breakerless system. **Fig. 16.9** shows a relatively common change brought about, namely the retardation of the spark at an increasing rate as the rpm goes up. This particular example is not very great, but I have on occasions seen electronic ignition systems that retarded about 7 or 8°. **If you intend to use a breakerless ignition system, don't fit it after you have had your distributor curved, buy it at the time you are getting a distributor built or modified for your particular** engine.

Ignition For Forced Induction

When an engine is supercharged the amount of ignition advance, under boosted conditions, is less

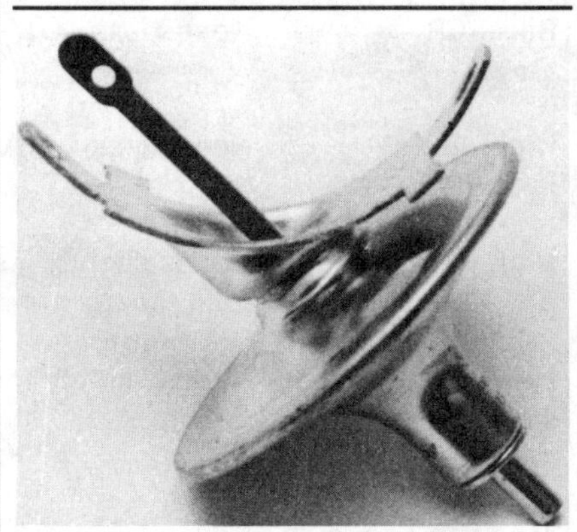

A vacuum advance unit. Don't try to build your high performance street motor without this unit on. It's use means mileage. To find out what the characteristics of the vacuum advance can are, take a look at the numbers stamped on the cannister just by the vacuum take-off-point. Somewhere in this area there should be a number, then a zero, then a number. This indicates the number of inches of Mercury at which the vacuum starts advancing and at which it finishes.

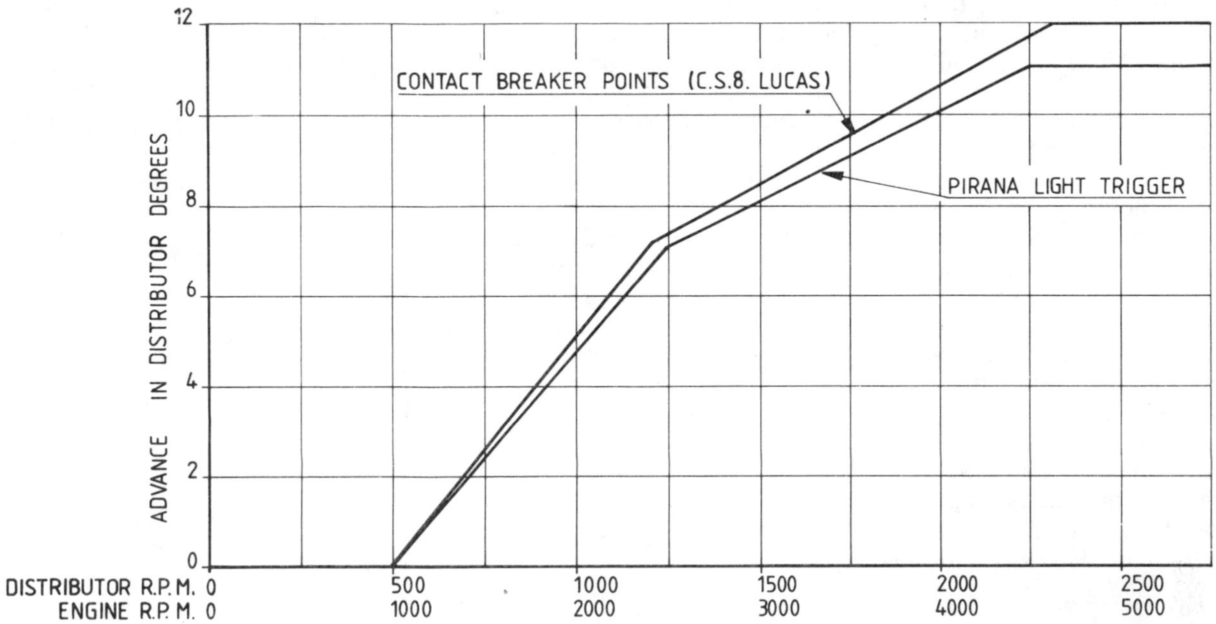

Fig. 16.9 Advance curve change with change in trigger used to initiate spark

than is required for a normally aspirated engine. Assuming 97 octane fuel, a typical normally aspirated engine requires around 35° of total advance. As a rule of thumb, you can figure that for every pound of boost the engine will require about 1.1° less total advance. In other words, an engine boosted to 10 psi requires 20 to 24° total advance, and this amount of advance generally needs to be 'all in' by about 3000rpm. Up to 3000, most engines, as long as they haven't too high a compression, tend to run with a fairly normal advance curve. The easiest way to stop the distributor over-advancing is, of course, to put a dab of weld on the end of the stop where it contacts the stop pin and then file it back until it gives the desired total advance. The only snag to this is in the case of insufficient advance at part throttle. Here, it would be necessary to have a distributor which gives boost retard. This particular problem can, in fact, be attacked two ways. You can either have a double diaphragm distributor which has vacuum advance on one side and boost retard on the other, or you can resort to an all-electronic setup which retards the spark when boost comes on by physically delaying the spark delivery to the cylinders. Many of the turbo kit manufacturers use the **Microdynamics** turbo ignition module. This unit is factory set to deliver a certain number of degrees retard per 1000rpm, superimposed upon the distributor's normal advance curve. Note that it's retarding in relation to engine rpm, not boost. Although this is not necessarily the ideal method, it is highly functional because generally speaking, the boost increases with increasing engine rpm. However, once the boost hits the waste gate setting, the amount of boost increase that occurs is very little in proportion to the rpm increase. Notwithstanding the small disadvantages the Microdynamics unit is very practical and does provide very acceptable results, especially when you consider that the unit also contains an electronic ignition system and a rev limiter. With such a system, a pressure sensing switch, usually set at 2 or 3 pounds, senses when the engine reaches a boost condition before it initiates any ignition retard. This means under normal driving conditions, the ignition timing can be set much nearer optimum for best mileage. Although its simplicity indicates that pressure retard would seem the obvious choice, the electronic Microdynamics system with rpm induced retardation produces such good results that it has, by virtue of simplicity in use, much to recommend it.

I have assumed that the ignition system for a supercharged engine had to be compatible with an engine using a cam of relatively mild timing. If a racing long period cam is used, then the situation, believe it or not, becomes simpler. For a low compression engine, equipped with a long period cam, the ignition advance required at idle and low rpm is far more advanced than that normally required. For instance, assume a supercharged engine utilizing, say, 16 psi boost. Under these conditions, the engine should have a CR around 6.5:1. To make an engine like this idle effectively will require 18 to 20° of initial timing. Under normal circumstances, as the rpm increase, so the engine requires more ignition advance, but in the case of a blown engine, as rpm goes up, so the boost starts to come on, thus reducing the advance over that normally required. On such an engine, under full throttle conditions and idle conditions, the engine requires in the region of 20° advance. This means that there is no advance curve as far as a centrifugal advance is concerned so from that point of view, the timing is fixed. However, to make the engine more responsive during part throttle operation, a vacuum advance should still be retained. Here a vacuum giving, say 8° of distributor advance will work fine, so that when the vehicle is cruising along at normal highway speeds, the ignition timing could have moved up to around 36 to 38°. Under such circumstances, you will see that it's very easy to build your own distributor. It's simply a question of fixing the distributor so that no centrifugal advance takes place, but still leaving the vacuum advance operative. The vacuum 'can' required though, needs to be sensitive because the amount of vacuum generated by, say, a 300° cam is not very great.

17 Bolt-On Mods

If you have read through the book so far, you should now have a relatively good understanding of the strong and weak points in the chain of power production events for the Series 'A' engine. Having acquired that background knowledge, the significance of various bolt-on goodies will be more apparent, and it will be more readily understood why certain bolt-on modifications are better value for money or better value for effort input than others.

Let's define what a bolt-on performance part is. We can say that anything from the cylinder head deck of the block upwards can be installed on the engine without having to tear into it. Actually going inside the engine usually means removal of the engine from the chassis. Anything that calls for engine removal falls outside the scope of a bolt-on performance part.

The purpose of a bolt-on performance part is to make the best of the new or good used engine. It is definitely not for pepping up an engine which is on its last legs and really needs an overhaul. If your engine has a good bottom end, namely the bores and bearings are in good condition, then it makes sense to use up at least the best part of the useful wear life of these components before going in and modifying the bottom end. This is especially true if you don't have money to throw around, and talking of money, brings me on to the next point of bolt-on mods. Essentially **bolt-on modifications are, or should be, a means of finding a useful measure of cheap, extra power, and the last point of bolt-on modifications is a means of finding easy extra power.** Let's face it, for many people, time is money, so if a modification which can be performed relatively quickly, although not necessarily dramatically cheaply, still may be worthwhile if it produces very positive results.

The purpose of this chapter is to look at what can be done either cheaply or quickly, or a combination of both, and to analyse just what is best for your particular needs.

Finding The Weak Link

To improve the power output of

Fig. 17.1 Bolt on performance priority guide

	5 Star	4 Star	3 Star	2 Star	1 Star
A 30	1, 2, 4, 6, 7, 10	8, 14	13	–	–
A 35	1, 2, 3, 4, 6, 7	8, 10	9, 11	13	14
A 40	1, 2, 3, 4, 6, 7	8, 10	9, 11	13	14
Austin Healey Sprite Mk. I and II and Mk. I MG Midget (948 cc)	1, 5, 6, 7	8, 9	11	13	14
Austin Healey Sprite Mk. III and Mk. II Midget (1098 cc)	1, 5, 6, 7, 12, 15	8, 9	11	13	14
Austin Healey Sprite Mk. IV and Mk. III Midget (1275 cc)	1, 5, 6, 7, 12, 15	8, 9	11	13	14
Austin and Morris 1100 saloon	1, 2, 4, 6, 7	8, 9	11	13	14
Austin and Morris 1300 saloon	1, 4, 6, 7, 12, 15	3, 8, 9	11	13	14
Austin America	1, 4, 6, 7, 12, 15	3, 8, 9	11	13	14
Austin 1300 GT	1, 5, 6, 7, 12, 15	8, 9, 10	11	13	14
Mini 850	1, 2, 4, 6, 7	8, 11	9, 10	13	14
Mini 998 (single carb)	1, 2, 4, 6, 7	8, 11	9, 10	13	14
Cooper 997	1, 5, 6, 7	8, 9	10, 11	13	14
Cooper 998	1, 5, 6, 7, 12, 15	8, 9	10, 11	13	14
Cooper 'S' (all sizes)	1, 5, 6, 7, 12, 15	8, 9	10, 11	13	14
Mini 1275 GT	1, 4, 6, 7, 12, 15	3, 8, 9	10, 11	13	14
Morris Minor (803 cc)	1, 2, 4, 6, 7, 10	8, 14	13	–	–
Morris Minor 1000 (948 cc)	1, 2, 4, 6, 7	8, 10	9, 11	13	14
Morris Minor 1000 (1098 cc)	1, 2, 4, 6, 7	8, 9	11	13	14
MG 1100	1, 5, 6, 7, 12, 15	8, 9	11	13	14
MG 1300	1, 5, 6, 7, 12, 15	8, 9	11	13	14
Morris 1098 Marina	1, 2, 3, 4, 6, 7	8, 9, 10	11	13	14
Morris 1275 Marina	1, 4, 6, 7, 12, 15	8, 9, 10	11	13	14
Morris 1275 Ital	1, 4, 6, 7, 12, 15	8, 9, 10	11	13	14
Metro 998	1, 4, 6, 7, 12, 15	8, 9, 10	11	13	14
Metro 1275	1, 4, 6, 7, 12, 15	9, 10	11	13	8, 14
MG Metro 1275	6, 7, 12, 15	9, 10, 11	–	–	8, 14
MG Turbo Metro	1, 6, 7, 11	15	–	–	–
Allegro 1100	1, 2, 4, 6, 7	8, 9	11	13	14
Allegro 1275	1, 4, 6, 7, 12, 15	3, 8, 9	10, 11	13	14

Modification	**Code No.**
Install performance air filter/s	1
Install 1½-inch S.U.	2
Install 1¾-inch S.U.	3
Install single carb. water-heated intake manifold	4
Modify existing twin S.U. manifold	5
Install free flow high efficiency silencers	6
Install modified cylinder head	7
Install high performance exhaust manifold	8
Install side draft carburettor	9
Install high performance induction system (i.e. 28/36 Weber and manifold)	10
Install higher ratio rockers	11
Install turbocharger/supercharger	12
Install electronic ignition	13
Install electric fan	14
Install N20 injection	15*

Not numbered above, but needs to be done in every case, is general engine tune-up. This will include checking and replacing as necessary, plugs, plug leads, points, condenser, carb, needle, jet and spring (replace if wrong spec or worn) distributor shaft wear, tappet settings (valve lash) and air cleaner element if paper type is being used.

** A five-star mod. in terms of performance gained for money/effort expended, but do not forget limited time of use before bottle needs refilling.*

The use of Nitrous Oxide or turbochargers/superchargers is only recommended where the original factory installed piston is able to withstand the temperatures and pressures involved. Power increases with such bolt-ons need to be done in moderation if acceptable reliability is to be retained.

any engine, it's necessary to find out where in the chain of power production events the greatest measure of inefficiency occurs. Just to make a point, let's assume you are modifying an engine which had double overhead cams and four valves per cylinder. If that engine operated on a small, single carb, it's fairly obvious to one and all that the first move on the engine wouldn't be a modified cylinder head. It should be carburation changes to improve the limited breathing in that area. Although the Series 'A' engine doesn't have a double overhead cam setup or four valves per cylinder, the same basic rule applies. We must **find the weakest link first and improve on that. By doing so, the greatest gains in power will be achieved per component change.** Mind you, there is one factor we need to take into account here, and that is the amount of effort required to change that component. It could be that the least effective component is a 'bear' to get at, and even though it may cost little or nothing to change, if it involves two days' work, it may be disqualified from being a 'practical' bolt-on mod.

Since there are so many variants in the Series 'A' engine, it is hardly surprising that the most inefficient aspect of the Series 'A' engine can be different with each variant. For instance the cylinder head on an A35 is far more inefficient than the cylinder head on a 1275 Cooper 'S', and modifying a Cooper 'S' head brings about far less a percentage improvement in power than a well modified head bolted on to an A35. So with this factor in mind, it's really necessary to look at each engine variant as an individual. To make this easier, I have made up a chart which lists all the common modifications and assigns them a star rating. A five-star modification is one which, in my opinion, gives good results in terms of both time and effort put into installing it and the amount of cash spent in relation to the power gained. A one-star modification is one which I feel requires the expenditure of either a lot of time or cash or both, but only gives limited returns.

As with any job, the workmanship involved needs to be done properly if you are going to get good results. Also as a final operation to any bolt-on modification, get the vehicle set up on a rolling road dynamometer. A good dyno tune is going to find all those little nit-picking things that you would otherwise miss. **Even if you have your engine set up exactly to the factory specification as per the overhaul manual, there is no guarantee that what your particular engine needs are anywhere near those particular settings.** This is especially so on earlier engines when production tolerances of virtually every part in the engine tended to be wider. Sometimes bolt-on modifications can be very disappointing, especially if they tend to exaggerate some fault already there. For instance, raising the CR on an engine which has a marginal ignition system for one reason or another, can often cause the engine to run worse, when in fact a high compression would make the motor go better if the short-comings in the ignition system were rectified.

Another factor which often crops up is the assumption that the needle in the carb is correct for the standard engine. Very often carb calibration (again, more so on earlier vehicles) can be a mile out, and doing nothing other than a straight dyno tune can make a difference in the vehicle's performance akin to night and day.

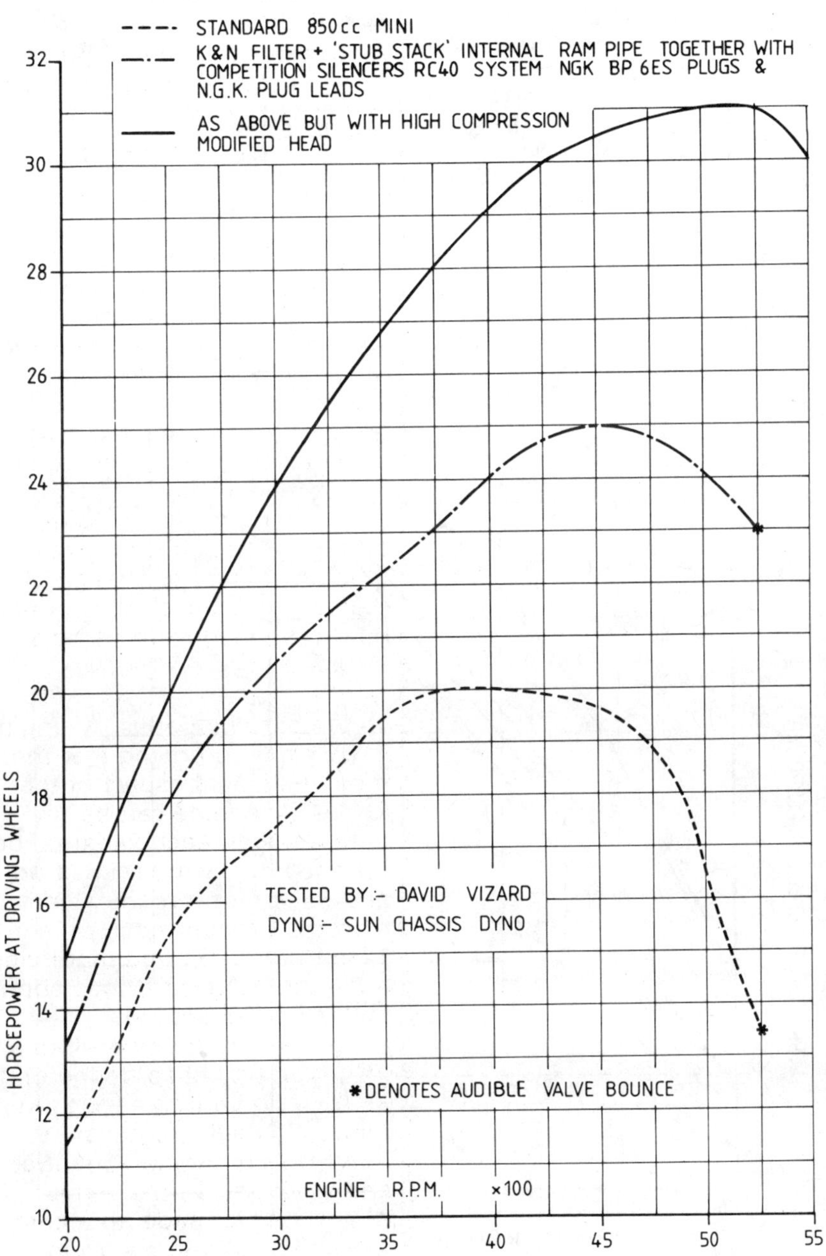

Fig. 17.2 Comparison of standard engine and engine modified by bolt-on performance equipment

Bolt-On Potential

If you had the opportunity to ask, no doubt the number 1 question at this point would be just how much power can we expect from bolt-ons? Well, that does depend on the vehicle you start off with and what deficiencies it may suffer. So let's look at examples from both ends of the scale. First, let's take an 850 Mini. The graph **Fig. 17.2** shows the baseline power figure at the wheels. You will note that in this particular example 20 hp was registered at the wheels. Not bad for an 850 Mini. The average tends to be between 16 and 18 when measured on a Sun rolling road dyno, which is a pretty common dyno.

Parts for this engine were bolted on in two stages. Firstly, a high efficiency filter with an internal ram pipe was installed, together with an E3 needle, and along with this went a high efficiency silencing system. At the same time a set of NGK leads, plus some NGK BP6 ES plugs were installed, and last but not least, a set of points with sufficient spring tension to avoid points bounce up to about 6500 rpm was fitted. The result of these modifications was an increase in hp to 25 maximum at the wheels. That's a straight 25% increase over what was there before. The curve 2 in **Fig. 17.2** shows just how the hp increased. You will note there was quite a bit of gain from low down, so top gear acceleration figures were markedly improved.

The next step was to install a modified cylinder head, having a relatively high compression. Standard compression on this particular engine was 8.3:1. The high compression head raised this to 9.4:1. The cylinder head was properly done in as much as flow bench tests showed it flowed well, but it was to a design I have subsequently stopped using. The shapes and designs shown in the cylinder head chapter earlier, will produce marginally better results, so if you are following this route, and you cut a cylinder head using standard valves as described in the cylinder head chapter, you can expect to see a little more power than is shown here. However, this particular cylinder head showed an increase of some 6 hp at the wheels over the standard head, thus bringing the power up to 31 hp at the wheels or 55% improvement over a standard 850. At that power level, it's still 10 to 15% better than a standard 998. Obviously if you are going to breathe on a 998, proportionally better results will be achieved, and even better results yet on the single carb 1100 engine.

Since the 1275 engine is by far the most popular engine to tune, it would be better if this was looked at in detail, so let's hit the bolt-on trail on a 1275 GT engine and see how the power climbs (**Fig. 17.3**).

In standard form, the particular test engine used

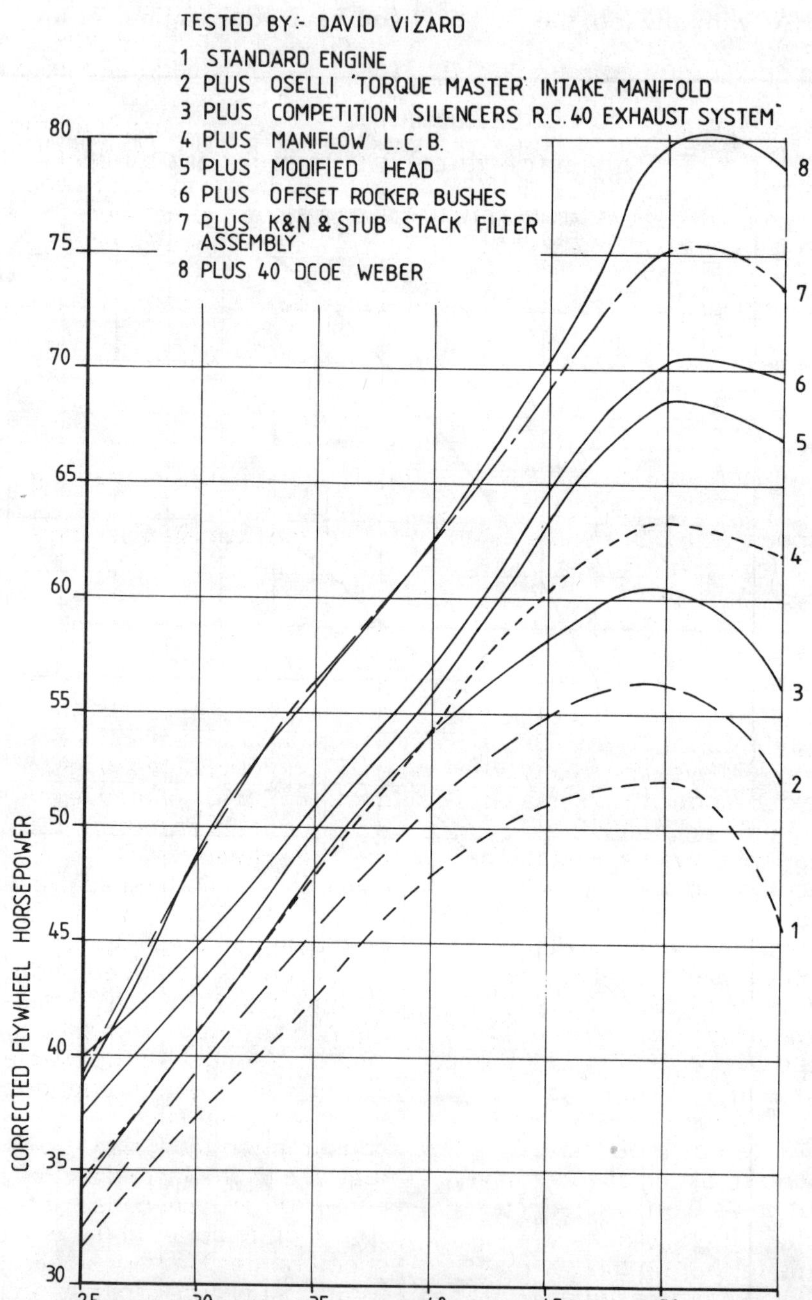

Fig. 17.3 Effect of bolt on performance equipment on a 1275 GT Mini engine

cranked out a whole 52.1 hp at 5000 rpm. The first move on this engine was to remove the standard intake manifold from the exhaust manifold by sawing it off, grinding clearance as necessary and installing an Oselli Torquemaster intake. Same carb, same air filter, no other changes, only recalibration of the carb to bring the mixture back right. This put the power output up from 52.1 to 56.3 hp and along with that was a torque increase of around 4 to 6 ft lb throughout the rev range.

For the next modification, the standard 1275 GT exhaust system was ditched in favour of a Competition Silencers RC 40 system. This added another 3 to 4 ft lb of torque and put maximum power up to 60.4 hp.

If you have ever spent time sawing an intake off an exhaust, you will realize it can be a pretty grim and sweaty job, especially in a hot climate. For this reason it makes a lot of sense, assuming the budget will stretch to it, to install an LCB at the same time that the intake manifold is changed, thus getting rid of the entire standard cast intake/exhaust manifold system. This was the next job done on this particular engine. The standard cast iron manifold was replaced by a medium bore LCB which dumped into the Competition Silencers exhaust system. As you can see from the power curve, the LCB didn't really produce any more hp until a little over 4000 rpm, but then it came into its own and it raised peak power from 60.4 hp to 63.3. Not as much power increase as the silencer gave, but still a worthwhile increase. Where it really did score was a little further up the rev range as it allowed the engine to turn much higher rpm and still produce power. At 5500 rpm the power output was up almost 6 hp.

At this point then, the intake and exhaust manifolds and silencers have been changed, nothing else, and this has resulted in some 11 hp increase over standard, that representing a 20% increase.

Changing the cylinder head is probably just about the biggest and most complex bolt-on mod, short of actually adding something like a supercharger to the engine. The particular modified head used sported a CR one point higher than the one that came off the engine. Also, an important point here is that the cylinder head the baseline figures were produced with was a new cylinder head, so it wasn't being affected by any carbon deposits or anything like that. I am making this point because the gains shown here represent the absolute minimum gains that are likely to be seen with a well modified head, still retaining standard valves.

If your vehicle has done more than, say, 15,000 or 20,000 miles, the bottom end could very well still be in perfect condition, but the breathing ability of the engine could be reduced by virtue of carbon deposits, so bolting on a head at this point could well add another 1 or 2 hp, maybe even 3 over what is seen here. This particular head, which essentially is nothing special in the way of cylinder heads, just employing all the simple modifications described in the cylinder head chapter, raised the power output from 63.3 to 68.6 hp. Along with this was an increase in torque throughout the rpm range. Fuel consumption tests also showed that due to the increase in compression, fuel consumption was slightly better in the higher speed ranges. At very small throttle openings accurate fuel consumption measurements actually showed the fuel consumption to be very slightly worse, i.e. about 1 to 2%.

Again I need to refer you to some of the details set forth in the cylinder head chapter. If you study the flow curves you will note that modified heads tend to go on increasing in flow up to much higher valve lift values than standard cylinder heads. The particular head used here was no exception, and to show that high lift is beneficial, a set of high lift bushes were installed in the standard rockers. This does two things for the valve lift. Firstly it increases the valve lift and secondly it increases the rate of valve lift so at any particular point in the lift curve the valve has reached at greater distance from the seat. Although the high lift bushes only give around 30 thousandths more lift, the effect on power and torque is immediately measurable. Essentially the power and torque were increased throughout the rev range, power generally going up by 1½ to 2½ hp between 2500 and 5500. On the bigger engines, going to high lift rockers which give a ratio of some 1.5, thus adding around 60 to 65 thousandths lift when used with a standard cam, will add more power than shown here. In fact they usually account for something in the region of double the power increase that is given by the high lift bushes. Although high lift rockers are more expensive, they are much easier to install than the high lift bushes, as pressing out bushes from pressed steel rockers is difficult and installing the new bushes is even more difficult, especially as they have to be reamed out to size after fitting in the rockers. The offset bushes are cheap, but they require many more man hours to install. High lift rockers are more expensive but they can be installed relatively quickly and they do produce more power, typically around a 4 to 5 hp increase at the top end.

The next move, curve 7 on

the graph of **Fig. 17.3**, is a modification which I would recommend very early on; that is if you don't mind an increase in induction roar. The reason it wasn't done earlier on in this test series is that parts simply weren't to hand when my allotted time on the dyno began, hence it is displaced to position 7. However, the test is still valid because it shows exactly what happens when the standard 1275 GT air filter is replaced with an Advanced Products/K & N filter and Stub Stack Ram Pipe. Power was increased throughout almost the whole rev range. Top end hp took a jump of some 5 and torque from 3000 to 5500 went up by around 4 to 6 ft lb. A very creditable increase considering that the removal of the old filter installation and the fitting of the new one involved only 10 minutes or so, and the cost in comparison with many of the other modifications, is relatively small.

At this stage the engine's power output has increased by some 50%. Many of the components now bolted on to the engine are highly conducive to improved breathing, and that raises the question of how much power would be gained using a side draft Weber. At this power level, a 40 DCOE is more than capable of supplying the air demand of the engine, and so I will make that the final mod to demonstrate bolt-on practicalities.

Curve 8 is the power output given by the use of a side draft 40 DCOE Weber on a 5 inch intake manifold. A point to note here is that the power curve didn't really start to show any improvement over the SU until over 4000 rpm, thus indicating that the 1½ inch SU on a Torquemaster intake with a K & N filter is completely adequate to supply the air demand for engines up to 65 to 70 hp or so. This particular engine still wanted to go more than that, and as you can see, a 40 DCOE was what it took to just make this 80 hp motor. A substantial improvement over standard, I'm sure you will agree, and all achieved without any loss of drivability.

Super Power Bolt-Ons

As you can see from the list **Fig. 17.1**, such items as nitrous oxide and turbochargers can and do often fall into the category of bolt-on modifications. However, I am not including them here because with a turbocharger for instance, the boost can be increased such that power outputs are quite phenomenal but the engine life can be incredibly short. The same goes for nitrous oxide. Both forms of tuning, though safe in themselves, can easily be overdone. If you want to know about these subjects, refer to the relevant chapters.

18 Example Engine Builds Part 1

Up to this point, I have dealt with the Series 'A' engine almost exclusively in terms of components, however, it must be stressed that the success of any engine does not depend upon having a few 'trick' components in it, but the right combination of components working in harmony. It's really a bit like an orchestra - the success of an orchestra depends not on a few brilliant musicians, but the combined effort of all of them, and so it is with engine building. If you are developing an engine, make only one change at a time, and then stop and think, after you have made a number of changes, whether the first thing you experimented with might not work in your subsequent specification of engine, although it didn't work to start with. This can often be the case, but above all, be logical in your concept of what your engine should be. In the following few pages I describe how to put together some competition engines, but do bear in mind that in racing, nothing stands still, least of all the regulations. The recommendations that follow are nothing but general guidelines. If you want to be competitive and you want to build your own engine, then it will be necessary not only to get your brain in high gear, because there are some sharp engine builders out there you are going to have to compete with, but also it will be necessary to spend quite a bit of time dyno testing. The chances of getting lucky and coming up with the right combination are pretty slim.

If you are a first time or inexperienced engine builder, avoid going overboard on engine specification. A little conservatism will often produce an engine you can live with when trying for the ultimate the first time around will not. I don't intend to try and cover every conceivable combination in this chapter because it's just not possible. The engine building hints put forward here will help with whatever combination you build.

Hot Road 850

Because it would be cheaper to go to a bigger engine, I can only conceive of any hot 850 for

Remember, your engine is the sum of the parts. It doesn't matter whether you are building an inexpensive 850 or a big, high cost 1600, the parts that go into it should be in perfect condition and your preparation should be as near perfect as possible.

road work as being one done on a very small budget, so with that in mind, I will assume that the operation will involve the use of as many standard parts as possible, some machining and whatever time it takes you, as the engine builder, to rework various components.

The golden rule with any engine building is that all parts must be up to scratch. It is better to spend money on new pistons and rings, even if they are of a relatively standard spec, than trying to build an engine with old pistons and rings and a racey cam. This being the case, the engine will almost certainly require a rebore. If it does, then look through the pistons available (discussed in the relevant chapter) and select a piston to suit your budget. If you can afford to have the cylinder block decked, then you can go for the flat topped 948 Sprite piston, rather than the standard Mini piston. Have the block skimmed so that the pistons are just a few thousandths proud of the top of the block, and then have the pistons skimmed so that each piston is flush with the deck of the block at TDC. If you can't afford to have balancing done, then leave the rods and crank as standard. Because the pistons have been skimmed, it will be necessary to check their weight. If the pistons are within a gram of each other, then that will be fine, though not perfect. If your budget runs to balancing, then I suggest you lighten the rods and rework the crank as detailed in a previous chapter.

Because 850 blocks don't have cam bearings for anything but the front cam journal, you should avoid using a cam that is anything approaching a race spec. I wouldn't recommend anything more than a 270° duration cam.

The choice of cylinder head is important. Remember, an 850 is going to want all the compression it can get. The more time you can put into the cylinder head the better. Bigger valves will work toward producing better top end power. A good 998 Cooper casting is a better basis than the typical 850 Mini type casting. However, be sure you can get the compression being sought, so check the cylinder head does have enough meat to produce a CR of at least 9.5:1. If you are using a relatively short cam (up to 250° period) then a 10:1 CR is perfectly acceptable. If you are going to use a longer cam than this, something such as a **Piper 270 Magnum**, try and go for

There's performance to be gained at virtually zero expense from the bottom end of any 850. It just requires hard work to lighten all the components as seen here.

about 11:1 CR, but in using these high compressions, be sure you do not use the standard exhaust heated intake manifold. If you have to use the standard manifold, separate it from the exhaust. 11:1 CRs and exhaust heated intake charges can produce detonation.

If you can afford to spend some money, then equip the carb with an efficient air filter. **Don't be tempted to run without an air filter**. It may give you a bit more power while your engine is new and unworn, but you will very quickly have a new, worn engine which won't give as much power.

The next thing to do if you can afford it is to change the intake manifold to one of the many, good, high performance manifolds available. The next mod should be an efficient silencing system and lastly, install a small bore LCB manifold.

850 Mini-Xer

A formula that has been gaining popularity hand over fist, is 850 Mini-X. It's relatively cheap, the regulations are such that engine reliability is pretty good, and since most of the racing is done on either grass tracks or tight asphalt tracks, the engines are powerful enough to produce close and exciting racing. In this class of racing, it is easy to be competitive but to be a front runner and a championship winner, like any form of racing, requires the very best.

One of the top engine builders in this category of racing is Dave Mountain of **Mountune Race Engines**, and at the time of writing, his championship winning engines were producing 47 hp at the wheels, on a rolling road dyno, where a standard 1275 'S' in really good running order produces 55. When you realize that the rules call for many limitations to be complied with it can be appreciated that 47 hp at the wheels is an exceptionally good figure.

First, the rules call for a single 1½ inch SU carb. The favoured carb is an HS 4 or an HIF 38. The air filter for this needs to be a deep K & N with a ram pipe between 2½ inch and 4 inch long inside. The carb and filter need to be mounted on a **Janspeed** 850 Mini racing intake manifold and for the exhaust, a small bore LCB seems to be the best of the conventional exhaust manifolds, but the CV exhaust seems to produce a little more in the mid-range. It would be well worth trying both the small and medium bore size CV exhausts, because the engines at 47 hp at the wheels are just on the cross-over between requiring a small bore and a medium bore.

The real trick to making an 850 Mini-X engine perform is knowing what to do within the block assembly. **For 850 Mini-X valve sizes are limited to standard, and lift limited by virtue of having to use a standard cam. As a result, the airflow of the cylinder head is all-important. This means that any old head just won't do.** The head used needs to be a well-developed item that flows plenty of air, and it needs to have as much compression as possible. This means it's necessary to select a good casting and then get it to pass all the air possible. This necessitates a really good porting job, which in turn makes almost essential a set of Rimflo valves. A compression ratio of 12.5 to 13:1 will be required; not an easy figure to achieve.

Regarding the pistons the hot favourite for a winning engine are **Carl Schmitt** pistons because they are strong and have a solid skirt. By the way, rules call for standard or standard replacement pistons, no hot rod pistons are allowed. The best oversize to go for is +40. For some inexplicable reason, engines at +60 don't seem to have quite the power of those at +40. As far as setting piston height is concerned, the block needs to be skimmed so that the pistons come out of the top of the block by 5 thousandths. When the

If you are putting together an 850 Mini-Xer, remember, the two most important criteria for success are finding a head so you can produce shallow enough combustion chambers for a high CR and lightening all the engine's moving components

head is torqued down, and the gasket compressed, this leaves static about 18 thousandths between the piston and cylinder head face. At 6500 rpm there is about 12 thousandths 'stretch' in the crank and rod assembly, so this brings the piston to within about 4 thousandths of the head. When setting the piston height, it's not a bad idea to skim the block a sufficient amount so that the pistons can have a fairly large amount taken off the top of them, i.e. 30 or 40 thousandths, because this will do two things. Firstly it will move the top ring nearer the top of the piston which is beneficial, and secondly it will lighten the piston, the **Carl Schmitt** piston not being the lightest 850 Mini piston available.

Now onto rods. These need to be lightened as much as possible, and the little end clamping bolt needs to be replaced with a high tensile socket-headed screw or one of the high tensile, multi-hex bolts. The same goes for the big end bolts. Because 850 Mini-X racing is done on tracks with many tight corners, acceleration out of a corner is of paramount importance and this means that the weight of reciprocating and rotating parts is very important. **Lightness within the engine should be considered a prime goal. Not only should the rods be lightened, but also the crankshaft needs to be lightened as much as possible.**

In preparing the crank, it's a good idea to have the crank ground -.040. Have your crank grinding man put a large fillet radius in the corner, i.e. 30 thousandths, and have all bearings sized to about ½ a thousandth under bottom limit. When building the block, use an 1100 centre main bearing shell, as these are narrower than the Mini ones, and more than adequate for the job. To make sure that a minimal amount of power is lost to the lubricating system, use one of the latest all-aluminium case oil pumps. Strip it and machine the whole unit so that it's thinner. You can reduce the thickness of the actual pump by about 20%. This means it will absorb about 20% less power in driving it. Attention should be given to the ports on the pump. All the sharp edges should be taken off so as to 'flow' the pump.

The hot cam for an 850 Mini-Xer is the scatter-pattern VP4 which can be ordered through Kent Cams. This cam gives a good 'over-rev' capability, necessary for the standard wide ratio box that are inevitably used. Additionally it comes on the cam as soon as do many cams of 280° or so. This means the minimum compromise of low-end power to top-end power. Apart from that it usually beats everything else by 3-5 hp.

As far as the valve train is concerned, there is scope for a bit of lightening here and there, as detailed in the cam chapter. The rockers you need to use are 1300 pressed steel rockers, but they must be used with standard bushes. Offset bushes are not allowed, simply because they will put the lift over the specified 280 thousandths for the type of cam being used.

As far as valve springs are concerned, a set of standard double valve springs from a 998 Cooper will get the job done.

The ignition system needs close attention. Obviously it must be up to par in every respect, and there are a few mods you can do to make it function better. These are mostly detailed in the chapter on ignition systems. However one thing you do need to take into account is that this motor has a very high compression, with a very short cam. As a result, it will need less ignition timing everywhere in the rev range in comparison with the standard motor, except right up the top end. It may also need the advance to come on a lot slower, otherwise low and mid rpm range detonation problems may occur.

It is possible to build a competition vehicle at a relatively inexpensive level. This machine, an 850 Mini-Xer, was a championship winning car and it didn't cost an arm and a leg. 18-4

850 Mini Rod

In many respects, the preparation of an 850 Mini Rod engine follows that of an 850 Mini-X engine including the use of a VP4 which proves an effective choice.

As far as intake and exhaust are concerned, rules call for a standard Leyland cast iron intake and exhaust manifold. This being the case, the best is the Metro cast iron LCB type system. However, it may be necessary to check the rules to see if this particular one can, in fact, be used. Remember, that if possible, the intake should be separated from the exhaust, but check that the rules allow it. If they don't then make saw cuts into the manifold to reduce the adjoining area as much as possible. If there is only a small connecting amount between exhaust and intake, the amount of heat transferred will be substantially reduced.

850 Mini 7 Racer

Mini 7 racing has been popular in England for many years. It has become a very sophisticated formula in respect of engine development. Originally the idea was to produce a formula that would go fairly fast but be cheap. In essence, the same sort of idea of a lot of formulas. The rules, however, are relatively liberal in respect of the actual engine, block and cylinder head, with the rule limitations being called for on the outside of the engine. This being that only one induction barrel is allowed and that the bore of the carb or a spacer between the carb and manifold must not exceed 1½ inch. In the early days, most racers used SU carbs. Some found the Reece/Fish beneficial, but the high airflow properties of the Weber carb proved its superiority and now the general approach is to use one barrel of a 45 Weber carb, the 1½ inch rule being satisfied by a spacer between the carb and the intake manifold.

Manifolding for such an engine is important. If the manifold is not a good one, power will be lost. At the time of writing (1984) **Maniflow** had one of the best intake manifolds for 850 Mini 7.

Topping off the induction system, **Advanced Products** has a K & N setup, specifically for one barrel of a Weber.

If you want to get the best exhaust manifold setup dyno testing will be required unless you copy an engine spec that is known to work. This is because the type of exhaust used is dependent on the cam, and over the last few years at least, different cam profiles have shuffled in and out of popularity and so many different profiles have been tried, that it's almost a question of using the profile of the week. Remember, cams much over 300° generally need 3-into-1 manifolds. Those around 300 or less need LCBs. A cam under 300° duration is likely to prove uncompetitive. A modern 850 Mini engine is a screamer and an intake duration of 300 to 310° is fairly common. Exhaust figures can range from 310 to 320, and sometimes as much as 330. Try a medium bore exhaust manifold of both LCB and 3-into-1 variety in both CV and non-CV configurations. I would test this variety simply because the cam situation for 850s is by no means static. In fact 850 Minis do have a problem in the cam area. The cylinder head used for an 850 Mini racer, is a 12G940 casting, the valve of which overhangs the block. This means a cutout must be put in the block so the valve doesn't come into contact with the block. The more lift the valve has, the deeper the valve pocket must be, and there comes a time when the pocket will undercut the position the top ring reaches. At this point, the limit of the pocket depth is also reached. The deeper the pocket is made, the more unfavourable the combustion chamber shape is, and the less CR there will be. However, CR is not quite such a heavyweight factor here, because cylinder heads generally have enough meat on them so that chamber volumes can be reduced as required. Because the exhaust valve lift is limited, usually to around 425 thousandths, it is more important to give it extra duration in comparison with the intake. Because of the limited valve lift, valve acceleration becomes more important. Fortunately the cylinder head's flow characteristics are quite good for an engine of only 850 cc displacement. This means that super high lifts may not really be necessary even if they could be used. The cam alone, however, will not necessarily provide all the breathing required for an 850 motor. First, we have to remember that even though we are dealing with an 850 Mini engine, its siamese port configuration will not respond well to over-long valve timing. Thus, if valve accelerations can be increased, the timing can be shortened slightly to produce higher trapping efficiencies, therefore producing the power without having to turn quite as many rpm. Also, although there is a limit of about 425 thousandths valve into the chamber without causing too much of a on exhaust lift, this does not mean that the intake lift must be limited similarly. On the other hand, if valve lift goes much above this, it is going to be necessary to pocket the pistons which won't help combustion chamber shape. Again, rapid valve accelerations with slightly later opening of the valves may well compensate for this, but valve accelerations are limited by the diameter of the cam follower. To open the valves faster, higher ratio rockers are needed, even if the actual valve lift required is within the span of the standard rocker ratio. By staying with a 1.25 rocker ratio, it is still possible to reach the same limiting lift as a 1.5 rocker, but there will be correspondingly less valve acceleration. So my formula for a valve train for an 850 would be to use 1.5 rockers along with a custom cam profile that opens the valves fast to about 0.425.

The cylinder head, like any race head, needs to have good flow capabilities, but bear in mind that this is a small motor in which valve lift need not necessarily be very high and, thus, the intake port needs to be kept down in size. Valve sizes most commonly used are 1.4 inch intake, 1.215 inch exhaust. The actual type of valves used are important. Use a Rimflo intake valve because of its improved flow capabilities, and its anti-reversionary qualities, as this should help cut the inherent camminess of such a small motor.

For the exhaust, either modify a Rimflo valve or use a conventional valve. The Rimflo valve has taller height from the seat to its combustion chamber face which is not going to help the valve-to-block clearance problem that occurs here. However, the Rimflo valve's good flow properties are an aid to performance, and so reproducing the shape of the valve but cutting down the land width should help cram this compromise between block fouling and shape for extra breathing.

As far as the bottom end is concerned, use a set of forged pistons mounted on fully prepared 998/1100 fully floating pin rods, coupled to a fully prepared crank. The block, of course, will need cam bearings, and drive to the cam should be via a duplex chain setup at least, but preferably via a belt drive.

Example Engine Builds Part 2

The 948 Engine

If you are preparing a 948 engine, whether it be for street or competition, the first consideration is to overbore it to accept 998 Cooper pistons. This will bring the capacity of the engine up to 998 cc and give you a wider range of pistons from which to choose. Rods or pins will need to be changed so that a compatible rod/pin combination is used. If the engine is for the street, preparation can follow the lines of the street 850. Since all 948 engines are of inline installation, the justification for using a sidedraft carb is often greater than for a transverse installation, simply because a good intake manifold can more often be fitted without any complications. From here on, treat the 948 engine as an 850 except, in most instances, if you are going to change the exhaust manifold, use a medium bore system. If you are building a competition 948 engine to SCCA rules for powering, say, an Austin Healey Sprite, then an overbore won't be possible and it will be necessary to use a standard cylinder head as the basis for the head on the engine. Because of the limited valve size, it becomes imperative that the head be reworked by someone who is competent on this type of work. Other than A.P.T. One of the best cylinder head modifiers for the Series 'A' engines in the States is David Taber of **Comptune.** I have seen enough of his work to be sure that he is an extremely competent engine man in every respect.

As far as cams are concerned, the choice is really between **Elgin, A.P.T.**, or a **Comptune** designed, **Iskenderian** ground cam from David Taber of **Comptune.** If the type of racing you intend to do does not limit you to utilizing or starting with the standard head, then a 998 Cooper head casting or a 12G 940 head casting are good options. I have had good results with both these head castings on racing 948 engines, and 998 Cooper heads produce nicer power curves, but do give away the odd 2 or 3 hp to a well ported 12G 940 head casting.

998s For The Road

There are probably more 998 Mini engines around now than any other type, so it's a popular one for modifications, especially in view of the fact that a 998 engine carries less of an insurance premium than a 1300. **If you want a high performance 998 without being heavily loaded, then it's an entirely practical proposition to make a 998 into a Cooper 'S' eater without necessarily having an insurance problem.** Most insurance companies like neither carbs with twin chokes, nor twin carbs. The carburettor type is associated more with performance than anything else, except maybe a turbo or a supercharger, so **if power can be obtained without actually resorting to twin carbs or a turbo, then the chances of your insurance being loaded are limited.**

Firstly, cylinder heads. Most single carb 998s have the small valve head as opposed to the 998 Cooper style head. A well ported cylinder head makes a lot of difference to these engines. However, I must emphasize 'well ported'. The best plan, if the budget will run to it, is to use a fully prepared 998 Cooper head. A high CR is needed and I would recommend 9.5:1 with cams up to 250 to 255°, and about 10.5:1 with cams around 270°, if the right induction system is used.

Attention to detail such as zeroing deck heights, getting crank clearances, etc. to recommended limits, and general preparation should be carried out as described in the relevant chapters. Fully preparing and lightening the crank plus the rods is certainly beneficial, especially as much redundant weight is carried in this area. My favourite cams for a road-going 1000 are the **Kent** 276 Megadyne or **A.P.T.** VP6 or VP7, or the **Piper** Magnum 270. These produce a *very* sporty power curve. Although it will run low down, the engine definitely picks up its heels and goes from 2500 up. If you don't like a power curve that's a little soft at the low end of the rpm range, then the **Piper** Magnum 255 or a Kent 256 or 256 Megadyne is a better bet. Any cams more than about 270° in a road-going 1000 tend to be a little too cammy. Sure, they'll make more hp to the top end, but there is a definite sacrifice down the low end and I don't think it's altogether justified, especially when you can get very acceptable power figures without going over 270°.

60 hp at the wheels from a 998 cc with minimal insurance loading. Here's what it looked like. Noisy from induction roar but fast (for a 1000) and totally street drivable, as it would pull right down to 20 mph in top gear.

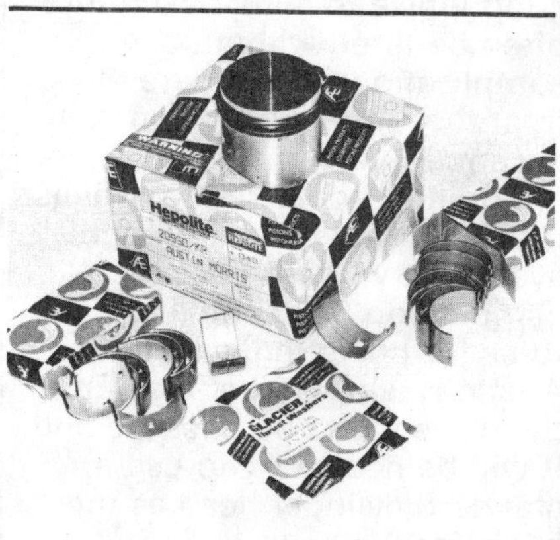

Pistons and bearings for a hot street engine are only as far away as your local AE dealer. Shown here is a Powermax flat topped piston for a 998 engine and the big end and main bearings for the same engine.

As an illustration of tuning a 998 engine I will use an example of an engine built by myself and Dave Mountain of **Mountune Race Engines** for *Auto Performance* magazine. This particular engine utilized an **Advanced Products/K & N** Stub Stack-equipped HIF6 carb, mounted on a **Howley** intake manifold, together with a **Maniflow** medium bore LCB plumbed through a **Competition Silencers** exhaust system. The engine itself had a selected 998 Cooper head, modified by Dave Mountain, equipped with Rimflo valves and cut to give a compression of 10:1. Valve sizes employed were 1¼ inch for the intake and 1$\frac{1}{16}$ inch for the exhaust. +40 Powermax pistons were used, and these were set flush to the deck of the block. The rods and crank were fully reworked as described in the relevant chapter, and the whole assembly balanced. The camshaft employed was a **Piper** Magnum 270 cam, set to give full lift at 105° on the intake. The distributor was a special **Aldon** unit which was equipped with an advance curve designed to keep the engine out of detonation in the lower rpm ranges where detonation is likely, and then to bring the advance in to the optimum in the mid and upper rpm ranges. This combination of parts resulted in a perfectly street driveable engine which delivered about 75 hp at the flywheel on the single HIF6 carb. It proved to be more than a match for any standard 1275 'S' engine in a straight line, and was in fact, as fast 0-60 as a turbo Metro.

If the HIF6 carb is replaced with a side draft 45 on a good intake manifold, then 80 bhp at about 6800 rpm can be expected.

Mini Miglia Engines

Racing Mini Miglia engines is another area of Series 'A' competition where the difference between being successful and an also-ran can be marked by a very fine dividing line. The success of the current winning engines is largely dependent on getting the right cylinder head and cam combination. This doesn't mean to say that things like intake manifolds, carbs and exhaust systems aren't important; they are, but the most efficient formula for those is well known by now.

If you are building an engine for this category, it never hurts to check what the current race-winning engines are using, though the builders are not likely to give away anything. For a competitive motor, here's what I would recommend:

First and foremost, the carb will be a 45 Weber. As a matter of course, I would equip it with a big **K & N** filter. This I would mount on a long **Maniflow** intake manifold of the type described in the intake manifold Chapter. The cylinder head would, of course, be a 12G 940 head casting, modified to the revelant drawings. The valve size would be 1.4inch intake, 1.215 inch exhaust and they would be Rimflo valves throughout. The CR would be about 13·5:1 but would be dependent on the cam spec used. As far as the cam is concerned, a good, safe proposition would be to use 649 profile, but to refine on that, I would use 1.5 rockers with a specially profiled cam to give around 425 exhaust, 450 intake lift, with 300 to 305° period on the intake and around 310 to 315 on the exhaust. I would pay particular attention to reasonably rapid exhaust opening rates. For pistons I would tend to go for the forged **Omega/Howley** items. Unlike some engine builders, however, I would run these pistons with either a flat top or with only a very shallow wedge dish to keep the depth of the valve cutout used with a high lift cam as small as possible.

There are several points of importance to note on this 998 Miglia racing engine. Firstly the cut-outs. The higher the cam lift used, the deeper these cut-outs must be and the less satisfactory the chamber shape becomes. A compromise must always be struck between these two factors. Note that the water pump bypass has been blocked off. The thermostat has been modified to compensate.

The rods and crank should be fully prepared and lightened items. To take the drive from the crank to the cam, I would use a belt drive setup rather than a chain, because at very high rpm less power is lost from the belt drive, and cam timing is more accurate throughout. For the exhaust, a medium bore LCB would be my choice as a starting point, but I would check the efficiency of the 3-into-1 system and a CV system of each type when used with the particular long period cam that would be the final choice.

For ignition, a contactless **Aldon** distributor would be used with an advance curve that came in fast to about 4500 rpm and then advanced a further 3 or 4° from 5000 to 8000 rpm.

A successful engine, built along the lines I have described, should produce between 95 and 98 hp. You may hear figures over 100 hp are sometimes quoted, but these are very rare, and in many instances can be due to slight differences in dynamometer read-outs.

1098 Engines

Essentially, the 1098 engine can be looked upon as a long stroke version of the 998 engine, so much of what has been said about the 998 engine applies equally. However, the longer stroke does affect the characteristics of the engine. First, the engine is more tolerant of longer period cams and if you are building a road motor, a 270° or so cam will run without any really apparent camminess low down, and a cam with as much as 280° produces quite acceptable results even down the low end of the rev range, and is worth about 5 hp over a 270 Magnum. To make a 285° work though it's necessary to have a well ported head, preferably with big valves. If the head is indifferent, the advantages of the bigger cam will be substantially reduced.

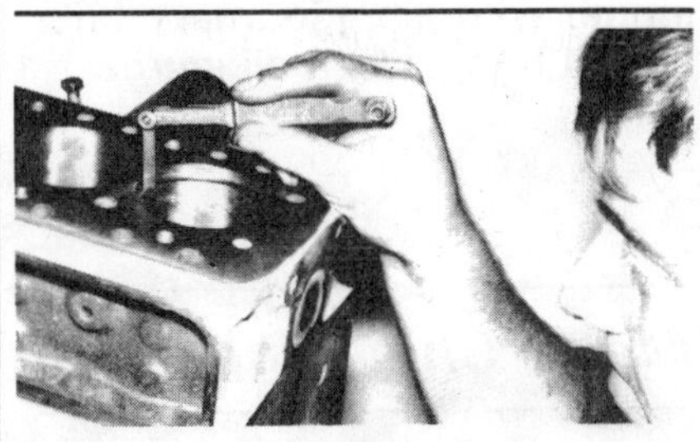

Careful engine building is always part of the secret of producing good results. If it takes you all day to gap the rings properly, then take all day. Just remember, top quality results take top quality efforts.

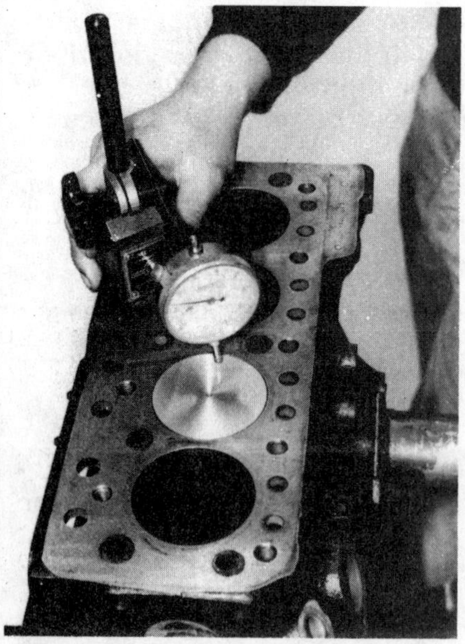

Almost irrespective of the engine you have, small bore or big bore, it's a good idea to deck the block and flush the pistons. This operation should only be omitted if doing it causes the CR to go too high, but this situation only crops up on bored and stroked big bore motors.

If you are building an inline engine for race purposes, remember to build it using the big journal crank as a basis, rather than the earlier, small journal ones. Although these engines will turn high revs eventually the crankshaft will fatigue and crank life is much longer with a big journal crank than it is with a small journal crank.

Metro Challenge

The Metro Challenge Series has been one of the more successful one-make series. Unlike some of these series which attempt to run standard or near-standard cars, the Metro Challenge cars are undoubtedly 'real' race cars. They are fast,

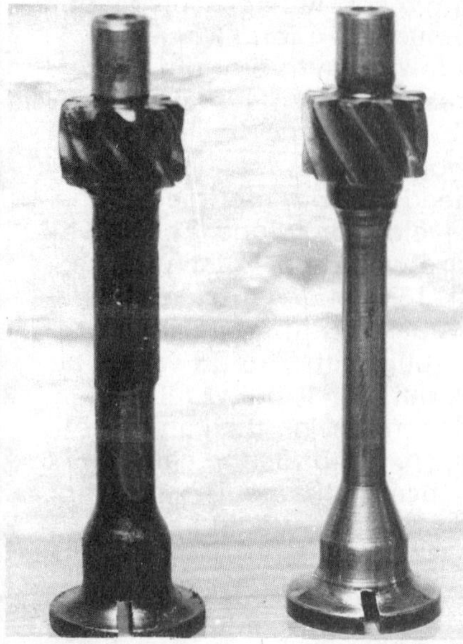

Slimming down the distributor drive shaft to the form shown on the right is vital on stroked, big bore engines so it clears the crank. However, reducing the shaft thickness on any engine seems to increase distributor bearing life and allows the distributor to hold the dwell angle longer.

Whatever engine you are building, it pays to time the cam accurately, and if you are building essentially the same type of engine time and time again, it pays to experiment with cam timing. Your particular combination of parts may run better with a cam advanced or retarded slightly from that recommended by the cam grinder. 18-12

and the racing produced is close.

This formula has changed since its inception and in 1986 the Metro Challenge became the Turbo Metro Challenge. Because of the way the rules are structured, power levels have not gone up significantly from those of the normally aspirated engine. A good normally aspirated engine was typically around 105 hp, though I had heard of race winning engines up as high as 115 hp. Talking to a few engine builders building turbo engines for the Turbo Metro Challenge the power levels seemed to be of a similar nature, 115 to maybe 120 hp. Personally I cannot make any comment on the Turbo Metro engines because I've been unsuccessful in trying to obtain an engine from Leyland to work on. Metro Challenge parts and technical information are available from Ripspeed which has been appointed by Austin Rover Motorsport as sole distributors for all Metro Challenge accessories.

SCCA 1275s

Without labouring the point, the whole secret to making an SCCA race engine work lies in good airflow preparation of the entire induction and exhaust system, and the use of the right cam. Because race fuels have higher octane in the States, most of the SCCA race machines are utilizing CRs at least a whole number higher than their European counterparts. Compressions around 14 to 14.4:1 are relatively common. Those people preparing successful engines such as Bob Griffith of **Bhp Developments**, David Taber and Dimitri Elgin, are all proponents of the high lift cam (just as I am). Most of the successful engines are using valve lifts of 460+ and these engines appear to be usefully revving to about 9000 rpm. As a result the engines are hard on valve trains and the cam often requires replacing or at least reprofiling by the cam manufacturer. Most cams are good only for a season in competitive racing.

Since this form of racing is so competitive and expensive to do, it is necessary to find out what the latest state-of-art is and deal with those people who are producing the parts equipping the winning cars.

For race engines, the best non-adjustable chain setup to use is a lightweight steel 'S' one, but if you can't afford that, the duplex 1300 scintered iron gears, still do a passable job.

Anytime you start building an engine for special or modified saloons you can reckon that nothing short of 'all-out' will get the job done. Everything from the crank upwards must be fully prepared to maximize the engine's output.

1300 Special Saloons

Building a 1300cc Mini for hill climbing, sprinting, modified saloons, etc., has got to the point of being a well tried and tested art. Thus, I will give you a straightforward specification of a motor that will produce working results. Starting at the carb, we have the inevitable sidedraft Weber. A 45 will produce good results, but a 48mm or 50mm side draft Weber may give a slight edge up at the top end.

For an intake manifold I would opt for the long tubular steel **Maniflow** item and this would feed into a fully ported head. If the engine is a short stroke unit, utilizing a 73½mm bore, then the cylinder head should employ 1.480 inch intakes with 1.250 inch exhausts, the exhaust valve being offset. If it's a standard bore engine, then I would use 1.4 intake and 1.215 exhaust. Again, Rimflo valves would be my choice.
Compression would be set at around 13 to 13.5:1, depending upon the cam finally chosen. A good all-round cam is, of course, the 649 profile, but it lacks lift. For a good, reliable motor, my choice of cam would be a 290 to 300° period on the intake, and a 305 to 310° one on the exhaust, operating through 1.5 or higher rockers to give around 500 thousandths lift.

For pistons I would opt for a 4cc dish in a **Cosworth** of similar forged pistons. If I were building a short stroke engine **Omega** pistons with a 6cc wedge dish would be my choice

There are times when motor racing can start to look very expensive and you know it had nothing to do with the combination of parts you build. Just a set of developing circumstances.

in a 73.5mm bore. For an exhaust system, I would use either a big bore LCB or a big bore 3-into-1 CV exhaust system. Again, the final choice would be dependent on the cam spec used.

For an unlimited racer, that is an engine from which maximum power is required, irrespective of how it's achieved, then in normally aspirated form I would use a 1480 cc engine based on a 74mm bore. If supercharging of any sort is permitted then, of course, it would have to be a turbocharged 1275.

If it's muddy out there, one way of keeping the windshield clear is to cut a hole in it.

Hot Street & Cheater Motors

Mini enthusiasts in Germany are in the unfortunate position of not legally being able to modify their vehicles because any modified vehicle will automatically fail the TÜV. This is the German equivalent of our M.O.T. However, the TÜV is much more stringent and modifications of the vehicles, if detected, simply invalidate the test. From this point of view it's necessary for Mini owners to be very discreet as to how they modify their vehicles. Externally at least they must appear standard. This means that the first modifications the engine should receive need to be in the form of a gas-flowed cylinder head with, of course, more compression. A hotter cam also helps the cause and if the car is equipped with a single carb, it is probably safe enough to assume that a Metro intake manifold, since it is a Leyland part, will go unnoticed. If the particular vehicle is a 1275 'S' then the intake manifold can be extensively internally modified, and an LCB exhaust system as opposed to the standard 3-into-1 system would not arouse suspicion.

One way to ease plotting distributor advance curves and to assist in timing the cam is to index the damper as shown here.

Regarding silencers any silencer that makes more noise than the standard one would probably come in for close scrutiny so, if power without noise is a criterion, then really the only choice is a **Competition Silencers** RC40.

If an engine is being overhauled and a rebore is necessary, take the opportunity to overbore a 1275 out to 73½ mm. If it's a later 1300 or Metro block, then 74mm is acceptable.

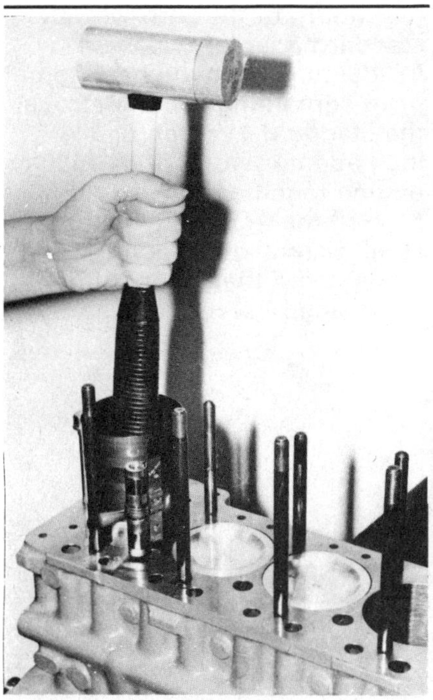

Always use the proper tools to install the piston and connecting rod assembly into the block. Those double rail oil control rings can easily be damaged during installation. Keeping the ring compressor hard down on the block is important.

There is evidence to suggest that the notched V-belt of the type shown here absorbs slightly less power than the plain V-belt. They also seem to have less tendency to fly off.

These big bore engines make excellent road engines for the increase in torque and revability is instantly noticeable. Of course any increase in displacement is undetectable on the outside and if power has to be achieved by virtue of extra

Here I am leaning against the Auto Performance magazine project Mini. The engine for this car was built by Nolan Pitts (right) and myself, with Nolan doing the rest of the car in his Plymouth-based workshop. On the left is my ex-race partner and good friend Colin Ashdown-Pogmore. Colin, who is an excellent chassis builder, now owns and runs a company that makes special batch production and short run gaskets much of which goes to the performance industry.

cc, then a longer stroke to stretch the engine to 1430 or even 1480 is to be recommended, but these long stroke engines are expensive, whereas stretching an engine to 1400 is not. Essentially for any road motor, it is my opinion that going to a sidedraft Weber should be considered the very last step to take to produce power. In the case of engines being modified for use in Germany, a sidedraft Weber is out, but careful selection of the right SU carb is not. At this point I would have to say that the best choice of carb would be an HIF 7 with an extensively reworked standard air filter assembly, and make sure that assembly does, in fact, flow. This would involve putting a larger pipe into the air filter case, a **K & N** element in the filter case itself and, a tweak I have not mentioned before, the use of about ¼ inch spacer plate between the right-angle adaptor on the front of the carb and the carb. The reason I am proposing this is because there is too much of a sharp turn into the mouth of the carb with the standard adaptor which, incidentally, you will have to get from something like a Metro, as the standard 1½ inch or 1¾ inch adaptor won't fit. A 1430 cc engine together with an HIF 7 plus all the 'good guy' internals should crank out around 115 to 118 hp, and that's enough to make any Mini move in style.

19 Nitrous Oxide Injection Part 1

Nitrous oxide is a chemical that was first 'discovered' by Sir Humphrey Davey, the same gentleman that invented the miner's safety lamp. Davey's discovery of nitrous oxide caused barely a stir, since first he was mostly unaware of its properties and, secondly, no really worthwhile applications presented themselves. During the mid-18th century, it was noted that nitrous oxide possessed a mild anaesthetic effect and, as such, was adopted for use by dentists as a pain killer. A side effect of nitrous oxide caused some patients to break out in mild fits of laughter, and as a result, the gas became popularly known as 'laughing gas'.

So far, none of the properties of nitrous oxide seem of any consequence to the performance enthusiast. However, an investigation of its chemical structure reveals nitrous oxide as one-third oxygen by weight and though the oxygen is combined with nitrogen, it can be broken down into its constituent elements at relatively low temperatures. This will release the oxygen for other purposes – such as for burning fuel. This characteristic was realized by chemists during the years between the two world wars and 'secret' projects were instigated by both allied and axis powers to determine if practical use could be made of the potential that existed for nitrous oxide to be used to boost performance of fighters and high altitude reconnaisance planes. Both sides managed to utilize nitrous oxide successfully, raising power levels of aircraft engines by as much as 600 hp. The irony is that both sides had nitrous oxide and each believed the other did not, each being capable of sending aircraft aloft to altitudes in excess of 50,000 feet.

The introduction of jet engines caused a dramatic increase in the performance of military aircraft. A jet engine provided such a huge margin of performance over piston engined aircraft both in terms of speed and altitude capability, that the use of piston engines and nitrous oxide became a dated technological field. As jet fighters started to enter squadron service in large numbers shortly after World War Two the use of nitrous oxide didn't re-emerge until the late '50s, in USA, and here again, it wasn't heralded with a fanfare of trumpets. It was once again used as 'the secret weapon', but now by those competing on the race track.

For more than ten years, nitrous oxide managed to stay under wraps, used by just a knowledgeable few to achieve an edge over the opposition. In the early '70s nitrous oxide emerged but it was not received at first with any great degree of enthusiasm even by the performance orientated enthusiast. It was probably deemed too different or too good to be true when compared with what is normally accepted as performance equipment. The claims for its power boosting capability always appeared a bit exaggerated. **A number of kits were produced in the mid-'70s mostly by people having no real idea of the technology required to build a successful and safe nitrous kit.** Many early kits caused dramatic engine explosions or melt-downs and as a result, public acceptance of nitrous oxide was retarded.

Toward the late '70s in the States, the use of nitrous oxide began to increase and by 1980, thousands of kits per month were being sold. Success was achieved by virtue of two factors. Firstly the kits worked. They produced huge amounts of extra power, and secondly, most nitrous oxide kit manufacturers had learned to be conservative with the power increase level supplied to customers. In other words, reliability figured higher on the priority list than power.

What Is Nitrous Oxide?

The factor that concerns us most is that nitrous oxide's symbol, N_2O, indicates that the gas is made up of nitrogen and oxygen. The important part, that is the oxygen content, is, by weight, 36.3% of the whole. At normal temperatures, nitrous oxide, in a free state, is a gas. If it is kept under sufficient pressure, it can be stored as a liquid in much the same way as can propane gas. When held in a pressure vessel, the self-generated pressure which keeps N_2O as a liquid, is known as the vapour pressure. At 75°F. the vapour pressure of N_2O is about 800 psi. At 95°, the vapour pressure is up to around 1030 psi. **If a pressure gauge is connected to a cylinder of N_2O, the gauge reading is the result of the temperature of the contents, not the amount the bottle contains.** The pressure gauge will read the same whether the bottle is three-quarters full or three-quarters empty. Not until the last drop of liquid N_2O is gone from the bottle, will the pressure start to drop.

When N_2O is released from the cylinder in which it is contained, it can be ejected in two forms, depending on the attitude of the bottle. Take a look at the drawing **Fig. 19.1**. When the bottle is in position A, the pickup pipe is submerged in N_2O liquid. Opening the valve causes the high pressure gas above the liquid to force the liquid out through the pickup pipe. In this instance, the discharge of N_2O will be in liquid form right up to the point at which it is ejected into the atmosphere and for a few tenths of a second after that. In drawing B the pickup pipe is above the surface of the liquid, so opening the valve will discharge gaseous N_2O. In either case, the action of releasing N_2O from the cylinder causes the pressure in the cylinder to drop. The boiling point of any liquid, N_2O included, is dependent on the pressure it is subjected to. The resultant lowering of the pressure due to releasing N_2O from the cylinder, causes the liquid to boil and the vapour given off from boiling tends to re-establish the pressure in the cylinder at the vapour pressure temperature the bottle is at.

Okay, so we've got around to temperatures, so let's stick with that for while. At this point, two things should be considered; first when a liquid boils, it absorbs energy, and secondly allowing it to expand from a liquid state in the high pressure confines of the

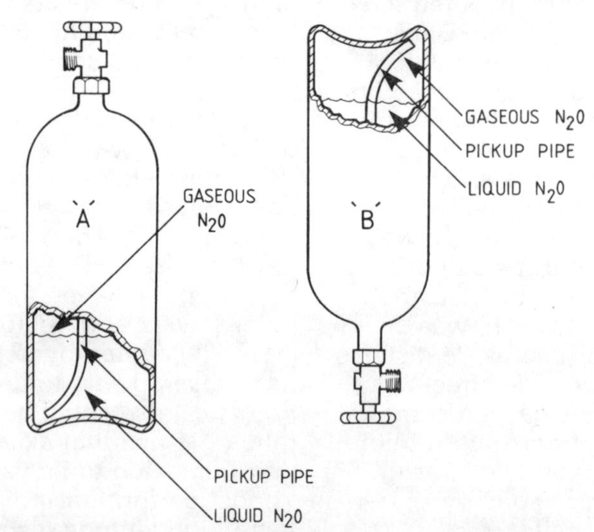

Fig. 19.1 Correct pick-up arrangement for N_2O bottle

cylinder to a gas in the open air causes it to become super-cooled. Not unexpectedly, this causes two things to happen. The cylinder and its contents will drop in temperature (typically 10° to 15° for N_2O) and the liquid released through the jet almost instantly turns to a gas and its temperature drops to as low as −128° F. I say 'almost instantly' because this statement really is dependent on the amount of liquid N_2O issuing from the jet. For the sort of applications being considered here, the bulk of the N_2O exists as a liquid for only a few tenths of a second. The tremendous temperature drop when discharged through the jet as a liquid, can be very important from the point of view of developing extra power, an aspect I intend to build on later.

Power Production Practicalities

No doubt most of you are wondering whether or not N_2O can produce the dramatic results often claimed, and if so, how and why. Let's go back to an earlier statement: N_2O is 36.3% by weight oxygen. By comparison, the atmosphere we breathe is just short of 20% oxygen. The implication here is that if an engine were run on an atmosphere composed solely of N_2O, it would produce proportionally more power, about 81% more in fact. Putting things another way, it should be equivalent to having about 12lb of supercharger boost pressure. **Tests on aircraft engines during World War Two indicated that when run on N_2O gas, either partially or completely, the power figures produced just about fell in line with those expected from simple theory.** The only snag to the use of N_2O in gaseous form for anything but relatively small increases, is its inconvenience. Handling a large number of cubic feet of gas is difficult and presents its own unique set of problems when a simple changeover from air to N_2O and back is required. The logical step then, is to inject liquid N_2O. This, as it happens, turns out to have many advantages over the use in the gaseous form.

To liquid N_2O's primary advantage I have attached a really fancy term, that is the **displaced charge factor**, or DCF for short. If an engine is run totally on N_2O in gas form, it will achieve around an 81% power increase on a 100% concentration of N_2O, the original atmospheric charge being completely replaced by a N_2O gas charge. Thus, a normal atmospheric charge will be contributing zero hp to the engine's total power output. In other words, the DCF is 100%.

At this point let us assume that N_2O is injected into the engine at exactly the same rate but this time in liquid form. **A pound of liquid N_2O only takes up 1/660th of the space that a pound of gaseous N_2O does. This means that if the N_2O stays liquid right up until the time it's in the cylinder, the DCF is only 0.15%. In other words, the engine can go on producing 99.85% of the power produced from the normally induced charge, plus the power the N_2O is going to contribute.** Putting numbers to it for those of you who care to know, injecting the maximum N_2O charge in liquid form will give an hp increase proportional to 181% minus 0.15% which equals 180.85%. In practice, the N_2O won't stay as a liquid for very long. Just how much turns to a gas before it hits the cylinder depends on a number of criteria, like how close the jet is to the intake valve, how hot the engine is, how many rpm it's turning, etc. These and other factors increase the N_2O vaporization rate, thereby increasing the displaced charge factor, but there is an action taking place within the manifold which has a tendency to reverse this. Remember that as the N_2O issues from the jet, it becomes super-cooled. This cooling action causes the atmospheric charge contained in the manifold, to contract and become denser. This tends to increase the weight of the induced charge which in turn beneficially reduces the displaced charge factor. Under normal conditions, the DCF will be between 3% and 15% when a well-designed N_2O system is used. Once in a while a system will, as a result of luck, an educated guess and, in rare cases, design, have everything at optimum. Under these 'ideal' conditions, it is possible to achieve a negative DCF. When working with some of my own experimental N_2O systems, I have found on occasion, that carburettor airflow would increase by as much as 10% when the N_2O system was in operation. No doubt a portion of this effect was due to super cooling, but also I surmised part of the increased volumetric efficiency could be due to the high speed nitrous oxide stream literally 'dragging' extra air into the engine.

Having outlined the reasons, let's summarize the case for injecting liquid N_2O. Because of the small amount of space taken up, substantial quantities of liquid N_2O can be injected into an engine to produce hp additional to its original output. **So long as distribution problems are not encountered, it is entirely possible to inject enough N_2O to achieve outputs of 200 bhp per litre if no other limitation on hp exists.**

No doubt many of you are impressed with the possibility of 200 hp per litre, but don't overlook the final part of the preceding statement. By injecting N_2O, the engine can get more than enough oxygen into its cylinder to produce almost unimaginable hp numbers, but to keep in touch with reality it's necessary for us to consider the 'if no other limitations exist' criterion. Any action involving combustion is brought about by a chemical reaction of two components, namely an oxidant and a fuel. **The N_2O is the oxidant. It will not burn on its own, so fuel must be added to go with the oxygen content in the N_2O charge.** By using the relevant equations the chemically correct N_2O to fuel ratio can be determined. A typical fuel consists of a blend of numerous compounds, making the use of any chemical equations exceedingly difficult. Fortunately, such a fuel can be well represented by iso-octane, whose chemical composition is C8 H18. This compound has a specific gravity of 0.72 and belongs to the paraffin, i.e. hydrocarbon, series. If you feel inclined to wade through chemical equations, **Fig. 19.2** will show how the chemically correct N_2O to fuel ratio is arrived at. However, it is not necessary for you to know how it's done. It will suffice to appreciate the fact that the correct fuel to N_2O ratio is approximately 1 lb of fuel to 9.6 lb of N_2O. Such a ratio will produce the complete combustion of the oxygen and fuel available, so there will be neither excess fuel nor excess oxygen. However, the use of a chemically correct mixture of fuel and N_2O is not the way to go. To see why, it's necessary to look at a basic property of combustion. As stated earlier, the atmosphere contains about 20% oxygen and apart from a few small amounts of impurities, the remaining 80% is nitrogen. Nitrogen is a relatively stable compound. It doesn't readily form chemical combinations except under high temperatures and pressures when it can break down and form oxides of nitrogen (NOX). The most important contribution nitrogen makes during the combustion cycle, is that it acts as a temperature damper. If the nitrogen content of the total charge is reduced, the combustion temperature will climb very rapidly. Nitrous oxide is some 63.7% nitrogen, so in a N_2O boosted charge there's quite a bit less nitrogen present to act as a temperature damper. If an attempt is made to double the power output of an engine by injecting N_2O, then the nitrogen content of the charge will drop by some 15%. This causes combustion temperatures to climb substantially. Unfortunately, the cycle of events doesn't stop there. To double the power, the amount of oxygen put into the engine has to be more or less doubled. Not only does this mean twice the potential power, but also the engine must dump approximately twice the amount of waste heat. The presence of this extra heat means that, if no additional means of cooling is employed, the engine will run substantially hotter than previously, and in so doing, become far more prone to detonation. **Unless steps are taken to the contrary, overheating and detonation will limit the power output possible by N_2O injection long before any limitations brought about by the amount of N_2O that can physically be crammed into the cylinders.**

The excess heat problem looks a bit daunting, but fortunately it's one that can, to

Fig. 19.2 Equation to determine N_2O to fuel ratio

The atoms of both carbon and hydrogen, of which the fuel (Iso-octane) is comprised will, when combined with oxygen, from carbon dioxide (CO_2) and water (H_2O), so a balanced chemical equation will look like this:

Fuel — Nitrous Oxide

$$C8\,H18 + 25\,N_2O \longrightarrow 8CO_2 + 9H_2 + 25N_2$$

The hydro-carbon fuel (Iso-octane) is combining with the oxygen content of the nitrous oxide and during combustion, forms 8 molecules of carbon dioxide, 9 molecules of water and 25 diatonic molecules of nitrogen (N_2)

The atomic weights of the elements we are dealing with are:

Hydrogen	*1*
Carbon	*12*
Nitrogen	*14*
Oxygen	*16*

The N_2O to fuel ratio by weight is equal to $\dfrac{\text{mass oxidant}}{\text{mass fuel}}$

Or

$$\frac{\underset{\text{(nitrogen)}}{(25 \times 14 \times 2)} + \underset{\text{(oxygen)}}{(25 \times 16)}}{\underset{\text{(carbon)}}{(8 \times 12)} + \underset{\text{(hydrogen)}}{(1 \times 18)}}$$

This equals $\dfrac{700 + 400}{114} = \dfrac{1100}{114} = 9.645$

So the chemically correct N_2O to fuel ratio by weight is 9.645:1.

a large degree, be overcome. The key to the whole problem can be summed up in three words: Excess Fuel Factor (EFF). Essentially the detonation and overheating problem is the principal reason why a chemically correct N_2O to fuel ratio should not be used. By using a fuel rich mixture, the excess fuel acts both as a means to cool the combustion chamber and piston and also to suppress detonation. This is why a properly developed N_2O kit with any power potential doesn't end up melting the engine.

The knowledgeable engine builders amongst you may, at this point, raise a point of real significance: it's a known fact than an air/fuel mix that is far too rich, causes a substantial drop in power and, therefore, using an over-rich mixture to cool the engine and eliminate detonation will rob the engine of much of its potential power increase. If a mixture is too rich in a normal situation, power does drop substantially, but fortunately, when N_2O is injected, it takes a large amount of extra fuel by comparison before the power really starts to drop off. It's possible to run mixtures 80% to 100% rich without losing much more than 5% to 10% of the N_2O kit's potential power increase. By running mixtures this rich, the engine does survive. In respect of this tolerance to very rich mixtures, petrol (gasoline) and N_2O are similar to alcohol and air.

If, when the N_2O is injected, the mixture becomes lean, then the increased heat plus the surplus oxygen will literally burn the chambers, plugs, valves and piston crowns to a crisp. What will, in effect, be happening, is that the engine will become a self-consuming petrol (gasoline) N_2O cutting torch. The amount of warning given that this may be happening, is very limited. On one occasion I saw an engine highly loaded with N_2O digest its innards in about two seconds from the time N_2O injection was initiated.

There comes a point when increasing the excess fuel factor does little or nothing further to reduce an engine's tendency to detonate. It's at about this point that the end of the line is reached as regards possible power increases when the only means of additional cooling is the excess fuel factor. Assuming at present that this is the limitation, it's worthwhile posing the question: how much extra power can be expected from a Series 'A' engine using N_2O injection? Well, let's establish one criterion first: any N_2O injection system used should NOT seriously impair the engine's reliability due to the additional power produced. Assuming your Series 'A' engine has a good set of pistons, then for road use, I would advise for small bore engines, power increases up to 25 hp and for big bore engines, power increases up to around 40 hp. For reliable competition application, a further 10 hp can be added to these two figures. If engine life is not a critical factor, then greater power increases than this can be used, but cylinder pressures can become exceedingly high, especially in the lower rpm ranges, and so for greater power outputs, it is often necessary to build two-stage N_2O systems, but more of that later.

Power In Practice

So far I have talked about power increases in terms of abstract figures, but these alone fail to convey one very important point. The use of N_2O causes the power to increase due to a substantial increase in engine torque. Also, **because the N_2O is injected at a relatively constant rate, the torque increase in the lower rpm range is greater than it is in the higher rpm range.** To a degree it's easier to look at the change of engine output in terms of hp. **Because of the constant injection rate, N_2O tends to produce a uniform hp increase irrespective of the rpm involved. Since hp is a function of torque and rpm this means that production of a given hp increase at low rpm involves a huge torque increase.** It's this very phenomena which causes a N_2O injected motor to seem like it has a far bigger power increase than it actually has. I have found when demonstrating an N_2O injected vehicle to people who run dynamometers, and as a result can fairly accurately estimate power increases, that they always over-estimate the hp increase given by the N_2O injection unit. For instance, on a Metro engine I was working for '*Cars & Car Conversions*', an experimental 25 hp kit was installed. I let three experienced dyno operators drive the car with and without the N_2O in operation. I asked each to estimate the power increase given by the N_2O. The lowest estimated power increase was 40. If you have to make a comparison between 25 hp gained by conventional tuning techniques and 25 hp gained by N_2O, you will find that if your only criterion is riding in the car and experiencing the sensation of the extra power, you will have great difficulty in equating the fact that two similar power increases can produce two such dramatically different effects on the vehicle's performance. When conventional tuning techniques are employed, the power increases normally occur up in the higher rpm ranges, so it's not until the engine really starts to turn some rpm that the

The N.O.S/A.P.T. kit being tested here on the A.P.T. dyno was finally developed to give a 90 hp increase over the engine's existing output. 19/62

benefit of the extra power is actually felt. On the other hand, with N_2O, 'punching the button' causes the extra power to occur instantaneously. The graph **Fig. 19.3** should amply illustrate the point I am trying to put over. Curve 1, is the output at the wheels of a Metro which is standard apart from the fact that the intake manifold has been separated from the exhaust. Curve 2 shows what happened to the hp output at the wheels when a modified cylinder head with a standard CR was fitted, plus a high flow intake manifold and a high performance **Janspeed** exhaust system. These modifications raised the power output from 56 to 68 hp at the wheels, a very creditable increase for the modifications done. However, the point to note here is that down at 3500 rpm, these modifications only raised the power from 38 to 39 hp.

Curve 3 is the power curve produced when a 25 hp N_2O kit is brought into action. Peak power goes up from 68 to 88 hp. This, on its own can be considered to be a very healthy jump in power, but take a look at what the N_2O does to the power output lower down in the rev range. At 3500 rpm the power is increased from 39 to 60 hp, an increase of 21 hp, or to look at it another way, a 33⅓% increase in power at that point in the rpm range. To give you an idea of what this feels like to drive, if you were to stand on the throttle of your 1275cc injected engine at 3500 rpm, it would feel as if it had an extra 2 cylinders delivering power, so it's going to pull like a 1900cc engine rather than a 1275.

From the dramatic change in the shape of the power curve shown in Fig. 19.3, it's a lot easier to appreciate the fact that nitrous oxide is powerful stuff in terms of the increased acceleration it produces, at least until the cylinder of N_2O runs out. Not unexpectedly, the consumption of N_2O is proportional to the amount of extra power developed by the system. Fortunately for a well

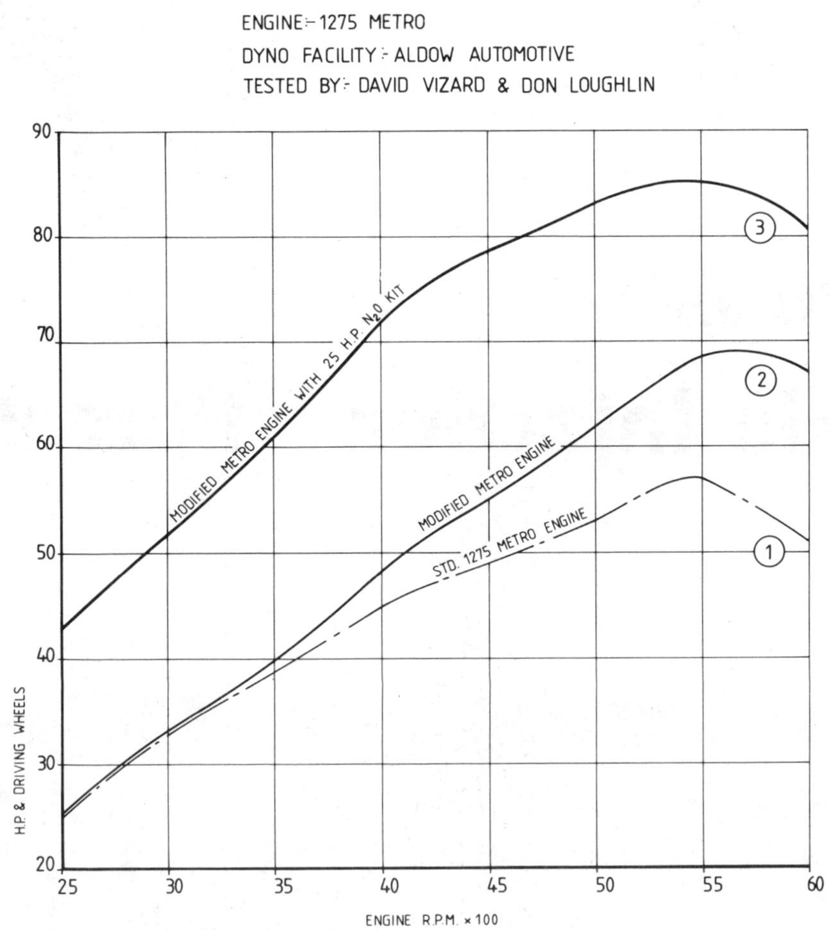

Fig. 19.3 Comparison of power produced with and without N_2O injection

calibrated N_2O system, an easy-to-remember relationship gives a good guide as to how long a cylinder of N_2O will last. This relationship indicates that 1 lb of N_2O will produce 100 hp for 10 seconds. Half the power increase will make 1 lb of N_2O last twice as long, and so on.

Although bottles can be had in a wide range of sizes, 10 or 15 lb bottles seem to be the most popular. This means a 50 hp system will produce about 5 minutes running time on a 15 lb bottle. That may not sound very long, but it must be realized that N_2O is not a power augmentation system for long, cross-country dashes. It's not really convenient for such applications. **N_2O's real forte lies in the fact that it will give your car super sprint capabilities.** In other words, for drag racing, hill climbing, or sprinting a car so equipped can bring to bear enough additional acceleration to quickly dispose of some pretty potent machines.

Nitrous Oxide Injection Part 2

Systems Operation

So far the discussion of N_2O use has been essentially theoretical and it is now time to consider the mechanical application of N_2O. I am going to centre my discussion on a system which was developed jointly by **Aldon Automotive, Oselli Engineering and me.** This system has not, however, reached production mainly because of the availability of low-cost mass produced parts available from the US. UK-produced N_2O kits are available but most UK-available kits actually come from California.

At A.P.T. we've worked closely with N.O.S., which is just a few miles down the road from our dyno shop. We've developed a kit using their components specifically for the Series 'A' engine. To show you the basic operating principles of N_2O, I'm still going to use the system I worked on with Alden and Osselli because in many ways it was more sophisticated. Suffice it to say, at the moment an off-the-shelf kit is available from A.P.T. to give power increases of up to 50 hp in a 1275 engine with no untoward effect. As a matter of record, we've used this kit calibrated at higher power levels to produce power figures in the region of 200 hp from a 1380 cc engine. If you find the subject of N_2O a fascinating one and would like more information, I would suggest you read, *Nitrous Oxide Injection* by David Vizard published by S-A Designs.

Fig. 19.4 shows an idealized basic layout of an N_2O system. The distribution blocks shown are for an engine with four cylinders and four intake ports. As we know, the Series 'A' engine has only two intake ports, so a monojet nozzle will go into each intake port (these are best situated in the intake manifold about an inch away from the cylinder head intake port) and the two spare outlets of the distribution blocks will simply be plugged. The solenoid control valve block is unusual in that most N_2O kits have as separate items all the components that make up the valve block. Basically the solenoid control valve block houses two solenoids and a fuel pressure regulator. One solenoid controls the N_2O flow, the other the additional flow of fuel required. To keep the

Shown here is a basic nitrous oxide kit. This particular example was an experimental one, hence some of the rather agricultural looking parts. 19-1

calibration of the system reasonably constant from vehicle to vehicle, it's necessary to have a pressure regulating valve so that whatever the fuel pump pressure may be, the pressure at the monojet is controlled to give the correct fuel delivery. If a fuel blockage occurred, for instance, the engine would overheat when the nitrous oxide was injected. To minimize the chances of a destructive fuel blockage, an

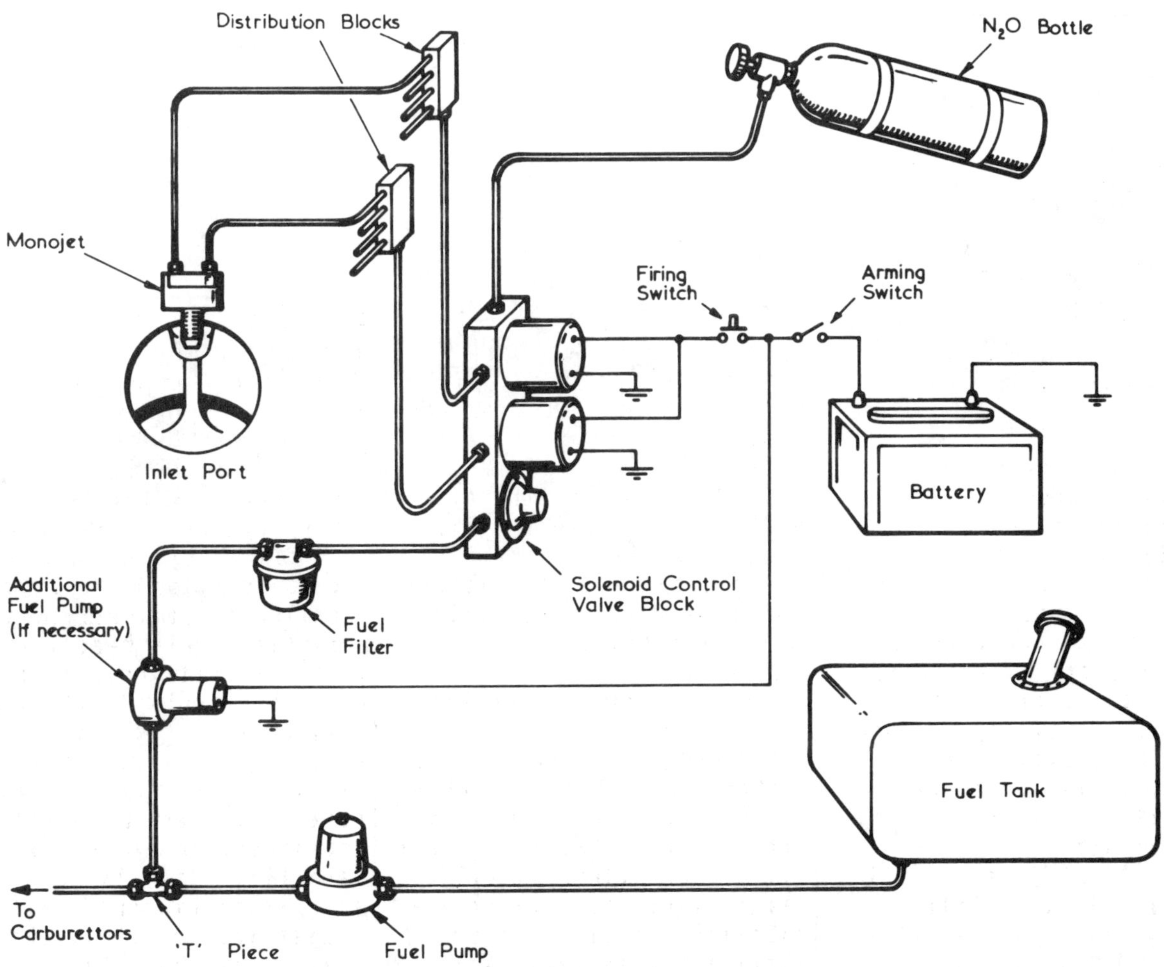

Fig. 19.4 An N_2O injection system

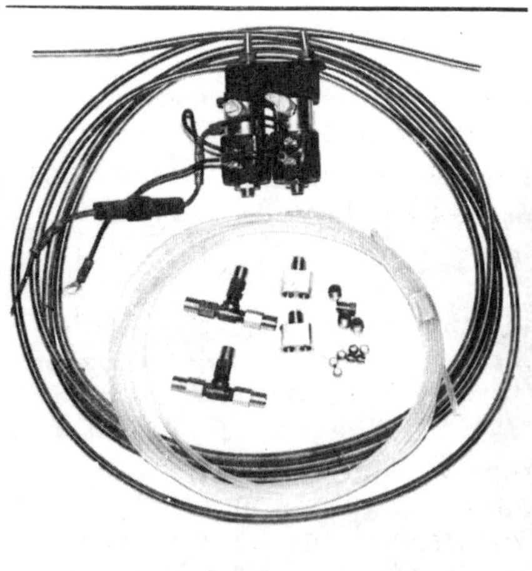

The solenoids for the N_2O and the fuel look identical but they are, in fact, of an entirely different specification to handle the specific needs of each substance. The special plastic hose and fittings are capable of handling over 2000 psi.

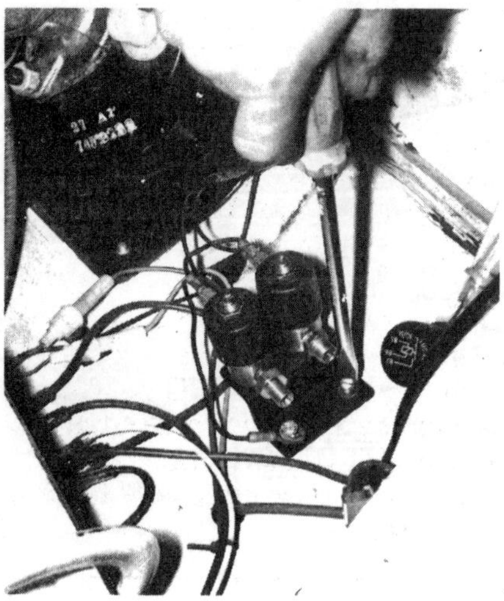

The solenoids that actuate the N_2O system can be mounted anywhere that is convenient and away from exhaust heat.

efficient fuel filter is put in the line. Supplying fuel to the monojet can either be achieved with a standard fuel pump (if it delivers enough pressure) or by the addition of a second pump. Most competition vehicles are likely to have large fuel pumps which can deliver adequate fuel, but the pressure variation of all the fuel pumps it's possible to use successfully can be wide. Fuel pump pressures can range from about 2½ to 3 psi, as provided by a big SU pump, to 16 psi as produced by a Holley competition fuel pump. The sole purpose of the pressure regulating valve is to set the fuel pressure so that the required fuel delivery rate is attained. At this point, it's well worth making an issue over the fact that you must not confuse capacity with pressure. Some fuel pumps can deliver a high pressure but only at low flow levels. As soon as the required flow rate goes up, the pressure will drop dramatically. On the other hand, there are many high capacity pumps which produce only 3½ psi stalled, that is with the outlet blocked up, but still deliver 3 psi when they are flowing 100 gallons an hour. **An important point to appreciate is that the fuel pump must always be capable of delivering a greater pressure than is required at the monojet nozzle at the maximum fuel flow needed by the N_2O system, plus the original engine requirements.**

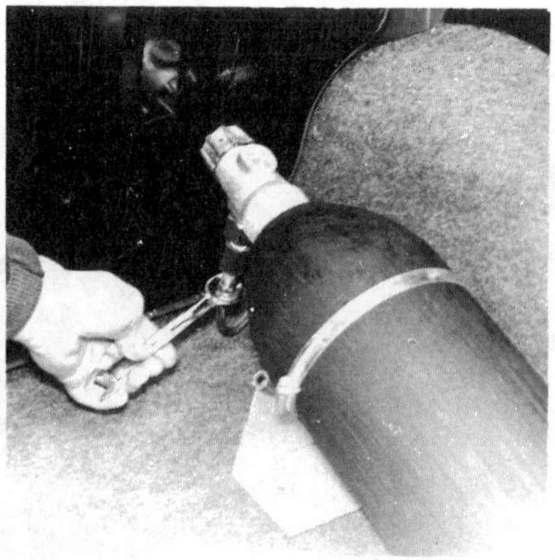

The largest item to mount is, of course, the nitrous oxide bottle. This particular item is a 15 pounder, and it does take up a little of the boot space. The fore and aft installation and the angle on the bottle is necessary to allow as much liquid nitrous oxide as possible to pass out of the bottle. Acceleration forces will push the liquid over the pickup pipe rather than the reverse.

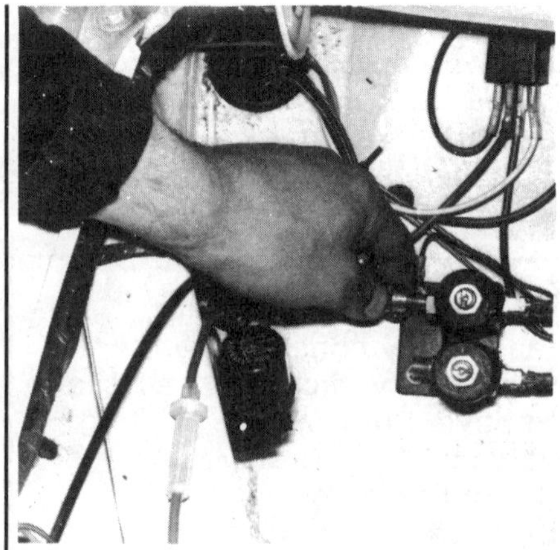

The N_2O is conveyed from the tank by an armoured hose, which can be seen in the top left-hand side of the picture passing through the rubber grommet. Here, I am just connecting the N_2O hose up to the input side of the N_2O solenoid.

Shown here is the N_2O and fuel plumbing close to the solenoid. The two translucent plastic pipes leading from the solenoids are conveying fuel and N_2O to the distribution block mounted on the manifold. Fuel to the distribution block goes via a pressure regulating valve which sets the fuel pressure to a constant 3 psi.

Actual triggering of the N_2O system is quite simple; pressing the firing switch will start the flow of N_2O and fuel to the engine. There are, however, a number of installation points worth noting. First the actuating solenoids should be mounted, as close to the injection nozzles as possible, although they need to be away from a source of great heat. The need for close mounting relates to line lengths from the solenoids to the injection nozzle. With 750 lb pressure behind it the N_2O can travel from the actuating solenoid valve to the nozzle much faster than the fuel with only 3 psi pushing it. Remembering that a lean mixture is something to be avoided at all times, it follows that the fuel should enter the cylinder ahead of the N_2O. So, a short fuel line is a help in this instance. A further factor, which is very important with a front wheel drive car, is how hard the system 'hits' when it's initiated. If a very short line between the N_2O solenoid valve and the injection nozzle is used, the system will hit hard. In other words, system initiation will cause the power output to increase from the base level to the new level in a few engine revolutions. Often this can 'shock' the tyres loose, and instead of producing dramatic acceleration, it produces large quantities of tyre smoke. To avoid this situation it is better to have the system hit 'soft'. This can be achieved by having a relatively long line length between the injection nozzle and the solenoid. Of course, on 25 hp kits, with vehicles with relatively grippy tyres, shocking the tyres loose is not a profound problem, but at 50 hp it most certainly is. To a degree, some heat in the line from the N_2O solenoid to the jet can actually help in this respect. As the initial slug of liquid leaves the solenoid valve, it turns to a gas within the line because of a relatively high line temperature. This builds pressure in the line but the content is gaseous rather than liquid. Because of the pressure in the line, the flow of liquid N_2O from the cylinder is, for a moment, impeded, at least until the gas is expelled from the line. By having the right line length combined with a line the temperature of which has been raised by the engine there can be a one-quarter to half-a-second of flow build-up time which N_2O will prevent the onset of uncontrollable wheelspin. Mind you, **there comes a time when no matter how slow the N_2O system hits, the extra power it produces will break the tyres loose. At that point, any further increase in power is pointless.**

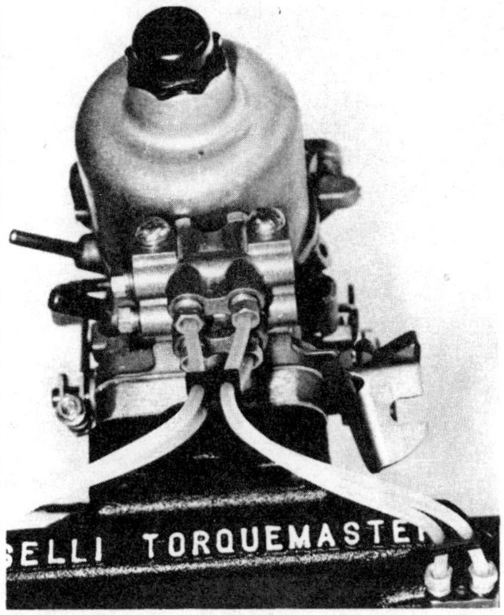

Shown here is a prototype N_2O setup that I built on an Oselli Torque-Master inlet manifold. Input of the fuel and N_2O is to the distribution blocks mounted on the carb plate immediately behind the SU bell chamber. Fuel and N_2O come into the distribution block on the right-hand side. The two screws sticking out of the distribution block on the left-hand side are blanking screws.

Another factor to consider in system design is that calibration of the nitrous flow should always be allowed to occur at the jet. If flow regulation occurs at the solenoid valve due to inadequate sizing, there will be a large pressure drop across the solenoid valve and as a result its temperature will drop. This can cause the solenoid valve to freeze. If it freezes open, (the most likely case) you *will* have a problem.

Here's a side view of the prototype N_2O Kit. This particular example is the one I use for most of the dyno testing involved in this chapter.

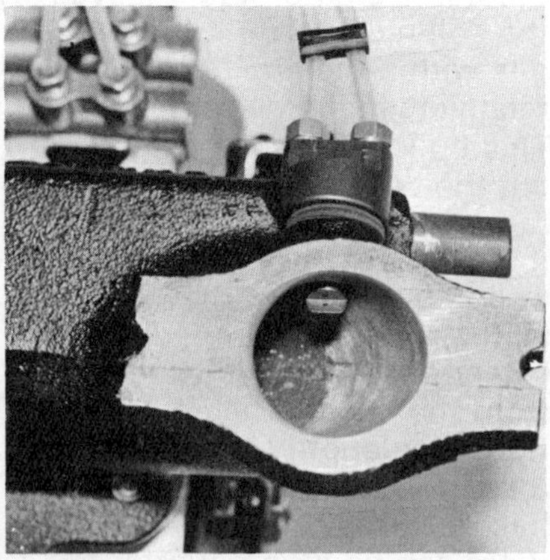

A view down the port of the intake manifold shows how the monojet nozzle is positioned in relation to the direction of flow into the intake manifold. The fuel squirts straight down into the port whereas the N_2O coming out of the lower hole at right angles to the fuel nozzle impinges into the fuel jet and mixes and breaks up the fuel.

In this shot we see the N.O.S. nozzles from the A.P.T. N_2O kit installed in an Oselli manifold. This particular Oselli manifold has two bosses on it in the appropriate position for N_2O nozzles installation.

Power & Efficiency

Selecting the power level an N_2O kit will produce is, in the first instance, a question of the amount of N_2O fed into the engine. Obviously this is dependent upon the jetting of the N_2O part of the system. With some kits such as those produced by **N.O.S.** in America, both N_2O and fuel is introduced through a dual spray bar system installed in a plate which fits between the carb and intake manifold. Whereas such a system works admirably on a typical V8 engine, it does tend to have a few shortcomings on a single carb Series 'A' engine. Some experiments I conducted showed that the point of injection on a Series 'A' engine can be quite critical. On a single carb engine, injecting at the carb manifold interface produces less power from the same amount of N_2O and fuel than injecting into the intake manifold just before the face of the cylinder head. To give some comparative figures, an identical flow of N_2O and fuel produced an 18 hp increase when injected at the carb to manifold face and a 25 hp increase when injected by the manifold cylinder head interface. The underlying factor here is that distribution of the mixture can be effected in a typical single carb intake manifold, and the amount of distance that the mixture of N_2O and fuel has to travel to get to the cylinders means that it has more opportunity to pick up heat when injected at the carb

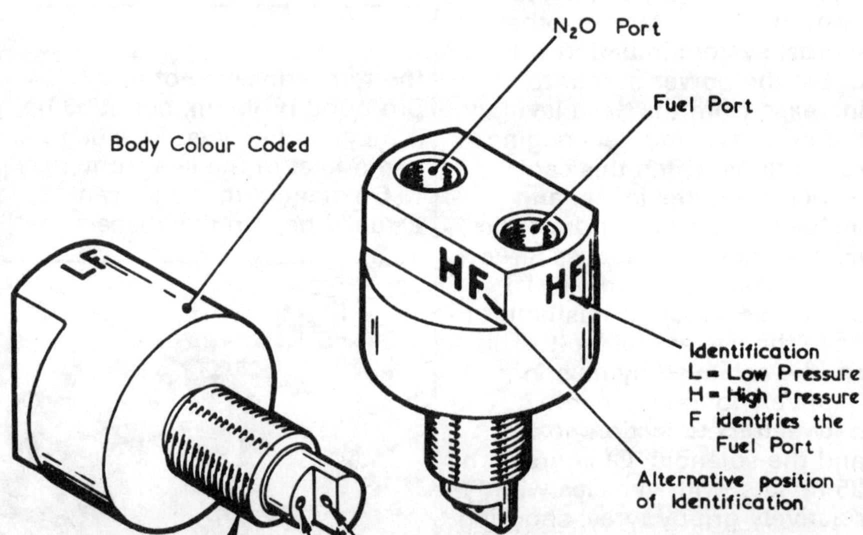

Fig. 19.5 N_2O/fuel injection nozzles

than when injected nearer the cylinder head port. Also, most of the N_2O issuing from a spray bar mounted at the carb face will impinge upon the back wall of the manifold, thus absorbing a great deal of heat from the manifold. On the other hand, if the N_2O stays in the air, at least it cools the air which performs a useful function rather than cooling and absorbing the heat of the intake manifold which does little to help power. If a twin choke, side draft carb is used, the plate and spray bar setup between carb and manifold works well, and **N.O.S.** can supply 'specials' for such applications. I do, however, believe that A.P.T.'s work in utilizing their 'fogger' nozzle (in effect their verslon of the monojet nozzle) obviates the need to use the plate setup.

The situation concerning power levels is straightforward. Most kit manufacturers will calibrate the kit to give, within reason, the power you require (don't be greedy). On the Aldon/Oselli type system, and any system using the 'fogger' nozzle changing the power level is simply a question of changing monojets. **Fig. 19.5** shows a typical monojet. Note that the N_2O nozzle is at 90° to the fuel nozzle. When installed in the engine, the stream of N_2O is, as would be expected, in the same direction as the airflow. As fuel issues from the fuel nozzle, it is impinged upon by the high speed stream issuing from the N_2O nozzle. This causes the fuel to be broken down to easily combustible droplet sizes and ensures thorough mixing of the N_2O and fuel. This last feature (shared by the 'fogger' nozzle) alone has proved to be worth a 10% power increase. Detailed attention to operational functions such as this makes the most of every lb of N_2O. In other words, if you are after a given power increase, an efficient system will use N_2O at a slower rate than an inefficient system. It's worth remembering that even an inefficient system can make enough hp, if calibrated to do so, to grenade your Series 'A' engine.

Returning to the port injection system for a moment, **Fig. 19.7a** shows the installation of the monojet in the manifold. The use of relatively soft fibre washers between the manifold and the base of the monojet not only allows a seal to be made against a rough, cast finish, but also the compressibility of the washer allows the monojet to be tightened so that it is pointing in the correct direction.

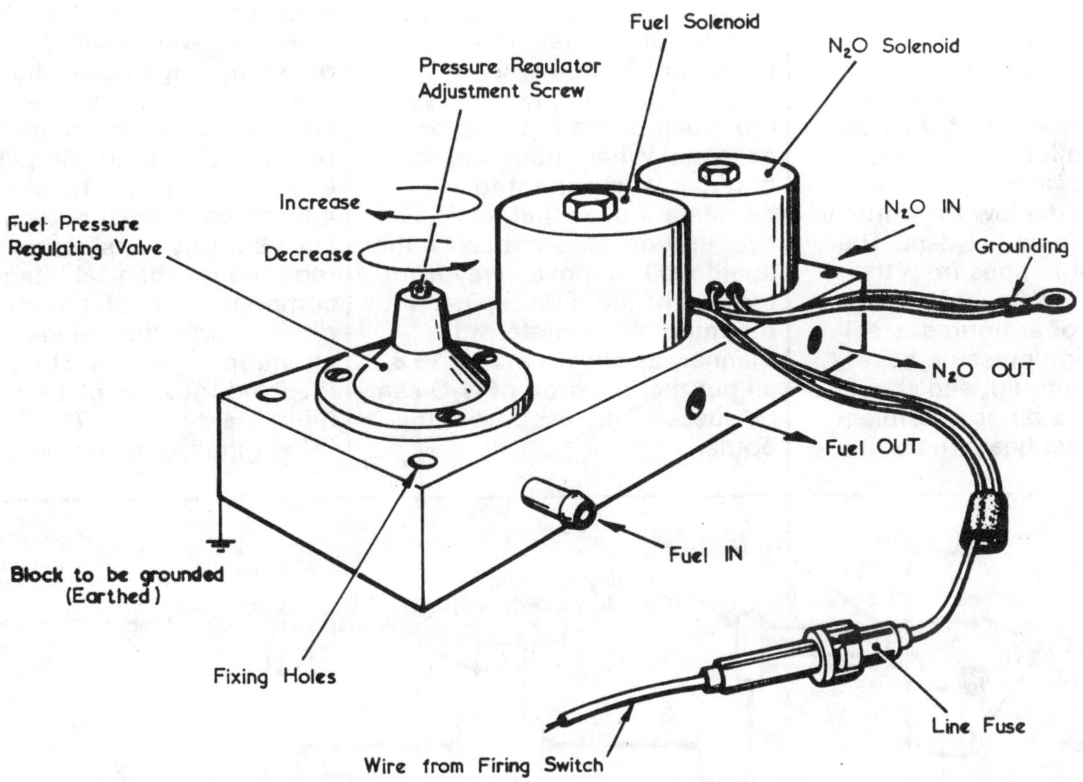

Fig. 19.6 Solenoid control valve block

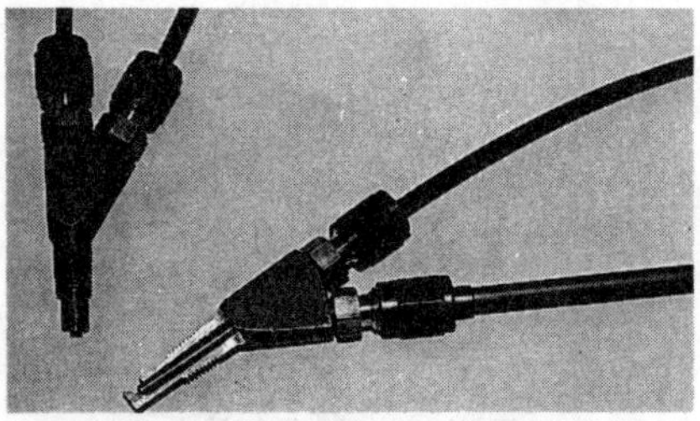

Here's the cut-away of the N.O.S. fogger nozzle. One of the beauties of this type of nozzle is that as the mass of N_2O going through the nozzle changes, so the amount of fuel drawn out changes roughly in proportion, therefore holding the N_2O to fuel mixture ratio at the desired level. 19/64

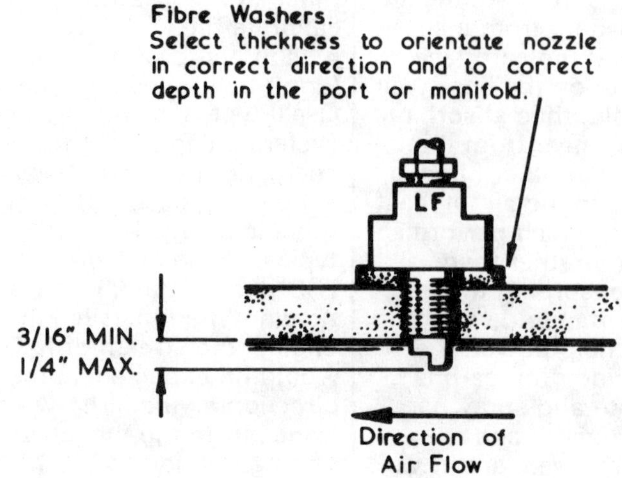

Fig. 19.7a Positioning of monojet to orientate orifice

Bottle Mounting

The N_2O bottle mounting attitude in the vehicle can have a significant effect on performance, also from the safety point of view, it's very important that the N_2O bottle be firmly fixed into the vehicle. Mounting brackets which are flimsy, could allow the bottle to tear loose in an accident. The high pressure lines from the bottle to the solenoid valves should be of armoured, steel braided, high pressure hose such as Aeroquip, and should be routed as far as possible from exhaust heat. The bottle itself should be kept out of direct sunlight, particularly in such places as Australia and the South West area of USA where the temperature of the bottle can be raised considerably. If the bottle temperature goes too high, a substantial drop in power production will be experienced, simply because too much of the N_2O will be gaseous. When mounting the bottle, it is important to orientate it such that acceleration does not cause the liquid N_2O to move away from the end of the pickup pipe. If it's mounted in the preferred manner, as shown in **Fig. 19.8**, all but the last drop of N_2O can be successfully used from the bottle.

Building A Nitrous Oxide Motor

Because N_2O is a relatively straightforward way of obtaining huge power figures from a Series 'A' engine, it's worthwhile looking at the practicalities of building an engine specifically to take advantage of such potential. Let's start by considering the required mechanical integrity of components. First, the most obvious area that needs attention is pistons. The pistons involved must be good, strong, high quality items. The Series 'A' engine has available a wide

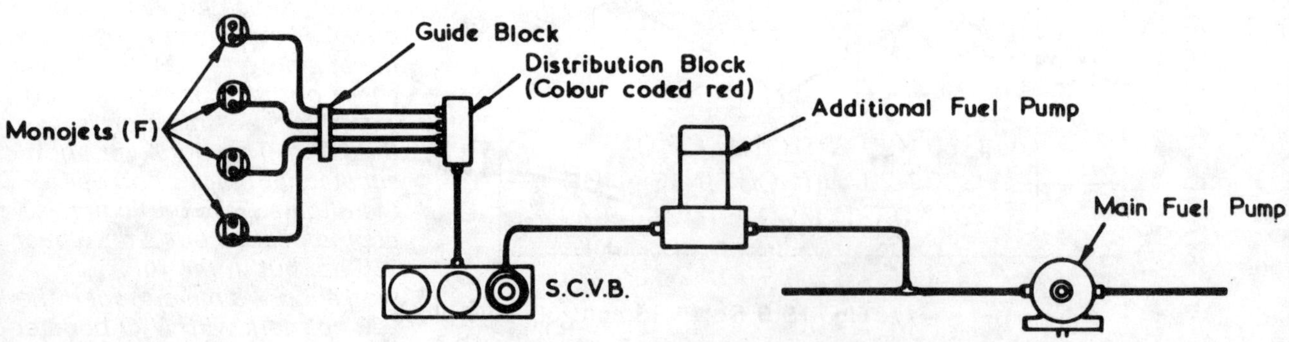

Fig. 19.7b Fuel connections

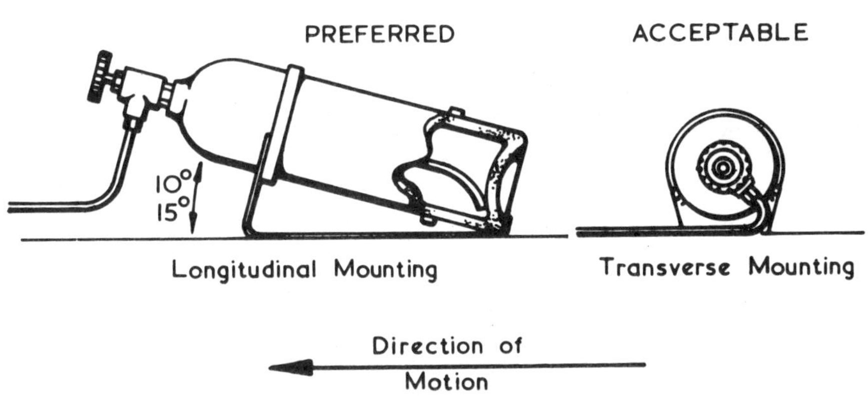

Fig. 19.8 Correct installation of N_2O bottle

range of pistons capable of withstanding relatively high N_2O loads. In this respect, the big bore motors are better catered for than the small bore motors. For a small bore engine, a forged piston is virtually a must if a high power N_2O system is to be fitted. On the other hand, on big bore motors, the standard Metro piston, as opposed to any of the other pistons in the range, have shown good reliability in coping with up to 50 N_2O-developed hp. The criterion to consider for an 'N_2O piston' is the position of the top ring. This needs to be well down from the piston crown, i.e. 5⁄16 inch to 3⁄8 inch, otherwise ring life is very short, due to the high temperatures.

Another critical area is the top ring land, that is the distance from the ring to the crown of the piston. This needs

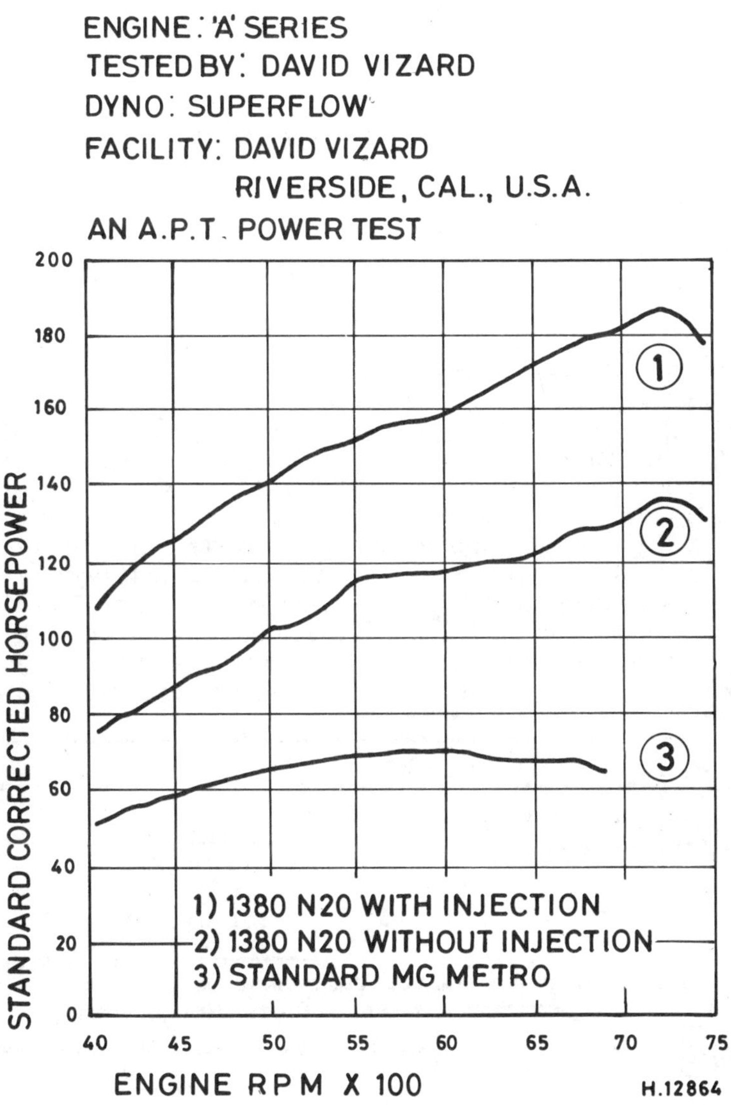

Fig. 19.8a MG Metro N_2O/standard output

Here is a slightly different way of looking at a power increase from normal. This 3D graph starts off with the horsepower scale at zero to give a true picture of the extra power delivered by the engines in question. As can be seen, the 1380 produces virtually double the power Leyland's highest output normally aspirated engine, the MG Metro, produces. Also of significance is the area under the curve. With the 1380, this is substantially greater, indicating the usable power is significantly more than the standard MG Metro. Now check out the curve produced when the N_2O injection is triggered. This adds yet another substantial block of power to the engine. This particular engine peaked out a little over 180 hp, but in the A.P.T. workshop we have topped the 200 hp mark with N_2O boosted Series 'A' engines.

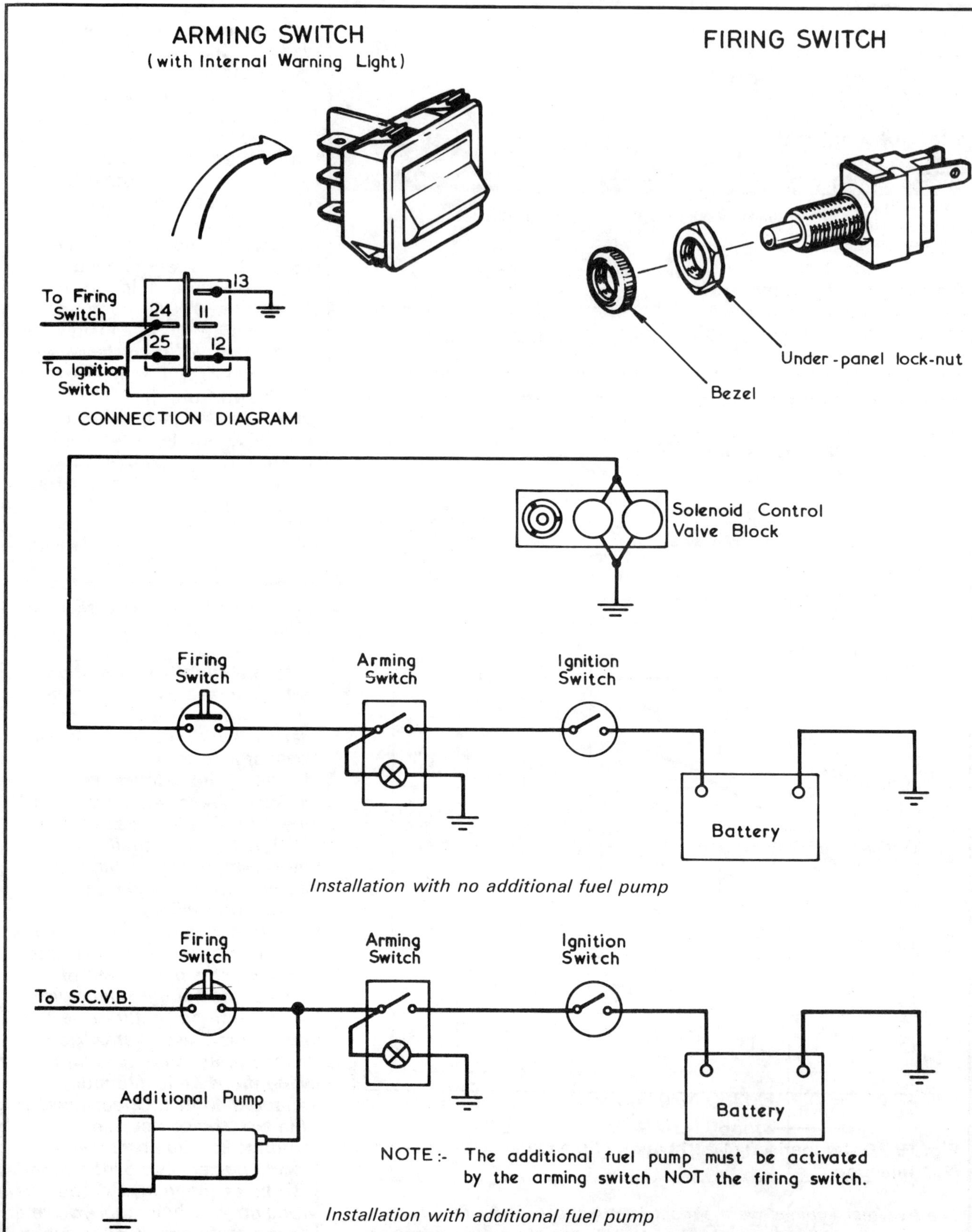

Fig. 19.9 Typical wiring diagrams for N_2O injection installations

to be thick (as will be the case if the top ring is well down the piston) solely to stop the edge of the piston collapsing and trapping the top ring. In standard or normal overbore sizes, most of the forged pistons available for the 1275 engine, appear to be okay for use with N_2O. Overbored, big bore engines, i.e. those utilizing a 73.5 or 74mm bore, present more of a special case. The **Omega** cast piston has, after many, many hours of dyno testing, proved to be perfectly adequate with N_2O loads up to about 50 hp. In fact in all the N_2O testing done on 1400 engines, I have experienced no problems with the Omega cast pistons until power levels in excess of 170 hp were reached. However, and here's where the compromise comes in, these big overbores cause the stiffness of the bore wall to be reduced. Heavy loads of N_2O produce high cylinder pressures which cause bore flexure, and this in turn leads to excessive ring blow-by. As a result, ring life, especially that of the top ring, is limited. On a sprint motor, I would say that the engine would need re-ringing at least once during a typical race season.

The entire induction system seen here was successfully used in a 1400 hillclimber racer. The car was competitive but could be driven to the supermarket with no great drama.

Because of the high cylinder pressures involved, I would not recommend the use of N_2O injection on engines with bore sizes greater than 74mm. Even at 25 hp levels they tend to have more blow-by and less ring life than is acceptable. For a 1400 sprinter, I would say around a 40 hp nitrous kit bringing the power level up to around, say, 160 hp or so total, should be regarded as the limit. An engine with 73½mm bore is in a little better shape than this. Underscoring the bore flexure problem is the ironical fact that N_2O injection works better on big bore, short stroke engines, than it does on long stroke engines. If you refer to the chart **Fig. 19.10** you can see the effect that the same N_2O kit has on two engines which

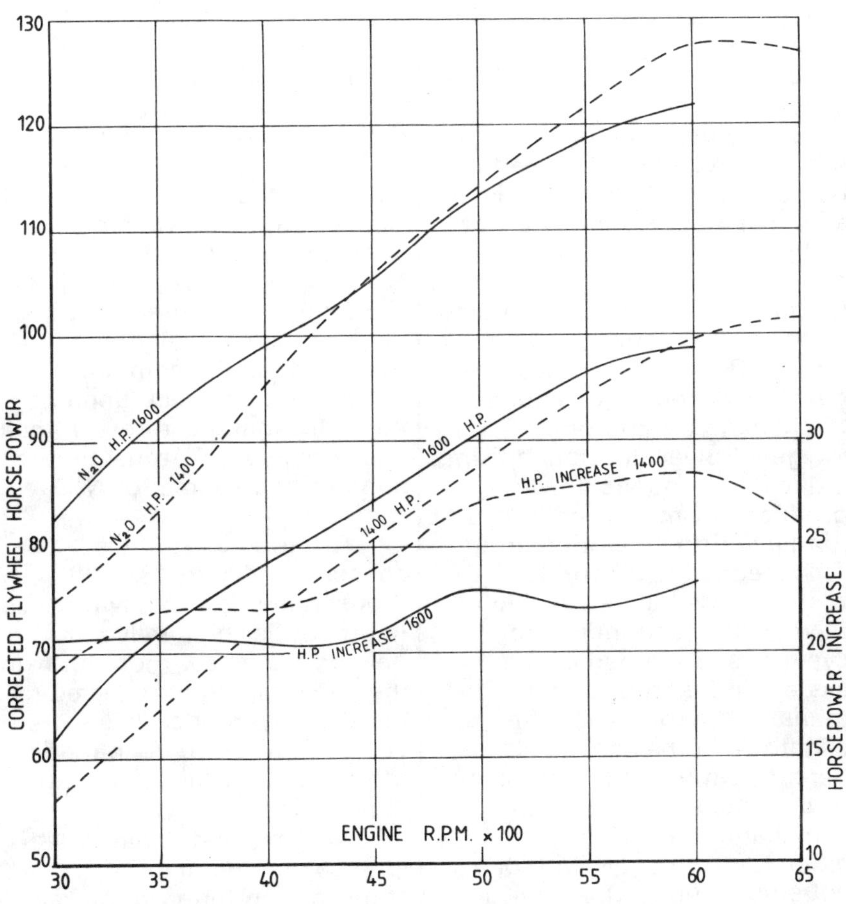

Fig. 19.10 Comparison of 1400 and 1600 engines with and without N_2O injection

The two test engines were almost identical except for the shorter stroke of the 1400. It is interesting to note that the power increase was greater on the shorter stroke engine.

Port injecting direct into a 45 DCOE manifold is very convenient and effective. The N_2O jets used here were set to boost power at the wheels by 25 hp though this increase could easily have been doubled.

differ basically only in stroke length. The 1400 engine produced more top end hp than the 1600 engine, in spite of having less capacity. On the other hand, the 1600 engine produced much better power up to about 4500 rpm. Under racing conditions, of course, power up to 4500 is quite academic, since the most used power band is going to be above 4500, even for a long stroke, relatively low revving engine, like the 1600.

Another point to note is that the power increase given by the N_2O on the 1400 is better at virtually all points tested. It could well be that because high cylinder pressures are involved, ring friction against the bore is increased, and on the long stroke engine, the necessity to drag the piston further up and down the bore means a greater proportion of the power produced in the cylinder is lost to friction. The bottom line then, for an 'all out' N_2O motor is to use forged pistons at no more than, say, +0.040 inch oversize. Because of its inherent stiffness, the best block to use as a basis would be the latest Metro block. If really big hp numbers are contemplated, this is best used in conjunction with the big journal crank, preferably a crank machined from a billet, and of EN40B. Because the power levels are likely to be very high, you need to take some hints on the style of crank to use from that described in the Chapter on big motors.

As far as rods are concerned, go for the best of the big journal rods and rework them as necessary to produce maximum reliability.

For rod bolts, there are no requirements for extra strength over and above a good standard item, because most of the loads taken by the connecting rod will be compressive. There is no need to use a lot of rpm with a N_2O motor. Power is developed by virtue of high torque rather than high rpm.

As far as developing hp is concerned, many of the requirements of an engine not normally N_2O injected need to be built into a N_2O motor. First, all the same rules apply as far as breathing is concerned. **The better breathing the engine has, the more economical it is going to be on N_2O for a given power output**. Remember, power developed from N_2O requires the hauling around of the oxygen. Power developed from inducted air means power produced from oxygen that is not having to be carried around. This means an 'all out' N_2O motor needs big valves, high flow ports, good induction system, and a tuned exhaust system in just the same manner. Where it can differ slightly is in the camshaft department. It is not necessary to be quite so radical in the camshaft area with a nitrous oxide motor, as it is with one not employing N_2O. A typical race N_2O cam will need about 280 - 290° max for the intake with 15-20° more timing on the exhaust. Lobe centreline angles need to be 110-112° and the cam needs to be 6-8° advanced. More cam timing than this does not seem to increase power to any great extent. However, the motor will become far more cammy when N_2O is not being used if a longer cam is used. Incidentally, one N_2O system characteristic is that when the button is punched, if the motor was off the cam, it will instantly come 'on cam', not because of an rpm increase, but because N_2O reduces the 'camminess' of an engine when it's in operation. This is fairly good proof that it is the velocity of the exhaust producing uni-direction flow out of the exhaust port which is the critical factor in deciding whether an engine is running 'on' or 'off' the cam. One aspect about an N_2O cam that is very useful is its very sedate manners and good driveability when the N_2O **isn't** being used. In this respect the cam spec just outlined makes a good high performance road cam. The power balance between an N_2O cam and a non-N_2O cam is about 10%, ie, it makes about 10% less power when the N_2O is off and 10% more when it's on than the equivalent non-N_2O cam. With all the cam and N_2O testing that has been done at A.P.T. for manufacturers in their businesses, A.P.T. must be the number one source for N_2O cams.

As far as valve lift is concerned, 400 to 425 thousandths is a totally practical figure. Super high valve lifts will only extract a few more hp from the engine in either injected or non-injected mode. This extra power could have been made more easily by injecting a little more N_2O.

Non-injected engines with long cams need a lot of compression to make the cam work. One of the reasons a 'big' cam is neither needed nor desirable in a N_2O motor is because the CR needs to be lower. This stems from the fact that not only does N_2O make the fuel more detonation prone, but also a heavy load of N_2O

injection can increase cylinder compression just prior to ignition. This inevitably moves things nearer the detonation limit. Using a cam of around 290° period, and a CR around 9.5 to 10:1 works well with about a 50hp N_2O kit. Using a cam of around 290° period, then a CR around 9.5 to 10:1 works well with about a 50 hp N_2O kit.

If the octane values are available, up to an 80hp increase on a 10:1 CR is feasible but figure 100+ octane is a necessity.

With very high loads of N_2O, there comes a time when the excess fuel content that's injected along with the N_2O, ceases to be effective at suppressing detonation. When this point is reached, water injection should be used to suppress detonation. In fact, any time you feel the N_2O power content of the engine may be too much in terms of thermal load for the pistons to withstand, water injection should be considered. One of the beauties of water injection used together with N_2O is that a lot of water can be injected before the power starts to drop due to 'water saturation.' If water injection is used, the fuel to N_2O ratio can be set nearer the chemically correct mixture because the duty of cooling pistons and suppressing detonation can be assigned solely to the water. With water injection rates equal to 50% of the fuel flow into the engine, power has not, in my experience, been seen to drop. Only when the water injection rate equals that of the fuel flow does power start to decrease at all. However, with a lot of water going into the engine like this, operating temperatures of the engine are brought down below those normally seen on non-nitrous injected engines, so from this aspect, water injection is a very good reliability enhancer. However, if the engine is properly set to water injection is not needed until power levels (in a big bore moter) of 180hp plus are reached, assuming 100 RON or more for the fuel.

This 1380 cc DV dyno test engine produced over 190 hp, 60 of which came from the N_2O oxide injection routed into the 45 DCOE Manifold.

Ignition Requirements

The timing of the ignition is dependent upon the rate of flame propogation. Whenever the percentage of oyxgen in the charge is increased, the burn rate is increased. Faster burning rates mean less ignition advance for optimum power. With injection rates much above 25 hp for a 1300 engine, ignition timing will need to be retarded from that giving best power without the N_2O. For instance, a 50 hp N_2O kit on a 1300 Series 'A' engine would generally require in the region of 24° of total timing. Rather than have a motor which, due to ignition retard, runs very soft when the N_2O is not being used, a 'two-stage' distributor can be easily made up using a vacuum distributor as a basis and have the vacuum plate swing the required amount to retard the ignition when the N_2O is used and advance it when it is not. It's quite simple to make up a solenoid actuated baseplate so that initiating the nitrous automatically retards the ignition.

It's also possible to achieve the same effects electronically, in fact many may favour this route, and probably the best company to contact for a special ignition system to achieve this would be **Microdynamics.**

Correct spark plug selection is very important on a N_2O injected engine. Because of the temperatures involved, it is easy to cook spark plugs accidentally. The combustion heat can be so high that not only is the centre electrode length, which normally governs the heat range of the plug, critical, but so is the side electrode. I have seen plugs on a N_2O injected engine which will melt the side electrode from the plug even though the centre electrode does not show signs of overheating. On very high powered engines, it's necessary to keep a watchful eye on the temperature of the side electrode. If, when the plugs are pulled, any signs of overheating are seen, steps must be taken to remedy the situation. Usually if the engine is water injected, plug overheating is not a problem, but sometimes water injection is not convenient, especially for a road motor. If plug side electrodes are overheating, the use of retracted nose plugs is often a simple cure. By keeping the side electrode out of the flame path, and by having it attached directly to the body of the plug, the temperature of the side electrode is reduced substantially. If the side electrode overheats the engine will detonate and typically take out the piston (usually numbers 2 and 3) in as little as 10 seconds!

The requirement for very cold running plugs in a N_2O engine does bring about a problem. Because the heat range of plugs required is toward the very cold end, it will be necessary to have an ignition system which will fire the spark

The selection of the correct heat range of spark plug is 19/66 important. A spark plug which burns out, such as those shown here, will cause that particular cylinder to detonate just prior to losing fire. The detonation usually takes the piston with it.

Any engine development project entails a lot of time on the dyno. I was a one man business until 1986, so it's been necessary to produce test results fast as time is money. My Superflow dyno allows me to produce results in six weeks that would normally take twelve months on a more normal dyno.

As you probably saw earlier in the book, the DV/A.P.T. dyno facility is a little more than your average dyno facility. This often leads prominent speed equipment companies with promising products to seek time on this dyno. Here, a Mountune engine has been shipped to the US to be put through its paces. Amongst other well-known companies that have had their products tested on this dyno are Cosworth, Kent Cams, Piper Cams Advance Product, plus a whole host of smaller, but significant component manufacturers. Each has appreciated the value of the technical feedback from such tests along with the commercial value of the exposure they will get if the product is successful. From the magazine reader's point of view, the facility supplies realistic numbers rather than the often inflated numbers given by the advertising-copy writer. 19/67

plugs under all conditions because the plugs are almost certainly going to cold foul during normal running. Because of the wide heat range the plugs are subjected to, it is necessary to have an electronic ignition on a N_2O injected engine simply to take care of the times when the plugs may be fouled. As such, employing an electronic ignition system with N_2O does not add to the hp. It just reduces the misfires when the N_2O isn't being used.

Firing Methods

Most N_2O kits on the market use a simple push-on, release-off button to fire the N_2O system. Although this is the simplest and cheapest method, it does present a few problems. If you are wrestling the car round a corner the last thing you want to do is let go of the steering wheel and push a button, which inevitably will add a whole chunk of hp and change the vehicle balance through the corner. By far the best bet is to have a throttle actuating switch to turn the N_2O on and off. The method I favour is to have a roller or leaf operated micro switch actuated by an arm on the carb throttle spindle. The micro switch should be set to operate just before the throttle reaches the wide open position. This means that feathering the throttle, even a small amount, will cause the N_2O system to switch out of operation and thus allow an easy way to regain control of the car if, due to the excess power, you are about to lose it all (not uncommon with N_2O).

When the throttle is opened, the lever end (arrow A) swings down and operates the leaf on the micro-switch (arrow B). Since micro-switches cannot handle a great deal of current, it is necessary for the micro-switch to operate ...

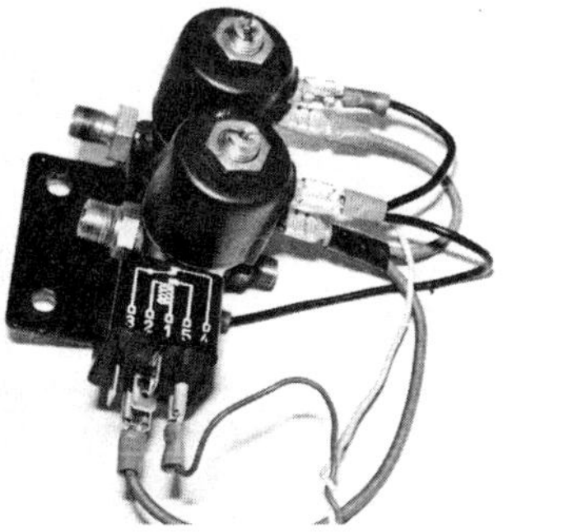

... the solenoids through a relay. The relay shown in the foreground here which can handle high currents, then switches the current to the solenoid valves, bringing the N_2O system into operation.

Nitrous Oxide Injection Part 3

Nitrous Oxide-Assisted Turbos

The most discussed shortcoming of the turbocharged engine is the so-called 'turbo lag'. Not only can the turbocharged engine suffer from slow throttle response, but it can lack low end punch, simply because no boost takes place at low rpm, even if the throttle is wide open. The turbocharger's forte is in producing extreme high power at the mid- and top end of the rpm range, more, in fact, than most engines can physically stand. It will be recalled that the shortcomings of N_2O are a high consumption rate and a tendency to produce less power at the top end of the rev range than at the mid- and bottom-end. If N_2O injection and turbocharging are combined, they compliment each other in a way that eliminates their respective shortcomings. To see how combining these two power augmentation systems works, let's go through a sequence of events, step-by-step

First, let us assume our turbocharged vehicle is cruising down the road at relatively slow speed and that engine rpm is, say, 1500. At this point, on a conventionally turboed engine, flooring the throttle is unlikely to cause any boost, because the mass of gas passing through the engine is not enough to drive the exhaust turbine sufficiently fast to produce boost at the intake turbine. As a result, initial acceleration from this point is lazy. When the engine speed gets to about 2500 rpm, the boost starts to come on and the car starts to perform. When N_2O injection is used at the same slow engine speed, flooring the throttle and adding a shot of N_2O causes power augmentation due to the N_2O injection. This will instantly raise the torque output of the engine, and to the driver, this will feel exactly the same as if the engine had come on boost. By virtue of injecting the N_2O, a large quantity of exhaust is produced, virtually instantly. This large quantity of exhaust impinging on the exhaust turbine is sufficient to drive the intake turbine into boost, even though the engine rpm may have only climbed 50 to 100. Because the engine is now

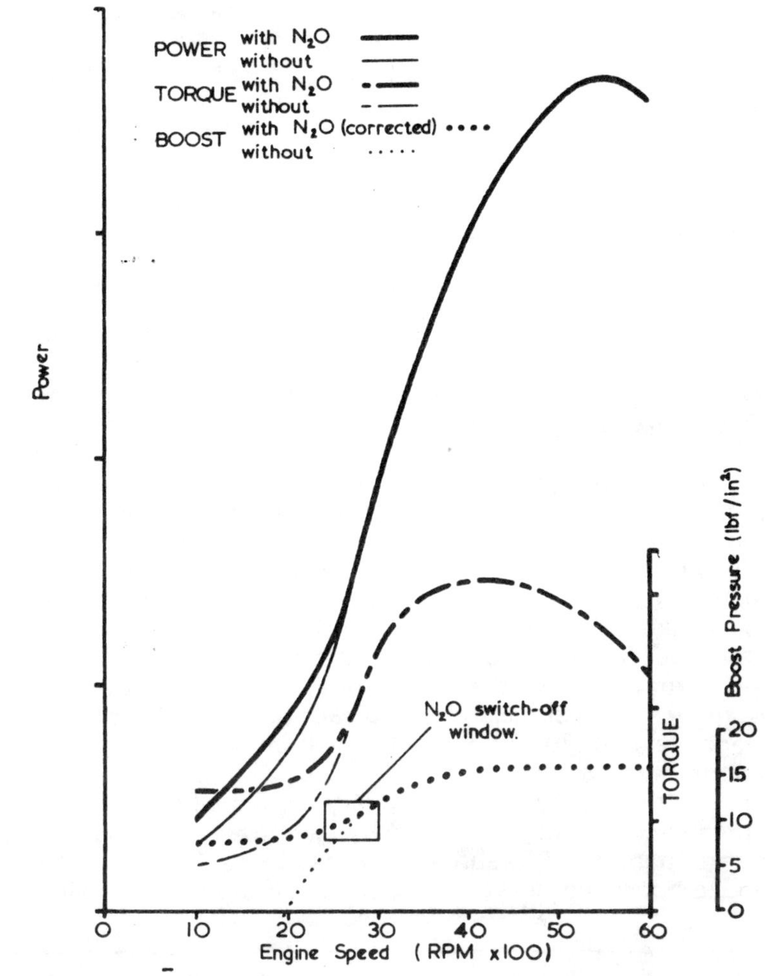

Fig. 19.11 N_2O-assisted turbocharged engine performance

driven into boost, exhaust gas propogation is increased proportionally by virtue of the airflow through the engine and the boost condition becomes self-sustaining. As soon as the boost pressure reaches a certain value, a pressure sensing switch turns the N_2O off, but if the boost starts to drop, the N_2O injection will come back on. Except at very low rpm, the N_2O tends not to cycle in and out. As a result, flooring the throttle at anything but the very lowest rpm causes a very short shot of N_2O to be injected into the engine, with a duration typically of ½ to 1 second.

Another aspect contributing to N_2O economy is that less N_2O is needed for a turbo engine than is needed when N_2O is the sole power augmentation method employed. A typical road N_2O system would be around 25 hp. That's enough to provide some exciting driving in, say, a Mini, but if N_2O is used to eliminate turbo lag, then an amount of N_2O equal to what it would take for about 10 hp in an unturboed engine, is all that is needed. As a result of minimal N_2O consumption, a single bottle charge can last a long time. Depending upon the style of driving, a 15 lb bottle can last anything from one to three months.

As far as peak power is concerned, a N_2O assisted turbo motor is not necessarily any more powerful than an unassisted example, simply because the nitrous is only used at low rpm. **Fig. 19.11** shows a typical change in the power curve due to N_2O assistance on a turboed engine. The particular engine these figures were taken from was highly boosted, hence the relatively large bulge in the torque curve at around 4000 rpm. Of course, this torque curve could have been levelled out more by increasing the amount of N_2O injection that occurred prior to the boost coming on. Even with the relatively small amount of N_2O injected into the engine, you will note that the torque increased by as much as one-third at 1000 rpm. You will note also that the boost is shown in a 'corrected' form when the N_2O system is activated. This is simply the easiest way to illustrate the situation as the N_2O content of the charge has the same effect as a certain number of pounds of boost. Whereas the boost normally doesn't start until 2000 rpm, the corrected boost on this particular engine is equivalent to 7 psi at 1000 rpm. Such boost figures are rarely achieved even with a belt-driven, positive displacement, mechanical supercharger. Probably the best way to describe a N_2O assisted turbo is to imagine an engine with the assets of both a belt-driven supercharger and a turbocharger. The price paid for such a power plant is a bottle refill once in a while.

Driving a nitrous assisted turbo powered vehicle, bears no relationship in terms of driver sensation to a conventional turbo car. The sensation produced can be likened to driving a vehicle with the engine capacity increased by a very substantial amount. For instance, a nitrous assisted turbo Metro with 10 psi boost would be like driving a Metro with an engine of around 2.3 litres. N_2O and a turbocharger is without a doubt the best way to achieve the monster motor feel.

Nitrous Oxide-Assisted Turbo Installation

The actual mechanics of building a nitrous assisted turbo installation are relatively simple. **Fig. 19.12** shows a typical layout. In this layout, the N_2O injection is activated by a vacuum switch. Depressing the throttle will cause the vacuum sensed by the switch to diminish. As soon as the intake manifold pressure upstream of the turbo, that is before the turbo, reduces to a sufficiently low negative value the N_2O will be brought into action. Another switch, this time a pressure sensing switch, is situated in the intake manifold. As soon as the intake boost pressure reaches a pre-determined level, say, 3 to 4 psi, then the N_2O system is switched off, the manifold switch overriding all other switches. This system works fine if the manifold vacuum switch turns the N_2O on only when the manifold vacuum is very low, say, 3 or 4 inch of Mercury at most (2 psi). However, if a lot of high altitude driving is to be done, the N_2O may come on when the driver does not require anything like full power. In such conditions, it is better to use a throttle-operated switch to turn the N_2O on, although the pressure switch for turning it off would remain the same.

If you take a look at the wiring diagram accompanying **Fig. 19.12**, you will see that a disarming relay is installed in the system. Without the disarming relay, the N_2O would switch on if the engine were stalled, which could cause a very big back-fire upon start-up. By having a disarming relay coupled into the oil pressure switch, if the engine stalls, oil pressure is lost; this operates the relay which, in turn, disarms the entire N_2O system. The oil pressure needs to be of a high pressure variety i.e. 20 psi, and screwed directly into the oil galleries, not at the end of a long line, otherwise the oil pressure may drop at such a slow rate that the engine may have been stalled for 2 or 3 seconds before the system is disarmed. During starting, the arming switch of the entire system must be in the OFF position. For this reason, it's a good idea to have a large and very obvious warning light which indicates when the N_2O system is armed.

The setup shown in **Fig. 19.12** employs a 'suck-through' carburation arrangement on the turbo, however, not all turbo setups are suck-through. Some, such as the **Janspeed** setup, are blow-through, in which case there will be differences in the way the N_2O system functions. **Fig. 19.13** shows the differences in layout. In a blow-through system, both the pressure and

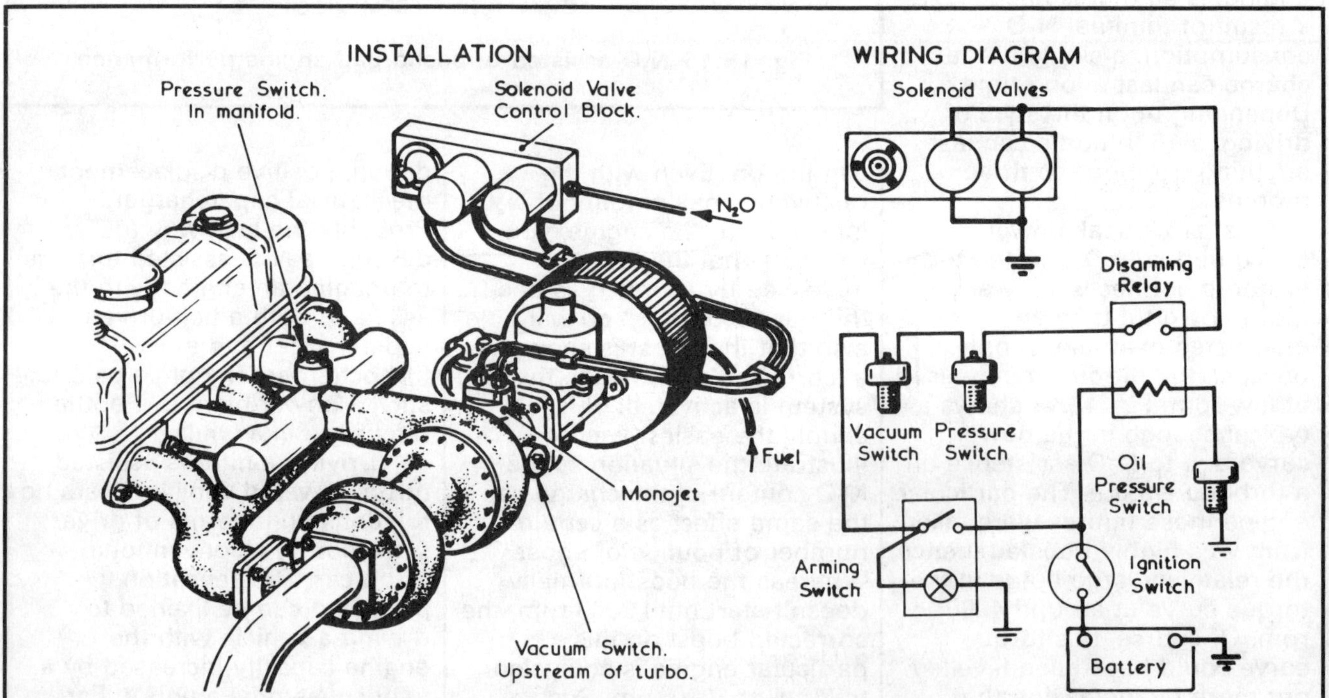

Fig. 19.12 N_2O-assisted turbocharger installation

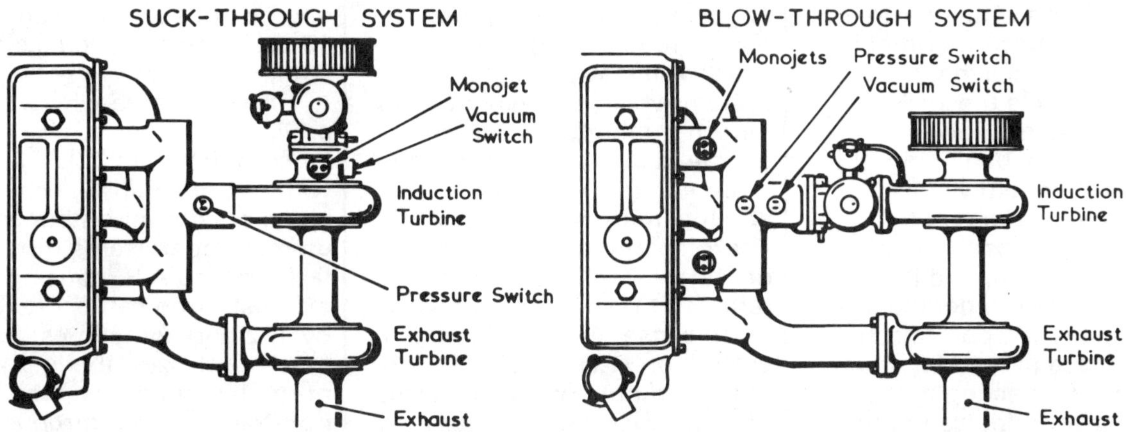

Fig. 19.13 Comparison of turbocharger suck-through and blow-through installations

vacuum switches will be in the intake manifold and although it is not essential, the preferred position of the monojets will be in the ports of the intake manifold rather than in the turbo itself. If the monojets are situated in the intake manifold on the boosted side, it will be necessary for them to be supplied with fuel which is always at some 3 psi (the calibrated pressure of a monojet) above the boost pressure. This is achieved using a pressure regulating valve which maintains the fuel pressure from a high pressure pump (providing at least 5 psi more than the highest boost pressure expected to be seen) at least 3 psi above the manifold pressure.

The experimental system shown here made provision for such a situation. This was catered for simply by supplying boost pressure to the diaphragm of the special fuel pressure regulating valve in the solenoid control valve block.

If, on a blow-through system, a throttle shaft actuating switch is employed rather than a vacuum switch to initiate N_2O injection, then the situation becomes a little simpler. If instead of 2 monojet nozzles, one in each manifold

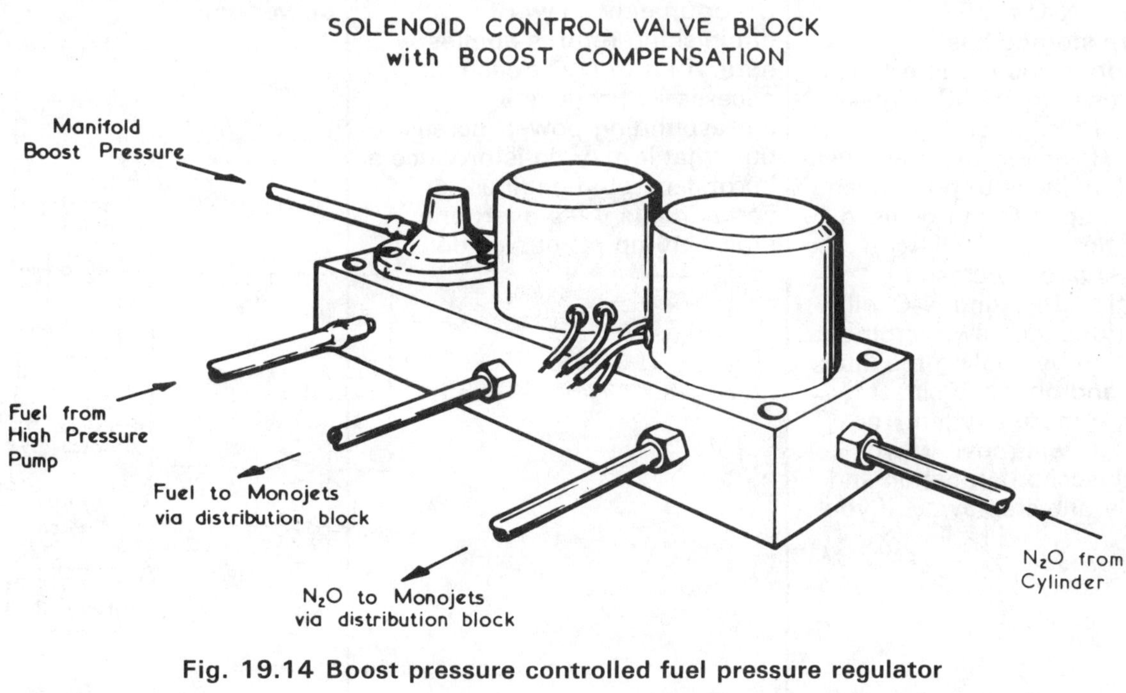

Fig. 19.14 Boost pressure controlled fuel pressure regulator

port runner, 1 nozzle located on the intake side of the turbo is used, things become simpler yet. In fact the system now closely parallels the suck-through system. However, injection at the beginning of the intake system allows much more time for the N_2O to expand as it goes through the induction system, and it takes a little more N_2O to get the job done than when port injected. The increase in N_2O for equal effect when injecting at the turbo entrance instead of the port runner, can be as much as one-third.

Nitrous Oxide Intercooler

Up to now we have discussed N_2O merely as a means of improving turbo response. There is, however, another area where N_2O can be usefully employed on a turbo installation; that is, where there is physically not enough room for an intercooler. Because release of N_2O from high-pressure storage has a tremendous cooling effect, it can be used to cool the intake charge of a supercharged engine. However, there are well-defined limits as to how much intercooling N_2O can be used. Remember, a supercharged engine is already close to detonation. Injecting N_2O will move it one step away from detonation by cooling the intake charge and one step closer by providing more oxygen. Too much N_2O will move it two steps closer to detonation and still only one step away. If you have equipped your turbo engine with an N_2O system to eliminate turbo lag, it is well worth while adding an override system which will allow you to 'punch the button' when the engine is on full boost, as most turbo engines can tolerate about a 10 hp addition to their intake charge without any detrimental effects. I use the term '10 hp addition' in a very loose sense. All through this section on N_2O I have referred to the quantity of N_2O in terms of how much extra power it will give. To define the rate of N_2O injection, apply the simple relationship of power to rate of N_2O injection, namely 1 lb for 10 seconds = 100 hp. From this, a kit rated at 10 hp would inject N_2O at a rate of 0.6 lb/min.

A supercharged engine given a moderate amount of N_2O produces a power increase out of all proportion to the quantity injected. As little as 5 hps worth of N_2O, that is a quantity that would produce a 5 hp increase on a normally aspirated engine, can, when used as an intercooler, produce as much as 15 extra hp from a turbocharged engine. Unfortunately a law of diminishing returns applies here. A lot of N_2O does not necessarily produce a corresponding power increase but what it may do is produce a lot of separated engine components if it's overdone. A typical 10 hp nitrous oxide injection system will, on an engine boosted to around 10 to 15 psi, normally produce a 15 to 20 hp increase.

Another area where additional power can be gained by combining an N_2O system with a turbocharger, especially when maximum output is the ultimate goal, stems from the fact that the A.R. housing of the exhaust turbo can be increased by 1, 2 and sometimes even as much as 3 sizes above that normally found best when a non-N_2O assisted turbo is used. This is because the size of housing selected is a compromise to produce the best power curve over the rpm range. For a given boost, the bigger the A.R. housing used on the exhaust side, the more top end power the engine will develop, simply because there is less back pressure in the exhaust manifold. **When an N_2O assisted system is used in conjunction with a turbo engine, the N_2O takes care of the low end power output, the turbo the top end. This means that the turbo exhaust housing can be selected with a view to producing the best top end power only.**

20 Water Pumps, Radiators & Ancillaries

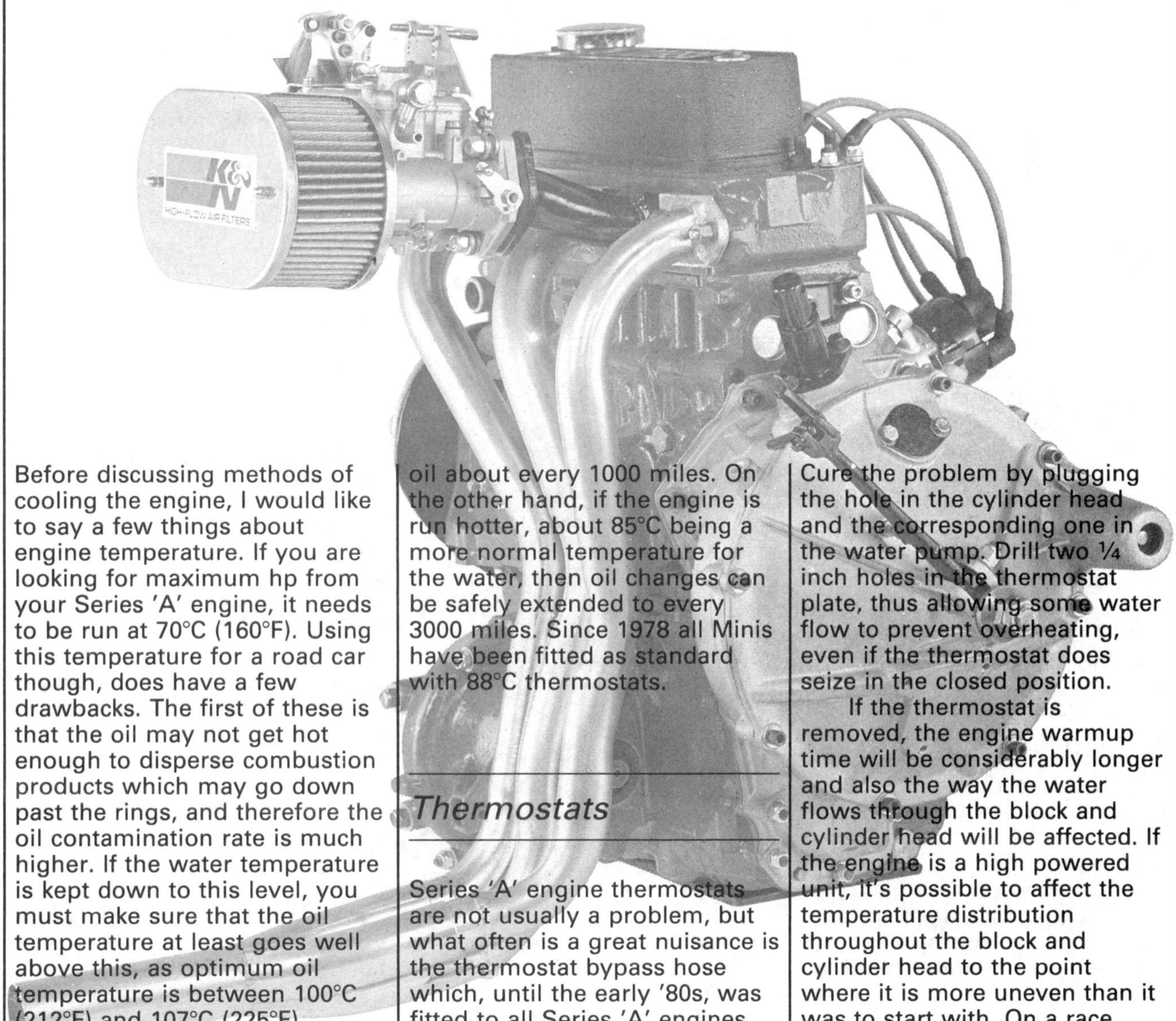

Before discussing methods of cooling the engine, I would like to say a few things about engine temperature. If you are looking for maximum hp from your Series 'A' engine, it needs to be run at 70°C (160°F). Using this temperature for a road car though, does have a few drawbacks. The first of these is that the oil may not get hot enough to disperse combustion products which may go down past the rings, and therefore the oil contamination rate is much higher. If the water temperature is kept down to this level, you must make sure that the oil temperature at least goes well above this, as optimum oil temperature is between 100°C (212°F) and 107°C (225°F). Unless the oil temperature can be maintained at that level, it will be necessary to change the oil about every 1000 miles. On the other hand, if the engine is run hotter, about 85°C being a more normal temperature for the water, then oil changes can be safely extended to every 3000 miles. Since 1978 all Minis have been fitted as standard with 88°C thermostats.

Thermostats

Series 'A' engine thermostats are not usually a problem, but what often is a great nuisance is the thermostat bypass hose which, until the early '80s, was fitted to all Series 'A' engines. The bypass hose connects the water jacket adjacent to the thermostat to the water pump. Cure the problem by plugging the hole in the cylinder head and the corresponding one in the water pump. Drill two ¼ inch holes in the thermostat plate, thus allowing some water flow to prevent overheating, even if the thermostat does seize in the closed position.

If the thermostat is removed, the engine warmup time will be considerably longer and also the way the water flows through the block and cylinder head will be affected. If the engine is a high powered unit, it's possible to affect the temperature distribution throughout the block and cylinder head to the point where it is more uneven than it was to start with. On a race engine, the use of the thermostat can cause the temperature to reach a higher

If the thermostat is removed because the engine is to be used for competition purposes, then a thermostat blanking sleeve like the one shown here will be needed to maintain the correct water cirulation through the head.

level than it would if it were absent, so it is a good idea to remove it. However, it's necessary to restore the circulation status quo. The simple answer is to use a thermostat blanking sleeve. This acts as a flow 'diverter', restoring the flow of water around the block and head to something resembling the original pattern. The blanking sleeve is a competition item and should be available at a modest price from any Series 'A' specialist.

If the thermostat is taken out and a blanking sleeve (or an equivalent) is not used, then the cylinders farthest away from the thermostat end of the engine will tend to run hotter, and those nearer the thermostat will tend to run cooler because some of the water is taking a short cut directly from the water pump to the thermostat housing area of the head and then out through the top hose. Finally, use of a thermostat or blanking sleeve causes slight restriction to flow, and is thus said to reduce pump cavitation at high rpm.

On highly tuned engines, big bore engines or turbocharged engines, water circulation at the far end of the engine may still be inadequate and something additional in the way of water circulation is often required. One solution is to install a water fitting in the freeze plug hole on the end face of the block by number 4 cylinder. From here, run a short length of pipe to route water directly into the heater take-off point on the cylinder head. I am assuming here that a very highly turboed engine is likely to be a competition engine and that the heater is not fitted. If the heater is fitted, then it may be necessary to drill and tap the end of the cylinder head.

This particular mod was passed on to me by Paul Ivey. Paul says that using this modification he has never had a problem with overheating on a race engine. However, he was quick to point out that all his race engines are built with new blocks and new cylinder heads; used blocks and heads have a rust build-up in the water jacket

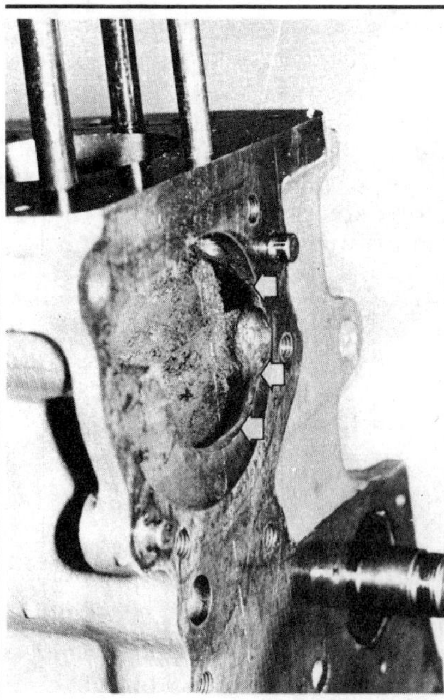

To promote a better flow of water to the back of the cylinders on big bore motors, grind the water jacket to give more clearance as indicated by the arrows in this photo.

which causes poor heat transfer, and this allied to restriction of the waterways leads to overheating. Also, displaced rust scale is transferred to the radiator which eventually becomes sludged up. **If you think your engine may produce enough hp to have a cooling problem and you are building it on a used block and cylinder head, be sure to have these two components cleaned out in a caustic hot tank to get them down to bare metal.** To stop them from rusting heavily again use a corrosion inhibiting anti-freeze.

The particular Paul Ivey mod just mentioned has been used successfully on turbo engines. However, it would appear that if the engine is very highly boosted, to much over 160 hp, yet more in the way of cooling may be needed. Assuming the radiator has ample heat dissipating capacity, then it's a question of making sure water circulation is brought up to a level which will actually transfer the heat from the engine to the radiator. This can often be done by matching all the holes in the block with the head, (it's more important to match the ones farthest from the water pump than those nearest) and then drawing water from the heater take-off on the cylinder head, routing it through an additional radiator and then back into the system either via the bottom hose, top hose or the header tank of the original cooling system. The most popular and easiest method, is to route the water from the additional radiator back into the bottom hose. However, there is a sort of thermal balance that it's necessary to achieve here, otherwise the whole idea may fall flat on its face. The additional radiator must be big enough to drop the temperature of the water going through it to a level comparable to or below that in the bottom hose, otherwise cooling efficiency may be reduced rather than increased. An additional radiator of the size shown in the accompanying photo is typically adequate for the job.

Overheating problems occur mostly on big bore engines, simply because they make the most hp and have less water jacket volume in relation to their power output. A block mod to aid cooling on big bore engines, is shown in the accompanying photo. It entails grinding away the outer case of the block on the inside so as to allow water from the pump to make its way more easily round the turn and into the area behind the cylinders. I have been doing this modification on all my big bore 1400 cc engines for a couple of years, and it seems to work.

Water Pumps

Water pumps and their inability to keep the engine cool, used to be a problem on the Series 'A' engine. Early small bore engines used a water pump of lesser impeller diameter and length than current production pumps. Since the introduction of the 'S', water pumps have been upgraded in a number of ways. Some mods produced a better performing water pump as an aside to the factory's original goal of reducing production costs. When 'S' engines were originally introduced, *the* mod to do for small bore engines was to shorten the impellor on an 'S' pump and install that.

If your Series 'A' engine is equipped with a cast iron pump, then ditch it. Use an aluminium pump like this. Use the smallest capacity pump that gets the job done.

When fitting an aluminium pump, be sure to use a good, thick washer between the head of the bolt and the pump.

On a high rpm engine, the pump can be running too fast, possibly causing cavitation. Slow the pump down by using larger pulley.

The highest capacity pump to date is the turbo Metro unit, but before rushing out and buying such a high volume device, note the following. Water circulation in the Series 'A' engine is at its most critical not at high rpm full throttle, but at low rpm full throttle. The pump designer has to produce a pump which will push sufficient water around the engine at 2000 rpm to keep it cool. Indeed, if a pump with a variable drive was to hand, you would find that over about 2000 engine rpm, the pump doesn't need to go any faster to keep the engine cool even if engine speed is up to 6000 rpm or more. This is partially due to the characteristics of the pump. Since it is essentially a centrifugal pump, its capability to move water goes up with the square of the rpm. This means that at low rpm, the pump is hardly moving any water at all, but by the time the engine reaches 2000 rpm, it's shifting all the water required for the function of cooling. Over approximately 2000 engine rpm, the pump is simply sapping unnecessary power from the engine. If the engine does not spend much time lugging down around the 3000 rpm mark, there is not a lot of point installing a pump big enough to cater for low rpm lugging. If it's a race engine, and the engine is revving between 6000 and 9000, then adequate pumping capacity may be achieved with either a smaller pump or a large pump running at reduced speed by virtue of a larger pulley, and in some cases, a combination of both.

The Golden Rule is to use the smallest and/or slowest rotating pump that will get the job done. Engines that will be most demanding on water pumps are those which make a lot of hp at low rpm, particularly overbored 1275s that are used for street applications and are cammed such that they will pull down around 2000 rpm.

As for the pump to use, well the latest pumps are good. The standard BL pump with the aluminium housing and cast iron rotor, works well. If you have an old cast iron water pump, this is the type you should replace it with. Not only does it save a few pounds, but also it is more effective at pumping water and it dissipates more heat.

Radiators

No amount of modification to the cooling system will prevent overheating if there is simply not enough radiator capacity. If overheating occurs, check first that the radiator is thoroughly clean internally. If the radiator still proves inadequate, take steps to install a radiator with greater heat dissipating capacity. Some vehicles, such as the Metro and Turbo Metro, have ample radiator capacity. Indeed, the turbo Metro has a radiator about 50% bigger than it actually needs. On the other hand, some vehicles such as the Cooper 'S' are very poorly catered for. It only takes a hot day plus a traffic jam to start an 'S' boiling in no time at all.

Here's the additional radiator setup in **Ripspeed's** *Stud Nicolao Mini. This particular engine displaces nearly 1600cc and since it's used for rallycross work where speeds, especially on slippery tracks, can be much slower than on road races, it needs all the cooling it can muster. A big 'S' radiator, plus the additional radiator, gets the job done. 20-8*

Sometimes an engine only overheats during certain modes of driving, when, it will blow a little water out of the radiator. Eventually the vehicle will overheat all the time simply because the coolant level has dropped so much. This condition is often cured by utilizing a radiator expansion tank. The tank will catch any water blown out of the header

This shot of one of the author's Minis depicts two things relevant to maintaining water temperature at the correct level. First, note the very large radiator, built from a Hillman Hunter core. By utilizing a radiator this size, no fan was needed. Secondly, note the expansion bottle mounted on the inner wing valance. Any water blown out of the radiator is returned during the cooling cycle. 20-2

tank and, as the engine cools down, it will draw the water back into the radiator, thus maintaining the level. To an extent, a high pressure radiator cap can be viewed as an item which performs a similar function. By holding the contents of the radiator under pressure, the boiling point of the water is raised. This means it will require a higher water temperature to actually start blowing the water out of the radiator in the first place, so for this purpose, I would always recommend the use of a 13 lb pressure radiator cap.

Race Minis have a severe cooling problem especially in the hotter climate of the South West U.S.A. Big radiators are a must, and this SCCA Mini shows that it is possible to install a big enough front-mounted radiator to make the fan redundant, and do it all without the radiator sticking out through the body. 20-9, 20-10

Fans

Series 'A' engines have been equipped with a variety of fans, ranging from two-blade to sixteen-blade items. Early Minis have probably accounted for a greater variety of fans than any other type of Series 'A' vehicle. The first 850s had a two-blade metal fan. This was quickly replaced with a four-blade metal fan which was found necessary to keep them cool in traffic. Later, the four-blade fan was replaced by a sixteen-blade fan, this being introduced to cut fan noise which was quite extensive with both the two- and four-blade fan. After the sixteen-blade fan, came the plastic fan era. The moulded plastic fans have more aerodynamically shaped blades and thus, they absorb less power and move more air than the early metal fans. Also, the plastic fans have staggered blades which has the effect of cutting fan noise. For high performance use, most plastic fans will move more air than is necessary. It is not practical to cut whole blades off the fan, simply because of the imbalance produced because no two blades are diagonally opposite. To cut the power absorption with this type of fan, trim the length of the blades.

Obviously a fan of any size driven by the engine is going to absorb hp. **The ideal situation is to have a sufficient radiator area to be able to dissipate heat without a fan, and for race applications where speeds are generally high, getting rid of the fan is not a difficult job.** It is at slower speeds that the problem can arise of overheating, due to an inadequate fan. **To improve cooling in traffic conditions, an electric fan is a good idea, but it will not aid cooling one iota over 30 mph. Above 30 mph, virtually all the cooling is**

achieved through the forward motion of the vehicle pushing air through the radiator. It follows that overheating at high speed will not be cured by an electric fan; it's more radiator that is needed.

Air dams can be useful, not only as aerodynamic devices but also as a means of housing a large oil cooler within the confines of a cramped engine compartment.

Ancillaries

It is probable that problems with overspeeding the dynamo or alternator will be encountered on really high speed engines. Dynamos in particular were not built to operate on an engine turning consistently at 8000 rpm. The solution is to run the dynamo at a lower speed by fitting a pulley about 25% larger. Although alternators can withstand more rpm than a dynamo, they simply absorb more power at these high rpm. Here again, slow them down by means of a larger pulley. The parts to do this are available from most Series 'A' performance specialists.

If you want to go really overboard on saving power in driving the ancillaries, then by far the most sophisticated ancillary drive system is the cogged belt setup produced by Bill Quine at **Manx Racing Developments**. Depending upon the rpm involved, this setup will contribute between 1 and 2 hp to the engine's output.

If you want the ultimate in ancillary drives, then this cog belt setup from Manx Racing Developments is the one for you.

21 Economy

Mention has been made of the economy aspects of various modifications throughout the book. In this Chapter I intend to condense much of what has been said before and deal with economy exclusively so that if economy is your main goal, you won't be distracted by the power attributes of previously described modifications.

Building an engine which has good economy is really a question of efficiency. Some of the mods done to produce power don't necessarily produce good economy. With that in mind, I'll make a start at the induction end of the engine and work through to give you the rundown on what works and what doesn't.

In the days when the 50 mpg/50 mph car was still just all talk, this car was built and went better than 55/55. This car went all around the U.S.A. come rain, come shine, come hail or snow ...

Here is the Mini dwarfed by 'Big Boy', an example of the world's largest locomotive. At around 7000 hp and designed to draw 4000 tons it's hardly in the economy class.

Air Filters

The important consideration concerning air filters is whether or not they should be fed with hot air. The simple answer is that in the summer hot air intakes are mostly of academic value. In the winter they are worth their weight in gold. The break-even temperature seems to be about 40°F. Many Series 'A' engines are equipped with air filter cases that will draw heated air during part throttle cruise conditions, but full throttle causes them to switch to a cold air duct. This type of filter case has much to commend it because it does give the best of both worlds. However, it tends not to be super-efficient in terms of airflow, so it's necessary to make the most of the induction system in other respects, such as installing high flowing filter elements.

If the filter case is one that has a winter and summer setting, then it's simply a question of turning the filter case round at the appropriate time. However, if you are looking for performance with economy, you will find this type of filter case does not give good airflow characteristics. They can often be improved substantially by increasing the size of the tube feeding into the case, as this is normally the restriction. During the summer, or any part of the world where the climate is warm year round, a performance air filter will get the job done both from the power and economy point of view, but for good economy, remember, the engine needs a carburettor induction temperature of at least 80°F, but preferably nearer 100. A lot of carburettor heat will, of course, be supplied by the engine heating the surrounding air. In very cold climates, the under-bonnet temperature is insufficient. Add to this the fact that the fuel evaporating further cools the carb, and you can see that it won't take too much during cold, damp conditions, to literally freeze the carb into an ice ball.

Which Carb?

At one time or another I have done economy tests on almost all the popular carbs. **To date I have found nothing to beat a correctly calibrated SU carb** and in this respect, my number 1 choice is an **SU from the HIF range, closely followed by an HS type SU.** Now you'll note I said 'a correctly calibrated SU'. Although an SU would seem an easy carb to calibrate (you simply change the needle) there is more to it than meets the eye; it is almost certainly going to be necessary to have a lengthy session on a rolling road dyno to calibrate that carb spot-on. I have assumed that the choice of carb will involve only one SU rather than two. As I have said before, with the airflow capability of the bigger SUs and the quality of the single carb manifolds available, there really is not a lot of point in using twin SUs, especially if the ultimate goal is economy. An HIF 4 or 6 on a good intake manifold really does get the job done from both the power and economy point of view.

If you feel inclined to sidestep, to a degree, some of the calibration difficulties of the SU carb, then it is possible to get very good results from a 28/36 Weber. The economy figures produced by a 28/36 Weber can literally be within the odd percent or so of that given by an SU carb. However, many enthusiasts feel more at home calibrating a Weber carb, where each calibration change is a finite step, due to a jet change here or an emulsion tube change there, and as a result, Weber proponents are likely to get better results from a 28/36 Weber than an SU. However, choose your manifold carefully. Read through what I have to say about Weber intake manifolds in the manifold Section.

Intake Manifolds

The standard exhaust heated manifold used on most Series

'A' engines would, at first sight, seem to be a good choice. With all the heat applied to the manifold, the fuel must vapourize well and distribution should be pretty good at part throttle. However, **extensive part throttle dyno tests have indicated that exhaust heat put to the intake manifold causes too much heat in the induction system and better part throttle economy figures can be obtained with less heat**. The exact reason why this should be so is not clear, but a water heated intake manifold of identical geometry to an exhaust heated intake manifold produces slightly better economy figures, and also results in a useful increase in torque, due to the reduced charge temperature. This in itself can be an aid to economy because it will allow the driver to maintain a higher gear ratio longer. It may only be a small point, but it can be worth 1 or 2% improvement in mileage achieved whilst driving as opposed to tests conducted on the dyno.

When choosing an intake manifold where economy is the number one concern, always select one with the smallest bore and least volume. Again, reference to the manifold Section will help in this respect. As far as the water heating facility is concerned, this is best used all the year round because in really hot weather conditions, a non-water heated intake manifold picks up heat from the exhaust and from the surrounding air to heat it to the same temperature that it would be anyway if it were water heated. So as the water heating is effective in the winter yet is of no disadvantage in the summer, you may as well leave it connected 100% of the time.

One more point concerning exhaust heated versus water heated intake manifolds. During highly controlled dyno testing, it was shown that in the first couple of minutes from start up, an exhaust heated intake manifold did provide better economy, but after about two minutes of running, the water heated manifold equalled or bettered that of the exhaust heated manifold. Let me impress upon you, however, that the difference is small enough to be merely academic.

Cylinder Heads

When economy is the sole criterion for your engine, breathing ability on the intake side of the cylinder head becomes far less important. Total airflow capability gives way to port velocity and what is needed is a cylinder head with small ports having a rough surface finish. Also, compact combustion chambers are an aid to good economy. I have never tried putting a small bore cylinder head onto a big bore motor to see what it would do for economy, but I have tried reworking both small bore and big bore heads to go on their respective blocks for just the purpose of improving economy. On big bore heads, the technique for an economy motor is to increase the exhaust valve and exhaust port flow as much as possible. Increasing exhaust valve size is usually an aid to economy so long as the cam timing is kept short. A 12% increase in exhaust port flow usually results in around a 1% increase in the part throttle fuel economy. Also, a worthwhile gain in hp is achieved by just going up on the exhaust valve size. Of course it is nothing like as large as the head that has the intake side improved as well, but a power gain is always a nice bonus when economy was originally the sole intention of the exercise.

Regarding valves, it is important to make sure they have as much uni-directional flow bias as possible. This means the use of a Rimflo intake valve because of its one-way bias and a Rimflo exhaust valve because it has very good flow capabilities at low valve lift values.

If you are working with a small bore engine, then combustion chamber modifications should be limited to removing the minimum possible amount of metal consistent with the valve size used and giving the combustion chamber a good polish. Polishing achieves two things. First the shiny surface can help reflect heat into the cylinder, and secondly a smooth finish has less surface area than a rough one and therefore less heat is conducted to the water jacket.

Compression Ratio

If power is the sole consideration, we find that no further power increases occur over about 13 to 13.5:1 if the fuel used is typical service station, low octane stuff, as opposed to special racing fuel. At compressions above 13 or so to 1, full throttle brake specific fuel consumption improvement levels out, but at part throttle, a different situation exists. Because the engine is throttled considerably, and a cylinder does not receive a full charge of air, the effective CR is a lot lower than the static or measured CR. Under these conditions, a static CR raised to 15, 16 even 17:1, will help the part throttle economy. Unfortunately, this makes the engine almost unusable at full throttle, so a compromise has to be struck to allow the engine to be driven at full throttle. From this point of view, it is a good idea to raise the CR as much as possible. If you retain a relatively standard distributor, then 10 or 11:1 is as high as the CR can be raised using fuel of currently available octane levels. On the other hand, if you are prepared to spend the money and have a special distributor built, or a detonation sensing system installed, then there is much more scope for improving the part throttle economy by raising the CR to around 13 or maybe even 14:1. Here's how a high economy system like this will work:

At part throttle the engine is not receiving a full cylinder charge so the ignition timing can be optimized at these part throttle, high manifold vacuum operating conditions. Under these conditions the fuel has a sufficiently high octane value to satisfy the engine's requirements. As the throttle is opened and the cylinder receives a denser charge, so the effective CR will go up. As full throttle is approached, so the likelihood of detonation will increase. Under these conditions, the ignition timing has to be artificially retarded from optimum, solely to keep the engine out of detonation. There are two ways this can be done: first you can have **Aldon Automotive** build a special distributor which has vacuum advance curve characteristics for optimum timing at part throttle but will, when the throttle is depressed and the manifold vacuum drops, retard the ignition more than normal to prevent detonation. Another way this goal can be achieved is to use a detonation sensing ignition retard system such as made in the U.S.A. by the **Carter Carburettor Company**. In the U.K. it would be available through **John Woolfe Racing**. It's not an inexpensive item by any means but it does get the job done. It functions by 'listening' to the engine combustion noise and as soon as detonation is heard, it will retard the ignition to keep the engine just on the point of detonating. There is a difference here between incipient detonation and hard detonation. Incipient detonation is light detonation and is just about audible on the typical car. Hard detonation is what ruins pistons. The detonation sensor detects incipient detonation and an engine run at this level suffers no damage. There is an advantage to using the detonation sensing device because on cold, humid days the engine can stand more advance and hence produce more power than it can on hot, dry days, so the detonation sensing ignition retard device makes the most of the performance aspect of the car, whereas the distributor method has to set ignition timing for the worst situation, and, therefore when more advance could be used, that advance is not available. The distributor from **Aldon Automotive** is designed for an engine of the type under discussion and it's about a quarter of the price of the detonation sensing system. You have to go a lot of miles to make that price difference up in fuel savings. In fact, the price difference may not be made up for in 100,000 miles, so this still leaves the distributor as a very viable route to go.

If we are to look at the CR in terms of how much part throttle economy can be gained, then it will quickly be seen to be one of the most effective ways of producing extra mileage.

Most of the modifications I discussed contribute only 1, 2, 3 or maybe 5% to economy when driving. It's easy to fall into the trap of believing the advertised figures claimed by the economy device manufacturers. You will notice they often say 'up to 10%' or 'up to 15%' or even 'up to 20%'. I am trying to avoid that situation by giving you realistic numbers.

The Series 'A' engine is fundamentally economical and in the last few years, Leyland have done much to improve even that. Thus, you can consider it a great success if you perform a modification which produces 2 or 3% more economy.

Bottom End Mods

If you are building an engine from scratch, then there are mileage gains to be made in correct parts selection and assembly for the engine bottom end. First, pistons should always be flush to the block deck. This allows more compression to be used before detonation sets in. Secondly, the hydrodynamic three point support skirt pistons that **A.E. Autoparts** currently sell are worth 1 to 2% improvement in economy by virtue of reduced skirt friction. Thirdly, pistons with a multiplicity of rings cause more internal friction; a good three ring piston is all that's necessary. Bore friction increases as piston skirt clearance is reduced, so have the bores sized so that the maximum recommended factory clearance exists. Also, cast pistons seem to fare better than forged pistons when relatively close clearances are involved. To ensure that bearing losses are kept to a minimum, have the crank journals sized to the factory recommended bottom limit. If you want to do anything special in the bottom end, then re-working the crank and rods as for a performance engine, will contribute slightly. Again, the sort of improvements we are talking about are minimal but, by reworking these parts, performance will be gained without any expense in terms of reduced fuel economy. I would, simply because it's a very quick and easy job to do, 'flow' the ports of the oil pump, and make sure the pressure release valve sets the oil pressure to 40 psi and no more. Remember, the higher the oil pressure, the more power it takes to drive the lubrication system.

Exhaust System

In terms of actual bolt-on economy, the exhaust system, next to the CR changes, represents the best area for potential economy gains. Read throughoughly the chapter on exhaust manifolds and silencers to appreciate the full extent of what's possible. If you have a budget for bolt-on parts and you are looking for something which will improve the performance and the economy, the first place you should spend your money is on the silencer system. This is especially so if you have an engine with the simple, short, 3-into-1 cast iron exhaust manifold. If you have a Metro with a cast iron LCB, then changing the manifold is not so important as getting a free flow exhaust silencing system on the car. Depending upon the exhaust system fitted, a good straight-through, free flow silencer system can contribute up to 8% in mileage gains as well as adding a power bonus.

We find, that the type of exhaust manifold to give the best results, is dependent on the cam. However, the situation regarding economy in relation to exhaust manifolds is quite different from that regarding power. When short period cams are used, the 3-into-1 exhaust manifold has the edge on an LCB exhaust manifold up to about 3500 rpm. Above that, the LCB manifold shows slightly better economy figures. This means that for all normal purposes, the 3-into-1 system will, on a normally geared Series 'A' car, show better mileage at speeds up to about 60 - 70 mph. Since this is the most important range to consider, because most driving is done under 70 mph, then a 3-into-1 would be the choice. However, the 3-into-1 exhaust manifold often doesn't produce any more power on a short cammed engine, by comparison with the standard cast iron 3-into-1 system. On the other hand, an LCB does tend to produce more power, but only at the top end of the rev range. When we consider the situation with longer cammed engines, that is, 250° or more, then we find that the 3-into-1 and the LCB system are much more closely matched, and to get the best economy from a hot street cammed motor, then a 3-into-1 often has the advantage because it allows the engine to produce better bottom end torque, thus reducing the number of gear changes required during normal driving, though it may not produce the top end power. This means that if your criterion is to have power and economy, the choice of manifold is much more difficult. You must make a decision for yourself whether the addition of a few extra foot pounds of torque that is given by the 3-into-1 system, is worthwhile or the extra power that usually results from using an LCB system on engines with cams between 270 and 300° . By the way, it's worth pointing out that the economy gains available from an exhaust manifold can be totally annihilated by a restrictive exhaust system, so all that is being discussed here makes the assumption that you have first changed your silencing system to a low back pressure, free flowing installation.

Cams For Economy

This is going to be a short paragraph because the practical position regarding cams and economy is relatively simple: the shorter the cam, the better the economy. The shortest standard Leyland cams produce the best mileage. If however, you are looking for power and economy, a Piper Magnum HR 255 or the Kent Cams 256 Megadyne will produce quite a bit more power and a little more economy. A.P.T.'s own VP1 also comes into this category though with 1.5 rockers the bias is towards power rather than economy. These cams help primarily by increasing torque and reducing the need to change gear as often. With Series 'A' development at its present level, I am inclined to regard 255° of cam timing as the maximum that should be used if economy is a prime consideration.

Ignition Systems

The ignition system is one of those 'obvious' areas which affect the economy of an internal combustion engine.

Don't forget that if you super-tune your ignition system for the high output contactless installation, plug gaps, ignition timing, etc. will need to be adjusted to make use of the potential of the ignition system's output. This will be especially so if ... 21-4

... you use a multi-spark system like this MSD 7A unit which gives about 10 sparks per firing stroke. The same thing would also apply if you used a Plasma ignition system. 21-5

Simply, if the ignition system isn't up to par, the economy won't be either. The fundamental rule for economy (and power) is that the ignition system must produce an *adequate* spark at the *right* time, or the performance will suffer. The adequate spark aspect is easily appreciated, but the at 'the right time' aspect is not. This is why it is important to have vacuum advance on the distributor functioning correctly and have the right advance characteristics for the engine.

Most Series 'A' engines are equipped with a contact breaker type ignition system and the spark is delivered by a conventional coil. There are on the market numerous electronic ignition systems which either augment the contact breaker setup or replace it. Just installing one of these systems is unlikely to give worthwhile gain in economy. It is only by utilizing the extra spark capability of an electronic ignition, that any significant gains in economy can be achieved, and even then, these gains are, for the most part, relatively small, about 2 to 3%.

The first move when an electronic ignition is installed is to open up the plug gaps. Typically 30 to 35 thousandths will help but plug gaps up to 50 thousandths can be beneficial as long as no problems occur with arcing in other areas. To stop arcing, it may be necessary to shield the plug leads, etc., yet all the time, effort and trouble to run 50 thousandths wide plug gaps, may not be worth the very small gain that it may produce on your particular engine.

Utilizing larger plug gaps improves the lean burn capability of the engine, therefore mileage gains can be had by leaning out the mixture slightly. Just how much it can be leaned out does depend on how effectively the mixture can be ignited. There comes a time when the ignition system cannot ignite the mixture properly. Although it doesn't misfire, the economy will drop slightly.

For greatest economy the leaner the mixture that can be wholly ignited, the better. For instance, if we could build an ignition system which would ignite a 100:1 air/fuel ratio mixture, we would be very closely approaching a diesel-like cycle, i.e. the amount of manifold vacuum would be near zero and the effective CR even at very light throttle openings would be near that achieved at full throttle. I am using the word 'throttle' here in a context for which it was not really designed because, remember, control of engine output on a spark ignition engine is by 'throttling' the air supply. On a diesel engine the 'throttle' is wide open all the time. The power output is controlled solely by adjusting the amount of fuel injected. If our lean-burn system could operate at such extreme ratios, then rather than have a 'throttle' in the true sense of the word, it would have a fuel control device which regulated the amount of fuel going into an engine which, to all intents and purposes, operated at wide open throttle.

If you understand this, you will be able to comprehend that the leaner the mixture the ignition system can ignite, the more economy the engine is likely to produce under part-load. As a help toward achieving this goal, multi-spark and plasma-type ignitions need to be considered. However, neither type of system is cheap. Multi-spark ignition systems in the form of **Autotronics** MSD system is available, but plasma systems are not readily so, **Microdynamics** being one of the few companies producing such a system. If you look at the ignition system in terms of extending the engine's lean burn capability, then you will begin to see how electronic ignition systems can improve mileage. Thus, if the ignition capability is of a high enough order, then the carb must be modified to deliver the very wide mixture range required. Again, a system for leaning out the mixture under cruise conditions was detailed in the carburation chapter.

Operating Conditions

Not only does an engine need to have a quick warmup time, but also the temperature at which it operates for maximum economy can be substantially different from that for maximum power. Basically, maximum power water temperature is around 70°C, (160°F) but for maximum economy, the water temperature needs to be around 100 to 105° (212 to 220°F). Temperatures above the normal boiling temperature are, of course, achieved because the coolant is under pressure. Typically a 13 lb psi radiator cap would be used on the radiator.

Another fluid temperature

that is important to good fuel economy is oil temperature. This, in fact, needs to be hotter than the water, and 110° to 115° (230 to 240°F) seems to be about the mark. However, if oil film breakdown is to be avoided, it is necessary to use a good quality oil. The best oils in terms of temperature and film strength, are the new synthetic oils. However, although they do increase mileage due to their more slippery nature and lower viscosity, they are expensive, and you may find it's cheaper to use a good quality conventional oil changed regularly and suffer the cost of extra oil changes and slightly increased fuel consumption.

At present, the only synthetic oils I have tried have been the French Motul synthetic oil and Mobil 1. I have used these both as a 50-50 mix with conventional oil and in totally synthetic form and I have been impressed with the result achieved. They appear to be highly functional, both for competition engines and for road engines. Certainly if price were not a factor, these are the oils I would chose.

When lubricant viscosity goes up, the economy achieved is likely to come down. If you use a high quality oil, it will, in all probability, do everything you expect of it and additives will not be required.

In my experience oil additives generally fall into the 'snake oil' category. Latterly I have dyno tested no additives that showed any advantage over high quality oil in a sound engine. However, I am reluctant to condemn all additives, as some are technologically proven lubricants. However, even the best of these additives will do little to improve the currently available high grade oils. They will improve a low grade oil considerably, but if you care about your engine, you have no business using a low grade oil.

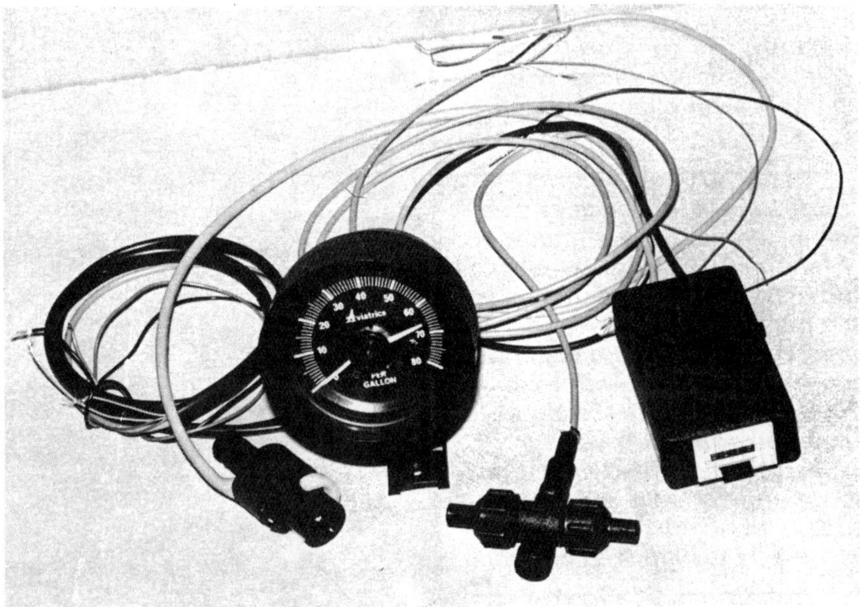

On my round U.S.A. economy Mini, I used a mpg gauge such as the one you see here, together with a 'gallons-gone' indicator which showed the total amount of fuel used. 21-6

Monitoring MPG

It's all very well to build an engine for economy, but real savings will not be made until the driver applies himself to driving the vehicle in a fuel efficient manner. Most car enthusiasts have a competitive spirit of some form or another and the problem in driving economically is that it tends not to be much fun. However, if you have a goal you are trying to achieve or surpass, then it does put an element of fun back into the driving. To this end, I have found that fitting of a trip computer to the car has improved the mpg I have achieved. This is because it tells me two things; first it tells me when I am driving in an

The heart of any mpg indicating device, the fuel-flow sensor. Friction is always present in the bearings, but over a certain level of fuel flow, the brain of the trip computer or mpg gauge can compensate. 21-7

Here's the fuel flow sensor. It goes with the trip computer that Leyland sell as an add-on for the Metro. In case you are finding difficulty locating its position, the camera is looking straight down on the engine and the unit is mounted by the bottom edge of the subframe.

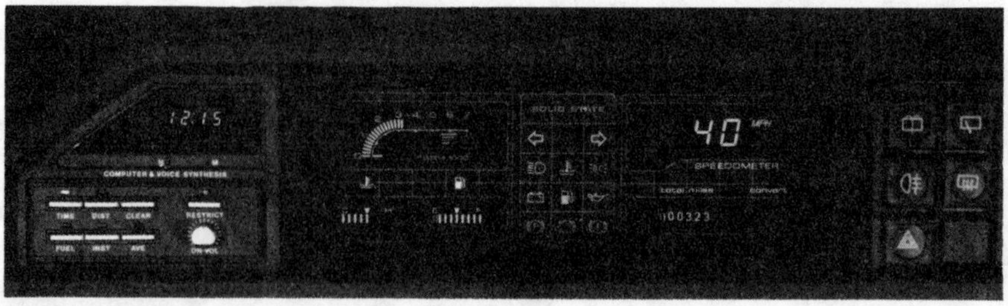

Leyland's trip computer is a relatively simple device with only a few functions, but some ...

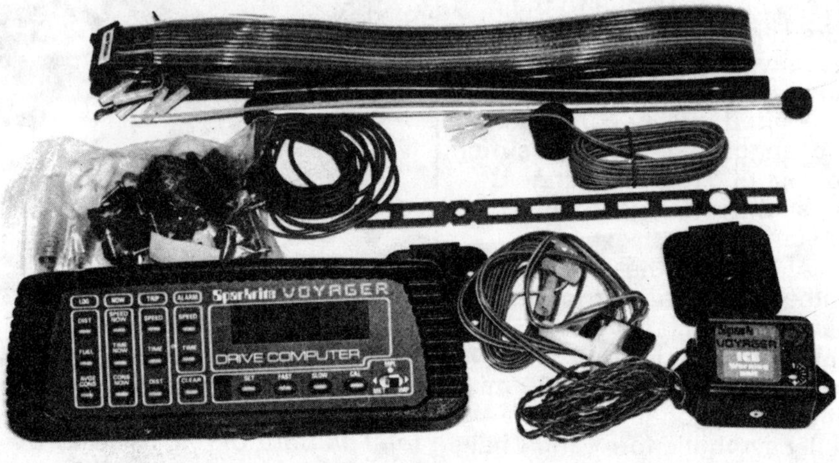

... trip computers such as this Sparkrite one, have a multitude of functions. Not only will it show average mpg, but also will give you your average speed for the trip, the distance you have travelled, warn you if freezing conditions are about to arise, tell you when you are exceeding the speed limit, and so on and so forth. In fact this is a unit I can, from my own experiences, highly recommend.

uneconomical fashion and by just how much I am being uneconomical, and secondly it has allowed me to improve my light foot economy driving technique by finding the driving modes where I thought I was being economical, but was not. After a while of driving with a trip computer, one finds efforts are continually being made to better previous best mpg readings. This puts the competition spirit and some fun back into economy driving. Effective trip computers are available from a number of the big electronics companies. Over the years I have tried many trip computers I didn't like and some I tried were downright unreliable. Indeed one example actually set fire to the vehicle in which it was installed impressing neither me nor, for that matter, the friend whose car it was. One unit I can recommend, however, is that manufactured by **Sparkrite**. Over a two-year period, up to the time of writing this book, this unit has proved to be very reliable and totally functional. If you drive by the trip computer, you are almost certainly going to pick up 5 mpg. At that sort of saving, it shouldn't be too long before you will recoup the cost of the trip computer and apart from fuel savings, it does have many other uses.

22 Dyno Tuning

I discussed dynamometers at length at the beginning of the book, from the point of view of the numbers they develop, and the comparisons between one dynamometer and another. Here, I want to impress upon you the value of dyno tuning the engine after you have built it and run it in.

In the course of my work I get lots of 'phone calls (I wish I didn't) from people who call up and tell me they have modified their Series 'A' engine and it is equipped with a so-and-so carburettor. What jets, etc. should it have in it? At this point, I get all confidential and ask them if they would like to know the secret of really making their setup work. **Since most of the recipients of this conversation are eager to get the best and latest speed secrets, they inevitably answer 'Yes' and I tell them that the way I do every one of my motors is to set it up on a dyno, because guessing is strictly for amateurs.**

Let's face it, if you have spent many careful hours building your engine and installing it in the car, it stands to reason that you will want to get the best from it, and have that hairy cam, big valve head, and fancy carburation working to their best advantage. If you have carried out even some of the simple modifications described you will have quite a bit of time and money invested in your engine. In either case the overall performance of parts and modifications will only be as good as the final integrated adjustment of the various aspects like carburation and ignition which make up the whole.

In previous chapters I have given you an idea of what various settings would need to be, but these are only a starting figure. Every engine differs slightly, and there are so many combinations that it's just not possible to cover them all. Each engine must be set up to suit it's own specification and this is where the rolling road dynamometer comes in.

This device, as you would expect, measures hp at the driving wheels. More than that, it allows the dyno operator to make a change of engine settings to see whether or not more hp is produced. No guesswork, just a positive read-out. Couple the dyno to an electronic engine analyser which can trace the smallest

Even though the engine of the economy Mini project spent a lot of time on the engine dyno, all the calibrations were checked and verified on the chassis dyno. Here, David Ray, co-conspirator on the project, checks out the gauge readings on the Sun 1120 engine analyser. Whether it's a race car, road car or economy car, the dyno tune should be part of your engine project. If you are serious about building race engines ... 22-1

faults in your ignition or set your distributor dwell correctly, or analyse the mixture strength and 101 other things, and you have the means of getting the very best from your engine. Don't think you can substitute seat-of-the-pants tuning for a rolling road dynamometer session. I have spent many years setting up carburation and ignition, and a host of other factors on high performance engines and, take my word for it, you have to be incredibly lucky to get anywhere near the best by guessing. One hour on a rolling road is worth weeks of testing up and down the road. Mind you, the dyno isn't the answer to everything unless it has a good operator. Here, you are best advised to go to the people who are used to setting up high performance machinery. It definitely helps if the dyno operator is in possession of sufficient expertise in fine-tuning the type of carburation with which your engine is equipped. Apart from that, a passing knowledge of the quirks of a Series 'A' engine is also a definite asset. Since the engine is so popular, at least in England, most dyno operators in a performance orientated shop, are pretty switched-on.

If you have enough time and patience and you know what it's all about, you could set your car up reasonably well with just a stopwatch. It could take you all day, maybe more, to do it, and even when you had finished, you could never really be sure that you were any closer than 10% off the optimum, and it is that last 10% that is so important. By utilizing a dyno, you can get the motor set up and know what it's producing in the way of hp. Where else can you find out just what effect is produced from changing an emulsion tube on a carb. It may only make 2 hp difference at some particular point in the rev range, but in each case, the dyno will give a positive indication as to whether a change has produced results in the right direction or not. For instance, you can go on increasing jet sizes or maybe coming down on air corrector sizes, until the best possible hp curve is produced, and this necessitates taking power measurements at a numerous number of points in the rev range. In a similar manner, the engine's ignition timing can be set to give maximum power and, if you are not sure the engine has the right ignition advance curve, it's very simple to plot out the optimum advance curve the engine requires. Although you may leave the dyno shop with the wrong advance curve in your distributor, at least you now know what advance curve is required. From here it's a simple job to have a distributor built by a company like **Aldon**

... then the engine dyno is a far better development tool, as it's more sensitive, and of course access is better for component changes. The particular dyno shown here is one I used to use at the old TJ injection facility in Plymouth, but for some years now I have had my own dyno facility.

Automotive having the right advance curve. When your new distributor arrives, you will know exactly what to set the static timing to, based on the figures produced on the dyno.

A dynamometer with a good operator is worth at least the value of a big valve cylinder head in terms of extra power.

Dyno Test Procedure

Let's have a brief run-through of a typical basic dyno tune on a Series 'A' powered vehicle. Obviously the first thing that will be done is to put the vehicle's driving wheels onto the dyno rollers. At this point, the operator will then make the various connections from the electronic engine analyser to the ignition system and an exhaust sampling tube will be positioned in the exhaust pipe. A preliminary power run will be taken, noting the power all the way up the rev range. At the same time, the operator will keep an eye on the various dials, the scope and gauges on the electronic engine analyser. Power at each point will be noted, but if anything appears to be amiss on the dials, such as the mixture starting to become lean, the operator will terminate the test. On the road, you wouldn't know the mixture was becoming lean, possibly until something untoward happened. Usually this wouldn't be anything worse than a melted plug or a bad misfire, but on occasion it could result in a burnt out piston.

Faults apparent on the first run would now be remedied. Such faults need not necessarily be due to worn parts even on a used engine, more usually being due to incorrect settings. For instance, the points dwell can easily be incorrect. On a high performance Series 'A' engine, it makes a lot of difference to top end power. A seemingly simple factor like having a few thousandths too wide on the points gap can make as much as 4 or 5 hp difference on an engine revving to 8000 rpm.

Having fixed obvious things at fault, the operator will then take another power run, noting any increases or decreases as may be the case. On this run, the operator will generally pay greater attention to mixture, noting at each rpm whether it's rich or lean, or about right. For maximum power, a Series 'A' engine likes to run around 12.5 to 13:1 fuel/air ratios. For good economy it needs to be slightly leaner than this, and 14:1 at full throttle is quite acceptable for a road-going economy engine. However, for part throttle use for an economy engine, mixtures of 18:1 or more, depending upon the type of ignition system, will produce the best economy.

At each speed and throttle opening at which the vehicle is driven the mixture analyser will show the air/fuel ratio being delivered by the carburettor. Most of these devices aren't super accurate, but they do at least give you a clear indication of what's going on. Because most operators know their equipment, they can make allowances for its errors. Remember, it's no good setting the idle mixture on an SU, Weber or Dellorto and expecting the mixture to be right throughout the rev range. It won't be. It will only be right at idle because that's where you set it. Once the mixture has been set to what is required, the ignition timing should be looked at. It should not be done in the reverse sequence because the ignition timing for optimum power can be heavily dependent upon the air/fuel ratio used. This isn't to say the air/fuel ratio changes the ignition timing. It simply means the engine requires a different timing. So getting back to the point in question, once the air/fuel ratio has been set, then it's the turn of the ignition system to be timed for optimum output. Typically the engine will be run at 4000 rpm or more if it's a race engine, and the distributor swung this way and that until the maximum power is shown. For this method of ignition setting to be successful, the engine must be at full operating temperature, otherwise an over-advanced state can be achieved. This is because a cool engine has slower mixture burning characteristics. As a result of the increased burn time, more advance is needed to get the power. When the engine gets hot, the ignition timing that was optimum for the cool condition, is now too far advanced. In the worst case, this could result in detonation. This, in turn, will lead to even more heat, and still more detonation and so the circle tightens, leading eventually to a burnt out motor.

The moral here is to set the ignition timing when the temperatures are right up to what they ought to be. This usually isn't a problem because most Series 'A' engines will run hotter on the dynamometer than on the road, simply because there is not the airflow past the engine to keep the temperature down. The difference between the dyno operating temperature and the track operating temperature can be looked upon as a safety buffer.

The second important point is that the engine must hold a steady power reading on the dyno for a significant period. At least 10 to 15 seconds. That may sound like a short period of time, but when you are sitting there, looking at the dial, with the engine turning 8000 rpm just in front of you, 10 seconds feels like an age. If the engine was up to operating temperature prior to the dyno run, 10 seconds should be an adequate time to hold a power reading to be sure it is a reliable number being produced.

Beware of ignition timings which give a high hp reading for a moment and then drops as engine temperature rises. This can be an indication of over-advance. A flash reading or a reading which can't be held for a reasonable period is not really a true reading. It could be that you have a dyno operator who is trying to prove he can get more hp than the guy running the last dyno shop you went to.

A dyno operator who is really on the ball can go one step further than just power-timing the engine at one particular rpm. This one step won't apply so much to a racing Series 'A' engine, but it does apply to a road–going one. What he will do is check that the distributor advance curve suits the engine requirements. I alluded to this earlier on and this is how it's done:

Firstly, the ignition advance on the distributor is locked out. Then the engine is power-timed, first at 1500 rpm or so, and for the ignition timing to give maximum power, this timing is noted. Then the ignition is re-timed with the engine running at 2000 rpm. Again, the ignition timing is noted. Then 2,500 and so on. At each stage, the amount of ignition timing required to give maximum power, is noted. The temperature the engine is running at is important for this particular plot to be successful. At least 30 seconds to 1 minute will be required to guard against producing an ignition timing figure that doesn't take into account the hottest combustion chamber conditions likely to be seen.

Having taken measurements, the dyno operator will then draw out the advance curve required by the engine. If the dyno shop you are frequenting has a distributor machine handy, the dyno operator can then adjust the weights and spring tension to tailor the distributor advance curve to give as near as is mechanically possible a curve which fits the engine requirements.

By now I hope you begin to see that you can't really do this on the road. You have no accurate means of telling exactly how much power increase or decrease each little change makes. Apart from that, 2 hp gained at the wheels is so small that the driver will never feel it. On the dyno, 2 hp change can be seen and 2 hp gained on half-a-dozen little things adds up to 12. This is a very real gain and don't think I am exaggerating by quoting 12 hp because that is typically the sort of extra power seen at the wheels after an expert dyno operator has 'tuned in' the engine built by a competent enthusiast.

To sum up the situation then, just as you wouldn't postpone fitting the fourth spark plug in your engine, don't keep putting off having your engine dyno tuned. As soon as it is run in sufficiently to be able to wind it up, then you get onto that rolling road, because you can bet it is the only way you will make the most of your engine.

Useful Addresses

The following addresses and telephone numbers were believed to be correct at the time of going to press. However, as these are subject to change, particularly telephone area codes, no guarantee can be given for their accuracy.

A.E. Autoparts Ltd P.O. Box 10, Legrams Lane, Bradford, Yorks BD7 1NQ, England
Tel: 01274-723481

AKM Racing, Sofievej 14. DK7700 Thisted, Denmark
Tel +45 97 92 01 82
Fax +45 97 92 29 32

Aldon Automotive, Breener Industrial Estate, Station Drive, Brierley Hill, W. Midlands DY5 3JZ, England
Tel: 01384-78508

Alexander Engineering Co. Ltd., Thame Road, Haddenham, Aylesbury, Bucks, HP17 8BZ, England
Tel: 01844-291345

Advanced Performance Technology (APT)
595 Iowa Avenue, Suite 'C' Riverside, CA 92507, U.S.A.
Tel: (001) 909-686-0260
Freephone: 800-278-3278
Fax: (001) 909-686-2831
E-mail: techinfo@aptfast.com
Web: www.aptfast.com
Contact: David Anton

Avonbar Racing Ltd, 219 New Haw Road, Addlestone, Weybridge, Surrey, England
Tel: 01932-42024/40058

Auto Electrical (Plymouth), Unit 2 Wall Park Close, Newnham Industrial Estate, Plympton, Plymouth, Devon, England
Tel: 01752-330011

Auto Power Services, South March, Long March Industrial Estate, Daventry, Northants, England
Tel: 013272-76161

Autotronic Controls Corp, 6908 Commerce, El Paso, TX 79915, U.S.A.
Tel: (915) 772-7431

Bhp Developments, 349 Bingen Road, RD # 4, Bethlehem, PA 18015, U.S.A.
Tel: (215) 838-0361

Caltrack Automotive Design, Unit 5, Caxton Road, St. Ives, Huntingdon, Cambs PE17 4LS, England

The Carburettor Centre, 389 Archway Road, London N5, England
Tel: 020-8348 4218

Carl Schmitt Pistons,
Winston & Pecks Ltd, Petrie House, Lisset Road, off Stafferton Way, Maidenhead, Berks SL6 IAZ, England
Tel: 01628 74009

Challenge Mini Supplies, 6 Kemp Court, Bagshot, Surrey, England
Tel: 01784 34167

Chris Montague Carburettor Co,
(Timavo) 380 Finchley Road, London NW2, England
Tel: 020-7794 7766

Competition Grinding,
2712 N. Columbus Blvd., Tucson, AZ 85712, U.S.A.
Tel: (520) 323-1578

Competition Silencers, 29-31 Friern Barnet Road, Southgate, London N11 1NE, England
Tel: 020-8368 6292

Comptune, 7096 Lafayette Road, Medina, OH 44252, U.S.A.
Tel: (330) 9483338

Contact Developments Ltd, 13 Boult Street, Reading, Berks RG1 4RD, England
Tel: 01734-598955/6

Cosworth Engineering Ltd, St. James' Mill Road, Northampton NN5 5JJ, England
Tel: 01604-52444

Cosworth Engineering Inc,
23205 Early Avenue, Torrance, CA 90505, U.S.A.
Tel: (310) 534-1390

Crane Cams Inc., 530, Fentress Blvd, Daytona Beach, Florida FL32014, U.S.A.
Tel: (904) 252 1151

Datum, 180 Hersham Road, Hersham, Walton on Thames, Surrey, KT12 5QE, England
Tel: 01932 221955

Elgin Cams, 55 Perry Street, Redwood City, CA 94063, U.S.A.
Tel: (650) 364-2187

Garrett Airesearch Ltd, West Pimbo, Skelmersdale, Lancs, England
Tel: 01695-22391

Garrett Airesearch,
6201 W. Imperial Highway, Los Angeles, CA, U.S.A.
Tel: (213) 646-2770

Hepworth & Grandage Ltd
(same address as A.E. Autoparts)

Huffaker Engineering,
28013 Arnold Drive, Sonoma, CA 94576, U.S.A.
Tel: (707) 935-0533

Iskenderian, Ed, 16020 S. Broadway, P.O. Box 30, Gardena, CA 90248, U.S.A.
Tel: (213) 770-0930

Jack Knight (Developments) Ltd, Butts Road, Woking, Surrey GU21 1JU, England
Tel: 014862 64326

Janspeed Engineering Ltd, Castle Road, Salisbury, Wilts SP1 2SQ, England
Tel: 01722-21833/6

Jeff Howe Exhausts, rear of Hollyville Cafe, Main Road A.20, West Kingsdown, Kent TN15 6ET, England
Tel: 0147 485 2347

John Cooper Garages Ltd, 50 Ferring Street, Ferring, Worthing, W. Sussex BN12 5JP, England
Tel: 01903-504455

John Woolfe Racing Ltd, Woolfe House, Norse Road, Bedford MK41 OLF, England
Tel: 01234-41441

Jonspeed, Galley Common Service Station, Galley Common, Nuneaton, Warks, England
Tel: 0247 6351495/6

Kent Cams Units 1-4 Military Road, Shorncliffe Industrial Estate, Folkestone, Kent, CT20 3SP, England
Tel: 01303-248666

K & N Engineering Inc, 561 Iowa Avenue, P.O. Box 1329, Riverside, CA 92502, U.S.A.
Tel: (714) 684-9762

K & N Europe, Wilderspool Causeway, Warrington, Cheshire WA2 7PA, England
Tel: 01925-636950

Leeds Mini Spares Ltd, 12 Westfield Road, Leeds, Yorkshire LS3 1DE, England
Tel: 01532 441041

Leonard Reece & Co. Ltd, Beeches Avenue, Carshalton, Surrey SM6 3LB, England
Tel: 020-8674-3646

Lincoln Mini and Metro Spares, Drake Street, Lincoln LN1 1P, England
Tel: 01522 26363

Lumention Ltd, 640 Ripple Road, Barking, Essex, IG11 0RU, England.
Tel: 020-8593 7434

Lynx Corporation Pty. Ltd, 166-176 Parramatta Road, Burwood, Sydney 2134, Australia
Tel: (02) 747-3333

Magard Ltd, 372 East Park Road, Leicester LE5 5AY, England
Tel: 01533-730831

Mainly Minis, 1964 Dalkeith Road, Edinburgh EH16 5DU, Scotland
Tel: 0131-668 2428

Maniflow (Exhausts), St. Paul's Road, Salisbury, Wilts, England
Tel: 01722-320834

Manx Racing Developments, Bratch Lane, Dinton, Wilts SP3 5EB, England
Tel: 0172-276 639

Metro Mini Spares, 126 Lower Richmond Road, Putney, London SW15, England
Tel: 020-8789 5307

Mini Spares Centre, 29-31 Friern Barnet Road, Southgate, London N11 1NE, England
Tel: 020-8368 6292

MiniSport, Thompson Street, Padiham Lancashire BB12 7AP, England
Tel: 01282 778731

Minisprint, 295 Avenue Road Extension, Leicester LE2 3ER, England
Tel: 0533 702996

Minnow-Fish Carburetters Ltd, Minnow House, Lochgilphead, Argyll, Scotland
Tel: 0546-2418

Moroso Performance Products Inc., Carter Drive, Guilford, CT 06437, U.S.A.
Tel: (203) 453-6571

Mountune Race Engines, Unit 3 Causeway Industrial Units, The Causeway, Maldon, Essex, England
Tel: 01621 854029

Mowatt, John Basildon, Essex, England
Tel: 01268-726225

N.O.S. Inc, 5930 Lakeshore Drive, Cypress, CA 90630, U.S.A.
Tel: (714) 821-0580

Omega Pistons, Oak Barn Road, Halesowen, W. Midlands B62 9DW, England
Tel: 0121-559-6778/9

Race Engine Components, 37 Beeches Road, West Bromwich, W. Midlands B70 6QP, England
Tel: 0121-553-1103

Race Techniques, Elim Works, Dunally Parade, Cheltenham, Glos., England
Tel: 01242 230403

Richard Longman & Co, 5 Airfield Road, Airfield Way Industrial Estate, Christchurch, Dorset, England
Tel: 01202-486569

Ripspeed International, 54 Fore Street, Edmonton, London, N18 2SS, England
Tel: 020-8803 4355

Seven Enterprises Ltd, 716 Blue Crab Road, Newport News, VA 23606, U.S.A.
Tel: (757) 873-0007

Shaun Wall, 104 Finian Park, Shannon, Co. Clare, Ireland

Slough Road and Race Parts, 33 Stoke Road, Slough, Berks, SL2 5AH, England
Tel: 01753 35717

Sparkrite, 82 Bath Street, Walsall, W. Midlands WS1 3DE, England
Tel: 01922-614791/6

Specialized Valves Ltd, 37 Beeches Road, West Bromwich, W. Midlands B70 6QP, England
Tel: 0121-553-1103

Steve Harris Motor Engineering, 3 Crane Street, Cranborne, Dorset BA21 5QD, England
Tel: 0172-54-376

Suhner, Finishing Aids & Tools Ltd., Little End Road, Eaton Socon, St. Neots, Cambs PE19 3JH, England
Tel: 01480-216060

Suhner Industrial Products Inc., P.O. Box 2938, Rome, GA 30161, U.S.A.
Tel: (404) 235-8047

Superflow Corp, 3512 N. Tejon, Colorado Springs, CO 80907, U.S.A.
Tel: (303) 471-1746

The 1275 GT Centre, 105 Brinksway, Stockport, Cheshire, SK3 0BZ, England
Tel: 0161-480 8808

Titan Motorsport & Manufacturing Co., The Harley Works, Paxton Hill, St. Neots, Cambridge PE19 4RA, England
Tel: 01480 474402

Turbotechnics, 17 Gallowhill Road, Brackmills, Northampton NN4 0EB, England
Tel: 01604-64005

TWM Induction, 325, Rutherford Street, Goleta, California 93117, U.S.A.
Tel: (805) 967 9478

Weber Concessionaires Ltd, Dolphin Road, Sunbury-on-Thames, Middx TW16 7HE, England
Tel: 0193-27818 805

Windsor, D., E.S.C., 72A Selsdon Road, S. Croydon CR2 6PZ, England
Tel: 020-8688 6809